THE
BIOCHEMISTRY
OF PLANTS

A COMPREHENSIVE TREATISE

Volume 7
Secondary Plant Products

E. E. Conn, editor

Department of Biochemistry and Biophysics
University of California
Davis, California

1981

ACADEMIC PRESS

A Subsidiary of Harcourt Brace Jovanovich, Publishers
New York   London   Toronto   Sydney   San Francisco

ACADEMIC PRESS, INC.
111 Fifth Avenue, New York, New York 10003

United Kingdom Edition published by
ACADEMIC PRESS, INC. (LONDON) LTD.
24/28 Oval Road, London NW1 7DX

Library of Congress Cataloging in Publication Data
Main entry under title:

The Biochemistry of plants.

  Includes bibliographies and indexes.
  CONTENTS: v. 1. The plant cell.--v. 2. Metabolism
and respiration.--v. 3. Carbohydrates.--[etc.]--v. 7.
Secondary plant products.
  1. Botanical chemistry. I. Stumpf, Paul Karl,
Date. II. Conn, Eric. E.
QK861.B48   581.19'2   80-13168
ISBN 0-12-675407-1 (v. 7)   AACR1

PRINTED IN THE UNITED STATES OF AMERICA

81 82 83 84   9 8 7 6 5 4 3 2 1

Contents

List of Contributors

Numbers in parentheses indicate the pages on which the authors' contributions begin.

Wolfgang Barz (35), Lehrstuhl für Biochemie der Pflanzen, Westfälische Wilhelms-Universität, D-4400 Münster, Federal Republic of Germany

E. A. Bell (1), Department of Plant Sciences, King's College London, University of London, London SE24, United Kingdom

Stewart A. Brown (269), Department of Chemistry, Trent University, Peterborough, Ontario, Canada K9J 7B8

V. S. Butt (627), Botany School, University of Oxford, South Parks Road, Oxford OX1 3RA, United Kingdom

E. E. Conn (479), Department of Biochemistry and Biophysics, University of California, Davis, California 95616

Otis C. Dermer (317), Department of Biochemistry, Oklahoma State University, Stillwater, Oklahoma 74074

Donald K. Dougall (21), W. Alton Jones Cell Science Center, Lake Placid, New York 12946

Heinz G. Floss (177), Department of Medicinal Chemistry, Purdue University, West Lafayette, Indiana 47907

L. Fowden (215), Rothamsted Experimental Station, Harpenden, Herts., United Kingdom

Hans Grisebach (457), Lehrstuhl für Biochemie der Pflanzen, Institut für Biologie II der Universität, D-7800 Freiburg/Br., Federal Republic of Germany

G. G. Gross (301), Universität Ulm, Abteilung Allgemeine Botanik, D-7900 Ulm, Federal Republic of Germany

Klaus Hahlbrock (425), Biologisches Institut II der Universität, D-7800 Freiburg/Br., Federal Republic of Germany

Kenneth R. Hanson (577), Department of Biochemistry and Genetics, The Connecticut Agricultural Experiment Station, New Haven, Connecticut 06504

Edwin Haslam (527), Department of Chemistry, University of Sheffield, Sheffield S3 7HF, United Kingdom

Evelyn A. Havir (577), Department of Biochemistry and Genetics, The Connecticut Agricultural Experiment Station, New Haven, Connecticut 06504

Wolfgang Hösel (725), Lehrstuhl für Biochemie der Pflanzen, Westfälische Wilhelms-Universität, D-4400 Münster, Federal Republic of Germany

Johannes Köster (35), Lehrstuhl für Biochemie der Pflanzen, Westfälische Wilhelms-Universität, D-4400 Münster, Federal Republic of Germany

C. J. Lamb (627), Department of Biochemistry, Oxford University, Oxford, OX1 3RA, United Kingdom

E. Leistner (403), Institut für Pharmazeutische Biologie und Phytochemie, Westfälische Wilhelms-Universität, D-4400 Münster, Federal Republic of Germany

Peder Olesen Larsen (501), Chemistry Department, Royal Veterinary and Agricultural University, DK-1871 Copenhagen, Denmark

Mario Piattelli (557), Department of Chemistry, University of Catania, Catania, Italy

Jonathan E. Poulton (667), Department of Botany, University of Iowa, Iowa City, Iowa 52242

David S. Seigler (139), Department of Botany, University of Illinois, Urbana, Illinois 61801

T. A. Smith (249), Long Ashton Research Station, University of Bristol, Long Ashton, Bristol BS18 9AF, United Kingdom

Helen A. Stafford (117), Department of Biology, Reed College, Portland, Oregon 97202

George R. Waller (317), Department of Biochemistry, Oklahoma State University, Stillwater, Oklahoma 74074

Rolf Wiermann (85), Botanisches Institut der Universität, D-4400 Münster, Federal Republic of Germany

General Preface

In 1950, James Bonner wrote the following prophetic comments in the Preface of the first edition of his "Plant Biochemistry" published by Academic Press:

> There is much work to be done in plant biochemistry. Our understanding of many basic metabolic pathways in the higher plant is lamentably fragmentary. While the emphasis in this book is on the higher plant, it will frequently be necessary to call attention to conclusions drawn from work with microorganisms or with higher animals. Numerous problems of plant biochemistry could undoubtedly be illuminated by the closer applications of the information and the techniques which have been developed by those working with other organisms . . .
>
> Certain important aspects of biochemistry have been entirely omitted from the present volume simply because of the lack of pertinent information from the domain of higher plants.

The volume had 30 chapters and a total of 490 pages. Many of the biochemical examples cited in the text were derived from studies on bacterial, fungal, and animal systems. Despite these shortcomings, the book had a profound effect on a number of young biochemists since it challenged them to enter the field of plant biochemistry and to correct "the lack of pertinent information from the domain of higher plants."

Since 1950, an explosive expansion of knowledge in biochemistry has occurred. Unfortunately, the study of plants has had a mixed reception in the biochemical community. With the exception of photosynthesis, biochemists have avoided tackling for one reason or another the incredibly interesting problems associated with plant tissues. Leading biochemical journals have frequently rejected sound manuscripts for the trivial reason that the reaction had been well described in *E. coli* and liver tissue and thus was of little interest to again describe its presence in germinating pea seeds! Federal granting agencies, the National Science Foundation excepted, have also been reluctant to fund applications when it was indicated that the principal experimental tissue would be of plant origin despite the fact that the most prevalent illness in the world is starvation.

The second edition of "Plant Biochemistry" had a new format in 1965 when J. Bonner and J. Varner edited a multiauthored volume of 979 pages; in 1976, the third edition containing 908 pages made its appearance. A few textbooks of limited size in plant biochemistry have been published. In addition, two continuing series resulting from the annual meetings and symposia of photochemical organization in Europe and in North America provided the biological community with highly specialized articles on many topics of plant biochemistry. Plant biochemistry was obviously growing.

Although these publications serve a useful purpose, no multivolume series in plant biochemistry has been available to the biochemist trained and working in different fields who seeks an authoritative overview of major topics of plant biochemistry. It therefore seemed to us that the time was ripe to develop such a series. With encouragement and cooperation of Academic Press, we invited six colleagues to join us in organizing an eight volume series to be known as "The Biochemistry of Plants: A Comprehensive Treatise." Within a few months, we were able to invite over 160 authors to write authoritative chapters for these eight volumes.

Our hope is that this Treatise not only will serve as a source of current information to researchers working on plant biochemistry, but equally important will provide a mechanism for the molecular biologist who works with *E. coli* or the neurobiochemist to become better informed about the interesting and often unique problems which the plant cell provides. It is hoped, too, the senior graduate student will be inspired by one or more comments in chapters of this Treatise and will orient his future career to some aspect of this science.

Despite the fact that many subjects have been covered in this Treatise, we make no claim to have been complete in our coverage nor to have treated all subjects in equal depth. Notable is the absence of volumes on phytohormones and on mineral nutrition. These areas, which are more closely associated with the discipline of plant physiology, are treated in multivolume series in the physiology literature and/or have been the subject of specialized treatises. Other topics (e.g., alkaloids, nitrogen fixation, flavonoids, plant pigments) have been assigned single chapters even though entire volumes, sometimes appearing on an annual basis, are available.

Finally, we wish to thank all our colleagues for their enthusiastic cooperation in bringing these eight volumes so rapidly into fruition. We are grateful to Academic Press for their gentle persuasive pressures and we are indebeted to Ms. Barbara Clover and Ms. Billie Gabriel for their talented assistance in this project.

P. K. Stumpf
E. E. Conn

Preface to Volume 7

Secondary compounds may be defined as those natural products, usually of plant origin, which do not function directly in the primary biochemical activities that support the growth, development, and reproduction of the organism in which they occur. Such compounds are usually restricted in their occurrence or distribution in nature. As others have observed, the choice of the adjective "secondary" to describe such "non-primary" compounds was unfortunate because it suggested that they were unimportant. That choice may have for a time even hindered progress in studying the function(s) of secondary compounds in nature. Fortunately, that obstacle, if it existed, has diminished, and today abundant evidence supports the hypothesis that many secondary (natural) products have played an indispensible role in maintaining a species during the course of evolution. To the extent that this volume informs the broader biochemical community of the research being carried out on these compounds, we shall have achieved one of our objectives. We also hope that this volume will prove useful to those who are already working on the biochemistry of secondary compounds.

The treatment of secondary plant products in this volume is divided into three sections. Chapters 1 through 7 discuss broader aspects of secondary (natural) products, such as their physiological roles, their metabolic turnover, or their relationship to plant taxonomy. The specific examples and relationships covered in these seven chapters, it is hoped, will find general application in future investigations carried out on other secondary compounds.

Chapters 8 through 19 deal with 12 classes of secondary compounds. In some chapters, emphasis is placed on the chemical nature and properties of the class; in other chapters, various aspects of their metabolism, including regulation, are stressed. In Chapter 12, the authors have confined their attention to the enzymes of alkaloid metabolism, an approach that has permitted the subject of alkaloids to be restricted to one chapter in the present volume. Whereas the chapters in this second section discuss 12 different classes of secondary plant products, other compounds that can properly be described as secondary plant products are dealt with elsewhere in this treatise (e.g., Chapters 5 and 15 in Volume 3; Chapters 13, 14, 17, 18, and 19 in Volume 4; and Chapters 8, 9, 10, and 11 in Volume 6). By discussing some secondary compounds elsewhere in the Treatise, we have maintained Volume 7 at a reasonable size.

Chapters 20 through 23 in the final section are concerned with enzymes that are of more than ordinary interest in the metabolism of secondary plant products. Thus, Chapter 20 discusses phenylalanine ammonia-lyase which catalyzes the conversion of the primary metabolite phenylalanine to cinnamic acid, the first reaction in the biosynthesis of phenylpropanoid compounds. Also treated in this section are oxygenases, methyl and glycosyl transferases, and glycosidases, all of which are intimately involved in the metabolism of secondary plant products.

I wish to thank the authors who contributed to this volume for their cooperation and their patience during its production. I also am grateful for the full cooperation of the publishers and for the secretarial assistance of Ms. Billie Gabriel and Ms. Barbara Clover.

E. E. Conn

The Physiological Role(s) of Secondary (Natural) Products

1

E. A. BELL

I. INTRODUCTION

Secondary compounds can be defined as compounds that have no recognized role in the maintenance of fundamental life processes in the organisms that synthesize them. This definition excludes the intermediates and end products of primary metabolic pathways and compounds such as the photosynthetic pigments of green plants and the oxygen-carrying pigments of mammalian blood.

Primary metabolic pathways are not necessarily the same in all species, however, and a given compound occurring in two species may have a primary role in one and not in the other. It is also certain that as our knowledge of biochemistry increases we shall have to revise our ideas concerning the role or roles of many biological compounds.

Whereas secondary compounds are in no way restricted to the plant kingdom, it is from plants and microorganisms that the great majority have been isolated. This difference between plants and animals in their commitment of genetic potential, nutritional resources, and storage capacity to the synthesis and accumulation of secondary compounds is, I believe, of relevance when

The Biochemistry of Plants, Vol. 7

1

considering their biological roles. Most animals are mobile and depend to a greater or lesser extent on their mobility to obtain food and evade predators. Most plants lack mobility and must necessarily have evolved alternative strategies for survival. Among such alternative strategies may be the synthesis of secondary compounds that serve to deter potential predators, to discourage competing plant species, to attract pollinators, to attract symbionts, or to further the interests of the plant in other ways.

We have of course no reason to believe that all secondary plant compounds have an ecological role. Some may merely be the end products of aberrant biosynthetic pathways, and others excretion products. A gene controlling the synthesis of a valueless secondary compound might conceivably be passed from generation to generation for a very long time if it was closely associated on the chromosome with a gene essential for the plant's survival.

In practice, however, we know so little about the biology of secondary compounds that any general statement concerning their significance would be meaningless. It is impossible to reconstruct past history and evaluate the selective advantage, if any, conferred by a particular secondary compound on a long dead plant occupying an unknown habitat and subject to environmental pressures of which we are ignorant. It may therefore seem equally impossible to decide whether such a compound has conferred a selective advantage on the species containing it. Fortunately another source of evidence is available to us, for no organism lives and evolves in isolation and each coevolving species influences its neighbors. Somewhat paradoxically then, the clearest evidence for the role of a secondary compound in a given species of plant may be found not in the plant itself but rather in the biochemistry of other plant species that compete with it, in the biochemistry of animals that eat it, or in the biochemistry of pathogens that invade it. In discussing the role of toxic compounds in seeds I have written (Bell, 1978):

> A seed does not exist in an ecological vacuum but in a complex living web which includes the seed's parent plant and all other living species with which the plant and seed interact. This implies that the presence of a secondary compound in a seed will be accompanied by other related adaptations (morphological, biochemical and physiological) in the seed, in the whole plant and in the interacting species. These adaptations will result from the progressive modification of the co-evolving species in a manner which will optimize any evolutionary advantage derived from the presence of the secondary compound and minimize any evolutionary disadvantage.

This statement is true not only of seeds but also of living organisms in general and, as I have suggested, the clearest indication of the significance of a secondary compound may well be provided by the biochemical responses it elicits in other organisms.

Although observations in the field suggest that chemical interactions between plants and plants and between plants and other organisms are widespread, the basic biochemistry of these interactions has rarely been investigated. In the following sections I have chosen a few examples that illustrate the ways in which some secondary compounds appear to serve the needs of plants, and I hope that these examples will awaken the interest of others in a fascinating and as yet little explored interface of biology and biochemistry.

For more comprehensive accounts than I am able to give here the reader is referred to the books of Florkin and Schoffeniels (1969), Sondheimer and Simeone (1970), and Harborne (1977a); to the multiauthor volumes edited by Harborne (1972, 1978), Gilbert and Raven (1975), Wallace and Mansell (1976), Luckner *et al.* (1976), Friend and Threlfall (1976), Marini-Bettolo (1977), and Rosenthal and Janzen (1979); and to the shorter review articles of Whittaker and Feeny (1971), Levin (1976), Seigler (1977), Harborne (1977b), and Swain (1977).

II. PLANT–VERTEBRATE INTERACTIONS

There is no tree more characteristic of the English churchyard than the yew (*Taxus baccata*). This tree was grown to provide wood for longbows, and it was planted within a churchyard wall to prevent village animals from grazing on the foliage and poisoning themselves.

This example reminds us that humans have been aware for a very long time that plant compounds can affect vertebrate biochemistry, yet they still know relatively little of such compounds and even less of the ways in which they act.

Most studies on the effects of plant secondary compounds in vertebrates have been concerned with compounds that are toxic to or physiologically active in humans or their domestic animals. Secondary compounds that have been the subject of such studies include alkaloids such as those of the yew tree, cyanogenic glycosides, glucosinolates, phenolics, nonprotein amino acids, amines, peptides, and compounds from a number of other miscellaneous chemical groups.

Cyanogenic glycosides, which are treated in detail elsewhere in this volume, are toxic to vertebrates and many other forms of life because they can be hydrolyzed enzymatically to give hydrogen cyanide. In the intact plant the enzyme and the glycoside remain separate, but if the plant tissue is damaged, the two are brought together and hydrogen cyanide is liberated (Fig. 1).

Hydrogen cyanide is extremely toxic to many organisms, and the lethal dose to humans has been variously estimated to be between 0.5 and 3.5 mg/kg body weight (Montgomery, 1969). This compound is an inhibitor of cytochrome oxidase and consequently of cellular respiration (Conn, 1973).

Several species of plant, including bird's-foot trefoil (*Lotus corniculatus*), clover (*Trifolium repens*), and bracken (*Pteridium aquilinum*), are known to be

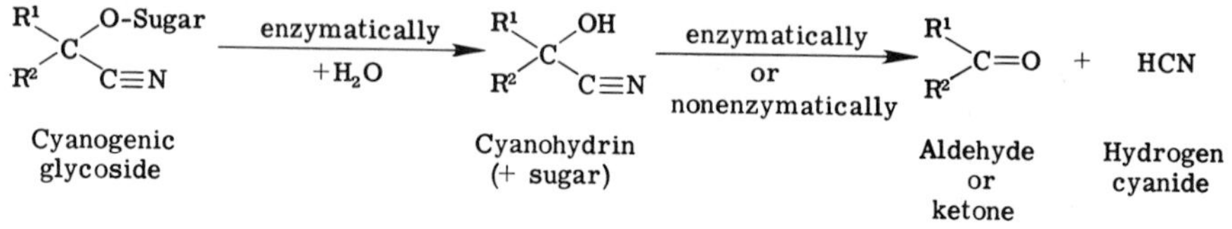

$$\underset{\substack{\text{Cyanogenic}\\\text{glycoside}}}{\overset{R^1}{\underset{R^2}{>}}\!\!C\!\!\underset{C\equiv N}{\overset{O\text{-Sugar}}{<}}} \xrightarrow[\substack{\text{enzymatically}\\+\,H_2O}]{} \underset{\substack{\text{Cyanohydrin}\\(+\text{ sugar})}}{\overset{R^1}{\underset{R^2}{>}}\!\!C\!\!\underset{C\equiv N}{\overset{OH}{<}}} \xrightarrow[\substack{\text{enzymatically}\\\text{or}\\\text{nonenzymatically}}]{} \underset{\substack{\text{Aldehyde}\\\text{or}\\\text{ketone}}}{\overset{R^1}{\underset{R^2}{>}}\!\!C\!\!=\!\!O} + \underset{\substack{\text{Hydrogen}\\\text{cyanide}}}{HCN}$$

Fig. 1. The liberation of hydrogen cyanide from cyanogenic glycosides.

polymorphic with respect to cyanogenesis, and it has been reported by Cooper-Driver and Swain (1976) that the acyanogenic form of bracken is grazed by deer and sheep although the cyanogenic form is not. This observation suggests that the cyanogenic forms may be less palatable as well as more poisonous to some animal species. One such species is undoubtedly the human one, for highly toxic cyanogen-rich varieties of cassava (*Manihot esculenta*) are referred to as "bitter cassavas" in West Africa and are avoided as a source of food. "Sweet cassavas" contain much lower concentrations of toxins, and these are further reduced by soaking the cassava flour in water before cooking.

Cyanogenic glycosides give rise to hydrogen cyanide because their aglycones are cyanohydrins, that is, molecules in which both nitrile and hydroxyl groups are attached to the same carbon atom. Some cyano compounds in plants do not have this structure, however, and even though they cannot liberate hydrogen cyanide, they may nevertheless be toxic to vertebrates. Two such nitriles are β-aminopropionitrile (I) and β-cyanoalanine (II).

The first of these, β-aminopropionitrile, was originally characterized from the seeds of the sweet pea (*Lathyrus odoratus*) as the γ-glutamyl derivative (Schilling and Strong, 1954) and has since been found in other species of the same genus.

When fed to experimental animals, β-aminopropionitrile and its γ-glutamyl derivative produce fundamental morphological changes which include gross skeletal deformation in young rats and aortic aneurysms in birds. These effects are due to the inhibition, by the nitrile, of enzymes responsible for the synthesis of cross-linking peptides in collagen (Levene, 1961) and elastin (O'Dell *et al.*, 1966).

Recent findings (M. F. Wilson and E. A. Bell, unpublished), to which reference is made later, indicate that β-aminopropionitrile is also liberated in the free form by seedlings of *L. odoratus,* suggesting that this compound may play more than one role in the plant.

β-Cyanoalanine, which occurs both free and as the γ-glutamyl derivative in seeds of *Vicia* species, produces convulsions and death in chicks and rats (Ressler *et al.*, 1963). It has also been shown (Ressler *et al.*, 1964) that the toxin acts as an inhibitor of cystathionase, the enzyme responsible for the conversion of cystathionine to cysteine in the mammalian liver. Cystathionase is a vitamin B_6-dependent enzyme, and the ability of vitamin B_6 to protect rats against the effects of β-cyanoalanine suggests that β-cyanoalanine is a vitamin B_6 antagonist.

$$NC-CH_2-CH_2-NH_2 \qquad NC-CH_2-CH(NH_2)-CO_2H$$

β-Aminopropionitrile　　　　　　β-Cyanoalanine

(I)　　　　　　　　　　　　　　　(II)

Cycasin (III), the β-glucoside of methylazoxymethanol, is sometimes referred to as a pseudocyanogenic glycoside. It occurs in the leaves and seeds of cycad species and has been implicated as a probable cause of amyotrophic lateral sclerosis in humans (Whiting, 1963). This neurological disease is particularly common among island peoples of the Pacific who use the seeds of *Cycas circinalis* as food.

In experimental animals cycasin has proved to be a powerful carcinogen when fed orally. It is nontoxic when injected intravenously, however, as the active carcinogen is the aglycone liberated by the β-glucosidase enzymes that occur in microorganisms of the animal gut. Methylazoxymethanol can act as a methylating agent, and it is probable that its ability to methylate cellular components including nucleic acids accounts for its carcinogenic properties (Matsumoto and Higa, 1966).

$$CH_3-N=N-CH_2-O-Gl$$
$$\overset{|}{O}$$

Cycasin

(III)

Among the nonprotein amino acids synthesized by plants there are a number that are toxic to mammals. These include 2,4-diaminobutyric acid, which was first reported in *Polygonatum* species by Fowden and Bryant (1958) and subsequently found to accumulate in high concentrations in seeds of *Lathyrus latifolius* (Ressler *et al.*, 1961). This amino acid was found to cause convulsions and death in rats when fed to them or injected intravenously. O'Neal *et al.* (1968) showed that 2,4-diaminobutyric acid (which is the lower homologue of ornithine) induced ammonia toxicity in the animals by inhibiting the action of the enzyme ornithine transcarbamylase and disrupting the urea cycle (Fig. 2).

We have seen that cycasin itself is inactive as a carcinogen but gives rise to the highly toxic methylazoxymethanol when enzymatically hydrolyzed by β-glucosidase. In the same way hypoglycin A (IV), which occurs in the unripe fruit of the akee (*Blighia sapida*) and can cause acute hypoglycemia and death in humans, is probably the precursor of a toxic compound rather than the toxin itself. Von Holt *et al.* (1964) have proposed that the toxin is an

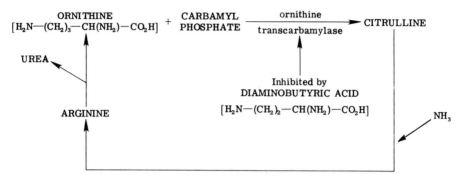

Fig. 2. The inhibition of ornithine transcarbamylase of the urea cycle by 2,4-diaminobutyric acid.

unsaturated carboxylic acid (V) formed from the amino acid in the mammalian system. This compound may block fatty acid metabolism and cause in turn the characteristic depletion of carbohydrates.

Hypoglycin A

(IV)

Methylenecyclo-
propylacetic acid

(V)

An even more interesting relationship between a nonprotein amino acid and its degradation product has recently been established with respect to mimosine (VI). This heterocyclic amino acid, which occurs in certain species of the Mimosoideae, notably *Leucaena leucocephala,* causes loss of hair and liver damage when ingested by animals. In ruminants these symptoms are also accompanied by enlargement of the thyroid gland. The goitrogenic effect has recently been shown (Hegarty *et al.,* 1979) to be caused not by mimosine itself but by 3-hydroxy-4(1*H*)-pyridone (VII) which is formed from mimosine by the action of microorganisms present in the rumen.

Mimosine

(VI)

3-Hydroxy-4 (1*H*)-
pyridone

(VII)

Among the plant amines are numerous compounds that produce physiological effects in vertebrates. Some, such as the methylated tryptamines of

Anadenanthera (*Piptadenia*)*peregrina*, have hallucinogenic effects in humans, whereas others, such as the tyramine (VIII) and *N*-methyltyramine (IX) of *Acacia berlandieri*, produce neurological disorders in cattle and goats that eat the foliage. It is of interest that both these *Acacia* amines, which are close structural analogues of noradrenaline (X), increase blood pressure in the anesthetized rat and increase both the force and rate of contraction of guinea pig right atrium by inducing the release of noradrenaline (Evans *et al.*, 1979).

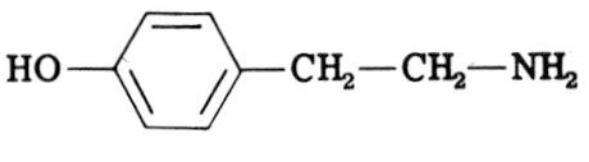

HO—⟨ ⟩—CH_2—CH_2—NH_2 HO—⟨ ⟩—CH_2—CH_2—$NH(CH_3)$

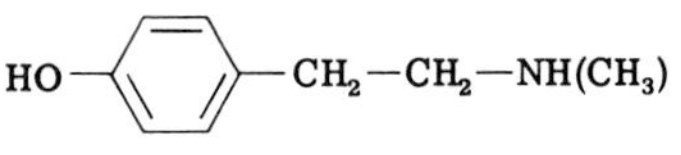

Tyramine   *N*-Methyltyramine

(VIII)   (IX)

HO
|
HO—⟨ ⟩—CHOH—CH_2—NH_2

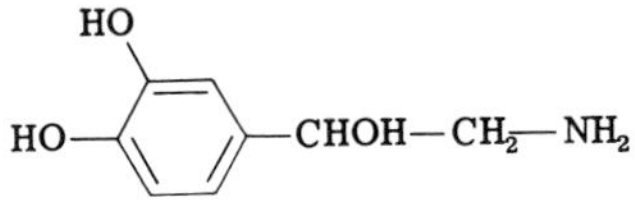

Noradrenaline

(X)

Physiologically active alkaloids from plants are so numerous and of so many structural types that I shall refer to one group only, pyrrolizidine alkaloids. These are primarily responsible for the poisoning of cattle and other domestic animals by various species of *Crotalaria* and *Senecio*. *Senecio jacobaea* (ragwort) causes the death of more domestic animals in Britain than all other British poisonous species taken together (Forsyth, 1968), and it contains several different hepatotoxic pyrrolizidine alkaloids. Mattocks (1972) has presented evidence that it is the metabolites of pyrrolizidine alkaloids rather than the alkaloids themselves that are toxic. In the rat pyrrolizidine alkaloids are metabolized to pyrrole derivatives, and the author states that "there are good reasons for believing that they are responsible for some or all of the toxic effects of the alkaloids." In sheep, however, it has been established that the pyrrolizidine alkaloid heliotrine is detoxified in the rumen (Dick *et al.*, 1963; Lanigan and Smith, 1970), the toxin being degraded to two nontoxic products (Fig. 3).

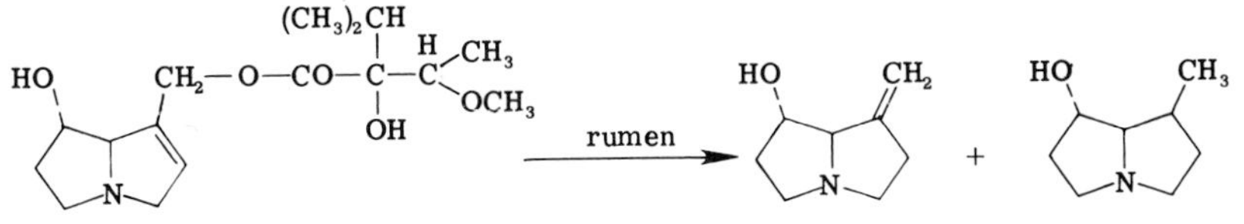

Heliotrine   Nontoxic metabolites

Fig. 3. The detoxication of heliotrine in sheep's rumen

The examples given in this section establish that secondary plant compounds may frequently be toxic to vertebrates. The preference of deer and sheep for acyanogenic bracken and the ability of sheep to detoxify heliotrine suggest that vertebrate predation may have been a significant factor in the natural selection of plants that accumulate toxic secondary compounds. Substantial evidence has yet to be obtained, however, as most studies on the effects of secondary compounds on vertebrates have been planned to provide results of medical and veterinary rather than ecological significance.

III. PLANT–INSECT INTERACTIONS

Insects, like vertebrates, are affected by a wide range of secondary compounds. Secondary compounds do not, however, appear to provide plants with total protection against insects, as almost any plant species one chooses to consider will be found to suffer the attack of one insect species or another. A lack of total protection does not necessarily mean that a secondary plant compound lacks a defensive role. Such a compound may successfully protect the plant from a great variety of potential predators in a given habitat and only fail to deter a specialist predator which, in the course of evolution, has developed biochemical modifications enabling it to avoid the consequences normally associated with the ingestion of the particular toxin.

Using the larvae of a seed-eating beetle, the cowpea weevil (*Callosobruchus maculatus*), Janzen *et al.* (1977) showed that many alkaloids and nonprotein amino acids were lethal to these larvae when added to their natural food [the meal of cowpeas (*Vigna unguiculata*)]. The lethal concentrations were in many instances lower than the concentrations at which the secondary compounds occur naturally in the seeds of other legume species. Clearly their presence in these species would be sufficient to explain the inability of *C. maculatus* larvae to feed on them. The transfer of *C. maculatus* eggs to legume seeds other than those of *V. unguiculata* (Janzen, 1977) again confirmed that the seeds of the great majority of species examined contained toxins that killed *C. maculatus* either in the egg or larval stage. The nonprotein amino acid canavanine (XI) proved lethal to the larvae of *C. maculatus*

$$H_2N-C(=NH)-NH-O-CH_2-CH_2-CH(NH_2)-CO_2H$$

Canavanine

(XI)

at a concentration of 5% and was markedly toxic at 1%. As this amino acid occurs in seeds of *Dioclea megacarpa* at concentrations of 8% and greater, it was of considerable interest to find that these seeds provided the larvae of a second seed beetle, *Caryedes brasiliensis,* with their only source of food. Because of its structural similarity to arginine the properties of canavanine as a possible arginine antagonist have been studied in numerous biological systems (Rosenthal, 1977). When canavanine is fed to the larvae of the tobacco hornworm (*Manduca sexta*), it is activated by the arginyl-tRNA synthetase of this insect and incorporated into the insect protein in place of arginine with toxic results (Rosenthal and Dahlman, 1975). Incorporation of canavanine into the proteins of *C. brasiliensis* does not occur, however, and it has been demonstrated that the arginyl-tRNA synthetase of this insect can discriminate against canavanine (Rosenthal *et al.,* 1976). This finding suggests that the chemical challenge provided by the plant has elicited a biochemical response in a coevolving insect. The need for such a response can be explained only if canavanine does in fact confer protection against other less specialized predators.

The toxicity of cycasin to mammals has been discussed in the previous section, but evidence that one of its roles may be the protection of cycad plants from insect attack is provided by the work of Teas (1967) who has shown that the larvae of the moth *Seirarctia echo,* which feed on the leaves of *Cycas* and *Zamia* species, are able to sequester cycasin (III). More surprising were two other findings. First, the larval gut contained β-glucosidase which must hydrolyze cycasin and, second, all the methylazoxymethanol in the insect was bound as the one glucose derivative (cycasin) whereas the plants contained a number of different glycosides of methylazoxymethanol. These findings suggest that the insect's response to these particular toxins is to hydrolyze them in the gut and then render the aglycone harmless by glucosylation or, in the case of cycasin, reglucosylation. It is of additional interest and perhaps of ecological significance that the presence of the sequestered cycasin renders the larvae of this insect highly toxic to potential vertebrate predators.

Another example of a predatory insect turning a potential plant toxin to its own use is provided by the cinnabar moth (*Tyria jacobaeae*). The distinctive black and yellow caterpillars of this insect feed on the leaves of ragwort (*Senecio jacobaea*) which, as we have already learned, is a major cause of poisoning among British farm animals.

The cinnabar moth shows even greater biochemical ingenuity than *S. echo*. Not only do its larvae sequester pyrrolizidine alkaloids but these compounds occur in the adult insect also, providing this sluggish, highly colored, day-flying moth with a chemical defense against potential predators of its own. Aplin *et al.* (1968) have shown moreover that the various pyrrolizidine alkaloids of ragwort do not occur in the same proportions in the plant, larvae, and adult moth.

One of the most interesting discoveries concerning plant–insect interrelationships at the biochemical level is the finding that some plants not only synthesize molecules that mimic insect hormones but also synthesize the hormones themselves.

Butenandt and Karlson (1954) isolated 25 mg of one insect molting hormone (α-ecdysone, XII) and 0.33 mg of a second (β-ecdysone, XIII) from a ton of silkworms. Both compounds were subsequently shown to be sterols, which indicated that each must be derived from a sterol precursor present in the food plant, as insects are unable to synthesize this class of compound. Twelve years after these original isolations it was found that some plants accumulated high concentrations of ecdysone-like substances and even β-ecdysone itself (Nakanishi *et al.*, 1966; Takemoto *et al.*, 1967). Jibza *et al.* (1967) isolated 25 mg of β-ecdysone from 2.5 g of dried rhizomes of the fern *Polypodium vulgare*, which is in striking contrast to the 0.33 mg obtained from the ton of silkworms. Although β-ecdysone itself proved relatively nontoxic when supplied orally to insects (possibly because it acts as a feeding inhibitor), other phytoecdysones seriously affected growth and metamorphosis.

Compounds with juvenile hormone activity have also been isolated from plants, and the reader is referred to the fascinating account provided by Williams (1970) of the discovery of and early research on these compounds in his and other laboratories.

Although we have no direct evidence that the accumulation of compounds with insect hormone activity has conferred a selective advantage on any plant species, it is clear that any insect dependent on a plant for a hormone precursor is very vulnerable to changes in the plant's biochemistry. It is quite possible that minor metabolic changes in the plant could lead to synthesis of the hormone itself or to analogues with hormone-like activity rather than to the precursor needed by the insect. Such changes could lead to the synthesis of compounds with either general or specific activity.

The juvenile hormone, juvabione (XIV), which was isolated from the wood of balsam fir (*Abies balsamea*) by Bowers *et al.* (1966), provides an example of a compound that is highly specific, as it affects only species of a single insect family, the Pyrrhocoridae.

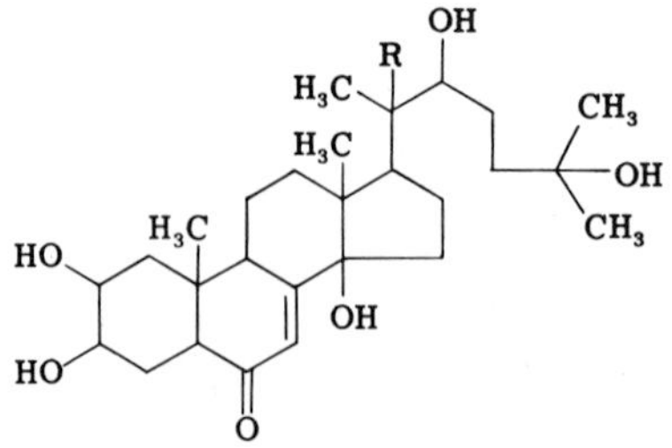

(XII) R = H α-Ecdysone
(XIII) R = OH β-Ecdysone

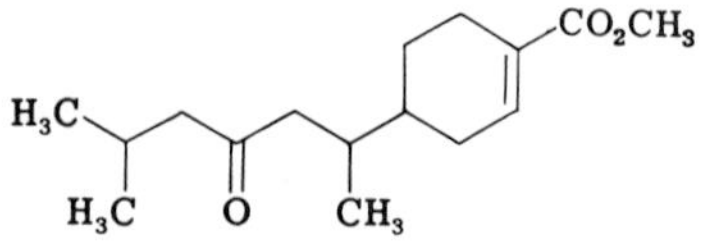

Juvabione
(XIV)

So far I have discussed the possible roles of secondary compounds as insect toxins and feeding inhibitors. If, however, an insect becomes adapted to the presence of a potentially toxic secondary compound, as *C. brasiliensis* has become adapted to canavanine and *S. echo* to cycasin, then it is clearly in the interest of the insect to stay with the host plant that synthesizes the compound to which it is adapted and so avoid competition from other less well-adapted insect species. To adopt this strategy the insect must be able to recognize the species to which it has restricted feeding rights, and the simplest recognition signal may often be the presence of the secondary compound in question. This evolutionary change in the role of a secondary compound is well illustrated in the interrelationship of the cabbage butterfly (*Pieris brassicae*) with the members of the cabbage family (Cruciferae). Cruciferae accumulate glucosinolates which are glycosides of mustard oils. These compounds are not only repellent but also toxic to most insects, yet they stimulate the adult female cabbage butterfly to lay her eggs on cabbages and act as feeding stimulants for the larvae. Other examples of potentially toxic secondary compounds used by specifically adapted insects for the recognition of their host plant are discussed by Harborne (1978).

IV. PLANT–PLANT INTERACTIONS

Some plants grow well together, others do not. The walnut tree (*Juglans nigra*) provides an often cited example of a plant that has an adverse effect on other plant species growing under its leaf canopy. The compound responsible for this allelopathic effect is juglone (XV) an hydroxyquinone formed in the soil by hydrolysis of the 4-glucoside of 1,4,5-trihydroxynaphthalene (XVI). This aromatic glucoside is synthesized by the walnut tree and washed

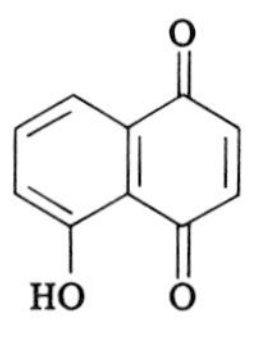

Juglone

(XV)

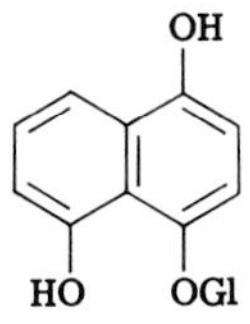

1, 4, 5-Trihydroxy-
naphthalene-4-glucoside

(XVI)

from the leaves onto the underlying soil by rain and dew. Juglone inhibits the growth of many plant species, but not all, and one species that grows well with the walnut is Kentucky bluegrass (*Poa pratensis*) which has clearly evolved a mechanism for circumventing the effects of juglone.

Other water-soluble phytotoxins from plants include salicylic acid (XVII) from the oak (*Quercus falcata*) and a variety of phenolic acids and quinones

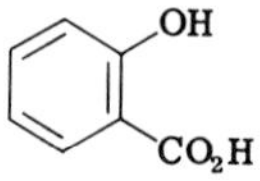

Salicylic acid

(XVII)

from the shrubs *Adenostoma fasciculatum* and *Arctostaphylos glandulosa*. These are species of California chaparral, which have been studied by Muller and Chou (1972). Where these shrubs are dominant they inhibit the growth of herbaceous species. If, however, their aerial parts are burned (as happens at intervals of about 20 yr in the California fire cycle), formerly dormant seeds of herbaceous species germinate and the herbs thrive under the postfire conditions. Gradually, however, the regenerating shrubs reestablish their dominance and the herbs disappear, but not before they have shed seed which will come into its own following the next fire.

Other species studied by Muller produce volatile rather than water-soluble phytotoxins. With the use of gas chromatography it has been possible, for example, to show that *Salvia leucophylla* liberates cineole (XVIII), camphor (XIX), and related compounds through its leaves into the surrounding atmo-

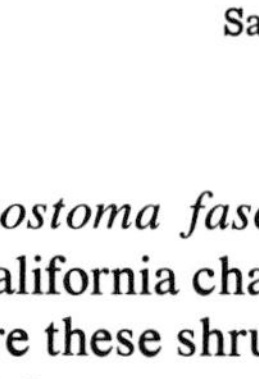

Cineole   Camphor

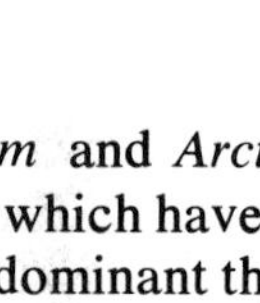

(XVIII)   (XIX)

sphere and that after absorption by the dry soil these compounds inhibit the germination and growth of grassland herbs, which would otherwise take place with the arrival of the winter rains.

In the examples given above at least some of the compounds responsible for the observed allelopathic effects have been identified. As yet, however, little is known of the biochemistry of these processes or of the ways in which the producer species protect themselves against their own phytotoxins.

Evidence as to how plants may be able to discriminate against potentially toxic compounds that they themselves synthesize has, however, been provided by Peterson and Fowden (1965) with respect to the nonprotein imino acid, azetidine-2-carboxylic acid (XX). This imino acid occurs in high concentrations in the green parts of lily of the valley (*Convallaria majalis*), in the

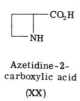

Azetidine-2-
carboxylic acid

(XX)

rhizomes of Solomon's seal (*Polygonatum multiflorum*), in the seeds of species of the legume genus *Bussea,* and in lower concentrations elsewhere in the plant kingdom.

This imino acid proved highly toxic to a number of plant species to which it is foreign, and this toxicity was shown to be related to the incorporation of azetidine-2-carboxylic acid into the plant's protein in place of its higher homologue, proline. Such incorporation does not take place in *C. majalis* or in *P. multiflorum*. The ability of *P. multiflorum* to exclude azetidine-2-carboxylic acid from its proteins depends on the ability of its prolyl-tRNA synthetase to discriminate against the nonprotein imino acid. The corresponding enzyme of *Phaseolus aureus,* a plant that does not synthesize azetidine-2-carboxylic acid, accepts both proline and the lower homologue as substrates.

The liberation of potentially toxic secondary compounds from the aerial parts of plants has already been discussed. There are, however, many examples of such compounds occurring in high concentrations in the seeds and roots of plants, and if any of these are to exercise a role in plant–plant interactions, they must be liberated by the seed or root in sufficient concentrations to influence the growth of neighboring species. This possibility has been examined recently using legume seeds known to be rich in nonprotein amino acids and related compounds. The ninhydrin-reacting compounds liberated by seeds during germination and by seedlings during growth have been monitored and the effects of these compounds on the germination and growth of the producer species and other plants determined. One of the species examined was *Neonotonia* (*Glycine*) *wightii* which accumulates high concentrations of canavanine and 3-carboxytyrosine in its seeds (Wilson and Bell, 1978a). During seed imbibition it was found that the principal constituent eluted from the seed was the aromatic amino acid. Immediately after emergence of the radical, the liberation of amino acids ceased for a period of about 30 hr; after this period canavanine was detected in the seedling exudate and continued to be the major ninhydrin-reacting compound liberated by the developing seedling during the remaining 300 hr of the experiment. Canavanine is a powerful antagonist of arginine in many living systems, and it was not surprising to find that seedlings of lettuce (*Lactuca sativa*) suffered inhibition of growth when exposed to the canavanine-containing eluate of *N. wightii* or to canavanine solutions of the same concentration. The

3-carboxytyrosine did not inhibit lettuce growth under the conditions of the experiment and what role, if any, this compound plays is a matter of conjecture.

The observation that the exudates of some plant species contain secondary products that are not accumulated in their seeds or seedlings is another indication that these secondary compounds may be of ecological significance. As mentioned in Section II, the exudate of sweet pea (*L. odoratus*) seedlings contains β-aminopropionitrile whereas the seeds contain its γ-glutamyl derivative. This nitrile, which failed to inhibit the growth of *L. odoratus* seedlings themselves, was extremely toxic to the seedlings of *L. aphaca,* a closely related species of the same genus, but not to seedlings of a totally unrelated species, lettuce (Wilson and Bell, 1978b). Findings such as these clearly indicate the biochemical variability that exists from one plant species to another, and the need to study species that have evolved together in a common environment if we are to learn more of the ways in which such species have adapted to one another biochemically during the course of evolution.

V. PLANT–MICROORGANISM INTERACTIONS

Some microorganisms, such as the nitrogen-fixing bacteria (*Rhizobia*) of leguminous root nodules, maintain a symbiotic relationship with higher plants. Others, including various viruses, bacteria, and fungi, are pathogenic. Of these pathogens the fungi are the most widespread and cause the greatest economic losses. The Irish famine in 1846 was caused by *Phytophthera infestans* which attacks the potato, and Dutch elm disease which has killed millions of trees in Europe and America in recent years is caused by the fungus *Ceratocystis ulmi* which is carried by an insect vector.

In discussing chemical mechanisms by which higher plants resist the invasion of pathogenic microorganisms Ingham (1973) has used the terms "preinfectional" and "postinfectional" to describe the antimicrobial toxins synthesized by plants.

The preinfectional compounds are antimicrobial toxins that occur in healthy plant tissue and contribute to the plant's ability to resist the invasion of pathogens. The resistance of the lima bean (*Phaseolus lunatus*) to the organism causing powdery mildew (*Phytophthera phaseoli*) is due to the presence of phenolic compounds, and tannins provide a number of other plant species with resistance to *Verticillium* wilt (Bell, 1972). The presence of fungitoxic diterpenes on the leaf surfaces of *Nicotiana glutinosa* (Bailey *et al.,* 1974) and of fungitoxic isopentylisoflavones on the leaf surfaces of *Lupinus albus* (Harborne *et al.,* 1976) suggests that the secretion of such compounds onto leaf surfaces may prove to be a widespread method of defense in plants.

Juglone, the phytotoxin of walnut, is also a powerful fungicide, and it too may be formed from its glycoside precursor on the leaf as well as in the soil beneath the tree.

In addition to the preinfection fungitoxins present in healthy tissues in concentrations sufficiently high to ward off attack, other preinfectional fungitoxins are known that are detectable at low concentrations before infection and in high concentrations after infection. These compounds, whose synthesis appears to be stimulated by infection, have been referred to as "inhibitins."

Postinfectional microbial toxins can be divided into two major categories. First, those that preexist in the plant in bound forms from which they are released after microbial invasion, and second, those that are synthesized by the plant only under stress such as that caused by fungal attack.

Secondary compounds that fall into the first category include cyanogenic glycosides and glucosinolates, which we have already considered in different contexts. Hydrogen cyanide is toxic to a broad range of fungi, and tissue damage by microorganisms no less than tissue damage by vertebrate or insect predators brings about the release of this compound in cyanogenic species. The leaf spot pathogen *Stemphylium loti,* which is able to invade the leaves of birds'-foot trefoil (*L. corniculatus*), a cyanogenic species, circumvents the plant's defense by producing an enzyme (formamide hydrolyase) which converts hydrogen cyanide to formamide (Fry and Millar, 1972). This biochemical response of *S. loti* can be reasonably explained only if cyanogenesis is effective protection against other fungi lacking a means of detoxifying hydrogen cyanide. That glucosinolates may have fungitoxic roles similar to those of cyanogenic glycosides has been shown by Greenhalgh and Mitchell (1976) who demonstrated that the invasion of *Brassica* species by the mildew pathogen *Peronospora parasitica* was inhibited by allylisothiocyanate formed enzymatically from sinigrin following tissue damage in the plant.

The second group of postinfectional inhibitors, those detected only in infected plants, are phytoalexins. Ingham (1973) has defined phytoalexins as "postinfectional metabolites whose formation involves either gene derepression or activation of a latent enzyme system." Cruickshank (1977) calls them "fungal-elicited host-metabolites with an antifungal activity formed in hypersensitive tissue which play a primary role in the inhibition of fungal growth *in vivo*."

The formation of these fungitoxic compounds whose synthesis is "turned on" by the invading fungus can be most simply demonstrated by placing a droplet of water containing nonpathogenic fungal spores on the leaf surface. As the spores germinate, the germ tubes penetrate the leaf cells and the plant responds by synthesizing a phytoalexin found not only within the leaf but also in the leaf droplet, which can be readily tested for fungitoxicity.

Phytoalexins show a variety of chemical structures. The first to be isolated and characterized was pisatin (XXI), a pterocarpan from *Pisum sativum*

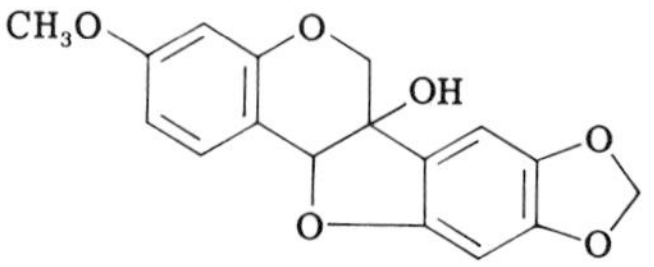

Pisatin

(XXI)

(Cruickshank and Perrin, 1960), and it has since been shown that many of the phytoalexins produced by legume species are isoflavonoids. Species of Solanaceae commonly synthesize terpenoid phytoalexins, and Orchidaceae produce phenanthrenes.

More comprehensive accounts of phytoalexins have recently been provided by Cruickshank (1977) and by Harborne and Ingham (1978).

VI. CONCLUSIONS

Although we classify secondary compounds according to their chemical structure, there is no reason to believe that this classification has very much meaning in ecological terms. It is quite possible that different plant species have developed different chemical solutions to the same biological problem. The accumulation of a cyanogenic glycoside may be one plant's response to predation by a particular herbivore, and the accumulation of a saponin the response of another. Similarly one species may synthesize a brightly colored anthocyanin which attracts pollinators, whereas another synthesizes a betalain pigment which has the same effect.

Some secondary compounds probably have more than one role. Juglone, for example, may protect the walnut tree from fungal pathogens and also inhibit the growth of competing plant species. Quinolizidine alkaloids are found in many species of the less advanced tribes of the Papilionoideae, but they are almost totally absent from the more advanced tribes of this leguminous subfamily (Kinghorn and Smolenski, 1980). In the more advanced tribes, however, it is common to find species that synthesize and accumulate high concentrations of nonprotein amino acids such as canavanine (Bell *et al.,* 1978). If the more toxic alkaloids and the less toxic nonprotein amino acids have a single common defensive role, one would have expected natural selection to favor the more toxic alkaloids, but apparently it has not. A possible explanation may be that nonprotein amino acids serve at least two functions in the plant. First, like the alkaloids, they may deter potential predators, and second, as suggested by their high concentrations in seeds and subsequent metabolism, they may act as storage compounds. We can then postulate that evolutionary advancement has been accompanied by a simplification of secondary biochemistry, more advanced papilionoid species using a single compound for the economical solution of two problems, storage and defense. The lower toxicity of the nonprotein amino acid is apparently offset by its higher concentration.

Secondary compounds, no less than the morphological features of plants, are subject to the selectionary pressures of the environment. It is probable that countless millions of secondary compounds have been synthesized by plants during the course of evolution. Some have survived because they increased the competitive fitness of the plant in which they arose. The others, and these probably constituted the great majority, conferred no such selectionary advantage or were even a liability. The plants containing such compounds (and the compounds themselves) are likely to have disappeared—replaced by better adapted forms.

The idea that all secondary compounds are merely of secondary importance to the plant is tenable only if the plant is considered in total isolation from its environment. Consider a species whose seeds contain 10% canavanine. In isolation the synthesis of this compound is difficult to explain. It may serve as a good source of readily available nitrogen for the germinating seedling, but then arginine could fill this role equally well without the need for specialist enzymes. In the wild, however, 10% canavanine can protect the seeds against most seed beetles, and the expenditure of resources and energy by the plant in providing the enzymes for canavanine metabolism would prove a very sound economic investment indeed if the expenditure were less than the cost to the plant of producing the additional seeds that would be otherwise needed to compensate for losses inflicted by beetles in the absence of canavanine.

Although it is possible that some secondary compounds are of no value to the plants that synthesize them, it has become increasingly clear that others are of prime importance in establishing and maintaining the relationships that exist between one plant and another and between plants and the other living organsims with which they share a common environment.

REFERENCES

Aplin, R. T., Benn, M. H., and Rothschild, M. (1968). *Nature (London)* 219, 747–748.

Bailey, J. A., Vincent, G. G., and Burden, R. S. (1974). *J. Gen. Microbiol.* 85, 57–64.

Bell, A. A. (1972). *In* "Verticillium Wilt of Cotton," pp. 34–37. U.S.D.A. Agric. Res. Serv., Washington, D.C.

Bell, E. A. (1978). *In* "Biochemical Aspects of Plant and Animal Coevolution" (J. B. Harborne, ed.), pp. 143–161. Academic Press, New York.

Bell, E. A., Lackey, J. A., and Polhill, R. M. (1978). *Biochem. Syst. Ecol.* 6, 201–212.

Bowers, W. S., Fales, H. M., Thompson, M. J., and Uebel, E. C. (1966). *Science* **154**, 1020–1021.

Butenandt, A., and Karlson, P. (1954). *Z. Naturforsch., B: Anorg. Chem., Org. Chem., Biochem., Biophys., Biol.* **9B**, 389–391.

Conn, E. E. (1973). *In* "Toxicants Occurring Naturally in Foods" (F. M. Strong, ed.), pp. 299–308. Nat. Acad. Sci., Washington, D.C.

Cooper-Driver, G. A., and Swain, S. (1976). *Nature (London)* **260**, 604.

Cruickshank, I. A. M. (1977). *In* "Natural Products and the Protection of Plants" (G. B. Marini-Bettolo, ed.), pp. 509–569. Scr. Varia 41. Pontifical Academy of Sciences, Vatican City.

Cruickshank, I. A. M., and Perring, D. R. (1960). *Nature (London)* **187**, 799–800.

Dick, A. T., Dann, A. T., Bull, L. B., and Culvenor, C. C. J. (1963). *Nature (London)* **197**, 207–208.

Evans, C. S., Bell, E. A., and Johnston, E. S. (1979). *Phytochemistry* **18**, 2022–2023.

Florkin, M., and Schoffeniels, E., eds. (1969). "Molecular Approaches to Ecology." Academic Press, New York.

Forsyth, A. A. (1968). "British Poisonous Plants," Bull. 161. HM Stationery Office, London.

Fowden, L., and Bryant, M. (1958). *Biochem. J.* **70**, 626–629.

Friend, J., and Threlfall, D. R., eds. (1976). "Biochemical Aspects of Plant-Parasite Relationships." Academic Press, New York.

Fry, W. E., and Millar, R. L. (1972). *Arch. Biochem. Biophys.* **151**, 468–474.

Gilbert, L. E., and Raven, P. H., eds. (1975). "Coevolution of Animals and Plants." Univ. of Texas Press, Austin.

Greenhalgh, J. R., and Mitchell, N. D. (1976). *New Phytol.* **77**, 391–398.

Harborne, J. B., ed. (1972). "Phytochemical Ecology." Academic Press, New York.

Harborne, J. B. (1977a). "Introduction to Ecological Biochemistry." Academic Press, New York.

Harborne, J. B. (1977b). *Pure Appl. Chem.* **49**, 1403–1421.

Harborne, J. B., ed. (1978). "Biochemical Aspects of Plant and Animal Co-evolution." Academic Press, New York.

Harborne, J. B., and Ingham, J. L. (1978). *In* "Biochemical Aspects of Plant and Animal Co-evolution" (J. B. Harborne, ed.), pp. 341–405. Academic Press, New York.

Harborne, J. B., Ingham, J. L., King, L., and Payne, M. (1976). *Phytochemistry* **15**, 1485–1487.

Hegarty, M. P., Lee, C. P., Christie, G. S., Court, R. D., and Haydock, K. P. (1979). *Aust. J. Biol. Sci.* **32**, 27–40.

Ingham, J. L. (1973). *Phytopathol. Z.* **78**, 341–335.

Janzen, D. H. (1977). *Ecology* **58**, 921–927.

Janzen, D. H., Juster, H. B., and Bell, E. A. (1977). *Phytochemistry* **16**, 223–227.

Jibza, J., Herout, V., and Sorm, F. (1967). *Tetrahedrom Lett.* pp. 1869–1891.

Kinghorn, A. D., and Smolenski, S. J. (1981). *In* "Advances in Legume Systematics" (R. M. Polhill and P. H. Raven, eds.). HMStationery Office, London (in press).

Lanigan, G. W., and Smith, L. W. (1970). *Aust. J. Agric. Res.* **21**, 493–500.

Levene, C. I. (1961). *J. Exp. Med.* **114**, 295–310.

Levin, D. A. (1976). *Annu. Rev. Ecol. Syst.* **7**, 121–159.

Luckner, M., Mothes, K., and Nover, L., eds. (1976). "Secondary Metabolism and Co-evolution," *Nova Acta Leopold, Suppl. 7.* Barth, Leipzig.

Marini-Bettolo, G. B., ed. (1977). "Natural Products and the Protection of Plants," Scr. Varia 41. Pontifical Academy of Sciences, Vatican City.

Matsumoto, H., and Higa, H. H. (1966). *Biochem. J.* **98**, 20C–22C.

Mattocks, A. R. (1972). *In* "Phytochemical Ecology" (J. B. Harborne, ed.), pp. 179–200. Academic Press, New York.

Montgomery, R. D. (1969). *In* "Toxic Constituents of Plant Foodstuffs" (I. E. Liener, ed.), pp. 143–157. Academic Press, New York.

Muller, C. H., and Chou, C. H. (1972). *In* "Phytochemical Ecology" (J. B. Harborne, ed.), pp. 201–216. Academic Press, New York.

Nakanishi, K., Koreeda, M., Sasaki, S., Chang, M. L., and Hsu, H. Y. (1966). *Chem. Commun.* pp. 915–917.

O'Dell, B. L., Elsden, D. F., Thomas, J., Partridge, S. M., Smith, R. H., and Palmer, R. (1966). *Nature (London)* **209,** 401–402.

O'Neal, R. M., Chen, C., Reynolds, C. S., Meghal, S. K., and Koeppe, R. E. (1968). *Biochem. J.* **106,** 699–706.

Peterson, P. J., and Fowden, L. (1965). *Biochem. J.* **97,** 112–124.

Ressler, C., Redstone, P. A., and Erenberg, R. H. (1961). *Science* **134,** 188–190.

Ressler, C., Nigam, S. N., Giza, Y.-H., and Nelson, J. (1963). *J. Am. Chem. Soc.* **85,** 3311–3312.

Ressler, C., Nelson, J., and Pfeffer, M. (1964). *Nature (London)* **203,** 1286–1287.

Rosenthal, G. A. (1977). *Q. Rev. Biol.* **52,** 155–178.

Rosenthal, G. A., and Dahlman, D. L. (1975). *Comp. Biochem. Physiol. A* **52A,** 105–108.

Rosenthal, G. A., and Janzen, D. H., eds. (1979). "Herbivores: Their Interaction with Secondary Plant Metabolites." Academic Press, New York.

Rosenthal, G. A., Dahlman, D. L., and Janzen, D. H. (1976). *Science* **192,** 256–258.

Schilling, E. D., and Strong, F. M. (1954). *J. Am. Chem. Soc.* **76,** 2848.

Seigler, D. S. (1977). *Biochem. Syst. Ecol.* **5,** 195–199.

Sondheimer, E., and Simeone, J. B., eds. (1970). "Chemical Ecology." Academic Press, New York.

Swain, T. (1977). *Annu. Rev. Plant Physiol.* **28,** 479–501.

Takemoto, T., Ogawa, S., Nichimoto, N., Arihara, S., and Buc, K. (1967). *Yakugaku Zasshi* **87,** 1414–1418.

Teas, H. J. (1967). *Biochem. Biophys. Res. Commun.* **26,** 686–690.

von Holt, C., Chang, J., von Holt, M., and Böhm, H. (1964). *Biochim. Biophys. Acta* **90,** 611–613.

Wallace, J. W., and Mansell, R. L., eds. (1976). "Biochemical Interaction Between Plants and Insects." Plenum, New York.

Whiting, M. G. (1963). *Econ. Bot.* **17,** 271–302.

Whittaker, R. H., and Feeny, P. P. (1971). *Science* **171,** 757–770.

Williams, C. M. (1970). *In* "Chemical Ecology" (E. Sondheimer and J. B. Simeone, eds.), pp. 103–132. Academic Press, New York.

Wilson, M. F., and Bell, E. A. (1978a). *J. Exp. Bot.* **29,** 1243–1247.

Wilson, M. F., and Bell, E. A. (1978b). *Phytochemistry* **197,** 403–406.

Tissue Culture and the Study of Secondary (Natural) Products

2

DONALD K. DOUGALL

I. INTRODUCTION

Significant advances in our understanding of the synthesis of natural products have been achieved through the use of plant cell cultures. For this reason, a discussion of the role of plant cell cultures in the study of natural or secondary products is included in this volume. Plant cell culture is a tool that is an adjunct to studies on whole plants. Like studies on whole plants, it has advantages and disadvantages. The advantages of plant cell culture lie in the uniformity, reproducibility, and control possible in suspension culture systems, the ready availability of large quantities of cells, and the decreased level of structural organization relative to that of the plant. The disadvantages lie in the continued effort required to maintain cell culture systems by serial passage and the possibility that cultures may change during serial passage over long periods of time. The effort required to maintain cultures by

The Biochemistry of Plants, Vol. 7

serial passage is proportional to the number of cell lines being maintained. The maintenance of cell lines requires preparation of media and subculturing at 1 to 4 week intervals. There is also a requirement for incubators, sterile working areas, and other equipment for the cultures. The alternative to serial passage for maintenance of cultures is some form of storage. This area of plant cell culture is showing signs of development and is reviewed by Dougall (1980), Withers (1978), and Withers and Street (1977).

Reports of cell cultures giving yields of secondary products approaching or exceeding those found in the plants of the producing species began to increase in about 1970. The total number now is in excess of 20 (Zenk, 1978; Dougall, 1979a,b). Prior to 1970 the reported yields from cell cultures were lower than those found in the plant by several orders of magnitude. This early work and its implications for the production of secondary products by cell cultures have been reviewed by Puhan and Martin (1971), Constabel *et al.* (1974), and Butcher (1977). The currently reported yields provide a basis for optimism concerning the possibility of commercially producing secondary products in cell cultures (Zenk, 1978; Dougall, 1979a,b). This is reinforced by observations that in some cases the yields in cell culture are several- to manyfold greater than those observed in the whole plant. The limit of these increases in yield relative to the yield from whole plants is unknown at present.

The 20 or so cultures producing secondary products in high yield do so using sugars as carbon and energy sources together with other inorganic nutrients, vitamins, and growth substances such as auxins and cytokinins. The cells in these cultures clearly contain and use the information for the accumulation of secondary products. In addition to the behavior represented by these 20 or so cultures several other classes of behavior, in terms of accumulation of secondary products, are displayed by plant cell cultures. Cultures that do not produce secondary products in simple media may be treated as two further classes. In these classes are cultures that contain the information but do not use it; they accumulate few or none of the secondary compounds characteristic of the plant. However, on the regeneration of plants from these cultures, the accumulation of characteristic compounds by the plant is found. These examples have been reviewed by Dougall (1979a,b) and presumably indicate that the complete information for accumulation is present in the cells but is not used under the conditions of cell culture investigated. In these cases there is no evidence to distinguish between the two additional possibilities that either none of the information is available for use or that only part of it is accessible. The second class of behavior thus is one in which some, but not all, of the steps of the biosynthetic pathway can be carried out in cell cultures. With these cultures, feeding of intermediates yields the desired secondary products or compounds further along the pathway of biosynthesis of the desired secondary product. Observations that appear to be of this class of behavior are the biotransformation of cardiac glycosides by cell cultures of *Digitalis lanata* (Alfermann *et al.*, 1977; Heins *et al.*, 1978), the conversion of pulegone into isomenthone by cell cultures of specific chemotypes from *Mentha* (Aviv and Galun, 1978), the conversion of geraniol but not geraniol pyrophosphate into geranial, neral, nerol, and citronellol by cell cultures from *Rosa* species (Macrae and Howe, cited in Jones, 1974), the conversion of *S*-(2-carboxypropyl)-L-cysteine into the sulfoxide in onion callus tissue (Selby *et al.*, 1980), and the conversion of (−)-codeinone into (−)-codeine but not into (−)-morphine by cell cultures of *Papaver somniferum* (Furuya *et al.*, 1978). The third possible class is cell cultures that cannot use the information for secondary product accumulation or have lost it. These classes are simply convenient ways in which to group the information available on the ability of plant cell cultures to accumulate secondary products.

Cultures that perform the latter steps of biosynthesis can be used to explore the pathway of biosynthesis of specific secondary products. This can be done by feeding compounds suspected to be intermediates and determining whether or not they are converted into the desired end product. In such experiments one can anticipate finding additional products, for example, hydroxylation at C-5 and C-7 in addition to the desired hydroxylation at C-12 of digitoxigenin described by Furuya (1978). In addition there have been experiments in which analogues of intermediates have been fed to cultures to yield analogues of the normal products (Steck and Constabel, 1974). In addition, cell cultures that produce secondary products when grown in simple media offer further possibilities for studies on biosynthesis and its regulation. These opportunities include identification of intermediates, studies on enzymes catalyzing steps in biosynthesis, studies on the synthesis and degradation of the mRNA coding for the enzymes, and regulation of these events. An example of the exploitation of some of these possibilities can be found in the work of Hahlbrock (1977, and this volume, Chapter 14), Hahlbrock *et al.* (1978), and Schröder *et al.* (1979), where various aspects of the regulation of the synthesis of flavonoids including measurements of the rate of synthesis of mRNA for phenylalanine ammonia lyase, flavanone synthetase, and UDP-apiose synthase have been investigated using parsley and soybean cell cultures. In the case of many other cell culture systems though, there is a need for much more information about the basic behavior of the systems before they can be explored at a similar level. As yet untouched are questions involving the reasons and mechanisms that allow either none or parts of biosynthetic pathways to operate in cell cultures although they do operate in whole plants.

Earlier in this chapter, the 20 or so cultures giving high yields of secondary products when grown on simple media were emphasized. The question, "Are cultures giving high yields of secondary products mandatory?" needs to be addressed. The answer is clearly "Yes," if a commercial objective exists. The answer is "Maybe" in a number of other situations. If the isolation of an enzyme is to be attempted, one will probably have less difficulty if the enzyme represents a high proportion of the proteins present in the cell. In this sense, high yields of secondary products in cell culture, which can be expected to depend on the presence of high enzyme capacities and thus high amounts of enzyme, may be an advantage. In studies on intermediates in biosynthesis, the use of sensitive methods (e.g., the use of radiolabeled compounds) can overcome the problems of yield or quantity of product. However, such investigations would be facilitated by using high-yielding cultures.

The achievement of high-yielding cultures has been discussed by Dougall (1979a,b) and Zenk (1978). There are three general requirements for achieving high yields. The first of these is to select from among a series of cultures a culture that grows well and produces the compound of interest. The series of cultures from which this selection is made must be initiated from a wide range of individuals within a producing species. The second is to determine the medium composition and culture conditions giving maximum yield. The third is to examine the heterogeneity within the culture by cloning and measuring the yield given by the various sublines or clones. A high-yielding clone or subline can then be selected. In retrospect, these requirements are not unexpected. They reflect the influence of plant genetics on yields of secondary products, the influence of environmental conditions on yield, and the widely recognized variability of plant cell cultures. However, they have been systematically applied in only one case (Zenk *et al.*, 1977).

II. SYNTHESIS OF SPECIFIC SECONDARY PRODUCTS

Plant cell cultures have been used in studies on biosynthesis and biotransformation of a number of classes of secondary products. The successes of these studies will be illustrated in the following sections which deal with studies on the synthesis of specific types of chemical compounds. In these sections recent papers will be cited and supplemented with reviews where available.

A. Mono- and Sesquiterpenes

Aviv and Galun (1978) showed that four of the suspension cultures established from six different chemotypes of *Mentha* species converted pulegone to isomenthone. They discussed the correlation between the capacities of the cell cultures to carry out this biotransformation and the chemotypes of the plants from which the cultures were initiated. Suspension cultures of rose oxidized geraniol to geranial, neral, nerol, and citronellal but could not convert geraniol pyrophosphate to these compounds (MacRae and How, cited by Jones, 1974). The oxidation of primary and secondary allylic alcohols (e.g., geraniol, nerol, verbenol) to the corresponding aldehydes and ketones by callus cultures of *Cannabis sativa* has been described (Itokawa *et al.,* 1977). The 2-*cis*, 6-*trans*-farnesyl diphosphate and *Z*-γ-bisabolene were shown to be intermediates in the synthesis of the sesquiterpene paniculide B by cell cultures of *Andrographis paniculata* (Overton and Picken, 1976; Mackie and Overton, 1977). Mackie and Overton (1977) have concluded that the cells of *A. paniculata* contain an enzyme that isomerizes *trans,trans*-farnesyl diphosphate to the cis–trans form. Overton (1977) has reviewed these studies and others dealing with mevalonate and its metabolism in cell cultures.

B. Steroids, Triterpenoids, and Cardiac Glycosides

Much of the work on the synthesis of these compounds has been reviewed by Overton (1977) and Stohs and Rosenberg (1975). In addition, Seo *et al.* (1978) have used [4-^{13}C]mevalonate and [1,2-^{13}C]acetate to study the labeling pattern of β-sitosterol in cell cultures of *Isodon japonicus*. The conversion of cholesterol into diosgenin by cell cultures of *Dioscorea deltoidea* has been examined by Stohs *et al.* (1969) and in *D. tokoro* by Tomita and Uomori (1971). The latter authors identified several intermediates and demonstrated the conversion of diosgenin into yonogenin and tokorogenin. Biotransformations of steroids and cardiac glycosides by plant cell cultures have been studied extensively and have been reviewed by Reinhard (1974), Alfermann *et al.* (1977), Furuya (1978), Stohs (1977), and Stohs and Rosenberg (1975). Additional papers on the reduction of cholesterol to 5α-cholestan-3β-ol by rape and soybean cell cultures (Weber, 1978) and on the hydroxylation of progesterone at C-6, C-11, and C-14 by cell cultures of *Vinca rosea* (Gallili *et al.,* 1978; Yagen *et al.,* 1978), of digitoxigenin at C-5 by cell cultures of *Daucus carota* (Jones *et al.,* 1978), and of β-methyldigitoxin at C-12 by cell cultures of *Digitalis lanata* (Heins *et al.,* 1978) have appeared.

C. Anthraquinones, Naphthoquinones, and Related Compounds

The anthraquinone production by cell cultures of *Morinda citrifolia* was stimulated twofold by the addition of *o*-succinylbenzoic acid to the culture medium (Zenk *et al.,* 1975), and ^{14}C-labeled *o*-succinylbenzoic acid labeled the anthraquinones (Leistner, 1975). *o*-Succinylbenzoic acid stimulated the formation of anthraquinones in cell cultures of *Gallium mollugo* (Bauch and Leistner, 1978a). The feeding of labeled mevalonate or labeled *o*-succinylbenzoic acid to *G. mollugo* suspension cultures yielded labeled anthraquinones (Bauch and Leistner, 1978b). Inoue *et al.* (1979) have shown that during the synthesis of anthraquinones by a cell culture of *G. mollugo,* the prenylation of *o*-succinylbenzoic acid occurs adjacent to the ketone in the succinyl portion of the acid.

Bauch and Leistner (1978b) found that labeled *o*-succinylbenzoic acid did not label the naphthols produced in cultures of *G. mollugo* and suggested that the lack of labeling may be due to the extremely high rate of turnover of the naphthols. Inouye *et al.* (1979) have shown that *p*-hydroxybenzoic acid, *m*-geranyl-*p*-hydroxybenzoic acid, and geranylhydroquinone are intermediates in naphthoquinone biosynthesis by callus cultures of *Lithospermum erythrorhizon*. In addition, they showed that a culture that did not produce naphthoquinones and others grown under nonproducing conditions did not convert *m*-geranyl-*p*-hydroxybenzoic to geranylhydroquinone. Thus they demonstrated the location of the interruption in the biosynthetic pathway in a culture that did not produce the naphthoquinones.

D. Coumarins, Furanocoumarins, and Furanochromes

The pathway of biosynthesis of the coumarin scopoletin and its glucoside, scopolin, from hydroxylated cinnamic acids by tobacco tissue cultures was elucidated by Fritig *et al.* (1970). Brown and Tenniswood (1974) examined hydroxylated cinnamic acids, coumarins, and furanocourmarins in normal tobacco and in tobacco tumor cell cultures. As a result, they suggested that in the tumor cell cultures the metabolism had been diverted from furanocoumarin biosynthesis to coumarin biosynthesis. Steck and Constabel (1974) showed that cell cultures of *Ruta graveolens* metabolized analogues of the furanocoumarin precursor umbelliferone to give the corresponding furanocoumarins. The enzyme transferring a dimethylallyl group from dimethylallyl pyrophosphate to the 6-position of umbelliferone has been purified and characterized from *R. graveolens* suspension cultures by Dhillon and Brown (1976). Brown and Sampathkumar (1977) concluded that xanthotoxin rather than bergapten was the major intermediate in the biosynthesis of isopimpinellin by cultures of *R. graveolens*. The purification of furanocoumarin *O*-methyltransferases using affinity chromatography on *S*-adenosylhomocysteine coupled to AH-Sepharose 4B has been described (Sharma and Brown, 1978,1979). Thompson *et al.* (1978) concluded the cell cultures of *R. graveolens* probably contained two *O*-methyltransferases involved in furanocoumarin biosynthesis. One of these methylated 5-hydroxyl groups and the other 8-hydroxyl groups of furanocoumarins. These two furanocoumarin *O*-methyltransferases were separated by affinity chromatography on AH-Sepharose 4B with 5-(3-carboxypropanamido) xanthotoxin as the bound ligand. Binding to the column took place in the presence of *S*-adenosyl-L-homocysteine or *S*-adenosylmethionine, and the two enzymes were separated by elution with xanthotoxol or bergaptol (Sharma *et al.*, 1979).

E. Alkaloids

The use of cell-free systems of *Catharanthus roseus* to achieve the biosynthesis of ajmalicine and other indole alkaloids from tryptamine and secologanin was described by Scott and Lee (1975) and confirmed by Stöckigt *et al.* (1976). The condensation of tryptamine and secologanin to yield strictosidine (Stöckigt and Zenk, 1977; Stöckigt, 1979; Treimer and Zenk, 1979a,b), which is converted into 4,21-dehydrocorynantheine aldehyde (Stöckigt *et al.*, 1978) and then cathenamine (Stöckigt *et al.*, 1977; Stöckigt, 1979) in the biosynthesis of ajmalicine and other indole alkaloids, has been documented using cell cultures of *C. roseus*. Scott *et al.* (1977) have also provided clear evidence for the role of strictosidine and the noninvolvement of vincoside in the biosynthesis of indole alkaloids. Lee *et al.* (1979) have concluded that geissoschizine is an intermediate in ajmalicine biosynthesis using cell-free extracts from *C. roseus* callus cultures. This compound appears to be on a route alternative to that via cathenamine for the synthesis of ajmalicine. No 19-epiajmalicine was synthesized by the cell-free system used in this investigation, in contrast to their previous report (Scott *et al.*, 1977) and those of Stöckigt and Zenk (1977), Stöckigt *et al.* (1976), and Treimer and Zenk (1978). This difference is intriguing in its own right. In addition, Lee *et al.* (1979) appear to be concerned that, although cathenamine has the same stereochemistry at C-19 as 19-epiajmalicine, its stereochemistry at C-19 is opposite that of ajmalicine and tetrahydroalstonine, thus leading to questions involving the role of cathenamine in the synthesis of the latter two compounds.

Treimer and Zenk (1979a,b) have surveyed the distribution of strictosidine synthase and shown it to be present in cell cultures of 15 members of the family Apocynaceae examined and not present in cell cultures of *Nicotiana tabacum* or *Trifolium pratense*. They purified the enzyme from extracts of *C. roseus* suspension cultures 50-fold and measured its pH optimum (6.8) and apparent K_m values for tryptamine (2.3 mM) and secologanin (3.4 mM). No evidence for isozymes was found by DEAE chromatography or disc electrophoresis, and the enzyme showed a high degree of specificity for its substrates. Mizukami *et al.* (1979) have purified strictosidine synthetase 740-fold from suspension cultures of *C. roseus* and described some of its properties including the lack of inhibition by ajmalicine, vindoline, and catharanthine.

In view of the role of tryptophan in indole alkaloid biosynthesis, Scott *et al.* (1979) have examined the production of indole alkaloids by a cell culture of *C. roseus* that is resistant to 5-methyltryptophan. This line overproduces tryptophan relative to the sensitive culture, accumulates tryptamine to approximately the same level as the sensitive culture, but does not accumulate ajmalicine which, however, is detectable in the sensitive line. The anthranillate synthetase from their resistant line was less sensitive to inhibition by tryptophan in cell-free extracts than the enzyme from the sensitive line. Schallenberg and Berlin (1979) also examined 5-methyltryptophan-resistant lines of *C. roseus* including 4 parental and 10 resistant lines. The resistant lines showed marked elevation of tissue tryptophan levels which did not correlate with the levels of serpentine accumulated by the cultures. The anthranillate synthetases from sensitive and resistant cell lines were all equally sensitive to tryptophan in cell-free extracts. In resistant lines, the capacity of cell-free extracts to decarboxylate tryptophan is not increased over that found in the sensitive lines. These observations may indicate that the availability of tryptamine limits the capacity to synthesize alkaloids. The possibility that the availability of secologanin, the second precursor of the alkaloids, limits synthesis or that there are other sites of limitation of synthesis must also be considered.

The pathway of biosynthesis of furoquinoline alkaloids and edulinine from the precursor 4-hydroxy-2-quinolone was investigated in detail in *R. graveolens* suspension cultures by Boulanger *et al.* (1973). Diversion of the precursor to edulinine apparently involves N-methylation as an initial step, whereas the formation of furanoquinoline alkaloids apparently involves the addition of mevalonate to the 3-position of the precursor. Steck *et al.* (1973) showed that increased yields of furoquinoline alkaloids resulted from feeding the precursor 4-hydroxy-2-quinoline to cell cultures of *R. graveolens*.

The biosynthesis of β-carboline alkaloids such as harman and norharman from tryptophan by cell cultures of *Peganum harmala* (Copeland and Slaytor, 1974; Nettleship and Slaytor, 1971, 1974) and *Phaseolus vulgaris* (Veliky, 1972; Veliky and Barber, 1975) has been described. It is to be noted that in *P. vulgaris* the compounds harman and norharman are not known as normal constituents. Pearson and Turner (1979) have shown that, when serine hydroxymethyltransferase from ox liver or methylene-tetrahydrofolate reductase from pig liver generates methylene tetrahydrofolate in the presence of tryptamine, 1,2,3,4-tetrahydro-β-carboline is formed. The implications of this observation, which is thought to involve a nonenzymatic reaction between tryptamine and the formaldehyde in equilibrium with methylene tetrahydrofolate, have not been investigated. On the basis of studies with ^{14}C-labeled compounds, Ogutuga and Northcote (1970) concluded that in tea tissue cultures caffeine was formed from purines released by degradation of nucleic acids rather than from a purine pool. Romeike (1975) showed that cell cultures of *Datura inoxia* esterified exogenously supplied tropine with exogenously supplied tropic acid to give hyoscyamine, whereas *D. metel* and *D. stramonium* esterified tropine with acetic acid from their metabolism. *Hyoscyamus niger, Scopolia lurida,* and *Atropa belladonna* cell cultures were not able to esterify the supplied tropine. Furuya *et al.* (1978) showed that cell cultures of *Papaver somniferum* converted (+)-(S)-reticuline to (−)-(S)-scoulerine and (−)-(S)-cheilanthifoline but did not convert (−)-(R)-reticuline to morphine-type alkaloids. However, (−)-codeinone was converted to (−)-codeine but not further to (−)-morphine.

F. Flavones and Flavanols

The biosynthesis of flavones and flavanols has been extensively studied, mainly by Hahlbrock and co-workers using cell cultures of parsley and soybean. These studies have been reviewed by Hahlbrock (1977, this volume, Chapter 14), Hahlbrock and Grisebach (1975), and Schütte (1978). The enzymatically catalyzed steps in the synthesis of flavonoids fall into two groups on the basis of their behavior on induction (Hahlbrock, 1977). The first group, consisting of the steps catalyzed by phenylalanine ammonia-lyase, cinnamic acid 4-hydroxylase and *p*-coumaroyl-CoA ligase, in addition to being involved in flavonoid biosynthesis, are also the first steps of lignin biosynthesis. In addition to the studies on these enzymes discussed in the reviews cited above, there are some more recent studies. Ibrahim and Phan (1978) measured the levels of these three enzymes in flax cell cultures. Pfändler *et al.* (1977) have shown that cinnamic acid 4-hydroxylase from parsley cell cultures is specific for *trans*-cinnamic acid. Postius and Kindl (1978) have concluded that phenylalanine ammonia-lyase and cinnamic acid 4-hydroxylase occur in the endoplasmic reticulum of cell cultures of soybean. The purification and properties of a single 4-coumarate:CoA ligase from illuminated parsley suspension cultures (Knoblock and Hahlbrock, 1977) have been described. The coordinated induction of these enzymes, which precedes induction of the remaining enzymes of flavonoid biosynthesis after illumination of parsley cell cultures, has been described by Hahlbrock *et al.* (1976) and Ebel and Hahlbrock (1977).

In the second group of enzymes of flavonoid biosynthesis, the first enzyme, which condenses *p*-coumaryl-CoA with 3 mol of malonyl-CoA to give the flavanone naringenin, has been purified and characterized from parsley cell cultures (Kreuzaler and Hahlbrock, 1975; Kreuzaler *et al.*, 1979), its substrate specificity and products have been examined (Hrazdina *et al.,* 1976), and it has been shown to catalyze CO_2 exchange and decarboxylation of malonyl-CoA (Kreuzaler *et al.,* 1978). The flavanone synthetase from cell cultures of *Haplopappus gracilis* has been purified and its properties shown to be similar to those of the same enzyme from cell cultures of parsley (Saleh *et al.,* 1978). Two enzymes that transfer methyl groups from S-adenosylmethionine to the 3-hydroxyl of 3,4-dihydric phenols were isolated from soybean suspension cultures (Poulton *et al.,* 1976a). On the basis of substrate specificity, one of these is thought to be involved in flavonoid biosynthesis. A UDP-glucose:flavanol 3-O-glycosyl transferase from cell cultures of soybean has been partly purified and characterized by Poulton and Kauer (1977). Stepwise methylation of quercitin by cell-free extracts of cultures from *Citrus mitis* has been shown (Brunet *et al.*, 1978). Gustine *et al.* (1978) have described the increase in the ability of cell-free extracts of jack-bean cell cultures to methylate isoflavonoids but not caffeic acid or flavones after stimulation of the tissue to produce the phytoalexin medicarpin.

G. Lignin

The synthesis of lignin appears to have intermediates and enzymes in common with flavonoid biosynthesis up to p-coumaric acid (Schütte, 1978). Studies on other enzymes of lignin biosynthesis have been reported. A caffeic acid O-methyltransferase has been reported in cell-free extracts of tobacco cultures (Yamada and Kuboi, 1976; Kuboi and Yamada, 1976, 1978; Sugano *et al.*, 1978), *P. vulgaris* (Haddon and Northcote, 1976), *Ruta graveolens* (Sharma and Brown, 1979), and soybean (Poulton *et al.*, 1976a). That from soybean has been purified and characterized (Poulton *et al.*, 1976b). A partial separation of two O-methyltransferases from tobacco has been achieved by Tsang and Ibrahim (1979a,b) on DEAE-cellulose. These two enzymes seem to have different substrate specificities and preferences for the meta and paraphenolic hydroxyl groups. Cinnamoyl-CoA:NAPH reductase from soybean cell cultures has been purified and characterized by Wengenmayer *et al.* (1976). UDP-glucose:coniferyl alcohol glycosyl transferase from cell cultures of rose has been purified and described by Ibrahim and Grisebach (1976). Hösel *et al.* (1978) examined β-glucosidases from the cell wall of chick-pea cells grown in culture and concluded from substrate specificity studies that these may be involved in lignin formation.

H. Anthocyanins

The synthesis of anthocyanins seems to be an alternative pathway to flavone and flavonol synthesis. These pathways both begin with naringenin, the product of flavanone synthase. Fritsch and Grisebach (1975) have described the hydroxylation of naringenin to dihydrokaempferol and eriodictyol, the hydroxylation of dihydrokaempferol to dihydroquercitin, and the conversion of dihydroquercitin to cyanidin by cell cultures of *Haplopappus gracilis* and microsomal preparations obtained from these cells. The hydroxylations have a requirement for oxygen and NADPH, suggesting that they are performed by mixed-function oxidases. The synthesis of phenylalanine ammonia-lyase and anthocyanin in *D. carota* cells is decreased when the cells are grown in the presence of gibberellic acid (Heinzmann and Seitz,

1977). In the same system, Heinzmann *et al.* (1977) showed that three isozymes of *p*-coumaric acid:CoA ligase were present in anthocyanin-producing cell cultures of *D. carota* but that only two were present in the same cultures treated with gibberellic acid, which did not produce anthocyanins. A UDP-glucose:cyanidin 3-*O*-glycosyltransferase from cell cultures of *H. gracilis* has been purified and its properties described by Saleh *et al.* (1976).

III. CONCLUSIONS

The investigations summarized above show that plant cell cultures have been successfully used for studies on biosynthetic pathways, for the identification of intermediates, as sources of enzymes, and in studies on the regulation of synthesis of secondary (natural) products. New questions have been raised by the studies performed, so that the use of plant cell cultures to further understand the metabolism of secondary products seems to be ensured. Such studies may also contribute to the commercial production of secondary products, a possibility that appears to have an increasing probability of being achieved (Zenk, 1978; Dougall, 1979b).

ACKNOWLEDGMENTS

The author wishes to thank Drs. Siu-Leung Lee, H. A. Collin, and Jochen Berlin for making copies of their manuscripts available prior to publication.

REFERENCES

Alfermann, A. W., Boy, H. M., Döller, P. C., Hagedorn, W., Heins, M., Wahl, J., and Reinhard, E. (1977). *In* "Plant Tissue Culture and Its Bio-technological Application" (W. Barz, E. Reinhard, and M. H. Zenk, eds.), pp. 125–141. Springer-Verlag, Berlin and New York.
Aviv, D., and Galun, E. (1978). *Planta Med.* **33**, 70–77.
Bauch, H.-J., and Leistner, E. (1978a). *Planta Med.* **33**, 105–123.
Bauch, H.-J., and Leistner, E. (1978b). *Planta Med.* **33**, 124–127.
Boulanger, D., Bailey, B. K., and Steck, W. (1973). *Phytochemistry* **12**, 2399–2405.
Brown, S. A., and Sampathkumar, S. (1977). *Can. J. Bot.* **55**, 686–692.
Brown, S. A., and Tenniswood, M. (1974). *Can. J. Bot.* **52**, 1091–1094.
Brunet, G., Saleh, N. A. M., and Ibrahim, R. K. (1978). *Z. Naturforsch., C: Biosci.* **33C**, 786–788.
Butcher, D. N. (1977). *In* "Plant Cell, Tissue, and Organ Culture" (J. Reinert and Y. P. S. Bajaj, eds.), pp. 668–693. Springer-Verlag, Berlin and New York.
Constabel, F., Gamborg, O. L., Kurz, W. G. W., and Steck, W. (1974). *Planta Med.* **25**, 158–165.
Copeland, L., and Slaytor, M. (1974). *Physiol. Plant.* **31**, 327–329.

Dhillon, D. S., and Brown, S. A. (1976). *Arch. Biochem. Biophys.* **177**, 74–83.

Dougall, D. K. (1979a). *In* "Proceedings of the Fourth Annual College of Biological Sciences Colloquium: Plant Cell and Tissue Culture—Principles and Applications." pp. 727–743. Ohio State Univ. Press.

Dougall, D. K. (1979b). *In* "Cell Substrates" (J. C. Petricciani, H. Hopps, and P. J. Chapple, eds.), pp. 135–152. Plenum, New York.

Dougall, D. K. (1980). *In* "Plant Tissue Culture as a Source of Biochemicals" (E. J. Staba, ed.). pp. 115–122. CRC Press, Boca Raton, Florida.

Ebel, J., and Hahlbrock, K. (1977). *Eur. J. Biochem.* **75**, 201–209.

Fritig, B., Hirth, L., and Ourisson, G. (1970). *Phytochemistry* **9**, 1963–1975.

Fritsch, H., and Grisebach, H. (1975). *Phytochemistry* **14**, 2437–2442.

Furuya, T. (1978). *In* "Frontiers of Plant Tissue Culture 1978" (T. A. Thorpe, ed.), pp. 191–200. The Bookstore, University of Calgary, Calgary, Alberta, Canada.

Furuya, T., Nakano, M., and Yoshikawa, T. (1978). *Phytochemistry* **17**, 891–893.

Gallili, G. E., Yagen, B., and Mateles, R. I. (1978). *Phytochemistry* **17**, 578.

Gustine, D. L., Sherwood, R. T., and Vance, C. P. (1978). *Plant Physiol.* **61**, 226–230.

Haddon, L., and Northcote, D. H. (1976). *Planta* **128**, 255–262.

Hahlbrock, K. (1977). *In* "Plant Tissue Culture and Its Bio-technological Application" (W. Barz, E. Reinhard, and M. H. Zenk, eds.), pp. 95–111. Springer-Verlag, Berlin and New York.

Hahlbrock, K., and Grisebach, H. (1975). *In* "The Flavonoids" (J. B. Harborne, T. J. Mabry, and H. Mabry, eds.), Part 2, pp. 866–915. Academic Press, New York.

Hahlbrock, K., Knobloch, K.-H., Kreuzaler, F., Potts, J. R. M., and Wellmann, E. (1976). *Eur. J. Biochem.* **61**, 199–206.

Hahlbrock, K., Betz, B., Gardiner, S. E. Kreuzaler, F., Matern, U., Ragg, H., Schäfer, E., and Schröder, J. (1978). *In* "Frontiers of Plant Tissue Culture 1978" (T. A. Thorpe, ed.), pp. 317–324. The Bookstore, University of Calgary, Calgary, Alberta, Canada.

Heins, M., Wahl, J., Lerch, H., Kaiser, F., and Reinhard, E. (1978). *Planta Med.* **33**, 57–62.

Heinzmann, U., and Seitz, U. (1977). *Planta* **135**, 63–67.

Heinzmann, U., Seitz, U., and Seitz, U. (1977). *Planta* **135**, 313–318.

Hösel, W., Surholt, E., and Borgmann, E. (1978). *Eur. J. Biochem.* **84**, 487–492.

Hrazdina, G., Kreuzaler, F., Hahlbrock, K., and Grisebach, H. (1976). *Arch. Biochem. Biophys.* **175**, 392–399.

Ibrahim, R. K., and Grisebach, H. (1976). *Arch. Biochem. Biophys.* **176**, 700–708.

Ibrahim, R. K., and Phan, C. T. (1978). *Biochem. Physiol. Pflan.* **172**, 199–212.

Inoue, K., Shiobara, Y., Nayeshiro, H., Inouye, H., Wilson, G., and Zenk, M. H. (1979). *J. Chem. Soc., Chem. Commun.* pp. 957–959.

Inouye, H., Ueda, S., Inoue, K., and Matsumara, H. (1979). *Photochemistry* **18**, 1301–1308.

Itokawa, H., Takeya, K., and Mihashi, S. (1977). *Chem. Pharm. Bull.* **25**, 1941–1946.

Jones, A., Veliky, I. A., and Ozubko, R. S. (1978). *Llyodia* **41**, 476–487.

Jones, L. H. (1974). *Proc. FEBS Meet.* **30**, pt. 2, 813–833.

Knobloch, K. H., and Hahlbrock, K. (1977). *Arch. Biochem. Biophys.* **184**, 237–248.

Kreuzaler, F., and Hahlbrock, K. (1975). *Eur. J. Biochem.* **56**, 205–213.

Kreuzaler, F., Light, R. J., and Hahlbrock, K. (1978). *FEBS Lett.* **94**, 175–178.

Kreuzaler, F., Ragg, H., Heller, W., Tesch, R., Witt, I., Hammer, D., and Hahlbrock, K. (1979). *Eur. J. Biochem.* **99**, 89–96.

Kuboi, T., and Yamada, Y. (1976). *Phytochemistry* **15**, 397–400.

Kuboi, T., and Yamada, Y. (1978). *Biochim. Biophys. Acta* **542**, 181–190.

Lee, S. L., Hirata, T., and Scott, A. I. (1979). *Tetrahedron Lett.* pp. 691–694.

Leistner, E. (1975). *Planta Med., Suppl.* pp. 214–224.

Mackie, H., and Overton, K. H. (1977). *Eur. J. Biochem.* **77,** 101–106.
Mizukami, H., Nordlöv, H., Lee, S. L., and Scott, A. I. (1979). *Biochemistry* **18,** 3760–3763.
Nettleship, L., and Slaytor, M. (1971). *Phytochemistry* **10,** 231–234.
Nettleship, L., and Slaytor, M. (1974). *Phytochemistry* **13,** 735–742.
Ogutuga, D. B. A., and Northcote, D. H. (1970). *Biochem. J.* **117,** 715–720.
Overton, K. H. (1977). *In* "Plant Tissue Culture and Its Bio-technological Application" (W. Barz, E. Reinhard, and M. H. Zenk, eds.), pp. 66–75. Springer-Verlag, Berlin and New York.
Overton, K. H., and Picken, D. J. (1976). *J. Chem. Soc., Chem. Commun.* pp. 105–106.
Pearson, A. G. M., and Turner, A. J. (1979). *FEBS Lett.* **98,** 96–98.
Pfändler, R., Scheel, D., Sandermann, H., Jr., and Grisebach, H. (1977). *Arch. Biochem. Biophys.* **178,** 315–316.
Postius, C., and Kindl, H. (1978). *Z. Naturforsch., C: Biosci.* **33C,** 65–69.
Poulton, J., Grisebach, H., Ebel, J., Schaller-Hekeler, B., and Hahlbrock, K. (1976a). *Arch. Biochem. Biophys.* **173,** 301–305.
Poulton, J., Hahlbrock, K., and Grisebach, H. (1976b). *Arch. Biochem. Biophys.* **176,** 449–456.
Poulton, J. E., and Kauer, M. (1977). *Planta* **136,** 53–59.
Puhan, Z., and Martin, S. M. (1971). *Prog. Ind. Microbiol.* **9,** 13–39.
Reinhard, E. (1974). *In* "Tissue Culture and Plant Science 1974" (H. E. Street, ed.), pp. 433–459. Academic Press, New York.
Romeike, A. (1975). *Biochem. Physiol. Pflanz.* **168,** 87–92.
Saleh, N. A. M., Fritsch, H., Witkop, P., and Grisebach, H. (1976). *Planta* **133,** 41–45.
Saleh, N. A. M., Fritsch, H., Kreuzaler, F., and Grisebach, H. (1978). *Phytochemistry* **17,** 183–186.
Schallenberg, J., and Berlin, J. (1979). *Z. Naturforsch. C: Biosci.* **34C,** 541–545.
Schröder, J., Kreuzaler, F., Schäfer, E., and Hahlbrock, K. (1979). *J. Biol. Chem.* **254,** 57–65.
Schütte, H. R. (1978). *Prog. Bot.* **40,** 126–149.
Scott, A. I., and Lee, S. L. (1975). *J. Am. Chem. Soc.* **97,** 6906–6908.
Scott, A. I., Lee, S. L., de Capite, P., and Culver, M. G. (1977). *Heterocycles* **7,** 979–984.
Scott, A. I., Mizukami, H., and Lee, S. L. (1979). *Phytochemistry* **18,** 795–798.
Selby, C., Turnbull, A., and Collin, H. A. (1980). *New Phytol.* **83,** 351.
Seo, S., Tomita, Y., and Tori, K. (1978). *J. Chem. Soc., Chem. Commun.* pp. 319–320.
Sharma, S. K., and Brown, S. A. (1978). *J. Chromatogr.* **157,** 427–431.
Sharma, S. K., and Brown, S. A. (1979). *Can. J. Biochem.* **57,** 986–994.
Sharma, S. K., Garrett, J. M., and Brown, S. A. (1979). *Z. Naturforsch., C: Biosci.* **34C,** 387–391.
Steck, W., and Constabel, F. (1974). *Lloydia* **37,** 185–191.
Steck, W., Gamborg, O. L., and Bailey, B. K. (1973). *Lloydia* **38,** 93–95.
Stöckigt, J. (1979). *Phytochemistry* **18,** 965–971.
Stöckigt, J., and Zenk, M. H. (1977). *FEBS Lett.* **79,** 233–237.
Stöckigt, J., Treimer, J., and Zenk, M. H. (1976). *FEBS Lett.* **70,** 267–270.
Stöckigt, J., Husson, H. P., Kan-Fan, C., and Zenk, M. H. (1977). *J. Chem. Soc., Chem. Commun.* pp. 164–166.
Stöckigt, J., Rueffer, M., Zenk, M. H., and Hoyer, G.-A. (1978). *Planta Med.* **33,** 188–192.
Stohs, S. J. (1977). *In* "Plant Tissue Culture and Its Bio-technological Application" (W. Barz, E. Reinhard, and M. H. Zenk, eds.), pp. 142–150. Springer-Verlag, Berlin and New York.
Stohs, S. J., and Rosenberg, H. (1975). *Lloydia* **38,** 181–194.
Stohs, S. J., Kaul, B., and Staba, E. J. (1969). *Phytochemistry* **8,** 1679–1686.
Sugano, N., Koide, K., Ogawa, Y., Moriya, Y., and Nishi, A. (1978). *Phytochemistry* **17,** 1235–1237.

Thompson, H. J., Sharma, S. K., and Brown, S. A. (1978). *Arch. Biochem. Biophys.* **188,** 272–281.

Tomita, Y., and Uomori, A. (1971). *J. Chem. Soc., Chem. Commun.* **7,** 284.

Treimer, J. F., and Zenk, M. H. (1978). *Phytochemistry* **17,** 227–231.

Treimer, J. F., and Zenk, M. H. (1979a). *FEBS Lett.* **97,** 159–162.

Treimer, J. F., and Zenk, M. H. (1979b). *Eur. J. Biochem.* **101,** 225–233.

Tsang, Y. F., and Ibrahim, R. K. (1979a). *Z. Naturforsch. C: Biosci.* **34C,** 46–50.

Tsang, Y. F., and Ibrahim, R. K. (1979b). *Phytochemistry* **18,** 1131–1136.

Veliky, I. A. (1972). *Phytochemistry* **11,** 1405–1406.

Veliky, I. A., and Barber, K. M. (1975). *Lloydia* **38,** 125–130.

Weber, N. (1978). *Z. Pflanzenphysiol.* **87,** 355–363.

Wengenmayer, H., Ebel, J., and Grisebach, H. (1976). *Eur. J. Biochem.* **65,** 529–536.

Withers, L. A. (1978). *In* "Frontiers of Plant Tissue Culture 1978" (T. A. Thorpe, ed.), pp. 297–306. The Bookstore, University of Calgary, Calgary, Alberta, Canada.

Withers, L. A., and Street, H. E. (1977). *In* "Plant Tissue Culture and Its Bio-technological Application" (W. Barz, E. Reinhard, and M. H. Zenk, eds.), pp. 226–244. Springer-Verlag, Berlin and New York.

Yagen, B., Gallili, G. E., and Mateles, R. I. (1978). *Appl. Environ. Microbiol.* **36,** 213–216.

Yamada, Y., and Kuboi, T. (1976). *Phytochemistry* **15,** 395–396.

Zenk, M. H. (1978). *In* "Frontiers of Plant Tissue Culture 1978" (T. A. Thorpe, ed.), pp. 1–13. The Bookstore, University of Calgary, Calgary, Alberta, Canada.

Zenk, M. H., El-Shagi, H., and Schulte, U. (1975). *Planta Med., Suppl.* pp. 79–101.

Zenk, M. H., El-Shagi, H., Arens, H., Stöckigt, J., Weiler, E. W., and Deus, B. (1977). *In* "Plant Tissue Culture and Its Bio-technological Application" (W. Barz, E. Reinhard, and M. H. Zenk, eds.), pp. 27–43. Springer-Verlag, Berlin and New York.

Turnover and Degradation of Secondary (Natural) Products

3

WOLFGANG BARZ
JOHANNES KÖSTER

I. INTRODUCTION

The history of plant secondary products is characterized by several discernible periods. Isolation and structural elucidation were a major effort of both natural product and pharmaceutical chemists for several decades. Then botanists and geneticists learned to use secondary products as suitable markers for physiological, morphological, and genetic studies. Syntheses of complex plant constituents later intrigued organic chemists. More recently studies on natural product biosynthesis with respect to the pathways employed and the enzymes involved have occupied the interest of plant

The Biochemistry of Plants, Vol. 7

biochemists. Only quite recently have systematic investigations been initiated to elucidate the dynamic role of secondary metabolites in plants. It has repeatedly been found that these plant constituents are not metabolically inactive, inert end products but that they are subject to turnover and degradation in the producing plant. Investigations on catabolic pathways of secondary products such as alkaloids, phenolics, and terpenoids represent a new field of plant biochemistry. For a long time it had been assumed that plants accumulated these products during their entire life cycle but did not degrade them. This belief is summarized in statements describing secondary products as accumulated waste material (Schwarze, 1958) with "no biochemically useable energy potential" (Reznik, 1960). This assumption was largely based on insufficient experimental data. Present concepts regard secondary product metabolism as a complex dynamic system where synthesis and degradation are two partially linked processes which are integrated into the complete network of plant metabolism (Robinson, 1974; Ellis, 1974; Barz and Hösel, 1975; Barz and Hösel, 1978). Many factors influence the two processes, and the actual rate of secondary product metabolism must be analyzed in each case.

This chapter will, whenever possible, place the main emphasis on catabolic pathways of selected classes of phenolic and alkaloid compounds. Physiological and morphological aspects of plant metabolism will only briefly be discussed, whereas the various metabolic forms of turnover will be outlined in more detail. Furthermore, many observations on the turnover of various secondary compounds that have not yet been substantiated by chemical analyses will not be presented. Attempts to present complete lists have not been made.

II. THE CONCEPT OF TURNOVER OF SECONDARY PRODUCTS

Although plants seemingly accumulate their products in specialized cells or tissues without further metabolism, sophisticated analytical techniques and the use of radioactive tracers permit measurements of the dynamics of secondary product metabolism. Such measurements have shown that synthesis and further metabolism (turnover, catabolism) occur simultaneously and have thus documented that there may be a permanent flow of carbon through metabolic pools of secondary constituents. Variations in this flow result from parameters that regulate secondary metabolism.

The following sections will describe the main aspects of metabolic turnover and integrate this phenomenon into the general metabolism and physiology of higher plants.

A. Evidence for Turnover of Secondary Compounds

Several methods have been used to determine that secondary products are not inert end products but rather are further metabolized and even degraded. In the absence of specific knowledge of the reactions involved, such further metabolism will be defined in this chapter as *turnover*. If the events of further metabolism result in partial or complete degradation of the compound, this form of turnover can be designated *catabolism*. The most simple method concerns cases where the complete disappearance of or the gradual decrease in a secondary compound in a tissue demonstrates that catabolic turnover occurs. The well-known phenomenon "youth anthocyanin" is a suitable example which must be explained by catabolic reactions (Zenk, 1967). Germination of seeds and growth of buds or seedlings are physiological stages often characterized by intense secondary metabolism. Thus various flavonoid compounds are known to disappear completely at times of intense growth (Barz, 1975a). Numerous examples can be cited where alkaloids in unfertilized ovules disappear after fertilization. Similarly, seeds that are rich in alkaloids undergo a net decrease in these compounds during germination (Waller and Nowacki, 1978). Previous reviews (Barz, 1975a; Barz and Hösel, 1975; Robinson, 1974) have listed various other examples.

Seasonal variation in the level of secondary constituents is a further indication of a dynamic role for these products. Fluctuations in total concentration and/or rate of turnover have been found for various alkaloids (Waller and Nowacki, 1978), phenols, and flavonoids (Ellis, 1974; Barz and Hösel, 1975), as well as other secondary products. The amount of primulaverin, a xyloglucoside of gentisic acid methyl ester in *Primula elatior* L., varies from 1.6% of the dry weight in March to 0.2% in June. The phenolglucosides in *Salix purpurea* show seasonal fluctuations in their concentration (Thieme, 1965) of one order of magnitude. The indole alkaloids in *Catharanthus roseus* reach a maximum the third week after germination. They then disappear almost completely and finally reappear at about eight weeks (Mothes *et al.*, 1965).

The numerous descriptions (see Waller and Nowacki, 1978) of diurnal variations in the concentration of morphine alkaloids in *Papaver somniferum* (Fairbain and Wassel, 1964) further document high rates of secondary metabolism. Half-lives in the range of 10–20 h have been observed (Robinson, 1974). Such data indicate great variability in rates of synthesis and subsequent catabolism. This is not surprising, because secondary metabolism is presumably linked to primary metabolism by the rates with which substrates can branch off from the primary pathways and funnel into the secondary biosynthetic routes. Obviously, all fluctuations in primary metabolites, variations in enzyme activity of the primary energy-yielding reactions (Deitzer *et al.*, 1978), and photoperiodic phenomena (Engelsma, 1979; see Section II,B) affect the rate of secondary metabolism, and analyses merely measuring accumulation do not show whether diurnal variation is due to changes in the rate of synthesis or the rate of catabolism.

The best and most versatile method for measuring turnover rates involves the use of isotopically labeled precursors. The usual procedure is to administer one pulse of a precursor and follow this by a metabolic period during which carbon of the precursor is incorporated into and eventually transported through the metabolic pool of the compound under investigation. During this metabolic period, analyses are made of the amount of the secondary product, the total radioactivity incorporated, and its specific radioactivity in order to calculate the turnover of the substance being investigated. First-order kinetics of turnover are in most cases assumed to be effective. A decrease in the total and specific radioactivity with time, although the secondary product remains constant or even increases, indicates turnover of the product. A full application of the kinetic principles (Reiner, 1953a,b) allows the determination of a variety of metabolic situations. Thus the turnover of a compound together with rates of synthesis and subsequent metabolism, a precursor–product relationship, the position of a compound in a metabolic network, and great differences in the turnover of chemically related compounds in a single plant have been determined using this technique (Brown and Wetter, 1972; Barz and Hösel, 1975; Molderez et al., 1978; Marcinowski and Grisebach, 1977).

Such experiments have revealed the following complex metabolic situation: Numerous secondary products appear to reach steady state concentrations (see Barz and Hösel, 1975) where the rate of synthesis equals the rate of further reactions. The steady-state situation clearly implies that the enzymes for subsequent reactions are formed in the plant soon after development of the biosynthetic pathways. Evidence in support of this view is available in the case of an isoflavone 7-O-glucoside-specific glucosidase which appears in Cicer arietinum roots slightly before maximum isoflavone levels have been reached (Hösel, 1976).

With the use of pulse-labeling techniques isoflavones in Phaseolus aureus and C. arietinum (Section III,C,4), flavonols in various plants (Section III,C,3), several structurally different alkaloids (Waller and Nowacki, 1978), and cinnamoyl derivatives such as chlorogenic acid (Taylor and Zucker, 1966) and cinnamoyl glucose esters (Molderez et al., 1978) were shown to remain at constant levels as a result of simultaneous and equally rapid synthesis and subsequent consumption. In these and other cases, biological half-lives ranged from as low as 10–20 h up to 300–400 h, where a clear difference no longer existed between slow turnover and accumulation without any turnover. It must be pointed out, however, that the values for biological half-life vary greatly because of the many factors that may influence the rates of anabolism and catabolism (see the following section).

A potentially powerful tool for measuring turnover rates is the use of inhibitors that specifically block biosynthesis. During inhibition, the formation of secondary constituents is prevented, whereas further metabolism (turnover) should proceed unaltered. Using N-benzyloxycarbonyl-α-aminooxy-β-phenylpropionic acid as an inhibitor of phenylalanine deamination, Amrhein and Diederich (1980) have successfully demonstrated isoflavone turnover in *C. arietinum* roots. Their results with respect to biological half-life and to the metabolic differences between the isoflavones involved were identical to those from previous studies with labeled precursors (Barz, 1969). The potential of this method should be more rigorously investigated.

In general, studies during the last decade have conclusively shown that the turnover of secondary products in higher plants is much greater than previously recognized. Such studies have further revealed the existence of several separate pools of one compound having different rates of turnover. On the other hand, turnover cannot be expected to occur with all compounds. Even structurally closely related compounds (the isoflavones in *C. arietinum*; see Section II,C,4) may differ in their turnover rates. Differences in cellular or tissue localization, form of conjugation, physiological condition of the plant, and many other factors appear important.

Once demonstrated, turnover can be further investigated by feeding experiments with suitable substrates. Though successful, feeding experiments are confronted with the problem that endogenously formed and exogenously applied compounds may be metabolized very differently. Examples are found in Section III,B. A further problem of feeding experiments for determining turnover or further metabolism of secondary compounds is the, as yet unknown, specificity of many catabolic enzymes. It is obviously difficult to evaluate the significance of studies where (1) analogous or unnatural compounds have been degraded or (2) natural products have been degraded by plants or cell cultures that do not contain these products as normal constituents. Other pitfalls of feeding studies must also be noted (Harborne, 1967). Such experiments describe the range of possible reactions but must be supported by enzymological studies. Feeding experiments have further shown that in many cases the concentration of the catabolites formed is very low (Barz and Hösel, 1975; Meyer and Barz, 1978). This has hampered structural elucidation and requires the use of special techniques such as cell suspension cultures (see Hösel *et al.*, 1977; Barz, 1977b). Microbial contamination must be rigorously avoided, and casual handling of this problem in feeding experiments renders many results rather doubtful (see Waller and Nowacki, 1978; Barz and Hösel, 1975).

Pulse-labeling and feeding experiments have also shown that secondary compounds are often part of a complex metabolic grid (Sections II,C,4b and III,B,4), which means that a particular compound may react in more than one direction. This and the many physiological factors influencing secondary metabolism combine to make a rather complicated picture with many unsolved questions.

B. Physiological and Developmental Aspects of the Turnover and Accumulation of Secondary Products

Formation, accumulation, and the further metabolism (turnover) of secondary products are fully integrated into programs of differentiation and development in the producing plant. Turnover and degradation are thus seen as the last steps in the full development of secondary metabolism programs. Some aspects will be described here to illustrate the effectiveness of these principles in turnover and degradative reactions.

Wiermann (this volume, Chapter 4) has stressed that the tissues and cells in which secondary products are formed, together with their biosynthetic enzymes, are products of differentiation processes. He has further noted that these morphological processes appear as prerequisites for the full development of secondary product metabolism. The selective induction of these differentiation and synthetic processes is governed by the principles of differential gene expression (Luckner *et al.*, 1977; Luckner, 1980). Though not yet investigated to a great extent, it is reasonable to assume that the morphological, cytological, and biochemical prerequisites for formation and accumulation of secondary products must also be met in order for turnover and degradation to occur. All factors that participate in the regulation of biosynthetic enzyme activity should be considered for catabolic enzymes as well. Such catabolic enzymes are then turned on after the biosynthetic program has fully developed. [A sequential formation of biosynthetic and catabolic enzymes can be deduced from studies on isoflavone turnover in *C. arietinum* roots (Hösel, 1976; Amrhein and Diederich, 1980).] Whether catabolic enzymes are induced by their substrates is as yet unknown.

The well-known compartmentation of secondary metabolism on the level of the organ, tissue, and cell must be taken into consideration when studying turnover and degradative processes of secondary substances (see Wiermann, 1979; McClure, 1975; Luckner, 1980). With respect to sites of synthesis and accumulation, there are various possibilities (McClure, 1975; Waller and Nowacki, 1978), but of equal importance is the relation between the site of biosynthesis and the site of turnover.

The main question to be answered is whether degradation (or further metabolism) occurs at the sites of synthesis and accumulation or whether different sites for degradative reactions exist. Knowledge on this question is at present fragmentary, because most of the degradative pathways, as well as the enzymes involved and their location, are unknown. In *Sorghum bicolor* the cyanogenic glucoside dhurrin is located in the epidermal layer of leaves, whereas the catabolic dhurrin β-glucosidase and the hydroxynitrile lyase reside in the mesophyll tissue (Kojima *et al.*, 1979). In *C. arietinum* the glucosidase specific for isoflavone 7-glucoside is found in the same tissue (cortex) as the formononetin glucosides but is separated from the biochanin A glucosides which reside in the rhizodermis (Barz, 1977a). In these and other examples, the physical separation of secondary products and their catabolic enzymes leads to the conclusion that, in certain cases, the secondary products must be translocated for catabolism to occur. Though the full range of compartmentation of secondary metabolism is at present not known, these aspects will be of great importance in explaining turnover and its dynamics.

Secondary products are mainly stored in specially constructed reservoirs or storage organelles (glands, hair, storage cells, chromoplasts, vacuoles, membrane systems). Though stored in such reservoirs as vacuoles (Saunders and Conn, 1978; Sasse *et al.*, 1979), an established flow of secondary material in and out of such compartments must occur in order to explain cases where permanent turnover has been shown. This is indicated by the strictly first-order kinetics of turnover and the complete disappearance of radiocarbon from secondary products observed in several pulse-labeling experiments (see Barz and Hösel, 1975; Robinson, 1974). Though studies have been made only on a limited scale so far, it can be predicted that degradative enzymes such as specific glucosidases (see Hösel, this volume, Chapter 23) are not in the vacuole or associated with the inner tonoplast membrane and that the secondary products involved must be released from the vacuole to be turned over (further metabolized). Since nothing is known concerning the vacuolar efflux of secondary products, studies in this field will be of great importance for secondary metabolism.

Few enzymes associated with catabolic pathways are presently known. However, the endoplasmic reticulum (hydroxylases, oxygenases, lyases, phenolases, peroxidases), cytoplasm (peroxidases, glycohydrolases), chloroplasts (benzoic acid synthetase, hydroxylases, oxygenases), lysosomes (phenolases, hydrolases), and mitochondria (cinnamic acid-degrading enzymes) are the cellular organelles of major interest for catabolic processes (Section II,C,4,b). For polymerization processes, the cell wall must also be mentioned (Section II,C,3).

Visible light is of utmost importance for the full development of plants and their machinery of secondary metabolism (Luckner, 1980; McClure, 1975; Waller and Nowacki, 1978). The induction by light of the biosynthesis of phenylpropanoid compounds (see Hahlbrock, this volume, Chapter 14) or betalain pigments (see Piattelli, this volume, Chapter 19) has been intensively studied. The photocontrolled accumulation of various phenolic constituents and photoperiodic alterations of content are known (Lamb and Rubery, 1976; Engelsma, 1979). The pronounced and extremely variable effects of light on alkaloid accumulation have been repeatedly substantiated (Waller and Nowacki, 1978).

Many years of research have shown that basic differences exist between light- and dark-grown plants in nearly all aspects of primary and secondary metabolism. With regard to turnover and degradation of secondary products, several studies have shown that darkness–light changes also greatly influence this part of secondary metabolism. As one example, Fig. 2 (Section II,C,2) shows that the conversion of O-sinapoylglucose to O-sinapoylmalate is severely reduced in dark-grown seedlings. Another example is the light-induced degradation of anthocyanin in *Sinapis alba* (Bopp and Pop, 1975), though the target reaction cannot be specified.

When changes in light regimens have led to a decreased or increased accumulation of a secondary constituent, both the rate of synthesis and the rate of turnover (further metabolism) might have been affected. Such changes are illustrated by the results of pulse-labeling studies with the isoflavone formononetin and the flavonol kaempferol in the garbanzo bean (Barz, 1975b). Upon transfer of light-grown plants to darkness, the rates of synthesis of both compounds remained constant, while isoflavone turnover was reduced and flavonol turnover increased. This response was reversed upon reillumination. Diurnal fluctuations of other secondary products might also be explained by varying degrees of turnover. More detailed studies are required to elucidate the catabolic reactions and the enzymes behind such observations.

Decisive factors in the metabolism of secondary products are phytohormones, whose effect has especially been demonstrated in plant cell cultures (Alfermann *et al.*, 1975; Zenk *et al.*, 1975; Kleist, 1975). In intact plants, the promoting or inhibiting effects of phytohormones on alkaloids (Robinson, 1974) or plant pigments have also been demonstrated (McClure, 1975). Induction or repression of enzymes, changes in the storage capacities of cells, morphogenetic effects, and various other unknown responses might be results of the action of phytohormones. Studies of this kind again demonstrate that in plant systems both the rate of synthesis and the rate of turnover of secondary products can be altered.

Suitable examples stem from studies on the isoflavones formononetin and biochanin A in *C. arietinum* cell cultures (Barz, 1977b). In one situation, high levels of isoflavones resulted from a stimulation of synthesis, although the rate of turnover remained essentially unaltered. In another case, an accelerated rate of turnover with an only slightly changed rate of synthesis led to low levels of isoflavones. The published data on the biological half-life of biochanin A (Barz, 1977b) in various organs of plants and in phytohormone-treated cell cultures (values ranging from more than 320 h down to a low of 25 h) demonstrate the dynamic character of turnover and its dependence on different cellular conditions. The rate of turnover of the alkaloid ricinine in

Ricinus communis also varies greatly depending on physiological conditions (Waller and Lee, 1969). Various other examples can be cited to establish that further metabolism is an important parameter in regulating stationary concentrations of secondary compounds (Luckner, 1980).

Major physiological, morphological, and biochemical changes occur when plant tissues or fruits become senescent (Simpson *et al.*, 1976). Degradation of chlorophyll and carotenoids, conversion of chloroplasts to chromoplasts, esterification of xanthophylls, substantial alteration of pigmentation and of nitrogenous compounds, and eventually destruction of tissue structure are some of the basic changes occurring during this period. Prior to and especially after the peak of autumn coloration, the degradation of secondary products, predominantly phenolic pigments, becomes very important. Though the rate and extent of discoloration have often been measured, the precise structure of the breakdown products is not yet known. Polymerization reactions by liberated oxygenases and peroxidases (Sections II,C,3,C,3,a, and C,3,b) should be of major interest for further studies in this field. With respect to the degradation of anthocyanins, which also occurs in senescent tissues, model systems have been studied and their chemistry described (Jurd, 1972; Simpson *et al.*, 1976).

Many other factors (i.e., nitrogen availability, inorganic nutrients, temperature, metabolizable carbon sources) have an influence on secondary product accumulation and turnover. The exact effect of each factor on catabolic processes remains to be determined. A representative example for the effects of different environmental conditions, and thus different physiological conditions, on turnover is found in the studies by Weissenböck and his group on *C*-glycosylflavones in *A. sativa* (Popovici and Weissenböck, 1976, 1977). When grown under field conditions, the primary leaves of oat accumulate the four glycosides shown in Fig. 1.

These compounds first accumulate in the early phases of leaf development and then decrease dramatically in the second half of the life of the leaf. Degradative processes are therefore obviously occurring. In contrast,

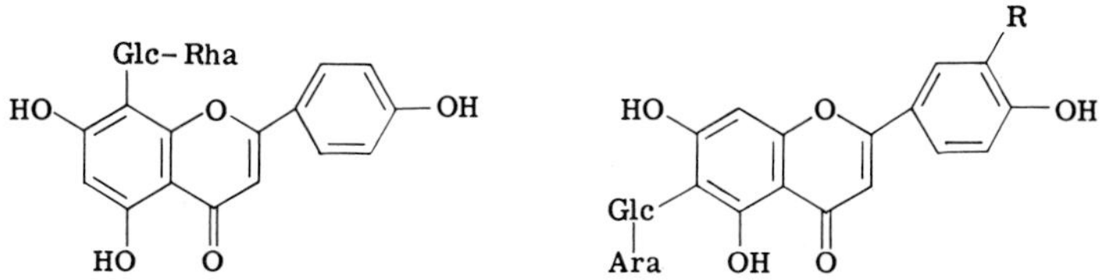

Fig. 1. *C*-Glycosylflavones from primary leaves of *Avena sativa*. Compound **IV** only accumulates under field conditions (Popovici and Weissenböck, 1976, 1977).

phytotron-grown plants accumulate only compounds I–III in Fig. 1, show sigmoid accumulation curves, and show no decrease in flavonoid material; turnover could not be measured by tracer studies.

These and other studies (see Barz and Hösel, 1975; Waller and Howacki, 1978) clearly demonstrate that the rate and extent of turnover of secondary products are dynamic values that have to be measured for each individual compound when physiological conditions change.

C. The Metabolic Forms of Turnover

Once the disappearance or turnover of a particular secondary product has been established, more detailed analyses must be conducted to determine the exact nature of the metabolic process. Turnover (or further metabolism) of a secondary plant product may involve four types of reactions.

1. Interconversion reactions involved in biosynthetic sequences.
2. Conjugation reactions involving other compounds to which the product under investigation is attached.
3. Oxidative polymerization reactions leading to insoluble structures of high MW.
4. Degradative reactions where the products are converted to structures of primary metabolism.

The main aspects of these four alternatives of aromatic metabolism will be discussed in the following sections in a general way.

1. Interconversion Reactions

Interconversion reactions mainly refer to anabolic processes in which compounds are transferred to monomers (or polymers) with a higher degree of substitution. The compounds turned over may therefore be regarded as biosynthetic intermediates. Such reactions are discussed in this section because, in several cases, the intermediates themselves either form comparatively large, measurable pools or accumulate for longer periods during plant development. In assessing such interconversions, (1) one can measure the permanent active turnover of the compound regardless of its rather constant level or (2) one can observe the transient appearance of the intermediate compound during a developmental stage.

Hydroxylation, methylation, cyclization, and rearrangement reactions are important interconversion reactions, and the following are examples.

In some stages during the development of *Tulipa* anthers first cinnamic acids and later chalcones accumulate, which eventually disappear. The transient appearance of these products is explained by their consumption for flavonol and anthocyanin biosynthesis in this organ (Wiermann, 1979).

The ripening processes of flowers and fruits are quite often accompanied by the transient accumulation and subsequent conversion of carotenoid intermediates to higher substituted or higher desaturated carotenoids or xanthophylls (Goodwin, 1976).

At all developmental stages lucerne (*Medicago sativa*) plants contain considerable amounts of the isoflavone daidzein which has been demonstrated to be an intermediate in coumestrol biosynthesis (Grisebach and Barz, 1969; Dewick and Martin, 1979) (Section III,C,4).

The isoflavone formononetin [Scheme 1 (V)] is a well-known constituent of many leguminous plants. It is also an intermediate in the formation of the rotenoids rotenone (VIa) and amorphigenin (VIb) in *Amorpha fruticosa* (Crombie *et al.*, 1971), of the phytoalexins demethylhomopterocarpin (VII), sativan (VIII), and vestitol (IX), and of 9-*O*-methylcoumestrol (X) in *M. sativa* (Dewick and Martin, 1979). The 7-hydroxy-4'-methoxyisoflavanone may be the more immediate precursor, though the latter compound is not accumulated to any extent.

A well-documented example from alkaloid biochemistry is the (transient) accumulation in various plants of the benzyltetrahydroisoquinoline compound norlaudanosoline which is known to be converted to several major classes of alkaloids (Waller and Nowacki, 1978). Similarly, turnover studies on the morphine alkaloids thebaine and codeine in young poppy plants are consistent with the known biosynthetic pathway: thebaine → codeine → morphine (Waller and Nowacki, 1978). Though interconversion reactions are sometimes quite evident from a detailed analysis of the constituents of a plant, more refined methods must be employed to determine if more complex structures and metabolic grids are involved.

2. Conjugation Reactions

Many secondary products do not occur in plants in the free form (aglycone) but rather are conjugated with a variety of different monomeric or oligomeric compounds. Aliphatic and phenolic hydroxyl groups, carboxyl functions, and amino and mercapto groups of aglycones are involved in conjugation reactions. *C*-glycosides must also be considered in this category, though their resistance to normal acid hydrolysis places them in a somewhat distinct group.

Occurrence in the form of conjugates may be regarded as a characteristic feature of secondary compounds such as phenols, aromatic acids, flavonoids, steroids, and many other secondary products. On the other hand, some classes of secondary products such as alkaloids and monoterpenes occur primarily in free form. Conjugates of these compounds are, however, also known.

Conjugation reactions are of importance because (1) they drastically alter the chemical (solubility, pK values) or physiological (transport through cells or membranes, biological activity) properties of secondary products, (2) they

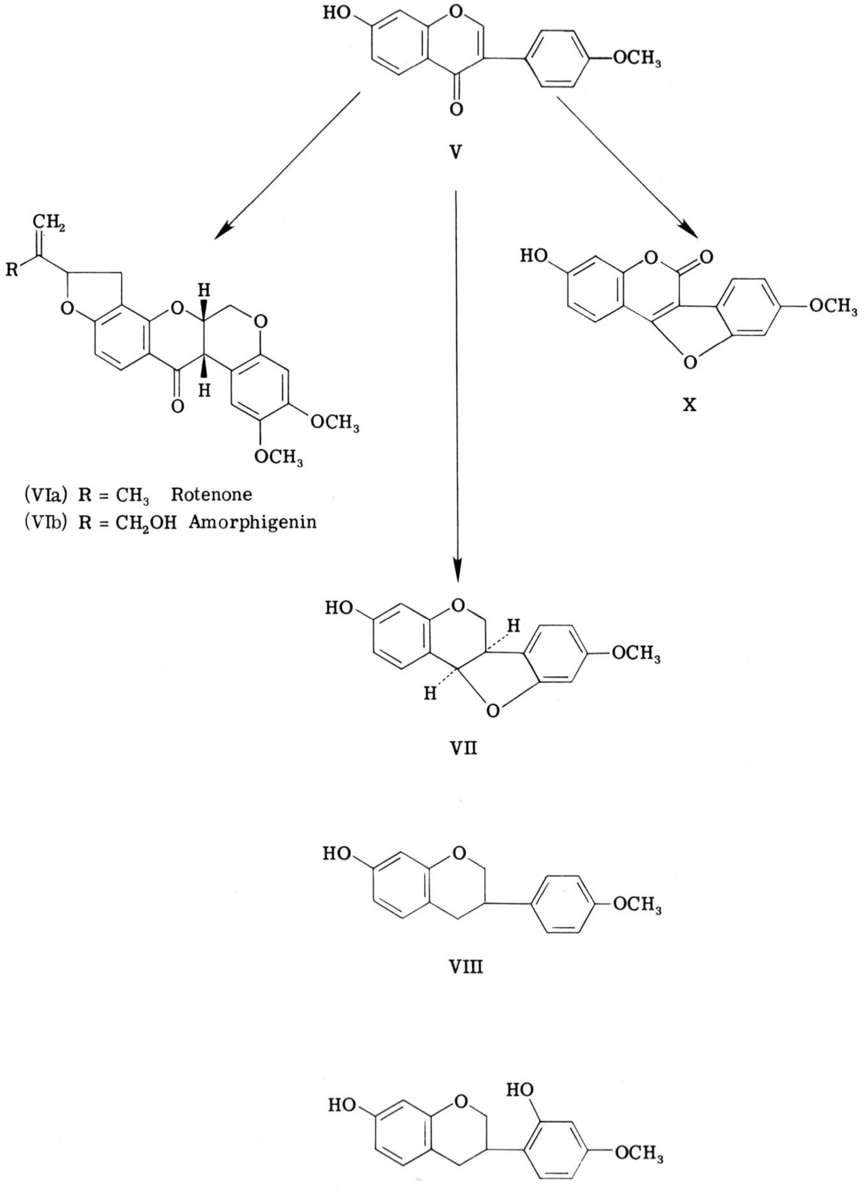

(VIa) R = CH₃   Rotenone
(VIb) R = CH₂OH Amorphigenin

V

X

VII

VIII

IX

Scheme 1.   Interconversion reactions depicting role of formononetin (V) as intermediate in rotenoid, pterocarpan, isoflavan, and coumestan biosynthesis.

may result in a site of accumulation different than that occupied by the aglycone, (3) the conjugated compound may enter a different metabolic pathway than the corresponding aglycone, and (4) they greatly determine whether a compound is a metabolically active species or a metabolically inactive (end) product. Therefore conjugation reactions represent an important way to store materials or to detoxify unwanted material.

Detoxification appears to be an essential aspect of conjugation reactions, both for exogenous material (Sandermann *et al.*, 1977) and for endogenous products. As an example, the excessive accumulation of *p*-coumaroylputrescine and other cinnamoylputrescine conjugates in a mutant line of tobacco cell cultures may be cited (Berlin and Widholm, 1978). An increased turnover of phenylalanine and a 10-fold higher phenylalanine ammonia-lyase (PAL) activity in these cells accelerate the influx of phenylpropanoid material into secondary metabolism, and this excess material is obviously trapped as conjugates.

The reservoir function of conjugates whereby metabolically active compounds can be temporarily stored should also be mentioned. As an example, nicotinic acid conjugates (*N*-glycosides and *N*-methyl derivatives) are reservoirs for nicotinic acid which is used for pyridine nucleotide synthesis (Barz, 1977b).

The biochemical and physiological properties of many secondary products cannot be understood unless their conjugated moieties are known. Analysis of plant extracts for aglycones alone will not reveal the full complexity and possible conjugate-caused compartmentation of secondary metabolism. This conclusion is supported by the interesting and complex metabolism of the caffeoyl moiety of chlorogenic acid (3-*O*-caffeoylquinic acid). This widely occurring depside, known to be rapidly turned over (Taylor and Zucker, 1966) and to fluctuate in amount in various plants (see Colonna, 1978), is intimately related to turnover of the caffeoyl unit, as the latter is converted to 3,5-dicaffeoylquinic acid (Mialoundama and Paulet, 1975), to lignin via ferulic acid (see Grisebach, this volume, Chapter 15), and to polymeric material by oxidative reactions (Section II,C,3), and as it is catabolized by degradative processes (Koeppe *et al.*, 1970; Barz and Hösel, 1975).

The great variety of conjugating compounds makes it difficult to predict what sort of conjugating moiety will be found for a given compound in a given plant. Furthermore, many cases are known where a mixture of conjugates from one particular aglycone is found. In these mixtures the individual compounds may possess very different turnover rates. This is exemplified by studies (Strack, 1973; Strack and Reznik, 1976) on flavonol glycosides in *C. maxima* Duchesne. The seedlings contain kaempferol, quercetin and isorhamnetin glucosides, or biosides and triosides. During seedling development the monosides have a biological half-life in the range of 30–36 h, and the biosides of approximately 48 h, whereas the triosides show no turnover at all.

Such data, which indicate a biosynthetic sequence leading from monosides to triosides, point to an important role for the degree of conjugation in this and many other cases found in the literature. Various aspects of conjugation reactions have previously been summarized (Harborne, 1964; Miller, 1973; Towers, 1964; Furuya, 1978; Herrmann, 1978). Table I lists some of the more important conjugating moieties found in natural secondary products or upon addition of exogenous substrates to plant tissues or plant cell cultures. A great variety of conjugated products can obviously be derived from the molecules, and in some complex structures it is often difficult to classify the aglycone and the conjugating moiety. This is demonstrated by the structure of the caffeoylrhamnosyl O-glucoside of 3,4-dihydroxyphenylethanol (acteoside from *Syringa vulgaris*) (XI). Although the phenylethanol and the caffeic acid moieties may be called aglycones, the caffeoylglucose part can also be considered a conjugating moiety.

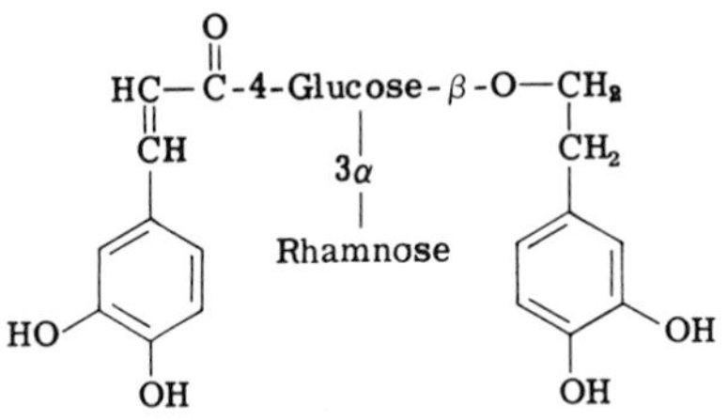

Acteoside

XI

High-MW conjugates exist in addition to the low-MW conjugates formed with the molecules shown in Table I. Various phenolics (for instance, chlorogenic acid, ferulic acid) have been shown to occur covalently bound to proteins, cellulose, and hemicellulose (van Sumere *et al.*, 1975). The covalently bound phenolics in proteins are of particular interest as a trap in explaining the rapid turnover of soluble conjugates (Molderez *et al.*, 1978).

Secondary products are often metabolized while bound to a conjugating moiety. The rapid turnover or disappearance of a conjugate must, however, not be taken as evidence for degradation of the aglycone portion. Transconjugation reactions may also be involved, as illustrated (Fig. 2) by studies (D. Strack, personal communication) on sinapine (sinapic acid choline ester) metabolism in *R. sativus* (Strack, 1977) and *S. alba* (Bopp and Lüdicke, 1975). Sinapine, occurring together with other sinapoylesters in the seeds, completely disappears upon germination and growth of cotyledons as a result of the action of a sinapine esterase (Nurmann and Strack, 1979). The sinapic acid thus formed is converted to sinapic acid glucose ester and finally metabolized to form O-sinapoylmalate. The latter compound is considered an end product, whereas the intermediate sinapic acid glucose ester consti-

TABLE I

Main *Conjugating Moieties* of Secondary Products in Plants

Sugars
Monosaccharides
  Glucose
  Fructose
  Galactose
  Rhamnose
  Mannose
  Sorbose
  Fucose
  Glucosamine
  Glucuronic acid
  Arabinose
  Xylose
  Ribose
  Apiose
  Hamamelose
  Glycerol
Disaccharides
  Rutinose
  Gentiobiose
  Sophorose
  Sambubiose
  Primverose
  Glucosylarabinose
  Arabinosylglucose
  Glucosylxylose
  Rhamnosylarabinose
  Apiosylglucose
  Xylosylgalactose
Oligosaccharides

Aromatic acids
  Benzoic acid
  p-Hydroxybenzoic acid
  Protocatechuic acid
  Gallic acid
  3,4-Dihydroxyphenyllactic acid
  p-Coumaric acid
  Ferulic acid
  Caffeic acid
  Sinapic acid

Amines
  Tyramine
  Tryptamine
  Choline
  Putrescine
  Cadaverine
  Spermidine
  Spermine

Lipid-soluble conjugates
  N-Methyl
  O-Methyl
  Acetyl
  Dimethylallyl
  Long-chain alcohols
  Stearic acid
  Palmitic acid
Amino acids and peptides
  Aspartic acid
  Glutamic acid
  Alanine
  Glycine
  Valine
  Phenylalanine
  DOPA
  Tryptophan
  Glutathione

Aliphatic acids
  Malic acid
  Tartaric acid
  Citric acid
  Malonic acid

Cyclic hydroxy compounds
  Shikimic acid
  Quinic acid
  Myoinositol
  cis,-1,2-Dihydroxycyclohexane
Inorganic acids
  Sulfuric acid
  Phosphoric acid

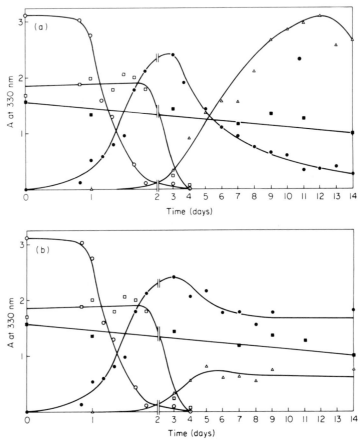

Fig. 2. Turnover of sinapoylcholine (sinapine) and disinapoylsaccharose in cotyledons of light-grown (a) and dark-grown (b) seedlings of *Raphanus sativus* (D. Strack, personal communication). Data are given as absorption units at λ_{max} 330 nm of sinapic acid per pair of cotyledons. Time in days after germination. ○, Sinapoylcholine (sinapine); □, sinapoylglucoraphenin; ■, disinapoylsaccharose; ●, sinapoylglucose; △, *O*-sinapoylmalate

tutes a metabolically reactive species. Similar observations on the metabolic activity of glucose esters and other glucoside intermediates have repeatedly been described (Kojima and Uritani, 1973; Köster *et al.*, 1978; Schlepphorst and Barz, 1979; Molderez *et al.*, 1978). The sequence described in Fig. 2 for sinapic acid is not only typical of illuminated *Raphanus* seedlings but also explains why conjugate mixtures of one compound are found depending on the time of analysis. In dark-grown plants (Fig. 2b) the final accumulation of *O*-sinapoylmalate is severely inhibited, again showing how much secondary metabolism is dependent on the general metabolism of a plant (Section II,B).

Conjugation reactions must also be considered in feeding experiments aimed at elucidating catabolic reactions. The applied substrates may either be trapped into conjugates that prevent them from undergoing further metabolism (Hösel *et al.*, 1977), or they may be converted into conjugates that differ significantly from the derivatives formed when the same compounds are endogenously produced (Barz and Hösel, 1975; Furuya, 1978).

3. Oxidation Polymerization Reactions

Numerous studies have involved the polymerization of various aromatic and heterocyclic compounds (Ellis, 1974; Barz, 1977a; Barz and Hösel, 1978; Waller and Nowacki, 1978; Patzlaff and Barz, 1978). In cases where such secondary products disappear from living tissue without the obvious formation or accumulation of other metabolites (Waller and Nowacki, 1978), their incorporation into lignin-like material or into cell walls has been assumed to occur. Other observations where highly oxidized metabolites are formed from secondary products (Barz and Hösel, 1975; Nowacki and Waller, 1975) also support the concept of oxidative polymerization reactions via hydroxylated intermediates (see also Sections III,C,1 and C,4). The ethanol-insoluble, lignin-like material formed from phenolic substrates has so far resisted detailed chemical analysis; typical properties of lignin were, however, shown to be absent (Berlin and Barz, 1971).

In some cases, polymer formation greatly depends on the substitution pattern of the substrates, as illustrated by the isoflavone structures in Fig. 3 (Barz, 1975b). The readily polymerized compounds (XII–XIV) possess a 4'-hydroxyl or a 3', 4'-dihydroxyl group in ring B and will eventually produce polymers via ortho-quinoid structures. A 4'-methoxy group (XV–XVII) or a 6,7-dihydroxy group cannot undergo polymer formation via quinoid intermediates. Another interesting example is the protoalkaloid hordenine (*N,N*-dimethyltyramine) (Section IV), which is polymerized without demethylation (Meyer and Barz, 1978) or cleavage of the side chain.

The numerous observations on secondary product polymerization and the incorporation of xenobiochemicals into polymeric structures (Lamoureux *et al.*, 1973) suggest that this process constitutes a powerful mechanism for inactivating certain endogenous and exogenous materials. Cell wall and membrane systems seem to be preferred sites of polymerization reactions, perhaps because their polysaccharide and/or protein structures serve as a matrix on which the polymers can form (Section II,C,2). Inactivation by polymerization may constitute an important process in plants that lack uniform excretory systems and have to cope with a great variety of organic compounds. Root tissue, which is especially exposed to large amounts of soil phenolic material, is an important site of polymerization reactions, and studies with root tissue clearly support this concept (Brandner, 1962). On the other hand, more work is required to prove conclusively that *endogenous*

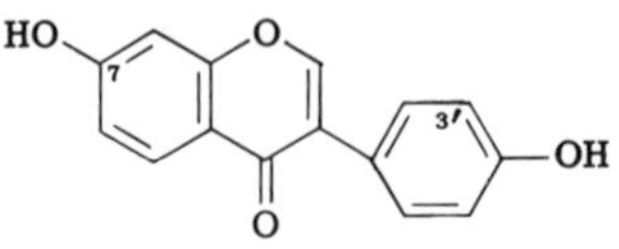

XII

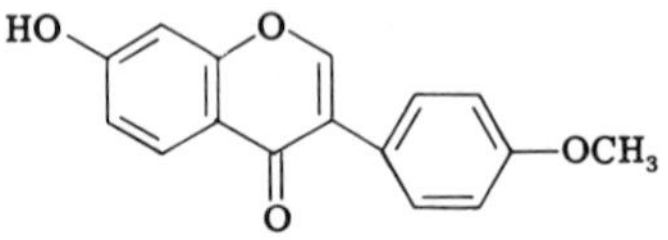

XV

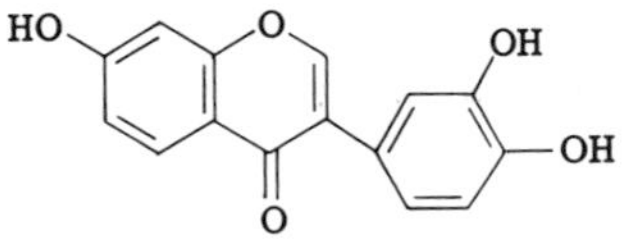

XIII

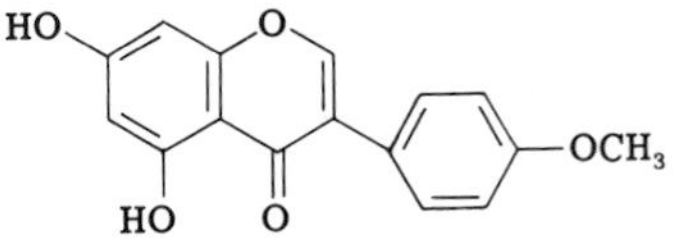

XVI

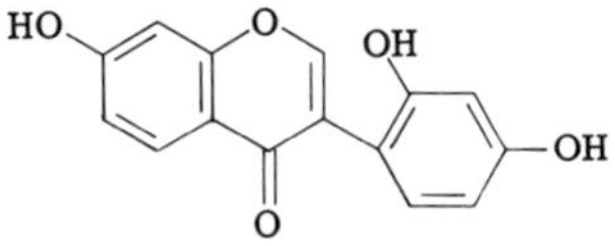

XIV

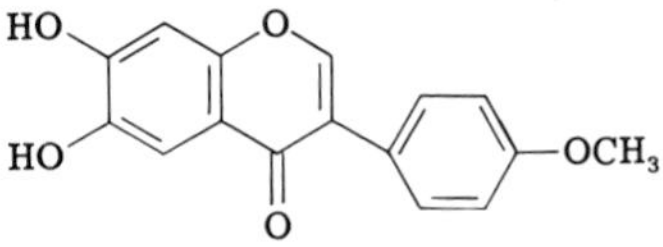

XVII

Fig. 3. Structures of readily (XII–XIV) and poorly (XV–XVII) oxidatively polymerized isoflavones as investigated in *Cicer arietinum* (Barz, 1975b).

secondary products are constantly funneled into polymers under normal conditions and in unwounded plant material. Phenolases and peroxidases, the enzyme systems mainly involved in oxidative polymerization reactions, will briefly be discussed in the following sections.

a. Phenolases. Phenolases and polyphenol oxidases (for reviews, see Butt, 1978; Mayer and Harel, 1979) catalyze two reactions [Eqs. (1) and (2)].

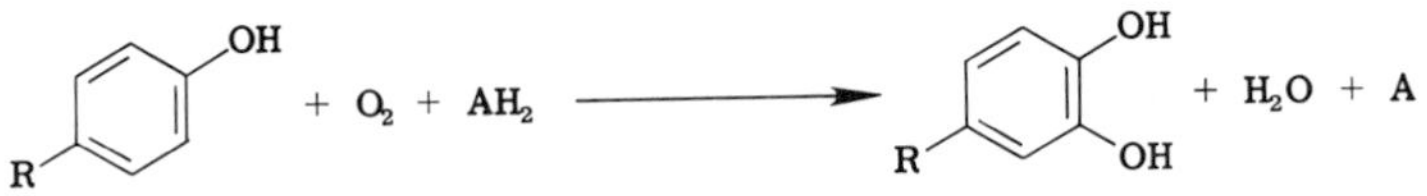

(1)

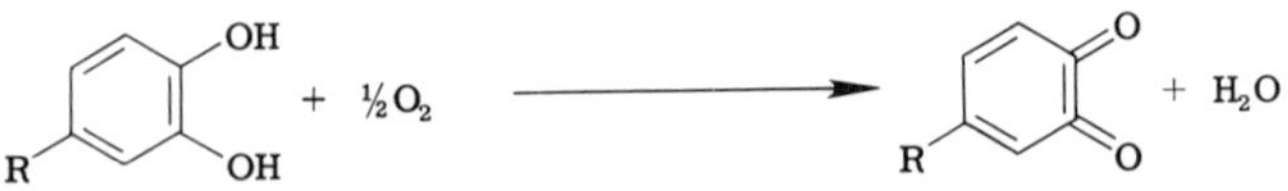

(2)

Reaction (1), the hydroxylation of a *para*-substituted phenol to an *o*-diphenol, requires molecular oxygen and an electron donor (which may be the product itself). Reaction (2) is the oxidation of an *o*-diphenol to an *o*-quinone by molecular oxygen; *p*-diphenols are likewise oxidized to *p*-quinones. Many phenolases occur as isozymes that show a differential ability to catalyze these two reactions. The highly reactive quinones produced give rise to insoluble polymers by self-polymerization or by condensation reactions with compounds carrying —NH or —SH groups (i.e., proteins, amino acids) (van Sumere et al., 1975). The polyphenol-derived browning reactions in damaged, infected, or senescent tissue are the obvious result of phenolase-catalyzed, stress-induced polymerization reactions. However, phenolases are often implicated as the enzymes responsible for hydroxylation and polymerization of aromatic secondary products occurring under normal physiological conditions.

b. Peroxidases. Peroxidases (donor: H_2O_2-oxidoreductase, E.C. 1.11.1.7) catalyze a variety of different reactions (with or without H_2O_2, with or without oxygen) of significance in the metabolism of natural products. These include (1) oxidation of substrates with H_2O_2 (peroxidase reactions), (2) introduction of oxygen into a substrate (oxygenase reaction), (3) electron transfer reactions (oxidase reaction), (4) transalkylation, and (5) halogenation reactions.

Specific examples of interesting peroxidase-catalyzed reactions have been described for flavonols, flavanones, and aurones (Barz and Hösel, 1975; Patzlaff and Barz, 1978) (Sections III,C,1 and C,3), p-hydroxybenzoic acids (Berlin and Barz, 1975) (Section III,B), ferulic acid (Pickering et al., 1973), the coumarin scopoletin (Reigh et al., 1973), and the protoalkaloid hordenine (Meyer and Barz, 1978) (Section IV).

The peroxidase-catalyzed polymerization reaction involves the formation of phenoxy radicals as an initial step (Harkin, 1967); the unpaired electron is stabilized by delocalization throughout the molecule. These radicals can either condense to form a polymer (lignification or polymerization) or react with H_2O_2 or oxygen to form higher oxidation products. Figure 4 shows the phenoxy radicals of some of the peroxidase substrates mentioned above and demonstrates that only a few structural requirements are necessary to render an aromatic compound sensitive to peroxidative attack. Because of the various possible linkages, structural elucidation of the polymers formed will be difficult and may require special techniques such as ^{13}C-nmr spectroscopy (Nimz, 1974). Protein and carbohydrate material may furthermore act as a backbone for the polymerization process.

The participation of various peroxidase isozymes has unequivocally been demonstrated under in vitro conditions (Sections III,C,3 and B). It is, however, not clear how many of these in vitro reactions represent in vivo functions. The formation of lignin is clearly catalyzed by peroxidases (Gross, 1978; Grisebach this volume, Chapter 15). An intensively studied physiological function of peroxidases is the peroxidative destruction of indoleacetic acid (IAA) and other growth regulators (Stafford, 1974).

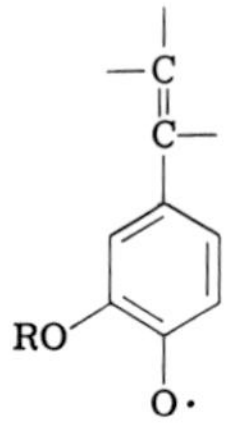

Fig. 4. Examples of peroxidase-generated phenoxy radicals showing ferulic acid (coniferyl alcohol) (XVIII) (Pickering *et al.*, 1973), vanillic acid (XIX) (Berlin and Barz, 1975), scopoletin (XX) (Reigh *et al.*, 1973), and (XXI) naringenin (Patzlaff, 1978).

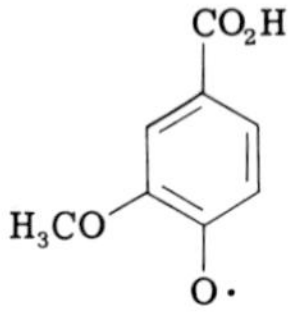

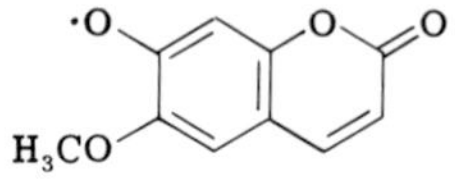

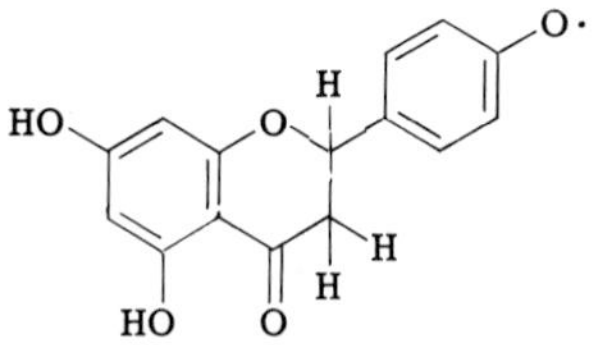

Peroxidases are found in soluble or bound form (Stafford, 1974; Parish, 1975; Henry, 1975; Barz, 1977a) in the cytoplasm, endoplasmic reticulum, tonoplast, chloroplast, plasmalemma, and cell walls. Some tissues such as rhizodermis, endodermis, and xylem are especially rich in peroxidases. Various cellular or tissue sites of polymerization can thus be expected. The cell wall, with its bound peroxidases and its capacity to generate H_2O_2 with the aid of a specific malate dehydrogenase (Gross, 1978; Mäder and Schloss, 1979), appears to be a preferential site of polymerization of suitable substrates. Any exogenous reactive substrate penetrating the cell will most likely be subject to such a peroxidative attack (see Section II,C,3 for studies with root tissue). The extracellular occurrence of peroxidases is also responsible for the polymerization of various polyphenols, as shown with cell cultures (Barz *et al.*, 1978a).

4. Degradative Reactions

The further metabolism (turnover) of secondary plant products may involve degradative (catabolic) reactions in which larger molecules are broken down to simpler cellular constituents. Such catabolic reactions may lead to primary constituents of cell metabolism or may constitute degradative processes in which smaller secondary products are formed. The term "catabolism" will be used regardless of whether or not free energy is being released that can be used by the plant cell.

Since degradation of secondary products by the producing plant is a new aspect of plant biochemistry, numerous apparently unrelated observations have to be brought together in order to construct a unifying picture. A comparison with microbial pathways might be helpful in this process.

a. The Rationale of Microbial Catabolic Pathways. The oxidative metabolism of secondary plant products in microorganisms is governed by the principle that these substrates will be used as a source of carbon, nitrogen, sulfur, and/or energy. The substrates must be modified and degraded in such a way that the intermediates formed can be funneled into the energy-yielding primary metabolism of the organism. It is important to recall that only the terminal cycles of metabolism (i.e., glycolysis and the Krebs cycle) are energy-yielding reactions and that they are catalyzed by constitutive enzymes. The introductory reactions (e.g., N-demethylation, O-demethylation, hydroxylation, deamination) and the intermediate steps (e.g., ring fission, chain shortening) are in most cases carried out by induced enzymes formed only in the presence of the substrate. These principles of catabolism dictate that the substrates being degraded must be metabolized in such a way that only a few compounds of primary metabolism (e.g., acetate, formate, propionate, pyruvate, succinate, and other dicarboxylic acids) will be formed. The molecular organization of catabolic pathways that degrade complicated molecules to a few simple aliphatic compounds in microorganisms has been studied extensively (Dagley, 1971). With respect to the chemical principles involved, such microbial pathways should serve as excellent examples for studies on plant systems.

b. Degradative Pathways in Plants. The ability of higher plants to degrade secondary products of very different chemical structures has repeatedly been described. However, the available information stems mainly from feeding experiments and a limited number of enzymatic investigations. Therefore only a rough picture of degradative pathways for secondary products, of the enzymatic reactions involved, and of the extent of catabolism in plants may be given.

Table II summarizes the reactions that participate in catabolic processes of secondary products. Appropriate examples are listed to illustrate the type of reaction. Whenever possible, the enzyme involved is also given. The reader will note that a certain overlap occurs with reactions involved in biosynthetic sequences because a stringent separation of anabolism and catabolism is not possible.

A few important differences between catabolic pathways for secondary products in plants and the unidirectional catabolic routes in microorganisms should be described.

While microorganisms use degradable compounds as substrates for

TABLE II

Classification of Reactions Involved in the Degradation of Secondary Products in Plants Together with Relevant Enzymes and Suitable Examples

Reaction	Enzyme involved	Examples		
		Substrate	Product(s)	Reference
O-Demethylation	Monooxygenase (?)	Veratric acid	Vanillic acid + CO_2	Barz, 1975b
N-Demethylation	Monooxygenase (?)	Nicotine	Nornicotine + C_1 unit (HCHO?)	Barz et al., 1978b
		N-Methyl-nicotinic acid	Nicotinic acid + C_1 unit (HCHO?)	Barz, 1977b
Decarboxylation	Decarboxylase	Phenylpyruvate	Phenylacetaldehyde + CO_2	Berlin and Barz, 1975
Oxidative decarboxylation	Peroxidase	Syringic acid	2,6-Dimethoxyquinone + CO_2	
Amino acid decarboxylation	Amino acid decarboxylase	Ornithine	Putrescine + CO_2	Mizusaki et al., 1973
Amino acid oxidation	Phenolase	Tyrosine	p-Hydroxyphenylacetic acid + NH_3	Butt, 1978
Aliphatic mono-hydroxylation	Mixed-function oxygenase with cytochrome P-450	Geraniol	Hydroxygeraniol	Madyastha et al., 1976
Aliphatic C-hydroxylation	Monooxygenase (without cytochrome P-450?)	p-Hydroxy-benzylnitrile	p-Hydroxymandelonitrile	Conn, 1979
Aromatic mono-hydroxylation	Mixed-function oxygenase with cytochrome P-450	t-Cinnamic acid	p-Coumaric acid	Potts et al., 1974
N-Hydroxylation	Oxygenase	L-tyrosine	N-Hydroxytyrosine	Conn, 1979
Nitrile lyase	FAD lyase	Mandelonitrile	Benzaldehyde + HCN	Gerstner and Kiel, 1975

N-Oxide formation	Microsomal oxygenase	Thebaine	Thebaine N-oxide	Phillippson et al., 1976
Epoxidation	Mixed-function oxygenase with cytochrome P-450	18-Hydroxyoleate	18-hydroxy-cis-9,10-epoxystearic acid	Croteau and Kolattukudy, 1975
Aromatic ortho-hydroxylation	Phenolase	p-Coumaric acid	Caffeic acid	Schill and Grisebach, 1973
	Peroxidase	Naringenin	Eriodictyol	Patzlaff and Barz, 1978
Cleavage of C—C double bonds, ring fission	Dioxygenase	Protocatechuic acid	3-Carboxy-cis,cis-muconic acid	Mohan et al., 1979
	Peroxidase	Catechol	Unidentified + CO_2	Prasad and Ellis, 1978
	Peroxidase	Phloroglucinol derivatives	Muconic acid derivatives	Patzlaff and Barz, 1978
Chain cleavage	Lipoxygenase followed by aldehyde lyases	Linoleic acid	Via 13-hydroperoxide to trans-hexen-2-al-1 and 12-oxododecen-(10 trans) acid, or via 9-hydroperoxide to nonodien-(2-trans, 6 cis)-al-1 and 9-oxononanic acid	Axelrod, 1974
Chain shortening	"p-Hydroxybenzoate synthase" (not involving ATP and CoA)	p-Coumaric acid	p-Hydroxybenzoic acid (via p-hydroxybenzaldehyde) + C_2 unit	Hagel and Kindl, 1975
Deamination	Monoamine oxidase	Tryptamine	Indole acetaldehyde	Alibert et al., 1979
	Diamine oxidase	Putrescine	Pyrroline	Smith, 1980
	Aldehyde amino-transferase	Primary amine	Aldehyde	Smith, 1980
				Unger and Hartmann, 1976
Aldehyde oxidation	Dehydrogenase	Phenylacetaldehyde	Phenylacetic acid	Nurmann and Strack, 1979
Ester hydrolysis	Esterase	Sinapine	Sinapic acid + choline	
Glucohydrolysis (as specific case for glycohydrolysis)	Glucosidase	Formononetin-7-O-glucoside	Formononetin + glucose	Hösel, this volume, Chapter 23
Amide hydrolysis	Amidase	Feruloylputrescine	Ferulic acid + putrescine	Smith, 1980

growth, plants, in most cases, do not rely on secondary compounds as energy sources or to sustain growth. The catabolism of secondary compounds in plants can be observed to occur simultaneously with their synthesis of these products. Substrate induction of catabolic enzymes, observed with microorganisms, has not yet been conclusively shown to occur in tissues of higher plants, although conjugation reactions of aromatic acids seem to be substrate-induced (Venis and Stoessl, 1969). Because of these differences, the principles under which unidirectional catabolic pathways operate in microorganisms may apply to plants only with significant restrictions. Such principles are, however, useful for demonstrating the chemical reactions to be possibly expected in plants.

Finally, it should be noted that degradative processes in plants appear to be integrated into a metabolic grid. In such a grid the substrate to be degraded and the intermediates formed may be simultaneously subjected to further degradation, conjugation, interconversion, and/or polymerization reactions (Sections II,C,1–3). Suitable examples are encountered during metabolism of the protoalkaloid hordenin (Meyer and Barz, 1978) (Section IV) and the isoflavone daidzein (Barz, 1977a) (Section III,C,4).

III. DEGRADATIVE REACTIONS OF PHENOLIC COMPOUNDS

Phenolic compounds form one of the most prominent classes of natural products in plants. They are very reactive products chemically and easily subjected to oxidation, substitution, and coupling reactions. In this section some recent aspects of the complex metabolism and degradation of phenolic constituents will be described. In particular, attention will be given to phenolic compounds whose catabolic sequences are at least partly established. A complete listing of all information on phenolic turnover in plant systems is beyond the scope of this chapter. Thus, several interesting groups of plant phenolics (i.e., coumarins, xanthones, and quinones) are not being dealt with. Previous data on plant degradative reactions are available in several reviews (Grisebach and Barz, 1969; Ellis, 1974; Barz and Hösel, 1975).

A. Aspects of Ring Fission Reactions

It was formerly believed (Schwarze, 1958; Reznik, 1960) that plants could form aromatic rings but that they did not possess the potential to cleave these structures. However, recent studies with plants and plant cell cultures, as well as enzymatic investigations, have unequivocally demonstrated the capability of plant systems for the fission of aromatic structures (Barz, 1977a,b; Ellis, 1974). A high degree of analogy with microbial pathways appears to exist. The fundamental mechanisms of oxygen-dependent ring

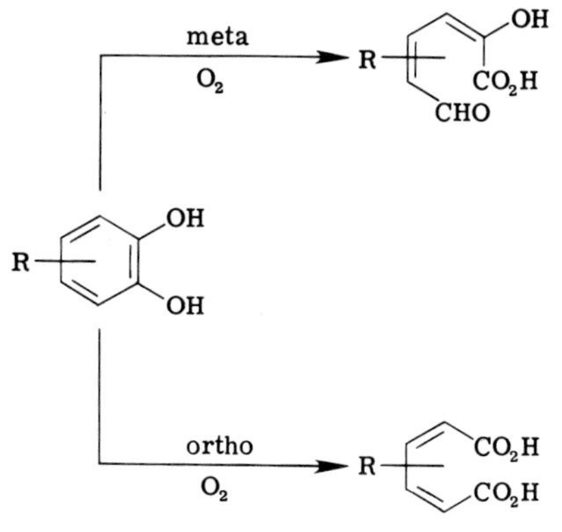

$$(3)$$

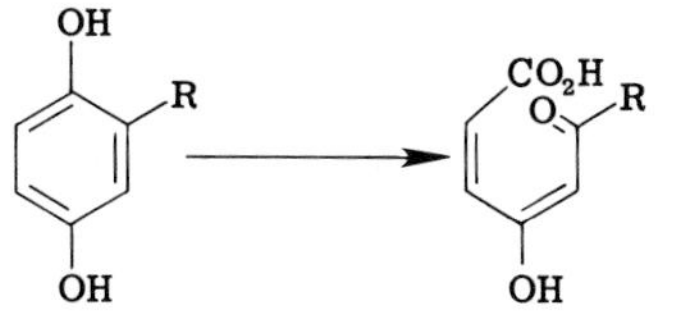

$$(4)$$

cleavage reactions [Eqs. (3) and (4)], namely, meta or ortho fission of 1,2-dihydroxy compounds and the rupture of 1,4-dihydroxy aromatic structures seem to occur in both microorganisms and plants.

Evidence for aromatic cleavage reactions in plants is available from tracer studies on the biosynthesis of the betalain precursor betalamic acid (Fischer and Dreiding, 1972; Impellizzeri and Piatelli, 1972), of stizolobic and stizolobinic acids (Ellis, 1976; Saito and Komamine, 1976), and of triglochinin (Sharples *et al.*, 1972) (Scheme 2). The meta cleavage of 3,4-dihydroxyphenylalanine (DOPA) (XXII) has clearly been demonstrated in the case of stizolobic acid (XXIII), betalamic acid (XXIV), and stizolobinic acid (XXV), and ortho cleavage of a 3,4-dihydroxyphenyl cyanogenic glucoside may be involved in the case of triglochinin (XVI) biosynthesis. The DOPA-cleaving enzymes involved in the formation of stizolobinic and stizolobic acids were isolated from *Stizolobium hassjoo* (Saito and Komamine, 1978) seedlings. They were shown to be specific for the carbon–carbon aromatic bond, which is cleaved, and for DOPA as the sole *o*-dihydroxy substrate.

Another example of a plant dioxygenase was reported by Durand and Zenk (1974a,b) who found in various plant cell cultures the homogentisic acid pathway long known from animals and microorganisms for the catabolism of tyrosine (Scheme 3). Homogentisate, which is used for the biosynthesis of prenylquinones in chloroplasts (Thomas and Threfall, 1974), is preferentially catabolized when given to cell cultures (Ellis, 1973) or when formed from exogenously applied tyrosine (Durand and Zenk, 1974a). The regulation of the homogentisate pathway and the properties of the dioxygenase are presently still unknown.

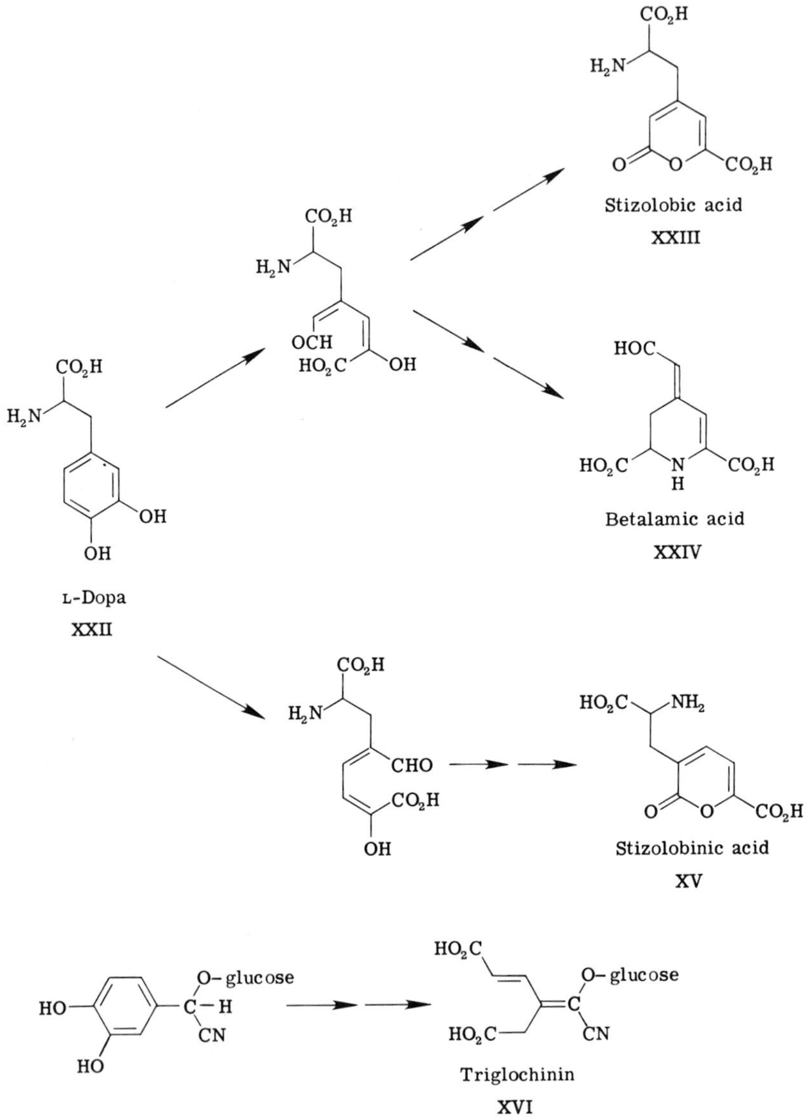

Scheme 2. Examples of ring fission reactions of *o*-dihydric phenols in plants. L-DOPA is a demonstrated precursor of three nitrogenous acids, and 3,4-dihydroxymandelonitrile is the expected precursor of triglochin (see text for references).

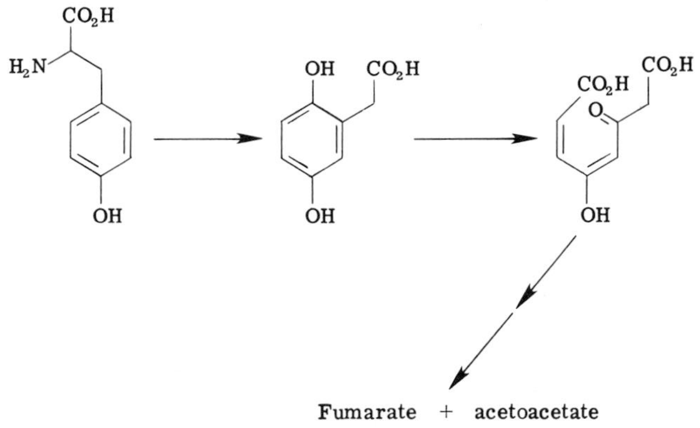

Fumarate + acetoacetate

Scheme 3. The homogentisate pathway (Durand and Zenk, 1974a,b).

Two cases of ortho fission by a dioxygenase were recently reported. Sharma and Vaidyanathan (1975) succeeded in isolating 2,3-dihydroxybenzoate 2,3-dioxygenase from *Tecoma stans*. Using protein extracts from the same plant, Mohan *et al.* (1979) purified protocatechuate 3,4-dioxygenase and characterized the product of the enzyme reaction as 3-carboxy-*cis,cis*-muconic acid. The enzyme required ferrous ions and was highly specific for protocatechuate because no substrate analogue was attacked.

More circumstantial evidence is available for the cleavage of such *o*-dihydric phenols as catechol, vanillic acid, 3,4-dihydroxyphenylacetic acid, and caffeic acid (Ellis, 1971; Barz and Hösel, 1975; Barz, 1977b). Though these compounds all appear to be better substrates for degradative reactions than the equivalent monophenols in intact plants (as judged by $^{14}CO_2$ formation from suitably labeled compounds), such *in vivo* studies should be interpreted with caution. The ring cleavage of phenols catalyzed by peroxidases has recently been observed (Section III,C,1) (Prasad and Ellis, 1978; Patzlaff and Barz, 1978). Aerobic fission of aromatic structures therefore may not necessarily be a result of the action of dioxygenases. The physiological significance of peroxidative ring cleavage reactions remains to be established, and further studies on ring fission in plants are required. The results reported above, however, show that most promising studies can be carried out in this field.

B. Simple Phenols and Aromatic Acids

Numerous independent observations have shown that simple phenols (i.e., catechol, hydroquinone), hydroxybenzoic acids, phenylacetic acids, and cinnamic acids are actively metabolized and finally either polymerized or catabolized by the producing plant. Aspects of ring fission reactions of these substrates can be found in Section III,A. Most of these compounds occur in the form of various glycosides or other conjugates (Section II,C,2), so that enzymes for the hydrolytic formation of the aglycones are an essential link in the postulated catabolic sequences (see Hösel, this volume, Chapter 23).

Turnover of catechol and hydroquinone that arise by hydroxylation-induced decarboxylation of salicylic acid and p-hydroxybenzoic acid (Kindl, 1971) or of salicin and picein is indicated by rapid diurnal or seasonal variation in various plants (i.e., Thieme, 1965; Wehnert, 1970). Turnover of the hydroquinone moiety of various conjugates is best explained by oxidative polymerization without ring fission (Reuter, 1970).

β-Oxidation of the side chain of cinnamic acids is the main biosynthetic pathway leading to benzoic acids. The substitution pattern of the latter is mainly determined at the cinnamic acid stage (Kindl, 1971). This conversion is one catabolic step in the degradative pathway of phenylalanine and tyrosine (Kindl, 1971; Barz and Hösel, 1975) and appears to be an important route for these C_6–C_3 compounds.

Various flavonoids (Section III,C,1) have also been shown to be catabolized via cinnamic acids to benzoic acids (Barz and Hösel, 1978). Recent enzymatic investigations have shown that the enzyme for p-hydroxybenzoate formation from p-coumaric acid appears to be associated with glyoxysomal membranes and/or with the mitochondrial membrane (Hagel and Kindl, 1975). Though this chain-shortening reaction is viewed as being similar to fatty acid β-oxidation, the intermediate formation of cinnamic acid–CoA esters has not been demonstrated.

Hydroxyphenylacetic acids that would arise from phenylalanine and tyrosine via the appropriate α-ketoacids have occasionally been demonstrated to be metabolically active intermediates in catabolic sequences (Section IV). The earlier literature on phenylacetic acid degradation has been summarized by Barz and Hösel (1975).

Recent studies have shown that in *Iris* species m-carboxyphenylalanine is degraded to m-carboxylphenylacetic acid (Larson and Wieczorkowska, 1978). With the exception of 2,5-dihydroxyphenylacetic acid (homogentisic acid), however (Section III,A), details about the subsequent degradative routes and the ring fission mechanisms still have to be sorted out.

Benzoic acid degradation has been especially investigated in connection with decarboxylation and O-demethylation reactions. Hydroxylation-induced decarboxylation of hydroxybenzoic acids constitutes a well-known route to simple phenols. The formation of hydroquinone, methoxyquinone, and catechol from p-hydroxybenzoic acid, vanillic acid, and salicylic acid are good examples (see Gross, this volume, Chapter 11). Studies with aseptically grown wheat seedlings and various cell suspension cultures (Harms *et al.*, 1971, 1972) demonstrated that benzoic acids bearing an hydroxyl group para to the carboxyl were most actively decarboxylated. Further investigations with purified enzymes (Berlin and Barz, 1975; Kamel *et al.*, 1977; Krisnangkura and Gold, 1979) showed that the oxidative decarboxylation of *p*-hydroxybenzoic acids was catalyzed by peroxidases using stoichiometric amounts of H_2O_2. Depending on the substitution pattern (Scheme 4) either *p*-quinones, dimers, or polymers were identified as main products. With respect to specificity all isoenzymes of peroxidase from various sources carried out this decarboxylation reaction equally well. The physiological significance of this peroxidative reaction again remains to be determined (see Barz and Hösel, 1978).

O-demethylation, indicating the ability of plant cells to split alkylaryl–ether bonds has been studied rather carefully with specifically labeled benzoic acids in plant cell suspension cultures (Harms *et al.*, 1972). The oxidizing system (see Poulton this volume, Chapter 22) appears to be very specific for the para position because *m*-methoxy groups are oxidized with almost two orders of magnitude less efficiently. Anisic acid, veratric acid, and 3,4,5-trimethoxybenzoic acid yield *p*-hydroxybenzoic acid, vanillic acid, and syringic acid, respectively, and in soybean cell cultures vanillic acid thus formed is immediately transformed to vanillic acid *O*-glucoside. As such it is protected against subsequent oxidative decarboxylation (Barz *et al.*, 1978a). These *in vivo* findings have not yet been followed up by enzymatic studies.

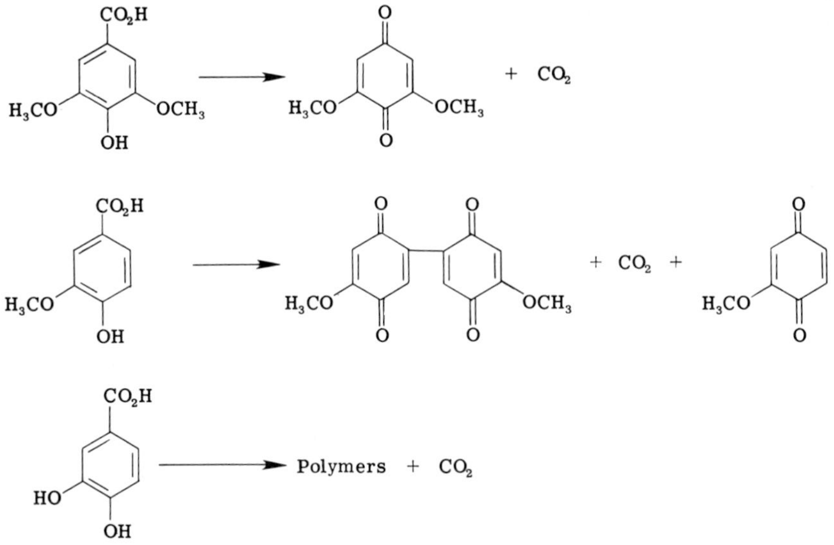

Scheme 4.   Peroxidative reactions of syringic, vanillic, and protocatechuic acid.

C. Flavonoid Constituents

Flavonoid turnover as such and in connection with various developmental processes is well documented (Barz, 1975b). Compounds of this group are therefore highly suitable for investigating the general principles of catabolism of secondary plant products. The various classes of flavonoid constituents have, however, been investigated to differing degrees. Thus anthocyanins, proanthocyanins, and catechins, which are among the most widely occurring plant phenols, have been rather poorly investigated with respect to degradative reactions (see review by Barz and Hösel, 1975). In this chapter, the more recent results on flavonoid degradation will be outlined. Emphasis is again placed on degradative pathways.

1. Chalcones, Flavanones, and Aurones

Chalcones and flavanones are the first C_{15} intermediates in flavonoid biosynthesis (Grisebach and Barz, 1969; Grisebach, 1978). They occur not only as transient intermediates in biosynthesis but also as major phenolic constituents in petals and anthers (see Wiermann, this volume, Chapter 4), grapefruit (Raymond and Maier, 1977), and peach buds. In the last-mentioned case the flavanone glucoside prunin has been shown to be rapidly metabolized prior to bud opening (Erez and Lavee, 1969).

In plant tissues chalcones and flavanones are readily interconvertible with the aid of chalcone-flavanone isomerases (see Hahlbrock, this volume, Chapter 14), so that catabolic reactions involving either group eventually affect both.

Feeding experiments with seedlings and cell cultures with ^{14}C-labeled flavanones or chalcones possessing either a phloroglucinol or a resorcinol type of substitution pattern in ring A demonstrated catabolic reactions to occur (Patschke et al., 1964a; Janistyn et al., 1971; Berlin et al., 1974). CO_2 from ring A and p-coumaric acid and p-hydroxybenzoic acid from ring B were isolated (Scheme 5). p-Coumaric acid itself was partly refunneled into flavonoid biosynthetic sequences (Patschke et al., 1964b).

Intensive enzymatic studies on the catabolism of 5,7,4'-tri-hydroxyflavanone (naringenin) have shown that the flavanone-degrading enzyme from soybean and chick-pea cell cultures and buckwheat plants is a peroxidase requiring H_2O_2 (Patzlaff, 1978; Patzlaff and Barz, 1978). Peroxidative attack on flavanones requires a free hydroxyl substituent in position 4' and occurs in a complex fashion so that a great number of catabolites are formed. Most of these are in turn sensitive to subsequent oxidation, so that further degradation or oxidative polymerization of the intermediates occurs. The structures elucidated so far (Scheme 6) allow us to explain the formation of primary metabolites (i.e., acetic acid, succinic acid.) or of CO_2

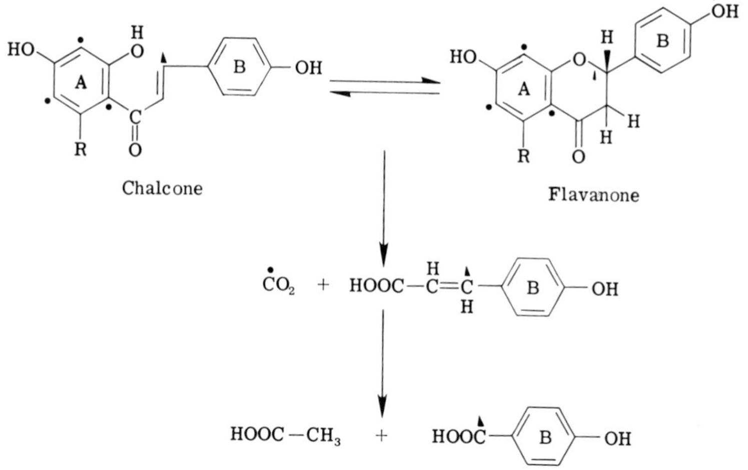

Scheme 5. Chalcone-flavanone degradation in plants where both phloroglucinol (R = OH) and resorcinol (R = H) substituted rings are split (Barz and Hösel, 1978).

from practically every carbon atom of the flavanone skeleton. Some products may well have arisen from nonenzymatic reactions with H_2O_2.

Four obviously parallel metabolic routes can be discerned in Scheme 6, indicating that flavanones are attacked at several positions of the heterocyclic ring and rings A and B.

1. Meta hydroxylations by peroxidases (conversion of I to XLI) have repeatedly been demonstrated with various substrates (Halliwell and Ahluwalia, 1976).
2. Formation of C_6–C_3 (XXXV) and C_6–C_1 (XL) compounds provides enzymatic evidence for the results of the feeding experiments described in Scheme 5. The pathway proposed for the formation of p-coumaric acid (XXXV) via XXXIII and XXXIV with the aid of peroxidases (Patzlaff and Barz, 1978) suggests that C-acylhydrolases, which are presumed to participate in the degradation of other compounds (Section III,C,2), are not involved in flavanone degradation.
3. Peroxidative formation of phloroglucinol carboxylic acid (XXX) from ring A is highly reminiscent of similar results obtained with various flavonols (Schreiber, 1975; Zaprometov, 1977; e.g., Section III,C,3). Isolation of the unsaturated lactone–carboxylic acid (XXXII) is of great interest because it provides further unequivocal evidence that dihydroxy aromatic rings can be split by peroxidases (Prasad and Ellis, 1978). Such data are in accord with chemical analogies where the fis-

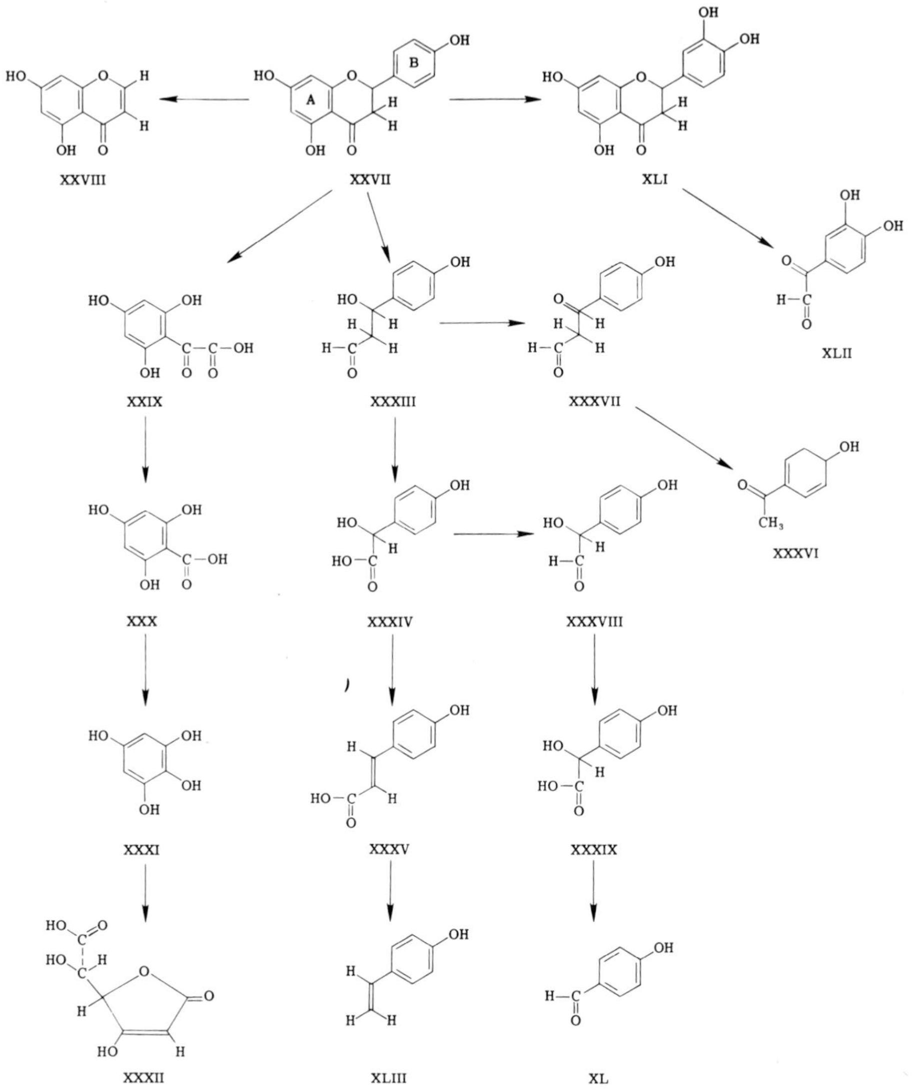

Scheme 6. Catabolites formed during peroxidative degradation of the flavanone naringenin (Patzlaff and Barz, 1978).

sion of o-diphenols by peracids (Schulz and Hecker, 1973; Walter and Hecker, 1973) or by metal dioxides (Tsuji and Takayanagi, 1978), yielding muconic acid-type compounds, has been demonstrated. These data on peroxidases offer interesting alternatives to the widely accepted dioxygenases as ring-cleaving enzymes (Section III,A).

4. Elimination of ring B with the formation of a chromone (XXVIII) further supports the proposal that chromones unsubstituted at the 2 and 3 positions (Chiji et al., 1978) arise by flavonoid degradation (Bourwieg and Pohl, 1973). This proposal was first substantiated by Janistyn and Stocker (1976) who, using an enzyme preparation from *Mentha longifolia*, observed the formation of a chromoneglycoside from the flavanone eriodyctyol 7-O-glycoside with liberation of the B ring as a substituted hydroquinone [Eq. (5)].

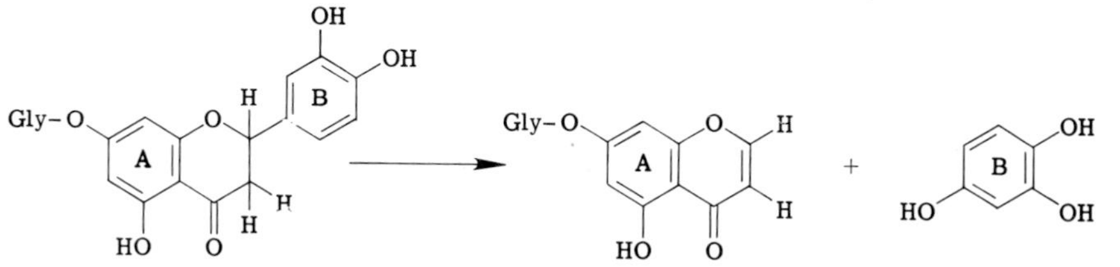

$$(5)$$

When naringenin ethers with methyl groups in position 5 and/or 7 are subjected to peroxidative degradation, formation of the equivalent chromones and hydroxylation at the 3' position of ring B become the most important reactions. Therefore free hydroxyl groups in ring A are essential for peroxidative degradation of the phloroglucinol structure. Mechanistic considerations of the pathways depicted in Scheme 6 (Patzlaff, 1978) visualize that peroxidative degradation of flavanones begins with the formation of a phenoxy radical in position 4' (Fig. 4), which can be stabilized by several resonance forms. Subsequent reactions of these isomers with either OH radicals or water explain the seemingly parallel sets of catabolic pathways proposed in Scheme 6. One likely intermediate would be a 2-hydroxyflavanone. Such compounds have recently been isolated as natural products from various plant sources (Chopin et al., 1978). Future studies are required to establish whether these 2-hydroxyflavanones are intermediates in flavanone degradation both in the plants concerned and in enzymatic studies.

The degradation of aurones such as 4',6-dihydroxyaurone (hispidol) was investigated in soybean cell suspension cultures (Barz et al., 1974). These cultures, which accumulate hispidol in low concentrations, degraded the applied ^{14}C-labeled substrate with the liberation of ring B as p-hydroxybenzoic acid [Eq. (6)]. The latter compound was only transiently

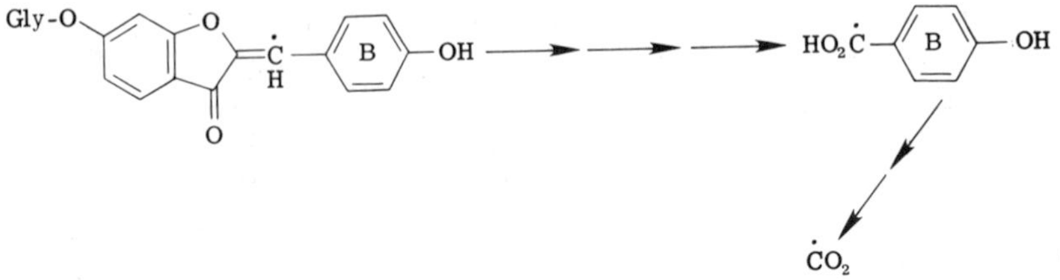

$$(6)$$

accumulated, whereas the bulk of the radioactivity of hispidol was incorporated into ethanol-insoluble material. Later enzymatic studies demonstrated that aurone catabolism in these cell cultures was catalyzed by peroxidases (Teufel, 1974).

The catabolic sequences described are remarkable because one enzyme, namely, peroxidase, catalyzes the complete degradation of the flavanone molecule. This degradative power of peroxidases, which has been observed with various other substrates (chalcones, flavonols), should be further studied and evaluated. Similarly the participation of peroxidases in biosynthetic reactions involving chalcones should be included in the evaluation (Wong and Wilson, 1976; Rathmell and Bendall, 1972).

The great similarity between the peroxidative degradation of flavanones/chalcones and the peroxidative destruction of aurones and flavonols (Section III,C,3) indicate that flavonoids are a new class of substrates for peroxidases. In all cases of 4'-hydroxy-substituted flavonoids, ring B is converted to the corresponding substituted benzoic acid. It is also of interest that this general pattern of flavonoid disintegration closely resembles microbial catabolism of these plant products (Barz and Hösel, 1975).

2. Dihydrochalcones

Though very few dihydrochalcones have been isolated from plants, some species such as apple accumulate rather high concentrations of these phenols. In apple leaves the dihydrochalcone glucoside phloridzin (Scheme 7) undergoes degradation. Three phloridzin-specific glucosidases have been purified and shown to be involved in phloridzin metabolism (Podstolski and Lewak, 1970). The resulting aglycone phloretin was further hydrolyzed to phloroglucinol and p-hydroxydihydrocinnamic acid (Grochowska, 1967) by protein extracts from the sap of apple spurs. Presumably a C-acylhydrolase is involved in this reaction. Further degradation of these two breakdown products has not yet been studied.

An alternative reaction of phloretin is catalyzed by apple phenolase and leads to polymerization via 3-hydroxyphloretin. Thus phloretin is another example where both catabolism and polymerization can occur.

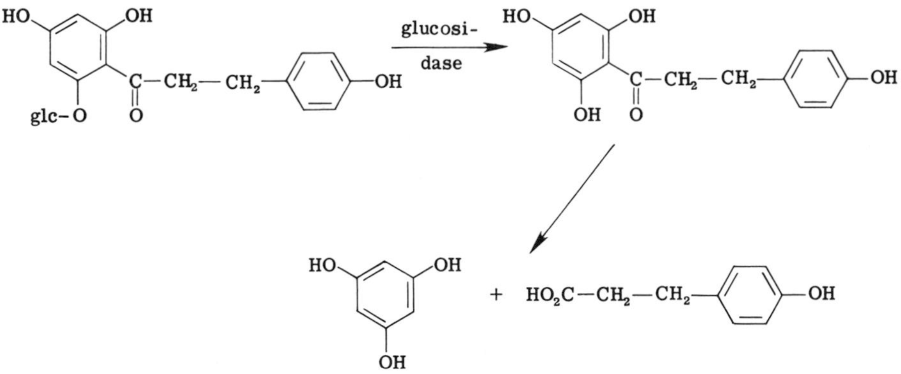

Scheme 7. Catabolic sequence for phloridzin in apple tissue (Grochowska, 1967).

3. Flavonols

Flavonols such as kaempferol (XLIVa), quercetin (XLIVb), and isorhamnetin (XLIVc) occur widely as glycosides in petals, leaves, and stem tissues. Many studies indicate that these compounds are actively metabolized in plants and that they are subject to catabolic reactions. Flavonol metabolism is most pronounced at times of intensive plant growth or during differentiation (cf. Wiermann, this volume, Chapter 4). The earlier literature on this subject has been reviewed (Barz and Hösel, 1975; Sections II,A and B).

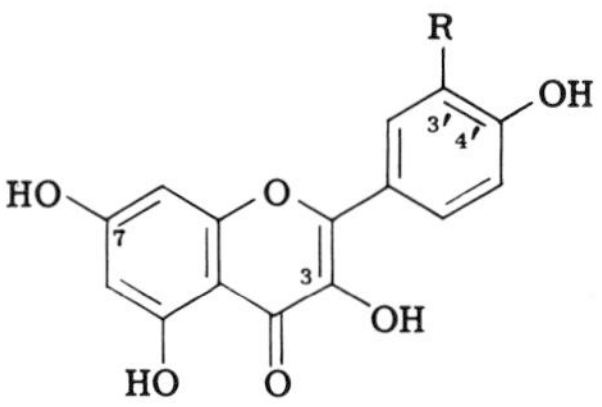

(XLIVa) R = H
(XLIVb) R = OH
(XLIVc) R = OCH$_3$

All data indicate that the constant concentrations of flavonols result from a balance between synthesis and degradation. Furthermore, environmental factors affecting the general metabolism of a plant ultimately influence the steady state concentrations of these phenolic constituents. The relationship between the physiology of flavonol-containing cells and the rate of flavonol turnover in these cells is indicated by studies on light-grown parsley cell suspension cultures (D. Hüsemann and W. Barz, unpublished; Barz, 1977b).

In these cell cultures flavonol accumulation and flavonol turnover are associated with growth, and both processes cease when the cultures enter the stationary phase.

It is generally accepted (cf. Hösel, this volume, Chapter 23) that degradation of polyphenol glucosides is initiated by glycosidase-catalyzed removal of sugar moieties, the liberated aglycones and carbohydrates being channeled separately into catabolic routes. Comparative feeding experiments with ^{14}C-labeled 3-O- and 7-O-glucosides of kaempferol (XLIVa) in plant cell suspension cultures have further supported this concept (Muhle *et al.*, 1976). However, kaempferol 3-O-glucoside was more rapidly oxidized than the 7-isomer. Such data are matched by similar observations on different turnover rates of these isomers in intact plants (Dittrich and Kandler, 1971). The preferential oxidation of the 3-isomer seems to indicate an active mechanism for the hydrolysis of flavonol 3-glucosides. However, intensive screening for flavonol 3-glycoside-specific β-glycosidases using a newly devised, sensitive assay has so far failed to detect any glycosidase activity with the required specificity (Surholt and Hösel, 1978). One may therefore postulate that the removal of sugars from position 3 of flavonols is catalyzed by a transferase-like reaction because these glycosides are energy-rich compounds (Sutter and Grisebach, 1975). The mechanism of glucose removal from flavonol 3-O-glucosides requires further study.

Tracer studies (Hösel *et al.*, 1972; Muhle *et al.*, 1976) with cell cultures and enzymatic investigations (Hösel *et al.*, 1975; Zhanayeva *et al.*, 1977; Zaprometov, 1977) with protein preparations from various plants and plant cell suspension cultures have demonstrated that peroxidases in the presence of H_2O_2 catalyze flavonol degradation (Scheme 8). Flavonols with free hydroxyl groups in positions 3 and 4' are first converted to the corresponding 2,3-dihydroxyflavanones. All peroxidase isoenzymes obtained from various plants (*S. alba, C. arietinum, Armoracia rusticana, Bupleurum multinerve*) possess this property of flavonol degradation without any quantitative difference in specificity.

In a second enzyme step these unstable 2,3-dihydroxyflavanones, which have recently become available by chemical synthesis (Hauteville *et al.*, 1979), are further degraded by peroxidases to substituted benzoic acids derived from ring B. Catabolites of ring A such as phloroglucinol carboxylic acid (Scheme 8) have so far only been found in studies on *B. multinerve* and *B. aureum* protein preparations (Zhanayeva *et al.*, 1977; Zaprometov, 1977). Results in the authors' laboratory have shown that CO_2 and a great number of other compounds still awaiting structural elucidation were formed in the second and further steps. In general, peroxidative degradation of flavonols leads to complete decomposition of the molecule and is thus comparable to peroxidative flavanone destruction (cf. Scheme 6). The catabolic sequence in Scheme 8 is substantiated by studies on the peroxidative degradation of the

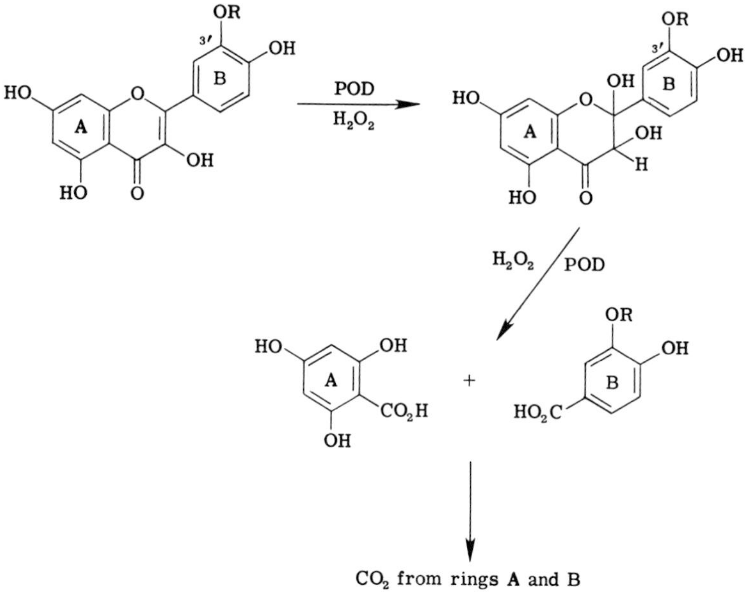

Scheme 8. Catabolites of peroxidative degradation of flavonols (Barz and Hösel, 1978). POD, peroxidase.

unsubstituted flavonol (Schreiber, 1975), where benzoic acid (from ring B) together with salicylic acid and 2-(o-hydroxyphenyl)glyoxylic acid (both from ring A and the heterocyclic ring) could be isolated.

Peroxidative degradation of flavonols that is H_2O_2-dependent is catalyzed by true peroxidases that possess ferriprotoporphyrin IX as a prosthetic group and also show β-indolylacetic acid (IAA) oxidase activity (see also Section II,C,3,b). The latter reaction has been described as an important pathway for auxin destruction (Hamilton et al., 1976; McClure, 1975). Quercetin (XLIVb) has repeatedly been shown to inhibit IAA oxidase activity (Shinshi and Noguchi, 1975; Sano, 1971; Frey-Schröder and Barz, 1979).

In contrast to the results with true peroxidases (Scheme 8) recent studies (Frey, 1977; Frey-Schröder and Barz, 1979) on Mentha piperita plants and M. arvensis cell suspension cultures resulted in the characterization of a distinctly different set of peroxidase isoenzymes. Though they converted quercetin to the corresponding 2,3-dihydroxyflavanone (Scheme 8), these enzymes differ from true peroxidases in various properties. The Mentha enzymes require hydroxyl substituents in positions 3, 3′, and 4′ in the flavonol, use molecular oxygen (0.5 mol O_2/mol quercetin) and not H_2O_2, possess no ferriprotoporphyrin moiety or IAA oxidase activity, and are strongly inhibited by reducing agents such as NADH and ascorbic acid.

These data and various other inhibitor studies suggest that the *Mentha* enzymes appear to be a new class of enzymes ("flavonol oxygenases") for flavonol degradation in plants. The mechanism of oxygen action with these flavonol oxidases requires further investigation.

As shown with flavonols (Scheme 8) and flavanones (Section III,C,1; Scheme 6) peroxidases catalyze the oxidation, degradation, and polymerization of numerous compounds from higher plants. It is by no means clear, however, that these *in vitro* reactions represent *in vivo* functions. Despite various attempts to correlate peroxidases and their different substrate specificities with physiological functions (Reigh *et al.*, 1973; Mäder *et al.*, 1977), a large number of unsolved problems remain. It is difficult to visualize that degradative reactions involving so ·many catabolites, as shown in Schemes 6 and 8, will be easily linked with the primary metabolism of a plant.

4. Isoflavonoids

The isoflavones biochanin A (XLV), formononetin (XLVIa), and daidzein (XLVIb) in the garbanzo bean, *C. arietinum* L., and in the mung bean, *Phaseolus aureus* Roxb., were among the first phenolic plant constituents to be systematically analyzed for turnover and degradation (see review by Barz and Hösel, 1975). These compounds are actively metabolized both in the intact plant and in plant cell cultures (Barz, 1977b). With formononetin low

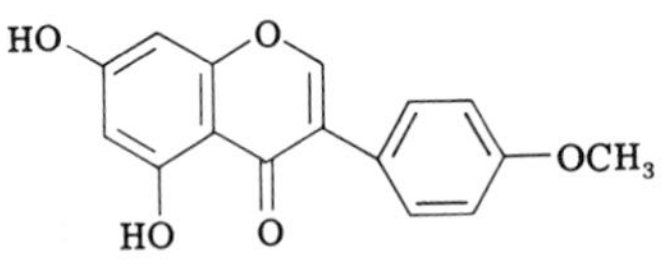

XLV

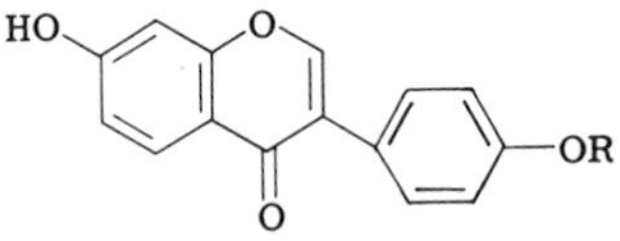

(XLVIa) R = CH$_3$
(XLVIb) R = H

values of approximately 50–70 h for the biological half-life indicated rapid turnover, and large variations in turnover rates resulted upon changes in the environmental or physiological conditions (Barz, 1975b; see Section II,B). These studies also demonstrated that turnover and degradation of a compound such as biochanin A depended greatly on the tissue site of accumulation. Rather rapid metabolism of biochanin A 7-O-glucoside could be measured in leaves and in cell cultures, although the same compound seemed to be metabolically inactive in the rhizodermis (Section II,B). The metabolic inactivity of this pool of biochanin A may be due to the fact that rhizodermis tissue lacks an isoflavone glucoside-specific glucosidase (cf. Hösel, this volume, Chapter 23). Such a glucosidase appears to be required to initiate the catabolic sequence since formononetin 7-O-glucoside in the cortex tissue is rapidly metabolized by a specific glucosidase found in the same tissue.

In recent experiments with potent inhibitors of PAL that block subsequent phenylpropanoid biosynthetic sequences (Amrhein and Holländer, 1979), Amrhein and Diederich (1980) independently determined formononetin and biochanin A turnover in roots of *C. arietinum*. These authors also measured an appreciable turnover of formononetin (72 h half-life), whereas no turnover could be measured for biochanin A under conditions of steady state concentrations. The systematic tracer studies on isoflavones help to explain the great variability observed in the amount of estrogenic isoflavones in important fodder plants such as clover and alfalfa (see review by Barz and Hösel, 1975).

Tracer and enzymatic studies on daidzein (XLVIb) metabolism in mung bean plants and cell cultures led to the formulation of a metabolic grid of four concomitant reactions (Barz, 1977a). Daidzein is (1) an intermediate in the biosynthesis of 3-phenylcoumarins, (2) glucosylated in the 7-position, (3) oxidatively polymerized via 7,3',4'-trihydroxyisoflavone, and (4) degraded via unknown intermediates. This metabolic grid demonstrates the full complexity of phenolic metabolism where the extent and ratio of the simultaneous reactions depend on the availability of enzymes and many other factors (Section II,B).

Isoflavone metabolism is further complicated by the fact that a rapid conversion to the corresponding isoflavanone seems to occur (Barz and Hösel, 1975; Dewick and Martin, 1979); the resulting isoflavanones may even be regarded as the more reactive species.

Although the degradation of various isoflavanones and isoflavones together with aromatic ring cleavage has unequivocally been shown to occur in sterile plants and plant cell suspension cultures, intermediates in these catabolic sequences have not yet been isolated (Barz and Hösel, 1978; Berlin *et al.*, 1974).

Great progress has recently been made in the isolation and characterization of various glucosidases specific for isoflavone O-glucosides (cf. Hösel, this volume, Chapter 23). The occurrence, specificity, and localization of these enzymes are in line with the assumption that such glucosidases represent the first step in funneling a glucosylated secondary compound into catabolic sequences (Hösel *et al.*, 1977). As measured with cell suspension cultures, a substantial portion of the glucosidase activity is located in cell wall structures. Therefore, in feeding experiments with labeled isoflavone O-glucosides, most of the substrate is hydrolyzed during uptake. Exogenously applied isoflavone glucosides thus do not reach the natural site of isoflavone metabolism unaltered (Hösel *et al.*, 1977; see also Section II,B).

IV. DEGRADATIVE REACTIONS OF NITROGENOUS COMPOUNDS

Higher plants synthesize a vast array of nitrogen-containing secondary products whose structures have long intrigued the natural product chemist. Since most of these compounds are derived from protein amino acids, the link between primary and secondary metabolism is especially obvious.

A. Early Studies of Alkaloid Metabolism

Alkaloids were among the first secondary compounds in plants for which turnover and degradation were predicted. It is well known that alkaloid-containing plants have optimum periods for harvest and that stages in plant development exist where the tissue is low in or devoid of alkaloids (Neumann and Tschoepe, 1966; Mothes and Schütte, 1969; Robinson, 1974; Waller and Nowacki, 1978; see also Section II,A). Modern studies with *Atropa belladonna* (atropine), *Conium maculatum* (coniine, γ-coniceine), *Papaver somniferum* (morphine, codeine, thebaine), and *R. communis* (ricinine) have clearly demonstrated rapid seasonal variations and even diurnal fluctuations in alkaloid constituents. Such fluctuations in alkaloid levels may be explained in part by rapid shifts between free and bound forms, with the latter yielding the free form upon hydrolysis. Waller and Nowacki (1978) summarize the details as found for hemlock and poppy alkaloids. On the other hand, it is clear that certain plants have the ability to degrade their alkaloidal constituents completely. The following section will present some specific examples.

B. Modern Aspects of Alkaloid Catabolism

Waller and Dermer (this volume, Chapter 12) describe the substantial evidence for catabolism of the pyridine alkaloids nicotine, anabasine, and ricinine. In this connection, recent studies on nicotinic acid degradation in sterile plant cell suspension cultures are of interest (Barz, 1977b). The ability to degrade nicotinic acid seems to be confined to several orders of higher plants. Contrary to expectations based on microbial pathways, 6-hydroxynicotinic acid is not an intermediate during catabolism. The exact catabolic sequence involved is under investigation.

As pointed out in Sections II,C,3 and C,4 catabolism under aerobic conditions involves higher oxidized intermediates where they serve as substrates for ring fission, chain cleavage, or polymerization reactions. This principle is demonstrated for alkaloids by the results shown in Figs. 5 and 6.

During feeding experiments with labeled sparteine (XLVII), lupanine (XLVIII), and thermopsine (XLIX) in *Thermopsis macrophylla* and *Baptisia*

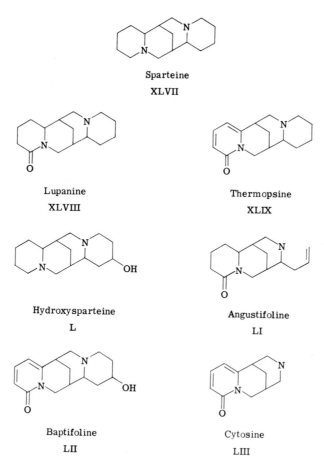

Sparteine

XLVII

Lupanine

XLVIII

Thermopsine

XLIX

Hydroxysparteine

L

Angustifoline

LI

Baptifoline

LII

Cytosine

LIII

Fig. 5.   Structures of lupine alkaloids derived from sparteine (XLVII), lupanine (XLVIII), and thermopsine (XLIX) in species of Leguminosae. (after Nowacki and Waller, 1975).

leucopheya, the corresponding higher oxidized alkaloids shown in Fig. 5 were derived from these substrates (Nowacki and Waller, 1975). The distribution of radioactivity in these derivatives supported the aforementioned principle and proved that alkaloid catabolism did occur.

In *Cephalotaxus harringtonia* trees (Delfel and Rothfus, 1977; Delfel, 1980) free and esterified alkaloids of the homoerythrina and cephalotaxine (LIV) type gradually disappear. 11-Hydroxycephalotaxine (LV) and drupacine (LVI) are believed to be early intermediates in this catabolism. Various other examples can be cited (Mothes and Schütte, 1969), although many of the proposed oxidized intermediates do not accumulate in large amounts.

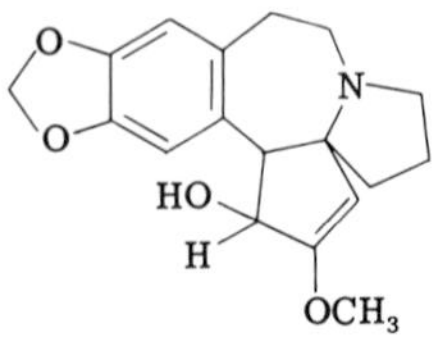

Cephalotaxine

LIV

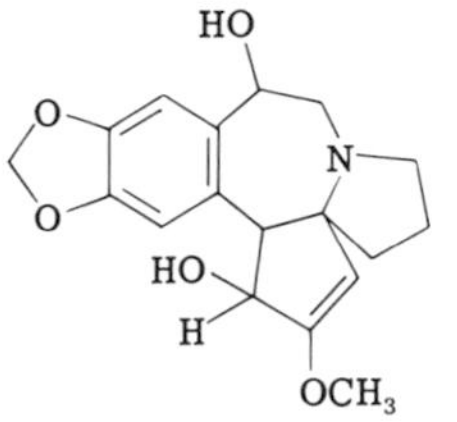

11-Hydroxycephalotoxin

LV

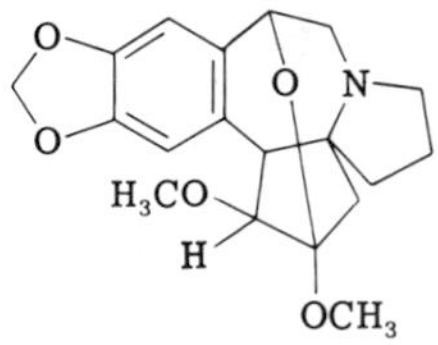

Drupacine

LVI

Fig. 6. Higher oxidized alkaloids (LV and LVI) derived from cephalotaxine (LIV) in *Cephalotaxus harringtonia*. (after Delfel, 1980).

Various other feeding experiments such as those with morphine (Miller *et al.*, 1973), atropine (Neumann and Tschöpe, 1966), and ricinine (Waller and Nowacki, 1978), have shown the conversion of substrate carbon into nonalkaloidal material such as sugars and amino acids. Except for the important and often observed early steps of O- or N-demethylation (morphine into normorphine, hyoscyamine into norhyoscyamine) (see Waller and Nowacki, 1978), such studies showed that degradation occurred but did not contribute to the knowledge of catabolic pathways. To overcome the difficulties caused by the low concentrations of catabolites, studies with suitably selected (Tabata *et al.,* 1978) plant cell cultures might be an alternative (Barz, 1977b; Meyer and Barz, 1978). Negative reports obtained with this approach Griffin, 1979) should not be considered representative.

In roots of young barley (*Hordeum vulgare*) plants the protoalkaloids hordenine (*N,N*-dimethyltyramine) and *N*-methyltyramine occur only during a few weeks after germination (Rabitzsch, 1959). Their accumulation is correlated with the presence of the enzymes tyramine and *N*-methyltyramine methyltransferase (Mann *et al.*, 1963). Although unsuccessfully studied in intact plants (Frank and Marion, 1956), recent investigations with barley cell suspension cultures and labeled hordenine, tyramine, and dopamine have demonstrated the catabolic sequence (Scheme 9) leading to 3,4,-dihydroxyphenylacetic acid (Meyer and Barz, 1978). This latter compound appeared to be the central catabolite in ring fission reactions (Section II,C,4).

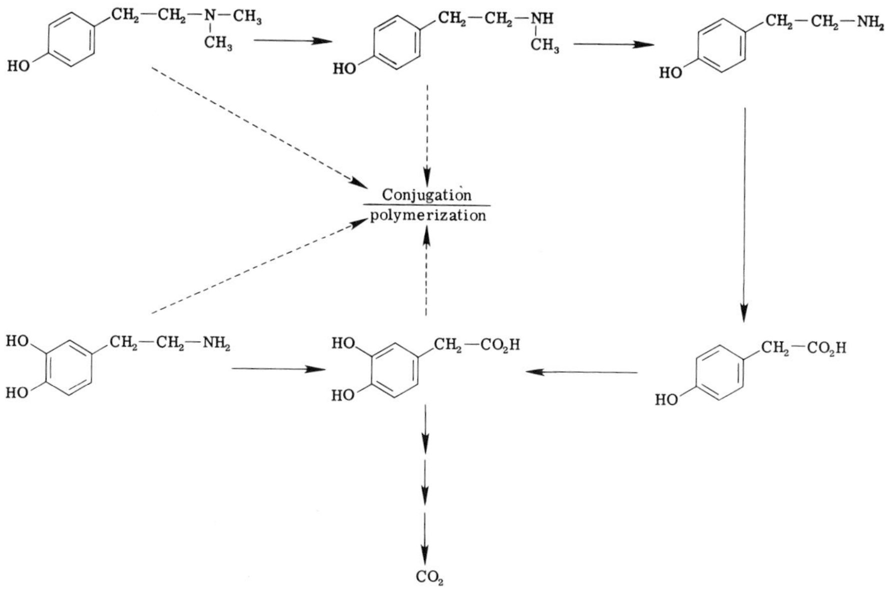

$$HO-\text{(ring)}-CH_2-CH_2-N(CH_3)-CH_3 \longrightarrow HO-\text{(ring)}-CH_2-CH_2-NH-CH_3 \longrightarrow HO-\text{(ring)}-CH_2-CH_2-NH_2$$

Conjugation
polymerization

$$HO,HO-\text{(ring)}-CH_2-CH_2-NH_2 \longrightarrow HO,HO-\text{(ring)}-CH_2-CO_2H \longleftarrow HO-\text{(ring)}-CH_2-CO_2H$$

$$\downarrow$$

$$CO_2$$

Scheme 9. Catabolic sequence of hordenine in cell suspension cultures of *Hordeum vulgare* (Meyer and Barz, 1978).

The additional isolation of p-hydroxybenzoic acid and p-hydroxymandelic acid as catabolites of p-hydroxyphenylacetic acid demonstrates that α-oxidation reactions also occur (see Galliard, Vol. 4, Chapter 3). As noted in Scheme 9, each intermediate also could undergo conjugation as glycosides (Section II,C,2) and polymerization reactions (Section II,C,3). Hordenine metabolism can therefore be described in terms of a metabolic grid (Sections II,A and III,C,4). Unpublished studies (E. Meyer, personal communication) with specifically labeled substrates have revealed that hordenine polymerization can occur without removal of the methyl groups and cleavage reactions of the side chain. This points to the aromatic ring as a site of polymerization reactions (Section II,C,3).

A puzzling observation involving hordenine, nicotine, and trigonelline N-demethylation (Barz, 1977b) is the high incorporation of isotope from ^{14}C-labeled N-methyl groups into insoluble cell residues. Such reactions have so far not been observed in O-demethylation studies because O-methyl groups were quantitatively oxidized to CO_2 (Barz, 1975a). This difference between N- and O-methyl groups requires enzymatic studies to assess the possible role of intermediate N-oxides (Phillipson and Handa, 1978) in N-dealkylation reactions. The results of studies with mammalian systems (Gescher et al., 1979) seem to support this difference (see also by Poulton, this volume, Chapter 22).

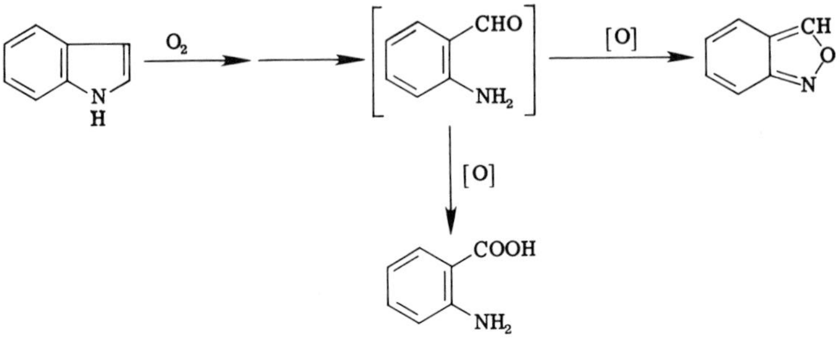

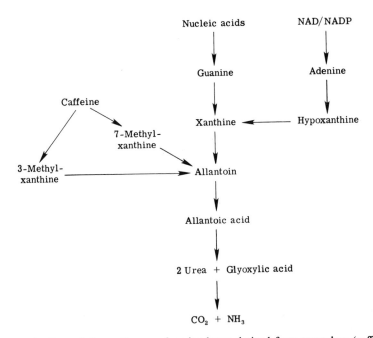

Scheme 10. Enzymatic cleavage of indole with an enzyme preparation from *Zea mays* L. (after Chauhan *et al.*, 1978).

Early studies with barley shoots (Digenis *et al.*, 1966; Digenis, 1969) on the tryptophan-derived alkaloid gramine (3-dimethylaminomethylindole) had already presented evidence for a side-chain oxidation sequence somewhat similar to the pathway shown in Scheme 9. 3-Hydroxymethylindole and indole-3-carboxylic acid were identified as the main metabolites. The important question remains whether the aromatic or heterocyclic rings of the indole system can also be cleaved by plant systems. As a first step in answer-

Scheme 11. Degradative pathways of purine bases derived from secondary (caffeine) or primary metabolism.

ing this question, recent studies by Chauhan and associates (1978) have succeeded in isolating from maize (*Zea mays* L.) leaves an enzyme preparation which in the presence of oxygen rapidly oxidizes indole as depicted in Scheme 10. Carbon atom 2 of indole is lost in this process. The metabolism of anthranilic acid has been thoroughly investigated in plants (see Köster *et al.*, 1978). With respect to the benzene ring of tryptophan early studies with plant cell cultures can be cited (Ellis and Towers, 1970). Although oxidation of [phenyl-^{14}C]tryptophan to $^{14}CO_2$ could be demonstrated, details of the reactions involved are unknown.

As a final example of alkaloid catabolism in plants, Scheme 11 summarizes findings on the degradation of caffeine and other purine bases such as guanine and adenine. This scheme for purine catabolism is based on studies by Wanner and Kolberer (1966), Wanner *et al.* (1975), Suzuki and Takahashi (1975), and Neuhann *et al.* (1979) on *Thea Sinensis* L. and *Coffea arabica*. It demonstrates that secondary (caffeine) and primary (adenine and guanine) compounds are introduced into the same catabolic pathway. Caffeine degradation is known to occur in older leaves, and degradation via di- or monomethylpurines is an important part of purine metabolism in this plant.

V. CONCLUSIONS

In contrast to earlier beliefs, numerous observations have unequivocally shown that secondary products are subject to turnover and degradation in plants. The metabolism of these compounds appears to be fully integrated into the complex morphological and biochemical organization of plants, so that many endogenous and exogenous factors influence the catabolism of secondary products. These factors have to be analyzed in each case to understand the rate and extent of secondary reactions. In most cases turnover is not an unidirectional pathway but rather consists of a considerable number of simultaneous reactions organized into metabolic grids. Degradation is one alternative on the grid, but it has to be measured in each case. Quite often degradation will be associated with particular developmental stages. Though many secondary compounds can be expected to be funneled into degradative pathways, the elucidation of a complete catabolic route has so far not yet been reported. Not even one example is known where the degradative pathway of a secondary compound has been worked out that shows the conversion of all carbon atoms into primary compounds *together* with the enzymes involved. Based on the sometimes complicated structures of secondary products, one can predict interesting biochemical reactions and long pathways. We are therefore justified in saying that this new field of research dealing with the degradation of secondary products in plants is at its very beginning.

ACKNOWLEDGMENTS

The authors thank Dr. D. Strack, Cologne, for the communication of unpublished data. The studies of the senior author have generously been supported by Deutsche Forschungsgemeinschaft.

REFERENCES

Alfermann, A. W., Merz, D., and Reinhard, E. (1975). *Planta Med., Suppl.* pp. 70–78.

Alibert, G., Bouyssou, H., and Ranjeva, R. (1979). *C. R. Hebd. Seances Acad. Sci.* **288**, 681–684.

Amrhein, N., and Diederich, E. (1980). *Naturwissenschaften* **67**, 40.

Amrhein, N., and Holländer, H. (1979). *Planta* **144**, 385–389.

Axelrod, B. (1974). In "Food Related Enzymes" (J. Whitacker, ed.), pp. 324–348. Am. Chem. Soc., Washington, D.C.

Barz, W. (1969). *Z. Naturforsch,* **24B**, 234–239.

Barz, W. (1975a). *Planta Med., Suppl.* pp. 117–133.

Barz, W. (1975b). *Ber. Dtsch. Bot. Ges.* **88**, 71–81.

Barz, W. (1977a). *Physiol. Veg.* **15**, 261–277.

Barz, W. (1977b). In "Plant Cell Cultures and its Bio-technological Application" (W. Barz, E. Reinhard, and M. H. Zenk, eds.), pp. 153–171. Springer Verlag, Berlin and New York.

Barz, W., and Hösel, W. (1975). In "The Flavonoids" (J. B. Harborne, T. J. Mabry, and H. Mabry, eds.), pp. 916–969. Chapman & Hall, London.

Barz, W., and Hösel, W. (1978). *Recent Adv. Phytochem.* **12**, 339–369.

Barz, W., Mohr, F., and Teufel, E. (1974). *Phytochemistry* **13**, 1785–1787.

Barz, W., Schlepphorst, R., Wilhelm, P., Kratzl, K., and Tengler, E. (1978a). *Z. Naturforsch., C: Biosci.* **33C**, 363–367.

Barz, W., Kettner, M., and Hüsemann, W. (1978b). *Planta Med.* **34**, 73–78.

Berlin, J., and Barz, W. (1971). *Planta* **98**, 300–314.

Berlin, J., and Barz, W. (1975). *Z. Naturforsch., C: Biosci.* **30C**, 650–658.

Berlin, J., and Widholm, J. M. (1978). *Phytochemistry* **17**, 65–68.

Berlin, J., Kiss, P., Müller-Enoch, D., Gierse, D., Barz, W., and Janistyn, B. (1974). *Z. Naturforsch., C: Biosci.* **29C**, 374–383.

Bopp, M., and Lüdicke, W. (1975). *Z. Naturforsch., C: Biosci.* **30C**, 663–667.

Bopp, M., and Pop, H. (1975). *Biochem. Physiol. Pflanz.* **168**, 101–111.

Bourwieg, D., and Pohl, R. (1973). *Planta Med.* **24**, 304–314.

Brandner, G. (1962). Doctoral Thesis, University of Freiburg/Brsg., Germany.

Brown, S. A., and Wetter, L. R. (1972). *Prog. Phytochem.* **3**, 1–46.

Butt, V. S. (1978). *Recent Adv. Biochem.* **12**, 433–456.

Chauhan, Y. S., RaShore, V. S., Garg, G. K., and Bhargava, A. (1978). *Biochem. Biophys. Res. Commun.* **83**, 56–59.

Chiji, H., Aiba, T., and Izawa, M. (1978). *Agric. Biol. Chem.* **42**, 159–164.

Chopin, J., Hauteville, M., Joshi, B. S., and Gawad, D. H. (1978). *Phytochemistry* **17**, 332–334.

Colonna, J.-P. (1978). Doctoral Thesis, University of Toulouse, France.

Conn, E. E. (1979). *Naturwissenschaften* **66**, 28–34.

Crombie, L., Dewick, P. M., and Whiting, D. A. (1971). *Chem. Commun.* pp. 1183–1184.

Croteau, R., and Kolattukudy, P. E. (1975). *Arch. Biochem. Biophys.* **170**, 61–72.

Dagley, S. (1971). *Adv. Microb. Physiol.* **6**, 1–46.

Deitzer, G. F., Hopkins, D. W., Haertle, U., and Wagner, E. (1978). *Photochem. Photobiol.* **27**, 127–131.

Delfel, N. E. (1980). *Planta Med.* **39**, 168–179.

Delfel, N. E., and Rothfus, J. A. (1977). *Phytochemistry* **16**, 1595–1598.

Dewick, P. M., and Martin, M. (1979). *Phytochemistry* **18**, 597–602.

Digenis, G. A. (1969). *J. Pharm. Sci.* **58**, 39–45.

Digenis, G. A., Faraj, B. A., and Abou-Charr, C. J. (1966). *Biochem. J.* **101**, 27c.

Dittrich, P., and Kandler, O. (1971). *Ber. Dtsch. Bot. Ges.* **84**, 465–472.

Durand, R., and Zenk, M. H. (1974a). *Phytochemistry* **13**, 1483–1492.

Durand, R., and Zenk, M. H. (1974b). *FEBS Lett.* **39**, 218–220.

Ellis, B. E. (1971). *FEBS Lett.* **18**, 228–230.

Ellis, B. E. (1973). *Planta* **111**, 113–118.

Ellis, B. E. (1974). *Lloydia* **37**, 168–184.

Ellis, B. E. (1976). *Phytochemistry* **15**, 489–491.

Ellis, B. E., and Towers, G. H. N. (1970). *Phytochemistry* **9**, 1457–1461.

Engelsma, G. (1979). *Plant Physiol.* **63**, 765–768.

Erez, A., and Lavee, S. (1969). *Plant Physiol.* **44**, 342–346.

Fairbain, J. W., and Wassel, G. M. (1964). *Phytochemistry* **3**, 253–258.

Fischer, N., and Dreiding, A. S. (1972). *Helv. Chim. Acta* **55**, 649–658.

Frank, A. W., and Marion, L. (1956). *Can. J. Chem.* **34**, 1641–1646.

Frey, G. (1977). Doctoral Thesis, University of Münster, Germany.

Frey-Schröder, G., and Barz, W. (1979). *Z. Naturforsch., C: Biosci.* **34C**, 200–209.

Furuya, T. (1978). *In* "Frontiers of Plant Tissue Culture 1978" (T. A. Thorpe, ed.), pp. 191–200. The Bookstore, University of Calgary, Calgary, Alberta, Canada.

Gerstner, E., and Kiel, U. (1975). *Hoppe-Seyler's Z. Physiol. Chem.* **356**, 1853–1857.

Gescher, A., Hickmann, J. A., and Stevens, M. F. G. (1979). *Biochem. Pharmacol.* **28**, 3235–3238.

Goodwin, T. W. (1976). *In* "Chemistry and Biochemistry of Plant Pigments" (T. W. Goodwin, ed.), 2 ed., Vol. 1, pp. 225–261. Academic Press, New York.

Griffin, W. J. (1979). *Naturwissenschaften* **66**, 58.

Grisebach, H. (1978). *Recent Adv. Phytochem.* **12**, 221–248.

Grisebach, H., and Barz, W. (1969). *Naturwissenschaften* **56**, 538–544.

Grochowska, M. J. (1967). *Bull. Acad. Pol. Sci.* **15**, 455–461.

Gross, G. G. (1978). *Recent Adv. Phytochem.* **12**, 177–220.

Hagel, P., and Kindl, H. (1975). *FEBS Lett.* **59**, 120–124.

Halliwell, B., and Ahluwalia, S. (1976). *Biochem. J.* **153**, 513–518.

Hamilton, R. H., Meyer, H. E., Burke, R. E., Feung, C. S., and Mumma, R. O. (1976). *Plant Physiol.* **58**, 77–81.

Harborne, J. B. (1964). *In* "Biochemistry of Phenolic Compounds" (J. B. Harborne, ed., pp. 129–170. Academic Press, New York.

Harborne, J. B. (1967). "Comparative Biochemistry of Flavonoids," p. 267. Acadmic Press, New York.

Harkin, J. M. (1967). *In* "Oxidative Coupling of Phenols" (W. J. Taylor and A. R. Battersby, eds.), pp. 95–116. Decker, New York.

Harms, H., Söchtig, H., and Haider, K. (1971). *Z. Pflanzenphysiol.* **64**, 437–445.

Harms, H., Haider, K., Berlin, J., Kiss, P., and Barz, W. (1972). *Planta* **105**, 342–351.

Hauteville, M., Chadenson, M., and Chopin, J. (1979). *Bull. Soc. Chim.* No. 3–4, pp. 125–131.

Henry, E. W. (1975). *J. Ultrastruct. Res.* **52**, 289–299.

Herrmann, K. (1978). *Fortschr. Chem. Org. Naturst.* **35**, 73–132.

Hösel, W. (1976). *Planta Med.* **30**, 97–103.

Hösel, W., Shaw, P. D., and Barz, W. (1972). *Z. Naturforsch, B: Anorg. Chem., Org. Chem. Biochem., Biophys., Biol.* **27B**, 946–954.

Hösel, W., Frey, G., and Barz, W. (1975). *Phytochemistry* **14**, 417–422.

Hösel, W., Burmeister, G., Kreysing, P., and Surholt, E. (1977). *In* "Plant Tissue Cultures and Its Bio-technological Application" (W. Barz, E. Reinhard, and M. H. Zenk, eds.), pp. 172–177. Springer-Verlag, Berlin and New York.

Impellizzeri, G., and Piatteli, M. (1972). *Phytochemistry* **11**, 2499–2502.

Janistyn, B., and Stocker, M. (1976). *Z. Naturforsch., C: Biosci.* **31C**, 408–410.

Janistyn, B., Barz, W., and Pohl, R. (1971). *Z. Naturforsch., B: Anorg. Chem., Org. Chem., Biochem., Biophys., Biol.* **26B**, 973–974.

Jurd, L. (1972). *In* "The Chemistry of Plant Pigments" (C. O. Chichester, ed.), pp. 123–142. Academic Press, New York.

Kamel, M., Saleh, N. A., and Ghazy, A. M. (1977). *Phytochemistry* **16**, 521–524.

Kindl, H. (1971). *Naturwissenschaften* **58**, 554–563.

Kleist, B. (1975). Die Hormonabhängigkeit des Sekundärstoffwechsels in pflanzlichen Zellkulturen. Staatsexamensarbeit, University of Münster, Germany.

Koeppe, D. E., Rohrbaugh, L., Rice, E. L., and Wender, S. H. (1970). *Phytochemistry* **9**, 297–301.

Kojima, M., and Uritani, I. (1973). *Plant Physiol.* **51**, 768–771.

Kojima, M., Poulton, J. E., Thayer, S. S., and Conn, E. E. (1979). *Plant Physiol.* **63**, 1022–1028.

Köster, J., Ohm, M., and Barz, W. (1978). *Z. Naturforsch., C: Biosci.* **33C**, 368.

Krisnangkura, K., and Gold, M. H. (1979). *Phytochemistry* **18**, 2019–2021.

Lamb, C. J., and Rubery, P. H. (1976). *Phytochemistry* **15**, 665–668.

Lamoureux, G. L., Stafford, L. E., Shimabukuro, R. H., and Zaylskie, R. G. (1973). *J. Agric. Food Chem.* **21**, 1020–1030.

Larsen, P. O., and Wieczorkowska, E. (1978). *Biochim. Biophys. Acta* **542**, 253–262.

Luckner, M. (1980). *In* "Secondary Plant Products" (E. A. Bell and B. V. Chalwood, eds.), pp. 23–63. Springer-Verlag, Berlin and New York.

Luckner, M., Nover, L., and Böhm, H. (1977). "Secondary Metabolism and Cell Differentiation." Springer-Verlag, Berlin and New York.

McClure, J. W. (1975). *In* "The Flavonoids" (J. B. Harborne, T. J. Mabry, and H. Mabry, eds.), pp. 970–1055. Chapman & Hall, London.

Mäder, M., and Schloss, P. (1979). *Plant Sci. Lett.* **17**, 75–80.

Mäder, M., Nessel, A., and Bopp, M. (1977). *Z. Pflanzenphysiol.* **82**, 247–260.

Madyastha, K. M., Meehan, T. D., and Coscia, C. J. (1976). *Biochemistry* **15**, 1097–1102.

Mann, J. D., Steinhart, C. E., and Mudd, S. H. (1963). *J. Biol. Chem.* **238**, 676–681.

Marcinowski, S., and Grisebach, H. (1977). *Phytochemistry* **16**, 1665–1667.

Mayer, A. M., and Harel, E. (1979). *Phytochemistry* **18**, 193–215.

Meyer, E., and Barz, W. (1978). *Planta Med.* **33**, 336–344.

Mialoundama, F., and Paulet, P. (1975). *Physiol. Plant.* **35**, 39–44.

Miller, L. P. (1973). *In* "Phytochemistry. The Process and Products of Photosynthesis" (L. P. Miller, ed.), Vol. I, pp. 297–375. Van Nostrand-Reinhold, Princeton, New Jersey.

Miller, R. J., Jollès, C., and Rapoport, H. (1973). *Phytochemistry* **12**, 597–603.

Mizusaki, S., Tanabe, Y., Noguchi, M., and Tamaki, E. (1973). *Plant Cell Physiol.* **14**, 103–110.

Mohan, V. P., Kishore, G., Sugumaran, M., and Vaidyanathan, C. S. (1979). *Plant Sci. Lett.* **16**, 267–272.

Molderez, M., Nagels, L., and Parmentier, F. (1978). *Phytochemistry* **17**, 1747–1750.

Mothes, K., and Schütte, H. R. (1969). "Biosynthese der Alkaloide." VEB Deutscher Verlag der Wissenschaften, Berlin.

Mothes, K., Richter, I., Stolle, K., and Gröger, D. (1965). *Naturwissenschaften* **52**, 431.

Muhle, E., Hösel, W., and Barz, W. (1976). *Phytochemistry* **15**, 1669–1672.

Neuhann, H., Leienbach, K.-W., and Barz, W. (1979). *Phytochemistry* **18**, 61–64.
Neumann, D., and Tschoepe, K. H. (1966). *Flora (Jena), Abt. A* **156**, 521–542.
Nimz, H. (1974). *Angew. Chem., Int. Ed. Engl.* **13**, 313–320.
Nowacki, E. K., and Waller, G. R. (1975). *Phytochemistry* **14**, 161–164, 165–171.
Nurmann, G., and Strack, D. (1979). *Z. Naturforsch., C: Biosci.* **34C**, 715–720.
Parish, R. W. (1975). *Planta* **123**, 1–13.
Patschke, L., Hess, D., and Grisebach, H. (1964a). *Z. Naturforsch., B: Anorg. Chem., Org. Chem., Biochem., Biophys., Biol.* **19B**, 1114–1117.
Patschke, L., Barz, W., and Grisebach, H. (1964b). *Z. Naturforsch., B: Anorg. Chem., Org. Chem., Biochem., Biophys., Biol.* **21B**, 201–205.
Patzlaff, M. (1978). Doctoral Thesis, University of Münster, Germany.
Patzlaff, M., and Barz, W. (1978). *Z. Naturforsch., C: Biosci.* **33C**, 675–684.
Phillipson, D. J., and Handa, S. S. (1978). *Lloydia* **41**, 385–431.
Philippson, D. J., Handa, S. S., and El-Dabbas, S. W. (1976). *Phytochemistry* **15**, 1297–1301.
Pickering, J. W., Powell, B. C., Wender, S. H., and Smith, E. C. (1973). *Phytochemistry* **12**, 2639–2643.
Podstolski, A., and Lewak, St. (1970). *Phytochemistry* **9**, 289–296.
Popovici, G., and Weissenböck, G. (1976). *Ber. Dtsch. Bot. Ges.* **89**, 483–489.
Popovici, G., and Weissenböck, G. (1977). *Z. Pflanzenphysiol.* **82**, 450–454.
Potts, R. J. M., Weklych, R., and Conn, E. E. (1974). *J. Biol. Chem.* **249**, 5019–5026.
Prasad, S., and Ellis, B. E. (1978). *Phytochemistry* **17**, 187–190.
Rabitzsch, G. (1959). *Planta Med.* **7**, 268–297.
Rathmell, W. G., and Bendall, D. S. (1972). *Biochem. J.* **127**, 125–132.
Raymond, W. R., and Maier, V. P. (1977). *Phytochemistry* **16**, 1535–1539.
Reigh, D. L., Wender, S. H., and Smith, C. (1973). *Phytochemistry* **14**, 1715–1717.
Reiner, J. M. (1953a). *Arch. Biochem. Biophys.* **46**, 53–79.
Reiner, J. M. (1953b). *Arch. Biochem. Biophys.* **46**, 79–99.
Reuter, E.-W. (1970). Ph.D. Dissertation, University Freiburg/Brsg., Germany.
Reznik, H. (1960). *Ergeb. Biol.* **23**, 14–46.
Robinson, T. (1974). *Science* **184**, 430–435.
Saito, K., and Komamine, A. (1976). *Eur. J. Biochem.* **68**, 237–243.
Saito, K., and Komamine, A. (1978). *Eur. J. Biochem.* **82**, 385–392.
Sandermann, H., Diesperger, H., and Scheel, D. (1977). *In* "Plant Tissue Culture and its Bio-technological Application" (W. Barz, E. Reinhard, and M. H. Zenk, eds.), pp. 179–196. Springer-Verlag, Berlin and New York.
Sano, H. (1971). *Biochim. Biophys. Acta* **227**, 565–575.
Sasse, F., Backs-Hüsemann, D., and Barz, W. (1919). *Z. Naturforsch., C: Biosci.* **34C**, 848–853.
Saunders, J. A., and Conn, E. E. (1978). *Plant Physiol.* **61**, 154–157.
Schill, L., and Grisebach, H. (1973). *Hoppe-Seyler's Z. Physiol. Chem.* **354**, 1555–1562.
Schlepphorst, R., and Barz, W. (1979). *Planta Med.* **36**, 333–342.
Schreiber, W. (1975). *Biochem. Biophys. Res. Commun.* **63**, 509–514.
Schulz, G., and Hecker, E. (1973). *Z. Naturforsch., C: Biosci.* **28C**, 662–674.
Schwarze, P. (1958). *In* "Handbuch der Pflanzenphysiologie" (W. Ruhland, ed.), Vol. 10, p. 523. Springer-Verlag, Berlin and New York.
Sharma, H. K., and Vaidyanathan, S. (1975). *Eur. J. Biochem.* **56**, 163–171.
Sharples, D., Spring, M. S., and Stoker, J. R. (1972). *Phytochemistry* **11**, 2999–3002.
Shinshi, H., and Noguchi, M. (1975). *Phytochemistry* **14**, 1255–1258.
Simpson, K. L., Lee, T.-C., Rodriguez, D. B., and Chichester, C. O. (1976). *In* "Chemistry and Biochemistry of Plant Pigments" (T. W. Goodwin, ed.), 2 ed., Vol. 1, Chapter 17. Academic Press, New York.

Smith, T. A. (1980). *In* "Secondary Plant Products" (E. A. Bell and B. V. Charlwood, eds.), pp. 433–460. Springer-Verlag, Berlin and New York.

Stafford, H. A. (1974). *Annu. Rev. Plant. Physiol.* **25**, 459–486.

Strack, D. (1973). Ph.D. Thesis, University of Cologne.

Strack, D. (1977). *Z. Pflanzenphysiol.* **84**, 139–145.

Strack, D., and Reznik, H. (1976). *Z. Pflanzenphysiol.* **79**, 95–108.

Surholt, E., and Hösel, W. (1978). *Phytochemistry* **17**, 873–877.

Sutter, A., and Grisebach, H. (1975). *Arch. Biochem. Biophys.* **167**, 444–447.

Suzuki, T., and Takahashi, E. (1975). *Biochem. J.* **146**, 79–85.

Tabata, M., Ogino, T., Yoshioka, K., Yoshikawa, N., and Hiraoka, N. (1978). *In* "Frontiers of Plant Tissue Culture 1978" (T. A. Thorpe, ed.), pp. 213–222. University of Calgary, The Bookstore, Calgary, Alberta, Canada.

Taylor, A. O., and Zucker, M. (1966). *Plant Physiol.* **41**, 1350–1359.

Teufel, E. (1974). Diplomarbeit, University of Freiburg/Brsg., Germany.

Thieme, H. (1965). *Pharmazie* **20**, 570–573.

Thieme, H. (1965). *Pharmazie* **20**, 688–691.

Thomas, G., and Threlfall, D. R. (1974). *Biochem. J.* **142**, 437–440.

Towers, G. H. N. (1964). *In* "Biochemistry of Phenolic Compounds" (J. B. Harborne, ed.), Chapter 7. Academic Press, New York.

Tsuji, J., and Takayanagi, H. (1978). *Tetrahedron* **34**, 641–644.

Unger, W., and Hartmann, T. (1976). *Z. Pflanzenphysiol.* **77**, 255–267.

van Sumere, C. F., Albrecht, J., Dedonder, A., de Pooter, H., and Pe, I. (1975). *In* "The Chemistry and Biochemistry of Plant Proteins" (J. B. Harborne and C. F. van Sumere, eds.), pp. 211–264. Academic Press, New York.

Venis, M. A., and Stoessl, A. (1969). *Biochem. Biophys. Res. Commun.* **36**, 54–56.

Waller, G. R., and Lee, J. L.-C. (1969). *Plant Physiol.* **44**, 522–526.

Waller, G. R., and Nowacki, E. K. (1978). *In* "Alkaloid Biology and Metabolism in Plants" (G. R. Waller, and E. K. Nowacki, eds.), Chapters 1 and 6. Plenum, New York.

Walter, G., and Hecker, E. (1973). *Z. Naturforsch, C. Biosci.* **28C**, 675–684.

Wanner, H., and Kolberer, P. (1966). *Abh. Dtsch. Akad. Wiss. Berlin, Kl. Chem., Geol. Biol.* p. 607.

Wanner, H., Pesakova, N., Baumann, T. W., Charubala, R., Guggisberg, A., Hesse, M., and Schmid, H. (1975). *Phytochemistry* **14**, 747–750.

Wehnert, H.-U. (1970). Ph.D. Dissertation, University of Hamburg, Germany.

Wiermann, R. (1979). *In* "Regulation of Secondary Product and Plant Hormone Metabolism" (M. Luckner and K. Schreiber, eds.), pp. 231–239. Pergamon, Oxford.

Wong, E., and Wilson, J. M. (1976). *Phytochemistry* **15**, 1325–1332.

Zaprometov, M. N. (1977). *Flavonoids Bioflavonoids, Proc. Hung. Bioflavonoid Symp. 5th, 1977* pp. 257–269.

Zhanayeva, T. A., Minayeva, V. G., and Zaprometov, M. N. (1977). *Proc. USSR Acad. Sci.* **233**, 722–725.

Zenk, M. H. (1967). *Ber. Dtsch. Bot. Ges.* **80**, 573–591.

Zenk, M. H., El-Shagi, H., and Schulte, U. (1975). *Planta Med. Suppl.* pp. 79–101.

Secondary Plant Products and Cell and Tissue Differentiation

4

ROLF WIERMANN

The Biochemistry of Plants, Vol. 7

I. INTRODUCTION

Plants exhibit the extraordinary ability to produce an abundant number of secondary products. Synthesis and accumulation of these compounds occur in organisms at very different evolutionary stages. Certain secondary products can appear only sporadically or are distributed widely throughout the plant kingdom; they can be specific for orders, families, species, and sometimes even intraspecific taxa. Also, within the same plant there can be variation in the pattern of secondary product composition; i.e., even organs or parts of organs can differ in the expression of their biosynthetic potential.

Coordination and cooperation among various cells are prerequisites for the existence of highly developed organisms and require structural and functional specialization of certain cells. This cell differentiation follows specific biochemical and morphological principles. The formation of morphologically specialized cells in multicellular organisms and the formation of secondary plant products are integrated parts of a differentiation process. Morphological–anatomical differentiation often coincides with the expression of secondary metabolism.

The connection between the expression of secondary metabolism and the differentiation program of an organism has been pointed out repeatedly. Especially impressive correlations are found in microorganisms (for review, see Luckner and Nover, 1977). For example, cultures of the fungus *Penicillium cyclopium* exhibit three developmental stages: germination, hyphal growth, and cell specialization (see Nover and Luckner, 1976, and references cited therein; Luckner, 1977). The phase of cell specialization is characterized not only by the formation of penicilli and conidiospores but also by an intensified formation of different secondary products, the most prominent of which are the benzodiazepine alkaloids cyclopenin and cyclopenol. The beginning of alkaloid production is associated with a marked increase in some enzymatic activities involved in alkaloid biosynthesis. No attempt will be made to cover this area within this chapter, since our focus shall be entirely on higher plants.

In higher plants and other more complex organisms, the occurrence of certain paths of secondary metabolism can depend on the general development of the organism and/or the development of single organs, tissues, and particular specialized cells. Synthesis and accumulation can be an endogenously controlled, development-dependent differentiation process and/or can be regulated by exogenous factors such as light, temperature, and wounding. Which tissues and cells are able to synthesize a distinct secondary compound is a manifestation of "primary differentiation." This primary differentiation determines the specificity of the cellular response to exogenous factors, e.g., induction of anthocyanin biosynthesis by light ("secondary differentiation") (cf. Wagner and Mohr, 1966, Mohr and Sitte, 1971; Mohr, 1972; Bopp and Capesius, 1973; Steinitz and Bergfeld, 1977).

This chapter is an attempt to describe the relations among the general development of the plant, the individual differentiation of organs, tissues, and cells, and the production of secondary compounds. Because of the great variety of chemical and morphological structures, a complete survey is not intended and only a few examples can be cited. With respect to the high number of secondary compounds, one has to expect a very complex "chemical" differentiation depending on organogenesis. This complexity, as well as fundamental similarities and differences, will be emphasized.

II. ORGANOGENESIS AND ACCUMULATION OF SECONDARY COMPOUNDS

Biosynthesis and accumulation of a number of secondary compounds take place during certain definite developmental stages of an organism. This is obvious in the differentiation and pigmentation of petals, but the same is also true for secondary compound formation during the development of entire plants and organs of plants. Some examples will show the different relations between ontogenetic development of plants and organs and the accumulation of secondary products.

A. Development of the Whole Plant

There are a number of cases showing that a definite developmental stage of a plant influences the expression of secondary metabolism and the actual potential for synthesis. In a race of *Papaver bracteatum,* Böhm (1967) demonstrated that three of the four major latex alkaloids disappeared during ontogenetic development of the plant and only thebain remained. In lupins, alkaloid synthesis begins not earlier than 2 weeks after germination and continues until the plants start to flower (Wiewiorowski *et al.,* 1966). In general, the beginning of flowering seems to influence alkaloid biosynthesis strongly; alkaloid production often slows or even stops at this stage of development.

Nicotine, which is formed in tobacco roots, can be demethylated to nornicotine in the shoot. In *Nicotiana silvestris* this reaction is at a maximum during stem elongation and flower induction. If the plant remains in the rosette growth stage (induced by short-day conditions), a significant decrease in demethylation is observed. The differences have not been attributed to direct dependence on photoperiods but rather considered to be a consequence of aging (Mothes *et al.,* 1957).

Popovici and Weissenböck (1976) investigated the differentiation of flavonoid metabolism during the development of oats grown under field conditions. They found a succession of B-ring substitution of flavonoids from 4'- via 3',4'- to 3'4',5'- which paralleled a differentiation of mono-C- to di-C-glycosyl substitution and O-glycosylation of flavones.

Germination and seedling development are characterized by a pattern of special physiological and biochemical processes. Therefore, the relations between ontogenesis of an organism and accumulation of secondary compounds (e.g., phenylpropanoid substances) have been intensively studied with respect to these developmental stages (Walton, 1968; Stafford, 1969;

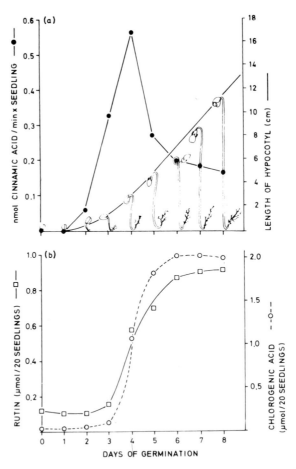

Fig. 1. (a) Hypocotyl lengthening in etiolated buckwheat seedlings (*F. esculentum* Moench) and development of PAL activity. (b) Time course of the accumulation of rutin and chlorogenic acid in dark seedlings of buckwheat. (Modified after Amrhein and Zenk, 1970.)

Amrhein and Zenk, 1970; Weissenböck and Reznik, 1970; Hahlbrock *et al.,* 1971; Weissenböck, 1971).

Experiments with etiolated seedlings of *Fagopyrum esculentum* have clearly shown that intensive accumulation of secondary products takes place at definite stages of development (Amrhein and Zenk, 1970; see Fig. 1). During the first day after sowing no visible morphological changes occurred, but on the second day the radicule and the hypocotyl appeared, and the latter constantly elongated from the fourth day to at least the eighth day. With regard to the accumulation of secondary compounds it was found that, after a lag period of 3 days, the accumulation rate of chlorogenic acid and rutin reached its maximum on the fourth day after germination was initiated.

From the sixth day on, no further increase in the pigment content of seedlings was observed. In addition, these studies clearly demonstrated that the rapid accumulation of these phenylpropanoid compounds was well correlated with an increase in the activity of phenylalanine ammonia-lyase (PAL), an enzyme involved in phenylpropanoid biosynthesis. The time of maximal PAL activity coincided with the time of maximal accumulation rate of the phenylpropane derivatives.

Studies with other plants have led to similar results. Thus, in seedlings of *Impatiens balsamina* and *Cucurbita maxima,* flavonol accumulation begins during the *early phase* of radicule and hypocotyl elongation (Weissenböck, 1971; Strack, 1973). Such correlations appear to be valid for the accumulation of a number of secondary compounds. Thus Nahrstedt *et al.* (1979) demonstrated that cyanogenic glycosides (especially triglochinin) accumulated extensively in early stages of seedling development of *Triglochin maritima* (see Conn, this volume, Chapter 16).

Secondary metabolism includes biosynthesis and catabolism as well as transport, interconversion, and accumulation. Three aspects of the relationships between organogenesis and the accumulation of secondary compounds have to be kept in mind.

1. At a given developmental stage of an organ, a large increase in the accumulation rate does not necessarily indicate synthesis of a product *de novo* but could be the result of translocation of a compound (which might undergo secondary alterations itself) from another organ. In other words, the site of synthesis of the compound may not be organs and tissues with a high rate of accumulation (cf. Müller, 1976). If accumulation is the result of transport, it is a storage problem rather than a biosynthetic problem for the organ in question. In this case "biochemical" differentiation means development of the ability to store secondary compounds. The accumulation of nicotine at different stages of leaf development following its formation in young parts of the roots is a well-known example (Romeike, 1956, 1959; Mothes and Romeike,

1958). Leaves of various *Ipomoea* species are the primary site of ergot alkaloid (or precursor) synthesis, whereas the seeds only accumulate and cannot synthesize alkaloids (Gröger *et al.*, 1963; Mockaitis *et al.*, 1973).

2. Changes in the accumulation rate may reflect interconversion of bound forms rather than *de novo* synthesis. During seedling development of *Raphanus sativus* and other Brassicaceae (Tzagoloff, 1963; Bopp and Lüdicke, 1975; Strack, 1977; Strack *et al.*, 1978), sinapine, which accumulates together with disinapoylsucrose and sinapoylgluco-raphenine in *R. sativus* seeds, rapidly disappears in the early stages of germination with a concomitant accumulation of sinapoylglucose (Linscheid *et al.*, 1980). This accumulation is not affected by addition of the *in vivo* inhibitor of PAL, α-aminooxy-β-phenylpropionic acid (Amrhein *et al.*, 1976; Amrhein and Gödecke, 1977; Amrhein and Holländer, 1979), although the formation of anthocyanins, flavonols, and a ferulic acid derivative, components that normally accumulate together with sinapoylglucose in *Raphanus* cotyledons, is severely depressed. Therefore the accumulation of sinapoylglucose at this stage of development does not seem to be the result of *de novo* synthesis of the C_6—C_3 skeleton but must be regarded as an interconversion of sinapine (Strack *et al.*, 1978; Linscheid *et al.*, 1980). This metabolism is catalyzed by a sinapine esterase (Nurmann and Strack, 1979; Strack *et al.*, 1980) and a glucosyltransferase (Strack, 1980). In later stages of *Raphanus* seedling development sinapoylglucose is inter-converted to sinapoylmalate by a direct transfer of the sinapic acid from glucose to malate (Tkotz and Strack, 1980). Figure 2 in the previous chapter illustrates the changes in sinapic acid esters in cotyledons of *R. sativus* (Barz and Köster, Chapter 3).

3. The accumulation rates observed are the sum of anabolism and catabolism at this stage of development and therefore represent a dynamic value. Investigations with *C. maxima* seedlings have shown that even during intensive accumulation of flavonol glucosides a significant turnover is measurable (Strack and Reznik, 1976). Comparable data were obtained by Barz and Hösel (1971) with *Cicer arietinum* (Barz *et al.*, 1971; see Barz and Hösel, 1975, and references cited therein; cf. Barz and Köster, this volume, Chapter 3). The process of differentiation creates the competence for synthesis and turnover at the same time.

B. Development of Leaves, Leaf Homologues, and Other Organs

1. Leaves and Cotyledons

Leaves not only exhibit great morphological diversity but are also sites of the synthesis and accumulation of a number of secondary compounds. Synthesis can vary in the different leaf organs formed by a plant, as demonstrated with seedlings of *F. esculentum.* Cotyledons accumulated both flavone *C*-glycosides and flavonol *O*-glycosides, whereas all leaves formed later produced only the latter compounds (Margna *et al.,* 1967; Rumpenhorst, 1968; Krause, 1971).

The correlation between differentiation of secondary metabolism and leaf development has been investigated repeatedly. Studies with *Petroselinum hortense* revealed a close relationship between the development of cotyledons and leaves and the accumulation of flavone glycosides. Based on four enzymes, all involved in the biosynthesis of apiin (7-*O*-[β-D-apiofuranosyl $(1 \rightarrow 2)$ β-D-glucosyl]-5,7,4'-trihydroxyflavone), it was demonstrated that *very young* organs contained the highest specific activities of the four enzymes involved in apiin biosynthesis. The specific activity then dramatically decreased with aging. With the individual organ as the reference, there was a close relationship between the development of enzymatic activities and the accumulation of flavone glycosides (Fig. 2a and b; see Hahlbrock *et al.,* 1971).

In another study there was a dramatic increase in *C*-glycosylflavones in primary leaves of *Avena sativa,* grown under field conditions, during the differentiation and growth phase (Popovici and Weissenböck, 1977). The flavone content there remained constant in the following mature phase but became markedly reduced with increasing senescence of the leaves (Fig. 3). The various flavone components showed specific differences in accumulation kinetics (Fig. 3). Among the major compounds isovitexin 2"-arabinoside reaches its maximum during the early growth and differentiation phase, whereas isoorientin 2"-arabinoside reaches its highest concentration in later phases of organ development.

These results are in good argeement with those obtained from experiments with leaves of *Corylus avellana* grown under natural environmental conditions (Staude and Reznik, 1973).

Tracer experiments with needles of *Picea abies* demonstrated dynamic changes in phenolic metabolism during organogenesis. Thus, flavonol glucosides are formed only in very young needles, whereas picein (4-hydroxyacetophenone glucoside) is produced only in fully differentiated needles (Dittrich and Kandler, 1971; cf. Ishikura, 1972; Andersen *et al.,* 1969; Tissut, 1972).

A detailed knowledge of organ growth is a prerequisite for correlating organogenesis with the often complex changes in secondary metabolism. Flavonoid accumulation in primary leaves of *Avena* seedlings is an instructive example (Effertz and Weissenböck, 1976; Weissenböck and Sachs, 1977). The primary leaf of these seedlings exhibits mainly basiplastic growth; i.e., all leaf tissues originate in the same basal meristematic zone. Growth rates of different leaf segments were determined by marking the leaves after definite time intervals (Fig. 4d). Based on the movement of these marks, three well-

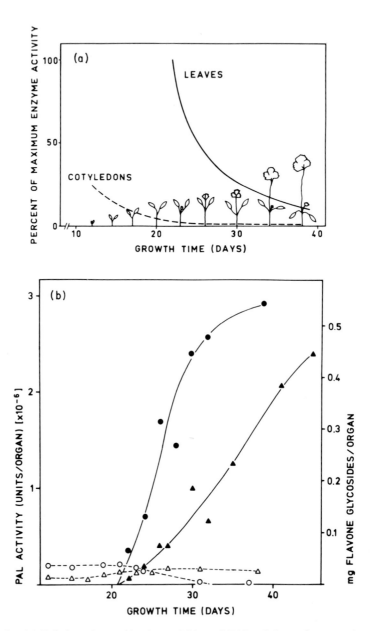

Fig. 2. (a) Relative changes in the activities of PAL, chalcone-flavanone isomerase, UDP-glucose:flavonoid 7-*O*-glucosyltransferase, and UDP-apiose:flavone glycoside apiosyltransferase in cotyledons (----) and leaves (———) of developing parsley plants (*P. hortense*) (after Grisebach and Hahlbrock, 1974). (b) Changes in PAL activity and accumulation of flavone glycosides per organ in developing parsley plants. All other enzymes mentioned above gave curves of identical shape and can be substituted for PAL. Cotyledons: ○, PAL; △, flavone glycosides. Leaves: ●, PAL; ▲, flavone glycosides. (Modified after Hahlbrock *et al.*, 1971; Hahlbrock and Grisebach, 1974.)

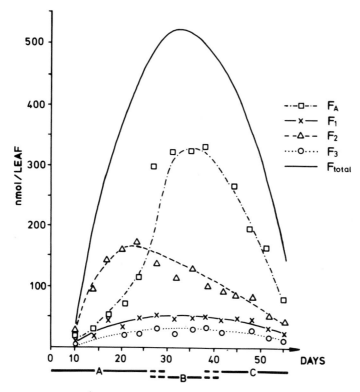

Fig. 3. Dynamics of the accumulation of *C*-glycosylflavones during the life cycle of primary leaves of oat (*A. sativa* L. 'Gelbhafer Flämingskrone'). F_A, Isoorientin 2″-arabinoside; F_1, isoswertisin 2″-rhamnoside; F_2, isovitexin 2″-arabinoside; F_3, vitexin 2″-rhamnoside; F_{total}, total flavonoid content as sum of single compounds. (A) Differentiation and growth phase, (B) mature phase, (C) Senescence phase. (Modified after Popovici and Weissenböck, 1977.)

defined leaf segments were recognized: (1) a basal segment (B), about 1 cm long, which represented the meristematic zone, (2) a top section (S), about 1 cm long, which did not show any growth during the period of investigation, and (3) a middle section (M) which elongated intensively during leaf development.

Under standard conditions, the kinetics of flavone accumulation and development of PAL activity are different in each particular segment. In the rapidly elongating section M, a large increase in flavone content occurs together with increasing PAL activity. Whereas the PAL activity decreases in section S continuously, its activity remains constant in the meristematic basal section B at a relatively high level after reaching a maximum on about the fifth day of seeding development (cf. Fig. 4a and b). Segment B exhibits a high specific PAL activity which remains constant throughout leaf development (Fig. 4c). This constant high specific activity of PAL in the basal meristematic zone—a significant decrease was noted in segments S and M—is probably responsible for a permanent high rate of formation of cinnamic acid and its derivatives in all leaf tissues originating in the meristematic zone (Weissenböck and Sachs, 1977).

2. Petals

The development of pigmentation as petals unfold is perhaps the most obvious example of the relation between definite stages of organogenesis and

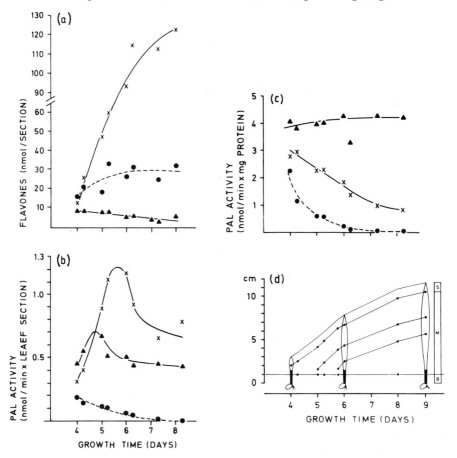

Fig. 4. Changes in total flavone contents (a) and PAL activity (b) in primary leaf sections of oat (*A. sativa* L. 'Gelbhafer Flämingskrone') during leaf growth. (c) Specific PAL activity in primary leaf sections during growth. (d) Displacement of ink marks successively applied during oat leaf development above the coleoptile (●—●). Mark placed at the coleoptile is not displaced. ●--●, Section S, 1-cm top section; ×, section M, middle section, increasing in length during growing time; ▲, section B, 1-cm basal section. (Modified after Weissenböck and Sachs, 1977.)

the accumulation of secondary compounds. Petals of *Primula obconica* produce various flavonol (kaempferol) glycosides and anthocyanins (malvidin, pelargonidin, and delphinidin glycosides). The accumulation of flavonol glycosides is observed in the youngest buds still in the cell multiplication phase. Anthocyanin accumulation, however, can be observed only in expanded flowers in which cell elongation is largely complete (Reznik, 1961). Moreover, it was shown that the less glycosylated kaempferol glycoside was accumulated first, followed by the more highly glycosylated compound, during development of these petals (Reznik, 1961).

The red-flowered petals of *I. balsamina* (genotype *LL HH PʳPʳ*) provide another example in that their accumulation sequence progresses from the simple anthocyanin pelargonidin 3-monoglucoside to more complex structures (e.g., acylated pelargonidin 3,5-diglucoside) (Strack and Mansell, 1977, 1979; cf. Hess, 1963; Hagen, 1966; Billot, 1974). Thus the chemical structures of anthocyanins present at different stages of flower development reflect the presumed biosynthetic sequence.

However, this relationship between anthocyanin pattern and petal development cannot be observed in all cases (see Hess and Endress, 1973). Unlike *P. obconica* and *I. balsamina* (see also *Petunia hybrida,* Hess, 1963), flower-color mutants of both *Pisum* and *Lathyrus odoratus* var. Chloe accumulate anthocyanins in an order that was the reverse of that expected from the biosynthetic pathway (Statham and Crowden, 1974).

3. Other Organs

Large amounts of flavanone glycosides rapidly accumulate in grapefruit during early growth. Maier and Hasegawe (1970) have shown that young grapefruits possess appreciable PAL activity and that a linear relationship exists between PAL activity and the rate of naringenin glycoside accumulation. Similar observations have been reported for a variety of other fruits (peach, Aoki *et al.,* 1971; strawberry, Hyodo, 1971; mandarin, Hyodo and Asahara, 1973; *Pyrus malus* L., Macheix, 1974; *Lycopersicum esculentum* var. Cerasiforme, Fleuriet, 1976; *Prunus avium* L. var. Bigarreau Napoléon, Melin *et al.,* 1977; *Pyrus communis* L. var. Passe-crassane, Billot *et al.,* 1978).

III. TISSUE- AND SEGMENT-SPECIFIC ACCUMULATION OF SECONDARY PRODUCTS

A. Tissue-Specific Accumulation

It has been noted previously in this chapter that single organs have a stage-dependent capacity for synthesizing and accumulating secondary products. But not all parts of an organ necessarily function in the same way;

various tissues of the same organ may differ significantly in their synthesizing ability. This is well demonstrated by experiments with separated leaf tissues of *Sinapis alba* L. (Wellmann, 1974). In general, flavonoids occur mainly in the epidermis of *Sinapis* cotyledons; a detailed study showed flavonols to be present in the upper epidermis and anthocyanins in the lower epidermis. In the same study almost all of the PAL and chalcone-flavone isomerase was located in the epidermis (see Table I and Steinitz and Bergfeld, 1977).

The localization of flavonols in leaves and leaf homologues follows a similar pattern in onion scales (*Allium* sp.; Tissut, 1974) and in leaves of *Bryophyllum crenatum, Allium porrum* (Tronchet, 1968, 1971), and *Tulipa* 'Apeldoorn' (Brüggen, 1974).

Effertz and Weissenböck (1976), however, observed a different distribution. These authors found that there was little difference in the flavone content of the upper epidermis, mesophyll, and lower epidermis of the primary leaves of *A. sativa*. The concentrations in these three tissues varied according to the developmental stage of the leaf. The pattern of accumulated flavones also developed in a chacteristic way.

As demonstrated, flavonoids are frequently accumulated preferentially in the epidermis. In other systems they occur in both epidermis and mesophyll (see Onslow, 1925; Nozzolillo, 1972). Tissue-specific accumulation is also important for other secondary compounds. For example, the cyanogenic glucoside dhurrin has been found exclusively in the epidermal layers of the leaf blade (Kojima *et al.*, 1979), whereas lupine alkaloids are found only in chloroplast-containing tissues (White and Spencer, 1964; see also Wink *et al.*, 1980).

The principle of tissue-specific accumulation of secondary compounds in

TABLE I

Distribution of Anthocyanin, Quercetin, Phenylalanine Ammonia-Lyase (PAL) and Chalcone-Flavanone Isomerase (CFI) in Mustard Cotyledons (*Sinapis alba* L.) After Continuous Far-Red Illumination[a]

Section	Anthocyanin (%)	Quercetin (%)	PAL (%)	CFI (%)
1 (upper epidermis)	12	80	65	42
2 (mesophyll)	1	6	4	3
3 (mesophyll)	0	4	2	4
4 (mesophyll)	5	1	3	6
5 (lower epidermis)	82	9	26	45

[a] After Wellmann, 1974.

leaf organs extends to other organs in higher plants. Anthraquinones of *Morinda citrifolia* L. (Rubiaceae) are found only in the root system, where they are not uniformly distributed but are preferentially localized in the root bark (Zenk *et al.*, 1975; see Thomson, 1971, too).

Even closely related secondary compounds can vary greatly in their location—and therefore in their biochemical differentiation—within an organ. In roots of *C. arietinum* (Fabaceae) the two isoflavones biochanin A (5,7-dihydroxy-4'-methoxy isoflavone) and formononetin (7-hydroxy-4'-methoxy isoflavone) are accumulated in high concentrations (Barz and Adamek, 1970) but are distributed differently. Histochemically, Gierse (1975) demonstrated that biochanin A was mainly accumulated in the rhizodermis and perhaps to some extent in the subepidermal cell layer, whereas formononetin was distributed evenly throughout the total cortical tissue (Fig. 5). Similar studies with other Fabaceae clearly indicated a tendency for the accumulation of 5-hydroxylated isoflavones in peripheral tissues, while the 5-deoxy compounds were detected in the central cortical tissues of roots (Isenberg, 1975; Käsekamp, 1976; Wendel, 1977).

B. Segment-Specific Accumulation

Individual segments of an organ may exhibit varying differentiation patterns in their ability to form secondary products. Maximum differentiation in the accumulative topography is observed in the characteristic pigmentation of petals. In the differentiation of petals, a segment-specific accumulation

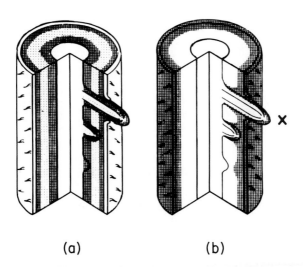

(a)   (b)

Fig. 5.   Block diagram of a root of *C. arietinum* and the distribution of formononetin (a, dotted) and biochanin A (b, dotted). x, Lateral root. (After Isenberg, 1975.)

pattern of secondary products, especially of flavonoid compounds and carotenoids, in many cases leads to the formation of highly characteristic "nectar guides," such as the uv-absorbing petal zone (Thompson *et al.,* 1972; cf. Bloom and Vickery, 1973; Kevan, 1978, and references cited therein).

The petals of *Torenia baillonii* in the final stage of development exhibit two distinctly colored zones (Fig. 6). The reddish-violet basal zone contains both carotenoids and anthocyanins, whereas the yellow apical zone is pigmented exclusively by carotenoids.

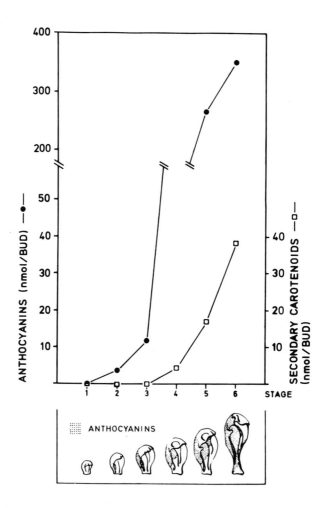

Fig. 6. Top: Concentration of anthocyanins and secondary carotenoids in developing buds of *T. baillonii.* Bottom: Stages of *Torenia* bud development with distinguishing marks of anthocyanin pigmentation. (Modified after Lang and Potrykus, 1971.)

The mixed-pigmented petals of *T. baillonii* arise from meristematic tissue containing chloroplasts. During development, differences in pigmentation arise in the apical and basal parts of the petals. In the region of the corolla tube, a high level of anthocyanins appears at an early stage of development. Initially, the anthocyanins appear exclusively in the lower and upper epidermis and later to some extent also in the mesophyll cells.

After formation of the anthocyanins, the differentiation of chromoplasts starts almost simultaneously in all tissues of the petals. This event marks the beginning of the accumulation of secondary carotenoids which are represented by five xanthophyll esters (Lang, 1970; Lang and Potrykus, 1971). The major phase of the formation of these secondary carotenoids is associated with a decline in chlorophyll content. Corresponding phases of differentiation in the formation of secondary carotenoids have been observed during the development of the petals of *Viola tricolor* (Lichtenthaler, 1969).

A highly segment-specific accumulation of secondary products may also occur in leaves. Weiler and Zenk (1976) have examined the distribution pattern of digoxin equivalents in leaves of *D. lanata* by radioimmunoassay. In the region of the veins, especially near the central vein and also toward the leaf base, low concentrations of digoxigenin glycosides were found, whereas high concentrations were recorded in green parenchymatous parts of the distal third of the lamina and their marginal segments.

IV. TISSUE-SPECIFIC CONTROL OF THE LEVELS OF ENZYMES INVOLVED IN SECONDARY METABOLISM

In Section III, it was shown that individual tissues and segments of particular organs differ in their ability to accumulate secondary products. Furthermore, studies by Wellmann (1974) have shown that secondary metabolism at separate sites within an organ can respond differently to the same external factor such as light.

In mustard seedlings, anthocyanin accumulation is induced to a similar degree both by continuous far-red (FR) light and by a short exposure to red (R) light (Lange *et al.,* 1971). In the same system, however, very little flavonol accumulation is induced by R-irradiation, whereas continuous FR irradiation intensifies the flavonol concentration (Table II). Thus the tissue levels of both classes of flavonoids are regulated to different degrees. Corresponding to these different rates of accumulation, phenylalanine ammonialyase, and chalcone-flavanone isomerase exhibit very little activity in the upper epidermis (the region of flavonol accumulation) after R irradiation. Under identical conditions, these enzymes show more than 50% of the activity found after irradiation with continuous FR light in the lower epidermis (the

TABLE II

Comparison of the Effect of Phytochrome, after Continuous Far-Red (FR) Irradiation and Short-Time Irradiation with Red Light (R), on the Accumulation of Anthocyanin and of Quercetin, and the Increase in Phenylalanine Ammonia-Lyase (PAL) and Chalcone-Flavanone Isomerase (CFI) Activity[a,b]

	Lower epidermis, FR/R (%)	Upper epidermis, FR/R (%)
Anthocyanin	$100 \pm 9 : 65 \pm 8$	
Quercetin		$100 \pm 12 : 13 \pm 6$
PAL	$100 \pm 13 : 58 \pm 9$	$100 \pm 18 : 11 \pm 8$
CFI	$100 \pm 24 : 75 \pm 15$	$100 \pm 21 : 8 \pm 5$

[a] After Wellmann, 1974.
[b] Results are for the upper and lower epidermis of mustard cotyledons (*Sinapis alba* L.). Dark values have been subtracted.

region of anthocyanin accumulation). Thus there is a meaningful correlation between the extent of accumulation and enzyme induction (Wellmann, 1974).

These results show that the enzymes in the upper and lower epidermis are regulated to varying extents by phytochrome. During primary differentiation (Wagner and Mohr, 1966) in the individual cells a different competence for the specificity of the photoresponse may be established.

V. THE INTEGRATION OF SECONDARY METABOLISM INTO THE DEVELOPMENTAL PROGRAM

The integration of secondary metabolism into the organization of differentiation programs is realized not only in many microorganisms but also in higher plants. Studies on anthers have shown a phase-specific formation of secondary constituents during pollen differentiation and pollen ripening (Wiermann, 1970; Quast and Wiermann, 1973). The pollen grains are formed in individual pollen sacs by meiotic division of the pollen mother cells. They exist in the free space of the loculus after breakdown of the tetrads, surrounded by metabolically active tapetum cells. These cells play an important nutritional role in the further development of the pollen grains (Vasil, 1973).

The outer pollen wall or exine is characterized by numerous structures and sculptures mainly composed of sporopollenins. From the chemical viewpoint, these are polymers derived from carotenoids and carotenoid esters

(Brooks and Shaw, 1971). Therefore this wall, with its complex structure, can be considered a product of secondary metabolism. With the various types of structures involved, it represents a system in which different secondary products (mainly pigments such as carotenoids and flavonoids) can be accumulated (Heslop-Harrison, 1968; Heslop-Harrison and Dickinson, 1969; Wiermann, 1970, 1979; Dickinson, 1973; Stanley and Linskens, 1974; Hesse, 1978). Interesting correlations are also found between morphological differentiation and the formation and accumulation of phenols. The systems so far studied differ noticeably in the process of accumulation during the entire phase of microsporogenesis. However, in all cases the accumulation of such phenols is a postmeiotic event.

A highly differentiated, phase-specific accumulation of different phenylpropanoid compounds occurs during pollen development in *Tulipa* 'Apeldoorn' (Fig. 7). After degradation of the callose wall (i.e., after disintegration of the tetrads), the accumulation of simple phenylpropanes (I), chalcones (II), and flavonols together with anthocyanins (III) takes place consecutively at various stages of postmeiotic microsporogenesis. Thus the anthers of *Tulipa* are the only known case in which the various compounds in flavonoid biosynthesis accumulate separately and sequentially during morphological differentiation.

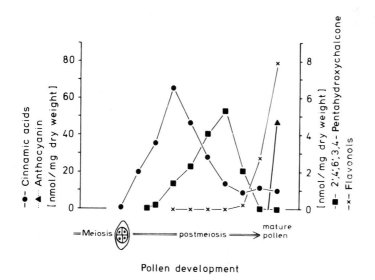

Fig. 7. Kinetics of phenylpropanoid accumulation during microsporogenesis of *Tulipa* 'Apeldoorn'. (Modified after Quast and Wiermann, 1973.)

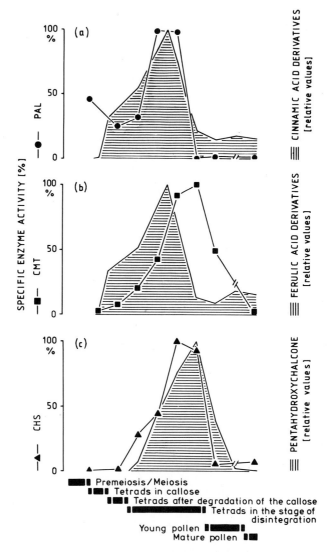

Fig. 8. Phase-dependent accumulation of secondary products and development of enzymes involved in their biosynthesis in the contents of anthers of *Tulipa* 'Apeldoorn'. (a) Hydroxycinnamic acid derivatives (ferulic and *p*-coumaric acids); PAL (R. Wiermann, unpublished). (b) Ferulic acid derivatives; *S*-adenosyl-L-methionine:caffeic acid 3-*O*-methyltransferase (Sütfeld and Wiermann, 1978, 1980). (c) 2′,3,4,4′,6′-pentahydroxychalcone; chalcone synthase (Sütfeld *et al.*, 1978; Sütfeld and Wiermann, 1980). Stages of microsporogenesis are shown at the bottom.

In stage I, *p*-coumaric acid and ferulic acid derivatives were detected (Quast and Wiermann, 1973; Wiermann, 1970).

In stage II, 2′,3,4,4′,6′-pentahydroxychalcone, as a major component, 2′,4,4′,6′-tetrahydroxychalcone, and 2′,4,4′,6′-tetrahydroxy-3-methoxy-chalcones as minor components were observed.

Kaempferol, quercetin, and isorhamnetin glycosides, and a delphinidin 3-rhamnoglucoside accumulated during stage III.

This phase-dependent accumulation of various secondary metabolites correlates to a considerable extent with the development of enzymes required in the different parts of the biosynthetic pathway (Fig. 8). Thus the activity of PAL increases during meiosis to a high level at the stages in which an intensive accumulation of hydroxycinnamic acids occurs. The development of *S*-adenosyl-L-methionine: caffeic acid 3-*O*-methyltransferase activity (CMT) (Sütfeld and Wiermann, 1978) correlates well with the accumulation sequence for ferulic acid (R. Sütfeld and R. Wiermann, 1980). Similarly, a high level of chalcone synthase activity is found during the stages of intensive chalcone formation (Sütfeld *et al.*, 1978). These results obviously suggest that a stage-specific accumulation of various phenylpropanoid compounds is due to the sequential formation (or activation) and subsequent inactivation of the corresponding enzymes.

VI. LIGNIFICATION, A PROCESS INTEGRATED IN CELL AND TISSUE DIFFERENTIATION

The formation of lignin is one of the most important achievements of secondary plant metabolism (see Grisebach, this volume, Chapter 15). Lignin formation takes place only in certain cells at a particular phase of development. Thus lignification should be considered a process that occurs during the morphogenesis of undifferentiated cells into xylem or other characteristic cell elements (e.g., sclerenchyma cells).

It is axiomatic that gymnosperms contain lignin of the guaiacyl type (4-hydroxy-3-methoxyphenyl), whereas dicotyledonous angiosperms have a higher content of the syringyl type (3,5-dimethoxy-4-hydroxyphenyl). The fact that lignin from the total wood of beech or birch trees contains both guaiacyl and syringyl phenylpropane units (Nimz, 1966; Larsson and Miksche, 1971) does not necessarily mean that such lignin is a uniform copolymer of the two forms. It has been shown that the lignin in different cells and cell complexes is not necessarily uniform in composition; instead it varies with the kind of cell. Certain cells or organ parts can contain guaiacyl lignin, whereas others are abundant in syringyl lignin (Fergus and Goring, 1970; Surholt, 1979). Thus in birch the transport vessels contain mainly guaiacyl lignin, whereas the secondary walls of fibers and ray cells contain primarily syringyl lignin.

In callus cultures of *Populus tremuloides,* which produce only vessels and parenchymatous cells, little or no syringyl lignin was detected, in contrast to the xylem of the intact plant (Wolter *et al.,* 1974). These results were interpreted as indicating that the guaiacyl and syringyl lignin in angiosperm wood is compartmentalized, with guaiacyl lignin in vessels and syringyl lignin in fibers and ray cells (see also Venverloo, 1969).

An extremely complicated picture of lignification in various tissues and/or various organs of *Phleum pratense* at different stages of development has been provided by Stafford (1962) in a critical study. This study revealed that the formation of guaiacyl lignin was preferred over that of syringyl lignin during differentiation of the metaxylem. The opposite preference was shown by the sclerenchyma of the bundle caps and subepidermal layer.

These conclusions, based on histochemical, chemical, and spectroscopic test procedures, have recently been confirmed by enzymatic analysis of lignin formation. Grand *et al.* (1977; Grand and Ranjeva, 1979) have examined the distribution of different coenzyme A (CoA) ligases in various tissues of poplar stems. They have found that feruloyl-CoA ligase is located mainly in the xylem, whereas sinapoyl-CoA ligase activity is found mainly in the fraction containing sclerenchymal elements (Table III). This characteristic tissue-specific distribution of CoA ligase activity agrees well with the observed ratio of syringyl to guaiacyl residues found in the lignin.

In addition to this tissue-specific synthesis of different lignins, the composition of lignin can vary in certain plants depending on the stage of development. The young wood of *Robinia* and *Populus* formed in spring contains predominantly guaiacyl lignin, whereas the late wood contains mainly syringyl lignin. Obviously, the biosynthesis of lignin in such tissues is regulated by the season and/or the stage of development.

TABLE III

Distribution of Coenzyme A Ligase Activity in Different Tissues of Poplar Stems and Comparison of the Syringyl/Guaiacyl Ratio[a]

	Enzyme activity (picokatals/g dry wt)			
Tissue	*p*-Coumaroyl-CoA ligase	Feruloyl-CoA ligase	Sinapoyl-CoA ligase	Syringyl/Guaiacyl ratio
Sclerenchyma	126	70	*400*	1.60
Xylem	190	*140*	50	1.14

[a] After Grand *et al.,* 1977.

VII. THE DIFFERENTIATION OF SPECIALIZED CELLS AND ACCUMULATION OF SECONDARY PRODUCTS

According to the results available, two principles determine the accumulation of secondary compounds: (1) In certain plant tissues, each cell potentially is able to synthesize and accumulate these products. (2) In other tissues there are only a few cells that specialize in the storage of secondary products; in such cells, known as idioblasts, the accumulation correlates with a specific morphological differentiation. Examples are seen in the accumulation of alkaloids in distinct alkaloid-accumulating cells, as well as in the excretion of terpenoids and flavonoids by special glandular cells.

A. Excretion Cells

Species of the Papaveraceae contain high concentrations of alkaloids in special cells called laticifers. These cells originate from laticifer primordial cells which are—as exemplified by *Chelidonium majus*—recognizable as well-defined cell elements located between the parenchyma cells of young leaves. In this species, laticifer development is marked by an increase in the number of vacuoles and a simultaneous degeneration of the other cell organelles, so that in the later stages the laticifers are composed mainly of small vacuoles which do not fuse with each other. Different procedures have shown that the alkaloids are localized in these small vacuoles. However, very young laticifers, which appear in the meristematic region, do not contain such vacuoles. These appear during further differentiation of the laticifers. Therefore the accumulation of alkaloids depends on a specific stage of differentiation of this compartment (Matile *et al.,* 1970; Neumann, 1976; Nessler and Mahlberg, 1978 and references cited therein; see also Section VIII).

Similar phenomena are found in the isodiametric alkaloid cells in the rhizome of *Sanguinaria canadensis;* here the alkaloids are localized in the large central vacuole (Neumann, 1976; Neumann and Müller, 1972).

Studies on alkaloid formation in *C. majus* have shown that the parenchymal cells surrounding the laticifers are the site of biosynthesis and that after formation the alkaloids are transferred to the laticifers for storage. These represent "sinks" for alkaloids (Müller, 1976, Müller *et al.,* 1976). In this case, therefore, the total process is the result of coevolution and codifferentiation of biosynthetic abilities and storage sites.

Tannins, too, may be stored in specialized, as well as normal, cells of a tissue (Swain, 1965), and there may be a similar correlation between cell differentiation and accumulation (Ginzburg, 1967; Amelunxen and Gronau, 1969; Mueller and Greenwood, 1978; Chafe and Durzan, 1973; see also Böhm, 1977).

Electron microscope studies on tannin vacuoles in *Oenothera* have shown that there are cells in the young tissues of the shoot apex with densely staining regions described as tannin inclusions. This material is localized in small vacuoles which, in the course of further differentiation, increase and finally fill the whole cell (Diers *et al.*, 1973; Mueller and Beckman, 1976; Parham and Kaustinen, 1977; see also Baur and Walkinshaw, 1974; Lang *et al.*, 1977). The endosplasmic reticulum (ER), Golgi bodies (Amelunxen and Gronau, 1969), and plastids (Mueller and Beckman, 1976) may play a role in the synthesis of tannins or their precursors.

B. Glandular Cells

In the case of glandular cells, plants have developed a highly specialized system for synthesis and elimination of secondary products. Such glands are often epidermal structures occurring in the form of hairs. [There are also internal glands that excrete into cavities (Kisser, 1958; Schnepf, 1969a,d; Reinhard and Nagel, 1976).] The morphology of glandular cells is as variable as the products they accumulate. They produce essential oils and resins, other terpenoids, several flavonoid compounds, and fatty oils or waxes, and they often excrete a mixture of these. However, in spite of their diversity in biosynthetic and anatomical properties, there are similar cytological aspects (Schnepf, 1976).

A striking feature of the glandular cells is the differentiation of a tubular, smooth ER which often becomes the predominating cell organelle and fills much of the cytoplasmic space. Thus there appears to be a correlation between the differentiation of this cell organelle and the production of secondary substances in the case of both essential oils (Schnepf, 1969b,c,e, 1972; Heinrich, 1973, 1977) and flavonoids (Wollenweber and Schnepf, 1970; Wollenweber *et al.*, 1971; Schnepf and Klasova, 1972; see Table IV and Fig. 9).

Electron microscope investigations of buds from *Alnus* have shown that the ER in young glandular cells is relatively undeveloped. However, with the beginning of the excretory phase, the gland cells show extended tubular, agranular ER. This cytological change is closely coordinated with the excretion of flavonoid aglycones and/or other lipophilic terpenoid substances. When secretion is completed, the ER decreases (Wollenweber *et al.*, 1971).

The substances produced often are accumulated in a subcuticular space of the glands (Schnepf, 1969b; Heinrich, 1973) or, in the case of flavonoids, penetrate the cuticula and crystallize on its surface (*Primula* species: Wollenweber and Schnepf, 1970; *Pityrogramma chrysoconia:* Schnepf and Klasová, 1972).

The role of the smooth tubular ER in the production of secondary products

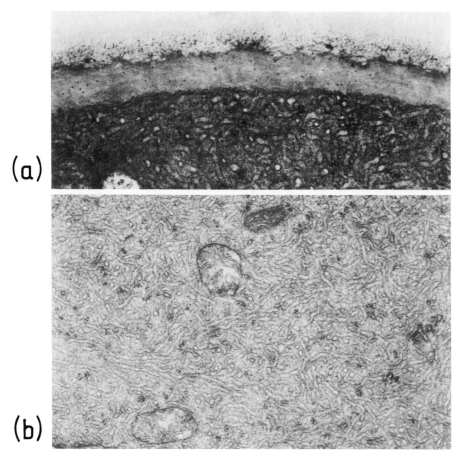

Fig. 9a. Farina-excreting gland of scape of *Primula denticulata*. Outer wall of the gland cell with cuticule and cutinized layer. The most prominent component of the gland cell cytoplasm is a smooth, tubular ER. Glutaraldehyde–osmium tetroxide plus potassium chromate. 31,000×.

Fig. 9b. Oil-excreting gland of an inflorescence axis of *P. obconica*. In the cell cytoplasm is smooth, tubular ER. Glutaraldehyde–osmium tetroxide plus potassium chromate. Both farina and oil are excreted through the cuticule and stored on the cuticular surface. 30,000×. (after Wollenweber and Schnepf, 1970).

in plant glands has repeatedly been pointed out (but see for *Viscaria vulgaris:* Tsekos and Schnepf, 1974). However, in the holocrine type of gland, there is some evidence for the function of plastids in the synthesis of terpenoids (Heinrich, 1966, 1969; Amelunxen and Arbeiter, 1967). The merocrine glandular cells of the resin canal cells of conifers are of this type too (Wooding and Northcote, 1965; Campbell, 1972; Schnepf, 1976).

TABLE IV

Types of Plant Glands That Excrete Secondary Products[a]

Anatomical gland type	Excretion products	Development of cell structures[b]					Species[c]
		Plastids	Tubular smooth ER	Rough ER	Periplast ER	Golgi apparatus	
Duct	Resins	+	+		+		*Pinus* species (1), *Picea abies* (1)
	Oils		+		+		*Solidago canadensis* (2)
	Essential oils, gums, slimes		++		+	+	*Heracleum* species (3), *Aegopodium podagraria* (3), *Dorema ammoniacum* (3)
Hair (peltate hair, hair with glandular head)	Oils		++		+		*Arctium lappa* (4), *Ledum palustre* (5), *Lonicera periclymenum* (6), *Ononis repens* (6), *Senecio viscosus* (6)
	Oil, resins		+	++			*Viscaria vulgaris* (7)
	Essential oils		++		+		*Cleome spinosa* (8)
	Flavonoids		++				*Primula* species (9)
	Flavonoids		++		+		*Pityrogramma chrysoconia* (10)
	Terpenes, flavonoids, slimes		++	+		+	*Alnus* species (11)
Epithelium	Fats, "wax"		++				*Ficus benjamina* (5)
	Terpenes, flavonoids	+	+				*Populus* species (12)
Cavity	Sesqui- and monoterpenes	+					*Citrus* species (13), *Poncirus trifoliata* (13)
	Essential oils	+					*Dictamnus albus* (14)

[a] Modified after Schnepf, 1976.

[b] +, Conspicuous; ++, predominating.

[c] Refs: (1) Wooding and Northcote, 1965; Campbell, 1972: (2) Schnepf, 1969c; (3) Schnepf, 1969e; (4) Schnepf, 1969b; (5) Schnepf, 1972; (6) Schnepf and Klasová, 1972; (7) Tsekos and Schnepf, 1974; (8) Amelunxen and Arbeiter, 1969; (9) Wollenweber and Schnepf, 1970; (10) Schnepf and Klasová, 1972; (11) Wollenweber *et al.*, 1971; (12) Charriere-Ladreix, 1973, 1979; (13) Heinrich, 1966, 1969; (14) Amelunxen and Arbeiter, 1967).

VIII. ACCUMULATION OF SECONDARY PRODUCTS AND THE DIFFERENTIATION OF STORAGE SPACES

Plant cells have developed two methods of storing secondary products. These may be classified as *intracytoplasmic* where vacuoles and plastids are the storage sites, and *extracytoplasmic* where the cell wall, pollen wall, subcuticular space, and cuticular surface are involved (Fig. 10). Thus the accumulation of secondary products should correlate with the differentiation of specific accumulation sites. This especially applies to the formation of vacuoles and the development of plastids, as well as to the wall systems mentioned above.

A. Vacuoles

The vacuoles are the most important, ubiquitously distributed storage spaces for secondary products (Küster, 1956; Blank, 1958). This was recently demonstrated in studies showing that isolated vacuoles of *Sorghum* seedlings represent the major site of accumulation of the cyanogenic glucoside dhurrin (Saunders and Conn, 1978; also Saunders *et al.,* 1977; see in this connection Wagner and Siegelman, 1975; Matile, 1976, 1978H; Leigh and Branton, 1976; Sasse *et al.,* 1979; Saunders, 1979). Sometimes the secondary products accumulate in such high concentrations that they even crystallize (e.g., anthraquinones in the cell suspension cultures of *Galium mullugo* L; Bauch and Leistner, 1978).

Anthocyanins often accumulate in the central vacuole of epidermal cells. In studies combining developmental physiology and electron microscope investigations, Steinitz and Bergfeld (1977) showed that there was a close correlation between the differentiation of vacuoles and the appearance of the ability to accumulate anthocyanins in cotyledons of *S. alba.* Degradation and mobilization of storage proteins in the aleurone vacuoles resulted in the formation of several separate vacuoles in the epidermal cells of mustard cotyledons which later coalesced so that finally a central vacuole was formed. It was demonstrated that the ability to accumulate anthocyanins was acquired during this transition phase. This process was interpreted as a

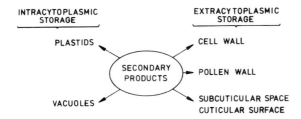

Fig. 10.   Scheme illustrating the different accumulation areas for secondary plant products.

transformation of aleurone vacuoles into a single central storage vacuole which stored anthocyanins. Similar correlations between the beginning of the production of secondary substances and the differentiation of vacuoles have previously been mentioned (Section VII,A).

B. Plastids

Chloroplasts of various species contain simple phenylpropanes and/or flavonoids (Zaprometov and Kolonkova, 1967; Weissenböck *et al.*, 1971, 1976; Monties, 1972; Kindl, 1971; Oettmeier and Heupel, 1972; Czichi and Kindl, 1975; Plesser and Weissenböck, 1977; Weissenböck and Schneider, 1974; Saunders and McClure, 1976a,b), as well as enzymes involved in phenolic metabolism (Sato, 1967; Barlett *et al.*, 1972; Löffelhardt *et al.*, 1973; Monties, 1974; Saunders and McClure, 1975; Czichi and Kindl, 1975; Gestetner and Conn, 1974; Weissenböck *et al.*, 1976; Ranjeva *et al.*, 1977a,b). In connection with the developmental and light-dependent production of *C*-glycosylflavones Weissenböck and Effertz (1974) have observed a correlation between flavone accumulation and the differentiation of both the primary leaf and the plastids. Because various flavones occur in high concentrations in the plastids of primary leaves and etioplasts and chloroplasts have their own specific flavone composition, it has been concluded that the differentiation of plastids is of importance for flavonoid metabolism during ontogenesis of the organ concerned (Weissenböck and Effertz, 1974; Luckner and Nover, 1977).

It has already been pointed out (Section III,B) that the differentiation of chromoplasts is important in the production of secondary carotenoids. In maturing fruits and plasmochromous petals the development of chromoplasts is accompanied by a marked increase in secondary carotenoids (Lichtenthaler, 1969). The transformation of chloroplasts into chromoplasts is initiated by the destruction of thylakoids, concurrently, the number and size of the plastoglobules serving as storage sites for the secondary carotenoids are greatly increased (Lichtenthaler, 1969, 1970a,b; Spurr and Harris, 1968; Dodge, 1970). While plastoglobules are found in numerous chromoplasts, other structural components occasionally predominate as the major sites of chromoplast pigments (e.g., tubular bundles, see Sitte, 1974, and literature cited therein; Smith and Butler, 1971; Falk, 1976; Winkenbach *et al.*, 1976; Wuttke, 1976; Sitte, 1977). Wooding and Northcote (1965) refer to a close correlation between the development of plastids and resin synthesis in the resin canal cells of *Pinus pinea* (see also Campbell, 1972).

C. Cell Wall

Secondary products may accumulate within the cell wall. This applies to various phenolic compounds (e.g., sphagnorubin, see Rudolph and Vohwinkel, 1969; Vohwinkel, 1975, and sphagnum acid (p-hydroxy-β-(carboxymethyl)cinnamic acid; see Tutschek, 1975; Tutschek *et al.*, 1973), but especially to lignins (see Grisebach, this volume, Chapter 15, Section VI). It has often been demonstrated that the formation of lignin as a typical incrustation of the cell wall is closely connected with the ontogenetic development of the individual wall layers. Thus the lignification process originates within the middle lamella and primary wall in the region of the cell corners about the time of the onset of secondary wall production. Successively, lignification then extends along the middle lamella and continues centripetally into the individual layers of the secondary wall (see Hepler *et al.*, 1970; Wardrop, 1971, and references cited therein). A similar picture was given for the incrustation of wall pigments in *Sphagnum magellanicum* (Tutschek *et al.*, 1978).

In all three cases described (Section VII,A–C), relationships between the development of specific accumulation sites and commencement of the accumulation of secondary plant compounds were demonstrated. It must nevertheless be emphasized that the subcellular location of the actual biosynthesis of these compounds is still unknown.

IX. CONCLUDING REMARKS

In a number of studies it has been demonstrated that the formation of secondary products as well as their storage site may be restricted to certain developmental stages of the plant, specific organs, tissues, or specialized cells. In the *intact plant* it is generally accepted that a close correlation exists between the expression of secondary metabolism and morphological and cytological differentiation (Luckner, 1971; Böhm, 1977). Secondary metabolism is therefore considered an aspect of the developmental process. It is not yet clear to what extent secondary metabolism depends on the development of specific structures, and it is unknown whether these two processes are genetically and/or physiologically linked. Results obtained from experiments with *cell cultures* have shown that specific morphological differentiation is not necessarily a prerequisite for the formation of secondary products (but see Neumann and Müller, 1974; Ellis, 1978).

The capacity for synthesis and accumulation of secondary compounds can vary greatly within an organ or tissue. This can be related to a specific and dynamic primary differentiation pattern which itself is under control of the genotype of each individual cell. That all cells of a plant show omnipotency is demonstrated particularly well by studies with cell cultures. And yet only part of the genetic potential is realized. The factors that determine the dynamics of the primary differentiation pattern as a manifestation of differential gene activity are unknown.

REFERENCES

Amelunxen, F., and Arbeiter, H. (1967). *Z. Pflanzenphysiol.* **58**, 49–69.
Amelunxen, F., and Arbeiter, H. (1969). *Z. Pflanzenphysiol.* **61**, 73–80.
Amelunxen, F., and Gronau, G. (1969). *Cytobiologie* **1**, 58–69.
Amrhein, N., and Gödecke, K.-H. (1977). *Plant Sci. Lett.* **8**, 313–317.
Amrhein, N., and Holländer, H. (1979). *Planta* **144**, 385–389.
Amrhein, N., and Zenk, M. H. (1970). *Z. Pflanzenphysiol.* **63**, 384–388.
Amrhein, N., Gödecke, K.-H., and Kefeli, V. I. (1976). *Ber. Dtsch. Bot. Ges.* **89**, 247–259.
Andersen, R. A., Lowe, R., and Vaughn, T. A. (1969). *Phytochemistry* **8**, 2139–2147.
Aoki, S., Araki, C., Kaneko, K., and Katayama, O. (1971). *Agric. Biol. Chem.* **35**, 784–787.
Barlett, D. J., Poulton, J. E., and Butt, V. S. (1972). *FEBS Lett.* **23**, 265–267.
Barz, W., and Adamek, C. (1970). *Planta* **90**, 191–202.
Barz, W., and Hösel, W. (1971). *Phytochemistry* **10**, 335–341.
Barz, W., and Hösel, W. (1975). *In* (J. B. Harborne, T. J. Mabry, and H. Mabry, eds.), "The Flavonoids" pp. 916–969. Chapman & Hall, London.
Barz, W., Hösel, W., and Adamek, C. (1971). *Phytochemistry* **10**, 343–349.
Bauch, H.-J., and Leistner, E. (1978). *Planta Med.* **33**, 105–123.
Baur, P. S., and Walkinshaw, C. H. (1974). *Can. J. Bot.* **52**, 615–619.
Billot, J. (1974). *Physiol. Veg.* **12**, 189–198.
Billot, J., Hartmann, C., Macheix, J.-J., and Rateau, J. (1978). *Physiol. Veg.* **16**, 693–714.
Blank, F. (1958). *In* "Handbuch der Pflanzenphysiologie" (W. Ruhland, ed.), Vol. 10, pp. 300–353. Springer-Verlag, Berlin and New York.
Bloom, M., and Vickery, R. K. (1973). *Phytochemistry* **12**, 165–167.
Böhm, H. (1967). *Planta Med.* **15**, 215–220.
Böhm, H. (1977). *Mol. Biol., Biochem. Biophys.* **23**, 103–123.
Bopp, M., and Capesius, I. (1973). *Ber. Dtsch. Bot. Ges.* **86**, 257–270.
Bopp, M., and Lüdicke, W. (1975). *Z. Naturforsch, C: Biosci.* **30C**, 663–667.
Brooks, J., and Shaw, G. (1971). *In* "Pollen: Development and Physiology" pp. 99–114. Butterworth, London.
Brüggen, J. (1974). Diplomarbeit, University of Münster, Germany.
Campbell, R. (1972). *Ann. Bot. (London)* [N.S.] **36**, 711–720.
Chafe, S. C., and Durzan, D. J. (1973). *Planta* **113**, 251–262.
Charrière-Ladreix, Y. (1973). *J. Microsc. (Paris)* **17**, 299–316.
Charrière-Ladreix, Y. (1979). *Phytochemistry* **18**, 43–45.
Czichi, U., and Kindl, H. (1975). *Hoppe-Seyler's Z. Physiol. Chem.* **356**, 475–485.
Dickinson, H. G. (1973). *Cytobios* **8**, 25–40.
Diers, L., Schötz, F., and Meyer, B. (1973). *Cytobiologie* **7**, 10–19.
Dittrich, P., and Kandler, O. (1971). *Ber. Dtsch. Bot. Ges.* **84**, 465–473.
Dodge, J. D. (1970). *Ann. Bot. (London)* [N.S.] **34**, 817–824.
Effertz, B., and Weissenböck, G. (1976). *Ber. Dtsch. Bot. Ges.* **89**, 473–481.
Ellis, B. E. (1978). *Can. J. Bot.* **56**, 2717–2729.
Falk, H. (1976). *Planta* **128**, 15–22.
Fergus, B. J., and Goring, D. A. I. (1970). *Holzforschung* **24**, 113–117.
Fleuriet, A. (1976). *Fruits* **31**, 117–126.
Gestetner, B., and Conn, E. E. (1974). *Arch. Biochem. Biophys.* **163**, 617–624.
Gierse, H.-D. (1975). Dissertation, University of Freiburg, Germany.
Ginzburg, C. (1967). *Bot. Gaz. (Chicago)* **128**, 1–10.
Grand, C., and Ranjeva, R. (1979). In press.
Grand, C., Ranjeva, R., Boudet, A. M., and Alibert, G. (1977). *Phytochem. Soc. Eur. Phytochem. Soc. North Am., 1st J. Symp.*

Gröger, D., Mothes, K., Floss, H.-G., and Weygand, F. (1963). *Z. Naturforsch B: Anorg. Chem., Org. Chem., Biochem., Biophys., Biol.* **18B,** 1123–1124.

Hagen, C. W. (1966). *Am. J. Bot.* **53,** 54–60.

Hahlbrock, K., and Grisebach, H. (1974). *In* "The Flavonoids" (J. B. Harborne, T. J. Mabry and H. Mabry, eds.), pp. 866–915. Chapman and Hall, London.

Hahlbrock, K., Sutter, A., Wellmann, E., Ortmann, R., and Grisebach, H. (1971). *Phytochemistry* **10,** 109–116.

Heinrich, G. (1966). *Flora (Jena), Abt. A* **156,** 451–456.

Heinrich, G. (1969). *Oesterr. Bot. Z.* **117,** 397–403.

Heinrich, G. (1973). *Planta Med.* **23,** 154–166.

Heinrich, G. (1977). *Biochem. Physiol. Pflanz.* **171,** 17–24.

Hepler, P. K., Fosket, D. E., and Newcomb, E. H. (1970). *Am. J. Bot.* **57,** 85–96.

Heslop-Harrison, J. (1968). *New Phytol.* **67,** 779–786.

Heslop-Harrison, J., and Dickinson, H. G. (1969). *Planta* **84,** 199–214.

Hess, D. (1963). *Planta* **59,** 567–586.

Hess, D., and Endress, R. (1973). *Z. Pflanzenphysiol.* **68,** 441–449.

Hesse, M. (1978). *Plant Syst. Evol.* **129,** 13–30.

Hyodo, H. (1971). *Plant Cell Physiol.* **12,** 989–991.

Hyodo, H., and Asahara, S. (1973). *Plant Cell Physiol.* **14,** 823–828.

Isenberg, P. (1975). Staatsexamensarbeit, University of Münster, Germany.

Ishikura, N. (1972). *Phytochemistry* **11,** 2555–2558.

Käsekamp, C. (1976). Staatsexamensarbeit, University of Münster, Germany.

Kevan, P. G. (1978). *In* "The Pollination of Flowers by Insects." (A. J. Richards, ed.), Linn. Soc. Symp. Ser., No. 6, pp. 51–78. Academic Press, New York.

Kindl, H. (1971). *Hoppe-Seyler's Z. Physiol. Chem.* **352,** 767–768.

Kisser, J. K. (1958). *In* "Handbuch der Pflanzenphysiologie" (W. Ruhland, ed.), Vol. 10, pp. 91–131. Springer-Verlag, Berlin and New York.

Kojima, M., Poulton, J. E., Thayer, S. S., and Conn, E. E. (1979). *Plant Physiol.* **63,** 1022–1028.

Krause, J. (1971). Dissertation, University of Münster, Germany.

Küster, E. (1956). "Die Pflanzenzelle." Fischer, Jena.

Lang, E., Hörster, H., Friedrich, H., Themann, H., and Amelunxen, F. (1977). *Cytobiologie* **15,** 372–381.

Lang, W. (1970). *Z. Pflanzenphysiol.* **62,** 299–301.

Lang, W., and Potrykus, J. (1971). *Z. Pflanzenphysiol.* **65,** 1–12.

Lange, H., Shropshire, W., and Mohr, H. (1971). *Plant Physiol.* **47,** 649–655.

Larsson, S., and Miksche, G. E. (1971). *Acta Chem. Scand.* **25,** 647–662.

Leigh, R. A., and Branton, D. (1976). *Plant Physiol.* **58,** 656–662.

Lichtenthaler, H. K. (1969). *Ber. Dtsch. Bot. Ges.* **82,** 483–497.

Lichtenthaler, H. K. (1970a). *Planta* **90,** 142–152.

Lichtenthaler, H. K. (1970b). *Planta* **93,** 143–151.

Linscheid, M., Wendisch, D., and Strack, D. (1980). *Z. Naturforsch., C: Biosci.* **35C,** 907–914.

Löffelhardt, W., Ludwig, B., and Kindl, H. (1973). *Hoppe-Seyler's Z. Physiol. Chem.* **354,** 1006–1012.

Luckner, M. (1971). *Pharmazie* **26,** 717–724.

Luckner, M. (1977). *In* "Cell Differentiation in Microorganisms, Plants and Animals" (L. Nover and K. Mothes, eds.), pp. 538–558. North-Holland Publ., Amsterdam.

Luckner, M., and Nover, L. (1977). *Mol. Biol., Biochem. Biophys.* **23,** 1–102.

Macheix, J. J. (1974). *Physiol. Veg.* **12,** 25–33.

Maier, V. P., and Hasegawe, S. (1970). *Phytochemistry* **9,** 139–144.

Margna, U., Hallop, L., Margna, E., and Tohver, M. (1967). *Biochim. Biophys. Acta* **136,** 396–399.

Matile, P. (1976). *Nova Acta Leopold., Suppl.* **7**, 139–156.

Matile, P. (1978). *Annu. Rev. Plant Physiol.* **29**, 193–213.

Matile, P., Jans, B., and Rickenbacher, R. (1970). *Biochem. Physiol. Pflanz.* **161**, 447–458.

Melin, C., Moulet, A. M., Dupin, J. F., and Hartmann, C. (1977). *Phytochemistry* **16**, 75–78.

Mockaitis, J. M., Kivilaan, A., and Schulze, A. (1973). *Biochem. Physiol. Pflanz.* **164**, 248–257.

Mohr, H. (1972). "Lectures on Photomorphogenesis." Springer-Verlag, Berlin and New York.

Mohr, H., and Sitte, P. (1971). "Molekulare Grundlagen der Entwicklung." BLV, München.

Monties, B. (1972). *Int. Congr. Photosynth., 2nd, 1971* pp. 1681–1691.

Monties, B. (1974). *C. R. Hebd. Seances Acad. Sci., Ser. C* **278**, 1465–1467.

Mothes, K., and Romeike, A. (1958). *In* "Handbuch der Pflanzenphysiologie" (W. Ruhland, ed.), Vol. 8, pp. 989–1049. Springer-Verlag, Berlin and New York.

Mothes, K., and Schütte, H. R. (1969). "Biosynthese der Alkaloide." VEB Dtsch. Verlag Wiss., Berlin.

Mothes, K., Engelbrecht, L., Tschöpe, K.-H., and Hutschenreuter-Trefftz, G. (1957). *Flora (Jena)* **144**, 518–536.

Mueller, W. C., and Beckman, C. H. (1976). *Can. J. Bot.* **54**, 2074–2082.

Mueller, W. C., and Greenwood, A. D. (1978). *J. Exp. Bot.* **29**, 757–764.

Müller, E. (1976). *Nova Acta Leopold., Suppl.* **7**, 123–128.

Müller, E., Neumann, D., Nelles, A., and Bräutigam, E. (1976). *Nova Acta Leopold., Suppl.* **7**, 133–138.

Nahrstedt, A., Hösel, W., and Walther, A. (1979). *Phytochemistry* **18**, 1137–1141.

Nessler, C. L., and Mahlberg, P. G. (1978). *Am. J. Bot.* **65**, 978–983.

Neumann, D. (1976). *Nova Acta Leopold., Suppl.* **7**, 77–81.

Neumann, D., and Müller, E. (1972). *Biochem. Physiol. Pflanz.* **163**, 375–391.

Neumann, D., and Müller, E. (1974). *Biochem. Physiol. Pflanz.* **165**, 271–282.

Nimz, H. (1966). *Chem. Ber.* **99**, 469–474.

Nover, L., and Luckner, M. (1976). *Nova Acta Leopold., Suppl.* **7**, 375–385.

Nozzolillo, C. (1972). *Can. J. Bot.* **50**, 29–34.

Nurmann, G., and Strack, D. (1979). *Z. Naturforsch., C: Biosci.* **34C**, 715–720.

Oettmeier, W., and Heupel, A. (1972). *Z. Naturforsch., B: Anorg. Chem., Org. Chem., Biochem., Biophys., Biol.* **27B**, 177–183.

Onslow, M. W. (1925). "The Anthocyanin Pigments of Plants," 2nd ed. Cambridge Univ. Press, London and New York.

Parham, R. A., and Kaustinen, H. M. (1977). *Bot. Gaz. (Chicago)* **138**, 465–467.

Plesser, A., and Weissenböck, G. (1977). *Z. Pflanzenphysiol.* **81**, 425–437.

Popovici, G., and Weissenböck, G. (1976). *Ber. Dtsch. Bot. Ges.* **89**, 483–489.

Popovici, G., and Weissenböck, G. (1977). *Z. Pflanzenphysiol.* **82**, 450–454.

Quast, L., and Wiermann, R. (1973). *Experientia* **29**, 1165–1166.

Ranjeva, R., Alibert, G., and Boudet, A. M. (1977a). *Plant Sci. Lett.* **10**, 225–234.

Ranjeva, R., Alibert, G., and Boudet, A. M. (1977b). *Plant Sci. Lett.* **10**, 235–242.

Reinhard, E., and Nagel, M. (1976). *Nova Acta Leopold., Suppl.* **7**, 335–343.

Reznik, H. (1961). *Flora (Jena)* **150**, 454–473.

Romeike, A. (1956). *Flora (Jena)* **143**, 67–86.

Romeike, A. (1959). *Flora (Jena)* **148**, 306–320.

Rudolph, H., and Vohwinkel, E. (1969). *Z. Naturforsch., B: Anorg. Chem., Org. Chem., Biochem., Biophys., Biol.* **24B**, 1211–1212.

Rumpenhorst, H. J. (1968). Dissertation, University of Münster, Germany.

Sasse, F., Backs-Hüsemann, D., and Barz, W. (1979). *Z. Naturforsch., C: Biosci.* **34C**, 848–853.

Sato, M. (1967). *Phytochemistry* **6**, 1363–1373.

Saunders, J. A. (1979). *Plant Physiol.* **64**, 74–78.

Saunders, J. A., and Conn, E. E. (1978). *Plant Physiol.* **61**, 154–157.
Saunders, J. A., and McClure, J. W. (1975). *Phytochemistry* **14**, 1285–1289.
Saunders, J. A., and McClure, J. W. (1976a). *Phytochemistry* **15**, 805–807.
Saunders, J. A., and McClure, J. W. (1976b). *Phytochemistry* **15**, 809–810.
Saunders, J. A., Conn, E. E., Lin, C. H., and Stocking, C. R. (1977). *Plant Physiol.* **59**, 647–652.
Schnepf, E. (1969a). *Protoplasmatologia* **8**, 8.
Schnepf, E. (1969b). *Protoplasma* **67**, 185–194.
Schnepf, E. (1969c). *Protoplasma* **67**, 195–203.
Schnepf, E. (1969d). *Protoplasma* **67**, 205–212.
Schnepf, E. (1969e). *Protoplasma* **67**, 375–390.
Schnepf, E. (1972). *Biochem. Physiol. Pflanz.* **163**, 113–125.
Schnepf, E. (1976). *Nova Acta Leopold., Suppl.* **7**, 23–44.
Schnepf, E., and Klasová, A. (1972). *Ber. Dtsch. Bot. Ges.* **85**, 249–258.
Sitte, P. (1974). *Z. Pflanzenphysiol.* **73**, 243–265.
Sitte, P. (1977). *Biol. Unserer Zeit* **7**, 65–74.
Smith, M., and Butler, R. D. (1971). *Protoplasma* **73**, 1–13.
Spurr, A. R., and Harris, W. M. (1968). *Am. J. Bot.* **55**, 1210–1224.
Stafford, H. A. (1962). *Plant Physiol.* **37**, 643–649.
Stafford, H. A. (1969). *Phytochemistry* **8**, 743–752.
Stanley, R. G., and Linskens, H. F. (1974). "Pollen: Biology, Biochemistry, Management." Springer-Verlag, Berlin and New York.
Statham, C. M., and Crowden, R. K. (1974). *Phytochemistry* **13**, 1835–1840.
Staude, M., and Reznik, H. (1973). *Z. Pflanzenphysiol.* **69**, 409–417.
Steinitz, B., and Bergfeld, R. (1977). *Planta* **133**, 229–235.
Stràck, D. (1973). Dissertation, University of Köln, Germany.
Strack, D. (1977). *Z. Pflanzenphysiol.* **84**, 139–145.
Strack, D. (1980). *Z. Naturforsch., C: Biosci.* **35C**, 204–208.
Strack, D., and Mansell, R. L. (1977). *Z. Pflanzenphysiol.* **85**, 243–252.
Strack, D., and Mansell, R. L. (1979). *Z. Pflanzenphysiol.* **91**, 63–67.
Strack, D., and Reznik, H. (1976). *Z. Pflanzenphysiol.* **79**, 95–108.
Strack, D., Tkotz, N., and Klug, M. (1978). *Z. Planzenphysiol.* **89**, 343–353.
Strack, D., Nurmann, G., and Sachs, G. (1980). *Z. Naturforsch., C: Biosci.* **35C**, 963–966.
Sütfeld, R., and Wiermann, R. (1978). *Biochem. Physiol. Pflanz.* **172**, 111–123.
Sütfeld, R., and Wiermann, R. (1980a). *Arch. Biochem. Biophys.* **201**, 64–72.
Sütfeld, R., and Wiermann, R. (1980b). *Z. Pflanzenphysiol.* **97**, 283–288.
Sütfeld, R., Kehrel, B., and Wiermann, R. (1978). *Z. Naturforsch., C: Biosci.* **33C**, 841–846.
Surholt, E. (1978). Dissertation, University of Münster, Germany.
Swain, T. (1965). *In* "Plant Biochemistry" (J. Bonner and J. E. Varner, eds.), pp. 552–580. Academic Press, New York.
Thompson, W. R., Meinwald, J., Aneshansley, D., and Eisner, T. (1972). *Science* **177**, 528–530.
Thomson, R. H. (1971). "Naturally Occurring Quinones," 2nd ed. Academic Press, New York.
Tissut, M. (1974). *C. R. Hebd. Seances Acad. Sci., Ser. D* **279**, 659–662.
Tissut, M., and Egger, K. (1972). *Phytochemistry* **11**, 631–634.
Tkotz, N., and Strack, D. (1980). *Z. Naturforsch., C: Biosci.* **35C**, 835–837.
Tronchet, J. (1968). *C. R. Hebd. Seances Acad. Sci., Ser. D* **266**, 882–884.
Tronchet, J. (1971). *Bull. Soc. Bot. Fr.* **118**, 173–184.
Tsekos, I., and Schnepf, E. (1974). *Biochem. Physiol. Pflanz.* **165**, 265–270.
Tutschek, R. (1975). *Z. Pflanzenphysiol.* **76**, 353–365.
Tutschek, R., Rudolph, H., Wagner, P. H., and Kreher, R. (1973). *Biochem. Physiol. Pflanz.* **164**, 461–464.

Tutschek, R., Rudolph, H., Asmussen, L., and Altena, U. (1978). *Rev. Bryol. Lichénol* **44**, 319–330.

Tzagoloff, A. (1963). *Plant Physiol.* **38**, 202–206.

Vasil, J. K. (1973). *Naturwissenschaften* **60**, 247–253.

Venverloo, C. J. (1969). *Acta Bot. Neerl.* **18**, 241–314.

Vohwinkel, E. (1975). *Chem. Ber.* **108**, 1166–1181.

Wagner, E., and Mohr, H. (1966). *Planta* **71**, 204–221.

Wagner, G. J., and Siegelman, H. W. (1975). *Science* **190**, 1298–1299.

Walton, D. C. (1968). *Plant Physiol.* **43**, 1120–1124.

Wardrop, A. B. (1971). *In* "Lignins–Occurrences, Formation, Structure and Reactions" (K. V. Sarkanen and C. Ludwig, eds.), pp. 19–41. Wiley (Interscience), New York.

Weiler, E. W., and Zenk, M. H. (1976). *Phytochemistry* **15**, 1537–1545.

Weissenböck, G. (1971). *Z. Pflanzenphysiol.* **66**, 73–81.

Weissenböck, G., and Effertz, B. (1974). *Z. Pflanzenphysiol.* **74**, 298–326.

Weissenböck, G., and Reznik, H. (1970). *Z. Pflanzenphysiol.* **63**, 114–130.

Weissenböck, G., and Sachs, G. (1977). *Planta* **137**, 49–52.

Weissenböck, G., and Schneider, V. (1974). *Z. Pflanzenphysiol.* **72**, 23–35.

Weissenböck, G., Tevini, M., and Reznik, H. (1971). *Z. Pflanzenphysiol.* **64**, 274–277.

Weissenböck, G., Plesser, A., and Trinks, K. (1976). *Ber. Dtsch. Bot. Ges.* **89**, 457–472.

Wellmann, E. (1974). *Ber. Dtsch. Bot. Ges.* **87**, 275–279.

Wendel, I. (1977). Staatsexamensarbeit, University of Münster, Germany.

White, H. A., and Spencer, M. (1964). *Can. J. Bot.* **42**, 1481–1485.

Wiermann, R. (1970). *Planta* **95**, 133–145.

Wiermann, R. (1979). *In,* "Regulation of Secondary Product and Plant Hormone Metabolism" (M. Luckner and K. Schreiber, eds.), pp. 231–239. Pergamon, Oxford.

Wiewiórowski, M., Podkowińska, H., Bratek-Wiewiórowska, M. D., Kuhn-Orzechóroska, M., and Boczán, W. (1966). *Abh. Dtsch. Akad. Wiss. Berlin,* 3rd. No. 3, pp. 215–233.

Wink, M., Hartmann, T., and Witte, L. (1980). *Z. Naturforsch., C: Biosci.* **35C**, 93–97.

Winkenbach, F., Falk, H., Liedvogel, B., and Sitte, P. (1976). *Planta* **128**, 23–28.

Wollenweber, E., and Schnepf, E. (1970). *Z. Pflanzenphysiol.* **62**, 216–227.

Wollenweber, E., Egger, K., and Schnepf, E. (1971). *Biochem. Physiol. Pflanz.* **162**, 193–202.

Wolter, K. E., Harkin, J. M., and Kirk, T. K. (1974). *Physiol. Plant.* **31**, 140–143.

Wooding, F. B. P., and Northcote, D. H. (1965). *J. Ultrastruct. Res.* **13**, 233–244.

Wuttke, H. G. (1976). *Z. Naturforsch., C: Biosci.* **31C**, 456–460.

Zaprometov, M. N., and Kolonkova, S. V. (1967). *Dokl. Akad. Nauk SSSR* **176**, 470–473.

Zenk, M. H., El-Shagi, H., and Schulte, U. (1975). *Planta Med., Suppl.* pp. 79–101.

Compartmentation in Natural Product Biosynthesis by Multienzyme Complexes

5

HELEN A. STAFFORD

The Biochemistry of Plants, Vol. 7

I. INTRODUCTION

A. Definitions and Terminology

Multienzyme systems have been described as aggregates of different but functionally related enzymes bound together by *noncovalent* forces into a highly organized structure (Ginsburg and Stadtman, 1970). The function of such a system is to increase catalytic efficiency by channeling the flow of carbon through the complex. Although these complexes may be soluble or membrane-associated, the organizational aspects leading to catalytic efficiency seem more dramatic in the case of soluble complexes. Compartmentation by either free or membrane-associated multienzyme complexes (the "surface model") should be distinguished from compartmentation due merely to restrictions by membranes (the "compartment model") (Davis, 1967).

The first well-studied soluble multienzyme complex was the complex of seven enzymes noncovalently linked in yeast and capable of fatty acid synthesis (Sumper *et al.*, 1969). Other evidence indicates that this enzyme system consists of multifunctional units, since the *in vivo* catalytic units are *covalently* linked in two identical or different polypeptide chains with multiple catalytic properties (Stoops *et al.*, 1975; Wieland *et al.*, 1979; Wood *et al.*, 1978, Lynen, 1980). During extraction procedures without added protease inhibitors, the catalytic units are "nicked" by specific proteases and give rise to an apparent multienzyme complex of still tightly bound but separate polypeptide chains. A similar result has been obtained with the *arom* gene cluster (Gaertner and Cole, 1977). Since the methods of detection described below do not distinguish between a multifunctional enzyme and a multienzyme complex, the former will be considered a variant of a complex for the purposes of this chapter. In most cases, no attempt has been made to distinguish between the two possibilities. Pyruvate dehydrogenase of *Escherichia coli* or of muscle mitochondria is another classical example of a tightly bound multienzyme complex. Such a complex has been recently isolated from plant tissues (Rubin *et al.*, 1978). Other membrane-embedded complexes involve the well-known electron transport systems of mitochondria and chloroplasts.

Complexes (including multifunctional polypeptides) can be separated into different groups based on the degree of binding of the catalytic units: tight complexes as in fatty acid synthetase, less strong ones typified by tryptophan synthetase and the *arom* complex, and very weak ones such as the glycolytic sequence (Masters, 1977; Reed and Cox, 1970; Mowbray and Moses, 1976). The intermediates formed may be either covalently bound as in the case of fatty acid synthetase, or associated by noncovalent attachments as in tryptophan synthetase (Mosbach and Mattiasson, 1976). Examples of weakly bound complexes are less well known but could be important as systems capable of binding reversibly to membranes or as self-aggregating systems as postulated for the glycolytic sequence (Ottaway and Mowbray, 1977; Mowbray and Moses, 1976). Loose complexes capable of association–dissociation might have physiological importance as a form of metabolic control (Reed and Cox, 1970; Masters, 1977). Since the forces that bind such enzymes to each other or to cytoplasmic structures are loose enough to be easily broken during extraction procedures, some and possibly all enzyme sequences found as separate soluble enzymes may be artifacts of extraction procedures.

Some of the evolutionary aspects of both multienzyme complexes and multifunctional enzymes, and their relationship to separate enzymes catalyzing the same reactions, have been discussed (Bloch, 1977; Kirschner and Bisswanger, 1977; Wood et al., 1978). It has been argued that the evolutionary sequence was from separate enzymes to multienzyme complexes to multifunctional polypeptide chains. However, one might also argue that some multienzyme complexes evolved from multifunctional ones if one assumes that looser connections are sometimes advantageous. In either case, various degrees of enzyme association would be found in living forms. However, one would need to differentiate between truly independent systems and artifacts produced by extraction procedures.

A correlation between a multienzyme complex and a cluster of genes in one linkage group, as in the case of the *arom* complex in *Neurospora,* has been emphasized by Ahmed and Giles (1969). However, this dependence is no longer crucial and might actually occur only in rare cases in eukaryotes, since genes for related enzymes of a pathway are frequently scattered on different chromosomes. The important requirement for multifunctional enzymes and tight multienzyme complexes might be the presence of a single mRNA made possible through mRNA splicing in eukaryotes (Darnell, 1978).

B. Major Advantages of Complexes

Four major advantages of multienzyme complexes and multifunctional enzymes are frequently cited (Reed and Cox, 1970; Stadtman, 1970; Mosbach and Mattiasson, 1976):

1. An increase in catalytic efficiency through metabolic channeling due to decreased diffusion times or direct transfer of intermediates, and a reduction in the lag phase prior to the steady state production of the final product.
2. The segregation of competing pathways (metabolic channeling).
3. The potential for a finer tuning of metabolic control by coordinate activation or inhibition, a change in pH optimum, etc.
4. Protection of unstable intermediates.

C. Kinetic Aspects of Multienzyme Complexes

In a multienzyme complex intermediates may remain enzyme-bound without equilibration with the surrounding medium, or their diffusion path may be limited to varying degrees within the microenvironment of the complex. The absence or limitation of free intermediates endows the complex with greater "catalytic facilitation" (Gaertner *et al.,* 1970). In a tight complex such as fatty acid synthetase of yeast, in which the intermediates are covalently bound, the rate of the overall reaction is determined only by the rate of transfer between the catalytic sites, not by the K_m for the various intermediates (Bloch, 1977). With less tightly bound enzymes such as those in the *arom* complex, K_m and V_{max} values for each site determine the extent of escape of an intermediate or the degree of "unchanneling" (Davis, 1967; Welch and Gaertner, 1976). It is considered important that the V_{max} of the second step of the complex be significantly greater than that of the first step (or that the first step be rate-limiting) to prevent saturation of the second step. Otherwise, some intermediates would become free in the solution and therefore would be available for equilibration with any added intermediate (Davis, 1967).

The activity of the first enzyme is postulated to be less in the diffusion-limited microenvironment of a complex than it would be if it acted independently in solution (Ottaway and Mowbray, 1977). However, a complex and its microenvironment, in which intermediates do not equilibrate with the surrounding medium, provide a finer system of metabolic control. For instance, the binding of only one allosteric effector to one enzyme component of a complex might also affect the activity of all the other members of the complex even though they themselves are not affected by the effector (Welch and Gaertner, 1976).

D. Criteria and Methodology Used to Detect Multienzyme Complexes

Evidence of the presence of multienzyme complexes can be based on genetic studies, tissue level isotopic tracer experiments, or cell-free studies. An example of genetic evidence is the demonstration of a cluster of genes for a sequence in one linkage group or the simultaneous loss of more than one catalytic function by a single point mutation (Kirschner and Bisswanger, 1977). Tissue level studies could involve isotopic evidence of separate metabolic pools (metabolic channeling) or isotopic evidence (via lower dilution values) that a precursor at the beginning of a sequence is a better precursor than one of the intermediates of the sequence (catalytic facilitation).

However, the bulk of the evidence must come from cell-free studies. An appropriate sequence of steps might be:

Step 1: Demonstration of multisteps for at least part of a sequence and determination of optimal conditions for the overall rate, for the rate starting with each intermediate for partial reactions, and for independent steps if feasible (Section II,A,C).

Step 2: Copurification to show a constant rate of activity of the overall reaction and of at least one independent step or partial reaction (Kirschner and Bisswanger, 1977).

Step 3: Demonstration of catalytic facilitation. The best method probably varies with the system.

a. The overall rate is greater than when starting with an intermediate. This may not be possible at substrate saturated levels; subsaturated levels may be necessary (see discussion in Section II,B,C).

b. Addition of an intermediate to limiting concentrations of the initial substrate does not increase the overall rate of the reaction (Gaertner et al., 1970).

c. Comparison of $^3H/^{14}C$ ratios of product and intermediate after incubation with substrate-limiting amounts of ^3H-labeled initial substrate and ^{14}C-labeled exogenous intermediate. Ratios greater than 1.0 show nonequilibration of the exogenous ^{14}C intermediate with that produced on the surface of the complex (^3H-labeled) (Sections II,B and C).

Other methodology has been used to study the kinetics of complexes. A comparison of K_m and K_{max} values for individual steps within the intact complex has been made in nonactivated and activated states. Coordinate activation of the intact *arom* complex is considered a key attribute of this complex and an indication of increased catalytic efficiency (Section II,A).

In some cases, the complex can be dissociated and the individual enzymes assayed independently in their nonaggregated form. Kinetic studies with these separate enzymes can be compared with those on the overall reaction of the intact complex. When the data for separate enzymes are not available, a computer can be used to simulate the dynamic behavior of the complex. A reduction in the extent of the lag phase by preincubation with the initial substrate is considered an important consequence of the physical state of the intact complex (Welch and Gaertner, 1975).

A useful technique for use in conjunction with the above studies is comparison of the kinetics of the intact complex with that of a model system made by the microencapsulation of solubilized enzymes in gels (Mosbach and Mattiasson, 1976; Srere and Mosbach, 1974). Anderson (1974), effectively used this technique to show that a three-enzyme system of urease–carbonic anhydrase–carboxylase was a more efficient CO_2 fixing enzyme in such a gel microenvironment than the soluble carboxylase assayed independently. In addition, she showed that the K_m value for the entire model multistep system was the same as that for the first enzyme, as long as the levels of the first enzyme were lower than that of the carboxylase and diffusion was limited. Such microencapsulation techniques should be useful along with computer-simulated data in studying the efficiency of a loose sequence such as the C_6-C_3 sequence under conditions of limited diffusion in microenvironments that might be comparable with *in vivo* reactions occurring in vesicles.

II. SPECIFIC EXAMPLES

A. The Soluble *Arom* Complex in *Neurospora* as a Model System

Although the products of the *arom* pathway are not secondary, this example of a relatively tight complex serves as an excellent model of methodology for demonstrating channeling and catalytic facilitation. The complex isolated from *Neurospora* is responsible for the initial five steps, starting with the C_7 sugar 3-deoxy-D-*arabino*-heptulosonate-7-phosphate (DAHP), which leads to the aromatic amino acids via chorismic acid. It was originally thought to consist of five constitutive enzymes catalyzing steps 2–6 of the aromatic pathway, encoded by five structural genes of the *arom* gene cluster (Giles *et al.*, 1967). However, recent evidence indicates that it is a multifunctional polypeptide consisting of a dimer of identical subunits, each of about 150,000 MW, controlled by one cluster gene (Gaertner and Cole, 1977). This biosynthetic pathway competes for two substrates held in common with a second pathway, consisting of three inducible separate enzymes (encoded in the *qa* gene cluster) which catalyze the catabolism of quinic acid (Giles and Case, 1975).

The primary function of the *arom* complex is to channel intermediates [dehydroquinate (DHQ) and dehydroshikimate (DHS), Table I] common to the two pathways. Since some cross-feeding of DHS between the biosynthetic and the catabolic pathways occurs under certain conditions, three methods of regulating this flow are postulated. One involves the greater lability of one of the enzymes of the catabolic pathway; a second involves the favoring of the biosynthetic pathway by a 20-fold greater K_m for the second substrate in the common pool by the enzymes in the catabolic sequence; the third involves the effectiveness of the channeling mechanism of the biosynthetic pathway (Strøman *et al.*, 1978). Their major evidence of channeling originally was the demonstration of a 10-fold increase in the rate of the overall reaction starting with DAHP, in contrast to that of one of the later intermediates (shikimate) (Gaertner *et al.*, 1970). However, in a study with different assay conditions that measured the rates of independent steps rather than multisteps, Welch and Gaertner (1976) emphasized a different type of evidence in demonstrating catalytic facilitation, i.e., an increase in catalytic efficiency due to coordinate activation by the first substrate. The

TABLE I

Comparison of K_m and V_{max} Values in the Prearomatic Pathway (*Arom* Complex) (Steps 2–6) with That of the Dhurrin Pathway (Steps 1-4)[a]

A. Prearomatic pathway, assayed as individual steps (data from Welch and Gaertner, 1976)

Step	1 mM Substrate	K_m (μM)		V_{max}[b]	
		Not activated	Activated	Not activated	Activated
2 (first of complex)	DAHP	60	<12	1	2
3 (I_1)	DHQ	100	20	20	20
4 (I_2)	DHS	40	20	4	4
5 (I_3)	SA	100	100	2	2
6 (I_4)	SAP	100	10	5	5

B. *Dhurrin pathway*, assayed as multiple steps (data from Conn, 1978)

Substrate	Product	K_m (μM)	V_{max}[c]
Tyrosine → 4-OH-benzaldehyde		30	145
I_1 : N—OH → 4-OH-benzaldehyde		110	345
I_2 : oxime → 4-OH-benzaldehyde		50	400
I_3 : nitrile → 4-OH-benzaldehyde		100	50

[a] Abbreviations: I, intermediate; DAHP, 3-deoxy-D-arabinoheptulosonate-7-phosphate; DHQ, dehydroquinate; DHS, dehydroshikimate; SA, shikimate; SAP, shikimate-5-phosphate; N-OH = N-hydroxytyrosine; oxime, p-hydroxyphenylacetaldoxime; nitrile, p-hydroxyphenylacetonitrile.

[b] V_{max} with DAHP, assayed as multiple steps, was >10 times that starting with SA in Gaertner et al. (1970).

[c] Nanomoles HCN hr^{-1} mg^{-1} protein.

values of K_m and V_{max} were determined for each step before and after activation with DAHP. Lower K_m values for four of the five enzymes were obtained after incubation with the initial substrate. The V_{max} values were unchanged except for a doubling for the first step (Table IA). The coordinate activation by the first substrate also shortened the lag phase for the overall reaction (Welch and Gaertner, 1975). The shikimate kinase step (step 5) was rate-limiting and is postulated to be involved in a novel regulatory device (Welch and Gaertner, 1976). Both the aggregate state of the complex and the 10- to 20-fold increase in the V_{max} of the second enzyme relative to the first are considered vital to the channeling mechanism.

Similar aggregates have been demonstrated in a variety of fungi and in *Euglena* (Ahmed and Giles, 1969). However, in extracts of prokaryotes, these five enzymes are always found separately, and no gene cluster is involved (Berlyn and Giles, 1969). In higher plants, two are associated with a bifunctional enzyme, 3-dehydroquinate hydrolase and shikimate : NADP$^+$ oxidoreductase (Boudet *et al.*, 1977; Koshiba, 1978; Polley, 1978), whereas the others are isolated as separate enzymes. It should be noted that the first enzyme in this polyaromatic pathway in *Neurospora* that catalyzes the condensation of phosphoenolpyruvate (PEP) and erythrose-4-phosphate (E-4-P) to DAHP is not tightly associated with the five enzymes of the *arom* complex. The first step is catalyzed by three isozymes encoded by three unlinked genes; each isozyme is feedback-inhibited by the three final products, tryptophan, phenylalanine, and tyrosine. Lineweaver–Burk plots indicated nonlinear kinetics (upward curvature or positive cooperativity). However, linear plots were obtained in the presence of the above inhibitors (Doy, 1967).

B. A Two-Step Sequence in the Phenylpropane (C$_6$–C$_3$) Pathway

A two-step sequence converting phenylalanine to *p*-coumarate (4-hydroxycinnamate) was demonstrated in an alga (Czichi and Kindl, 1975a) and in higher plants (Czichi and Kindl, 1975b, 1977). It involved two enzymes, phenylalanine ammonia-lyase (PAL) and cinnamate-4-hydroxylase (C4H). These studies also demonstrated the two-step sequence to coumarate in the same preparations, but only the first example will be discussed here. In potato extracts, this multiple step was associated with a defined subcellular fraction containing endoplasmic reticulum. The product, *p*-coumarate, was formed from phenylalanine at a faster rate than from exogenous cinnamate, as long as the latter was added in a concentration comparable to the amount of cinnamate formed by PAL activity. The above data implied that channeling occurred. However, Kindl obtained more effective evidence of channeling by using a double-labeling technique to determine the extent of equilibration between pools of bound or endogenously produced intermediate with that of the added or exogenous intermediate.

In this double-labeling technique, enzyme mixtures containing the suspected multienzyme complex were incubated with the starting substrate, ^{3}H-labeled phenylalanine, and the exogenous intermediate, ^{14}C-labeled cinnamate. After an incubation period, the cinnamate and the product (*p*-coumarate) were isolated and the ^{3}H/^{14}C ratio in both compounds was determined. The most useful way to compare the data is to compute the ratio of these labels in the product to that of the intermediate. If the ratio is less than 0.1, it indicates the presence of two independent consecutive enzymes with complete equilibration of the ^{3}H label from the initial substrate with the ^{14}C label of the added intermediate. A ratio greater than 1.0 means that only limited equilibration of the ^{3}H label with the ^{14}C label has occurred. A ratio of

5 or greater demonstrates the presence of a relatively tight complex, showing considerable catalytic facilitation (Czichi and Kindl, 1975a,b, 1977).

Some of the data from Kindl's laboratory are summarized in Table II with comparable data for the dhurrin pathway to be described below. Channeling was detectable only when both the phenylalanine and added cinnamate were kept at subsaturating levels. Four to 50 times as much phenylalanine (50–500 μM) was added, in contrast to the amount of cinnamate (about 10 μM). Czichi and Kindl (1977) reported K_m values of 350 and 3–6 μM for PAL and C4H, respectively, in the microsomal complex isolated from cucumber cotyledons. However, Havir and Hanson (1968) have reported non-Michaelis kinetics for PAL from potato, with a K_m of only 38 μM at low phenylalanine concentrations (10 μM), and a higher one of 260 μM at phenylalanine concentrations of 1 mM (see also Hanson and Havir, 1979, also this volume). Since Kindl used phenylalanine concentrations spanning this range, the percentage of PAL saturation may have varied in his different experiments. Nevertheless, the $^3H/^{14}C$ ratio data shown in Table II clearly indicate nonequilibration of the metabolic pools and therefore of channeling or catalytic facilitation. A plateau of channeling ("coupling") was reached when the [^3H]cinnamate concentration derived from the [^3H]phenylalanine reached 0.7 μM, which is below the K_m value. The dependency on subsaturating levels of both initial substrate and intermediate is explained by Czichi and Kindl (1977) as being due to the ease of saturation of the complex or microcompartment. The loss of channeling, however, could also be due to a too high V_{max} of the PAL step relative to that of the hydroxylation step, or

TABLE II

Comparison of $^3H/^{14}C$ Ratios of Product/Intermediate in the Two-Step C_6–C_3 Sequence (A) and in the Four-Step Dhurrin Pathway (B)

Sequence	$\dfrac{^3H/^{14}C \text{ product}}{^3H/^{14}C \text{ intermediate}}$
A. Phenylalanine → p-coumarate (two steps)	
^3H-Phe + ^{14}C cinnamate	
Alga (*D. marina*) (thylakoids) (Czichi and Kindl, 1975a)	14
Potato (ER) (Czichi and Kindl, 1975b)	2
Cucumber (ER) (Czichi and Kindl, 1977)	
Green	6
Etiolated	0.4
B. Tyrosine → 4-OH-benzaldehyde (four steps) (+ one nonenzymatic)	
(Conn, 1978)	
[^3H]Tyr + ^{14}C–I_1 (N—OH)	120
[^3H]Tyr + ^{14}C–I_2 (oxime)	3
[^3H]Tyr + ^{14}C–I_3 (nitrile)	80

to an excess of PAL relative to C4H. In the case of the potato preparations, the channeling effect was lost upon purification and could be demonstrated only in extracts of green cucumbers but not from etiolated cucumbers. These results might imply that the association between PAL and C4H was ionic and that the bonds were easily broken, or that proteases were active. In these experiments, Kindl used very high specific-activity isotopes in order to obtain enough counts for statistically significant data. Incubation times were kept short (6–10 min). Laborious techniques for isolating the product were also employed, such as paper chromatography, repeated recrystalization, gas-liquid chromatography, prior to scintillation counting (Czichi and Kindl, 1975a,b, 1977).

C. A Four-Step Sequence in the C_6–C_2 Pathway Leading to Dhurrin in Higher Plants (Table IB, IIB)

The same $^3H/^{14}C$ ratio method has also been used by Conn's laboratory to demonstrate the presence of a relatively tight complex involving four steps in the dhurrin pathway isolated from *Sorghum* tissues. Dhurrin biosynthesis consists of five steps starting with tyrosine. All but the final glucosidation step are bound to endoplasmic reticulum (ER) membranes. The latter enzyme is found in the "soluble" portion of a plant extract after centrifugation at 100,000 g. Whether the solubilization is an artifact of extraction has not yet been demonstrated. However, it has been shown that the glycosylation enzyme is not associated with intact vacuoles or the tonoplast as might be suspected (Saunders and Conn, 1975).

Multiple lines of evidence of a multienzyme complex of four enzyme activities have been presented by Conn's group (Conn, 1978; Conn *et al.*, 1979, also this volume). These include the demonstration of multisteps, evidence of the preferential utilization of tyrosine (initial substrate) in contrast to that of the later steps, the independent demonstration of four individual steps requiring different assay conditions, the determination of K_m and V_{max} values for multisteps starting with the initial substrate and with the intermediates, and the demonstration by $^3H/^{14}C$ ratios of the nonequilibration of enzymatically produced intermediates with added intermediates.

The $^3H/^{14}C$ ratios obtained by Conn's group are summarized in Table IIB. The product in this case was not dhurrin, since the glucosylation enzyme was not a part of the isolated complex. Instead, 4-hydroxybenzaldehyde was isolated, as it is the nonenzymatic decomposition product of the intermediate 4-hydroxymandelonitrile (I_4). These ratios demonstrate that, at subsaturated substrate levels, I_1 (*N*-hydroxytyrosine) and I_3 (*p*-hydroxyphenylacetonitrile) had the least access to the complex, whereas I_2 (*p*-hydroxyphenylacetaldoxime) had the greatest. The latter, however, still shows a significant amount of channeling. All three cases of added intermediates,

therefore, showed preferential conversion of tyrosine to 4-hydroxybenzalde-
hyde at subsaturated substrate levels. Two of the ratios indicated a complex
tighter than that studied by Czichi and Kindl in the PAL–C4H system. The
apparent tightness of the complex may account for the difficulty in detecting labeled intermediates in earlier trapping experiments (Conn, 1978).

Kinetic determinations of K_m and V_{max} values for multistep reactions,
starting with either tyrosine or one of the intermediates at saturated substrate levels, gave values similar in part to those obtained for the *arom*
complex in *Neurospora* (Table IB). It should be remembered that the activities of the *Neurospora* complex were assayed as a series of onestep reactions, whereas the sorghum dhurrin complex was assayed as multistep reactions. In both cases, the V_{max} for the second step was greater than that for the
first, an expected requirement of a complex showing channeling according to
Davis (1967). Tyrosine was a better substrate for the final product than the
nitrile, I_3. The second intermediate, I_2, which showed the greatest access to
the complex according to the ratio method, gave the highest V_{max}. In contrast, I_1, which showed much less access to the complex, gave a V_{max} almost
as high. Apparently, at saturation substrate levels of the intermediate I_1,
access to the complex was not a major limiting factor. The nitrile, I_3, gave
the slowest rate, consistent with the limited access to the complex measured
by the ratio method. The K_m values did not differ greatly, but the initial
substrate, tyrosine, had the lowest K_m value.

While the ratio method is an extraordinarily useful tool for detecting channeling or nonmixing of pools of an intermediate, it can be misleading if a
suspected component is not the *in vivo* intermediate but only an "unnatural"
precursor. It also cannot differentiate between compartmentation on the surface of a multienzyme complex and that involving membranes of an intact
organelle or vesicle. In such membrane-associated systems, compartmentation might be produced merely by the differential permeability characteristics of the membranes involved. Davis (1967) called this a "compartment
model" in contrast to the "surface model" of a multienzyme complex.

D. Flavanone Synthetase, a Multifunctional Enzyme

This is the first enzyme in a sequence (or series of sequences) leading to a
varied group of C_{15} secondary products called flavonoids (Hahlbrock, this
volume). The enzyme catalyzes the stepwise addition of three acetate units
from malonyl-coenzyme A (CoA) to the acyl moiety of a starter molecule,
p-coumaroyl-CoA at pH 8, and caffeoyl-CoA at pH 6.5–7 (Saleh *et al.*, 1978).
The initial C_{15} products are naringenin (5,7,4'-tetrahydroxyflavanone) and
eriodictyol (5,7,3'4'-tetrahydroxyflavanone), respectively. In the presence of
2-mercaptoethanol, three types of short-chain intermediates, lacking one or
two malonyl-CoA units, are released. None of these have been detected *in*

vivo. A mechanism for their production has been proposed (Hrazdina *et al.*, 1976). Product inhibition was observed, and the approximate K_m value for the starter molecule was 1.6 μM. The enzyme was easily solubilized.

Flavanone synthetase and the succeeding enzymes in the C_{15} pathway, as well as the acetyl-CoA carboxylase (a precursor of malonyl-CoA), show a coordinate induction by light in parsley cell suspension cultures, which is preceded by a coordinate increase in the C_6–C_3 pathway enzymes converting phenylalanine to 4-coumaroyl-CoA (Ebel and Hahlbrock, 1977). The above phenomenon, the evidence of tannins in vesicles associated with the ER (Parkham and Kaustinin, 1977), and the similarity of pH requirements suggest the existence of a loose multienzyme complex as part of the C_{15} sequence (Hrazdina *et al.*, 1978). In addition, a chloroplast localization has also been suggested (McClure, 1979). However, the presence of ER vesicles attached to plastids could account for the above results. The bulk of the flavonoids accumulate in the large central vacuole. The amount remaining in smaller vesicles in the cytoplasm is not known, and the concentrations in chloroplasts may be only transitory (McClure, 1979).

A tissue level study by Haslam (1977, 1979) can be interpreted as further suggestive evidence of metabolite compartmentation or channeling. [^{14}C]cinnamic acid was fed to various plants that accumulate simple condensed tannins (procyanindins), dimers of catechin. The two halves of the dimer were differentially labeled, with two to three times the amount of label appearing in one of the two identical monomer units. This implies that the two halves were derived from metabolically distinct or nonequilibrating pools, a result not expected from the postulated pathway of dimerization. In addition, trapping experiments indicated the absence of detectable amounts of the flavan-3,4-diol postulated as a precursor in this part of the C_{15} sequence. These results suggest the presence of bound intermediates or nonequilibration of metabolic pools.

Furthermore, tissue level studies with *Haplopappus* cell suspension cultures and some cell-free data imply at least a loose association of a multienzyme complex with membranes (presumably ER) in anthocyanin biosynthesis. The hydroxylation system converting naringenin to dihydrokaempferol and then to dihydroquercetin may be membrane-embedded. Subsequent steps to a cyanidin have not yet been demonstrated in cell-free extracts (Fritsch and Grisebach, 1975).

E. Summary of Evidence of Complexes in Higher Plants and Algae

The above suggestive evidence cited for the existence of a multienzyme complex associated with the multifunctional enzyme flavanone synthetase is similar to many fragments of evidence for other pathways leading to secondary products in higher plants. These are summarized in Table III, along with

TABLE III

Recent Examples of Natural Product Compartmentation by Multienzyme Complexes (or Multifunctional Enzymes) in Higher Plants and Algae[a]

	Reference
Cell-free evidence	
Multifunctional enzymes	
1. C_{15}—flavonoid pathway	Hrazdina et al., 1976
3 malonyl-CoA + p-coumaryl-CoA $\longrightarrow$ naringenin	Hahlbrock, this volume
Flavanone synthetase, "soluble"	
Multienzyme complexes—multiple evidence	
2. C_6–C_2—dhurrin pathway	Conn, 1978
Tyrosine $\xrightarrow{\text{4 steps}}$ p-hydroxyphenylacetonitrile	Conn, this volume
ER	
3. C_6–C_3—phenylpropane sequence	Czichi and Kindl,
Phenylalanine $\xrightarrow{\text{2 steps}}$ cinnamate	1975a,b, 1977
ER or algal thylakoids	
plus caffeic acid $\longrightarrow$ ferulic acid	Charriere-Ladreix, 1979
ER	
Suggestive cell-free evidence—multiple-step evidence only	
4. C_6–C_3—phenylpropane sequence (p-hydroxylation)	Ranjeva et al., 1977b
Phenylalanine $\xrightarrow{\text{3 steps}}$ caffeate	Alibert et al., 1977
ER + chloroplast, 27,000-g pellet	Stafford and Lewis, 1977
5. C_6–C_3 coumarin pathway (ortho-hydroxylation)	Ranjeva et al., 1977a
Phenylalanine $\xrightarrow{\text{2 steps}}$ o-coumarate	Czichi and Kindl, 1975a
Chloroplasts	
6. C_6–C_2—phenylacetate sequence	Löffelhardt, 1977
Phenylalanine $\xrightarrow{\text{2-3 steps}}$ phenylacetate	Stafford and Lewis, 1977
Algal thylakoids, 37,000-g pellet	
7. C_6–C_1—benzoate sequence	
Phenylalanine $\xrightarrow{\text{2 steps}}$ benzoate	Löffelhardt and Kindl,
Chloroplast thylakoids	1975
8. $(C_6$–$C_3)_n$—lignin polymerization sequence	Gross et al., 1977
2 malate + $2O_2$ + 2 coniferyl alcohol $\xrightarrow{\text{3 steps}}$ lignin	(-Localized in cell wall)
NAD+, H_2O_2	
in cell wall	
9. C_{15}—Flavonoid pathway	Fritsch and Grisebach,
Naringenin $\xrightarrow{\text{2 steps}}$ dihydroquercetin	1975
ER	
10. Indole alkaloids	
Geraniol $\longrightarrow$ 10-OH-geraniol $\longrightarrow$ indole alkaloids	Madyastha et al., 1977
cyt P_{450} in 20,000-g vesicles (provacuoles)	

(*continued*)

TABLE III (Continued)

	Reference
11. Terpenoid biosynthesis	

11. Terpenoid biosynthesis
  (a) Mevalonate ⟶ squalene ⟶ sterols,
        "Soluble"   ER + GA_3 Hartmann-Bouillon and
                  "soluble"   Benveniste, 1978

  (b) Mevalonate ⟶ kaurene ⟶ GA_3 Hedden *et al.,* 1978
        "Soluble"   ER in seeds,
                   seedlings

  (c) Mevalonate ⟶ kaurene ⟶ GA_3 Hedden *et al.,* 1978
        Proplastids,   Proplastids,
         etiolated   etiolated
          plants   plants
    CO_2 ⟶ Kaurene ⟶ GA_3
     Chloroplasts   Chloroplasts

  (d) Prenyltransferase-isomerase   Bauthorpe *et al.,* 1978
    Mevalonate ⟶ isothujone
    (Geraniol and nerol are intermediates that cannot be
    converted to isothujone *in vitro*)

Indirect cell-free evidence—coordinate induction
  12. Phenylalanine ⟶ C_6–C_3 phenolics, group I
    Light-mediated coordinate inductions, group I followed by   Ebel and Hahlbrock,
    group II   1977
  13. Phenylalanine ⟶ C_{15} flavonoids, group II

Tissue level studies
  14. Prenylquinone synthesis
    Two metabolic pools in thylakoid membranes   Grumbach and Lichten-
                                       thaler, 1975
  15. Flavonoid pathway: procyanidin biosynthesis   Haslam, 1977
    (condensed tannins)
    Asymmetric labeling of procyanidins

[a] See Stafford, 1974b, for earlier references.

the few concrete examples of multienzyme complexes discussed above. Most of the suggestive evidence for complexes is based merely on multiple-step demonstrations. Some of these results may not be due to actual complexes but only to soluble enzymes sequestered in vesicles or an organelle such as the chloroplast. It will be necessary to isolate these potential complexes from these organelles or vesicles to differentiate between these possibilities.

The secondary products cited in Table III include alkaloids, terpenoids, and a wide range of phenolic compounds. Many of the synthetic pathways leading to these products involve hydroxylation reactions. They all appear to be membrane-bound, either to ER or to vesicles that sediment at lower centrifugal speeds. The other enzymes tend to be easily solubilized. While this could be due to the *in vivo* state in the cytosol or in the matrix of an organelle, this soluble state more likely results from broken ionic bonds or from protease activity.

III. DISCUSSION AND CONCLUSIONS

A. Sparsity of Evidence for Complexes in Higher Plants. Why?

Although there are many indications in the literature of potential multienzyme complexes, soluble or membrane-associated, involving pathways leading to secondary products, definitive evidence in higher plants is sparse. Data for terpenoids and alkaloid products are suggestive, but the cell-free enzymology has only just begun to be published. The lack of progress in the area of the phenylpropane (C_6–C_3) hydroxylation–methoxylation sequence has been disappointing.

What is the reason for such a dearth of good examples? One obvious reason is that there are only a few complexes *in vivo* in higher plants, presumably because they have retained a more primitive organization (see Bloch, 1977, for a discussion of this in relation to fatty acid synthetase). However, complexes might be harder to detect in higher plants because they are associated with organelles such as chloroplasts and are destroyed upon disruption of the organelle. Proteases might nick tight complexes, producing loose ones as artifacts of extraction. One might argue that tight complexes are a disadvantage, in the sense that a more generalized (flexible) pathway is more important than a highly specialized (restricted) one, since the former would permit interchanges more easily. Therefore relatively weak ionic bonds may be more important in secondary product pathways of higher plants to permit a more flexible "grid" type of interrelated pathways involving a series of closely related compounds. Loose complexes capable of reversible association–dissociation, modulated by pH, light, etc., may be physiologically more important in higher plants. Such weak bindings might be easily altered in plants upon extraction, because of the well-known problems in grinding due to the cell wall and the release of vacuolar contents.

B. The C_6–C_3 Phenylpropane Pathway and the Special Case of Phenylalanine Ammonia-Lyase

The sequence of five reactions leading to a series of hydroxylated and methylated compounds, some of which act both as intermediates and as final products after esterification, is a good example of what might be a valid loose or weakly bound complex *in vivo*. Except for the still unidentified third hydroxylation step producing 5-hydroxyferulic acid, these reactions have been fairly extensively investigated as cell-free systems. However, although several laboratories have evidence of multiple-step sequences (Table III), the only evidence for a multienzyme complex based on a more complete demonstration is the data of Kindl's laboratory on the two-step microsomal reaction discussed earlier. Theoretically, this sequence should be a prime source of a series of complexes, since it competes with several chain-shortening pathways (Stafford, 1974a,b). On the other hand, it might consist of a single loose complex requiring regulation at each step to permit variations in the amounts of final products accumulated. Alibert supports such a hypothesis and implicates participation of more than one intracellular site in such a sequence (Alibert *et al.*, 1977). A similar interorganellar relationship has been postulated in flavonoid glandular cells by Charrière-Ladreix (1977, 1979). The two intermediates of the pathway that generally do not accumulate as free acids or esters are cinnamate and 5-hydroxyferulate, possibly because they are associated *in vivo* with a very tight membrane-bound complex. This might explain the difficulty in demonstrating C4H activity in some cell-free extracts and the inability so far to demonstrate the third hydroxylase step in any extract, although inactivation by proteases and oxygenases is also a prime suspect.

The first enzyme of the C_6–C_3 sequence, PAL, is highly soluble and exists as a series of isozymes. The next enzyme, C4H, is almost universally embedded in membranes of ER, while the third step, *p*-coumarate- or 4-hydroxycinnamate-3-hydroxylase (HCH) is frequently at least loosely membrane-associated. It also is either a part of, or aggregates with, a diphenol oxidase and has a much lower pH requirement than the other known enzymes of the sequence. Whereas some of the activities are generally soluble (transmethylases and esterifying enzymes), others have multiple intracellular sites (soluble, microsomal, peroxisomal, and chloroplastic) (Stafford, 1974a,b). Evidence indicates that O-methyltransferase activity can be found along with PAL and C4H in an ER fraction isolated from cells capable of secreting flavonoids. Since the methyltransferase activity was easily removed from these membranes upon addition of $(NH_4)_2 SO_4$ in the presence of EDTA, the binding was probably ionic (Charrière-Ladreix, 1977, 1979).

Is PAL truly soluble or is it just easily solubilized? Many classical soluble sequences, such as the glycolytic pathway, are now being found in loose aggregates (Mowbray and Moses, 1976; Knull, 1978). It is quite possible that none of the enzymes in a sequence of reactions are merely randomly distributed in the aqueous environment of the cytosol (Srere and Mosbach, 1974; Mosbach and Mattiasson, 1976). For instance, PAL may be loosely attached via ionic bonds to a membrane containing C4H and, in some cases, the subsequent enzyme HCH. Cinnamic acid is sufficiently lipid-soluble that it could be easily transported through the membrane. Both the chloroplast envelope and the thylakoid membranes are negatively charged above pH 4.3 (Nakatani *et al.*, 1978), an indication that a positively charged grouping of an enzyme such as PAL would be necessary for ionic attachment purposes. This enzyme is frequently extracted at high pH in borate buffer, which may effectively solubilize a large majority of its activity. Recent evidence, for example, indicates that glycolic acid oxidase is loosely attached to peroxisomes, an attachment controlled by both pH and light; the binding is easily lost during sucrose density gradients analysis (Roth-Bejerano and Lips, 1978). The ease of solubility or reversible ionic attachment to membranes or organelles might be typical of the first enzyme of a sequence. It also might account for a varied particulate localization, but the problem of artifacts during extraction is worrisome. Recall, however, that the first enzyme of the prearomatic pathway in *Neurospora* is not a part of the *arom* complex (Section II,A) (Ahmed and Giles, 1969). The low K_m of 38 μM at low phenylalanine concentrations may be important to counteract the postulated disadvantage of the first enzyme in a complex (Ottaway and Mowbray, 1977). However, it is still only about one-third that obtained for the first enzyme of the *arom* complex (K_m of less than 12); the latter was obtained after activation by preincubation with the substrate (Welch and Gaertner, 1976). Even the lowest K_m values for PAL are still about 10 times higher than that for C4H.

C. Special Case of the Chloroplast

Chloroplasts or leukoplasts have been postulated to play a special role in the synthesis of phenolic compounds, either in the stroma or tightly associated with thylakoid membranes (Table III) (McClure, 1979). Except for Czichi and Kindl's data (1975a), which place C4H in the thylakoids of an alga, this hydroxylase activity is generally limited to ER or microsomal preparations. Alibert *et al.* (1977) reported that [^{14}C]phenylalanine labeling experiments indicated the absence of C4H in chloroplasts isolated from *Petunia*, even though a two-step reaction to *o*-coumarate occurred. Although the localization of phenolic pathways in chloroplasts is a very attractive hypothesis, none of the data distinguish clearly between an enzyme or complex within the chloroplasts and activity due to ER vesicles or enzymes bound ionically to the outside of the chloroplast envelope. Such an external association could have all the benefits of the photosynthetic product NADPH via a shuttle system. The problem of artifacts during extraction and the possible ease with which soluble stromal enzymes leach out from the inside make it difficult to assess the localization of phenolic pathway enzymes in chloroplasts.

Protoplast preparations might be expected to provide the best chloroplasts and therefore a better answer to this question of intracellular localization.

However, at least two laboratories found that protoplast preparations from light-pretreated monocots, barley and sorghum, had no detectable PAL activity, even though whole-cell extracts from the same tissue were active. Either PAL is inactivated during the preparation of the protoplasts or is not present in the mesophyll cells that provide the best protoplasts (D. E. Blume and J. W. McClure, private communication; H. A. Stafford and K. Beckwith, unpublished data). PAL activity, however, has been reported in protoplasts from a dicot, tobacco (Kopp *et al.*, 1977).

D. The Importance of Membranes

The "problem" of the ease of solubility of PAL and of the subsequent transmethylase and transesterase, as well as the difference in the pH requirements of the two known hydroxylases in the C_6–C_3 sequence (pH 6 vs 8–9), might be accommodated by the complexes acting vectorially across membranes, the final product being accumulated in the cisternae of ER or in vesicles derived from them or the Golgi apparatus for transport to the site of accumulation (Table IV). The easily solubilized enzymes might be attached ionically to either side of the membrane. A sequence spanning a membrane would have two great advantages: (1) It could sequester the final product from the initial enzymes of the sequence, and (2) it would permit differences in microenvironments needed for optimal catalysis, such as pH. If the pH optima of a sequence are similar, the enzymes could be oriented toward the

TABLE IV

Interactions of Sites of Synthesis with Transport Channel and Sites of Accumulation

Site of synthesis	Transport channel	Site of accumulation	Example
ER membranes	Cisternal channels or vesicles	Vacuole	Dhurrin
ER or Golgi-derived vesicles	(Provacuolar?) vesicles	Vacuole	Tannins
ER membranes, chloroplast	Vesicles	Vacuole	C_6–C_3 esters, flavonoids
Chloroplasts	?	Vacuole	Coumarins
ER-derived vesicles with alcohols of cinnamic acid derivatives	Vesicles		
Golgi-derived vesicles with peroxidase and H_2O_2-generating system	Vesicles	Wall	Lignin
ER membranes, chloroplast	ER cisternae	Wall (epidermal surface)	Secreted flavonoids

same side of the membrane, as postulated for the two-step sequence discussed in Section II,B (Czichi and Kindl, 1977) and for the flavonoid pathway (Hrazdina *et al.,* 1978).

There are two sites of accumulation of secondary products in higher plants, the vacuole and the cell wall, which require secretion across the tonoplast and plasmalemma membranes, respectively. The cell wall, but not the central vacuole, is a probable site of synthesis as well. If the cell wall is the site of the final steps in lignin biosynthesis, the required enzymes must be packaged for secretion, in addition to substrates such as cinnamyl alcohol. (Gross *et al.,* 1977). *Populus* cells exuding flavonoids onto the epidermal surface represent another specialized case of secretion across the plasmalemma (Charrière-Ladreix, 1977). The secretory process and collagen synthesis in animal cells can serve as a useful analogy (Palade, 1975).

E. Future Approaches

Protease inhibitors should be added during extraction to determine whether proteases are the cause of the lack of complexes or whether a complex is multienzymatic or multifunctional. If the dearth of examples of multienzyme complexes is due to the prevalence of the loose type held together by ionic linkages, extraction methods should be devised to protect these ionic linkages. Light may be an important modulating agent in the case of *in vivo* dissociation–reassociation types of complexes (Montagnoli, 1977). Kinetic studies for determining K_m and V_{max} values of the intact complex and its isolated components assayed individually and in multistep assays should be compared with either computer-simulated or encapsulation models. The powerful $^3H/^{14}C$ ratio tool first used by Czichi and Kindl (1975a) for plant systems needs to be extended to other systems.

The nature of the microenvironment of these complexes should be investigated further. More information and in particular new methodology are necessary in order to determine the intracellular location and size of synthetic, catabolic, and storage pools in the microenvironments of the cytosol or of organelles, because it is within these microenvironments, and not within the large central vacuole, that regulation of these and other pathways by these secondary products might occur (Subramanian *et al.,* 1973).

F. Speculative Conclusions

I predict that all sequences producing secondary products are aggregated as complexes at least by loose ionic bonds. Predominantly soluble enzymes such as PAL and transmethylases are artifacts of extraction, either because the ionic bonds are broken or because of protease action. Hyroxylases, including HCH and the unknown 5-hydroxyferulate hydroxylase, are membrane-embedded *in vivo*. The presumed complexes act vectorially across membranes, permitting varied microenvironments and compartmentalization of final products. The products are released internally into vesicles or cisternal channels of ER for transport to and secretion at the site of accumulation either in the large central vacuole or the wall.

ACKNOWLEDGMENTS

Supported by National Science Foundation Grant PCM-7684392.

REFERENCES

Ahmed, S. I., and Giles, N. H. (1969). *J. Bacteriol.* **99**, 231–237.
Alibert, G., Ranjeva, R., and Boudet, A. M. (1977). *Physiol. Veg.* **15**, 279–301.
Anderson, L. E. (1974). *Plant Physiol.* **54**, 791–793.
Bauthorpe, S. V., Ekundayo, O., and Rowan, M. G. (1978). *Phytochemistry* **17**, 1111–1114.
Berlyn, M. G., and Giles, N. H. (1969). *J. Bacteriol.* **99**, 222–230.
Bloch, K. (1977). *Adv. Enzymol.* **45**, 2–84.
Boudet, A. M., Boudet, A., and Bonyssou, H. (1977). *Phytochemistry* **16**, 919–922.
Charrière-Ladreix, Y. (1977). *Physiol. Veg.* **15**, 619–642.
Charrière-Ladreix, Y. (1979). *Phytochemistry* **18**, 43–45.
Conn, E. E. (1978). *Naturwissenschaften* **66**, 28–34.
Conn, E. E., McFarlane, I. J., Møller, B. L., and Shimada, M. (1979). *Proc. FEBS Meet.* **55**, 63–71.
Czichi, V., and Kindl, H. (1975a). *Hoppe-Seyler's Z. Physiol. Chem.* **356**, 475–486.
Czichi, V., and Kindl, H. (1975b). *Planta* **125**, 115–125.
Czichi, V., and Kindl, H. (1977). *Planta* **134**, 133–143.
Darnell, J. E. (1978). *Science* **202**, 1257–1260.
Davis, R. H. (1967). *In* "Organizational Biosynthesis" (H. J. Vogel, J. O. Lampen, and V. Bryson, eds.) pp. 302–325. Academic Press, New York.
Doy, C. H. (1967). *Biochem. Biophys. Res. Commun.* **28**, 851–856.
Ebel, J., and Hahlbrock, K. (1977). *Eur. J. Biochem.* **75**, 201–209.
Fritsch, H., and Grisebach, H. (1975). *Phytochemistry* **14**, 2437–2442.
Gaertner, F. H., and Cole, K. W. (1977). *Biochem. Biophys. Res. Commun.* **75**, 259–254.
Gaertner, F. H., Ericson, M. C., and DeMoss, J. A. (1970). *J. Biol. Chem.* **245**, 595–600.
Giles, N. H., Cage, M. E., Partridge, C. W. H., and Ahmed, S. I. (1967). *Proc. Natl. Acad. Sci. U.S.A.* **58**, 1453–1460.
Giles, N. Y., and Case, M. E. (1975). *In* "Isozymes" (C. L. Markert, ed.), Vol. 2, pp. 865–876. Academic Press, New York.
Ginsburg, A., and Stadtman, E. R. (1970). *Annu. Rev. Biochem.* **39**, 429–472.
Gross, G. C., Janse, C., and Elsner, E. F. (1977). *Planta* **136**, 271–276.
Grumbach, K. H., and Lichtenthaler, H. K. (1975). *Planta* **141**, 253–258.
Hanson, K. R., and Havir, E. A. (1979). *Recent Adv. Phytochem.* **12**, 91–137.
Hartmann-Bouillon, M., and Benveniste, P. (1978). *Phytochemistry* **17**, 1037–1042.
Haslam, E. (1977). *Phytochemistry* **16**, 1625–1640.
Haslam, E. (1979). *Recent Adv. Phytochem.* **12**, 475–523.
Havir, E. A., and Hanson, K. R. (1968). *Biochemistry* **7**, 1904–1914.

Hedden, P., MacMillan, J., and Phinney, B. O. (1978). *Annu. Rev. Plant Physiol.* **29**, 149–192.
Hrazdina, G., Kreuzaler, F., Hahlbrock, K., and Grisebach, H. (1976). *Arch. Biochem. Biophys.* **175**, 392–399.
Hrazdina, G., Wagner, G. J., and Siegelman, H. W. (1978). *Phytochemistry* **17**, 53–56.
Kirschner, K., and Bisswanger, H. (1977). *Annu. Rev. Biochem.* **45**, 143–166.
Knull, H. R. (1978). *Biochim. Biophys. Acta* **522**, 1–9.
Kopp, M., Fritig, B., and Hirth, L. (1977). *Phytochemistry* **16**, 895–898.
Koshiba, T. (1978). *Biochim. Biophys. Acta* **522**, 10–18.
Löffelhardt, W. (1977). *Z. Naturforsch., C: Biosci.* **32C**, 345–350.
Löffelhardt, W., and Kindl, H. (1975). *Hoppe-Seyler's Z. Physiol. Chem.* **356**, 487–493.
Lynen (1980). *Evr. J. Biochem.* **112**, 431–442.
McClure, J. W. (1979). *Recent Adv. Phytochemistry* **12**, 525–588.
Madyastha, K. M., Ridgway, J. E., Dwyer, J. A., and Coscia, C. J. (1977). *J. Cell Biol.* **72**, 302–313.
Masters, C. J. (1977). *Curr. Top. Cell. Regul.* **12**, 75–105.
Montagnoli, G. (1977). *Photochem. Photobiol.* **26**, 679–683.
Mosbach, K., and Mattiasson, B. (1976). *In* "Methods in Enzymology" (K. Mosbach, ed.), Vol. 44, pp. 453–478. Academic Press, New York.
Mowbray, J., and Moses, V. (1976). *Eur. J. Biochem.* **66**, 25–36.
Nakatani, H. Y., Barber, J., and Foster, J. A. (1978). *Biochim. Biophys. Acta* **504**, 215–55.
Ottaway, J. H., and Mowbray, J. (1977). *Curr. Top. Cell. Regul.* **12**, 107–208.
Palade, G. (1975). *Science* **189**, 347–358.
Parkham, R. A., and Kaustinin, H. M. (1977). *Bot. Gaz. (Chicago)* **138**, 465–467.
Polley, L. D. (1978). *Biochim. Biophys. Acta* **526**, 259–266.
Ranjeva, R., Alibert, G., and Boudet, A. M. (1977a). *Plant Sci. Lett.* **10**, 225–234.
Ranjeva, R., Alibert, G., and Boudet, A. M. (1977b). *Plant Sci. Lett.* **10**, 235–242.
Reed, L. J., and Cox, D. J. (1970). *In* "The Enzymes" (P. A. Boyer, ed.), 3rd ed., Vol. 1, pp. 213–240. Academic Press, New York.
Roth-Bejerano, N., and Lips, S. H. (1978). *Phytochem. Photobiol.* **27**, 171–177.
Rubin, P. M., Zahler, W. L., and Randall, D. D. (1978). *Arch. Biochem. Biophys.* **188**, 70–77.
Saleh, N. A. M., Fritsch, H., Kreuzaler, F., and Grisebach, H. (1978). *Phytochemistry* **17**, 183–186.
Saunders, J. A., and Conn, E. E. (1978). *Plant Physiol.* **61**, 154–157.
Srere, P. A., and Mosbach, K. (1974). *Annu. Rev. Microbiol.* **28**, 61–83.
Stadtman, E. R. (1970). *In* "The Enzymes" (P. D. Boyer, ed.), 3rd ed., Vol. 1, pp. 397–459. Academic Press, New York.
Stafford, H. A. (1974a). *Annu. Rev. Plant. Physiol.* **25**, 459–486.
Stafford, H. A. (1974b). *Recent Adv. Phytochem.* **8**, 53–79.
Stafford, H. A., and Lewis, L. L. (1977). *Plant Physiol.* **60**, 830–834.
Stoops, J. K., Arslanian, M. J., Oh, Y. H., Aune, K. C., and Vanaman, T. C. (1975). *Proc. Natl. Acad. Sci. U.S.A.* **72**, 1940–1944.
Strøman, P., Reinert, W. R., and Giles, W. H. (1978). *J. Biol. Chem.* **253**, 4593–98.
Subramanian, K. N., Weiss, R. L., and Davis, R. H. (1973). *J. Bacteriol.* **115**, 284–290.
Sumper, M., Oesterheit, D., Riepertinger, C., and Lynen, F. (1969). *Eur. J. Biochem.* **10**, 377, 387.
Welch, G. R., and Gaertner, F. H. (1975). *Proc. Natl. Acad. Sci. U.S.A.* **72**, 4218–4222.
Welch, G. R., and Gaertner, F. H. (1976). *Arch. Biochem. Biophys.* **172**, 476–489.
Wieland, F., Renner, L., Uerfüth, C., and Lynen, F. (1979). *Eur. J. Biochem.* **94**, 189–197.
Wood, W. I., Petersen, D. O., and Bloch, K. (1978). *J. Biol. Chem.* **253**, 2650–2656.

Secondary Metabolites and Plant Systematics

6

DAVID S. SEIGLER

I. THE RELATIONSHIP OF CHEMICAL AND BOTANICAL DATA

A. Introduction

Many scientists, both chemical and biological, have sought to correlate chemical characters (i.e., the presence of certain types of compounds) with various taxonomic entities (De Candolle, 1804, 1816; Greshoff, 1893; McNair, 1965; Alston and Turner, 1963b). In the past, several factors have limited the success of such efforts, and only in recent years have such studies been done for many plant groups. In this chapter I will review several of these previous efforts, as well as examine current thinking in this area of endeavor. Several new ideas concerning the placement of selected plant

The Biochemistry of Plants, Vol. 7
Copyright © 1981 by Academic Press, Inc.
All rights of reproduction in any form reserved.
ISBN 0-12-675407-1

groups within taxonomic systems will be discussed, as well as certain enigmatic problems that can as yet not be clearly resolved. As a background to these discussions I will first describe the nature and goals of plant systematics to provide the reader with the necessary perspective to understand the needs of this science.

B. What Is Plant Systematics?

Systematics is the scientific study of the kinds and diversity of organisms and of the relationships among them (Raven *et al.*, 1976). In former times, much systematic work was based on the examination of preserved herbarium specimens in an effort to describe and classify various plant taxa (a term indicating taxonomic entities of unspecified rank). These studies frequently involved an examination of the morphology of relatively small numbers of specimens. Although this approach is still viable in many tropical areas of the world where rich and unstudied floras are in immediate danger of destruction or extreme modification (Turner, 1970a; Raven *et al.*, 1970), it is largely being supplemented by the examination of larger numbers of plants from living populations in temperate areas of the world, where the floras are better known.

By means of this latter method, often called biosystematics, one attempts to study as much of the biology of the plant as possible and to utilize these data to clarify the taxonomic and evolutionary relationships of the taxa involved. The information derived from both approaches is normally utilized in two ways: to prepare floras of a particular region (often a state or large natural geographical region) and to account for all the species within a given group, for example, a genus or a family, regardless of where the plants grow (Radford *et al.*, 1974). Although each of the above aspects of systematics assists in the identification and location of plant materials, this information may also be invaluable to workers in many other fields because of its predictive nature.

In this chapter, reference will be made to several contemporary systems of classification. Among them are those of Takhtajan (1966), Cronquist (1968, and in Jones and Luchsinger, 1979), Thorne (1976), Dahlgren (1975, 1977), Hutchinson (1973), and Melchior (1964). Several of these have been summarized by Becker (1973).

In summary, the principal goals of plant systematics are to (1) provide a convenient method of identifying, naming, and describing plant taxa, (2) provide an inventory of plant taxa via local, regional, and continental floras, and (3) provide a classification scheme that attempts to express natural or phylogenetic relationships and to provide an understanding of evolutionary processes and relationships (Radford *et al.*, 1974).

C. Phytochemical Data

1. Origin of Chemical Characters

As both morphological and chemical features are determined by the genetic makeup of an organism, the structure of a molecule must be as much a character as any other (Birch, 1974). Further, all the characters of a plant must be related and self-consistent. Thus it is scarcely surprising that new cytological and chemical data have provided valuable complementary information about the placement of groups within the taxonomic system rather than upsetting the results of extensive morphological investigations. How did these two types of characters arise and how do they differ?

In the course of evolution the fate of any change in the genetic material of an organism in large part depends on the function of the products produced. For example, changes in respiratory proteins, such as cytochromes, are unlikely to survive, whereas changes in the enzymes that produce secondary metabolic products are more likely to persist. The evolution of morphological and chemical features of an organism must be interrelated but, significantly, the forces of natural selection do not have the same effect on each type of genetic expression. These differences in selection are very important from a systematic standpoint, because the evolution of chemical constituents differs from morphological evolution, making the examination of both morphological and chemical characters an extremely valuable approach to the study of evolutionary problems (Turner, 1970b; Fairbrothers et al., 1975). Because the structure of any compound is determined by a series of biosynthetic steps, each of which is under different selective forces, not only may the structure of the compound itself be useful but the biochemical pathway by which it has arisen may be of systematic significance.

2. Rationale for Using Chemical Data

The application of macromolecules, in particular proteins, and micromolecules, mostly secondary metabolic compounds such as terpenes, flavonoids, alkaloids, cyanogenic and other glycosides, amino acids, and lipids of various types to taxonomic problems involves basically different approaches, and the data appear to be useful in different manners.

When one utilizes macromolecules, one examines the primary products of plant DNA, and changes in amino acids within the protein reflect changes in the base sequence of the DNA. Initial studies on protein sequences, especially those involving cytochrome c, indicate that these data provide valuable information about phylogeny and relationships at the higher taxonomic categorical levels (families, orders, classes). Cytochrome c, which occurs in both animals and plants, has been sequenced in several animal species

(Nolan and Margoliash, 1968). The fossil record for animals generally confirms information derived from these phylogenetic studies. The number of similarities in amino acids in particular positions in cytochrome c molecules from different animals makes it statistically improbable that they could have arisen from more than a single ancestral type with an ancestral cytochrome c molecule. By tracing the differences in amino acid substitutions it is possible to relate various groups of animals, as successive groups carry the changed cytochrome c molecule after a modification.

In plants, especially flowering plants, there is no extensive fossil record, and much of the current knowledge of relationships and phylogeny in this group is based on extrapolation from studies on morphological data. To date, relatively few plant cytochromes have been studied but, in the few that have been investigated, it is apparent from the number of similarities among amino acid sequences that plant and animal cytochromes are related. It is also evident that the sequences of amino acids in genera of the same family are more similar to each other than to those of other families, and that families thought to be closely related by morphological evidence generally resemble each other more closely than less related families (Fairbrothers *et al.,* 1975). The evolutionary history of plant groups, as well as of animals, appears to be recorded in this and other proteins.

Much recent work has established that micromolecular chemical data can also provide valuable insight into evolutionary processes (Turner, 1970b). Chemical studies on secondary products have proved useful in resolving many problems of speciation and evolution but, in contrast to protein sequencing data, have generally been applied to the study of lower taxonomic categories, i.e., problems at the species and genus level (Turner, 1967, 1969). However, as will be pointed out, they may also be of value at higher taxonomic levels.

To understand how secondary compounds can be useful in the study of systematic problems, it is necessary to consider how and why they arose. Plants have a multitude of proteinaceous materials, many of which have enzymatic functions. In primitive organisms these compounds were and are largely active in synthesizing primary metabolic components of cells. As these organisms evolved, genetic material and its derived proteins were duplicated and increased both in amount and in redundancy. Mutations occurred that subsequently produced changes in the proteins and their products. The forces of natural selection operated on all such products (Stebbins, 1974), selecting them for value to and compatibility with parental organisms and the ecological systems in which they occur. Many of these compounds were of a less critical nature than primary metabolites and were less widely distributed. Complications are introduced because one does not observe the primary gene products, but rather the pools of compounds they produce, the concentrations of which are partially functions of the relative amounts and activities of enzymes, the availability of certain precursors, and compart-mentalization and translocation within the cell (Seigler, 1974). Biosynthesis and accumulation of secondary products are distinct processes (Hegnauer, 1976). Subsequent mutations may affect steps in a biosynthetic sequence that we observe as the accumulation or disappearance of an altered prod-uct. These mutations usually involve the loss, gain, blockage, or altera-tion of the specificity of an enzyme system. Loss of synthetic ability is presumably more common than gain or alteration, since it merely implies destruction or blocking of a process instead of setting up a new one (Birch, 1974). This is partially confirmed by the observation that in several groups of species from the related genera *Parthenium, Hymenoxys,* and *Ambrosia* of the Asteraceae, more highly evolved members have simplified patterns of secondary compounds (Mabry, 1974). A one-gene loss may also block an entire pathway.

Determination of the homologous origin of similar compounds in different taxonomic groups is one of the fundamental problems inherent in the taxonomic application of secondary compounds (Hegnauer, 1967). Two taxa may synthesize or accumulate the same products by different pathways; therefore the mere presence of a compound is not necessarily an indication of a relationship; i.e., similarities in the chemistry of plant taxa (or mor-phological features) may reflect an evolutionary or phyletic similarity but may also be the result of convergent evolutionary processes (Seigler, 1974, 1977).

With a knowledge of the biosynthetic pathways of secondary compounds in plants, it should be possible to determine at what point in a sequence divergence occurred and what subsequent changes have come to pass (Birch, 1974). In reality this is rarely realized because of several factors; several classes of compounds do not appear to have specific structural re-quirements, whereas in others less variation can be tolerated. For example, most phenolic substances can serve as antioxidants, and many lipid com-pounds for surface coatings, as long as the necessary physical properties are met; but attractants for specific pollinators and diterpenes with hormonal activities must be precisely synthesized (Birch, 1974). Many plant products arise by simple processes such as the removal of an activating group [such as phosphate or coenzyme A (CoA)] or from oxidations, reductions, or meth-ylations of easily modified groups (Birch, 1974). In some cases the relative amounts of products produced may simply reflect the rates of two enzymes operating on a common precursor. Highly probable reactions, such as the introduction of a hydroxyl group ortho or para to an existing one in a phenol, occur frequently in nature. These types of changes are usually of only minor importance in considering the taxonomic significance of secondary com-pounds.

Other reaction sequences are reversible or are controlled by feedback inhibition controls such that, when a given compound disappears, it disappears without a trace or causes accumulation of a compound far removed in the sequence. For example, polyketide chains, probably as CoA esters, are rapidly degraded to their initial units unless some chemically irreversible stage is reached such as reduction or cyclization (Birch, 1974). In the fungus *Penicillium islandicum,* which produces polyketide anthraquinones, mutation simply leads to the complete absence of these compounds.

We have limited knowledge as to what pathways may be available in advanced plant groups, as we can see only the products of pathways that the plant utilizes at a particular time. By examination of amino acid fractions from sugar beets, Fowden (1972) demonstrated the presence of a number of amino acids not previously known from that source. The presence of a given constituent can be demonstrated only if an appropriate method for its detection is available (Hegnauer, 1976).

Several lines of work suggest that many plants are capable of carrying out complex reactions or reaction series but lack precursors or particular enzymes under normal situations. For example, when plants of *Nicotiana* are fed thebaine and certain other precursors of morphine, they are able to perform several biosynthetic steps and produce morphine (Mothes, 1966) which is not known to occur naturally in the genus. Interestingly, this conversion cannot be made by some species of *Papaver,* although other species of the genus contain thebaine and morphine.

In assessing the importance of a particular change as an evolutionary step it is necessary to decide on the probability of its occurrence. As a general rule, the more difficult the reactions and the less available the building blocks, or the more reaction steps required in a definite sequence to give rise to a compound, the rarer will be its convergent formation (Mothes, 1966).

3. Botanical and Chemical Literature

Many earlier publications were based on mass collections of materials often gathered from large geographical areas and/or of uncertain origin. Frequently, only the major constituents—those that were poisonous, crystallized readily, or had other easily detectable properties—were examined. These facts must be considered by those who intend to apply the information to a taxonomic problem. Another difficulty in utilizing chemical data from the literature is a lack of reliability of certain structure determinations and in particular the identification of plant products by such physical properties as gas–liquid chromatography retention time, paper and thin-layer chromatography R_f values, color reactions, and spot tests. Misidentification of compounds by wet chemical methods was not uncommon in the older literature before advanced spectral methods became available and must always be considered (Seigler, 1977).

One of the most serious problems in utilizing literature data is that almost no chemical reports are supported by adequately vouchered plant materials. Proper vouchering records would make it possible to examine the original materials and allow comparison with other collections in order to ascertain whether (1) the material was correctly identified and (2) certain phenomena, such as hybridization, introgression, and subspecific variations, exist. It would also permit subsequent workers to determine the presence of fungi, lichens, algae, insects, etc., that may be involved in the production of certain secondary compounds. If a small portion of the actual material utilized for the research were also preserved, it would permit later analysis for foreign contaminants.

In other cases, careful perusal of the botanical literature will reveal that taxonomists have placed taxa of various ranks incorrectly. These incorrect placements may range from questionable or aberrant species in a genus to the realignment of entire orders of plants. Chemical data can assist in resolving problems of this type, but they sometimes provide enigmatic results until sufficient information is available to allow a reassignment of the taxa involved.

One must look carefully and critically at *all* reported data to be sure that both chemical and botanical portions of the work have been done and interpreted correctly before applying the data to a problem under investigation.

4. Documentation of Plant Materials

As mentioned in the preceding section, many early reports of secondary compounds are suspect because accurate techniques required for the assignment of complex structures were not available. Nonetheless, the major problem in using these data for systematic studies is not the reliability of the chemical data but the identity of the plant materials that were examined (Hegnauer, 1967; Mears and Mabry, 1971).

To document the materials used, the investigator should always have a competent person identify the plant materials and a portion should be dried or otherwise preserved as a voucher specimen so that further examination of the specimen is possible should it be desirable. The selected plant should be typical of the population and, when possible, should have mature reproductive organs. Full collection or acquisition data (data, location, collector, habitat, etc.) should be provided, and the specimen deposited in a recognized herbarium. Taxonomists usually are willing to assist with the necessary details of voucher specimen preparation. Most major universities have collections of dried plant specimens (a herbarium) that provide a wealth of data about ranges, flowering times, uses, soil preferences, and other information about particular species as well as preserve materials for future study or reinvestigation.

In publications describing chemical results, one should record the locations and dates of plant collections, the parts of the plants used in the study, the name of the herbarium where the voucher specimens are deposited, and the name of the taxonomist who identified the plants. With this information and with the possibility of comparing specimens collected at other times with the original vouchers, later investigators can usually determine the relationship of the plants concerned to the original collection (Mears and Mabry, 1971; Ettlinger and Kjaer, 1968).

II. NATURE AND SOURCES OF VARIATION

Until sensitive separation techniques (column, paper, thin-layer, and gas chromatography; countercurrent distribution; etc.) and sensitive methods of instrumental analysis (ir, nmr, uv, and mass spectrometry) became available, it was not feasible to undertake the analysis of secondary plant constituents from single plants of most species in naturally occurring populations. These new microtechniques permit the chemist or botanist to obtain chemical data from single plants rapidly, allowing extension of the biosystematic approach to chemical as well as morphological characters.

When phytochemical workers began to examine single plants, they were often frustrated by apparently uninterpretable variations in chemical constituents. Many of these investigators did not do adequate sampling, ignored the significance of these variations, and came to conclusions based on a meager amount of data in comparison to what was actually needed. Recent combined chemical and morphological investigations have used this information more fully and proved that, instead of being troublesome, the study of chemical and morphological variation actually provides a key to the solution of many problems of biological speciation, hybridization, and introgression.*

A relationship between plant taxa is established by summarizing the similarities among groups of organisms and contrasting their differences. We consider two plants to be closely related if they have many common characters, and only distantly so (or at higher categorical levels) if the differences outweigh the similarities. In contrast to this, the important factors in evolution are change and the ability to maintain variability. Few natural populations are without measurable variation; that is, plants from interbreeding groups that share a gene pool have phenotypic and genotypic differences that can be seen even by inexperienced observers. How do these variations arise and how are they maintained?

---

* Introgression is the process by which the genes of one taxon are mixed with the genes of another by hybridization of the two taxa followed by backcrossing of the hybrid plants with either of the two parents. Even when hybrids are not significant in relative numbers, they can allow gene flow and mixing, producing increased variability of the two parental types.

Each individual plant must possess the ability to respond to its environment, but this variation must remain within the limits set by the genetic makeup of the taxon (Stebbins, 1974; Davis and Heywood, 1963). Thus phenotypic expression is determined by both genotypic composition and reaction to a specific environment. Some characters are little changed by environment—e.g., leaf arrangement or floral structure—and these have been considered "good characters" or to be "genetically fixed." Other characters are known to vary radically and are said to be "phenotypically plastic." Examples of characters of this type are leaf shape, stem height, and time of flowering. The effects of environment are superimposed on and may obscure genotypic variability; further, it is the phenotype produced by both that is exposed to the pressures of natural selection. Davis and Heywood (1963) have listed a number of important physical factors in determining the appearance of a plant in nature. Among these are light, seasonal variation, elevation differences, terrestrial vs epiphytic state, photoperiodism, temperature, temperature periodic effects, water (heterophylly), wind, soil (e.g., halophytes), and biotic factors such as fungal and bacterial infection, ant habitation, galls, grazing and browsing, fire, and trampling.

The population is considered by many to be the basic evolutionary unit and, when we discuss speciation and concomitant chemical change, it is necessary to understand something of the nature of variation both within and among populations of a given taxon or group of taxa. Populational variations are a function of the variation in individual plants and in the common gene pool they possess. Morphological and chemical features enable us to recognize the population, but they do not define it. It must also be remembered that the population is a dynamic entity. It changes in numbers of plants and, even in some perennials, in the particular individuals present in a given year. A population may occupy a much larger geographical area in some years than in others. It may separate into two or several new populations under some conditions that may be maintained, or later merge with the parental population. Taxonomic descriptions are sometimes based on a single plant specimen which may not reflect the nature of the species or its populations.

Several factors are important in determining genetic variation. Mutations usually produce a one-gene change, but these changes may have profound effects. Such changes as zygomorphic corollas to actinomorphic corollas in *Antirrhinum,* the gamosepalous to the polysepalous condition in *Silene,* spurred to nonspurred flowers in *Aquilegia,* and an annual to a biennial condition in *Atropa* are all known to be controlled by one gene (Davis and Heywood, 1963). Most mutations affect several characteristics of the phenotype. Thus a species may differ from another in several characters but still may be separated by only a one-gene difference. Characters that have no selective advantage in themselves can become established through the secondary effects of genes that have been selected as valuable to the organisms for completely different reasons (Davis and Heywood, 1963). Certain genetic variants coexist in temporary or permanent equilibrium within a single population in a single spatial region in a phenomenon known as polymorphism (Davis and Heywood, 1963).

Recombination of genetic variability in populations is largely determined by the breeding system. Cross-fertilized populations contain a large store of variability hidden in the form of recessive genes in the heterozygous condition. This variability serves as insurance in the presence of a constantly changing environment. In sexual populations breeding tends to take place principally between neighboring individuals.

In summary, the three factors that largely control variation in populations are (1) external environmental modification, (2) mutation, and (3) genetic recombination (Davis and Heywood, 1963). Populations rarely stay the same over a period of time but are affected by the process of natural selection in a stabilizing, disruptive, or directional manner. Populations separated by geographical, ecological, or reproductive barriers tend to differentiate into a series of populations that may have gradually accrued differences (clinal variation) or stepwise variations associated with ecological differences (ecotypic variation) (Davis and Heywood, 1963).

If the differences among populations increase sufficiently, and especially if reproductive barriers arise, these differentiating populations may be recognized as species. Stebbins (1974) considers four major factors in speciation: (1) mutation, (2) genetic recombination, (3) natural selection, and (4) isolation. In small, often peripheral populations, chance may play a greater role in speciation because the probability of loss of a particular character is greater; recessive genes are more likely to appear and become homozygous, and the genetic nature of the population may be determined by the founders or survivors of a period of catastrophic selection. These phenomena explain many of the variational patterns observed in the distribution and occurrence of secondary plant compounds, especially in the lower taxonomic ranks, and although they have mostly been examined by means of morphological characters, much evidence suggests that evolution and speciation may be studied or measured by chemical characters as well.

In the preceding discussion, variation in morphological characters has been considered. There is no reason to think that variation in chemical characters has not occurred and is not maintained in a similar manner. In contrast to morphological features, however, the specific structures and steps of biosynthetic pathways are easier to quantify and generally simpler in terms of genetic control (at least in principle).

Secondary compounds are affected by environmental as well as genetic factors (Tétényi, 1970; Flück, 1963; Adams, 1979). The inheritance and variation in several classes of secondary compounds in plants have been discussed (Smith, 1976; Alston and Turner, 1963a,b; Irving and Adams, 1973; Harborne, 1975; Mabry, 1970; Alston *et al.*, 1965; Crawford, 1972; Adams, 1972; Jones, 1973).

Flavonoids have been investigated more than any other group of plant constituents for taxonomic purposes. They are essentially ubiquitous, easily separated and characterized, and have relatively simple patterns of inheritance (Harborne, 1975; Smith, 1976). In studies on species of *Baptisia* and their hybrid progeny, Alston and Turner (1959) demonstrated that, in general, hybrids possessed patterns that were additive combinations of the two parental types, with occasional missing "spots" and extra or "hybrid" compounds. Numerous species of *Tragopogon* and their F_1 and F_2 hybrids were examined by Belzer and Ownbey (1971). Inheritance was generally additive in F_1 plants, whereas considerable segregation and recombination were observed in F_2 plants. In contrast, however, Dement and Mabry (1972) found that 44 populations of 13 species of *Thermopsis* all had practically the same pattern. Most genera probably lie somewhere in between these extremes.

Variations in the chemical composition of essential oils (composed mostly of monoterpenes and sesquiterpenes) have been studied in several plants. In *Mentha*, chemistry is controlled by genetic factors that reflect typical Mendelian ratios (Murray, 1960). Mirov (1961) extensively studied the turpentines of the genus *Pinus*. Adams (1972) identified putative hybrid plants of *Juniperus pinchotii* Sudw. and *J. monosperma* (Engelm.) Sarg. but concluded that introgression did not occur.

In the genus *Eucalyptus,* Pryor and Bryant (1958) found that they were not regularly able to predict the chemistry of F_1 hybrids and that in F_2 hybrids there was little correlation between morphological characters and terpenoid content.

Genetic and biochemical factors that control the synthesis of linamarin and lotaustralin in white clover (*Trifolium repens* L.) have been examined (Hughes and Conn, 1976). Non-cyanoglucoside-producing plants lacked at least two steps in the biosynthetic pathway.

In a study on alkaloids of the genus *Baptisia* (Fabaceae), Cranmer studied the variation in lupine alkaloids during the development of individual plants in different populations of *Baptisia leucophaea* Nutt. (Cranmer and Turner, 1967). Individual plants in each population exhibited considerable quantitative variation, whereas plants from different populations were similar at similar stages of development. However, there was a striking variation in the specific alkaloids produced, in the relative amounts of each, and in the total quantity of alkaloids present at any given time in development. Nowacki encountered similar variations in lupine alkaloids in the genus *Lupinus* (Nowacki, 1963).

A number of workers have examined the genetics of alkaloid production by the study of hybrid plants (Mothes, 1960, 1966; Nowacki, 1963; Smith, 1965; Dános, 1965, 1966). These results indicate that the genetic mechanisms that control alkaloid synthesis are complex and that hybridization and introgression can produce significant variations in the alkaloid content of plants within a population. Many past workers have been unaware of natural hybridization and, because these plants are occasionally indistinguishable from the parental species, have not been able to interpret the alkaloid patterns observed (Mothes, 1966; Mears and Mabry, 1971). Hybridization and introgression in the genus *Baptisia* has been extensively studied by workers at the University of Texas. Several populations that contained all possible hybrid combinations, plants derived from backcrossing these plants with the parental plants, and the parental plants were examined. The status of these plants was established by independent methods; subsequently the alkaloid chemistry was examined. The data indicated that the hybrid plants not only failed to exhibit the alkaloid chemistry of the parent species either singly or combined but also showed striking quantitative variation among individual hybrid plants. Mabry concludes that this variation is extremely useful and represents one of the best available techniques for detecting and documenting natural hybridization and introgression (Mabry and Mears, 1970).

Extensive variation can occur in the different parts of an individual plant (Hughes and Genest, 1973). Changes associated with the reproductive parts of a plant are often striking; these organs also exhibit the greatest amount of morphological change during a plant's growth and development.

The pigments of flowers and fruits (anthocyanins and other flavonoids, carotenoids, anthraquinones, betalains, etc.) are usually different than the compounds found in the vegetative portions of the plant. Compounds from roots and from wood also differ in most cases. Lipids from seeds and fruits sometimes contain unusual fatty acids, whereas those of leaves are usually common fatty acids found in all plants. Numerous examples are known in which flowers and fruits are chemically protected by the accumulation of large quantities of poisonous compounds. These compounds may play other roles in the plant as well, for instance, as sources of nitrogen or other elements or as regulatory compounds (Seigler and Price, 1976; Jones, 1979; Swain, 1976). Compounds that impart flavor or taste to fruits or leaves may differ from one plant part to another. Essential oils (which may contain terpenes, phenylpropanoids, esters, alcohols, aldehydes, ketones, hydrocarbons, etc.) frequently serve as attractive or repellant substances in plant parts, especially fruits and flowers. The complement of compounds that makes up an oil of this type is usually distinct from that of the rest of the plant.

The patterns of distribution of compounds other than those obviously involved in attractancy, repellency, or energy or nutrient storage may also be linked to functions within the plant (McClure, 1975; Levin, 1971; Harborne, 1977).

Cranmer and co-workers (Cranmer and Turner, 1967; Cranmer and Mabry, 1966) observed that in *Baptisia* species alkaloids often showed greater variation among organs of plants from a single species than among the same organs for different species. The total yield of alkaloids from different organs was also shown to vary significantly. The most thoroughly investigated plants in this regard are medicinally important ones such as *Papaver somniferum* L. and solanaceous plants of the genera *Nicotiana, Atropa, Hyoscyamus,* and *Datura* (Hughes and Genest, 1973).

At present, our lack of knowledge of the specific enzymology of the synthesis of secondary metabolites prevents direct comparison of many of the pathways involved in various taxa. Examination and comparisons must frequently be restricted to systems ascertained to be related by other reasoning, such as a knowledge of the structures of other compounds derived from and part of the biosynthetic pathways in the same and related species of plants.

Secondary compounds have classically been viewed as waste or excretion products (Mothes, 1966), but a body of information is accumulating that suggests that many have important coevolutionary defensive and attractive roles (Whittaker and Feeny, 1971; Freeland and Janzen, 1974; Van Emden, 1973) as well as primary metabolic importance (Seigler and Price, 1976; Robinson, 1974; Swain, 1976; Jones, 1979).

The forces of natural selection seldom operate on a single organism but on a total biological system. This is undoubtedly one reason convergence in the evolution of both morphological and chemical characters is observed.

It is well known, for example, that certain habitats are occupied by plants that possess similar morphological features (Stebbins, 1974; Davis and Heywood, 1963; Mabry, 1973; Went, 1971; Tétényi, 1974). It has not been definitely established, but it appears that various chemical components of plants can be selected to produce convergence of chemical types. Three species of the genus *Hymenoxys* (Asteraceae), *H. scaposa* (DC.) K. F. Parker, *H. acaulis* (Pursh) K. F. Parker, and *H. ivesiana* (Greene) K. F. Parker, contain more than 30 flavonoids. The patterns of distribution of these compounds are correlated more strongly with population positions along an east-west gradient extending from Arizona to Texas than with the diagnostic morphological features of different species in the same region, suggesting the action of common selective forces (Seeligman and Alston, 1967). It has been observed that small, isolated island populations of mainland taxa usually have fewer and simpler compounds than their mainland ancestors. This may be because of lowered selection by predation or because island habitats have different environmental requirements (Mabry, 1973).

III. BASIC PATHWAYS OF SECONDARY METABOLITE BIOSYNTHESIS

In the preceding section I have surveyed some of the ways in which variation originates and is maintained in plants. A knowledge of these variations is extremely important in systematic studies at the lower taxonomic levels (genus and species), but when one wishes to establish relationships at higher ranks, e.g., at the family, order, and subclass level, it is necessary to survey as many taxa and individuals as possible to reduce the effects of these variations. That is, one needs to know what morphological features are produced and what biosynthetic pathways exist in a particular group of taxa to compare them (Table I). This is made more difficult by an imperfect knowledge of biosynthetic pathways, but by careful observation of their products one can establish certain relationships.

A number of chemical reactions are known to occur in virtually all plants. Among these reactions are those that lead to the production of activated compounds such as phosphates and thioesters of CoA. Many enzymes involved in oxidative deamination, other oxidation and reduction processes

TABLE I

Major Groups of Secondary Compounds in Plants

Group	Biosynthetic origin
Amino acids	Citric acid cycle, shikimic acid, others
Alkaloids	Amino acids, aromatic amino acids often with other pathways such as terpenoids, acetate, etc.
Alkanes, alkenes, aldehydes, alcohols, ketones, waxes (long chain)	Acetyl-malonyl CoA (through fatty acids)
Acetylenic compounds	Acetyl-malonyl CoA (through fatty acids)
Anthraquinones, etc.	Polyketides or shikimic acid and mevalonic acid and others
Carbohydrates	Pentose cycle, photosynthesis
Cinnamic acids	Shikimic acid
Coumarins	Shikimic acid (through cinnamic acids)
Cyanogenic glycosides and lipids	Amino acids and carbohydrate or fatty acids
Fatty acids	Acetyl-malonyl CoA
Flavonoids	Aromatic amino acids and acetyl-malonyl CoA
Glucosinolates	Amino acids and carbohydrate
Phenylpropanoids	Cinnamic acids
Phenols (simple types)	Shikimic acid, acetylenic compounds, mevalonic acid, acetyl-malonyl CoA
Polyketides	Acetyl-malonyl CoA
Terpenes (steroids, carotenoids, etc.)	Acetyl-malonyl CoA (through mevalonic acid)

(such as flavin enzymes, phenol oxidases, peroxidases, dioxygenases, and mixed-function oxidases), and enzymes involved in oxidative demethylation appear to be universal. One carbon metabolism involves carboxylases, transcarboxylases, transformylase, hydroxymethyltransferases, methyltransferases (most commonly S-adenosylmethionine), and α-ketoacid decarboxylases. Other important and widespread enzymes are transaminases and amino acid decarboxylases (Luckner, 1972; Geissman and Crout, 1969). Although these reactions are common to most if not all plants, the enzymes are often found to be specific for a particular substrate or group of substrates. These reactions are not generally useful for taxonomic purposes above the specific or generic level. It is probable that pathways of this type had a common origin and have been inherited by most extant plant groups.

Many of the compounds considered secondary in plants are present in all plants and are essential to their survival. In order to assess the evolutionary importance of a particular compound, it is necessary to consider this information. Because fatty acids, alkanes and their relatives, certain mono- and sesquiterpenes, diterpenes, triterpenes (Luckner, 1972), carotenoids (Goodwin, 1974), cinnamic acids, flavonoids (Harborne, 1975), and possibly alkaloids and cyanogenic glycosides are ubiquitous in plants, all plants contain the pathways leading to them.

As previously discussed, the compounds most useful taxonomically are those that are restricted in distribution. Products that arise by removal of activating groups, oxidation, reduction, or methylation of easily modified groups are generally less useful than those that arise via more complex pathways. As a general rule, the more difficult the reactions and the less available the precursors, or the more reaction steps required in a definite sequence to give rise to a compound, the rarer its convergent formation (Mothes, 1966).

Space precludes outlining the pathways that produce the tens of thousands of secondary compounds known to occur in plants. Proposed biogenetic pathways leading to many of these have been reviewed (Luckner, 1972; Geissman and Crout, 1969; Mann, 1978; Haslam, 1974; Bernfield, 1967; *Spec. Period. Rep.: Biosynthesis,* 1972–present). Only a few of these pathways have been studied in a reasonably thorough manner, notably those leading to cyanogenic glycosides (Conn, 1979), glucosinolates (Ettlinger and Kjaer, 1968; Tapper and Reay, 1973), and flavonoids (Hahlbrock and Grisebach, 1975).

Several attempts have been made to quantify chemotaxonomic information based on the number of steps and complexity of pathways. Although these have been useful at the generic and familial levels, they have seldom been extended to higher taxonomic ranks, generally because of major differences that make estimation of the degree of relationship difficult (for examples, see Rezende *et al.*, 1975; Smith *et al.*, 1977; Cagnin and Gottlieb, 1978; Levy and Levin, 1975).

IV. A CONSPECTUS OF THE APPLICATION OF CHEMICAL DATA TO THE CLASSIFICATION OF VASCULAR PLANTS

In the following section, I will comment on the chemical characters that are unique or distinctive for each superorder, order, or major family (Table II). I will correlate the available chemical data with a currently accepted taxonomic treatment, that of Thorne (1974, 1976) as slightly modified by Young and Seigler (1981). Thorne's system has been updated and outlined diagramatically in Fig. 1. Much of the chemical data included in this chapter has been taken from Hegnauer (1962, 1963, 1964, 1966, 1969a, 1973a) and Gibbs (1974).

A. Lower Vascular Plants

Although the secondary compounds of a number of lower vascular plants have been examined, in only a few cases have these data been applied to taxonomic problems (Swain and Cooper-Driver, 1973). It is apparent that ferns do not synthesize the variety of compounds found in gymnosperms or angiosperms. Common fern metabolites are triterpene hydrocarbons, occasionally cyanogenic glycosides and, in the genus *Dryopteris,* complex polyketide-derived phenols (Berti and Bottari, 1968).

In a now classical study on chemical systematics, Smith and Levin (1963) and Harborne *et al.* (1973) found that flavonoid and mangiferin (a xanthone) glycosides were inherited in an additive manner in a number of hybrid taxa derived from *Asplenium montanum* Willd., *A. rhizophyllum* L., and *A. platyneuron* (L.) Oakes, and the origin of these hybrid taxa was elucidated.

Flavonoid aglycones occur as exudates on the fronds of many species of *Pityrogramma Cheilanthes,* and *Notholaena* (Smith *et al.,* 1971; Wollenweber, 1978).

Flavonoid chemical data do not support inclusion of the fern allies *Psilotum* and *Tmesipteris* in the ferns. The former taxa contain biflavonoids, whereas ferns typically contain flavonols and proanthocyanidins (Cooper-Driver, 1977; Wallace and Markham, 1978).

B. Gymnosperms

Extant gymnosperms consist of four classes: Coniferopsida, Cycadopsida, Ginkoöpsida, and Gnetopsida. In total there are only about 700 species and

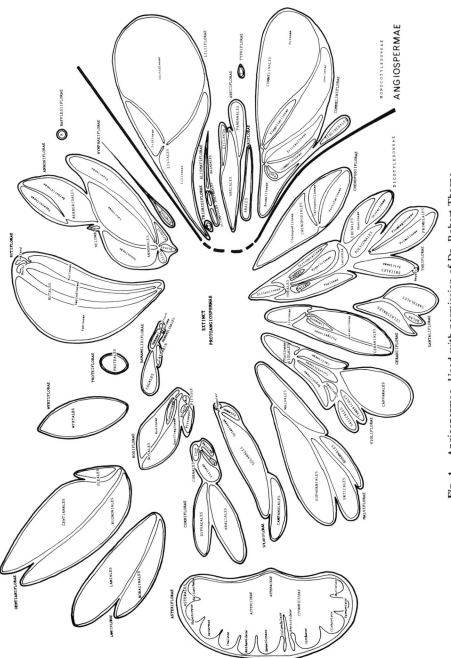

Fig. 1. Angiospermae. Used with permission of Dr. Robert Thorne.

TABLE II

Principal Chemical Constituents of Systematic Interest

Chemical constituent	Distribution	Reference
	Annoniflorae	
Benzylisoquinoline alkaloids	Annonales, Nelumbonales, Berberidales	Seigler, 1977
Sesquiterpene alkaloids	Nympheales	Seigler, 1977
Diterpene alkaloids	Ranunculaceae (*Aconitum* and *Delphinium*)	Seigler, 1977
Sesquiterpene lactones	Annonales (Illiciaceae, Magnoliaceae, Lauraceae), Berberidales (Menispermaceae)	Yoshioka *et al.*, 1973
Cardiac glycosides	Ranunculaceae (*Adonis* and *Helleborus*)	Hegnauer, 1973a
2-Pyrones	Anonales (Lauraceae, Piperaceae)	Gottlieb, 1972
	Theiflorae	
Iridoid monoterpenes	Theales (Actinidiaceae, Icacinaceae, Sarraceniaceae, Ericales (Ericaceae)	Hegnauer and Kooiman, 1978
Alkaloids	Actinidiaceae, Symplocaceae, Icacinaceae, Phellinaceae	Seigler, 1977
Triterpenes and polyterpenes	Sapotaceae	Gibbs, 1974; Hegnauer, 1973a
Naphthoquinones, benzoquinones, and related compounds	Plumbaginaceae, Ebenaceae, Ericaceae (Pyroloideae) Myrsinaceae, Primulaceae, Hypericaceae, Sapotaceae	Gibbs, 1974; Hegnauer, 1966, 1969a, 1973a
Myricetin	Many taxa of the Theiflorae	Gornal *et al.*, 1979; Young, 1981
5-Methoxyflavonols	Dilleniaceae	Gornal *et al.*, 1979; Young, 1981
	Cistiflorae	
Glucosinolates	Capparales	Vaughan *et al.*, 1976
Cyclopentene cyanogens	Flacourtiaceae, Passifloraceae, Turneraceae	Gibbs, 1974; Hegnauer, 1966, 1969a, 1973a
Erucic acid and other longer than normal fatty acids	Brassicaceae	Vaughan *et al.*, 1976
Volatile terpenes	Cistaceae	Hegnauer, 1964
Curcurbitacins	Begoniaceae, Cucurbitaceae	Hegnauer, 1964; Gibbs, 1974
Cardiac glycosides	Brassicaceae (*Erysimum* and *Cheiranthus*)	Hegnauer, 1964; Gibbs, 1974

Nonprotein amino acids	Cucurbitaceae	Hegnauer, 1964
Alkaloids	Brassicaceae (*Lunaria*)	Seigler, 1977
Flavones and their derivatives	Many members of the superorder	Gornal et al., 1979; Young, 1981
	Malviflorae[a]	
Cyclopropenyl fatty acids	Malvales (Sterculiaceae, Malvaceae, Bombacaceae, Tiliaceae)	Hegnauer, 1964, 1969a, 1973a
Caprylic acid	Ulmaceae	Smith, 1970
Numerous unusual fatty acids and lipids	Euphorbiaceae	Smith, 1970
Liquid wax esters	Simmondsiaceae	Smith, 1970
Fluorofatty acids	Dichapetalaceae	Seigler, 1977
Alkaloids	Eleagnaceae (unique type), Rhamnaceae (peptide and benzylisoquinoline), Eleagnaceae (harmine), Urticaceae (simple types), Moraceae (simple types), Cannabaceae (simple types), Euphorbiaceae (no less than eight types, at least two unique to the family), Pandaceae (peptide)	Seigler, 1977
Accumulation of terpenes	Dipterocarpaceae, Cannabaceae, Moraceae, Ulmaceae, Euphorbiaceae	Hegnauer, 1964, 1966, 1969a, 1973a
Anthraquinones and anthrones	Rhamnaceae	Gibbs, 1974; Hegnauer, 1969a
Cyanogens	Euphorbiaceae (in many taxa)	van Valen, 1978
Myricetin and its derivatives	Common in the superorder	Gornal et al., 1979; Young, 1981
	Santaliflorae	
Acetylenic acids	Olacaceae, Santalaceae, Loranthaceae, Viscaceae, Opiliaceae	Bohlman et al., 1973
Peptide alkaloids	Celastraceae	Seigler, 1977
Naphthoquinones	Celastraceae	Gibbs, 1974
Mono- and sesquiterpenes	Santalaceae	Hegnauer, 1973a
	Geraniiflorae	
Alkaloids (tropane type)	Erythroxylaceae	Hegnauer, 1966
Cyanogens	Linaceae, Tremandraceae	Gibbs, 1974; Hegnauer, 1966, 1973a
Glucosinolates	Tropaeolaceae, Limnanthaceae	Hegnauer, 1966, 1973a
Nitrocompounds	Malpighiaceae, Tropaeolaceae	Stermitz and Yost, 1978
Naphthoquinones	Balsaminaceae	Gibbs, 1974; Hegnauer, 1964

(continued)

TABLE II (Continued)

Chemical constituent	Distribution	Reference
	Chenopodiiflorae	
Betalains	All families of the superorder except the Caryophyllaceae Molluginaceae	Mabry, 1977
C-Glycosylflavones	Common within the superorder	Gornal et al., 1979; Young, 1981
Isoquinoline alkaloids	Chenopodiaceae	Seigler, 1977
Mesembryanthemum alkaloids	Aizoaceae	Seigler, 1977
Cactus alkaloids	Cactaceae	Seigler, 1977
	Hamamelidiflorae	
Cyanogens	Platanaceae, Trochodendraceae	van Valen, 1978b; Fikenscher and Ruijgrok, 1977
Iridoid monoterpenes	Hamamelidaceae, Eucommiaceae	Hegnauer, 1966
	Rutiflorae[b]	
Mono-, sesqui-, and diterpenes	Rutaceae, Burseraceae, Anacardiaceae, Juglandaceae, Myricaceae	Hegnauer, 1964, 1966, 1969a, 1973a
Triterpenes	Rutaceae, Simaroubaceae, Meliaceae, Burseraceae, Anacardiaceae	Polonsky, 1973a,b; Dreyer et al., 1972;
Triterpene saponins	Aceraceae, Sapindaceae, Hippocastanaceae	Hegnauer, 1964, 1966, 1973a
Cardenolides	Fabaceae (Coronilla)	Gibbs, 1974
Sesquiterpene lactones (picrotoxins)	Coriariaceae	Yoshioka et al., 1973
Nonprotein amino acids	Fabaceae, Sapindaceae, Aceraceae, Hippocastanaceae	Hegnauer, 1964, 1966, 1973a
Cyanogens	Fabaceae, Sapindaceae, a few Rutaceae	van Valen, 1979
Alkaloids	Very common in the Rutaceae and Fabaceae	Seigler, 1977
Anthraquinones	Fabaceae (Cassia)	Gibbs, 1974
Cyanolipids	Sapindaceae	Mikolajczak, 1977
Coumarins	Hippocastanaceae, Fabaceae	Gibbs, 1974; Hegnauer, 1966
Furocoumarins	Rutaceae, Fabaceae	Gibbs, 1974; Hegnauer, 1973a
5-Deoxyflavonoids	Rutaceae, Anacardiaceae, Fabaceae	Young, 1981
Isoflavones and their derivatives	Common only in the Fabaceae	Young, 1981

Rosiflorae

Acetylenic compounds	Pittosporaceae	Hegnauer, 1969b
Triterpene alkaloids	Daphniphyllaceae, Buxaceae	Seigler, 1977
Volatile terpenes	Rosaceae (fruits and flowers), Pittosporaceae, Myrothamnaceae	Gibbs, 1974; Hegnauer, 1969a, 1973a
Iridoid monoterpenes	Saxifragaceae (*Deutzia, Philadelphus, Hydrangea,* and *Escallonia*), Daphniphyllaceae	Hegnauer and Kooiman, 1978
Cyanogens	Rosaceae (common), Proteaceae (common), Saxifragaceae (*Hydrangea* and *Ribes*), Crassulaceae	Gibbs, 1974; Hegnauer, 1964, 1969a, 1973a
Keto fatty acids	Chrysobalanaceae	Smith, 1970
Naphthoquinones	Droseraceae	Nahrstedt, 1980
5-Methoxyflavonoids	Eucryphiaceae	Gornal *et al.,* 1979

Myrtiflorae

Mono- and sesquiterpenes	Myrtaceae	Hegnauer, 1969a
Cyanogens	Myrtaceae, Melastomataceae	Gibbs, 1974; Hegnauer, 1969a
Alkaloids	Punicaceae (pseudopelletierine type), Lythraceae (unusual type of quinolizidine alkaloid), Rhizophoraceae (tropane type alkaloids)	Seigler, 1977
Anthraquinones and naphthoquinones	Sonneratiaceae, Lythraceae	Gibbs, 1974; Hegnauer, 1966, 1973a
Ellagic acid	Most families of the superorder	Gibbs, 1974; Hegnauer, 1964, 1966, 1969a, 1973a

Gentianiflorae

Indole alkaloids	Gentianales (except Asclepiadaceae)	Seigler, 1977
Monoterpene alkaloids	Gentianaceae, Apocynaceae, Loganiaceae	Seigler, 1977
Emetine alkaloids	Rubiaceae	Seigler, 1977
Cardiac glycosides	Asclepiadaceae, Apocynaceae Scrophulariaceae (*Digitalis*)	Gibbs, 1974; Hegnauer, 1964, 1973a
Naphthoquinones, xanthones, anthraquinones	Apocynaceae, Gentianaceae, Rubiaceae, Scrophulariaceae, Bignoniaceae, Gesneriaceae	Gibbs, 1974; Hegnauer, 1964, 1966, 1973a
6-Hydroxyflavones	Buddlejaceae, Scrophulariceae, Gesneriaceae	Gornal *et al.,* 1979; Young, 1981
Iridoid monoterpenes	Bignoniaceae, Acanthaceae	Hegnauer and Kooiman, 1978
Coumarins	Most families of the superorder	Gibbs, 1974; Hegnauer, 1964, 1966, 1973a
Sesquiterpenes	Rubiaceae, Loganiaceae, Apocynaceae, Asclepiadaceae	Hegnauer, 1969a
	Myoporaceae	

(*continued*)

TABLE II (Continued)

Chemical constituent	Distribution	Reference
Solaniflorae		
Iridoid monoterpenes	Fouquieraceae	Dahlgren et al., 1976
6-Methoxyflavones	Polemoniaceae	Smith et al., 1977
Alkaloids	Convolvulaceae (tropane and ergot types), Solanaceae (tropane, steroidal, nicotine types), Campanulaceae (Lobelioideae) (simple types)	Seigler, 1977
Acetylenic compounds	Campanulaceae	Lam and Kaufman, 1969
Corniflorae		
Alkaloids	Rhizophoraceae (tropane types), Nyssaceae (camptothecin), Alangiaceae (emetine type), Garryaceae (diterpene type), Dipsacaceae (monoterpene type), Apiaceae (coniine type, benzylisoquinoline type)	Seigler, 1977; Gupta et al., 1976
Iridoid monoterpenes	Nyssaceae, Garryaceae, Alangiaceae, Cornaceae, Valerianaceae, Caprifoliaceae, Dipsacaceae, Valerianaceae	Hegnauer and Kooiman, 1978; Bate-Smith, et al., 1975
Mono- and sesquiterpenes	Apiaceae, Araliaceae, Valerianaceae	Hegnauer, 1978
Lamiiflorae		
Mono-, sesqui-, and diterpenes	Lamiaceae, Verbenaceae	Hegnauer, 1966, 1973a
Iridoid monoterpenes	Lamiaceae, Verbenaceae, Callitricaceae	Hegnauer and Kooiman, 1978
6-Hydroxyflavonoids	Verbenaceae	Gornal et al., 1979; Young, 1981
Pyrrolizidine alkaloids	Boraginaceae	Seigler, 1977
Naphthoquinones	Boraginaceae	Gibbs, 1974; Hegnauer, 1964
Asteriflorae		

The biology and chemistry of the Asteraceae has recently been reviewed (Heywood et al., 1977).

Constituent	Taxon	Reference
Liliiflorae		
Alkaloids	Liliaceae (colchicine, steroidal, and Amaryllidaceae types), Dioscoreaceae (unique type), Stemonaceae (unique type), Orchidaceae (a type mostly restricted to this family)	Seigler, 1977
Anthraquinones and naphthoquinones	Liliaceae (*Aloe*), Iridaceae	Gibbs, 1974; Hegnauer, 1963
Cardiac glycosides	Several Liliaceae	Gibbs, 1974; Hegnauer, 1963
Polyketide-derived compounds such as eleutherinol	Iridaceae	Gibbs, 1974; Hegnauer, 1963
Alismatiflorae		
Cyanogens	Typhaceae, Araceae	Gibbs, 1974; Hegnauer, 1963
Mono- and sesquiterpenes	Araceae (*Acorus*), Typhaceae (*Sparganium*)	Hegnauer, 1963
Areciflorae		
Lauric, myristic, and palmitic acids in seed oil	Arecaceae, Araceae	Gibbs, 1974; Hegnauer, 1963
Simple alkaloids	Arecaceae	Gibbs, 1974; Hegnauer, 1963
Commeliniflorae		
Mono- and sesquiterpenes	Bromeliaceae, Poaceae (*Andropogon* and *Cymbopogon*), Cyperaceae, Zingiberaceae	Hegnauer, 1963
Aliphatic ketones	Zingiberaceae	Gibbs, 1974; Hegnauer, 1963
6-Hydroxyflavonoids	Eriocaulaceae	Gornal, *et al.*, 1979
8-substituted flavonoids	Restionaceae	Gornal *et al.*, 1979
Alkaloids	Poaceae (simple types), Cyperaceae (harmine types)	Gibbs, 1974; Hegnauer, 1963

[a] Numerous constituents known to occur in the Euphorbiaceae have been omitted. The reader should consult Hegnauer (1966) and Gibbs (1974) for a more complete list.

[b] Numerous constituents of the Fabaceae and Rutaceae have been omitted. The reader should consult Hegnauer (1973a) and Gibbs (1974) for a more complete list.

65 genera. In terms of individuals, gymnosperms are among the most abundant plants. Almost all are trees or shrubs.

Mono-, sesqui-, and diterpenes are common in the order Coniferales (Coniferopsida). Several chemosystematic studies have been conducted within this group, notably those of Mirov (1961), Flake *et al.* (1969), Zavarin and Snajberk (1975), Zavarin (1968), Lawrence *et al.* (1975), von Rudloff (1969, 1972, 1975a,b), and Adams (1977). Several of these workers made careful analyses of diurnal and seasonal variation as well as other environmental factors. The studies of Adams (1977 and references therein) represent one of the most thorough biochemical systematics examinations of any group. He has effectively used numerical methods to resolve problems within the genus *Juniperus,* and his studies should serve as a model for others. Pines of the group *Diploxylon* contain only stilbenes (pinosylvins), whereas those of the *Haploxylon* group contain these compounds and, in addition, flavones such as chrysin (Erdtman, 1963).

Diterpenes of the Coniferales have been reviewed by Erdtman (1963). A series of complex diterpene lactones (ginkgolides) are only known to occur in the genus *Ginkgo* of the Ginkgoöpsida (Okabe *et al.,* 1967).

Alkaloids are not commonly encountered in gymnosperms (Seigler, 1977). Species of *Taxus* and *Cephalotaxus* (Coniferopsida-Taxales) contain complex alkaloids of the diterpene and homoerythrina types, respectively.

The presence of biflavonoids is a characteristic feature of gymnosperms (Geiger and Quinn, 1975; Dossaji *et al.,* 1975). These compounds are known to occur in all gymnosperm taxa except the Gnetopsida and Pinaceae (Coniferopsida).

Several glycosides of methylazoxymethanol occur in genera of Cycadopsida (Yang *et al.,* 1972; Whiting, 1963; Nishida *et al.,* 1959). These compounds are not known to occur in other plants.

C. Dicotyledonous Angiosperms

1. Superorder Annoniflorae

The Annoniflorae consists of 4 orders (Annonales, Nelumbonales, Berberidales, and Nymphaeales), 33 families, and probably more than 12,000 species (Thorne, 1976; Young and Seigler, 1981). This group of predominately woody plants is considered to contain an assemblage of characters regarded as primitive by most investigators.

Most systematists would consider this superorder a fairly well-defined group based on morphological characters. It is also coherent from a phytochemical standpoint.

The superorder Annoniflorae as a group is united chemically by the presence of benzylisoquinoline alkaloids which are relatively uncommon in other plant taxa. They occur in many families of the orders Annonales, Nelumbonales, and Berberidales but are absent in the order Nympheales and the Annonales suborder Piperineae (Gibbs, 1974; Hegnauer, 1962, 1963, 1964, 1966, 1969a, 1973a; Seigler, 1977). R. F. Thorne (personal communication) has recently transferred the Piperineae to a position more closely related to the Nymphaeiflorae. The Nympheales contain sesquiterpene alkaloids of a unique type.

Essential oils of primarily terpenoid and phenylpropanoid origin are characteristic of the Annoniflorae, with the exception of the suborder Berberidineae.

Sesquiterpene lactones are known to occur in the Illiciaeae, Magnolicaceae, Lauraceae, Menispermaceae, and Winteraceae (Annonales). They are otherwise known mostly from the Asteraceae and certain liverworts. Inclusion of the Coriariaceae in the Berberidineae (Cronquist, 1968) is supported by the presence of picrotoxins (a peculiar type of sesquiterpene lactones found in several Menispermaceae [as well as in *Hyenanche* (Euphorbiaceae)] (Yoshioka *et al.*, 1973).

The origin and taxonomic significance of phenylpropanoids in the Lauraceae have been examined (Gottlieb, 1972). A number of 2-pyrones are also known to occur in the Lauraceae as well as in the Piperaceae and Fabaceae. Both phenylpropanoids and 2-pyrones appear to supplant alkaloids in plant taxa in which they occur (Gottlieb, 1972).

Cyanogenesis is common in this superorder; all compounds isolated to date are derived from tyrosine (Hegnauer, 1973b; van Valen, 1978b).

2. Superorder Theiflorae

The Theiflorae consist of 5 orders (Theales, Ericales, Ebenales, Primulales, and Polygonales), 42 families, and approximately 13,000 species. Dahlgren (1977) segregated the Actinidiaceae, Clethraceae, Ericaceae, Epacridaceae, Empetraceae, Sarraceniaceae, and Icacinaceae and placed them in his superorder Cornaniflorae, largely based on the presence of unitegmic (nearly always tenuinucellate) ovules, usually cellular endosperm formation, and the presence of iridoid monoterpenes.

In contrast to the superorder Annoniflorae, the Theiflorae are not well defined chemically. Within this taxon alkaloids are uncommon. They occur in the Actinidiaceae (monoterpene type), Icacinaceae (camptothecin—otherwise known only from the Nyssaceae), and Symplocaceae (apophine type). Homoerythrina alkaloids have been reported from the Phellinaceae (Hoang *et al.*, 1970). They are otherwise known from *Cephalotaxus* (a gymnosperm) and from the Liliaceae (a monocotyledonous plant family).

Flavonoids and the systematics of a number of species of *Rhododendron* have been examined (King, 1977).

3. Superorder Cistiflorae*

The Cistiflorae consists of 4 orders (Cistales, Salicales, Tamaricales, and Capparales), 22 families, and approximately 8500 species. It is basically a tropical group of families related to the Theiflorae. Two groups of families within the superorder are chemically distinctive.

The Capparales possess myrosin cells and glucosinolates (Vaughan et al., 1976). These are also known from the Caricaceae (Cistales).

The second group, the families Flacourticaceae, Turneraceae, Passifloraceae, and Malesherbiaceae, frequently contain species that are cyanogenic. The compounds responsible are cyclopentenyl cyanogenic glycosides which probably arise from the same biosynthetic precursor as the cyclopentenyl fatty acids of the Flacourtiaceae.

4. Superorder Malviflorae

The Malviflorae consists of 4 orders (Malvales, Urticales, Rhamnales, and Euphorbiales), 22 families, and approximately 11,750 species. They are a diverse group of primarily tropical woody plants.

The chemistry of the superorder is highly varied (Table II). One cluster of families within the Malvales (Tiliaceae, Sterculiaceae, Malvaceae, Bombacaceae) is characterized by the presence of cyclopropenyl fatty acids in the seed oils.

The Euphorbiaceae (about 5000 species) are among the most diverse of all plants. As only one example of this, many types of fatty acids are known from the seed oils of this family. Among them are hydroxy, epoxy, and allenic fatty acids, as well as acids with unusual positions of unsaturation and both cis and trans isomers.

The similarity of alkaloids in certain genera of the Euphorbiaceae, Rhamnaceae, and Pandaceae suggests relationships among these taxa.

The accumulation of terpenes in the superorder is restricted to several families. Among these are the Dipterocarpaceae (mono-, sesqui-, and diterpenes) (Ourisson et al., 1965), Cannabaceae (including cannabinoids and other terpenes), Moraceae (tri- and polyterpenes), Ulmaceae (sesquiterpenes), and Euphorbiaceae.

The terpenoid chemistry of the Euphorbiaceae is also quite complex (Ponsinet and Ourisson, 1965). Several genera such as *Croton* accumulate mono- and sesquiterpenes. Others accumulate triterpenes (*Euphorbia*), and still others polyterpenes (*Hevea*). Several genera make cocarcinogenic diterpenes which are otherwise only known from the Thymeleaceae. The rather com-

---

* Recent evidence suggests that the Cistaceae are more closely allied with the Malviflorae. Transfer of this family will require a name change for this superorder (D. A. Young, personal communication).

plex structures of these compounds suggest a common biosynthetic origin and a close relationship between the two families.

5. Superorder Santaliflorae

The Santaliflorae consists of a single order (Santalales), 12 families, and approximately 2800 species. Because of its great evolutionary depth, the Santaliflorae is hard to characterize as a whole. Basically, members of the superorder are tropical woody plants, many of which have become specialized as semiparasites or total parasites.

Acetylenic acids, which are known to occur in glycerides of the Olacaceae, Opiliaceae, Santalaceae, Loranthaceae, and Viscaceae (Bohlmann et al., 1973), confirm proposed relationships among these families.

6. Superorder Geraniiflorae

The Geraniiflorae, with a single order (Geraniales), consists of 14 families and approximately 5000 species of primarily herbaceous plants.

Dahlgren (1977) considered the Limnanthaceae and Tropaeolaceae part of his Capparales (Violiflorae). This is supported by the presence of glucosinolates in the two families and the presence of C_{22} fatty acids in the seed oils of the Limnanthaceae.

A large variety of compounds occurs in this superorder (Table II). They do not appear, however, to provide phylogenetic information at the ordinal or subordinal level.

7. Superorder Chenopodiiflorae

The Chenopodiiflorae (Centrospermae of other authors) consists of a single order (Chenopodiales), 12 families, and approximately 6700 species.

Betalain pigments replace anthocyanins in all families of the order except the Caryophyllaceae and Molluginaceae (Mabry, 1977). These alkaloidal pigments are known only from this superorder.

The use of betalain chemical data for delineation of the Chenopodiiflorae (Mabry, 1977) has answered a number of questions but at the same time created much controversy. These data provided strong support for placement of the Cactaceae, Didiereaceae, and certain other families within the superorder. [It is of interest, however, to note that Hallier (1912) had located many of the enigmatic families in the superorder based on purely morphological information.]

The Caryophyllaceae and Molluginaceae are considered members of the Chenopodiiflorae on morphological grounds, especially the presence of P-III-type sieve tube plastids despite the fact that they lack the characteristic betalain pigments (Mabry, 1977), whereas the Plumbaginaceae and

Polygonaceae, which lack betalain pigments and type P-III plastids, were excluded. *C*-Glycosylflavones are widespread in the superorder (Gornal *et al.*, 1979; Young, 1981).

Saponins are commonly encountered and occur in all but one or two families.

8. Superorder Hamamelidiflorae

The Hamamelidiflorae, which is one of the smaller superorders of dicotyledons, consists of 3 orders (Hamamelidales, Casuarinales, and Fagales), 10 families, and approximately 1100 species. The Hamamelidiflorae are all woody plants.

The phytochemistry of many plants of this group has been reviewed (Mears, 1973).

Although Dahlgren (1977) grouped most iridoid-producing families into related superorders, he retained the Hamamelidaceae (and Daphniphyllaceae) in his Hamamilidiflorae but transferred the Eucommiaceae to the Corniflorae.

9. Superorder Rutiflorae

The superorder Rutiflorae consists of a single order (Rutales), 22 families, and approximately 20,000 species with over half of the species belonging to a single family, the Fabaceae (=Leguminosae, 12,000–13,000 species). Members of the Rutiflorae are typically tropical or subtropical woody plants. Although most taxonomists have not closely associated the Rutinae with the Fabaceae, there are many chemical similarities that support this placement. It is of interest that Hallier (1912) also recognized the close relationship of these groups (and the Connaraceae and Fabaceae).

Accumulation of terpenes is common within this superorder (Table II). Many triterpenes from the Rutaceae, Meliaceae, and Simaroubaceae are extensively oxidized and rearranged (Polonsky, 1973a,b; Mondon and Callsen, 1975a,b; Mondon *et al.*, 1975; Dryer *et al.*, 1972).

The triterpenes of these families are closely related, and those of the last two families are derived from more simple types which predominate in the Rutaceae.

Several genera of the Fabaceae accumulate terpenes, although they are not common in the family. Many species of the subfamily Caesalpinoideae have diterpenes. Saponins are found in all three subfamilies.

Nonprotein amino acids are accumulated in unusual numbers in seeds of the Fabaceae, Sapindaceae, Aceraceae, and Hippocastanaceae (Bell, 1971; Evans and Bell, 1978; Hegarty, 1978). These compounds are not commonly accumulated in other families of dicotyledonous plants (except the Cucurbitaceae).

Cyanogenesis occurs in many species of the Fabaceae (all three subfamilies) (van Valen, 1979), the Sapindaceae, and a few species of the Rutaceae. The similarity in structure of the cyanogens suggests a relatively close relationship.

Alkaloids are commonly found only in the Rutaceae and Fabaceae. Each of these families possesses several types of alkaloids, including some unique to each family. Although the terpenes of the Meliaceae and Simaroubaceae suggest a close relationship to the Rutaceae, these two families are almost totally lacking in alkaloids (Seigler, 1977).

Simple coumarins are common in the Hippocastanaceae and Fabaceae. More complex furocoumarins known from the Fabaceae and Rutaceae are otherwise found only in the Apiaceae and Araliaceae.

Many taxonomic problems in the genus *Baptisia* (Fabaceae) have been studied by investigations of the morphology and chemistry. The genus, of approximately 17 species, represents one of the most complex groups of plants in the southeastern United States and is now one of the most thoroughly studied of all plant genera. The flavonoid, alkaloid, and isoenzyme chemistry of various species was studied and the data from these studies applied to the analysis of populations where extensive hybridization occurred (hybrid swarms). These works documented instances of hybridization among two, three, or even four species (Alston and Turner, 1963a,b). Although the detection of gene flow from one taxon to another (introgression) by hybridization and backcrossing is very difficult with morphological characters, this phenomenon has been demonstrated in hybrid swarms of *Baptisia leucophaea* Nutt. and *B. sphaerocarpa* Nutt. by a chromatographic study of the flavonoids of more than 1400 plants (Turner, 1967, 1969).

In studies on the inheritance of flavonoid pigments in the genus *Baptisia*, of the legume family, it has been shown that hybrid compounds occur in addition to those predicted on an additive basis (Alston *et al.*, 1965), but in a study on the alkaloids of this genus it was established that the alkaloid chemistry of the hybrid plants resembled neither the combined or single chemistry of the parental species. Mabry concludes that the careful investigator can use to an advantage the fact that hybridization and introgression alter the genetic makeup and thus the chemistry of certain populations of species (Mabry and Mears, 1970).

The chemosystematic significance of heartwood flavonoids of *Rhus* (Anacardiaceae) (Young, 1979) and alkaloids of *Zanthoxylum* and *Fagara* (Rutaceae) (Fish and Waterman, 1973) have been examined.

10. Superorder Rosiflorae

The Rosiflorae consist of 3 orders (Rosales, Pittosporales, and Proteales), 30 families, and approximately 8500 species. Members of the Rosiflorae are extremely diverse in terms of habit, foliage, and anatomy and, like members of the Annoniflorae and Hamamelidiflorae, have retained many primitive features. There are strong tendencies toward the herbaceous and annual habit.

Dahlgren (1977) placed the Pittosporaceae in his Araliiflorae in agreement with chemical data previously reviewed by Hegnauer (1969b). In addition he placed many taxa recognized as subfamilies of the Saxifragaceae (*sensu* Thorne) in the Corniflorae. Several of these contain iridoid monoterpenes which are absent from the Saxifragaceae *sensu stricto*.

The chemotaxonomy of the Rosaceae has been examined by Challice (1973, 1974), largely by means of phenolic compounds.

11. Superorder Myrtiflorae

The Myrtiflorae, with a single order (Myrtales) consists of 8 families and approximately 8000 species, mostly tropical woody plants. Members of several families are mangrove taxa.

Mono- and sesquiterpenes are accumulated, often in large quantities, in the Myrtaceae. A number of ketones are also found in essential oils of this family.

Phytochemistry does not appear to provide much useful data on this superorder above the familial level.

12. Superorder Gentianiflorae

The Gentianiflorae consists of 3 orders (Oleales, Gentiales, and Bignoniales), 19 families, and nearly 22,000 species.

The order Gentianales is characterized by the presence of iridoid monoterpene glycosides and indole alkaloids which are present in all families except the Asclepiadaceae and possibly Buddlejaceae (Hegnauer and Kooiman, 1978). Alkaloids are found in the Rubiaceae (emetine, quinine, indole types), Gentianaceae (monoterpene type), Loganiaceae, and Apocynaceae (indole and steroidal types). The Asclepiadaceae contain tylophorine alkaloids which are unique to that family (Seigler, 1977).

The Asclepiadaceae and Apocynaceae share a number of cardenolides that do not occur elsewhere.

In the order Bignoniales, iridoid monoterpenes have been reported from all families.

Chemically the Gentianales and Bignoniales are similar in that both contain iridoid compounds but distinct in that alkaloids are common only in the former order.

13. Superorder Solaniflorae

The Solaniflorae consists of 2 orders (Solanales and Campanulales), 7 families, and approximately 6500 species.

Dahlgren *et al.* (1976) placed the Fouquieriaceae in his Corniflorae. The Fouquieriaceae is the only family in the superorder Solaniflorae that contains iridoid terpenes. The Polemoniaceae possess 6-methoxyflavonols and other characters which suggest that they are not closely related to the Solanaceae and Convolvulaceae (Smith *et al.,* 1977) but may be viewed as transitional between these families and the order Bignoniales.

The Solanaceae and Convolvulaceae contain tropane alkaloids, saponins, and a number of triterpenes. The Solanaceae additionally contain steroidal alkaloids and nicotine types, whereas the Convolvulaceae contain ergot alkaloids (which otherwise are known to occur only in a fungus, *Claviceps*) (Seigler, 1977). The Polemoniaceae and Fouquieriaceae appear to lack alkaloids entirely.

The presence of acetylenic compounds in the Campanulaceae (Lam and Kaufman, 1969) suggests affinities with the Asteraceae.

14. Superorder Corniflorae

The superorder Corniflorae consists of 2 orders (Cornales and Dipsacales), 17 families, and approximately 4700 species, primarily woody plants.

Dahlgren (1977) segregated the Araliaceae and Apiaceae and placed them in his Araliflorae (along with the Pittosporaceae). The biology and chemistry of the Apiaceae have been reviewed (Heywood, 1971). These families have acetylenic compounds, petroselinic acid in their seed oils, furocoumarins, and accumulate volatile terpenes, characters not otherwise found in the superorder Corniflorae (Hegnauer, 1969b). Further, they lack iridoid monoterpenes. The presence of sesquiterpene lactones, benzylisoquinoline alkaloids (Gupta *et al.,* 1976) (Apiaceae), furocoumarins, and volatile terpenes suggests alliances with the Asteriflorae and Rutiflorae (Meeuse, 1970; Dahlgren, 1977; Kubitzski, 1969). Cronquist (1968) also separated these two orders.

R. Hegnauer (1969b, 1978) has pointed out that the Hippuridaceae are quite distinct from the Haloragaceae and Gunneraceae and most closely resemble the Bignoniales in overall chemistry.

The flavonoids of fruits of many Apiaceae have been examined (Harborne and Williams, 1972) in an effort to establish tribal affinities within the family.

15. Superorder Lamiiflorae

The Lamiiflorae, with a single order (Lamiales), consists of 7 families and approximately 8100 species.

The Verbenaceae and Lamiaceae are very similar chemically; both accumulate many types of mono-, sesqui-, and diterpenes in various taxa, and both contain iridoid monoterpenes, coumarins, saponins, and anthraquinones.

The Boraginaceae and Hydrophyllaceae are chemically distinct from the former families (Table II). Dahlgren (1977) considered them to be part of his Solaniflorae.

16. Superorder Asteriflorae

The Asteriflorae consists of a single family, the Asteraceae (=Compositae). This family is the second largest family of flowering plants, having nearly 20,000 species of subcosmopolitan distribution, and accounts for approximately 10% of flowering plants.

The Asteraceae is chemically diverse. The biology and chemistry of the family have recently been reviewed (Heywood *et al.*, 1977, 1978). Most members of the family accumulate terpenes. Mono-, sesqui-, di-, and triterpenes are common within the family. Sesquiterpene lactones are commonly found in this group of plants. Polyterpenes are known, especially from the tribe Cichorieae.

Acetylenic compounds are found throughout the family, with the exception of the tribe Senecioneae where they are rare (Sørensen, 1977).

Alkaloids are rarely encountered in the family, except within the tribe Senecioneae where pyrrolizidine alkaloids are common (Culvenor, 1978).

D. Monocotyledonous Angiosperms

1. Superorder Liliiflorae

The Liliiflorae consists of a single order (Liliales), 8 families, and over 29,800 species, over 20,000 of these being members of the largest family of flowering plants, the Orchidaceae.

The Liliaceae has often been subdivided into numerous families (Melchior, 1964; Hutchinson, 1973; Dahlgren, 1977). This reflects the complex chemistry as well as morphology of the group. Almost all members have saponins and steroidal glycosides. Many members have colchicine as the principal alkaloid, and others possess three types of alkaloids collectively known as Amaryllidaceae alkaloids. Some genera (e.g., *Veratrum*) possess steroidal alkaloids. Many members of the group contain nonprotein amino acids.

Both the Dioscoreaceae and Stemonaceae possess unique types of alkaloids (Seigler, 1977).

2. Superorder Alismatiflorae

The Alismatiflorae consists of 4 orders (Alismatales, Zosterales, Najadales, and Triuridales), 11 families, and only about 500 species of primarily aquatic plants.

Cyanogenesis is known from the Hydrocharitaceae, Scheuchzeriaceae, and Juncaginaceae; triglochinin, a cyanogenic glycoside derived from tyrosine, has been isolated from the last two families.

3. Superorder Ariflorae

The Ariflorae consists of a single order (Arales), 3 families (Araceae, Lemnaceae, and Typhaceae), and approximately 2100 species.

Cyanogenesis (triglochinin) is common among the Araceae. The report of a benzylisoquinoline alkaloid from this group suggests affinities with the Magnoliflorae (Seigler, 1977).

In the Lemnaceae morphological variations which might prove to be insignificant in more complex organisms assume greater importance. The only reliable way to identify certain species was by analysis of their flavonoid chemistry (McClure and Alston, 1966).

4. Superorder Areciflorae

The Areciflorae consists of 3 monofamilial orders [Arecales (Arecaceae = Palmae), Cyclanthales (Cyclanthaceae), and Pandales (Pandanaceae)], and approximately 4500 species. The largest group is the Arecaceae (palms), with nearly 3500 species. Seed oils of many Arecaceae contain large amounts of caprylic, lauric, and myristic acids.

5. Superorder Commeliniflorae

The Commeliniflorae consists of 2 orders (Commelinales and Zingiberales), 22 families, and nearly 20,000 species, over half of which are members of a single family, the Poaceae (=Gramineae).

Cyanogenesis is known from the Juncaceae, Commelinaceae (*Tinantia*), Poaceae (many taxa), and Maranthaceae. The compounds of most taxa are derived from tyrosine pathways.

Williams (1978) examined flavonoids of the family Bromeliaceae and concluded that the family occupies an isolated position with regard to other monocot families.

ACKNOWLEDGMENTS

I wish to thank S. G. Saupe and D. A. Young for reading this chapter.

REFERENCES

Adams, R. P. (1972). *Taxon* **21**, 407–427.
Adams, R. P. (1977). *Ann. Mo. Bot. Gard.* **64**, 184–209.
Adams, R. P. (1979). *Am. J. Bot.* **66**, 986–988, and references therein.
Alston, R. E., and Turner, B. L. (1959). *Nature (London)* **184**, 285–286.
Alston, R. E., and Turner, B. L. (1963a). *Am. J. Bot.* **50**, 159–173.
Alston, R. E., and Turner, B. L. (1963b). "Biochemical Systematics." Prentice-Hall, Englewood Cliffs, New Jersey.

Alston, R. E., Rösler, H., Naifeh, K., and Mabry, T. J. (1965). *Proc. Natl. Acad. Sci. U.S.A.* **54,** 1458–1465.

Bate-Smith, E. C., Ferguson, I. K., Hutson, K., Jensen, S. R., Nielsen, B. J., and Swain, T. (1975). *Biochem. Syst. Ecol.* **3,** 79–89.

Becker, K. M. (1973). *Taxon* **22,** 19–50.

Bell, E. A. (1971). *In* "Chemotaxonomy of the Leguminosae" (J. B. Harborne, D. Boulter, and B. L. Turner, eds.), pp. 179–206. Academic Press, New York.

Belzer, N. F., and Ownbey, M. (1971). *Am. J. Bot.* **58,** 791–802.

Bernfield, P. (1967). "Biogenesis of Natural Compounds." Pergamon, Oxford.

Berti, G., and Bottari, F. (1968). *Prog. Phytochem.* **1,** 589–685.

Birch, A. J. (1974). *In* "Chemistry in Botanical Classification" (G. Bendz and J. Santesson, eds.), pp. 261–270. Academic Press, New York.

Bohlmann, F., Borkhardt, T., and Zdero, C. (1973). "Naturally Occurring Acetylenes." Academic Press, New York.

Cagnin, M. A. H., and Gottlieb, O. R. (1978). *Biochem. Syst. Ecol.* **6,** 225–238.

Challice, J. S. (1973). *Phytochemistry* **12,** 1095–1101.

Challice, J. S. (1974). *Bot. J. Linn. Soc.* **69,** 239–259.

Conn, E. E. (1979). *Naturwissenschaften* **66,** 28–34.

Cooper-Driver, G. (1977). *Science* **198,** 1260–1262.

Cranmer, M. F., and Mabry, T. J. (1966). *Phytochemistry* **5,** 1133–1138.

Cranmer, M. F., and Turner, B. L. (1967). *Evolution* **21,** 508–517.

Crawford, D. J. (1972). *Taxon* **21,** 27–38.

Cronquist, A. (1968). "The Evolution and Classification of Flowering Plants." Houghton-Mifflin, Boston, Massachusetts.

Culvenor, C. C. J. (1978). *Bot. Not.* **131,** 473–486.

Dahlgren, R. (1975). *Bot. Not.* **128,** 119–147.

Dahlgren, R. (1977). *Flowering Plants, Symp., 1976* pp. 253–283.

Dahlgren, R., Jensen, S. R., and Nielsen, B. J. (1976). *Bot. Not.* **129,** 207–212.

Dános, B. (1965). *Pharmazie* **20,** 727–730.

Dános, B. (1966). *Abh. Dtsch. Akad. Wiss. Berlin, Kl. Chem., Geol. Biol.* **3,** 363–367.

Davis, D. H., and Heywood, V. H. (1963). "Principles of Angiosperm Taxonomy." Van Nostrand-Reinhold, Princeton, New Jersey.

De Candolle, A. P. (1804). "Essai sur les propriétés médicales des plantes, comparées avec leurs formes extérieures et leur classification naturelle," 1st ed. Paris.

De Candolle, A. P. (1816). "Essai sur les propriétés médicales des plantes, comparées avec leurs formes extérieures et leur classification naturelle," 2nd ed. Paris.

Dement, W. A., and Mabry, T. J. (1972). *Phytochemistry* **11,** 1089–1093.

Dossaji, S. F., Mabry, T. J., and Bell, E. A. (1975). *Biochem. Syst. Ecol.* **2,** 171–175.

Dreyer, D. L., Pickering, M. V., and Cohan, O. (1972). *Phytochemistry* **11,** 705–713.

Erdtman, H. (1963). *In* "Chemical Plant Taxonomy" (T. Swain, ed.), pp. 89–125. Academic Press, New York.

Ettlinger, M. G., and Kjaer, A. (1968). *Recent Adv. Phytochem.* **1,** 59–144.

Evans, C. S., and Bell, E. A. (1978). *Phytochemistry* **17,** 1127–1129.

Fairbrothers, D. E., Mabry, T. J., Scogin, R. L., and Turner, B. L. (1975). *Ann. Mo. Bot. Gard.* **62,** 765–800.

Fikenscher, L. H., and Ruijgrok, H. W. L. (1977). *Planta Med.* **31,** 290–293.

Fish, F., and Waterman, P. G. (1973). *Taxon* **22,** 177–203.

Flake, R. H., von Rudloff, E., and Turner, B. L. (1969). *Proc. Natl. Acad. Sci. U.S.A.* **64,** 487–494.

Flück, H. (1963). *In* "Chemical Plant Taxonomy" (T. Swain, ed.), pp. 167–186. Academic Press, New York.

Fowden, L. (1972). *Phytochemistry* **11**, 2271–2276.

Freeland, W. J., and Janzen, D. H. (1974). *Am. Nat.* **108**, 269–289.

Geiger, H., and Quinn, C. (1975). *In* "The Flavonoids" (J. B. Harborne, T. J. Mabry, and H. Mabry, eds.), Part 2, pp. 692–742. Academic Press, New York.

Geissman, T., and Crout, D. H. G. (1969). "Organic Chemistry of Secondary Plant Metabolism." Freeman Cooper, San Francisco, California.

Gibbs, R. D. (1974). "Chemotaxonomy of Flowering Plants." McGill-Queens University Press, Montreal.

Goodwin, T. W. (1974). *In* "Algal Physiology and Biochemistry" (W. D. P. Stewart, ed.), pp. 266–305. Blackwell, Oxford.

Gornall, R. J., Bohm, B. A., and Dahlgren, R. (1979). *Bot. Not.* **132**, 1–30.

Gottlieb, O. R. (1972). *Phytochemistry* **11**, 1537–1570.

Greshoff, M. (1893). *Ber. Dtsch. Pharm. Ges.* **3**, 191–204.

Gupta, B. D., Bannerjee, S. K., and Handa, K. L. (1976). *Phytochemistry* **15**, 576.

Hahlbrock, K., and Grisebach, H. (1975). *In* "The Flavonoids" (J. B. Harborne, T. J. Mabry, and H. Mabry, eds.), Part 2, pp. 866–915. Academic Press, New York.

Hallier, H. (1912). *Arch. Neerl. Sci. Exactes Nat., Ser. 3* **1**, 146–234.

Harborne, J. B. (1975). *In* "Flavonoids" (J. B. Harborne, T. J. Mabry, and H. Mabry, eds.), Part 2, pp. 1056–1095. Academic Press, New York.

Harborne, J. B. (1977). "Introduction to Ecological Biochemistry." Academic Press, New York.

Harborne, J. B., and Williams, C. A. (1972). *Phytochemistry* **11**, 1741–1750.

Harborne, J. B., Williams, C. A., and Smith, D. M. (1973). *Biochem. Syst.* **1**, 51–54.

Haslam, E. (1974). "The Shikimate Pathway." Wiley, New York.

Hegarty, M. P. (1978). *In* "Effects of Poisonous Plants on Livestock" (R. F. Keeler, K. R. Van Kampen, and L. F. James, eds.), pp. 575–585. Academic Press, New York.

Hegnauer, R. (1962). "Chemotaxonomie der Pflanzen," Vol. 1. Birkhaeuser, Basel.

Hegnauer, R. (1963). "Chemotaxonomie der Pflanzen," Vol. 2. Birkhaeuser, Basel.

Hegnauer, R. (1964). "Chemotaxonomie der Pflanzen," Vol. 3. Birkhaeuser, Basel.

Hegnauer, R. (1966). "Chemotaxonomie der Pflanzen," Vol. 4. Birkhaeuser, Basel.

Hegnauer, R. (1967). *Pure Appl. Chem.* **14**, 173–187.

Hegnauer, R. (1969b). *In* "Perspectives in Phytochemistry" (J. B. Harborne and T. Swain, eds.), pp. 121–138. Academic Press, New York.

Hegnauer, R. (1969a). "Chemotaxonomie der Pflanzen," Vol. 5. Birkhaeuser, Basel.

Hegnauer, R. (1973b). *Biochem. Syst.* **1**, 191–197.

Hegnauer, R. (1973a). "Chemotaxonomie der Pflanzen," Vol. 6. Birkhaeuser, Basel.

Hegnauer, R. (1976). *In* "Secondary Metabolism and Coevolution" (M. Luckner, K. Mothes, and L. Nover, eds.), pp. 45–76. Dtsch. Akad. Naturf. Leopold., Halle.

Hegnauer, R. (1978). *Dragoco Rep. (Ger. Ed.)* **25**, 203–230.

Hegnauer, R., and Kooiman, P. (1978). *Planta Med.* **34**, 1–33.

Heywood, V. H., ed. (1971). "The Biology and Chemistry of the Umbelliferae." Academic Press, New York.

Heywood, V. H., Harborne, J. B., and Turner, B. L., eds. (1977). "The Biology and Chemistry of the Compositae," Vol. 1. Academic Press, New York.

Heywood, V. H., Harborne, J. B., and Turner, B. L., eds. (1978). "The Biology and Chemistry of the Compositae," Vol. 2. Academic Press, New York.

Hoang, N. M., Langlois, N., Das, B. C., and Potier, P. (1970). *C. R. Hebd. Seances Acad. Sci., Ser. C* **270**, 2154–2156.

Hughes, D. W., and Genest, K. (1973). *In* "Phytochemistry" (L. P. Miller, ed.), Vol. 2, pp. 118–170. Van Nostrand-Reinhold, Princeton, New Jersey.

Hughes, M. A., and Conn, E. E. (1976). *Phytochemistry* **15**, 697–701.

Hutchinson, J. (1973). "The Families of Flowering Plants Arranged According to a New System Based on Their Probable Phylogeny," 3rd ed. Oxford Univ. Press (Clarendon), London and New York.

Irving, R. S., and Adams, R. P. (1973). *Recent Adv. Phytochem.* **6**, 186–214.

Jones, D. A. (1973). *In* "Taxonomy and Ecology" (V. H. Heywood, ed.), pp. 213–242. Academic Press, New York.

Jones, D. A. (1979). *Am. Nat.* **113**, 445–451.

Jones, S. B., Jr., and Luchsinger, A. E. (1979). "Plant Systematics." McGraw-Hill, New York.

King, B. L. (1977). *Am. J. Bot.* **64**, 350–360.

Kubitzki, K. (1969). *Taxon* **18**, 360–368.

Lam, J., and Kaufman, F. (1969). *Chem. Ind.* (*London*) p. 1430.

Lawrence, L., Bartschot, R., Zavarin, E., and Griffin, J. R. (1975). *Biochem. Syst. Ecol.* **2**, 113–119.

Levin, D. A. (1971). *Am. Nat.* **105**, 157–181.

Levy, M., and Levin, D. A. (1975). *Evolution* **29**, 487–499.

Luckner, M. (1972). "Secondary Metabolism in Plants and Animals." Academic Press, New York.

Mabry, T. J. (1970). *In* "Phytochemical Phylogeny" (J. B. Harborne, ed.), pp. 269–300. Academic Press, New York.

Mabry, T. J. (1973). *Pure Appl. Chem.* **34**, 377–400.

Mabry, T. J. (1974). *In* "Chemistry and Botanical Classification" (G. Bendz and J. Santesson, eds.), p. 63. Academic Press, New York.

Mabry, T. J. (1977). *Ann. Mo. Bot. Gard.* **64**, 210–220.

Mabry, T. J., and Mears, J. A. (1970). *In* "Chemistry of the Alkaloids" (S. W. Pelletier, ed.), pp. 719–746. Van Nostrand-Reinhold, Princeton, New Jersey.

McClure, J. W. (1975). *In* "The Flavonoids" (J. B. Harborne, T. J. Mabry, and H. Mabry, eds.), Part 2, pp. 970–1055. Academic Press, New York.

McClure, J. W., and Alston, R. E. (1966). *Am. J. Bot.* **53**, 849–860.

McNair, J. B. (1965). "Studies in Plant Chemistry Including Chemical Taxonomy, Ontogeny, Phylogeny, etc." (A collection in book form of papers of the author between 1916 and 1945). Published by the Author, Los Angeles, California.

Mann, J. (1978). "Secondary Metabolism." Oxford Univ. Press (Clarendon), London and New York.

Mears, J. A. (1973). *Brittonia* **25**, 385–394.

Mears, J. A., and Mabry, T. J. (1971). *In* "Chemotaxonomy of the Leguminosae" (J. B. Harborne, D. Boulter, and B. L. Turner, eds.), p. 73. Academic Press, New York.

Meeuse, A. D. J. (1970). *Acta Bot. Neerl.* **19**, 133–140.

Melchior, H., ed. (1964). Engler's Syllabus der Pflanzenfamilen," 12th ed., Vol. II. Borntraeger, Berlin.

Mikolajczak, K. L. (1977). *Prog. Chem. Fats Other Lipids* **15**, 97–130.

Mirov, N. T. (1961). *U.S., Dep. Agric., Tech. Bull.* **1239**.

Mondon, A., and Callsen, H. (1975a). *Tetrahedron Lett.* pp. 551–554.

Mondon, A., and Callsen, H. (1975b). *Tetrahedron Lett.* pp. 699–702.

Mondon, A., Callsen, H., and Epe, B. (1975). *Tetrahedron Lett.* pp. 703–706.

Mothes, K. (1960). *Alkaloids* (*N.Y.*) **6**, 1–30.

Mothes, K. (1966). *Lloydia* **29**, 156–171.

Murray, J. J. (1960). *Genetics* **45**, 925–929, 931–937.

Nahrstedt, A. (1980). *Phytochemistry* **19**, 2757–2758.

Nishida, K., Kobayashi, A., Nagahama, T., and Numata, T. (1959). *Bull. Agric. Chem. Soc. Jpn.* **23**, 460–464.

Nolan, C., and Margoliash, E. (1968). *Annu. Rev. Biochem.* **37**, 727–790.

Nowacki, E. (1963). *Genet. Pol.* **4**, 161–202.
Okabe, K., Yamada, K., Yamamura, S., and Takada, S. (1967). *J. Chem. Soc. C* pp. 2201–2206.
Ourisson, G., Bisset, N. G., Diaz, M. A., Ehret, C., Palmade, M., Pesnelle, P., and Streith, J. (1965). *Mem. Soc. Bot. Fr.* pp. 141–142.
Polonsky, J. (1973a). *Fortschr. Chem. Org. Naturst.* **30**, 101–150.
Polonsky, J. (1973b). *Recent Adv. Phytochem.* **6**, 31–64.
Ponsinet, G., and Ourisson, G. (1965). *Phytochemistry* **4**, 799–811.
Pryor, L. D., and Bryant, L. H. (1958). *Proc. Linn. Soc. N. S. W.* **83**, 55–64.
Radford, A. E., Dickinson, W. C., Massey, J. R., and Bell, C. R. (1974). "Vascular Plant Systematics." Harper, New York.
Raven, P. H., Berlin, B., and Breedlove, D. E. (1970). *Science* **174**, 1210–1213.
Raven, P. H., Evert, R. F., and Curtis, H. (1976). "Biology of Plants." Worth, New York.
Rezende, C. M. A. da. M., Gottlieb, O. R., and Marx, M. C. (1975). *Biochem. Syst. Ecol.* **3**, 63–70.
Robinson, T. (1974). *Science* **184**, 430–435.
Seeligman, P., and Alston, R. E. (1967). *Brittonia* **19**, 205–211.
Seigler, D. S. (1974). *Chem. Br.* **10**, 339–342.
Seigler, D. S. (1977). *Alkaloids (N.Y.)* **16**, 1–82.
Seigler, D. S., and Price, P. W. (1976). *Am. Nat.* **110**, 101–104.
Smith, C. R., Jr. (1970). *Prog. Chem. Fats Other Lipids* **11**, 137–177.
Smith, D. M., and Levin, D. A. (1963). *Am. J. Bot.* **50**, 952–958.
Smith, D. M., Craig, S. P., and Santarosa, J. (1971). *Am. J. Bot.* **58**, 292–299.
Smith, D. M., Glennie, C. W., Harborne, J. B., and Williams, C. A. (1977). *Biochem. Syst. Ecol.* **5**, 107–116.
Smith, H. H. (1965). *Am. Nat.* **99**, 73–79.
Smith, P. M. (1976). "The Chemotaxonomy of Plants." Am. Elsevier, New York.
Sørensen, N. A. (1977). *In* "The Biology and Chemistry of the Compositae" (V. H. Heywood, J. B. Harborne, and B. L. Turner, eds.), pp. 385–409. Academic Press, New York.
Spec. Period. Rep.: Biosynthesis (1972–present). Chemical Society, London.
Stebbins, G. L. (1974). "Flowering Plants. Evolution above the Species Level." Belknap Press, Cambridge, Massachusetts.
Stermitz, F. R., and Yost, G. S. (1978). *In* "Effects of Poisonous Plants on Livestock" (R. F. Keeler, K. R. Van Kampen, and L. F. James, eds.), pp. 371–378. Academic Press, New York.
Swain, T. (1976). *In* "Secondary Metabolism and Coevolution" (M. Luckner, K. Mothes, and L. Nover, eds.), pp. 411–421. Dtsch. Akad. Naturf. Leopold., Halle.
Swain, T., and Cooper-Driver, G. (1973). *J. Linn. Soc. Bot.* **67**, Suppl. 111–134.
Takhtajan, A. (1966). "Systema et Phylogenia Magnoliphytorum." Soviet Sciences Press, Moscow and Leningrad.
Tapper, B. A., and Reay, P. F. (1973). "The Chemistry and Biochemistry of Herbage" (G. W. Butler and R. W. Bailey, eds.), Vol. 1, pp. 447–476. Academic Press, New York.
Tétényi, P. T. (1970). "Infraspecific Chemical Taxa of Medicinal Plants." Chem. Publ. Co., New York.
Tétényi, P. T. (1974). *In* "Chemistry and Botanical Classification" (G. Bendz and J. Santesson, eds.), pp. 67–78. Academic Press, New York.
Thorne, R. F. (1974). *Aliso* **8**, 147–209.
Thorne, R. F. (1976). *Evol. Biol.* **9**, 35–106.
Turner, B. L. (1967). *Pure Appl. Chem.* **14**, 189–213.
Turner, B. L. (1969). *Taxon* **18**, 134–151.
Turner, B. L. (1970a). *Taxon* **20**, 123–130.

Turner, B. L. (1970b). *In* "Phytochemical Phylogeny" (J. B. Harborne, ed.), p. 187. Academic Press, New York.

Van Emden, H. F., ed. (1973). "Insect-Plant Relationships." Wiley, New York.

van Valen, F. (1978a). *Planta Med.* **34**, 408–413.

van Valen, F. (1978b). *Proc. K. Ned. Akad. Wet., Ser. C* **81**, 198–203.

van Valen, F. (1978c). *Proc. K. Ned. Akad. Wet., Ser. C* **81**, 492–499.

van Valen, F. (1979). *Planta Med.* **35**, 141–149.

Vaughan, J. G., MacLeod, A. J., and Jones, B. M. G., eds. (1976). "The Biology and Chemistry of the Cruciferae." Academic Press, New York.

von Rudloff, E. (1969). *In* "Recent Advances in Phytochemistry" (M. K. Seikel and V. C. Runeckles, eds.), pp. 127–162. Appleton, New York.

von Rudloff, E. (1972). *Can. J. Bot.* **50**, 1025–1040.

von Rudloff, E. (1975a). *Biochem. Syst. Ecol.* **2**, 131–167.

von Rudloff, E. (1975b). *Phytochemistry* **14**, 1319–1329.

Wallace, J. W., and Markham, K. R. (1978). *Am. J. Bot.* **65**, 965–969.

Went, F. W. (1971). *Taxon* **20**, 197–226.

Whiting, M. G. (1963). *Econ. Bot.* **17**, 271–302.

Whittaker, R. H., and Feeny, P. P. (1971). *Science* **171**, 757–770.

Williams, C. A. (1978). *Phytochemistry* **17**, 729–734.

Wollenweber, E. (1978). *Am. Fern J.* **68**, 13–28.

Yang, M. G., Kobayashi, A., and Mickelsen, O. (1972). *Fed. Proc., Fed. Am. Soc. Exp. Biol.* **31**, 1543–1546.

Yoshioka, H., Mabry, T. J., and Timmerman, B. N. (1973). "Sesquiterpene Lactones." Univ. of Tokyo Press, Tokyo.

Young, D. A. (1979). *Am. J. Bot.* **66**, 502–510.

Young, D. A. (1981). *In* "Phytochemistry and Angiosperm Phylogeny" (D. A. Young and D. S. Seigler, eds.). Praeger, New York (in press).

Young, D. A., and Seigler, D. S. (1981). *In* "Handbook Series of Biosolar Resources" (O. R. Zaborsky, ed.-in chief), Vol. 1. C. R. C. Press, West Palm Beach, Florida (in press).

Zavarin, E. (1968). *Bull. Int. Assoc. Wood Anat.* pp. 3–12.

Zavarin, E., and Snajberk, K. (1975). *Biochem. Syst. Ecol.* **2**, 121–129.

Stereochemical Aspects of Natural Products Biosynthesis

7

HEINZ G. FLOSS

I. INTRODUCTION: EVIDENT VS CRYPTIC STEREOCHEMISTRY

One of the characteristic features of enzymes and of biological systems in general is the stereospecificity they exhibit. Such stereospecificity manifests itself in many ways, for example, in the fact that in a reaction generating a chiral center only one enantiomer is usually produced, that in a reaction involving a chiral substrate only one enantiomer is bound to the enzyme, and that the conversion of one chiral center into another occurs in a particular stereochemical reaction mode, i.e., inversion or retention. The stereospecificity exhibited by enzymes is a consequence of the fact that an enzyme, in order to achieve the extraordinary rate enhancement and the high sub-

The Biochemistry of Plants, Vol. 7

strate specificity characteristic of the catalytic process, must orient the reaction partners very specifically within a relatively rigid chiral matrix. The importance of stereochemistry in biological systems is impressively documented by two monographs published a few years ago (Bentley, 1969, 1970; Alworth, 1972), which summarized the state of knowledge in this field at that time.

Many features of enzyme stereospecificity are due to chiral recognition, i.e., the fact that an enzyme, being chiral itself, can recognize stereochemical differences between other molecules or within other molecules and use them to discriminate between stereochemically nonequivalent species or between stereochemically nonequivalent groups within a given molecule. For example, the fact that muscle lactate dehydrogenase converts L-lactate, but not the D isomer, into pyruvate is a clear case of chiral recognition. The enzyme in such cases either cannot bind the "wrong" stereoisomer or it binds it in a manner that does not allow reaction to occur. There are other stereochemical features of enzyme reactions that are not due to chiral recognition. For example, it is found that in mixed-function oxygenase-catalyzed hydroxylations at saturated carbon atoms the hydrogen is replaced by the hydroxyl group in a retention mode. This steric course is seen regardless of whether the reaction takes place at a (chiral) methine carbon, in which case chiral recognition could be invoked, at a methylene group, in which the two hydrogens may still be distinct, or at a methyl group, in which there is no longer any intrinsic difference between the three hydrogens that could form the basis for chiral recognition.

Obviously, the stereospecificity exhibited by enzymes is not limited to the reactions of chiral molecules for which the steric course of the transformation is evident from the configurations of substrate and product. In fact, the majority of biochemical reactions take place at centers that are not chiral; yet these reactions are usually still stereospecific. However, their steric course is not evident to the observer, and they are called "stereochemically cryptic" (Hanson and Rose, 1975). The steric course of such stereochemically cryptic reactions can be deduced by stereospecific isotope labeling, which converts the reaction center into a chiral center of discernible stereochemical behavior, making the generally valid assumption that the isotopic substitution does not alter the reaction stereospecificity. The discussion in this chapter will focus mainly on this approach which has been practiced extensively and with increasing sophistication during the past 30 years.

II. CONCEPTS AND DEFINITIONS

The properties of a chiral center are well known. Its presence in a compound results in the existence of two enantiomers having identical physical properties except for their interaction with polarized light. Combination with another chiral center gives rise to two pairs of diastereomers which differ in physical properties. The configuration at a chiral center is designated R or S on the basis of the sequence rules developed by Cahn *et al.* (1966) and tentatively adopted as official nomenclature by IUPAC (International Union of Pure and Applied Chemistry, 1970). The stereochemical course of reactions at chiral centers can generally be analyzed by classical spectroscopic methods.

Replacement of one of the four ligands at a chiral center Cabcd by one already present generates a prochiral center Caabd. The term "prochirality" was coined by K. R. Hanson (1966) and describes in general the property of an achiral object having a pair of features that can be distinguished only by reference to a chiral object or to a chiral reference frame. The two chemically like groups at a prochiral center are not identical because they differ in the chirality of their environment. In a tetrahedral prochiral center Caabd (Fig. 1A), for example, the sequence a → b → d appears clockwise when viewed from one group a, but counterclockwise when viewed from the other group a. The two groups a are thus intrinsically different because they occupy different, non-superimposable positions in space; they are therefore called "heterotopic" (Mislow and Raban, 1967) or "stereoheterotopic" (Hirschmann and Hanson, 1971). They are called "enantiotopic" if their individual replacement by a group not yet present gives rise to a pair of enantiomers. If the prochiral center is combined with a chiral center, the two heterotopic groups are called "diastereotopic," because their individual replacement by a group not yet present gives rise to diastereomers. Diastereotopic groups can have intrinsically different reactivities because they can react through different transition states. Likewise, diastereotopic protons can give distinct nmr signals. Enantiotopic groups can react differently in reactions involving chiral reaction partners or catalysts or a chiral environment, and enantiotopic protons

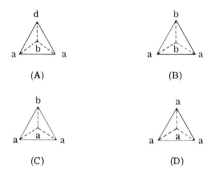

Fig. 1. (A) A prochiral center Caabd. (B) A pro-prochiral center Caabb. (C) A pro-prochiral center Caaab. (D) A pro-pro-prochiral center Caaaa.

are distinguishable by nmr when observed in a suitable chiral environment, e.g., in the presence of a chiral shift reagent. An enzyme may be viewed as a perfect asymmetric catalyst, and thus for either case an enzyme reaction to a prochiral center usually involves only one heterotopic group to the complete exclusion of the other.

The two heterotopic ligands at a prochiral center are designated "*pro-R*" and "*pro-S*" depending on whether giving the group to be named preference over the other produces a center of R or S chirality. Thus the two hydrogens at C-1 of ethanol are designated the *pro-R* (H_R) and the *pro-S* (H_S) hydrogens as indicated:

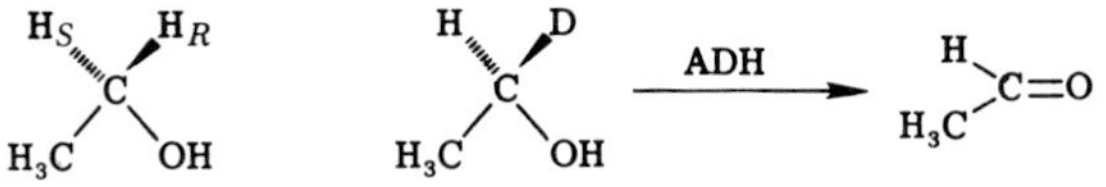

(R)-[1-D]-Ethanol

It is important to remember that isotopic substitution of heterotopic groups at the prochiral center (e.g., of H_R by deuterium) produces a genuinely chiral center. The difference between the prochiral and the corresponding stereospecifically labeled chiral species is reflected in the terminology used to describe their behavior. When stereospecifically deuterated ethanol is used to determine the steric course of the alcohol dehydrogenase (ADH) reaction, the results can be stated as indicating that acetaldehyde is formed with loss of deuterium from the R isomer and with retention of deuterium from the S isomer of [1-D]ethanol. In general terms, the reaction may be stated to involve loss of the *pro-R* hydrogen and retention of the *pro-S* hydrogen from C-1 of ethanol. It may also be described as proceeding with loss and retention, respectively, of the label from ethanol deuterated at C-1 in the *pro-R* and *pro-S* positions, respectively. However, it would be incorrect to speak of loss or retention of the *pro-R* or *pro-S* deuterium and loss of D_R and retention of D_S (unless one discusses the behavior of the species CH_3CD_2OH).

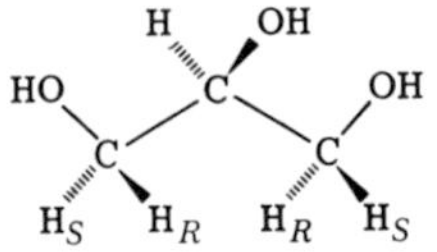

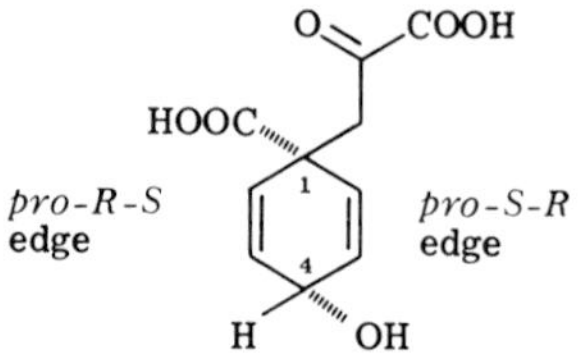

Heterotopic ligands may themselves again contain prochiral centers as in citric acid and in glycerol, in which all four methylene hydrogens are distinct. In such a situation it is necessary to specify not only H_R or H_S but also in which heterotopic ligand the hydrogen is located, for example, by referring to "the *pro-R* hydrogen of the *pro-R* CH_2OH group or H_R of the $(CH_2OH)_R$ group." Heterotopic ligands may be edges of a ring system as, for example, in prephenic acid. The "left" edge is designated the *pro-R* (relative to C-1)-*S* (relative to C-4) edge and the other one as the *pro-S-R* edge.

Isotopic substitution seldom changes the configurational designations in a molecule, because most simple molecules are fully characterized by application of subrule 1 of Cahn *et al.* (1966), which states that sequence rule priorities are established based on atomic number. Only when subrule 1 has been applied exhaustively are further distinctions made on the basis of subrule 2, which says that higher mass numbers have priority over lower mass numbers. Thus in an ^{18}O-labeled thiophosphate the priority sequence is

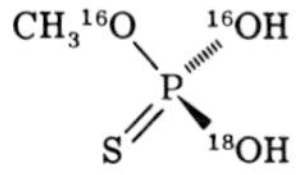

$$CH_3{}^{16}O \quad {}^{16}OH$$
$$\diagdown \underset{S}{\overset{\diagup}{P}} \overset{}{\diagdown} {}^{18}OH$$

(*R*)-Methyl-[^{18}O]-
thiophosphate

$S > CH_3{}^{16}O > {}^{18}OH > {}^{16}OH$, *not* $S > {}^{18}OH > CH_3{}^{16}O > {}^{16}OH$. In 2,4,5-tri-hydroxypentanoic acid the designation of the heterotopic hydrogens at C-3 does not change when C-4 is labeled with ^{13}C:

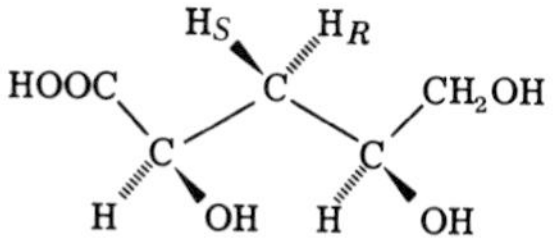

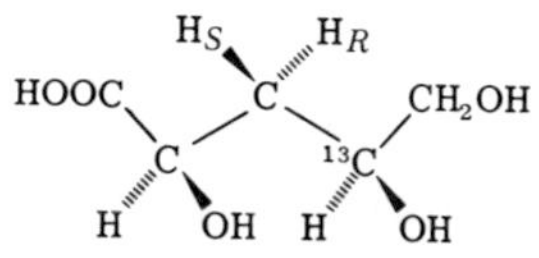

However, in the corresponding dicarboxylic acid, 2,4-dihydroxyglutaric acid, the situation is different because the configurational designations in the unlabeled compound in this case are based on a lower-ranking subrule (*R* chirality precedes *S* in enantiomeric ligands). Since with the labeled species

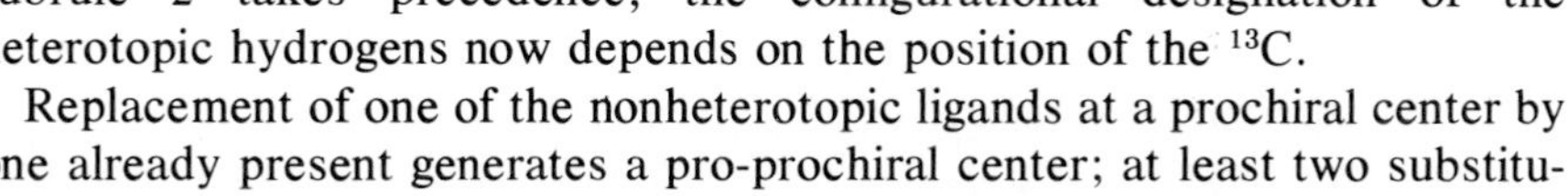

subrule 2 takes precedence, the configurational designation of the heterotopic hydrogens now depends on the position of the ^{13}C.

Replacement of one of the nonheterotopic ligands at a prochiral center by one already present generates a pro-prochiral center; at least two substitution steps are then necessary to convert the system into a chiral center. As shown in Fig. 1, two types of pro-prochiral centers exist in a tetrahedral system, one of type Caabb (Fig. 1B) containing two pairs of like substituents, and the other of type Caaab (Fig. 1C) containing three like and one different substituent. The simplest examples of biologically important Caaab systems are a methyl group attached to any other ligand different from hydrogen and the phosphate in a phosphate monoester. One of the simplest cases of a Caabb system is malonic acid. In either system the chemically like substitents are *not* heterotopic; there is no intrinsic difference between them because their positions in space can be interchanged by rotations of the system. For this reason, for example, the three hydrogens of a methyl group are completely equivalent and indistinguishable even to an enzyme. Despite the equivalence of the like substituents in a pro-prochiral center, however, the principle still applies that such a system can be converted into a true chiral center of observable stereochemical behavior by, in this case two, isotopic substitutions. Thus, by the introduction of the isotopes deuterium and tritium into a methyl group (Cornforth *et al.*, 1969; Lüthy *et al.*, 1969, cf. Floss and Tsai, 1979) or ^{17}O and ^{18}O into the phosphate group of a phosphate ester (Abbott *et al.*, 1978; Cullis and Lowe, 1978), two enantiomeric species can be generated:

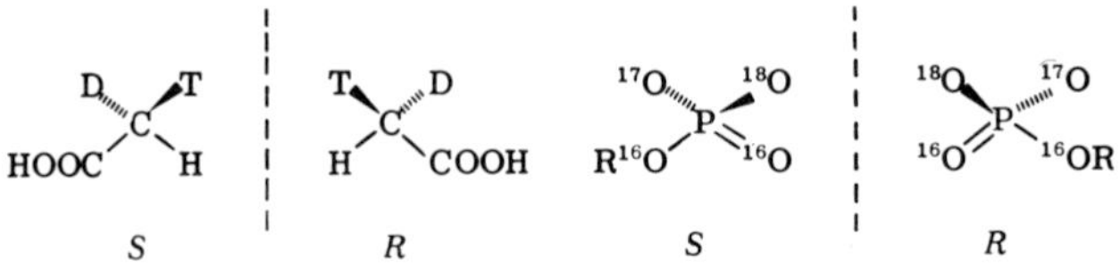

Alternatively, the stereochemistry of reactions involving a phosphate monoester group is frequently analyzed using the corresponding thiophosphate analogue (Eckstein, 1975, 1979), which of course simplifies the problem to one of analyzing a stereospecifically labeled prochiral center:

No example of a stereospecifically isotope-labeled pro-prochiral center of the Caabb type has been reported so far. However, Sedgwick *et al.* (1977) have prepared stereospecifically tritiated malonyl-CoA and studied its stereochemical fate in fatty acid biosynthesis; based on their work it is now possible to distinguish the R from the S isomer of a stereospecifically isotope-labeled sample of malonic acid.

Obviously, one further replacement of a different by an already present ligand is possible in a Caaab system, generating a pro-pro-prochiral center Caaaa. The biologically most important example of this system is inorganic phosphate. Again, the four ligands in this system are not heterotopic and are therefore indistinguishable; however, again, in principle a chiral center can be generated from such a system by, in this case three, isotopic substitutions. However, this possibility has not yet been realized, not the least because of the lack of a sufficient number of isotopic species of, for example, oxygen or hydrogen (See Note Added in Proof, p. 214).

Enzymes also exhibit stereospecificity in distinguishing between the two faces of a trigonal atom Cabx, most frequently encountered as part of a double-bond system. One of the classical examples of this of course is the reduction of pyridine nucleotides by different dehydrogenases, which occurs from the A side with some enzymes and from the B side with others:

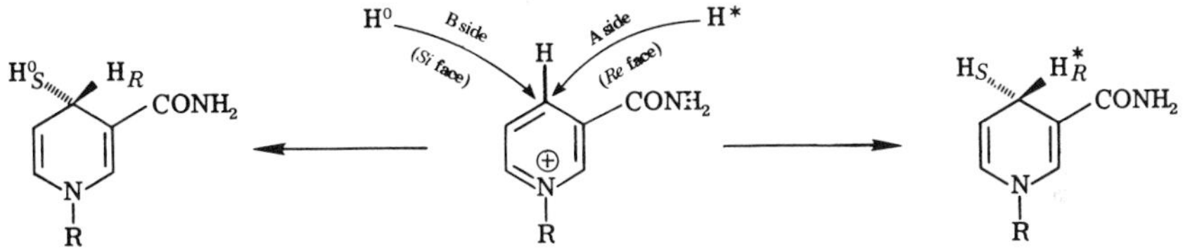

Other frequently encountered examples are reductions, or other addition reactions, at unsymmetrically substituted carbonyl groups and additions to carbon–carbon double bonds. If such a trigonal atom Cabx is part of an unsymmetrically 1,2-disubstituted double-bond system, an enzyme usually will not only distinguish between the E and Z isomers (for nomenclature, see Blackwood *et al.*, 1968), these are different compounds, but also between the two sides of the double bond in the isomer that is bound. The distinction between the two faces of an atom Cabx is of course maintained even if x is not another unsymmetrically substituted trigonal atom as, for example, the carbonyl carbon of acetaldehyde. Enzymatic reduction usually still occurs exclusively from one side only. On the other hand, the two faces of a trigonal system Caax are identical and indistinguishable if x is, for example, a carbonyl oxygen, as in formaldehyde or acetone, or another symmetrically substituted trigonal atom as, for example, in ethylene. The two faces of Caax are distinct, however, if x is an unsymmetrically substituted trigonal atom Cabx. Thus the two faces of the methylene group of phosphoenolpyruvate (PEP) are clearly different and are distinguished by enzymes. Of course many of the reactions of such trigonal systems are again stereochemically cryptic, but their steric course can again be revealed by stereospecific isotope labeling.

Convenient nomenclature has been devised by Hanson (1966) to designate the faces of trigonal atoms, based on the Cahn–Ingold–Prelog sequence rule priorities. Double bonds are saturated by replica atoms, as specified by the sequence rules, and the order of priority of the three ligands at the trigonal atom to be analyzed is then determined (ignoring the replica atom at this center). If the sequence from highest to lower to lowest priority is clockwise, the face is called the *Re* face (from *rectus,* to the right). The opposite face is called the *Si* face (from *sinister,* to the left); e.g.,

$$\underset{H}{\overset{H_3C}{>}}C=O \qquad \underset{H_3C}{\overset{H}{>}}C=O$$

Si face   Re face

The reduction of acetaldehyde by ADH, in which the hydride approaches the *Re* face, can be said to involve *Re* attack at the carbonyl carbon of acetaldehyde. When two trigonal atoms Cabx are joined together in a double bond, the face at each of them is specified separately. The faces of the double bond are then designated the *Re-Re* and *Si-Si* or the *Re-Si* and *Si-Re* faces:

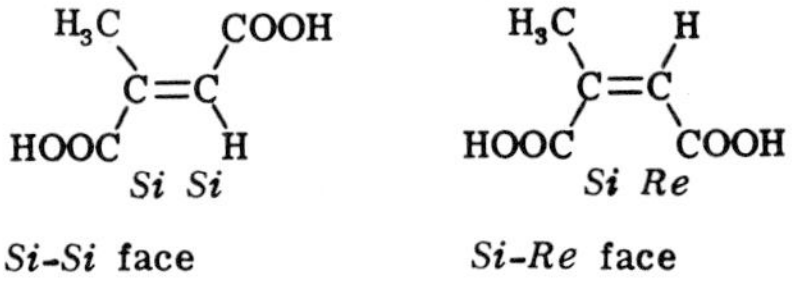

$$\underset{HOOC}{\overset{H_3C}{>}}C=C\underset{H}{\overset{COOH}{<}} \qquad \underset{HOOC}{\overset{H_3C}{>}}C=C\underset{COOH}{\overset{H}{<}}$$

Si Si   Si Re

Si–Si face   Si–Re face

If necessary, numbers are used to indicate which carbons the designations refer to. If a symmetrically substituted atom Caax is attached to an unsymmetrically substituted trigonal carbon Cabx in a double bond, it is defined that the designations of the faces at Caax correspond to those at the attached Cabx. Thus, for example, in PEP, the *Re* face at C-2 is also the *Re* face at C-3:

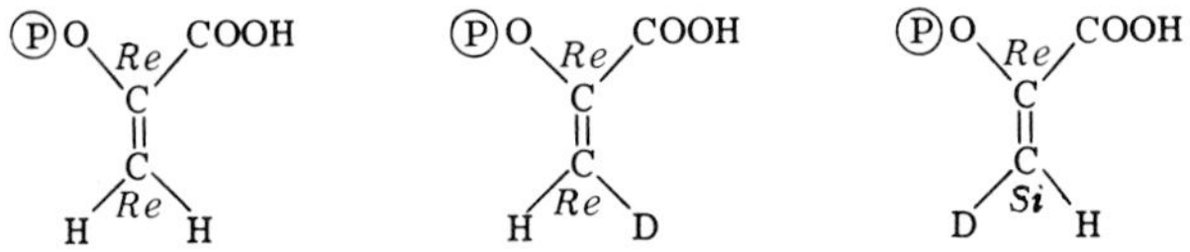

$$\underset{H}{\overset{\textcircled{P}O}{}}\underset{Re}{\overset{Re}{\underset{\parallel}{C}}}\overset{COOH}{\underset{H}{C}} \qquad \underset{H}{\overset{\textcircled{P}O}{}}\underset{Re}{\overset{Re}{\underset{\parallel}{C}}}\overset{COOH}{\underset{D}{C}} \qquad \underset{D}{\overset{\textcircled{P}O}{}}\underset{Si}{\overset{Re}{\underset{\parallel}{C}}}\overset{COOH}{\underset{H}{C}}$$

It is important to note that in this situation the designation of the faces may change upon introduction of an isotopic label at Caax because the faces at the formerly symmetrically substituted trigonal atom can now be designated independently. Thus the $3Re$ face in PEP is still $3Re$ in (E)-[3-D_1]phosphoenolpyruvate but becomes the $3Si$ face in the corresponding Z isomer.

III. EXPERIMENTAL APPROACHES

The task of unraveling the steric course of stereochemically cryptic reactions usually involves one or both of two unique steps, in addition to, obviously, the conversion of substrate into product with concomitant monitoring of the retention or loss of isotopic label. These are (1) the synthesis of stereospecifically isotope-labeled substrates of known absolute configuration at the labeled center and (2) the analysis of stereospecifically labeled enzyme reaction products to determine their configuration at the labeled center(s). The major principal approaches used to accomplish these tasks will be discussed in the following. The emphasis will be heavily on examples from studies with isotopes of hydrogen, since these make up the majority of the work carried out so far in this field.

A. Stereospecific Labeling

The preparation of stereospecifically labeled compounds may involve either suitable enzyme reactions or purely chemical reactions to introduce an isotope in a stereospecific manner. Not infrequently the initial stereospecific labeling is done enzymatically at the stage of a precursor, which is then converted into the desired substrate in a series of, usually chemical, reactions that do not destroy the chirality at the labeled center.

1. Stereospecific Labeling by Chemical Synthesis

Three main principles can be used in devising chemical syntheses of stereospecifically isotope-labeled compounds, and each of these will be illustrated by examples.

a. Simultaneous Generation of a Chiral and a Stereospecifically Labeled Achiral Center. This approach involves addition reactions of known steric course (syn or anti) to double bonds of known geometry, resulting in the simultaneous formation of two chiral centers, one of them the stereospecifically labeled analogue of an achiral center. In the process a racemate is formed in which the configurations of the two chiral centers are strictly correlated to each other. Resolution into the enantiomers based on the classical chiral center then also resolves the two enantiomeric forms of the stereospecifically labeled achiral center. An example is the general method for the preparation of amino acids tritiated or deuterated stereospecifically in the β position, which was developed in the laboratories of Battersby, Hanson and Kirby (Wightman *et al.,* 1972; Kirby and Michael, 1973):

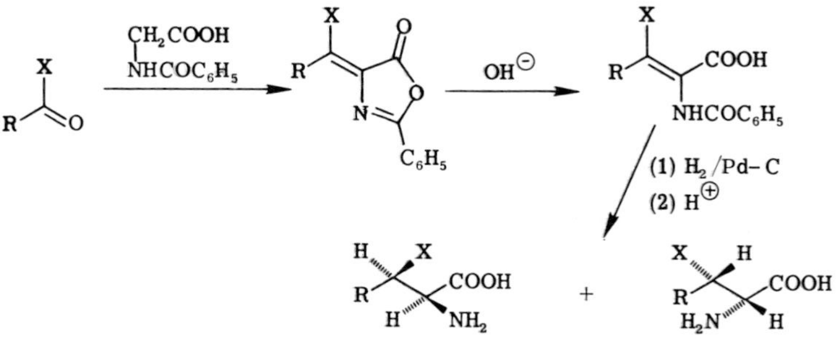

X = D or T
R = alkyl, phenyl, *p*-hydroxyphenyl, β-indolyl, or 4-imidazolyl
Resolve into 2*R* and 2*S* isomer, then racemize each at C-2

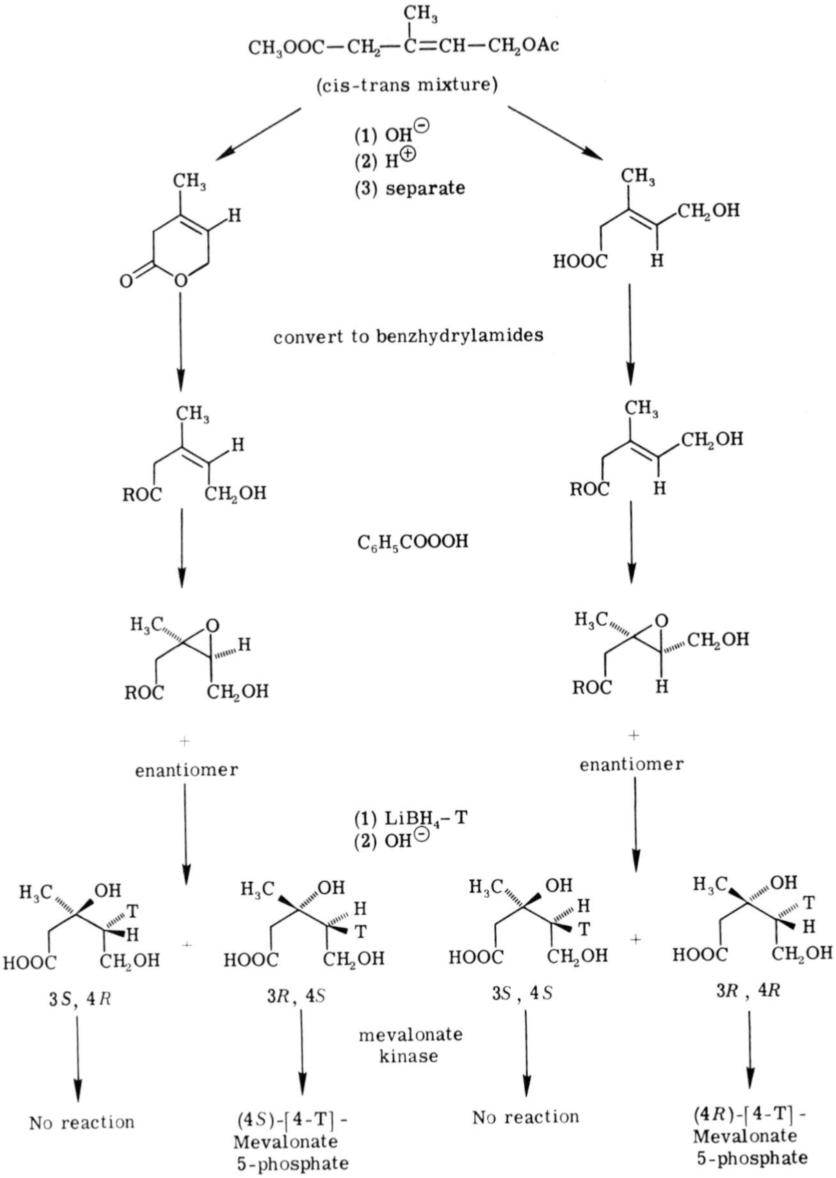

Fig. 2. Synthesis of mevalonates stereospecifically tritiated at C-4.

These stereospecifically β-deuterated or tritiated amino acids have been widely used to study the stereochemistry of plant enzymes and plant natural products biosynthesis.

The optical resolution step can be omitted in many cases, when it is known that the biological system will utilize only one of the enantiomers at the classical chiral center. This is the case, for example, with mevalonic acid, since the enzyme mevalonate kinase in every system examined shows complete specificity for the $3R$ isomer. Hence the two diasteromeric racemates of [4-T]mevalonate synthesized by Cornforth *et al.* (1966b) (Fig. 2) can serve as sources of $(3R,4R)$- and $(3R,4S)$-[4-T]mevalonate, respectively, in studies on isoprenoid biosynthesis.

The classical chiral center generated in such a sequence may serve only as an auxiliary center, allowing separation of the enantiomers, following which it may be destroyed in the further course of the reaction sequence as, for example, in a synthesis of chiral acetic acid in Cornforth's laboratory (Lenz *et al.*, 1971):

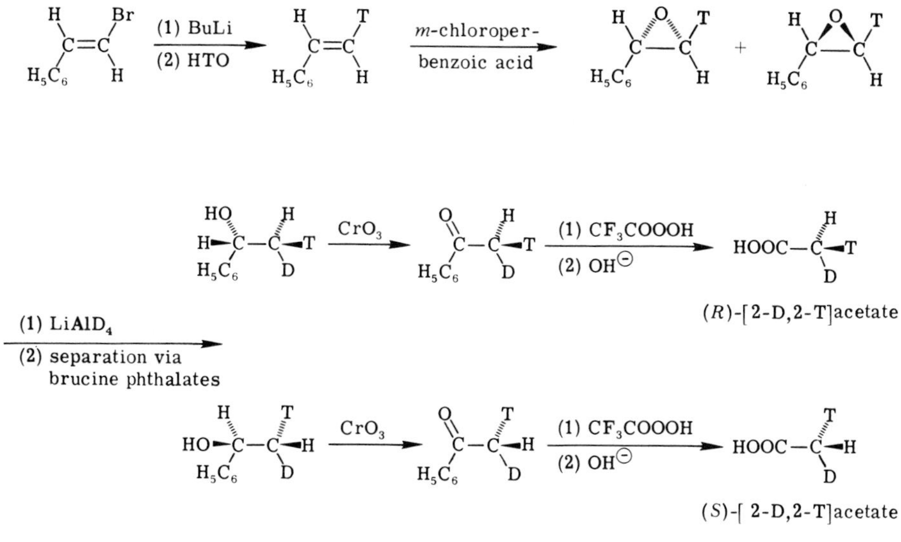

b. Conversion of a Chiral into a Stereospecifically Labeled Achiral Center.

In this approach the two enantiomers of a chiral compound are separately converted into the two enantiomers of a stereospecifically labeled achiral compound. A simple example is the synthesis of (R)- and (S)-[9-T]stearate from naturally occurring (R)-9-hydroxy-Δ^{12}-octadecenoate (Schroepfer and Bloch, 1965):

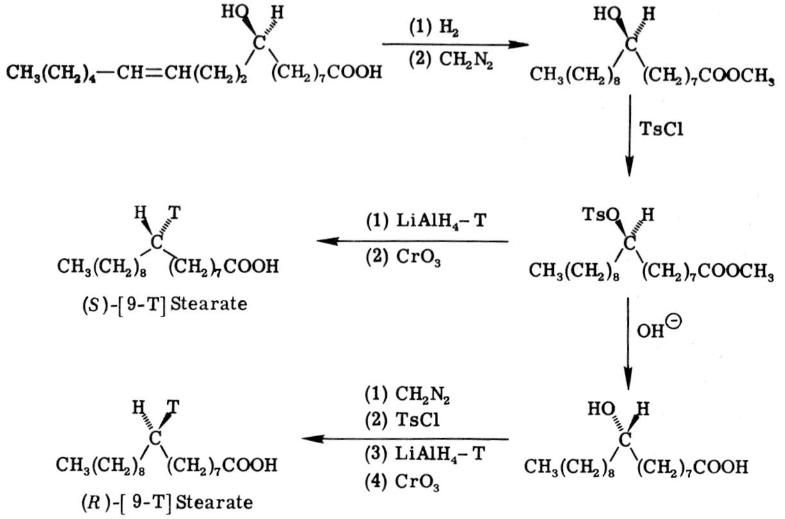

CH₃(CH₂)₄—CH=CH(CH₂)₂ ... (CH₂)₇COOH   →(1) H₂ / (2) CH₂N₂→   ... CH₃(CH₂)₈ ... (CH₂)₇COOCH₃

(with HO, H at chiral center)

| TsCl

(S)-[9-T] Stearate: CH₃(CH₂)₈ ... (CH₂)₇COOH (H, T at chiral center)   ←(1) LiAlH₄–T / (2) CrO₃—   TsO, H CH₃(CH₂)₈ ... (CH₂)₇COOCH₃

| OH⁻

(R)-[9-T] Stearate: CH₃(CH₂)₈ ... (CH₂)₇COOH (H, T at chiral center)   ←(1) CH₂N₂ / (2) TsCl / (3) LiAlH₄–T / (4) CrO₃—   HO, H CH₃(CH₂)₈ ... (CH₂)₇COOH

A variation of the same principle has been used in a number of syntheses of compounds labeled stereospecifically in one of two stereoheterotopic methyl groups, e.g., in a synthesis of isobutyric acid published by Aberhart (1975). The synthesis starts from optically active (2S,3R)-*trans*-2,3-epoxybutyric acid:

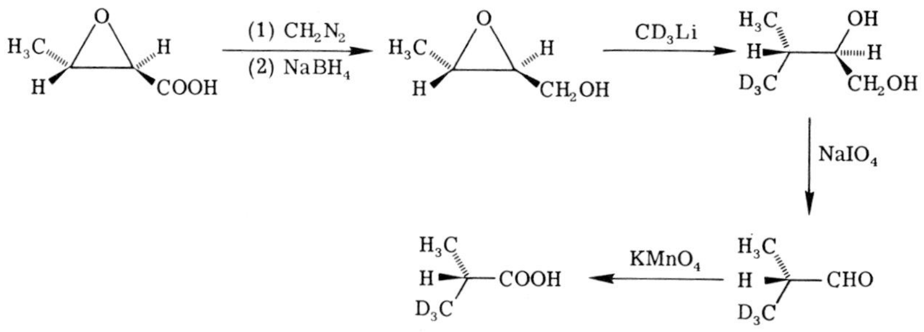

H₃C, H, H, COOH (epoxide)   →(1) CH₂N₂ / (2) NaBH₄→   H₃C, H, H, CH₂OH (epoxide)   →CD₃Li→   H₃C, OH, H, D₃C, CH₂OH

| NaIO₄

H₃C, H, D₃C —COOH   ←KMnO₄—   H₃C, H, D₃C —CHO

Obviously, in this case the ring-opening reaction with methyl lithium could have also been carried out on the racemic *trans*-epoxide, followed by separation of the enantiomers at the diol stage.

The principle of successive conversion of substituents at a chiral center into isotopic homologues of ones already present also forms the basis of recently published syntheses of chiral phosphate monoesters, such as the synthesis of [1(R)-¹⁶O,¹⁷O,¹⁸O]phospho-(S)-propane-1,2-diol in Knowles' laboratory (Abbott et al., 1978):

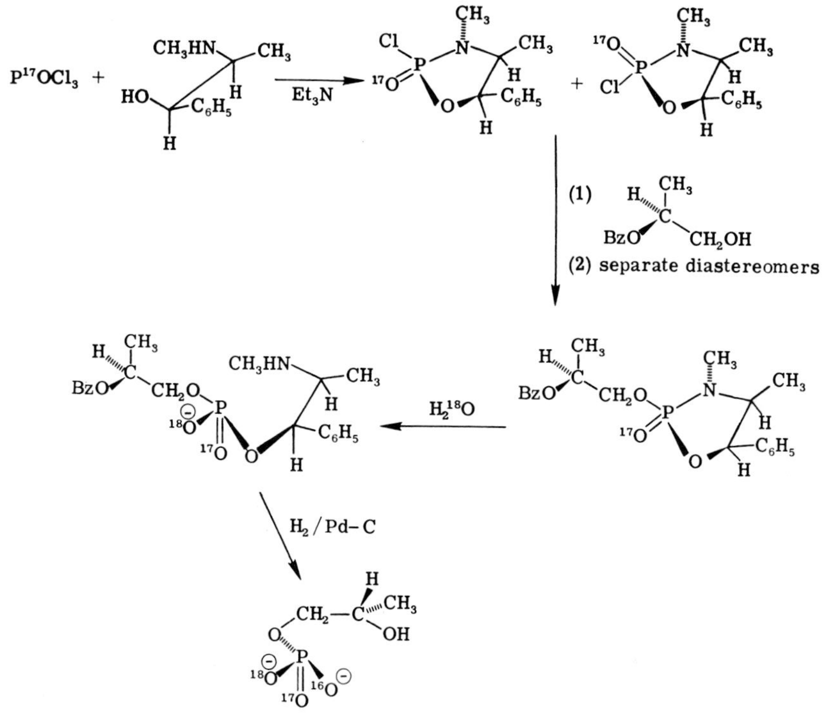

Another sophisticated application of the same general principle utilizes reactions involving chirality transfer. The synthesis of chiral acetate reported by Townsend *et al.* (1975) is a particularly elegant example:

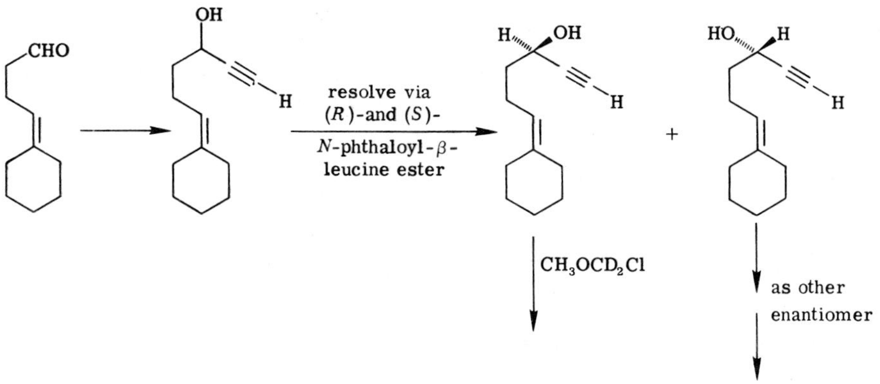

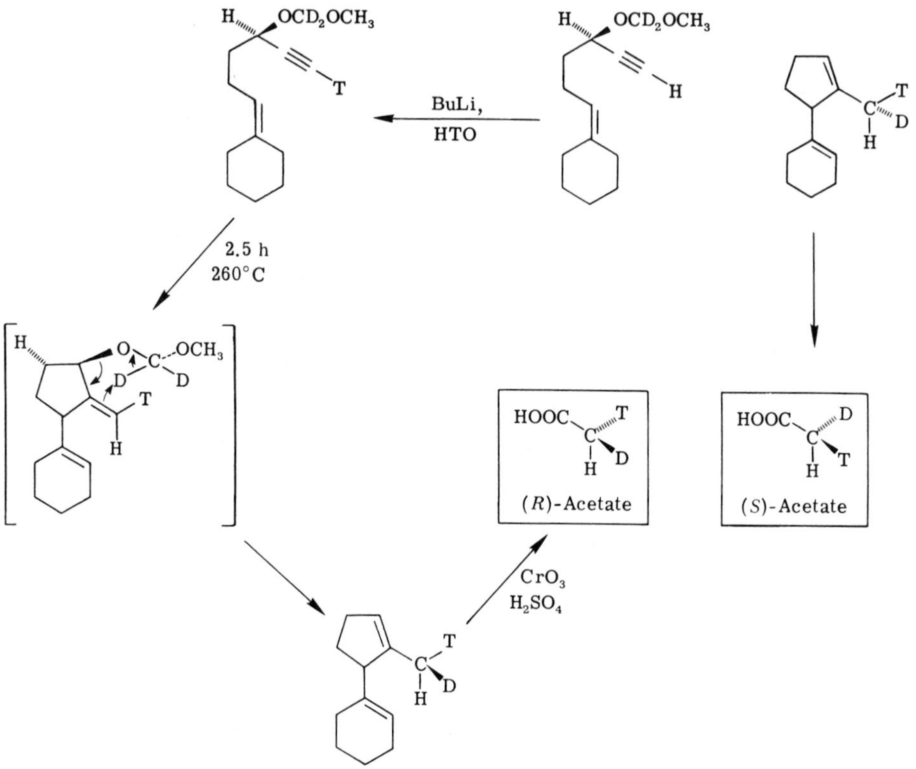

c. **Asymmetric Synthesis.** In this case one enantiomeric form of a stereo-specifically labeled chiral center is generated in great excess over the other either through asymmetric induction by one or more chiral centers present in the molecule or by using a chiral reagent. Asymmetric induction is particularly effective in conformationally rigid chiral molecules, as evidenced, for example, in the reduction of cholesta-5,7-dien-3β-ol with tritiated diimine, which results in the introduction of tritium exclusively at the 5α and 6α positions (Paliokas and Schroepfer, 1968):

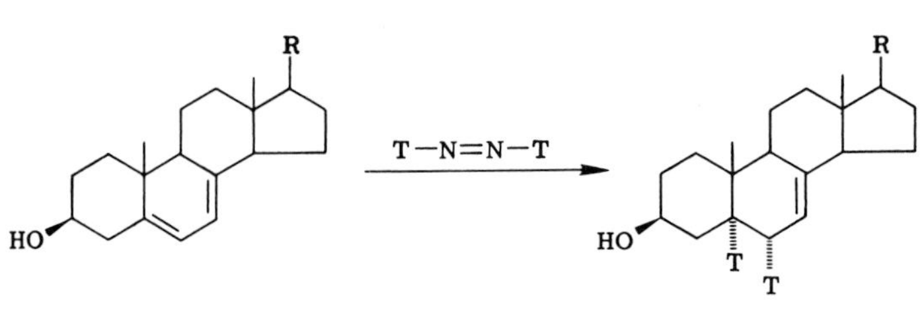

Considerable progress has been made in recent years in the development of asymmetric catalysts and chiral reagents for asymmetric syntheses, particularly asymmetric reductions of carbon–carbon and carbon–oxygen double bonds. For example, catalytic hydrogenations of α-acylamino acids with rhodium catalysts containing chiral phosphine ligands have been found to give as high as 74–100% enantiomeric excess (Vineyard *et al.*, 1977; Frydzuk and Bosnich, 1977). A particularly useful reagent for the reduction of aldehydes to give stereospecifically deuterated primary alcohols of high enantiomeric purity is Midland's reagent, B-3-pinanyl-9-borabicyclo[3.3.1]nonane, which is conveniently prepared from 9-BBN and (+)- or (−)-α-pinene (Midland *et al.*, 1979). With the reagent prepared from (+)-α-pinene, for example, practically enantiomerically pure (S)-[1-D]benzyl alcohol is obtained by the reduction of [1-D]benzaldehyde. This method of introducing a stereospecific isotope label was used in a recently published synthesis of (5′R)-[5′-D_1]adenosine (Parry, 1978). The synthesis starts with available [5′-D_2]2′-3′-isopropylideneadenosine.

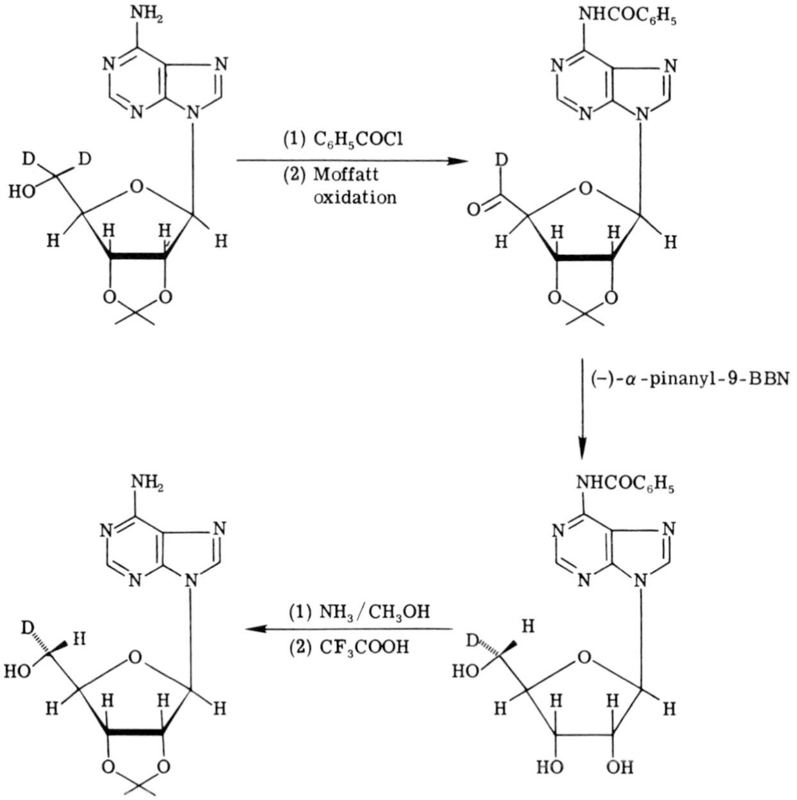

2. Enzymatic Synthesis of Stereospecifically Labeled Compounds

A multitude of enzymes can be used to introduce stereospecific isotopic labels into a variety of compounds of biochemical interest. These enzymes represent a number of different reaction types. The reduction of carbonyl groups in aldehydes by pyridine nucleotide-dependent dehydrogenases is a prominent type of reaction, and ADH is a particularly useful enzyme for this purpose (Loewus *et al.*, 1953). Both the yeast and the horse liver enzyme can be used, but the latter generally has a much broader substrate specificity. Both enzymes, with any substrate tested, catalyze transfer of the *pro-4R* hydrogen of NADH to the *Re* face of the aldehyde carbonyl group to give a primary alcohol carrying the newly introduced hydrogen in the *pro-R* position.

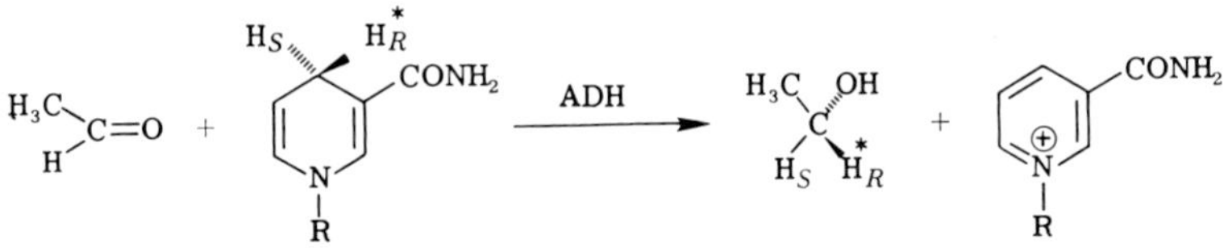

Labeling of alcohols with this enzyme can be carried out in a number of ways. One is the reduction of a labeled or unlabeled aldehyde with a stoichiometric quantity of unlabeled or labeled NADH, respectively. Alternatively, a catalytic amount of cofactor may be used in the presence of a regenerating system. When regenerating labeled NADH, it is not necessary for the regenerating system to have the same stereospecificity with respect to C-4 of the cofactor as ADH, provided the substrate/cofactor ratio is very large. If a higher-boiling alcohol is labeled, the regenerating system can also be excess ethanol (unlabeled or labeled) itself and ADH (Battersby *et al.*, 1976) or, better yet, cyclohexanol and ADH (Battersby *et al.*, 1975). In a further step, the reaction can also be carried out as an equilibration between two alcohols, the one to be labeled (e.g., 3-methylbuten-1-ol) and a cheap auxiliary alcohol present in excess, which can be easily removed (e.g., ethanol) (H. Hsu, J. M. Cassady, and H. G. Floss, unpublished). Finally, Günther *et al.* (1973a,b) have devised an exchange procedure for equilibrating the *pro-1R* hydrogen of alcohols with water protons in the presence of ADH, diaphorase, and catalytic quantities of NAD and NADH.

Addition reactions to double bonds can be used to label prochiral centers stereospecifically, as, for example, in the formation of L-malate from fumarate catalyzed by fumarase:

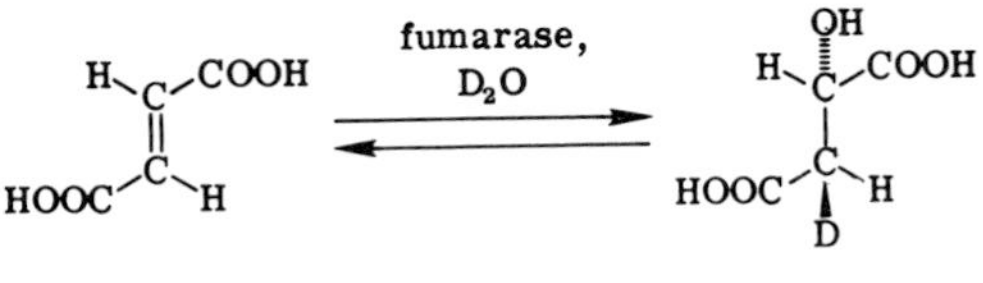

Decarboxylations (e.g., of amino acids or other carboxylic acids) usually result in stereospecific incorporation of a solvent proton into the product:

```
    COOH                                          COOH
     |                                             |
  H—C◄—OH              isocitrate                 C=O
     |              ───────────────►              |
  H—C""""COOH        dehydrogenase,          H—C"""T
     |                   HTO                       |
    CH₂                                           CH₂
     |                                             |
    COOH                                          COOH
```

Amino acid α-decarboxylations catalyzed by pyridoxal phosphate enzymes invariably seem to proceed in a retention mode; however, with various other decarboxylases both inversion and retention modes have been observed. Isomerases can be used for stereospecific labeling, e.g.,

```
   T—C=O                                        H
      |                                         ⫶
   H—C—OH                                   T—C—O
      |              glucose-6-                 |
  HO—C—H             phosphate                 C=O
      |             ─────────────►         HO—C—H
   H—C—OH            isomerase                  |
      |                                     H—C—OH
   H—C—OH                                       |
      |                                     H—C—OH
   H—C—O (P)                                    |
      |                                     H—C—O (P)
      H                                         |
                                                H

  [1-T] Glucose-                          (1S)-[1-T] Fructose-
   6-phosphate                               6-phosphate
```

In many cases such reactions can be carried out as exchange processes in which either the migrating hydrogen (e.g., glucose-6-phosphate isomerase, triose phosphate isomerase) or the hydrogen added or abstracted in the overall reaction (e.g., isocitrate dehydrogenase, aldolase, L-alanine transaminase) is equilibrated with solvent protons:

```
                                                         H
                         triose-phosphate               ⫶
                    ┌──────────────────►            T—C—OH
                    │   isomerase, HTO                   |
                    │                                   C=O
                    │                                    |
     H                                                 CH₂O (P)
     ⫶
  H—C—OH ───────────┤
     |
    C=O
     |
    CH₂O (P)
                    │
                    │                                    T
                    │                                    ⫶
                    └──────────────────►            H—C—OH
                        aldolase, HTO                    |
                                                        C=O
                                                         |
                                                        CH₂O (P)
```

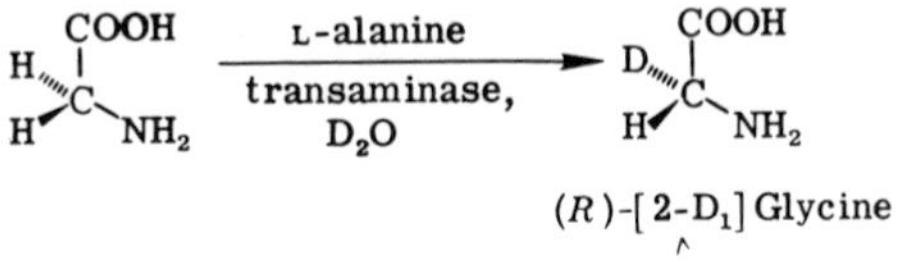

$$
\begin{array}{ccc}
\underset{H}{\overset{COOH}{\underset{|}{\overset{H_{\prime\prime\prime}}{C}}}}\overset{}{\underset{NH_2}{}} & \xrightarrow[\substack{\text{transaminase,}\\D_2O}]{\text{L-alanine}} & \underset{H}{\overset{COOH}{\underset{|}{\overset{D_{\prime\prime\prime}}{C}}}}\overset{}{\underset{NH_2}{}}
\end{array}
$$

$$(R)\text{-}[\,2\text{-}D_1]\,\text{Glycine}$$

A tabulation of enzyme reactions involving stereospecific hydrogen isotope transfer is given in a review by Simon and Kraus (1976).

In addition to isolated enzymes, intact microorganisms have been used to achieve stereospecific isotope labeling. The most prominent example is the reduction of [1-D]aldehydes to $(1S)$-[1-D]alcohols with actively fermenting yeast (Althouse *et al.*, 1966). More recently, stereospecific hydrogenations of α,β-unsaturated acids with hydrogen gas and a microorganism, e.g., *Clostridium kluyveri*, as a catalyst have been reported (Simon *et al.*, 1974).

Although in some cases the enzymatic reaction produces the desired stereospecifically labeled end product directly, more frequently a number of subsequent transformations are necessary to convert the initial enzyme product into the desired compound. These transformations may be either chemical or enzyme-catalyzed. For example, for studies on the stereochemistry of the amine oxidase from pea seedlings, Battersby *et al.* (1976) synthesized stereospecifically tritiated benzylamines from the corresponding benzyl alcohols which had been labeled with ADH:

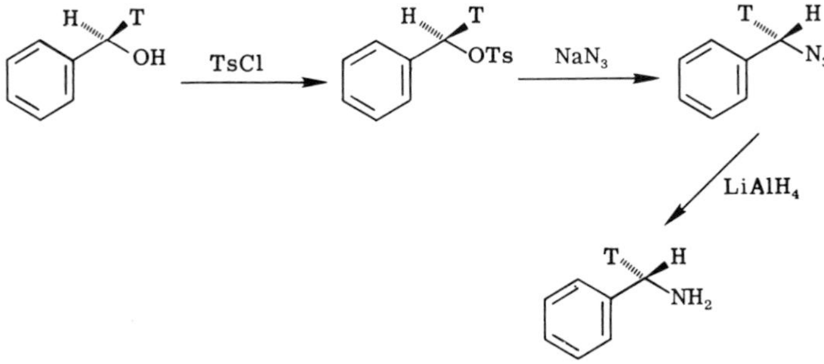

It is important in such a transformation sequence that either the stereospecifically labeled center is not involved in any of the reactions or, if it is, that reactions at this center proceed in a stereochemically homogeneous and predictable manner. Coupled enzyme reactions are often particularly useful for such transformation sequences. We have made use of this in a number of cases, for example, in a synthesis of $(3S)$- and $(3R)$-phospho-[3-T]glyceric acid from [1-T]glucose and [1-T]mannose, respectively (Floss *et al.*, 1972):

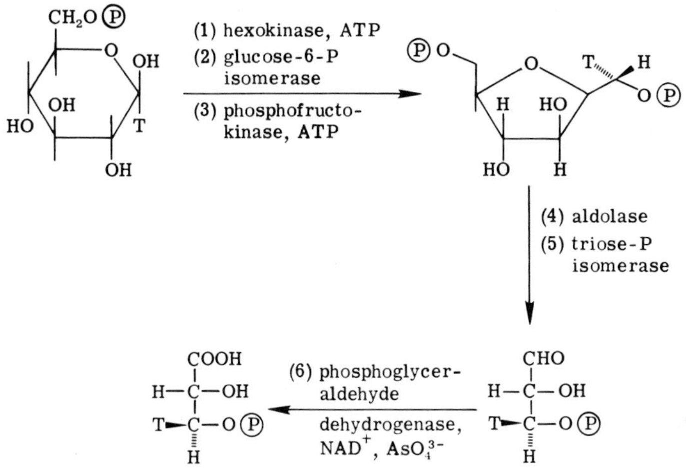

Since many enzymatic processes are equilibrium reactions, it is advantageous either to have an irreversible reaction at the end of the sequence or to push the last reaction to the product side by suitable manipulations (excess of cosubstrate or trapping of a product) in order to obtain maximum conversion. In the above case this was achieved by including arsenate instead of phosphate in the reaction mixture, which makes the phosphoglyceraldehyde dehydrogenase reaction irreversible. Such coupled reaction sequences can be used to construct efficient syntheses of many stereospecifically labeled enzyme substrates. For example, from the above phosphoglyceric acids, we have prepared by enzymatic syntheses (3R)- and (3S)-[3-T]serine (Floss *et al.*, 1976), (6R)- and (6S)-[4,D,6-T]glucose (Snipes *et al.*, 1977), (3R)- and (3S)-L-[3-D,3-T]alanine (L. Mascaro and H. G. Floss, unpublished), and (R)- and (S)-[2-D,2-T]acetate (Mascaro *et al.*, 1977).

B. Configurational Analysis of Stereospecifically Labeled Centers

1. Spectroscopic Methods

The two major spectroscopic techniques that are useful in characterizing and/or analyzing stereospecifically labeled compounds are polarimetry or optical rotatory dispersion (ORD) or circular dichroism (CD) spectroscopy and nmr spectroscopy. Measurement of rotations can serve to determine conveniently the configuration and chiral purity of stereospecifically deuterated prochiral compounds once the sign and degree of specific rotations have been established with an authentic sample. From a practical point of view the approach is limited to compounds containing no other chiral centers and, in view of the often low specific rotations, it is advisable to record the ORD or CD spectrum rather than to rely on a single wavelength measurement. In this fashion, rotational measurements have served to characterize optically active glycine (Battersby *et al.*, 1976, and references therein), α-deuterated fatty acids and aliphatic primary amines (LaRoche *et al.*, 1971, and references therein), and deuterated succinic acids (Cornforth *et al.*, 1966a). In the latter case, ORD in combination with mass spectrometry in determining the number of carbon-bound deuterium atoms allows an unequivocal characterization of each of the six nonexchangeably deuterated chiral succinic acids. The chiroptical method is not limited to deuterium-labeled compounds, or even to labeled prochiral compounds. Cullis and Lowe (1978) have recently reported that a chirally labeled pro-prochiral compound, methyl (R)-[^{16}O,-^{17}O,^{18}O]phosphate gives a measurable CD spectrum with a maximum at 208 nm ($\Delta\epsilon = +2.7 \times 10^{-3}$, corrected for 100% isotopic substitution). Obviously, however, when using such small rotations to characterize chirally labeled compounds, fair quantities of material are required and great care must be exercised to exclude any chiral impurities.

Proton nmr spectroscopy has been and continues to be used widely to determine the deuterium distribution between heterotopic hydrogens by observing either changes in the intensity of the signals of the heterotopic protons or changes in the coupling patterns of adjacent protons. Diastereotopic protons in conformationally rigid systems give separate nmr signals (unless they happen to have the same chemical shift), the integrations of which can be compared. For example, the 220-MHz proton nmr spectrum of the alkaloid elymoclavine biosynthesized from [3'-D$_3$]mevalonate shows a 38% decrease in the signal at δ3.65 (H$_{7S}$) but no significant decrease in the signal at δ3.08 (H$_{7R}$), indicating that one hydrogen from the methyl group of mevalonate is incorporated stereospecifically into the *pro-S* position at C-7 of the alkaloid (Floss *et al.*, 1974):

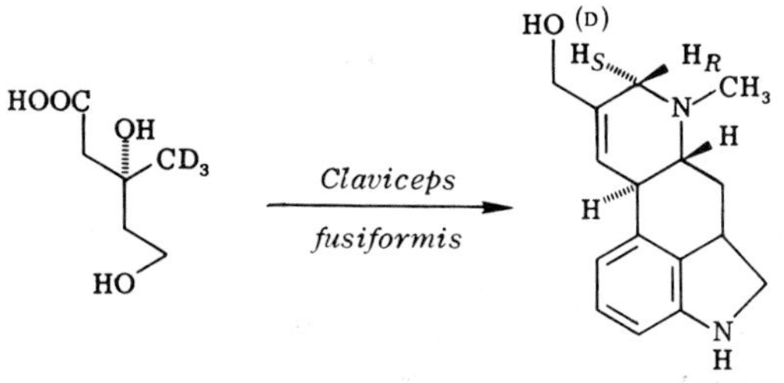

Not infrequently the heterotopic proton signals overlap with other signals and thus cannot be readily integrated. This is, for example, the case for the H-2' protons of the antibiotic granaticin. The location of deuterium in the *pro-S* position of a sample biosynthesized from [2-D]glucose could, however, be ascertained by observing the change in the coupling pattern of H-3'. This hydrogen appears as a doublet of doublets at δ4.06 ($J_{H_{3'}-H_{2'R}}$ = 8.5 Hz; $J_{H_{3'}-H_{2'S}}$ = 1.9 Hz). In the deuterated sample the large coupling is completely retained, whereas the small coupling is drastically reduced (Snipes *et al.*, 1979):

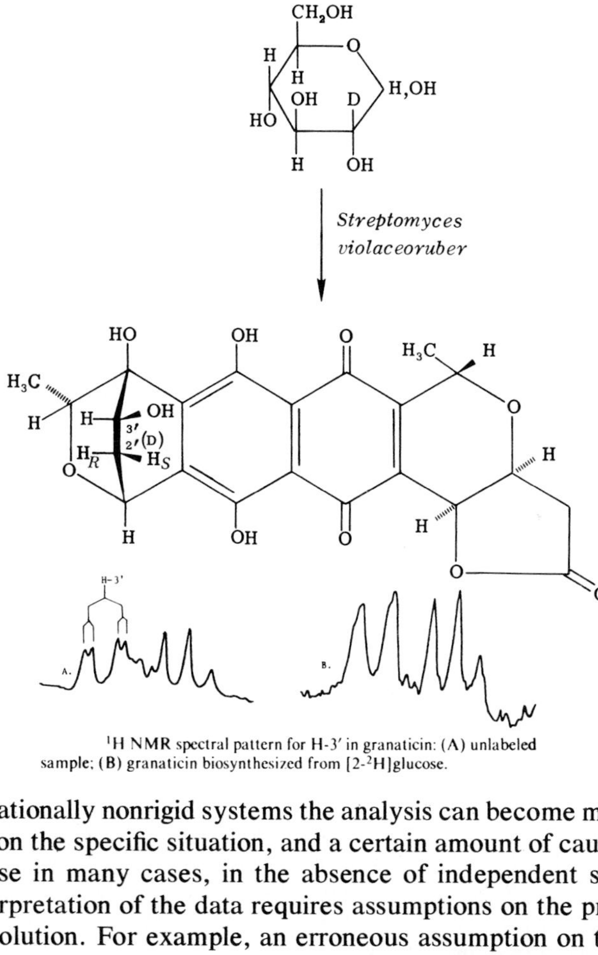

¹H NMR spectral pattern for H-3′ in granaticin: (A) unlabeled sample; (B) granaticin biosynthesized from [2-²H]glucose.

In conformationally nonrigid systems the analysis can become more complex depending on the specific situation, and a certain amount of caution is advisable because in many cases, in the absence of independent signal assignments, interpretation of the data requires assumptions on the preferred conformer in solution. For example, an erroneous assumption on the preferred conformation of malic acid at −196°C (Farrar *et al.*, 1957) initially led to an incorrect assignment of the configuration of deuterated malic acid, hence of the steric course of the fumarase reaction (Gawron and Fondy, 1959; Anet, 1960). Thus it is advisable either to analyze a conformationally rigid derivative or to confirm the signal or coupling constant assignments in an independent manner. However, with these caveats in mind the method is extremely useful and has served well in a vast number of studies (cf. Bentley, 1969, 1970).

Diastereotopic ligands other than hydrogens can also be distinguished by proton nmr spectroscopy, for example, the two diastereotopic methyl groups in valine. The *pro-R* methyl group of L-valine resonates at δ1.45 and the *pro-S* methyl group at δ1.38, as shown by analysis of authentic samples of valine carrying a trideuterated methyl group in one of the heterotopic positions (Hill *et al.*, 1973; Aberhart and Lin, 1973). Enantiotopic groups may be distinguishable by nmr either in the presence of chiral shift reagents (Goering *et al.*, 1974) or by analyzing a derivative with an optically active reagent, i.e., by converting the enantiotopic into diastereotopic groups. For example, the two methyl groups of isobutyric acid can be distinguished by analyzing the ester with $S(+)$-phenyltrifluoromethyl carbinol. In the ester, the two methyl groups appear at δ1.17 and 1.21; the ester prepared from $(2R)$-[3-D_3]isobutyrate shows only the signal at δ1.21 (Aberhart, 1975). In a similar way, the enantiotopic hydrogens at C-1 of a primary alcohol can be differentiated by analyzing the proton nmr spectrum of the (−)-camphanic acid ester in the presence of a shift reagent [Eu(dpm)$_3$] (Gerlach and Zagalak, 1973).

Observation of other nuclei by nmr can also be applied in the configurational analysis of stereospecifically labeled prochiral centers. For example, the two diastereotopic methyl groups in valine and penicillins have been clearly distinguished by ^{13}C-nmr (Neuss *et al.*, 1973; Kluender *et al.*, 1973). Deuterium nmr spectroscopy has recently come to some prominence in biosynthetic studies, including the analysis of stereochemical questions. For example, the steric course of the double-bond reduction in the conversion of 2′,7-dihydroxy-4′-methoxyisoflavone to the phytoalexin $(6aR,11aR)$-demethylhomopterocarpin has been determined by Dewick and Ward (1977) by feeding the deuterated isoflavone to seedlings of *Trigonella foenum-graecum* and analyzing the product by deuterium nmr. The spectrum showed a signal at δ3.92 (D-6R) but not at δ3.44 (D-6S):

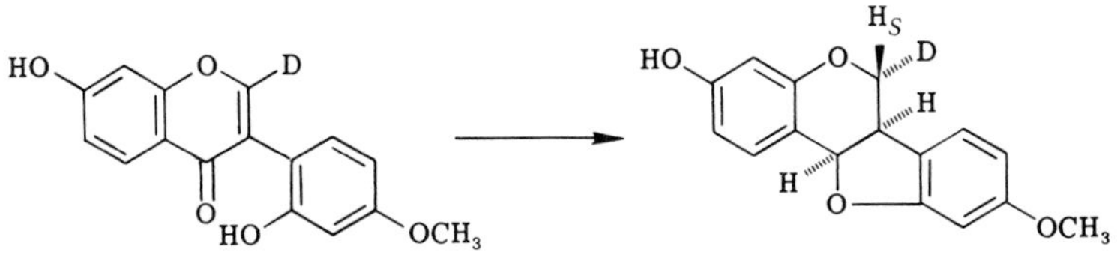

Biosynthetic applications of tritium nmr are still rare, mainly because of the relatively large amount of radioactivity required (at least 0.1 mCi). However, one intriguing stereochemical problem, the steric course of cyclopropane ring formation in cycloartenol biosynthesis, has been solved by tritium nmr in an elegant study in Altman's laboratory (Altman *et al.*, 1978).

2. Chemical Methods

Most purely chemical approaches to the configurational analysis of stereospecifically labeled centers have involved the synthesis of stereospecifically labeled compounds of known configuration, which were then compared with the enzymatically or biosynthetically generated samples. Particularly in the early days of studies on enzyme reaction stereospecificity this was the main approach. Examples of the configurational analysis of stereospecifically labeled compounds purely by chemical degradation are relatively rare. Basically, two of the three principles used in the chemical synthesis of stereospecifically labeled compounds, (1) simultaneous generation of a chiral and a stereospecifically labeled achiral center and (2) conversion of a chiral into a stereospecifically labeled achiral center, can be used in the reverse to analyze a stereospecifically labeled compound. A classical example from the early work on alkaloid biosynthesis is the degradation of hyoscyamine carried out by Bothner-By *et al.* (1962) and Leete (1962), which established the asymmetric incorporation of ornithine into the tropane ring system (Fig. 3). The original degradation only established that C-2 of ornithine labeled only one, and not both, of the bridgehead carbons. Subsequently, Leete (1964) demonstrated that the labeled carbon, C-1, had the *R* configuration.

3. Enzymatic Methods

Today a large body of knowledge exists on the steric course of many enzyme reactions, which forms a network of configurational correlations of stereospecifically labeled compounds. In many cases, therefore, the most expedient way to determine the configuration at a stereospecifically labeled center is to tie into this network via a suitable relay enzyme. This requires conversion of the compound to be analyzed by a series of chemical or en-

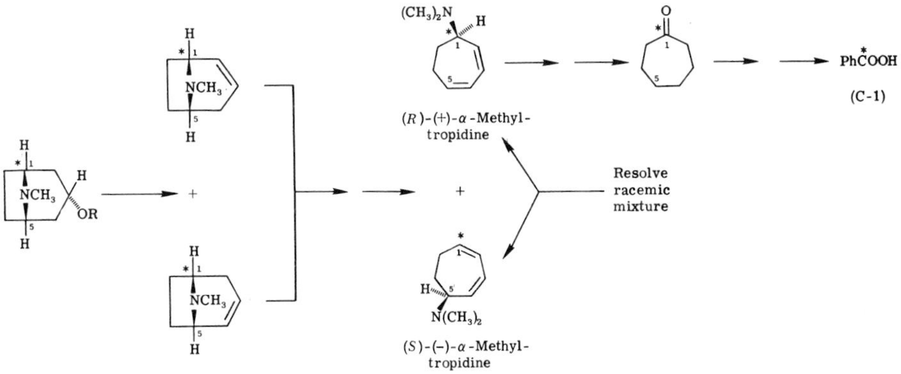

Fig. 3. Stereospecific degradation of hyoscyamine biosynthesized from [2-^{14}C]ornithine.

zymatic reactions into a substrate for one of the relay enzymes, followed by reaction with that enzyme. For example, to determine the configuration at C-6 of stereospecifically tritiated shikimic acid [obtained enzymatically from (E)- and (Z)-[3-T]phosphoenolpyruvate] the compound was degraded to DL-malate, which was analyzed for the tritium distribution between the diastereotopic hydrogen positions at C-3 by incubation with fumarase (Floss *et al.*, 1972):

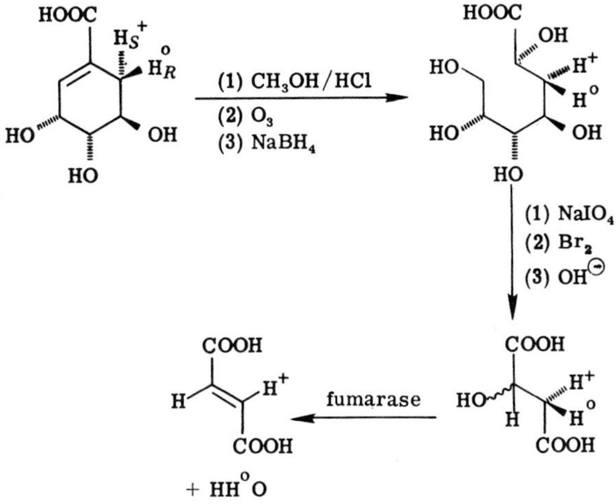

Some of the important relay enzymes are (cf. Simon and Kraus, 1976) ADH, fumarase, aspartase, glycolate oxidase, and L-lactate dehydrogenase (with glycolate as substrate), L-alanine transaminase (with glycine as substrate), and amine oxidase. In the degradations, it is obviously essential that the integrity of the stereospecifically labeled center is maintained. The reactions should preferably not involve cleavage of any bond at that center. Attention needs to be paid to possible racemization of the center to be analyzed, e.g., by enolization of an adjacent carbonyl function at an intermediate stage. For example, in a degradation of tryptophan to aspartate to determine the configuration at the stereospecifically tritiated methylene group it was found that ozonolysis of the N-acetyl derivative gave N-acetylaspartate in fair yield, but with extensive tritium loss and complete racemization of the tritiated methylene group (Tsai *et al.*, 1978a). Presumably, the known preferred ozonolytic cleavage of indoles at the 2,3-double bond led to the initial formation of a kynurenine derivative, which would have a pronounced tendency to enolize (double bond in conjugation to benzene ring):

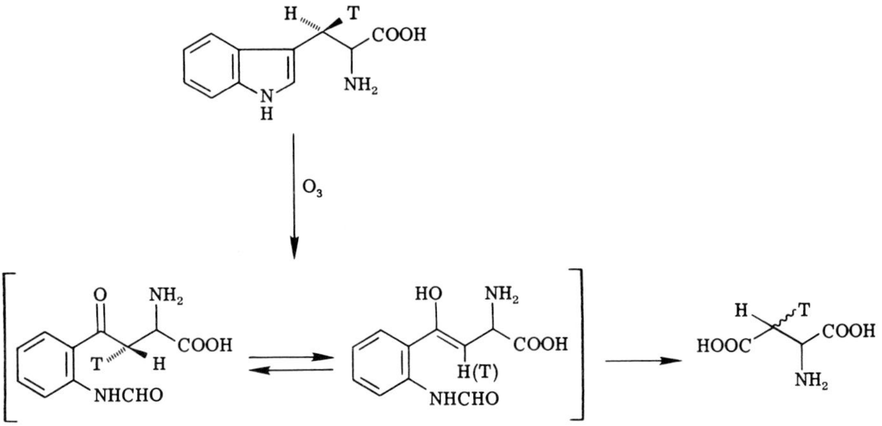

The problem was circumvented by first reducing tryptophan to the 4,7-dihydroderivative with lithium in liquid ammonia, rendering the six-membered ring more susceptible to cleavage by ozone. If the degradative reaction sequence involves bond cleavage and/or formation at the stereo-specifically labeled center, the reactions used must be stereospecific and their steric course must be unequivocally established. It is of particular importance to exclude any possibility that the reaction could proceed with the stereochemistry opposite that assumed (e.g., in a solvolytic reaction by a double inversion process), since the third possibility, a nonstereospecific reaction (e.g., in a solvolytic process by an S_N1 rather than an S_N2 reaction), is usually recognized in the final analysis.

4. Analysis of Pro-prochiral Centers

The configurational analysis of stereospecifically labeled pro-prochiral centers requires special methods, since the like substituents are not heterotopic and thus not distinguishable. The methods in use are based on replacement of one like ligand by a dissimilar group to generate a prochiral center, followed by isotopic analysis of the products of this reaction. The principles can be illustrated by examining the fate of a chiral methyl group in such a reaction.

When a chiral methyl group is converted into a methylene group in a stereospecific irreversible reaction, each enantiomer gives rise to a set of three products. Figure 4 illustrates this, assuming a reaction in which a methyl hydrogen is replaced by a group Y in an inversion mode. The remaining two hydrogens, assuming Y ≠ X, are now heterotopic and therefore distinguishable. The two sets of products, IIIa, IIIb, IIIc and IVa, IVb, IVc, are clearly different and can be distinguished in two principal ways. Since only

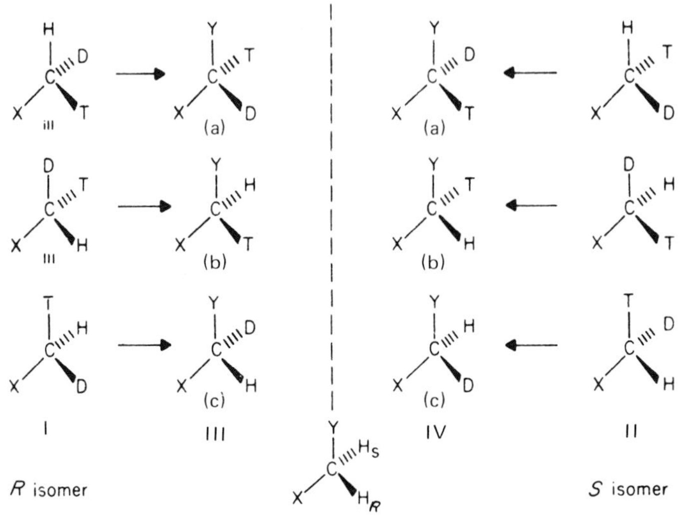

Fig. 4. Fate of the R and the S isomer of a chiral methyl group in a reaction replacing one methyl hydrogen by Y in an inversion mode. Group X is assumed to have sequence rule priority over group Y.

few molecules actually contain tritium, it is evident that species IIIc and IVc will be formed abundantly from X—CH₂D and are useless for the analysis; a distinction must thus be based on the sets IIIa + IIIb and IVa + IVb. This may be accomplished by determining whether the methylene species carrying tritium in the *pro-S* position contains a deuterium in the *pro-R* position, as in IIIa, or a normal hydrogen, as in IVb, and conversely, whether the species tritiated in the *pro-R* position carries D or H in the other heterotopic position. Alternatively, one can determine whether the species carrying tritium in the *pro-S* position has arisen from the replacement of H or from the replacement of D. This can be done by using a reaction for the conversion of I and II into III and IV, respectively, that exhibits a significant primary kinetic isotope effect in the C—H bond cleavage. Such an isotope effect is generally positive, thus favoring cleavage of the C—H bond over that of the C—D bond. Thus by generating the two sets of products in a reaction exhibiting a primary kinetic isotope effect and then determining the tritium distribution between the *pro-R* and *pro-S* positions, one can distinguish set III from set IV.

The most commonly used method for the chirality analysis of methyl groups is based on the second of these principles and involves condensation of the chiral methyl group, in the form of acetyl-coenzyme A (acetyl-CoA), with glyoxylate to form L-malate, catalyzed by malate synthase, followed by analysis of the tritium distribution in the malate by incubation with fumarase

(Lüthy *et al.*, 1969; Cornforth *et al.*, 1969). Retention of more than 50% of the tritium in the fumarase reaction indicates an *R* configuration of the methyl group (limit 79%), and retention of less than 50% (limit 21%) indicates an *S* configuration. Chiral methyl groups in other chemical environments have to be converted into acetate for chirality analysis by this method. For a more detailed discussion of the topic see the review by Floss and Tsai (1979).

The chirality analysis of a chiral phosphate group in a phosphate monoester worked out by Knowles' group (Abbott *et al.*, 1978) is based on the same principle. The chiral [^{16}O,^{17}O,^{18}O]phospho-(*S*)-propane-1,2-diol was converted into the cyclic diester with replacement of one of the oxygen isotopes by the oxygen of the hydroxyl group of the propanediol. The remaining two unsubstituted oxygens are then heterotopic. It was then determined whether, for example, the species containing ^{18}O in the *pro-R* position had arisen from loss of ^{16}O or ^{17}O in the cyclization reaction, i.e., whether it contained ^{17}O or ^{16}O, respectively, in the *pro-S* position. This was done by methylation of one of the oxygens and separation of the diastereomeric triesters, followed by metastable ion mass spectrometry, to determine which daughter ion in the mass spectrum had arisen from which parent.

IV. INFORMATION DEDUCIBLE FROM STEREOCHEMICAL STUDIES

The types of information that can be derived from studies on the steric course of stereochemically cryptic reactions fall into several categories. On the one hand, one can use stereochemical approaches to distinguish between different metabolic pathways and to probe for "cooperation" between enzymes; on the other hand, the steric course of the reaction catalyzed by a particular enzyme can give insight into the reaction mechanism and the structure and evolution of the active site. In the following a few examples will be given to illustrate the various points. For a more detailed discussion of the latter aspect, the reader is referred to the thought-provoking article by Hanson and Rose (1975) entitled, "Interpretation of Enzyme Reaction Stereospecificity."

A. Distinction between Metabolic Pathways

In certain instances alternative metabolic pathways can be distinguished on stereochemical grounds if the stereochemical data are incompatible with a particular metabolic route. For example, in the biosynthesis of tropane alkaloids it has been found that both ornithine and putrescine are efficient precursors, suggesting that the pathway may involve decarboxylation of ornithine to putrescine. However, the stereospecific degradation of hyoscyamine biosynthesized from DL-[2-^{14}C]ornithine (Leete, 1962) or from [1-^{14}C]acetate (Bothner-By *et al.*, 1962) mentioned earlier (Fig. 3) indicated that the molecule was labeled asymmetrically; all the ^{14}C from C-2 of ornithine resided at C-1, the bridgehead carbon of the R configuration. Since putrescine is a symmetrical molecule, its intermediacy in the pathway is ruled out, because a route via putrescine would have resulted in equal labeling of C-1 and C-5. It was subsequently established that methylation of one of the nitrogens of ornithine precedes the decarboxylation step, thus preserving the asymmetry of the molecule. These results on tropane alkaloid formation are in contradistinction to the situation in nicotine biosynthesis, where [2-^{14}C]ornithine labels C-2 and C-5 of the pyrrolidine ring equally and a symmetrical intermediate must be involved in the biosynthesis (Leete and Siegfried, 1957).

The situation is even more complex in the biosynthesis of the piperidine alkaloids, e.g., sedamine, which are derived from L-lysine. It has been shown that L-[2-^{14}C]lysine labels only C-2 of sedamine and that the corresponding diamine, cadaverine, is also incorporated, resulting of course in symmetrical labeling. Thus free cadaverine cannot be a biosynthetic intermediate. It was then suggested (Leistner and Spenser, 1973) that a bound form of cadaverine, e.g., a Schiff's base with pyridoxal phosphate, might be the actual intermediate. However, additional stereochemical data point up further complications. L-[2-T]lysine is incorporated into sedamine with retention of the tritium (Gupta and Spenser, 1970). Decarboxylation of L-[2-T]lysine by bacterial lysine decarboxylase proceeds in a retention mode to give (1S)-[1-T]cadaverine. When this material is fed to *Sedum acre* plants, the resulting sedamine shows loss of half of the tritium. The corresponding (1R)-[1-T]cadaverine, on the other hand, gives sedamine with complete retention of tritium (Leistner and Spenser, 1973).

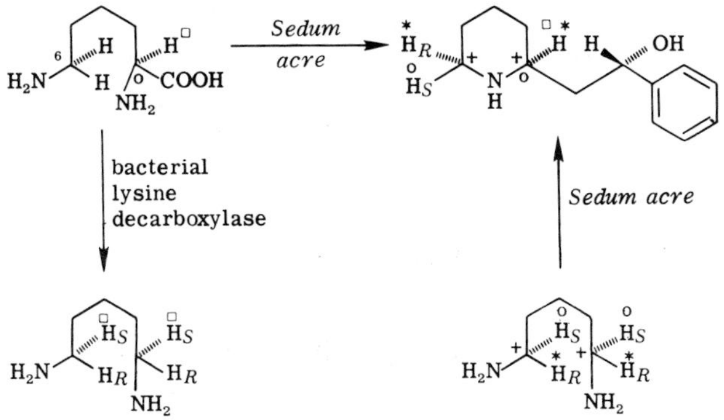

These paradoxical results must indicate that either (1) the lysine decarboxylase in *Sedum* plants has stereochemistry opposite that of the bacterial enzyme, i.e., it operates in an inversion mode, or (2) the fate of endogenous bound cadaverine is different from that of exogenously added free cadaverine; i.e., they are converted into sedamine by different enzymes with different stereospecificities. It is also conceivable, and in fact attractive as a hypothesis, that *Sedum* plants contain a pyridoxal phosphate enzyme which at the same active site decarboxylates lysine (with retention) and transaminates the intermediate to 5-aminopentanal:

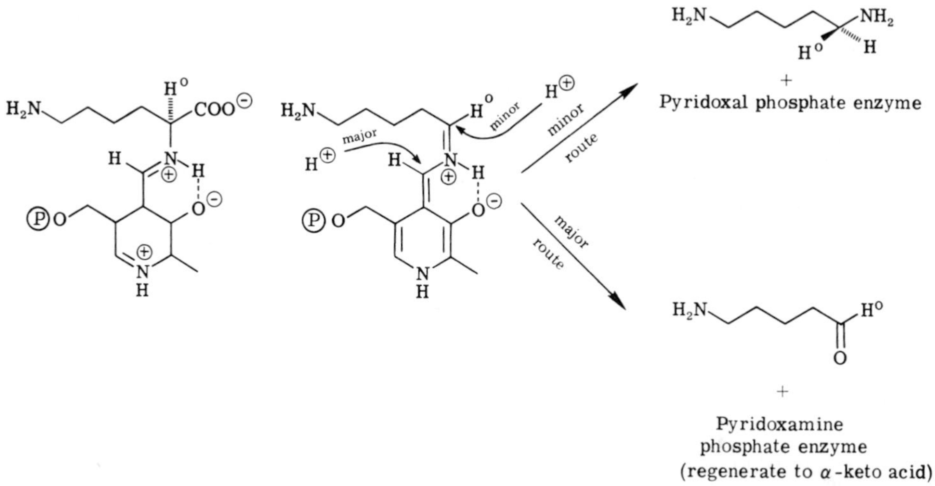

Pyridoxal phosphate enzyme

Pyridoxamine
phosphate enzyme
(regenerate to α-keto acid)

The stereochemistry of hydrogen removal in the conversion of free cadaverine to sedamine corresponds to that exhibited by diamine oxidase (Gerdes and Leistner, 1979), and it may be this enzyme that converts added free cadaverine into 5-aminopentanal and thus channels it into the biosynthetic pathway.

B. Enzyme Reaction Mechanisms

Stereochemistry has served as a powerful probe in the investigation of the mechanisms of organic reactions in free solution. It can also contribute to our understanding of the mechanisms of reactions that take place at the active site of an enzyme. However, different criteria need to be applied in the mechanistic interpretation of stereochemical data pertaining to biochemical processes, since enzyme reactions are almost always stereospecific. Thus the fact that a particular enzyme reaction proceeds in a stereochemical mode that is compatible with a particular mechanism does not prove this mechanism. In fact, it may not even rule out other mechanisms which in free solution would lead to a different stereochemical outcome. For example, Popják and Cornforth (1966) and their co-workers found that the prenyl transferase reaction, the enzymatic condensation of isopentenyl pyrophosphate with dimethylallyl pyrophosphate or geranyl pyrophosphate to give geranyl or farnesyl pyrophosphate, respectively, proceeds in an inversion mode with respect to the allylic center. This was interpreted to indicate that pyrophosphate is replaced by C-4 of isopentenyl pyrophosphate by an S_N2 mechanism. However, subsequent studies by Poulter and Rilling (1978) and their co-workers have provided ample evidence for the notion that the process is a stepwise one, involving initially ionic cleavage of the carbon–pyrophosphate bond. Presumably, therefore, the stereochemical identity of the two faces of the intermediate carbonium ion is maintained by preventing rotation around the C-1–C-2 axis, through ion-pairing forces, through conjugation with the double bond, or through both. Since the carbonium ion is held at the enzyme active site, the approach of the C-4 of isopentenyl pyrophosphate can occur exclusively at the face opposite that from which the pyrophosphate anion has departed.

There are, on the other hand, cases in which stereochemical information can clearly rule out a particular mechanism. An example is citrate synthase. This enzyme catalyzes proton exchange between water and the methyl group of acetyl-CoA in the presence of L-malate, but not in its absence. This finding was interpreted by suggesting (Eggerer, 1965) that the β-carboxyl group of oxalacetate (C-4) in its anion form may be the base that abstracts a proton from the methyl group of the second substrate, acetyl-CoA. L-Malate as a substrate analogue that cannot undergo the condensation reaction nevertheless confers acetyl-CoA enolase activity upon the enzyme. This interpretation would require the replacement of a methyl proton by C-2 of oxalacetate to occur in a retention mode, since C-2 and C-4 of oxalacetate cannot approach the methyl group from opposite sides. The subsequent experimental finding (Rétey et al., 1970); Lenz et al., 1971; Klinman and Rose, 1971) that the reaction proceeds in an inversion mode ruled out this mechanism. It is now assumed that the binding of L-malate causes a conformational change in citrate synthase, which alters the position of an essential base group.

As another example, the finding that the enzyme indolyl-3-alkane α-hydroxylase hydroxylates L-tryptophan methyl ester with stereospecific removal of the pro-S hydrogen from the β position, but produces the two epimers of β-hydroxy-L-trytophan methyl ester, strongly suggests that the reaction does not involve a concerted replacement of a β hydrogen by OH. The results instead indicate that only part of the overall process is mediated by the enzyme, namely, dehydrogenation to the corresponding alkylidene indolenine, whereas the subsequent oxygen insertion, by addition of water to the diene system, occurs outside the active site of the enzyme (Tsai et al., 1979). Conversely, it was observed that all the pyridoxal phosphate-catalyzed α,β-eliminations of β-substituted amino acids, including the reaction catalyzed by the plant enzyme S-alkylcysteine lyase (Tsai et al., 1978b, and references therein), involve stereospecific protonation of the β-methylene group to give the methyl group of pyruvate. Hence the protonation occurs at the active site, ruling out that, as had been proposed at one time, the intermediate pyridoxylidene α-aminoacrylate is hydrolyzed to free α-aminoacrylic acid which is released into the medium and there converted to pyruvate.

As pointed out earlier, the stereochemistry observed for a single enzyme usually cannot prove a particular mechanism. However, if a sizable group of enzymes of widely different structure, origin, and substrate specificity that catalyze the same reaction type do so with the same stereochemistry, then it is most likely that the steric course of the reaction is dictated by its mechanism. For example, the fact that all P-450-dependent aliphatic hydroxylases catalyze replacement of a hydrogen by OH in a retention mode (cf. Hanson and Rose, 1975) almost certainly is a consequence of the reaction mechanism; any mechanism proposed for these hydroxylations thus must account for this stereochemical observation. Six methyl transferases catalyzing transfer of the methyl group of S-adenosylmethionine (SAM) to various acceptors were studied (cf. Floss and Tsai, 1979), and in each case the transfer occurred with inversion of configuration at the methyl group. This again suggests a common (S_N2) mechanism and, in addition, indicates that in each case the methyl group is transferred directly from the sulfur of SAM to the acceptor group of the second substrate. On the other hand the fact that of several enzymes catalyzing β-oxidative decarboxylation reactions, some operate in a retention and some in an inversion mode, indicates that the CO_2 elimination and the enolization steps are independent of each other (cf. Hanson and Rose, 1975).

C. Active Site Conformation of Enzyme–Substrate Complexes

In certain situations one can derive from stereochemical studies a detailed picture of the conformation of the substrate or the coenzyme–substrate complex at the active site and of the position of essential groups of the enzyme in relation to it. A particularly instructive example is the enzyme TDP-glucose oxidoreductase. A member of the broader class of hexose-nucleotide oxidoreductases which initiate the various pathways to 6-deoxyhexose derivatives in plants, animals, and microorganisms, this *Escherichia coli* enzyme catalyzes the transformation of TDP-glucose into TDP-4-keto-6-deoxyglucose. In this process, H-5 of the sugar is exchanged with solvent protons and, mediated by a tightly enzyme-bound pyridine nucleotide, H-4 is transferred intramolecularly to C-6 of the hexose. We showed that this transfer leads to replacement of the 6-OH group by H-4 in an inversion mode. Since the transfer of the hydrogen is intramolecular, it must be suprafacial, defining the relative orientation of H-5 and the 6-OH group as

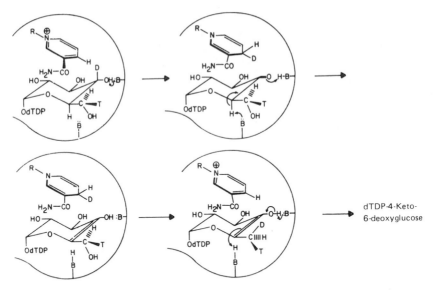

Fig. 5.   Stereochemical mechanism of the TDP-glucose oxidoreductase reaction.

syn (Snipes *et al.*, 1977). Based on the minimal motion rule of Hanson and Rose (1975), the results lead to a very compelling model for the molecular events at the enzyme active site during the catalytic process (Fig. 5).

Similarly, stereochemical studies on tryptophan synthetase have established in considerable detail the conformation of the coenzyme–substrate complex at the active site of the pyridoxal phosphate-containing β_2 subunit of this enzyme (Tsai *et al.*, 1978a). The demonstration of intramolecular proton transfer between C-α and C-β of the amino acid and, in a different reaction mode, between C-α and the *Si* face of C-4′ of the cofactor defines the conformation of the system as shown:

A very similar conformation is maintained at the active site of tryptophanase (Vederas *et al.*, 1978) and probably also at that of tyrosine phenol-lyase

H. Kumagai, Y. Yamada, E. Schleicher, and H. G. Floss, unpublished results).

D. Evolution of Enzymes and Enzyme Active Sites

In the above example of the pyridoxal phosphate enzymes catalyzing β replacement and α,β-elimination reactions, the occurrence of a common active site conformation of the coenzyme–substrate complex suggests that the observed quasi-six-membered ring conformation must have a catalytic advantage. In this case the evolutionary driving force seems to be the economy inherent in being able to carry out several sequential protonation and deprotonation steps using a single base and intrinsic proton recycling. This is reminiscent of situations in which in a series of enzymes a common stereochemical reaction path is dictated by a particular mechanism.

There are other cases, however, in which stereochemical homogeneity is observed in a series of enzymes, although there is no discernible advantage of one stereochemical reaction mode over the other. For example, the pyridoxal phosphate enzymes mentioned above could operate equally well on an alternative conformer of the coenzyme–substrate complex, which is rotated 180° around the C-α—N-α bond. Any proton transfer to C-4' would then occur at the Re rather than the Si face. The same holds for various other pyridoxal phosphate enzymes. The energy difference between these two conformations is relatively small, and there seems to be no mechanistic advantage of one operational mode over the other. Thus if a series of such enzymes evolves independently,   would expect some to operate on the Si and some on the Re face of the cofactor. Yet it has been found that, in all the enzymes studied so far, the addition of a hydrogen at C-4' of the cofactor–substrate complex occurs on the Si face. This has been interpreted (Dunathan and Voet, 1974) to indicate evolution of all these enzymes from a common ancestral protein, in which an arbitrary choice between the two faces of the cofactor has been made that has been preserved in present-day enzymes. A similar stereochemical consistency is displayed by a series of enzymes that catalyze the addition of electrophilic groups (H^+, CO_2, or C-1 of erythrose-4-phosphate) at C-3 of an active intermediate derived from PEP, possibly enolpyruvate. Although the two faces of PEP are completely equivalent, the attack in each case was found to be on the $3Si$ face, suggesting again an evolution of all these enzymes from a common ancestral protein (Rose, 1972).

E. Miscellaneous

There are numerous other ways in which stereochemical experiments can answer specific biochemical questions. In cases where potential reaction intermediates possess rotational symmetry one can probe for scrambling of a stereospecific isotope label to determine if such an intermediate is indeed involved. For example, in the first reaction of the shikimate pathway, the condensation of PEP and erythrose-4-phosphate catalyzed by DAHP synthetase, the two methylene hydrogens of PEP predominantly retain their identity (Floss *et al.*, 1972). Ruling out an earlier proposal, this finding indicates that the reaction cannot proceed via a structure with a freely rotating methyl group as an obligatory intermediate. In another, particularly intriguing application, Etemadi *et al.* (1969) have used this principle to determine whether squalene is a free intermediate in the biosynthesis of sterols or whether it may be passed from the active site of squalene synthetase to that of squalene oxidase in a concerted fashion without being released from the particulate enzyme complex. Squalene is a symmetrical molecule in which, for example, the two center carbons C-12 and C-13 are indistinguishable. However, as shown earlier (Popják and Cornforth, 1966), the molecule is formed from two molecules of farnesyl pyrophosphate in an unsymmetrical fashion, because one hydrogen from C-1 of one of the farnesyl residues is replaced by a hydrogen from NADPH. Thus, as long as squalene remains bound to a chiral environment, only one half of the molecule will carry a label introduced asymmetrically during its synthesis, e.g., from tritiated NADPH. The squalene oxidase epoxidizes the terminal double bond in one half of the squalene molecule. If the substrate has been "handed over" in a concerted fashion from the active site of squalene synthetase without ever leaving the chiral environment of the enzyme complex, either only the labeled half or only the unlabeled half will be epoxidized and tritium from NADPH-T will be located exclusively either at C-12 or C-13. However, if free squalene is bound to the oxidase, no distinction between the two halves is possible and the tritium will be distributed equally between C-12 and C-13. The tritium distribution can be determined by analyzing the sterols derived from the squalene epoxide. For the biosynthesis of lanosterol in a pig liver homogenate the experiment showed that most, if not all, of the sterol was formed via free squalene (Etemadi *et al.*, 1969).

As pointed out earlier, enzyme reactions are almost always stereospecific. It is therefore of particular significance if a biochemical reaction or reaction sequence is found to be nonstereospecific. Such an observation indicates that either a gyrosymmetric intermediate is involved or that one of the reactions or a part of a reaction is nonenzymatic. For example, in studies on the biosynthesis of papaverine in *P. somniferum* it was found that aromatization of the tetrahydroisoquinoline ring system involves stereospecific removal of the *pro-S* hydrogen from the C-3 of norreticuline. However, the removal of hydrogen from the C-4 of norreticuline was nonstereospecific and showed an isotope effect and only a slight preference for the removal of H_R. These results have been interpreted in terms of a two-stage process comprising first an enzymatic dehydrogenation to the 2,3-imine, followed by a nonenzymatic tautomerization to the enamine (Battersby *et al.*, 1977):

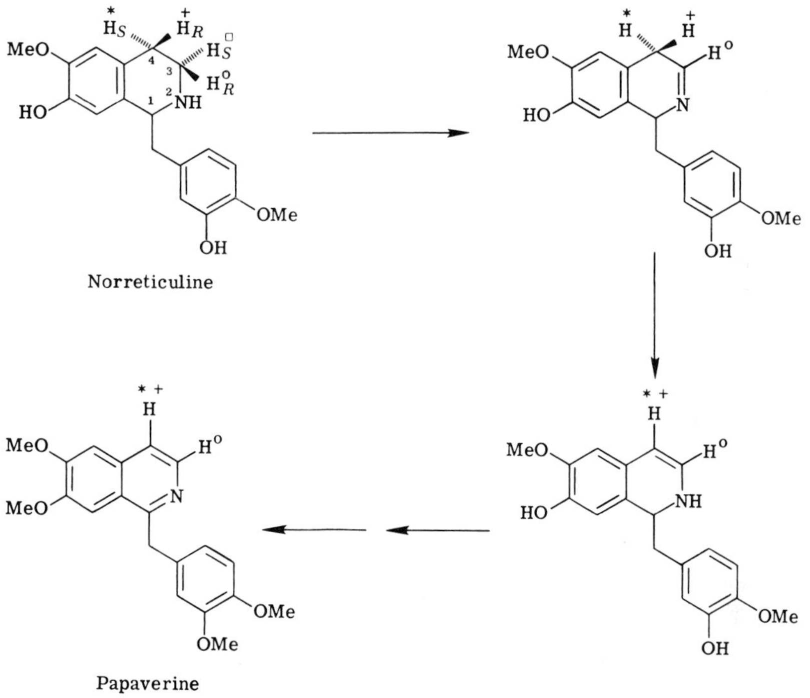

On similar grounds a nonenzymatic step has been implicated in the biosynthesis of camptothecin (Hutchinson *et al.*, 1979).

V. CONCLUSION

In the foregoing discussion of stereochemically cryptic reactions it has been attempted to outline the principles governing studies on the stereochemistry of such processes and to illustrate some of the methodology and applications. Results from stereochemical studies of this kind do not tell us everything about a particular reaction or reaction sequence. Knowledge of the stereochemistry is, however, an integral part of the totality of information about an enzyme or a biochemical reaction sequence. In this sense, the methodology discussed here occupies an important place in the hierarchy of experimental approaches used to study enzymes and metabolic pathways.

REFERENCES

Abbott, S. J., Jones, S. R., Weinman, S. A., and Knowles, J. R. (1978). *J. Am. Chem. Soc.* **100**, 2558–2560.

Aberhart, D. J. (1975). *Tetrahedron Lett.* pp. 4373–4374.

Aberhart, D. J., and Lin, L. J. (1973). *J. Am. Chem. Soc.* **95**, 7859–7860.

Althouse, V. E., Feigl, D. M., Sanderson, W. A., and Mosher, H. S. (1966). *J. Am. Chem. Soc.* **88**, 3595–3599.

Altman, L. J., Han, C. Y., Bertolino, A., Handy, G., Laungani, D., Muller, W., Schwartz, S., Shanker, D., DeWolf, W. H., and Yang, F. (1978). *J. Am. Chem. Soc.* **100**, 3235–3237.

Alworth, W. L. (1972). "Stereochemistry and Its Application in Biochemistry." Wiley (Interscience), New York.

Anet, F. A. L. (1960). *J. Am. Chem. Soc.* **82**, 994–995.

Battersby, A. R., Staunton, J., and Wiltshire, H. R. (1975). *J. Chem. Soc., Perkin Trans. 1* pp. 1156–1161.

Battersby, A. R., Staunton, J., and Summers, M. C. (1976). *J. Chem. Soc., Perkin Trans. 1* pp. 1052–1056.

Battersby, A. R., Sheldrake, P. W., Staunton, J., and Summers, M. C. (1977). *Bioorg. Chem.* **6**, 43–47.

Bentley, R. (1969). "Molecular Asymmetry in Biology," Vol. 1. Academic Press, New York.

Bentley, R. (1970). "Molecular Asymmetry in Biology," Vol. 2. Academic Press, New York.

Blackwood, J. E., Gladys, C. L., Loening, K. L., Petrarca, A. E., and Rush, J. E. (1968). *J. Am. Chem. Soc.* **90**, 509–510.

Bothner-By, A. A., Schutz, R. S., Dawson, R. F., and Solt, M. L. (1962). *J. Am. Chem. Soc.* **84**, 52–54.

Cahn, R. S., Ingold, C. K., and Prelog, V. (1966). *Angew. Chem., Int. Ed. Engl.* **5**, 385–415.

Cornforth, J. W., Cornforth, R. H., Donninger, C., Popják, G., Ryback, G., and Schroepfer, G. J. (1966a). *Proc. R. Soc. London, Ser. B* **163**, 436–464.

Cornforth, J. W., Cornforth, R. H., Donninger, C., and Popják, G. (1966b). *Proc. R. Soc. London, Ser. B* **163**, 492–514.

Cornforth, J. W., Redmond, J. W., Eggerer, H., Buckel, W., and Gutschow, C. (1969). *Nature (London)* **221**, 1212–1213.

Cullis, P. M., and Lowe, G. (1978). *J. Chem. Soc., Chem. Commun.* pp. 512–514.

Dewick, P. M., and Ward, D. (1977). *J. Chem. Soc., Chem. Commun.* pp. 338–339.

Dunathan, H. C., and Voet, J. G. (1974). *Proc. Natl. Acad. Sci. U.S.A.* **71**, 3888–3891.

Eckstein, F. (1975). *Angew. Chem., Int. Ed. Engl.* **14**, 160–166.

Eckstein, F. (1979). *Acc. Chem. Res.* **12**, 204–210.

Eggerer, H. (1965). *Biochem. Z.* **343**, 111–138.

Etemadi, A. H., Popják, G., and Cornforth, J. W. (1969). *Biochem. J.* **111**, 445–451.

Farrar, T. C., Gutowsky, H. S., Alberty, R. A., and Miller, W. G. (1957). *J. Am. Chem. Soc.* **79**, 3978–3980.

Floss, H. G., and Tsai, M. D. (1979). *Adv. Enzymol.* **50**, 243–302.

Floss, H. G., Onderka, D. K., and Carroll, M. (1972). *J. Biol. Chem.* **247**, 736–744.

Floss, H. G., Tcheng-Lin, M., Chang, C.-j., Naidoo, B., Blair, G. E., Abou-Chaar, C. I., and Cassady, J. M. (1974). *J. Am. Chem. Soc.* **96**, 1898–1909.

Floss, H. G., Schleicher, E., and Potts, R. (1976). *J. Biol. Chem.* **251**, 5478–5482.

Frydzuk, M. D., and Bosnich, B. (1977). *J. Am. Chem. Soc.* **99**, 6262–6267.

Gawron, O., and Fondy, T. P. (1959). *J. Am. Chem. Soc.* **81**, 6333–6334.

Gerdes, H. J., and Leistner, E. (1979). *Phytochemistry* **18**, 771–775.

Gerlach, H., and Zagalak, B. (1973). *J. Chem. Soc., Chem. Commun.* pp. 274–275.

Goering, H. L., Eikenberry, J. N., Koermer, G. S., and Lattimer, C. J. (1974). *J. Am. Chem. Soc.* **96,** 1493–1501.

Günther, H., Alizade, M. A., Kellner, M., Biller, F., and Simon, H. (1973a). *Z. Naturforsch., C: Biosci.* **28C,** 241–246.

Günther, H., Biller, F., Kellner, M., and Simon, H. (1973b). *Angew. Chem., Int. Ed. Engl.* **12,** 146–147.

Gupta, R. N., and Spenser, I. D. (1970). *Phytochemistry* **9,** 2329–2334.

Hanson, K. R. (1966). *J. Am. Chem. Soc.* **88,** 2731–2742.

Hanson, K. R., and Rose, I. A. (1975). *Acc. Chem. Res.* **8,** 1–10.

Hill, R. K., Yan, S., and Arfin, S. M. (1973). *J. Am. Chem. Soc.* **95,** 7857–7859.

Hirschmann, H., and Hanson, K. R. (1971). *Eur. J. Biochem.* **22,** 301–309.

Hutchinson, C. R., Heckendorf, A. H., Straughn, J. L., Daddona, P. E., and Cane, D. E. (1979). *J. Am. Chem. Soc.* **101,** 3358–3369.

International Union of Pure and Applied Chemistry (1970). *J. Org. Chem.* **35,** 2849–2867.

Kirby, G. W., and Michael, J. (1973). *J. Chem. Soc., Perkin Trans. 1* pp. 115–120.

Klinman, J. P., and Rose, I. A. (1971). *Biochemistry* **10,** 2267–2272.

Kluender, H., Bradley, C. H., Sih, C. J., Fawcett, P., and Abraham, E. P. (1973). *J. Am. Chem. Soc.* **95,** 6149–6150.

LaRoche, H.-J., Simon, H., Kellner, M., and Günther, H. (1971). *Z. Naturforsch., B: Anorg. Chem., Org. Chem., Biochem., Biophys., Biol.* **26B,** 389–394.

Leete, E. (1962). *J. Am. Chem. Soc.* **84,** 55–57.

Leete, E. (1964). *Tetrahedron Lett.* pp. 1619–1622.

Leete, E., and Siegfried, K. J. (1957). *J. Am. Chem. Soc.* **79,** 4529–4531.

Leistner, E., and Spenser, I. D. (1973). *J. Am. Chem. Soc.* **95,** 4715–4725.

Lenz, H., Buckel, W., Wunderwald, P., Biedermann, G., Buschmeier, V., Eggerer, H., Cornforth, J. W., Redmond, J. W., and Mallaby, R. (1971). *Eur. J. Biochem.* **24,** 207–215.

Loewus, F. A., Westheimer, F. H., and Vennesland, B. (1953). *J. Am. Chem. Soc.* **75,** 5018–5023.

Lüthy, J., Rétey, J., and Arigoni, D. (1969). *Nature (London)* **221,** 1213–1215.

Mascaro, L., Hörhammer, R., Eisenstein, S., Sellers, L. K., Mascaro, K., and Floss, H. G. (1977). *J. Am. Chem. Soc.* **99,** 273–274.

Midland, M. M., Greer, S., Tramontano, A., and Zderic, S. A. (1979). *J. Am. Chem. Soc.* **101,** 2352–2355.

Mislow, K., and Raban, M. (1967). *Top. Stereochem.* **1,** 1–38.

Neuss, N., Nash, C. H., Baldwin, J. E., Lemke, P. A., and Grutzner, J. B. (1973). *J. Am. Chem. Soc.* **95,** 3797–3798.

Paliokas, A. M., and Schroepfer, G. J. (1968). *J. Biol. Chem.* **243,** 453–464.

Parry, R. J. (1978). *J. Chem. Soc., Chem. Commun.* pp. 294–295.

Popják, G., and Cornforth, J. W. (1966). *Biochem. J.* **101,** 553–568.

Poulter, C. D., and Rilling, H. C. (1978). *Acc. Chem. Res.* **11,** 307–317.

Rétey, J., Lüthy, J., and Arigoni, D. (1970). *Nature (London)* **226,** 519–521.

Rose, I. A. (1972). *CRC Crit. Rev. Biochem.* **1,** 33–57.

Schroepfer, G. J., and Bloch, K. (1965). *J. Biol. Chem.* **240,** 54–63.

Sedgwick, B., Cornforth, J. W., French, S. J., Gray, R. T., Kellstrup, E., and Willadsen, P. (1977). *Eur. J. Biochem.* **75,** 481–495.

Simon, H., and Kraus, A. (1976). *In* "Isotopes in Organic Chemistry" (E. Buncel and C. C. Lee, eds.),Vol. 2, pp. 153–229. Elsevier, Amsterdam.

Simon, H., Rambeck, B., Hashimoto, H., Günther, H., Nohynek, G., and Neumann, H. (1974). *Angew. Chem., Int. Ed. Engl.* **13,** 608–609.

Snipes, C. E., Brillinger, G.-U., Sellers, L., Mascaro, L., and Floss, H. G. (1977). *J. Biol. Chem.* **252,** 8113–8117.

Snipes, C. E., Chang, C.-j., and Floss, H. G. (1979). *J. Am. Chem. Soc.* **101,** 701–706.

Townsend, C. A., Scholl, T., and Arigoni, D. (1975). *J. Chem. Soc., Chem. Commun.* pp. 921–922.

Tsai, M. D., Schleicher, E., Potts, R., Skye, G. E., and Floss, H. G. (1978a). *J. Biol. Chem.* **253,** 5344–5349.

Tsai, M. D., Weaver, J., Floss, H. G., Conn, E. E., Creveling, R. K., and Mazelis, M. (1978b). *Arch. Biochem. Biophys.* **190,** 553–559.

Tsai, M. D., Floss, H. G., Rosenfeld, H. J., and Roberts, J. (1979). *J. Biol. Chem.* **254,** 6437–6443.

Vederas, J. C., Schleicher, E., Tsai, M. D., and Floss, H. G. (1978). *J. Biol. Chem.* **253,** 5350–5354.

Vineyard, B. D., Knowles, W. S., Sabacky, M. J., Bachman, G. L., and Weinkauff, D. J. (1977). *J. Am. Chem. Soc.* **99,** 5946–5952.

Wightman, R. H., Staunton, J., Battersby, A. R., and Hanson, K. R. (1972). *J. Chem. Soc., Perkin Trans. 1* pp. 2355–2364.

Note Added in Proof

Two groups have now reported the synthesis of R- and S-[^{17}O,^{18}O]thiophosphate and have devised a method to distinguish the enantiomers [M. R. Webb and D. R. Trentham, *J. Biol. Chem.* **225,** 1775–1779 (1980); M. D. Tsai and T. T. Chang, *J. Am. Chem. Soc.* **102,** 5416–5418 (1980)].

Nonprotein Amino Acids

L. FOWDEN

I. INTRODUCTION

The characteristic properties of an amino acid are conferred by carboxyl and amino groups which are features of all compounds of this type (see Larsen, this series, Vol. 5, Chapter 6). The possession of an amino group places these compounds among the nitrogenous constituents of plants. Of the elements essential for plant growth, nitrogen tends to have a special position, for it is the element among those taken up by root systems whose partial deficiency most frequently causes stunted growth. The reasons for this must lie partly in the very ramified pathways by which nitrogen is assimilated to produce a great variety of cell constituents including proteins, nucleic acids, chlorophyll, and certain growth hormones—all indispensible components of

The Biochemistry of Plants, Vol. 7

living plants. It is then a little surprising that many plants channel nitrogen, sometimes in substantial amounts, into compounds that seem to be of secondary importance, and that they continue to do this when the plant's supply of nitrogen is less than optimal. Such compounds include alkaloids (the most numerous class of plant nitrogenous compounds), cyanogenic glucosides and glucosinolates, and about 200 amino acids which occur only free or as simple peptides in plants and not as constituents of protein. A discussion of this last group of compounds, commonly called nonprotein amino acids, forms the basis of this chapter.

More examples of nonprotein amino acids have been found in green plants than in nonchlorophyllous organisms, although many peptide antibiotics contain residues of one or more compounds of this type. Macrofungi are also sources of a number of compounds, but relatively few types have been identified in animal tissues. Many of the known compounds were first identified and isolated from seeds, in which they seem frequently to accumulate—leading to the suggestion that some may function as storage forms of nitrogen available to the succeeding generation of plants at the onset of growth. Particular compounds may be found widely distributed within the families of the plant kingdom—γ-aminobutyric acid and homoserine are such examples—but, more generally, compounds have a restricted distribution, frequently being found only within a closely related group of plants such as a genus or a few allied genera from a family. A few compounds, present in some but not all members of a family, are also encountered in quite unrelated families, or even in different orders of plants, in ways that appear random on the basis of present information. Azetidine-2-carboxylic acid, identified in liliaceous plants, some legumes, sugar beet, and a marine alga, provides an example of such apparent haphazard distribution.

The occurrence of individual amino acids in botanically allied species, and the coexistence of structurally similar and therefore probably biogenetically related substances in particular plants or groups of plants, suggest that this type of compound may be useful for classifying plants on the basis of chemical criteria, i.e., in chemotaxonomy. Many careful surveys of the distribution of these less common amino acids have confirmed this belief, and illustrations will be presented later in this chapter. The patterns of cooccurrence of amino acids exhibiting structural kinships also may suggest probable metabolic relationships, e.g., that compounds represent successive intermediates in a biosynthetic pathway that terminates at different points in different plants, or are products formed from a common precursor following a bifurcation in the pathway. It is important, however, to realize that the normal chromatographic methods used in survey work do not detect compounds whose concentrations fall below certain threshold values, and therefore substances can never be recorded as absent, merely not detected. This situation has been encountered during the fractionation of extracts of large quantities of plant material, when compounds not detected in the original extract have been recognized at later stages following their concentration in particular fractions (see, for example, Fowden, 1972). Many compounds then may have a wider distribution within members of the plant kingdom than we recognize at present. In turn, this focuses attention upon the extent to which the enzymatic complement necessary for particular biosyntheses may exist in different plants—a complement whose activity may be far from fully expressed.

Some nonprotein amino acids are simple homologues of those forming constituents of protein. Such compounds therefore have molecular sizes and conformations that do not differ too markedly from those of the corresponding protein amino acids, and so they may act as analogue molecules, sometimes mimicking the behavior of normal molecules and sometimes acting as metabolic antagonists or inhibitors. Structural differences other than those of homology are also consistent with analogue behavior. For example, a close approach to isosterism can result from the replacement of one atom (or group of atoms) by another of similar size and polar character, as seen in the behavior of canavanine as an analogue of arginine (replacement of $-CH_2-$ by $-O-$) and of S-aminoethylcysteine as a lysine analogue (replacement of $-CH_2-$ by $-S-$). Many such molecules were isolated and characterized initially merely as compounds illustrating the varied nature of the amino acid complex of plants, and only subsequently was their analogue behavior appreciated. However, other compounds resulted from searches for a chemical explanation of toxicities associated with certain plants eaten by humans or grazing animals—hypoglycins A and B and indospicine are examples of such nonprotein amino acids. Now a wider significance is attached to the occurrence of these toxic or antagonistic amino acids in plants, for there is increasing evidence, albeit largely indirect, suggesting that they may have an important role in influencing plant–pathogen or plant–pest insect relationships (see Bell, this volume, Chapter 1). They have also proved useful in achieving a better understanding of the pathways of intermediary nitrogen metabolism in plants by virtue of their ability to regulate selectively the rates of specific reactions. As a special case of their value in regulating metabolic reactions, there are indications that certain nonprotein amino acids, used to mimic the end products of biosynthetic pathways, may be applied in the selection of cell lines possessing altered regulatory control of earlier, intermediate steps in the pathway in ways leading to enhanced production of the normal end product. If such possibilities can be exploited in practice to provide new plant lines capable of further multiplication by the plant breeder, then a new approach to developing crops with enhanced nutritional attributes for humans and animals will have been created.

II. CHEMISTRY AND BIOGENESIS

The number of nonprotein amino acids isolated from plants and fully characterized now exceeds 200. Many other examples of this class of compound still remain to be identified, for it is common experience during the analysis of most plant extracts to encounter small quantities of uncharacterized substances that undoubtedly represent additional amino acids. The great majority of the present 200 or so compounds have been recognized within the past 30 yr, i.e., since paper and ion-exchange chromatographic methods became sufficiently facile to apply in a routine way to the survey of a wide range of plant species. But a few compounds predate this upsurge in chromatographic methodology, and canavanine, citrulline, and mimosine represent examples of compounds isolated considerably earlier by more classical procedures of natural product chemistry.

During the past 20 yr, the development of newer physicochemical techniques has tremendously simplified structural identifications. Mass spectrometry, and especially nmr spectroscopy, have become paramount in such studies, making identifications much more rapid and requiring smaller samples of isolates than the traditional methods relying on degradative chemical procedures.

The 20 amino acids constituting protein can be divided into structural classes. The least sophisticated possess an alkyl radical [R in $RCH(NH_2)COOH$] associated with one amino and one carboxyl group; others have an additional carboxyl in R and are termed acidic amino acids. A class of basic amino acids contains within R a further group capable of protonation. Amino acid amides possess a $CONH_2$ function, whereas the classes of hydroxy and sulfur amino acids are characterized by $-OH$ and $-S-$ in R. Aromatic amino acids are all examples of alanines in which an aryl moiety has been substituted on the β-carbon atom. A single imino acid, proline, occurs universally in plant proteins, and smaller amounts of 4-hydroxyproline are present in certain proteins exhibiting a structural function. On examination, nonprotein amino acids are seen to include many of the same chemical types. For example, a considerable number of additional acidic amino acids have been characterized, and some of these are found associated with the corresponding amino acid amides in the tissues of the plants from which they were isolated. The imino acid group is represented not only by further compounds based on the five-atom heterocycle typified by proline, but also by a range of other substances having either a four- or six-atom heterocycle. Many sulfur-containing amino acids are S-substituted derivatives of cysteine; the substituent may be a simple alkyl group or may contain a carboxyl or amino function or even a second sulfur atom, as in djenkolic acid. Nonprotein amino acids also exhibit structural features not encountered in their protein counterparts; for example, a considerable num-

ber of compounds contain ethylenic linkages, whereas only a small number
possess the acetylenic bond. The cyclopropane ring is encountered in about
10 nonprotein amino acids, and a variety of novel heterocycles (containing
nitrogen, oxygen, sulfur, or combinations thereof) occur in other compounds, usually as substituents attached to the β-carbon of alanine. Finally,
many plants contain γ-glutamyl derivatives of certain amino acids accumulating in large concentrations, typically in the seeds of certain species.

A. Acidic Amino Acids

The majority of such compounds normally possess two carboxyl groups
and a single amino function; the group as encountered in proteins is represented by aspartic and glutamic acids. A range of closely allied compounds
can be regarded as substituted glutamic acids (see review by Fowden, 1970).
The simplest types have hydroxy or methyl substituents on either the β- or
γ-carbon atom of glutamic acid, e.g., *threo*-γ-hydroxy-L-glutamic acid (Ia)
isolated from *Phlox decussata* and certain ferns and *erythro*-γ-methyl-L-
glutamic acid (Ib) obtained from *Phyllitis scolopendrium*. Or they may contain
both these substituents attached to the γ-carbon atom as in the $2S$, $4S$ form of
γ-hydroxy-γ-methylglutamic acid (Ic), which also occurs in *P. scolopendrium*. γ-Methylene-L-glutamic acid (IIa) was an earlier isolate in this class,
being obtained from *Arachis hypogaea* and *Tulipa gesneriana*. More recently
γ-methyleneglutamic acid has been identified in some ferns and fungi; in a
fungus (Hatanaka and Katayama, 1975) it is accompanied by the two higher
homologues γ-ethylidene- (IIb) and γ-propylidene-glutamic acids (IIc).
γ-Methylene-L-glutamine (III) usually coexists with γ-methyleneglutamic

$$HOOCC(R_1R_2)CH_2CH(NH_2)COOH$$

Ia. $R_1 = OH$, $R_2 = H$
Ib. $R_1 = CH_3$, $R_2 = H$
Ic. $R_1 = OH$, $R_2 = CH_3$

$$H_2NOCC(=CH_2)CH_2CH(NH_2)COOH$$

III

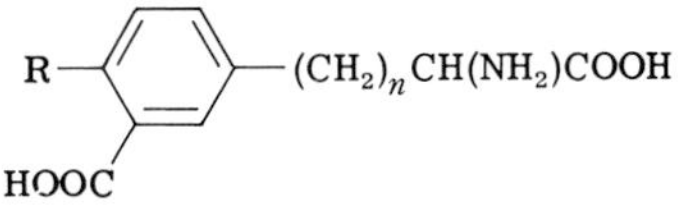

$$HOOC(=R)CH_2CH(NH_2)COOH$$

IIa. $R = CH_2$
IIb. $R = CHCH_3$
IIc. $R = CHCH_2CH_3$

IVa. $R = H$, $n = 1$
IVb. $R = OH$, $n = 1$
IVc. $R = H$, $n = 0$
IVd. $R = OH$, $n = 0$

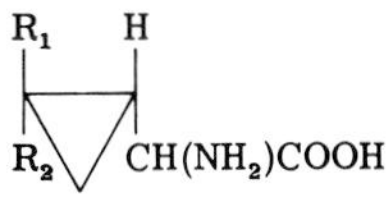

Va. $R_1 = H$, $R_2 = COOH$
Vb. $R_1 = COOH$, $R_2 = H$
Vc. $R_1 = CH_2COOH$, $R_2 = H$

$$RSCH_2CH(NH_2)COOH$$

VIa. $R = CH_2CH_2COOH$
VIb. $R = CH(CH_3)CH_2COOH$

acid and actually represents the principal nitrogenous constituent present in sap exuding from cut stems of *Arachis. threo-γ*-Hydroxy-L-glutamine occurs in *P. decussata*.

The compounds described so far possess the second carboxyl group attached to a simple linear carbon skeleton, but more heterogeneous types of dicarboxylic amino acids are known. For example, a group of amino acids characteristic of some members of the families Iridaceae and Resedaceae contain a carboxyl attached to a phenyl ring at the meta position: 3-carboxyphenylalanine (IVa), 3-carboxytyrosine (IVb), 3-carboxyphenyl-glycine (IVc), and 3-carboxy-4-hydroxyphenylglycine (IVd) (Larsen, 1967). Carboxyl groups attached to a cyclopropyl moiety are encountered in *cis*- (Va) and *trans-α*-(carboxycyclopropyl)glycines (Vb) isolated from certain *Aesculus* species (Fowden *et al.*, 1970) and, through an intervening —CH_2— group, in *trans-α*-(carboxymethylcyclopropyl)glycine (Vc) isolated from seed of *Blighia unijugata* (Fowden *et al.*, 1972a). A third type of dicarboxylic amino acid is found among S-substituted cysteines, of which S-carboxyethyl- (VIa) and S-carboxyisopropyl-L-cysteines (VIb) are structurally simple examples; both these compounds occur in seeds of some species of the genera *Acacia* and *Albizzia* (see Seneviratne and Fowden, 1968).

The biosynthetic mechanisms leading to nonprotein amino acids remain largely unknown. Many biogenetic hypotheses exist, based partly on structural relationships existing between compounds occurring in a single or related group of species, but these at best are mostly supported by data derived from isotope labeling experiments. Little enzymatic evidence is available. The pathway leading to the 3-carboxyphenylalanines perhaps represents the best documented example of biogenesis (Fig. 1). Doubly labeled (^{3}H and ^{14}C) shikimate (VII) was used to establish that these aromatic compounds were derived by an alternative of the normal shikimate aromatic pathway; after the formation of chorismate (VIII, a normal intermediate in the biosynthesis of phenylalanine and tyrosine) a different rearrangement of the C_3 moiety occurred to produce isochorismate (IX) and isoprephenate (X) and finally *m*-carboxyamino acids (Larsen *et al.*, 1975).

A simple biogenetic hypothesis explaining the formation of γ-substituted glutamic acids can be advanced on the basis of the known chemical condensation of two molecules of pyruvate to yield the keto (oxo)acid corresponding to γ-hydroxy-γ-methylglutamic acid; transamination could then be envisaged as the final step in the formation of the amino acid. γ-Methyleneglutamic acid could result from the loss of a molecule of water across the γ-carbon atom, and this in turn might be hydrogenated to yield γ-methylglutamic acid. This simple reasoning, however, has not been confirmed by definitive experiments, and the proposed pathway probably must be rejected as unlikely. A separate study using seedlings of the legume *Gleditsia triacanthos* provided evidence, based on ^{14}C labeling data, that

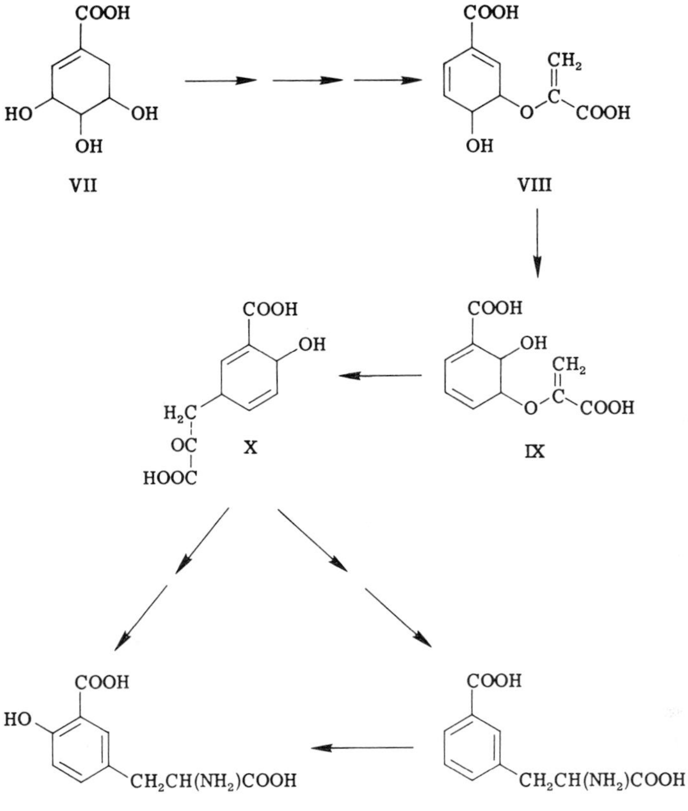

Fig. 1 The biogenetic pathway leading from shikimate to the m-carboxy-substituted aryl amino acids.

leucine was a precursor of γ-methylglutamic acid; seemingly one of the terminal methyl groups of leucine was oxidized to carboxyl (Peterson and Fowden, 1972). Although the seedlings contained γ-methyleneglutamic acid, no ^{14}C label was introduced into the unsaturated amino acid. The existence of a series of homologous unsaturated acids (γ-methylene-, γ-ethylidene-, and γ-propylidene-glutamic acids) is perhaps suggestive of a common biogenetic mechanism, in which C_1, C_2, and C_3 units are attached to a C_5 glutamic acid-like skeleton. The uncertainty still surrounding the biogenesis of these substituted glutamic acids, 30 yr after their initial characterization as plant products, typifies the general position regarding nonprotein amino acids—they are compounds that frequently accumulate in high concentrations but possibly by reactions proceeding relatively slowly and often in species rather intractable to experiment.

B. Imino Acids

In most plant species, proline (XIa) forms the principal imino acid in the soluble nitrogen fraction. Generally, it is not a predominant component in comparison with other amino acids, but substantial accumulation occurs in certain species and, more generally, in particular tissues (e.g., pollen) or under specific physiological conditions (e.g., water stress). Many ring-substituted derivatives of proline also exist (Fowden, 1976). *cis*-4-Hydroxy-L-proline (XIb) has been isolated from leaves of the sandal tree (*Santalum album*) and detected in extracts of other species, whereas *trans*-3-hydroxy-L-proline (XII) represents a major constituent of seed and seedlings of the legume tree, *Delonix regia*. *trans*-4-Methyl-L-proline (XIc) and *cis*-4-hydroxymethyl-L-proline (XId) occur in the fruits of apple and other members of the family Rosaceae. 4-Methyleneproline (XIe, apparently the DL racemate) occurs in seed of loquats (*Eriobotrya japonica*) and *cis*(*exo*)-3,4-methano-L-proline (XIII) in seed of *Aesculus parviflora*; the latter imino acid may be regarded as either a five- or six-atom heterocycle, and in the latter configuration resembles pipecolic acid.

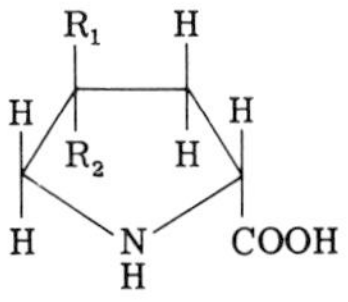

XIa. $R_1 = R_2 = H$
XIb. $R_1 = H$, $R_2 = OH$
XIc. $R_1 = CH_3$, $R_2 = H$
XId. $R_1 = H$, $R_2 = CH_2OH$
XIe. $R_1 + R_2 = CH_2$

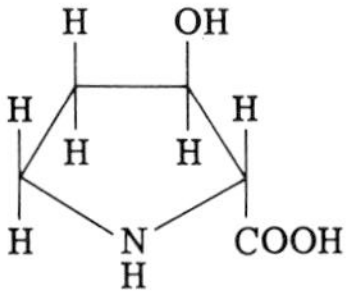

XII

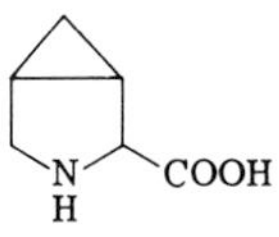

XIII

The lower homologue of proline, azetidine-2-carboxylic acid (XIVa), was isolated first from members of the family Liliaceae (Fowden, 1955). The imino acid represents the predominant soluble nitrogen compound in the tissues of plants such as *Convallaria majalis* and *Polygonatum multiflorum*. More recently, it has been found present in seeds or seedlings of some legumes (e.g., *D. regia* and *Bussea* spp.), in a red marine alga, in trace amounts in sugar beet, and probably in seed of *Fagus silvatica*. Although initially azetidine-2-carboxylic acid appeared to be a useful chemical marker for plants belonging to the Liliaceae and the closely allied Amaryllidaceae, later identifications have been from species widely scattered within the families and orders of the plant kingdom. In contrast to the situation recorded earlier for proline, there have been no reports of derivatives of azetidine-2-carboxylic acids occurring in chlorophyllous plants; 3-ethylidene-L-azetidine-2-carboxylic acid (XIVb), however, is a component of certain polyoxin antibiotics (Isono and Susuki, 1968).

The corresponding six-atom heterocyclic compound, pipecolic acid (XVa), is recognized as a constituent of many plants and is often clearly evident when extracts of legume seeds are examined chromatographically. In many legume plants, it is accompanied by *trans*-4- (XVb) and *trans*-5-hydroxy-L-pipecolic acids (XVc) and, in some, the related unsaturated compound, L-baikiain (XVI, 4,5-dehydro-L-pipecolic acid), coexists (Fowden, 1976). *trans*-5-Hydroxy-L-pipecolic acid also forms a characteristic constituent of date fruits (*Rhapis* spp.), and 4,5-dihydroxypipecolic acids (XVd) have been identified recently as plant constituents (Shewry and Fowden, 1976).

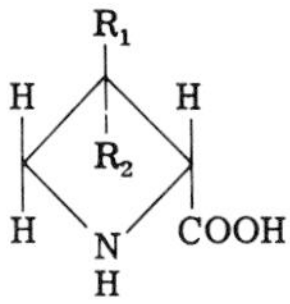

XIVa. $R_1 = R_2 = H$
XIVb. $R_1 + R_2 = CHCH_3$

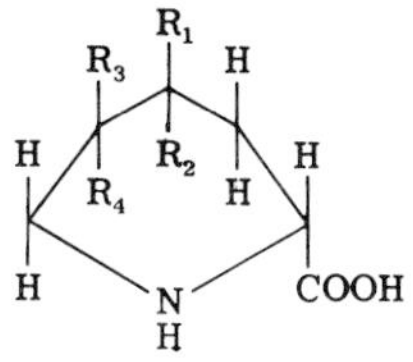

XVa. $R_1 = R_2 = R_3 = R_4 = H$
XVb. $R_1 = OH, R_2 = R_3 = R_4 = H$
XVc. $R_1 = R_2 = R_4 = H, R_3 = OH$
XVd. three isomers are recognized:
    (i) $R_1 = R_4 = H, R_2 = R_3 = OH$
    (ii) $R_1 = R_3 = OH, R_2 = R_4 = H$
    (iii) $R_1 = R_4 = OH, R_2 = R_3 = H$

As with acidic amino acids, structural likenesses existing among different imino acids, especially those forming families of compounds coexisting in a single species or in a group of allied species, suggest possible biogenetic relationships. These structural analogies extend to the dicarboxylic amino acids, for a number of imino acids may be produced in ways similar to that in which proline is derived from glutamic acid. For example, it is known that pipecolic acid can arise from α-aminoadipic acid, although normally lysine seems to be a better precursor. Similar reactions could relate γ-methyleneglutamic acid and 4-methyleneproline, γ-methyglutamic acid and 4-methylproline, γ-hydroxyglutamic acid and 4-hydroxyproline, and α-(carboxycyclopropyl)glycine and 3,4-methanoproline. These and other hypothetical conversions are shown in Fig. 2. The scheme suggests that the reverse reactions (imino acid to the corresponding dicarboxylic amino acid) are also likely, and isotope labeling studies have provided evidence in support of some of the individual reactions.

No single compound has been identified as the unique precursor of azetidine-2-carboxylic acid. Studies using specifically labeled methionine have established its conversion to the imino acid in *C. majalis*, the distribution of label in the final product indicating direct use of the C_4 portion of the methionine molecule probably via the activated *S*-adenosyl derivative (Leete *et al.*, 1974). In other experiments with *Convallaria*, α,γ-diaminobutyric acid

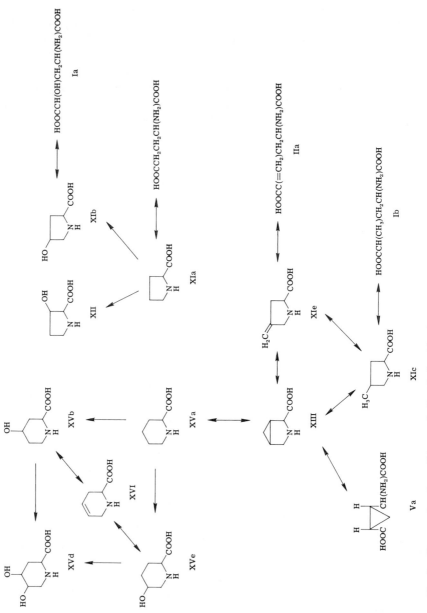

Fig. 2. Structural and possible metabolic relationships between C_5 and C_6 imino acids and corresponding dicarboxylic amino acids.

was tested as a precursor of azetidine-2-carboxylic acid, but no conclusive evidence in support of this conversion was obtained. A separate study on the biosynthesis of polyoxins in *Streptomyces cacaoi* has established that the C_6 skeleton of isoleucine is incorporated directly into 3-ethylidene-azetidine-2-carboxylic acid (Isono *et al.,* 1975).

C. Basic Amino Acids

The simplest type consists of the diamino acids based on a linear carbon skeleton. The series encountered in plants ends with the C_6 protein constituent, lysine (XVIIa). Three lower homologues, α,β-diaminopropionic acid (XVIIb), α,γ-diaminobutyric acid (XVIIc), and ornithine (XVIId), also are plant products. Of these ornithine is implicated as an essential intermediary metabolite in the formation of arginine, another protein constituent. Diaminopropionic and diaminobutyric acids are encountered less commonly, being principally recognized as components of the seeds of some legume genera, e.g., *Acacia* (Seneviratne and Fowden, 1968) and *Lathyrus* (Bell, 1962). Ornithine is formed metabolically from glutamic acid, and lysine can be produced from α-aminoadipic acid, but only in fungi and via intermediary stages different from those involved in ornithine biosynthesis. In green plants, lysine is formed by a quite distinct pathway originating from aspartic acid, in which α,ϵ-diaminopimelic acid is the penultimate compound. The mechanisms whereby diaminopropionic and diaminobutyric acids are produced are not known, but the latter may be derived from aspartic acid by steps similar to those involved in ornithine formation from glutamic acid.

This group of diamino acids gives rise to derivatives in many plants in which they occur. The ω-substituted acetyl derivatives (XVIIe) are found commonly to coexist with the parent compound, and all four ω-acetyl amino derivatives have been fully characterized. β- and γ-oxalyl derivatives (XVIIf) of diaminopropionic and diaminobutyric acids are present in seed of certain *Lathyrus* species, where they represent part of the complex of neurotoxic constitutents characteristic of some members of this genus.

A group other than a second amino can confer basic properties on α-amino acids. Examples are found among the protein amino acids in the guanidino group of arginine and the imidazole ring of histidine. Nonprotein amino acids containing the guanidino group include homoarginine (XVIIIa) and γ-hydroxyhomoarginine (XVIIIb), which are both components of some *Lathyrus* seeds; a basic guanidoxy group forms the ω-terminal portion of canavanine (XIX) that occurs widely in members of the Papilionoideae.

Enduracididine, isolated from *Lonchocarpus sericeus*, represents a strongly basic amino acid whose structure includes an aminoimidazole group (see next section). A similar basic β-substituted alanine, 3-aminomethyl-phenylalanine, occurs as part of a complex of unusual aromatic amino acids in seed of *Combretum* (Mwauluka *et al.*, 1975). Finally, basic compounds are found in S-substituted cysteines, where *S*-aminoethylcysteine from the fungus *Rozites caperata* has gained importance as a lysine analogue.

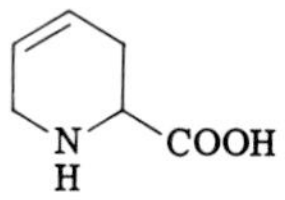

XVI

$$RHN(CH_2)_n CH(NH_2)COOH$$

XVIIa. R = H, n = 4
XVIIb. R = H, n = 1
XVIIc. R = H, n = 2
XVIId. R = H, n = 3
XVIIe. R = COCH$_3$, n = 1–4
XVIIf. R = COCOOH, n = 1 or 2

$$\underset{H_2N}{\overset{HN}{\diagdown}}CNH(CH_2)_2CHRCH_2CH(NH_2)COOH \qquad \underset{H_2N}{\overset{HN}{\diagdown}}CNHOCH_2CH_2CH(NH_2)COOH$$

XVIIIa. R = H   XIX
XVIIIb. R = OH

D. β-Substituted Alanines

This generic name is adopted to describe amino acids in which a phenyl or heterocyclic ring is attached to the β-carbon atom of alanine. Four protein amino acids may be defined as β-substituted alanines: phenylalanine, tyrosine, histidine, and tryptophan. The group displays a diverse range of structure and reactivity and includes a number of compounds mentioned previously.

Nonprotein amino acids containing a phenyl ring system include 3-hydroxyphenylalanine (XXa, *m*-tyrosine), 3,4-dihydroxyphenylalanine (XXb, DOPA), orcylalanine (XXI), 3-carboxyphenylalanine (IVa), 3-carboxy-4-hydroxyphenylalanine (IVb), 3-hydroxymethylphenylalanine (XXc), 3-hydroxymethyl-4-hydroxyphenylalanine (XXd), and 3-aminomethylphenylalanine (XXe). Compounds containing hydroxymethyl and aminomethyl substituents have been found together with carboxy-substituted amino acids, and it appears likely that after the formation of 3-carboxy compounds by the modified shikimate pathway (see earlier) reduction and amination reactions further convert $-COOH$ to $-CH_2OH$ and $-CH_2NH_2$. DOPA and probably *m*-tyrosine formation also follows from the primary products of the shikimate pathway, but available evidence suggests that orcylalanine, like a number of other β-substituted alanines next considered, results from condensation between the appropriate ring structure and a C_3 compound.

R$_1$—⟨ ⟩—CH$_2$CH(NH$_2$)COOH   HO—⟨ CH$_3$ / OH ⟩—CH$_2$CH(NH$_2$)COOH

R$_2$

XXa. R$_1$ = H,   R$_2$ = OH   XXI
XXb. R$_1$ = R$_2$ = OH
XXc. R$_1$ = H,   R$_2$ = CH$_2$OH
XXd. R$_1$ = OH, R$_2$ = CH$_2$OH
XXe. R$_1$ = H,   R$_2$ = CH$_2$NH$_2$

Heterocyclic β-substituted alanines are quite numerous, and the class includes compounds with ring systems having oxygen, nitrogen, or, in bacterial products, sulfur as the heteroatom. Known compounds include the following amino acids (the type of heterocyclic group and plant source are given in parentheses:) mimosine (XXIIa, pyridine, *Mimosa* and *Leucaena* spp.), β-pyrazol-1-alanine (XXIIb, pyrazole, Cucurbitaceae), willardiine (XXIIc) and isowillardiine (XXIId, pyrimidine, *Acacia* spp., peas), lathyrine (XXIIe, pyrimidine, *Lathyrus* spp.), enduracididine (XXIIf, imidazole, *Lonchocarpus* spp.), stizolobic (XXIIg) and stizolobinic acids (XXIIh, pyran, *Stizolobium* spp.), β-isoxazolin-5-one-4-ylalanine (XXIIi, isoxazoline, peas), and β-thiazolylalanine (XXIIj, thiazole, bacterial antibiotic). Experiments in which still other heterocycles were supplied to cucurbit seedlings resulted in the production of further types of β-substituted alanine based on triazole, aminotrizole, and aminotetrazole residues (Frisch *et al.*, 1967). Such experiments suggest that the natural occurrence of certain heterocyclic β-substituted alanines depends upon the ability of individual plant species to elaborate the required heterocycle and not upon the possession of specific enzymes effecting final amino acid production. Many amino acids are apparently derived by a condensation mechanism between the heterocyle and a C_3 moiety derived from serine. This mode of biosynthesis is supported by the results of ^{14}C labeling studies showing that the three-carbon chain of serine can be introduced as the alanine residue of mimosine, orcylalanine, willardiine and isowillardiine, and β-isoxazolin-5-one-4-ylalanine when [^{14}C]serine is supplied to seedlings of the producer plants. Other work has lead to the characterization of enzyme systems effecting condensations between O-acetylserine (an activated form of serine) and 3,4-dihydroxypyridine to form mimosine using *Leucaena* seedling extracts, and with β-isoxazolin-5-one to yield the β-substituted alanine derivative in pea seedling extracts (Murakoshi *et al.*, 1973). It is premature, however, to assume that this mechanism is general in the synthesis of heterocyclic β-substituted alanines; for example, lathyrine can be formed in a biosynthetic pathway commencing from homoarginine and involving γ-hydroxyhomoarginine as an intermediate, in which heterocyclic ring formation occurs only at a later stage.

$$RCH_2CH(NH_2)COOH$$

XXIIa. R =

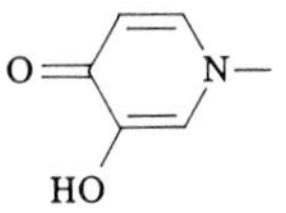

XXIIb. R =

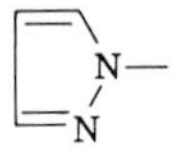

XXIIc. R =

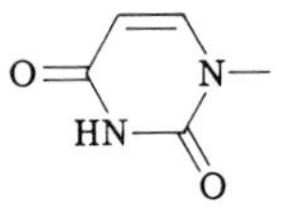

XXIId. R =

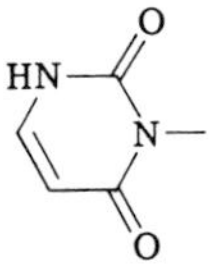

XXXe. R =

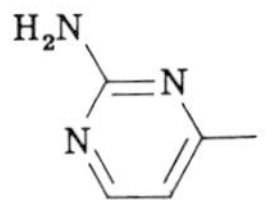

XXIIf. R =

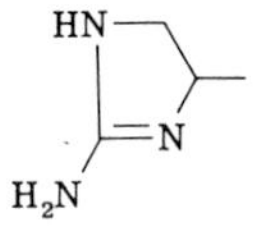

XXIIg. R =

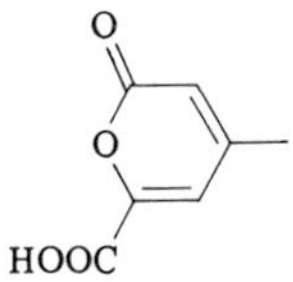

XXIIh. R = HOOC—

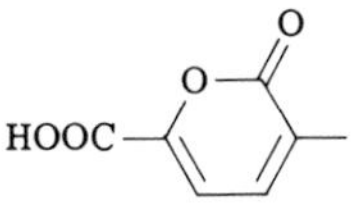

XXIIi. R =

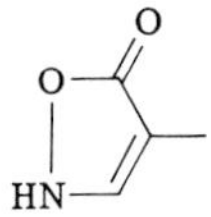

XXIIj. R =

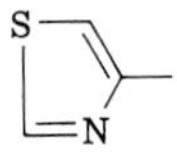

E. Branched-Chain and Cyclopropyl Amino Acids

Apart from the first cyclopropyl amino acid isolated from plants—1-aminoyclopropane-1-carboxylic acid (XXIII) from fruits of pear and apple—all other compounds of this type have been derived from members of the plant families Sapindaceae, Hippocastanaceae, and Aceraceae. The same plants often contain a range of C_6 and C_7 amino acids having noncyclic branched carbon skeletons, in which the positions of branching suggest possible biogenetic relationships with cyclopropane-containing acids. Many of these compounds are unsaturated, containing ethylenic or acetylenic linkages.

Interest in these compounds began when the toxic compound present in unripe akee fruits (*Blighia sapida*, Sapindaceae) was shown to be β-(methylenecyclopropyl)alanine (XXIVa, hypoglycin A); it is associated with even larger quantities of its γ-glutamyl derivative (XXIVb) in akee seeds (Ellington *et al.*, 1959). Not long afterward, the lower homologue, α-(methylenecyclopropyl)glycine (XXIVc), was obtained from seed of litchi (*Litchi chinensis*), another species in the family Sapindaceae. Later, all three compounds were demonstrated to coexist, together with the γ-glutamyl derivative (XXIVd) of α-(methylenecyclopropyl)glycine, in seed of *Billia hippocastanum* (Hippocastanaceae) and *Acer pseudoplatanus* (Aceraceae) (Fowden *et al.*, 1972b).

The most detailed survey of this class of amino acids has involved principally the genus *Aesculus*. Seed of *Aesculus californica* was shown to contain three C_7 amino acids possessing similarly branched carbon chains. One of these, 2-amino-4-methylhex-4-enoic acid (XXV), was the most dominant component of the free amino acid fraction extracted from the seeds; it was accompanied by its saturated form, homoisoleucine (XXVI), and its chain-terminal 6-hydroxy derivative. Although hypoglycin A (which has a closely allied C_7 skeleton) was not detected in the seeds, small amounts of β-(methylenecyclopropyl)-β-methylalanine (XXVII) were isolated, suggesting that hypoglycin A may act as an intermediate receptor in a C_1 transfer reaction. Several different species of *Aesculus* produce the cis and trans forms of α-(carboxycyclopropyl)glycine (Va and Vb); the trans isomer also occurs in seed of *B. sapida*, together with its γ-glutamyl derivative. Another species of *Blighia*, *B. unijugata*, contains *trans*-α-(carboxymethylcyclopropyl)glycine (Vc). The final cyclopropane-containing compound encountered in this group of plants has been mentioned previ-

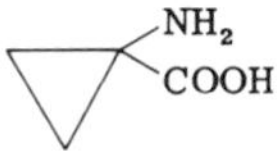

XXIII

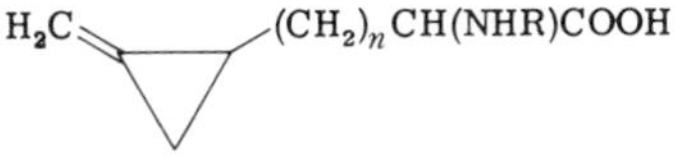

XXIVa. R = H, n = 1
XXIVb. R = OCCH$_2$CH$_2$CH(NH$_2$)COOH, n = 1
XXIVc. R = H, n = 0
XXIVd. R = OCCH$_2$CH$_2$CH(NH$_2$)COOH, n = 0

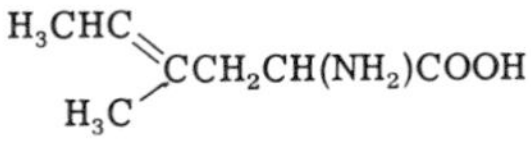

XXV

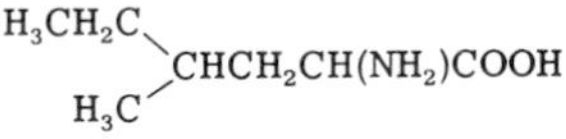

XXVI

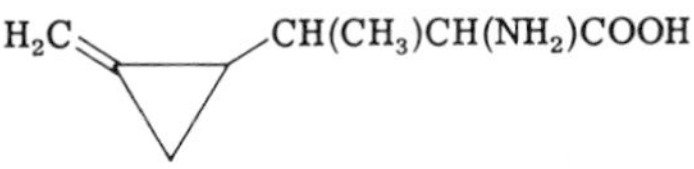

XXVII

ously as *cis*-3,4-methanoproline (XIII), an imino acid identified uniquely in *Aesculus parviflora* seed (Fowden *et al.*, 1970).

Two C_7 amino acids, with similarly branched carbon chains, were isolated from seed of *Euphoria longan* (Sapindaceae) and characterized as 2-amino-4-methylhex-5-ynoic acid (XXVIIIa) and 2-amino-4-hydroxy-methylhex-5-ynoic acid (XXVIIIb). A third acetylenic compound with an unbranched skeleton was shown to be 2-amino-4-hydroxyhept-6-ynoic acid (XXIX) (Sung *et al.*, 1969).

An amino acid, 2-amino-5-methyl-6-hydroxyhex-4-enoic acid (XXXa), also isolated from *B. unijugata*, differs from those previously described in having the branched point within the carbon skeleton one atom closer to the ω end; it represents a hydroxylated form of 2-amino-5-methylhex-4-enoic acid (XXXb) recognized only as a fungal constituent (in *Leucocortinarius bulbiger*) (Dardenne *et al.*, 1967).

$$HC \equiv C \diagdown$$
$$ CHCH_2CH(NH_2)COOH \qquad\qquad HC \equiv CCH_2CH(OH)CH_2CH(NH_2)COOH$$
$$RH_2C \diagup$$

XXVIIIa. R = H   XXIX
XXVIIIb. R = OH

$$RH_2C \diagdown$$
$$ C = CHCH_2CH(NH_2)COOH$$
$$H_3C \diagup$$

XXXa. R = OH
XXXb. R = H

The pattern of the carbon skeleton common to all C_6 cyclopropyl amino acids is such that it could be derived from either leucine or isoleucine by ring closure mechanisms. There is, however, no experimental evidence able to substantiate or refute this idea; the possibility that biosynthetic pathways lead directly to the cyclopropane-containing skeleton by condensation of smaller precursor molecules must be appreciated. A limited study on hypoglycin A biosynthesis in developing akee fruits has been undertaken using ^{14}C-labeled forms of several possible precursors, but no clear biogenetic hypothesis has emerged (R. J. Suhadolnik, personal communication).

More detailed information is available concerning the probable mechanisms responsible for 2-amino-4-methylhex-4-enoic acid formation in *A. californica*. By analogy with the pathway producing leucine from valine in plants, isoleucine can be envisaged as the probable precursor of the branched-chain C_7 amino acids in *Aesculus*. The biogenetic hypothesis (Fig. 3) requires chain lengthening at the carboxyl end of isoleucine, a process associated with acetyl-coenzyme A condensation coupled with loss of the

C-1 atom of isoleucine. Supporting evidence was gained from experiments in which various ^{14}C compounds were supplied to developing fruits of *A. californica* still subtended on the tree. Label from [U-^{14}C]isoleucine was incorporated significantly into 2-amino-4-methylhex-4-enoic acid; ^{14}C from either [1-^{14}C]- or [2-^{14}C]acetate was also incorporated, though in lower percentages. The simultaneous supply of unlabeled tiglate reduced the extent of ^{14}C incorporation from isoleucine—an expected result if the scheme in Fig. 3 is operative. In contrast, when [1-^{14}C]isoleucine was supplied, no radioactivity could be detected in the C_7 amino acid. Similarly, when [U-^{14}C]leucine, [3-^{14}C]serine, or [methyl-^{14}C]methionine was used, no ^{14}C entered 2-amino-4-methylhex-4-enoic acid. Therefore no support was obtained for an alternative biogenetic hypothesis in which the carbon chain of leucine would be extended by the addition of a C_1 unit from serine or methionine at one of the terminal methyl carbons (Fowden and Mazelis, 1971).

When [^{14}C]2-amino-4-methylhex-4-enoic acid itself was supplied to developing fruits of *A. californica*, label was introduced into homoisoleucine and, more actively, into 2-amino-4-methyl-6-hydroxyhex-4-enoic acid, indicating active reduction and hydroxylation processes. In contrast, no label was detected in 2-amino-4-methylhex-4-enoic acid after [^{14}C]homoisoleucine was provided (Boyle and Fowden, 1971). This evidence suggests that the unsaturated form of C_7 amino acids is first produced. The close similarity between the carbon skeleton of hypoglycin A and the branched-chain C_7 amino acids of *Aesculus* suggests the possibility of common biogenetic intermediates with isoleucine also acting as an initial precursor of hypoglycin A biosynthesis.

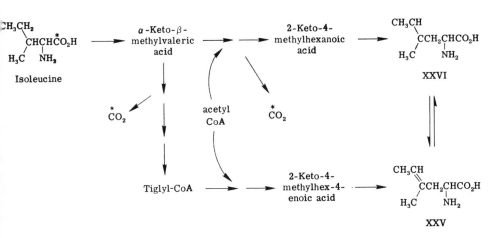

Fig. 3. Possible biosynthetic pathways leading to the formation of branched-chain C_7 amino acids from isoleucine.

III. NONPROTEIN AMINO ACIDS AS INDEXES FOR CHEMOTAXONOMY

Long before chemists were able to study the structure of molecules, it was recognized that related groups of plants often possessed similar chemical constituents. Early correlations relied upon the recognition of sweet, acidic, astringent, aromatic, pungent, or other organoleptic principles. Now that reliable and rapid methods exist for the isolation and structural characterization of plant constituents, interest in the distribution of certain plant products as taxonomic indexes has increased markedly. This interest is soundly based, since secondary metabolites such as nonprotein amino acids are more direct products of gene action than the anatomical and morphological features upon which much of the practice of plant classification depends.

We have already noted differences in the universality with which various amino acids are encountered in members of the plant kingdom, ranging from γ-aminobutyric acid, which is invariably present, to a substance like 3,4-methanoproline (XIII) which is recognized only as a product of the species (*A. parviflora*) from which it has been isolated. There are many other instances of compounds at present known to have a similar very restricted distribution. For example, the related 4-methyleneproline (XIc) was characterized initially from seed of *Eriobotrya japonica* and has been found subsequently only in a closely allied genus, *Raphiolepis*. Certain other compounds occur across a broad sweep of families constituting the plant kingdom, but their presence both within and across families may have a somewhat sporadic character, e.g., as in the case of azetidine-2-carboxylic acid.

Considerable attention has centered on the distribution of nonprotein amino acids in a large and economically important family, the Leguminosae. The members of this group have proven to be a rich source of "new" compounds, and canavanine may be cited as an example of an early isolate readily recognized by the use of a specific color reagent in association with paper chromatographic survey techniques. This basic amino acid is believed to be confined to species within the Papilionoideae; furthermore, refined and detailed examinations of seed extracts have permitted the construction of a probable phylogenetic tree indicating the relationships among the many tribes constituting the Papilionoideae (Fowden *et al.*, 1979).

At the finer level of relationship existing among species constituting a single genus, studies on nonprotein amino acid distribution have established subgeneric groupings within the genera *Lathyrus* and *Vicia* that show a close resemblance to the divisions established using more traditional taxonomic criteria. Within *Lathyrus*, subgeneric divisions were constructed on the basis of the distribution of two associated groups of amino acids (Bell, 1962). The first group contained the C_7 compounds homoarginine, γ-hydroxyhomoarginine, and lathyrine, a series of biogenetically related products. Lathyrine was synthesized by fewer species than homoarginine, presumptive evidence for a lack of the enzymatic complement necessary for the later stages of a biogenetic pathway in some species. The second group of nonprotein amino acids consists of α,γ-diaminobutyric acid, γ-oxalyl-α,γ-diaminobutyric acid, β-oxalyl-α, β-diaminopropionic acid, and γ-glutamyl-β-aminopropionitrile (XXXI); all show toxic properties if ingested by animals, the first three sometimes being described as neurolathyrogens, whereas the last is an osteolathyrogen. On the basis of occurrence of these and certain other uncharacterized seed compounds, Bell envisaged a primary subdivision in which ancestors of present-day *Lathyrus* species either elaborated α,γ-diaminobutyric acid or homoarginine (Fig. 4). Subsequently, the major subdivision producing diaminobutyric acid evolved to produce groups 1 and 5 of the present-day classification; group 1 species still contain α,γ-diamobutyrate as a major constituent, but now it is accompanied by its γ-oxalyl derivative and also β-oxalyl-α,β-diaminopropionic acid. Group 5 species are recognized as an offshoot of the main α,γ-diaminobutyrate stem line, having acquired the capacity to produce first β-aminopropionitrile (by decarboxylation of α,γ-diaminobutyric acid) and subsequently the γ-glutamyl derivative. Groups 2–4 are likely to have arisen from the ancestral stem line elaborating homoarginine. Species in group 2 now contain β-oxalyl-α,β-diaminopropionic acid in addition to homoarginine, whereas those in group 3 have acquired the enzymes necessary to covert homoarginine first into the γ-hydroxy derivative and finally into lathyrine. The presence of a neutral amino acid, γ-hydroxynorvaline, together with homoarginine distinguished species forming group 4. A related study (Bell and Tirimanna, 1965) on species assigned to the *Vicia* genus has relied upon information concerning the subgeneric distribution of the following amino acids: γ-hydroxyornithine (XXXIIa), γ-hydroxycitrulline (XXXIIb), γ-hydroxyarginine (XXXIIc), canavanine, β-cyanoalanine (XXXIIIa), and γ-glutamyl-β-cyanoalanine (XXXIIIb). Three distinct groups of species are recognized using these chemical indexes; one group containing about a third of the species examined possesses β-cyanoalanine

$$NHCH_2CH_2CN$$
$$|$$
$$COCH_2CH_2CH(NH_2)COOH$$

XXXI

$$RHNCH_2CH(OH)CH_2CH(NH_2)COOH$$

XXXIIa. R = H
XXXIIb. R = $CONH_2$
XXXIIc. R = $C{\underset{NH_2}{\overset{NH}{\big<}}}$

$$CH_2(CN)CH(NHR)COOH$$

XXXIIIa. R = H
XXXIIIb. R = $OCCH_2CH_2CH(NH_2)COOH$

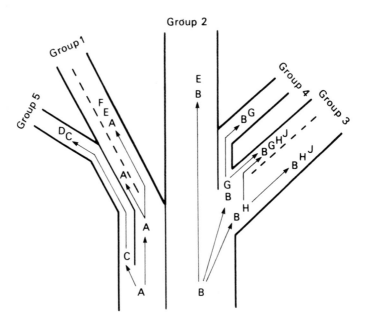

Fig. 4. A possible evolutionary scheme leading to the five groups of species forming the *Lathyrus* genus. Letters indicate the following amino acids: A, α,γ-diaminobutyric acid; B, homoarginine; C, β-aminopropionitrile; D, γ-glutamyl-β-aminopropionitrile; E, β-oxalyl-α, β-diaminopropionic acid; F, γ-oxalyl-α,γ-diaminobutyric acid; G, γ-hydroxynorvaline; H, γ-hydroxyhomoarginine; J, lathyrine.

and its γ-glutamyl peptide, perhaps because these species, unlike others, lack an active nitrilase hydrolyzing β-cyanoalanine to asparagine. The remaining species of *Vicia* may be divided into those synthesizing canavanine and those that do not. Species lacking both canavanine and β-cyanoalanine usually possess high concentrations of γ-hydroxyarginine and/or arginine in their seeds. There are then very clear differences in composition, not only among subgeneric groups within both *Lathyrus* and *Vicia* but also invariably among species from the different genera. Indeed, amino acid analysis has been used to establish unequivocally the generic affiliation of seed samples of uncertain status.

Cucurbitaceae represents another family producing characteristic amino acids (Dunnill and Fowden, 1965). β-Pyrazol-1-ylalanine and its γ-glutamyl peptide occur fairly widely among members of the family, but two substituted amides, N^4-ethylasparagine (XXXIVa) and N^4-hydroxyethylasparagine (XXXIVb), are restricted to the genera *Ecballium* and *Byronia*. *m*-Carboxyphenylalanine is also encountered within the family and serves to distinguish chemically between some otherwise closely allied species. The observed distribution of these, and certain other amino acids, in seed of species from 35 genera was generally in accord with the type of classification arrived at by Jeffrey (1961, 1964), i.e., a split into two major subfamilies consisting of eight tribes and one tribe, respectively. Where discrepancies were encountered between chemical and morphological criteria, the value of the chemical approach was confirmed by a realization that the anomalies almost always concerned species or genera whose morphological relationships remained in some doubt. This example then well illustrates what should be a general dictum of taxonomic work—that no one type of index is in itself adequate for taxonomic purposes and that the most worthwhile classification must depend upon the sifting and weighting of all types of data, including chemical and biochemical information.

$$RHNOCCH_2CH(NH_2)COOH$$

XXXIVa. R = CH_2CH_3
XXXIVb. R = CH_2CH_2OH

The distribution of branched-chain and cyclopropyl amino acids within the genus *Aesculus* is the last example of the chemotaxonomic approach. Information presented earlier indicated that compounds of these structural types occurred in seed of several genera within the allied families Sapindaceae, Hippocastanaceae, and Aceraceae. Hippocastanaceae is a small family consisting of just two genera, *Aesculus* and *Billia*. The 13 true species of *Aesculus* are grouped in five subgeneric sections; hybridization between species, either within a section or between members of different sections, is also not uncommon. The correlation between amino acid occurrence and the assignment of species to subgeneric sections based on morphological and cytological evidence is extremely good (Table I). A hybrid form, such as the ornamental tree *Aesculus carnea*, tends to reflect its parentage, in this case *A. hippocastanum* x *A. pavia*. On the basis of morphological evidence, the section Macrothyrsus is considered to be more closely related to the section Calothyrsus than to the section Pavia, although the present geographical distribution of the species does not support this conclusion (Hardin, 1957, 1960). The amino acid data are also contrary to this view, for although species assigned to the sections Pavia and Macrothyrsus all synthesize the C_6 group of compounds, those in the section Calothyrsus do not; instead they produce amino acids having a C_7 skeleton (Fowden *et al.*, 1970). Kinetic data gained from a survey of the properties of phenylalanyl-tRNA synthetases isolated from seed of members of the different sections serve clearly to distinguish the section Calothyrsus from the others (see Section IV,C for an account of aminoacyl-tRNA synthetase specificities).

TABLE I

The Distribution of Amino Acids in Seeds of *Aesculus* Spp. and Hybrids

Section and species[a]	Amino acids[b]					
	Va	Vb	XIII	XXV	XXVI	XXVII
Parryaneae						
A. parryi	S	W	0	0	0	0
Aesculus						
A. hippocastanum	0	0	0	0	0	0
A. turbinata	0	0	0	0	0	0
(*A. carnea*)	T	0	0	0	0	0
Calothyrsus						
A. californica	0	0	0	S	W	W
A. indica	0	0	0	0	W	0
Macrothyrsus						
A. parviflora	M	T	S	0	0	0
Pavia						
A. glabra var. *glabra*	M	W	0	0	0	0
A. glabra var. *arguta*	M	M	0	0	0	0
A. octandra	W	W	0	0	0	0
A. sylvatica	M	W	0	0	0	0
A. pavia	T	T	0	0	0	0
(*A. hybrida*)	T	T	0	0	0	0
(*A. arnoldiana*)	T	T	0	0	0	0

[a] Hybrids are shown in parentheses.

[b] Letters denote relative concentrations of amino acid: S, high; M, moderate; W, low; and T, trace. 0 indicates not detected.

IV. NONPROTEIN AMINO ACIDS AS ANALOGUES AND ANTIMETABOLITES

Instances in which nonprotein amino acids interfere with growth and/or metabolism in plants, animals, or microorganisms are now very numerous. Examples already cited involve hypoglycin A (XXIVa) and indospicine (XXXV), compounds identified following a search for the chemical basis of toxicities associated with akee fruits and *Indigofera* herbage. Other well-documented toxic amino acids include mimosine and canavanine, lathyrogens from *Lathyrus* and *Vicia* seeds, and selenium-containing amino acids, especially *Se*-methylselenocysteine (XXXVI), present in certain species of *Astragalus*. Most commonly, the highest concentrations of such toxic amino acids tend to occur in the seeds of producer species, and it is tempting to ascribe to them roles in protecting the seed against attack by predatory insects or as inhibitors of the growth of adjacent competing species after diffusion from the seed, especially during the early stages of germination (see Bell, this volume, Chapter 1).

$$\begin{array}{c} HN \\ \diagdown \\ H_2N \diagup \end{array} CCH_2(CH_2)_3CH(NH_2)COOH \qquad\qquad CH_3SeCH_2CH(NH_2)COOH$$

XXXV XXXVI

$$\begin{array}{c} H_2N \\ \diagdown \\ O \diagup \end{array} CNHCH_2CH(NH_2)COOH$$

XXXVII

Of the toxic amino acids just mentioned, two are close isosteres of constituents of protein. Indospicine and canavanine have molecular configurations differing little from that of arginine, and mimosine has the aromatic character and a shape similar to that of phenylalanine and tyrosine. Other compounds have been shown to behave as effective analogues and antimetabolites; these include azetidine-2-carboxylic acid as an analogue of proline, albizziine (XXXVII) of glutamine, and 2-amino-4-methylhex-4-enoic acid of phenylalanine. The fact that the analogue molecule possesses a shape and polarity very similar to that of a protein amino acid suggests that metabolic antagonism is due to the analogue mimicking the normal compound. In practice, antagonism is usually encountered in one or more of the following three principal ways:

1. The analogue molecule competes as an alternative substrate for permease systems responsible for amino acid uptake into cells.
2. The analogue acts as an inhibitor of amino acid biosynthesis, either as a direct competitor of the substrate for the reactive sites of the enzyme or by mimicking the role of the end product of a biosynthetic pathway in its feedback inhibitory action on earlier key enzyme(s) in the pathway.
3. The analogue competes with a normal amino acid in the activation process catalyzed by aminoacyl-tRNA synthetases, the initial reaction in a sequence responsible for the incorporation of amino acids into protein molecules; the analogue molecule may behave as a competitive inhibitor, but probably more often it is accepted by the synthetase as an alternative substrate and thereby itself becomes incorporated into an anomalous form of protein.

A. Amino Acid Uptake

There are few detailed studies on the uptake of analogue amino acids into higher plant cells or tissues, and precise kinetic measurements have not been made. Analogues are readily taken up by seeds during early water imbibition, germination, and seedling growth, and growth inhibition has often been recorded. The action of azetidine-2-carboxylic acid on the growth of seedlings of *Phaseolus aureus* can be cited in illustration (Fowden, 1963). Increased concentrations of the analogue produce progressively greater inhibition of radicle growth and ultimately cause the death of the developing seedling. The inhibition can be alleviated by supplying proline simultaneously with the analogue, suggesting that the two imino acids compete for transport across cell membranes; they also undoubtedly compete as substrates for prolyl-tRNA synthetase, because azetidine-2-carboxylate residues are found in the protein of the seedlings. Nevertheless, competition at the level of permease enzymes has been established far more clearly using bacterial systems. The study of proline uptake by *Escherichia coli* cultures as influenced by a range of imino acid analogues is a good example. The uptake of proline determined sensitively using ^{14}C-labeled material was inhibited by many analogues in the following descending order of effectiveness: 3,4-dehydroproline, thiazolidine-4-carboxylic acid, azetidine-2-carboxylic acid, and 4-methyleneproline; pipecolic acid did not inhibit the uptake of proline. The analogues reduced proline uptake by competing for sites on the permease and were themselves accumulated by the bacterial cells (Tristram and Neale, 1968).

When azetidine-2-carboxylic acid or dehydroproline is supplied to growing cultures of *E. coli*, growth retardation gradually occurs in comparison with the control cultures. It is possible, however, to select mutant cell lines unaffected by these analogues, whose proline permeases have undergone subtle structural changes associated with a much reduced affinity for the analogues but without marked effect upon the binding of proline. Similar observations have been made with other analogue molecules, especially with *o-, m-,* and *p*-fluorophenylalanines (see Fowden *et al.*, 1967).

B. Analogues and Amino Acid Biosynthesis

Analogue amino acids very commonly react with enzyme systems whose normal function is the catalysis of intermediary cellular metabolism. Certain classes of enzymes such as aminotransferases and amino acid decarboxylases and oxidases are not highly specific in their action, and many analogue molecules are accepted as substrates. In this section, however, more specific examples are selected to illustrate the manner in which analogue molecules can serve to regulate amino acid biosynthesis and, indeed, provide evidence concerning the pathways of nitrogen assimilation and incorporation into amino acids.

Since 1970, a dramatic reappraisal of the initial steps concerned in ammonia assimilation by higher plants and bacteria has occurred. For more than

30 years, glutamic dehydrogenase (GDH) was accepted as the key enzyme in this process, α-oxoglutaric acid (α-ketoglutaric acid) undergoing reductive amination to form glutamic acid. In 1970, however, a new enzyme (glutamine-α-oxoglutarate amidotransferase, GOGAT) was described (Tempest *et al.*, 1970); its function was to transfer the amide NH_2 group of glutamine to α-oxoglutarate in the presence of a suitable hydrogen donor [NAD(P)H or reduced ferredoxin]. Acting in conjunction with glutamine synthetase (GS) producing glutamine from ammonia and glutamic acid, it provided an alternative pathway for the assimilation of ammonia into glutamic acid. The alternative pathways may be summarized as follows

GDH pathway:

$$NH_3 + \alpha\text{-oxoglutarate} + NADH \rightarrow \text{glutamate} + NAD^+$$

GS-GOGAT pathway:

(i) $NH_3 + \text{glutamate} \xrightarrow{\text{ATP}} \text{glutamine}$

(ii) $\text{Glutamine} + \alpha\text{-oxoglutarate} + Fd(H_2) \rightarrow 2 \times \text{glutamate} + Fd$

Evidence favoring the GS-GOGAT pathway for ammonia assimilation in green plants under normal physiological concentrations was of three principle types: kinetic evidence resting upon determined K_m values for GDH and GS, which established that GDH must be inefficient at the low ammonia concentrations encountered in plant tissues; isotope labeling evidence indicating that ^{15}N from ammonia was incorporated far more rapidly into glutamine than glutamic acid; and evidence from the use of specific metabolic inhibitors, namely, amino acid analogues (see Miflin and Lea, this series, Vol. 5, Chapter 4).

Methionine sulfoximine (XXXVIII) inhibits GS but is without effect on GDH, and so the observation that the rate of glutamine and glutamate formation is reduced in the presence of this analogue is strong presumptive evidence for the GS-GOGAT pathway. Transfer of the amide group from glutamine to various acceptors, including α-oxoglutaric acid in the GOGAT-mediated reaction, is inhibited specifically by albizziine and azaserine (XXXIX), and these two analogues have been used extensively to show that glutamate production is dependent on GOGAT in a variety of plants and with different tissue and subcellular preparations (Miflin and Lea, 1977).

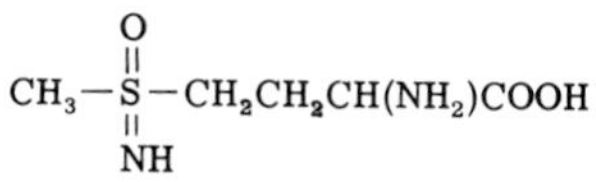

$$CH_3-\underset{\underset{NH}{\overset{\parallel}{\parallel}}}{\overset{\overset{O}{\parallel}}{S}}-CH_2CH_2CH(NH_2)COOH \qquad\qquad HOOCCH(NH_2)CH_2OCOCHN_2$$

XXXVIII   XXXIX

Albizziine has been studied as an inhibitor of asparagine biosynthesis, which results from another amidotransferase-type reaction catalyzed by asparagine synthetase and involving glutamine:

Glutamine + aspartic acid + ATP → glutamic acid + asparagine + ADP + P_i

The properties of asparagine synthetase preparations from seedlings of three legumes have been compared, especially the enzymes' sensitivity to albizziine as an inhibitor (Lea and Fowden, 1975). Two of the legumes (*Acacia farnesiana* and *Albizzia lophantha*) produce large quantities of albizziine, whereas the third (*Lupinus albus*) lacks this glutamine analogue. Presumably, asparagine synthetase has to function in *Acacia* and *Albizzia* in an environment containing at least some albizziine, and it is therefore interesting to note that the enzyme from these two species is far less readily inhibited by albizziine than the synthetase from *Lupinus* (Table II). The inference is that, during evolution, the protein possessing asparagine synthetase activity has been modified in species elaborating much albizziine in ways reducing the affinity of the enzyme's catalytic site for albizziine.

The metabolic pathway leading from aspartic acid to lysine provides a good example of the role of amino acid analogues as end product inhibitors. In this pathway, which has bifurcations leading to three protein amino acids (threonine, methionine, isoleucine), lysine behaves as a feedback inhibitor at two points shown in Fig. 5; at similar concentrations it inhibits dihydrodipicolinate synthetase more strongly than aspartate kinase, a not unexpected finding since the former enzyme functions in the later unbranched part of the pathway leading only to lysine. S-Aminoethylcysteine (XXXX), a molecule containing a sulfur atom in place of the C-4 methylene group of lysine mimics the normal amino acid in its role as end product inhibitor. The

TABLE II

Inhibition of Asparagine Synthetase Preparations from Three Legume Species by Albizziine

Glutamine concentration (mM)	Inhibition produced by 6 mM albizziine (%)		
	Lupinus	*Acacia*	*Albizzia*
0.012	97.3	88.7	100
0.12[a]	93.2	73.5	75
1.2[b]	79.3	33.6	2.8
6.0	71.3	3.6	0

[a] Concentration that saturates enzyme from *Lupinus*.
[b] Concentration that saturates enzymes from *Acacia* and *Albizzia*.

$$H_2NCH_2CH_2SCH_2CH(NH_2)COOH$$

XXXX

analogue then causes inhibition of growth in seedlings, e.g., barley, by limiting the amount of lysine synthesized and therefore available for introduction into cellular proteins. End product inhibition represents one of a series of regulating controls ensuring that a balanced production of intermediary metabolites is maintained within cells at levels appropriate to the physiological and environmental conditions under which the plant is developing at any one time. Genetic mutations may cause disturbances of this delicately regulated situation; for instance, mutant cell lines may elaborate lysine biosynthetic enzymes that either are not subject to feedback inhibition or are

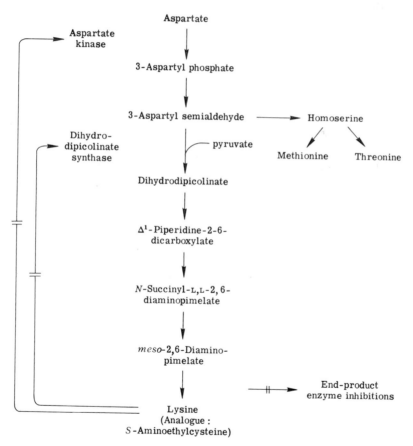

Fig. 5. The steps in the biosynthetic pathway forming lysine that are subject to end product inhibition.

far less strongly inhibited by the end product lysine. S-Aminoethylcysteine can be employed to screen for such mutant cell lines, for unlike normal cells whose growth would be strongly impeded by a much impaired lysine synthesis, those possessing altered forms of enzyme would continue to produce lysine and therefore grow normally; indeed, compared to normal cells, they would overproduce lysine. (It should be recognized that cell lines might exhibit resistance to growth inhibition by S-aminoethylcysteine for other reasons, e.g., an alteration in the kinetic parameters governing the analogue's uptake by cells). The use of analogues in this way provides a possible technique for the selection of cell lines having improved nutritional attributes for humans or animals; in practice, successful development depends upon an ability to screen readily very large numbers of individuals, and so priority must be given to research for developing suitable techniques for handling single cells, or better protoplasts, of agriculturally important crops such as cereals and legumes, including inducing them to grow and differentiate to produce normal plantlets.

C. Analogues and Aminoacyl-tRNA Synthetases

Proteins are built up from 20 amino acids which are initially activated by reaction with ATP catalyzed by 20 aminoacyl-tRNA synthetase enzymes, each specific for a single protein constituent. Aminoacyl residues are then transferred from the activated complexes formed in this first reaction to specific tRNA molecules. These transfers represent the second stage in the overall process of aminoacyl-tRNA formation. It is during these steps that the critical phase of amino acid selection occurs, since apparently little discrimination occurs in the later stages involving the incorporation of amino acid residues into the elongating polypeptides.

The exacting specificity of individual aminoacyl-tRNA synthetases ensures an almost perfect selection of substrates from within the group of 20 protein amino acids. Nevertheless, certain nonprotein amino acids can act as alternative substrates for certain enzymes, and it then follows that they can be introduced anomalously into protein molecules. Azetidine-2-carboxylic acid has provided an excellent example of this type of analogue behavior. The growth inhibition of bacterial cultures or seedlings caused by azetidine-2-carboxylate rests largely upon the fact that it is introduced into cellular proteins in place of proline residues. Since the presence of proline in a polypeptide is associated with an interruption of the α-helical structure and a change in the orientation of the chain axis, such residues play an important part in determining a protein's final tertiary structure. The substitution of an azetidine-2-carboxylate residue for a proline residue inevitably produces different bond angles at these important intramolecular sites, and extensive replacement of prolines within a single protein molecule must produce a marked change in its final conformation—and thereby an altered reactivity of enzymically active proteins.

Kinetic evidence relating to the use of azetidine-2-carboxylic acid as a substrate for prolyl-tRNA synthetase was obtained first with enzyme preparations from seed of *Phaseolus aureus*. This legume does not synthesize the imino acid but is a ready source of various aminoacyl-tRNA synthetases. The prolyl enzyme was considerably purified from seed meal, and its ability to activate proline, azetidine-2-carboxylic acid, dehydroproline, and pipecolic acid was assessed; substrates were used at enzyme-saturating concentrations. Under these conditions, dehydroproline was activated at about half the rate of proline, azetidine-2-carboxylate at 38% of the rate, and pipecolic acid was inactive. When prolyl-tRNA synthetase was prepared from an azetidine-2-carboxylic acid-producing species (*Polygonatum multiflorum*, Liliaceae), the enzyme showed no affinity for the analogue, although dehydroproline was still used as a substrate (Peterson and Fowden, 1965). This ability of a producer species to discriminate enzymatically against its own toxic product ensures that the analogue does not become incorporated into protein and so lead to the production of ineffective enzymes. As a phenomenon, it is now known to be of wider occurrence, as shown by a more comprehensive survey of the specificities of prolyl-tRNA synthetases from a variety of azetidine-2-carboxylate producer and nonproducer species (Table III). In this study, 3,4-methanoproline was used as

TABLE III

Activation of Proline Analogues by Prolyl-tRNA Synthetase Preparations from Various Plant Sources

	Analogue activation[a]	
Plant species	Azetidine-2-carboxylic acid	cis-3,4-Methanoproline
Parkinsonia aculeata[b]	0–5	42
Delonix regia[b]	0–5	22 (12 mM)
Convallaria majalis[b]	0–5	36
Beta vulgaris[c]	73	<3
Hemerocallis fulva[c]	75	<3
Phaseolus aureus[c]	55	<2
Ranunculus bulbosa[c]	66	0

[a] V_m determined for analogue expressed as percentage of V_m for normal substrate proline.

[b] Species synthesizing azetidine-2-carboxylic acid.

[c] Species in which azetidine-2-carboxylic acid is absent or present in only trace amounts.

an analogue having a slightly larger molecular size than proline. The results show very clearly that species elaborating azetidine-2-carboxylic acid possess a form of prolyl-tRNA that does not activate this imino acid, although 3,4-methanoproline is invariably activated. In distinct contrast, the enzyme from species in which azetidine-2-carboxylic acid cannot be detected use the imino acid as a substrate, but they cannot activate methanoproline. Sugar beet, a species known to contain trace amounts of azetidine-2-carboxylic acid, behaves enzymatically as though it were a nonproducer. From these observations, one may infer that in species producing azetidine-2-carboxylic acid the substrate-binding site of prolyl-tRNA synthetase can accommodate a somewhat larger molecule than that from nonproducer species; methanoproline therefore can bind to the former but not the latter type of enzyme, however, azetidine-2-carboxylic acid fits too loosely at the binding site of the former type of enzyme and thereby fails to be activated (Norris and Fowden, 1972).

A similar instance of positive discrimination against glutamic acid analogues by producer species was reported some years after the first studies with prolyl-tRNA synthetases. Glutamyl-tRNA synthetase was prepared from *P. aureus* and from a species (*Hemerocallis fulva*) elaborating *threo-γ*-hydroxy-L-glutamic acid and another (*Caesalpinia bonduc*) producing *erythro-γ*-methyl-L-glutamic acid. The substrate properties determined for the three preparations of glutamyl-tRNA synthetase are given in Table IV. The nonnatural disastereoisomers of each analogue were fortunately available and were included among the substrates used. The same conclusion was reached, namely, that the synthetase enzyme present in an analogue-producing species specifically (and sterically) discriminates against its own product, whereas nonnatural isomers are activated. The substrate specificity of the enzyme from the nonproducer species (*P. aureus*) is less absolute (Lea and Fowden, 1972).

A final example concerns phenylalanyl-tRNA synthetases from various *Aesculus* species. It is chosen partly to illustrate the unexpected analogue role of 2-amino-4-methylhex-4-enoic acid. A cursory examination of the molecular structure of this amino acid might suggest that it could act as an analogue of either leucine or isoleucine. But the presence of the $|-C\equiv C-$ group confers considerable planarity on the molecule about the C-4 atom, giving it a spatial conformation far more akin to that of phenylalanine; it can be regarded as having an incomplete phenyl ring (only four C atoms being present). It acts as an alternative substrate for many preparations of phenylalanyl-tRNA synthetase. Preparations of the enzyme from five species of *Aesculus* (one species from each of the five sections of the genus) all activated 2-amino-4-methylhex-4-enoic acid. When comparisons were made using enzyme-saturating concentrations of the substrates, the analogue was activated as effectively as phenylalanine itself by preparations from

TABLE IV

Activation of Glutamic Acid Analogues by Glutamyl-tRNA Synthetase Preparations from Three Plant Species

	Analogue activation[a]			
Plant species	*erythro*-γ-Methyl-L-glutamic acid	*threo*-γ-Methyl-DL-glutamic acid	*threo*-γ-Hydroxy-L-glutamic acid	*erythro*-γ-Hydroxy-DL-glutamic acid
Phaseolus aureus[b]	68	55	54	55
Hemerocallis fulva[c]	40	0	0	34
Caesalpinia bonduc[d]	0	20	23	0

[a] V_m determined for analogue expressed as percentage of V_m for normal substrate glutamic acid.
[b] Species in which none of analogues tested are found.
[c] Species producing *threo*-γ-hydroxy-L-glutamic acid.
[d] Species producing *erythro*-γ-methyl-L-glutamic acid.

Aesculus hippocastanum, *A. parryi*, and *A. parviflora*, and at 90% of the rate determined from phenylalanine by the synthetase from *A. glabra*; enzyme from *A. californica*, the only species of the five tested producing 2-amino-4-methylhex-4-enoic acid, utilized the analogue 30% as effectively as phenylalanine (Fowden *et al.*, 1970). Although discrimination by enzyme from the producer species is in this instance not absolute, it should be realized that reaction rates were based on measurements of ATP–^{32}PP$_i$ exchange rates (i.e., the first step, in reverse, of the two-stage activation process; see earlier). Possibly, further discrimination against the analogue occurs in the second step during transfer of the aminoacyl residue to tRNA^{Phe}

V. CONCLUDING REMARKS

Twenty-five years ago nonprotein amino acids were regarded generally as oddities of plant biosynthesis. Now, although we still know little detail about the manner in which many of them are elaborated by plants, they have as a group taken on a greater significance, especially as more and more individual compounds have been shown to possess toxic properties or to behave as metabolic analogues. There has been an awakening of interest in the possible physiological and ecological roles of many of these unusual substances, and opportunities seem to exist for their greater use in attempts to understand the regulation of biosynthetic processes and, indeed, to manipulate regulatory control mechanisms via mutant selection.

REFERENCES

Bell, E. A. (1962). *Biochem. J.* **83**, 225–229.
Bell, E. A., and Tirimanna, A. S.L. (1965). *Biochem. J.* **97**, 104–111.
Boyle, J. E., and Fowden, L. (1971). *Phytochemistry* **10**, 2671–2678.
Dardenne, G., Casimir, J., and Jadot, J. (1968). *Phytochemistry* **7**, 1401–1406.
Dunnill, P. M., and Fowden, L. (1965). *Phytochemistry* **4**, 933–944.
Ellington, E. V., Hassall, C. H., Plimmer, J. R., and Seaforth, C. E. (1959). *J. Chem. Soc.* pp. 80–85.
Fowden, L. (1955). *Nature (London)* **176**, 347–348.
Fowden, L. (1963). *J. Exp. Bot.* **14**, 387–398.
Fowden, L. (1970). *Prog. Phytochem.* **2**, 203–266.
Fowden, L. (1972). *Phytochemistry* **11**, 2271–2276.
Fowden, L. (1976). *Heterocycles* **4**, 117–130.
Fowden, L., and Mazelis, M. (1971). *Phytochemistry* **10**, 359–365.
Fowden, L., Lewis, D., and Tristram, H. (1967). *Adv. Enzymol.* **29**, 89–163.
Fowden, L., Anderson, J. W., and Smith, A. (1970). *Phytochemistry* **9**, 2349–2357.
Fowden, L., MacGibbon, C. M., Mellon, F. A., and Sheppard, R. C. (1972a). *Phytochemistry* **11**, 1105–1110.
Fowden, L., Pratt, H. M., and Smith, A. (1972b). *Phytochemistry* **11**, 3521–3523.

Fowden, L., Lea, P. J., and Bell, E. A. (1979). *Adv. Enzymol.* **50,** 117–175.

Frisch, D. M., Dunnill, P. M., Smith, A., and Fowden, L. (1967). *Phytochemistry* **6,** 921–931.

Hardin, J. W. (1957). *Brittonia* **9,** 173–195.

Hardin, J. W. (1960). *Brittonia* **12,** 26–39.

Hatanaka, S., and Katayama, H. (1975). *Phytochemistry* **14,** 1434–1436.

Isono, K., and Susuki, S. (1968). *Tetrahedron Lett.* No. 9, p. 1133.

Isono, K., Funayama, S., and Suhadolnik, R. J. (1975). *Biochemistry* **14,** 2982–2995.

Jeffrey, C. (1961). *Kew Bull.* **15,** 337–371.

Jeffrey, C. (1964). *Kew Bull.* **17,** 473–477.

Larsen, P. O. (1967). *Biochim. Biophys. Acta* **141,** 27–46.

Larsen, P. O., Orderka, O. F., and Floss, H. G. (1975). *Biochim. Biophys. Acta* **381,** 397–408.

Lea, P. J., and Fowden, L. (1972). *Phytochemistry* **11,** 2129–2138.

Lea, P. J., and Fowden, L. (1975). *Proc. R. Soc. London, Ser. B* **192,** 13–26.

Leete, E., Davis, G. E., Hutchinson, C. R., Woo, K. W., and Chedekel, M. R. (1974). *Phytochemistry* **13,** 427–433.

Miflin, B. J., and Lea, P. J. (1977). *Annu. Rev. Plant Physiol.* **28,** 299–329.

Murakoshi, I., Kato, F., Haginawa, J., and Fowden, L. (1973). *Chem. Pharm. Bull.* **22,** 918–921.

Mwauluka, K., Charlwood, B. W., Briggs, J. M., and Bell, E. A. (1975). *Biochem. Physiol. Pflanz.* **168,** 15–18.

Norris, R. D., and Fowden, L. (1972). *Phytochemistry* **11,** 2921–2935.

Peterson, P. J., and Fowden, L. (1965). *Biochem. J.* **97,** 112–124.

Peterson, P. J., and Fowden, L. (1972). *Phytochemistry* **11,** 663–674.

Seneviratne, A. S., and Fowden, L. (1968). *Phytochemistry* **7,** 1039–1054.

Shewry, P. R., and Fowden, L. (1976). *Phytochemistry* **15,** 1981–1984.

Sung, M. L., Fowden, L., Millington, D. S., and Sheppard, R. C. (1969). *Phytochemistry* **8,** 1227–1233.

Tempest, D. W., Meers, J. L., and Brown, C. M. (1970). *Biochem. J.* **117,** 405–407.

Tristram, H., and Neale, S. M. (1968). *J. Gen. Microbiol.* **50,** 121–137.

Amines 9

T. A. SMITH

I. INTRODUCTION

Amines may be considered to be derivatives of ammonia in which the three hydrogen atoms are replaced by one, two, or three alkyl moieties to give, respectively, primary, secondary, or tertiary amines. The carboxyl group is absent, and amines behave as cations at physiological pH values (the pK_a for amines lies between 9 and 11). Amines are widespread in plants and often serve as precursors of alkaloids. Many different structures of varying complexity are found. Plant amines have been reviewed by Werle (1955), Guggenheim (1958), Schütte (1969), Smith (1971, 1977a, 1980), and Smith *et al.* (1978a).

In many cases amino acid decarboxylases (listed by Smith, 1980) form the corresponding amine which may be subsequently modified. The generalized reaction for amino acid decarboxylation is

$$\underset{R-\overset{\overset{\displaystyle COOH}{|}}{CH}-NH_2}{} \longrightarrow R-CH_2-NH_2 + CO_2$$

The Biochemistry of Plants, Vol. 7
249

Metabolically, some amines may be quite important as precursors; for instance, the plant hormone indol-3-yl-acetic acid (IAA) is derived from tryptamine in some plants (Section V). However, amines may also be functional per se, as found for instance with the ubiquitous polyamines (Section III,C). Some amines may have a protective role in deterring predators; this is probably the reason for the accumulation of mescaline in cacti (Section V). However, as is well known, various ethnic groups have exploited mescaline for its hallucinogenic properties. More recently it has been suggested that some amine conjugates (aromatic amides, Section IV) are important as antifungal and antiviral agents.

II. ALIPHATIC MONOAMINES

These amines (listed by Smith, 1980) are especially common in the flowers of various members of the Rosaceae, in which they may simulate the smell of rotting meat. They therefore attract carrion-feeding insects which effect pollination. A similar function may be ascribed to amines found in the flowers of various members of the Araceae, where they form one component of a complex mechanism that has evolved to attract pollinating flies (reviewed by Smith, 1971, 1977a).

A nonspecific insoluble amino acid decarboxylase is responsible for formation of the amines found in many members of the Rhodophyceae (red algae). The best substrate was leucine, but norleucine, isoleucine, valine, norvaline, phenylalanine, methionine, cysteine, and homocysteine could also be decarboxylated. Incubation of the enzyme with substrate displaced pyridoxal phosphate, and centrifugation then yielded almost pure apodecarboxylase (Hartmann, 1972). The enzyme attacks free amino acids in seawater, and the resulting amines are adsorbed to the acidic polysaccharides of the cell walls. These amines may serve to protect the seaweed against bacteria (Hartmann and Aufermann, 1973).

In *Camellia sinensis* (tea plant) the ethylamine moiety of theanine (N^5-ethylglutamine) is produced from alanine by decarboxylation. Activity was much greater in the roots of seedlings grown in the light than in the dark. The enzyme was very labile in extracts, but it was partially stabilized in the presence of alanine (Takeo, 1978).

Although simple amines may be formed by amino acid decarboxylation in several higher plants, aldehyde amination appears to be the more common route of synthesis. The generalized reaction for aldehyde amination is

$$R-CHO + R'-\underset{\underset{NH_2}{|}}{\overset{\overset{COOH}{|}}{CH}} \longrightarrow R-CH_2-NH_2 + R'-\underset{\overset{COOH}{|}}{C}=O$$

An aldehyde transaminase from *Mercurialis perennis* aminated a wide range of aldehydes including the homologous series from ethanal to *n*-undecanal. Hexanal was the best substrate, with alanine, ε-aminocaproic acid, γ-aminobutyric acid, and glutamic acid as donors. The enzyme was inhibited by pyruvate, oxaloacetate, α-ketoglutarate, and glyoxalate. The reaction is thought to be catalyzed by a normal amino acid on ketoacid transaminase, with keto specificity that extends to the aldehydes (Hartmann *et al.*, 1972a,b; von Preusser, 1975; Unger and Hartmann, 1976). Aldehyde transamination also occurs in apple fruits. Amines were formed on infiltration of the corresponding aldehydes, but apple homogenates showed no activity (Hartmann, 1967).

In *Zea mays* seedlings transamination is a source of ethylamine and isoamylamine, but ethylamine at least may be formed additionally by decarboxylation of alanine. Methylamine, propylamine, and isoamylamine are formed in *Vicia faba* by transamination, but isobutylamine is produced by decarboxylation (von Preusser, 1975).

III. DI- AND POLYAMINES

A. Occurrence and Biosynthesis

The diamine putrescine and the polyamines spermidine and spermine (Fig. 1) have been found in all higher plants studied. They are known to stimulate the growth of several microbial, animal, and plant systems, and enhanced polyamine biosynthesis often occurs concurrently with growth (Jänne *et al.*, 1978). In view of their universal distribution and their ability to interact with nucleic acids, polyamines are considered to be of fundamental importance in growth processes. Polyamines stimulate many processes concerned with protein synthesis *in vitro*. Evidence at present links spermidine and spermine *in vivo* with ribosome function, probably in stabilizing nucleic acid (Algranati and Goldemberg, 1977). Spermine is probably also concerned with DNA synthesis in nuclei, and the absence of spermine in most prokaryotes supports this hypothesis (Stevens, 1970). Putrescine and polyamines may also have functions in membrane stabilization (Tabor and Tabor, 1976). In bacteria and plants, putrescine is also apparently involved in mechanisms for the control of cell pH (Smith, 1971).

In barley, putrescine is formed from arginine with agmatine and *N*-carbamylputrescine as intermediates (Fig. 1). The enzymes converting arginine to agmatine (arginine decarboxylase), and *N*-carbamylputrescine to putrescine (*N*-carbamylputrescine amidohydrolase), have both been partially characterized in the leaves of barley seedlings. Under conditions of

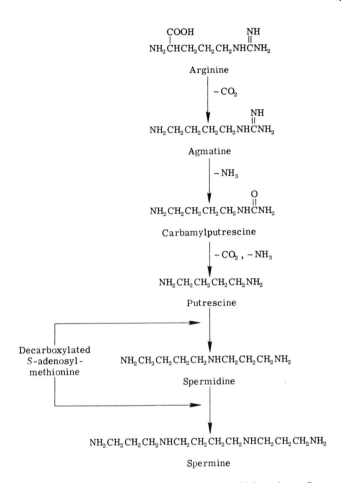

Fig. 1. Biosynthesis of putrescine and the polyamines in higher plants. Some plants (e.g., tobacco, see Fig. 3) utilize ornithine as the precursor, though this is probably unusual.

potassium deficiency, putrescine concentrations were increased up to 90-fold and agmatine was increased 5-fold. The activity of arginine decarboxylase and N-carbamylputrescine amidohydrolase was also increased 2- to 3-fold. Since feeding hydrochloric acid to barley seedlings also increased the activity of these enzymes, it appears that putrescine may be formed as an organic cation by the plant to control the intracellular pH. In black currant leaves with an extreme potassium deficiency, the potassium deficit may be compensated for to the extent of 30% by the formation of putrescine (reviewed by Smith, 1971, 1977a). N-carbamylputrescine does not occur in detectable amounts in barley leaves, even in potassium deficiency, but it is readily formed from agmatine in extracts of maize leaves (Smith, 1969). In *Sesamum indicum* citrulline decarboxylation appears to be the source of the *N*-carbamylputrescine that accumulates in potassium deficiency (Crocomo and Basso, 1974). Agmatine iminohydrolase has been purified 375-fold from groundnut cotyledons (Sindhu and Desai, 1979).

In *Lathyrus sativus* arginine is converted to putrescine via agmatine, and homoarginine is converted to cadaverine via homoagmatine (Fig. 2). Homoarginine decarboxylase, which was highly labile, was purified about 110-fold and resolved from the arginine decarboxylase. The homoarginine decarboxylase preparation also decarboxylated lysine; both reactions required Mn^{2+} and Fe^{2+} and showed the greatest activity at pH 8.4. If, as indicated by kinetic studies, these activities are both associated with the same enzyme, lysine is the preferred substrate with a K_m only 25% of that for homoarginine (Ramakrishna and Adiga, 1976). The arginine decarboxylase in *L. sativus* seedlings reaches maximal activity 5–7 days after germination. On purification 1000-fold, the enzyme appeared to be homogeneous and was shown to have a MW of 220,000. It appeared to be a hexamer with identical subunits. No differences could be detected in the properties of the enzyme isolated from the embryo or cotyledons. The optimum pH was 8.5, and 0.1 mM Mn^{2+} stimulated activity about 35%. The enzyme was sensitive to sulfhydryl inhibitors, and thiol reagents were required for maximum activity and stability. Inhibitor studies indicated that pyridoxal phosphate was involved as a cofactor. Moreover, after prolonged dialysis the activity could be stimulated 80% on adding 30 μM pyridoxal phosphate. Even so, spectral evidence for pyridoxal phosphate could not be obtained. The enzyme did not attack L-homoarginine, L-lysine, or L-ornithine. Activity was inhibited by spermine (K_i = 4 mM), indicating a product feedback control mechanism. L-Homoarginine, L-canavanine, L-lysine, L-ornithine, and D-arginine also inhibited the activity (Ramakrishna and Adiga, 1975). Arginine was 14 times

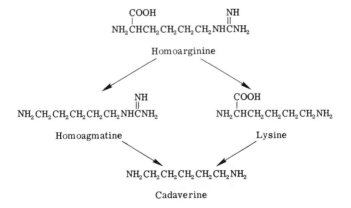

Fig. 2. Pathway of cadaverine biosynthesis in *L. sativus*.

more efficient than ornithine in the synthesis of agmatine, putrescine, spermidine, and spermine. Cadaverine was formed at least in part from homoarginine, though lysine was four times more effective as a precursor (Ramakrishna and Adiga, 1974).

The arginine decarboxylase from oats was specific for L-arginine, and it was inhibited by D-arginine. Activity was stimulated by pyridoxal phosphate, and the pH optimum was 7.0. The enzyme could be separated into two fractions (MW 195,000 and 118,000) by gel chromatography and electrophoresis (Smith, 1979a,b; Smith and Ng, 1979).

In cucumber cotyledons, benzyladenine, its riboside, and gibberellin A_3 increased the activity of arginine decarboxylase and the putrescine content. Abscisic acid inhibited cotyledon growth and reduced arginine decarboxylase activity and putrescine content, and this could be reversed by cytokinins. The *in vivo* half-life of the arginine decarboxylase, determined by treatment with putrescine or cycloheximide, was 3 or 3.7 h, respectively (Suresh *et al.*, 1978), considerably longer than that of ornithine decarboxylase, the analogous key enzyme in animal tissues, which has a half-life of about 10 min (Tabor and Tabor, 1976). Treatment with KCl decreased arginine decarboxylase and putrescine content, but hydrochloric acid feeding doubled arginine decarboxylase and putrescine despite reduced growth (Suresh *et al.*, 1978). Although nucleic acid biosynthesis, growth, and putrescine content increased on treatment with cytokinins, the polyamine concentration was reduced (Suresh and Adiga, 1978).

Plants grown in an ammonium-based nutrient solution have considerably higher concentrations of putrescine than those grown in nitrate-based solutions. This may be associated with the reduction in the pH of the medium known to occur in ammonium-based nutrient media as a result of the rapid uptake of the ammonium ion. Addition of solid calcium carbonate to the ammonium medium reduced the putrescine accumulation in pea roots (Smith and Wilshire, 1975). However, in soybean seedlings grown in an ammonium medium, putrescine concentration was about 100-fold greater than in seedlings grown in a nitrate medium, even when solid buffers were used to control the pH. Moreover, the potassium concentration was almost unaffected by the nitrogen source (Le Rudulier and Goas, 1971, 1975). Le Rudulier and Goas (1975) suggest that putrescine accumulation may result from direct competition between ammonium and potassium at a metabolic site within the cell.

Protoplasts isolated from oat leaves are unstable and undergo progressive senescence culminating in lysis. This can be delayed by the presence of di- and polyamines. Polyamines added to protoplasts during isolation also reduce RNase synthesis (Kaur-Sawhney *et al.*, 1978).

Loss of betacyanin by efflux from discs of beet root storage tissue is reduced by 1 mM spermidine and spermine, presumably as a result of binding to the phospholipid component of the membrane. The reversal of efflux by spermine occurs at 27° and 37°C, but not at 47°C, as a result of irreversible damage to the membrane. The destabilization induced by 8% ethanol is reversed by 0.2–0.6 mM spermine (Naik and Srivastava, 1978).

Turnip yellow mosaic virus and cowpea mosaic virus contain about 1% spermidine by weight, though several other plant viruses contain no detectable polyamines (Nickerson and Lane, 1977). The unusual polyamines diaminodipropylamine [$NH_2(CH_2)_3NH(CH_2)_3NH_2$] and homospermidine [$NH_2(CH_2)_4NH(CH_2)_4NH_2$] are found in several algal genera (*Scenedesmus, Coelastrum, Chlorella,* and *Euglena*) (Kneifel, 1977). *Euglena* also contains the tetraamine norspermine [$NH_2(CH_2)_3NH(CH_2)_3NH(CH_2)_3NH_2$] (Kneifel *et al.,* 1978; Aleksijevic *et al.,* 1979; Adlakha and Villanueva, 1980). Homospermidine has been detected in the symbiotic nitrogen-fixing bacteria in the genus *Rhizobium* found in the nodules of various legumes (Smith, 1977d), and it also occurs at up to 1.5% of the dry weight in the leaves of the sandalwood tree (*Santalum album*). In *Santalum,* arginine was a better precursor than ornithine of this polyamine by a factor of 7. In view of the molecular symmetry of homospermidine, it was suggested that biosynthesis may be effected by oxidation of putrescine to γ-aminobutyraldehyde, subsequent condensation with another putrescine moiety, and reduction of the resulting Schiff base (Kuttan and Radhakrishnan, 1972).

Methylated polyamines have also been found in plants. Tetramethylputrescine occurs in *Ruellia rosea'* (Acanthaceae) (Johne *et al.,* 1975), and in *Solanum carolinense* methylated homospermidine [solamine, $Me_2N(CH_2)_4NR (CH_2)_4Me$, R = H] occurs both free and as a urethane derivative (R = — COOEt) (Evans and Somanabandhu, 1977). Two derivatives of tetramethylhomospermidine (solapalmitine and solapalmitenine) occur in *Solanum tripartitum.* These inhibit the growth of animal tumors (Kupchan *et al.,* 1969) and bacteria (Silver and Kralovic, 1969).

In the biosynthesis of the polyamines from putrescine, the propylamino moiety is derived from S-adenosylmethionine (SAM) after decarboxylation (Fig. 1). Crude extracts of 3-day-old etiolated seedlings of *L. sativus* contained two components with SAM decarboxylase activity (Suresh and Adiga, 1977). One of these was artifactual and was dependent on the diamine oxidase present in the seedlings (Section III,B). Putrescine was oxidized by this enzyme with the formation of hydrogen peroxide which effected the decarboxylation. In the presence of putrescine, diamine oxidase decarboxylated lysine, arginine, ornithine, methionine, and glutamic acid in addition to SAM. As indicated by Suresh and Adiga, this phenomenon calls for caution in interpreting data on putrescine-stimulated SAM decarboxylase in other systems.

The biosynthetic SAM decarboxylase, which occurs primarily in the shoots, was detected after removing the diamine oxidase on an immunoaffinity column. This decarboxylase was Mg^{2+}-dependent and was not influenced by putrescine or catalase. The optimum pH was 7.6, and the enzyme was inhibited by methylglyoxal bis(guanylhydrazone) (MGBG). By utilizing the sulfhydryl nature of the enzyme, 40-fold purification was achieved on a p-hydroxymercuribenzoate AH-Sepharose affinity column which also resolved the spermidine synthase. Specific activity reached a peak in the shoots 3 days after germination. Activity was lower in the roots and negligible in the cotyledons (Suresh and Adiga, 1977).

With the use of a MGBG Sepharose affinity column SAM decarboxylase was purified 12,000-fold to homogeneity from yeast. The MW by gel filtration was 88,000, and 41,000 after sodium dodecyl sulfate (SDS) electrophoresis. Pyridoxal phosphate was not present, but one pyruvate residue was found per subunit. This is the first example of a eukaryotic enzyme containing a covalently linked pyruvoyl residue (Cohn *et al.*, 1977). Activity of purified yeast SAM decarboxylase was stimulated by putrescine, 1,3-diaminopropane, and cadaverine (Pösö *et al.*, 1975).

In tobacco, putrescine derived from ornithine by decarboxylation (Heimer *et al.*, 1979) is utilized in the biosynthesis of nicotine, which takes place in the roots (Fig. 3). *N*-Methylputrescine is formed first, with SAM as donor. The enzyme putrescine *N*-methyltransferase has been purified 30-fold. The MW is 60,000, and the pH optimum between 8 and 9. An amine oxidase specific for *N*-methylputrescine then forms *N*-methylpyrrolinium (Saunders and Bush, 1979). This enzyme, which contains copper, like the diamine oxidase from the Leguminosae (Section III,B), has been purified 150-fold (Mizusaki *et al.*, 1973).

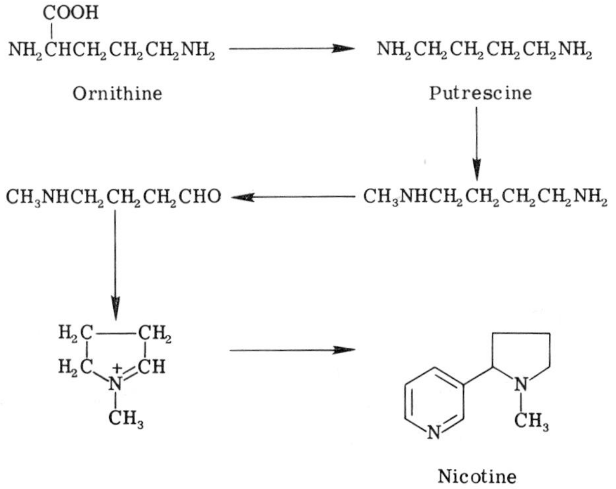

Fig. 3. Formation of nicotine from ornithine in tobacco roots.

B. Amine Oxidases

Enzymes capable of oxidizing amines occur in several families of higher plants (distribution reviewed by Smith, 1980). Diamine oxidase is widespread in the Leguminosae. The general equation for the oxidation of amines by this enzyme is

$$RCH_2NH_2 + O_2 + H_2O \rightarrow H_2O_2 + RCHO + NH_3$$

Putrescine and cadaverine are the best substrates for the enzyme isolated from pea seedlings, though many other amines are oxidized including aliphatic monoamines and aromatic amines, and also the amino acid lysine. The enzyme contains copper which can be removed by dialysis against chelating agents. Activity is restored specifically on adding Cu^{2+} (Hill and Mann, 1964). The oxidation state of the copper is not changed on adding substrate, although a shift in the epr spectrum indicates that the copper is bound differently under these conditions (Nylén and Szybek, 1974). When substrate was added under anaerobic conditions, the enzyme became yellow, and the pink color could be restored by oxygenation. The enzyme was sensitive to carbonyl reagents, though the presence of pyridoxal phosphate has not yet been demonstrated (Hill and Mann, 1964). Titration of the enzyme with phenylhydrazine indicates the presence of 1 mol of reactive carbonyl per mole of enzyme (Nylén and Szybek, 1974).

The K_m values for putrescine, spermidine, and spermine were 4×10^{-5}, 5×10^{-6}, and $9 \times 10^{-5}M$, with a relative V_{max} of 5.5 : 3.0 : 1, respectively, suggesting that at low substrate concentrations spermidine is a better substrate than putrescine (Smith, 1974). Calculated from electron microscopy, the MW was 96,000 (Hill and Mann, 1964), but sedimentation equilibrium centrifugation indicated a MW of 185,000 (McGowan and Muir, 1971). The amino acid analysis showed an unusually high content of ornithine. The major components were serine, glutamic acid, and glycine; cystine or cysteine could not be detected (Nylén and Szybek, 1974). The oxidative process appeared to involve conversion of the amine substrate to the corresponding aldehyde and a modified enzyme which was oxidized by oxygen in a separate step (Yamasaki et al., 1970). With benzylamine as substrate, activity was dependent on inorganic phosphate in glycine and borate buffers, but not in Tris buffer (McGowan and Muir, 1971). The phosphate requirement would be compatible with the possibility that the pea seedling enzyme is involved in the oxidation of the di- and polyamines associated with nucleic acids.

Diamine oxidase of *L. sativus* was purified to homogeneity and, like the pea enzyme, was inhibited by carbonyl reagents and metal chelators. The optimum pH was about 8.4 with putrescine and cadaverine, and activity was greater in Tris than in phosphate buffer (Suresh et al., 1976; Suresh and Adiga, 1979). The MW was 148,000 (dimer), and it had 2.7 g-atoms of copper

per mole, but ornithine could not be found in the amino acid residues, unlike the pea seedling enzyme.

The function of pea seedling amine oxidase is still obscure, but the high K_m for tryptamine $(3.2 \times 10^{-3} M)$ (McGowan and Muir, 1971) indicates that this enzyme is probably not primarily involved in the formation of IAA. However, the amine oxidase from pea seedling cotyledons differs from that in the axis. The K_m values for putrescine for the two enzymes were, respectively, 1.6×10^{-4} and $9 \times 10^{-5} M$, and on heating for 15 min at 70°C the enzymes retained about 90 and 20% of their activity, respectively. The electrophoretic mobility of the cotyledon enzyme purified 470-fold was about twice that of the enzyme from the embryo (purified 160-fold). In other respects the enzymes were similar. The differences in mobility were unlikely to be due to changes induced during purification, since the differences were also found with crude extracts (Srivastava and Prakash, 1977).

Cotyledon amine oxidase was detected 42 h after germination and increased to a plateau at 110 h. Activity appeared at 14 h in the embryo, again increasing up to 110 h. No activity was detected in the cotyledons on removal of the embryo before soaking. However, activity in the cotyledons was undiminished on removal of the embryo after soaking for 14 h. In seedlings where the embryo was retained after soaking for 14 h, amine oxidase appeared in the cotyledons later (after 28 h) than in seedlings from which the embryo was removed after soaking for 14 h (within 24 h). Activity was indistinguishable in these two groups after 32 h. The presence of the embryo later than 14 h after soaking therefore inhibits formation of the cotyledon amine oxidase, despite its earlier requirement for enzyme formation. After removal of the embryo, soaking the cotyledons in embryo extract induced 50% of the normal amine oxidase activity. The induction factor could be replaced by putrescine, spermidine, or ornithine. On soaking whole seeds in (2,4-dichlorophenoxy)acetic acid (2,4-D) or IAA, amine oxidase activity was reduced in both the cotyledons and the embryo. In the cotyledons this was observed only when the embryo was present during soaking. These hormones therefore act through the embryo (Srivastava et al., 1977a). Amine oxidase activity of the cotyledons and the embryo was reduced on feeding ethrel and chloroethanol, substances known to form ethylene in vivo (Srivastava et al., 1977b). Moreover, 2,4-D also causes ethylene formation in plants, and this function is blocked by red light (Burg, 1973). Since the 2,4-D inhibition of amine oxidase formation is reversed by red light, it seems likely that the effect of 2,4-D is mediated through ethylene (Srivastava et al., 1977b).

A polyamine-specific oxidase occurring widely in the Gramineae is associated with nucleic acid-containing particles which sediment in low centrifugal fields. The enzyme can be released from these particles by washing in 0.5 M sodium chloride and has been purified to apparent homogeneity

(1000-fold) from extracts of etiolated oat leaves. Gel chromatography indicates a MW of about 100,000. Unlike pea seedling diamine oxidase, this enzyme is relatively resistant to carbonyl inhibitors and chelating agents. No cofactor has yet been demonstrated. The polyamine oxidase from barley leaves showed the greatest activity with spermine as substrate at an optimum pH of 4.5 (Fig. 4). Spermidine was oxidized at only 7% of the rate with spermine at an optimum pH of 8. Even at pH 7.5 the enzyme is unstable, 50% of the activity being lost on incubation for 15 min at 25°C. It seems likely therefore that spermine is the natural substrate in barley. However, the oat enzyme oxidized both spermidine and spermine optimally at pH 6.5, and it showed greatest stability at this pH. Unlike the barley enzyme the enzyme from oats was inactive with diaminodipropylamine. With 0.1 M buffer alone, activity was about three times greater with spermidine as substrate than with spermine. On adding 0.5–1 M sodium chloride, phosphate, or citrate, activity with spermine was increased two- to threefold, but activity with spermidine was unaffected. The enzyme was inhibited 50% by Mg^{2+} or Ca^{2+} at 50 mM. The barley enzyme was strongly inhibited by spermidine at pH 4.5 and by spermine at pH 7.5 in the oxidation of spermine and spermidine, respectively. The spermidine analogue $NH_2(CH_2)_{10}NH(CH_2)_3NH_2$ was a very effective competitive inhibitor of the polyamine oxidase from both oats and barley. The K_i for the barley enzyme was 5×10^{-6} M. The barley and oat enzymes had unusually high K_m values for oxygen, similar to the concentration of oxygen dissolved in water in equilibrium with the air. This indicates that they might function as dehydrogenases *in vivo*, although no artificial acceptors could be found. In barley, the K_m for spermine oxidation was independent of oxygen concentration, suggesting a sequential mechanism (Smith, 1970, 1976, 1977e).

Diaminopropane and 1,3-aminopropylpyrroline appear to occur naturally in the leaves of barley seedlings (Smith, 1970). In maize, spermine is further

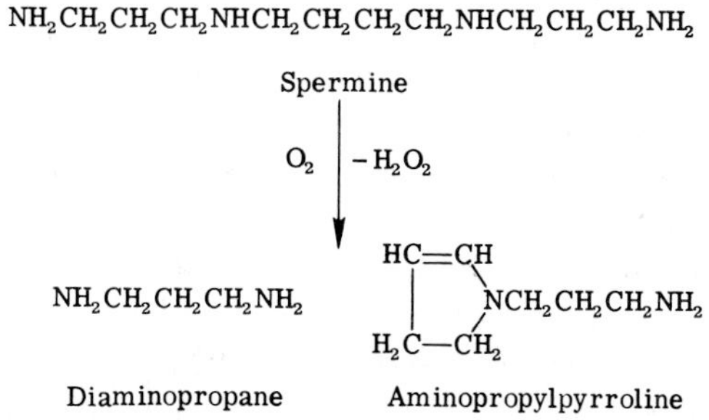

$$NH_2CH_2CH_2CH_2NHCH_2CH_2CH_2CH_2NHCH_2CH_2CH_2NH_2$$

Spermine

O_2 | $-H_2O_2$

$NH_2CH_2CH_2CH_2NH_2$

Diaminopropane

HC=CH
| NCH_2CH_2CH_2NH_2
H_2C—CH_2

Aminopropylpyrroline

Fig. 4. Oxidation of spermine by an enzyme found in the Gramineae. This enzyme also oxidizes spermidine with the formation of pyrroline. The position of the double bond has not yet been established.

metabolized (via diaminopropane) to β-alanine, and (probably via 1,3,-aminopropylpyrroline) to γ-aminobutyric acid (Terano and Suzuki, 1978a,b). In maize the polyamine oxidase was apparently associated with the vascular system (Smith, 1970).

C. Effects on Protein Synthesis and Growth

Spermidine and spermine stimulate the incorporation of amino acids into protein in the wheat germ cell-free system using exogenous mRNA. Addition of optimum concentrations of the polyamines reduced the optimum Mg^{2+} concentration needed for protein synthesis. In the translation of tobacco mosaic virus mRNA, the rate of chain elongation with optimum Mg^{2+} and polyamines was twice that found with the optimum Mg^{2+} concentration in the absence of polyamines. Moreover, the yield of full-length translation products was considerably increased. This may be due at least in part to the inhibition of RNase activity by the polyamines. The accelerated rate of chain elongation in the presence of polyamines increases the probability of synthesis of full-length protein products before the mRNA is degraded (Hunter *et al.*, 1977).

The effects of polyamines on the synthesis of turnip yellow mosaic virus by the wheat germ cell-free system have also been studied. Coat protein and three other polypeptides were synthesized. Spermine at $3 \times 10^{-5} M$ stimulated total protein synthesis twofold, and the optimal Mg^{2+} concentration was then reduced from 3.3 to $1.5 \times 10^{-3} M$. Spermine had an overall stimulatory effect on total protein synthesis at $10^{-3} M$ (optimal) Mg^{2+} concentrations, but at $2.5 \times 10^{-3} M$ Mg^{2+} spermine specifically stimulated coat protein synthesis (Benicourt and Haenni, 1976).

In the formation of polyphenylalanine in a barley ribosome cell-free system, putrescine, spermidine, and spermine showed discrete optimal concentration ranges for the partial replacement of Mg^{2+} (optimum $10^{-2} M$). Spermine was very efficient (optimum $5 \times 10^{-4} M$), and spermidine less so (optimum $5 \times 10^{-3} M$). Regardless of the polyamine concentration it was not possible to dispense with Mg^{2+}, though loss of activity was negligible with $5 \times 10^{-3} M$ Mg^{2+} in the presence of 0.1 mM spermine (Cohen and Zalik, 1978).

Polyamine concentrations are abnormally high in both animal and plant tumors. Habituated tumors of *Scorzonera hispanica* contain increased concentrations of spermidine (1.5-fold) and putrescine (10-fold) by comparison with normal tissue. In crown gall tumors induced by *Agrobacterium tumefaciens*, in which this bacterium is no longer present, spermidine increased 3-fold and putrescine 100-fold. However, spermine decreased (0.5-fold) in both types of tumors (Bagni *et al.*, 1972). A similar increase in putrescine was found in habituated *Nicotiana glauca* tumors, reflecting their enhanced growth rate (Audisio *et al.*, 1976).

The growth of *Helianthus tuberosus* (Jerusalem artichoke) explants was stimulated on treatment with polyamines or with IAA. Moreover, treatment with IAA induced the accumulation of polyamines. It is therefore possible that at least one of the processes by which IAA stimulates growth may be mediated by polyamines (Cocucci and Bagni, 1968). Inhibition of the formation of amine oxidase by IAA (Srivastava *et al.*, 1977a) may result in amine accumulation.

The addition of polyamines to cultures of Paul's Scarlet rose tissue did not stimulate growth, and only toxic effects were observed. On inoculation of cells into fresh medium, the putrescine concentration increased almost two-fold within the first 4 h (Smith *et al.*, 1978b).

On transferring carrot cells to both nonembryogenic and embryogenic media, putrescine increased about 7-fold to reach a peak after 5 h and then declined to a minimum after 20 h. After 50 h in the embryogenic medium putrescine concentration in the cells was about 30-fold greater than in the cells transferred to nonembryogenic medium (Montague *et al.*, 1978, 1979*).

Polyamine function has been studied in a strain of *Aspergillus nidulans* that requires polyamines for conidial growth. When conidia were placed in a minimal medium supplemented with putrescine, growth resembled that of the wild type. However, in the absence of putrescine, growth was limited 5–6 h after inoculation, a time when polyamines are rapidly accumulated in wild-type conidia. Putrescine deficiency in this mutant appears to be due to a lack of ornithine decarboxylase. Putrescine was by far the most active growth stimulator, but spermidine and spermine were also effective. Cadaverine, 1,3-diaminopropane, and ornithine had little activity (Stevens, 1975; Winther and Stevens, 1978).

S-Adenosylmethionine decarboxylase activity increased significantly during the first 2-h germination of yeast ascospores. Spermidine concentration increased 3-fold during the first 6-h germination, and during this time putrescine increased 2-fold and spermine 1.3-fold. The accumulation of spermidine and spermine was completely prevented by MGBG, but the putrescine concentration increased up to 2-fold. This treatment had no effect on the incorporation of labeled precursors into RNA or protein, or on the net DNA content, and germination was apparently normal. However, it is possible that the polyamine levels were already sufficient to support maximal rates of macromolecular synthesis. Alternatively the enhanced amount of putrescine or even MGBG, which is a polyamine analogue, may take over the functions of the polyamines (Choih *et al.*, 1978).

The addition of spermidine and spermine to a cell-free protein-

* On growth initiation tobacco cultures and developing tomato ovaries showed a considerable increase in ornithine decarboxylase, apparently associated with cell division (Heimer *et al.*, 1979).

synthesizing system from *Saccharomyces* increased phenylalanine polymerization 2-fold in 10 m*M* (optimum) Mg^{2+} and up to 10-fold in 2.5 m*M* Mg^{2+} (Wolska-Mitaszko *et al.*, 1976). Polyamines also increased the thermal stability of this protein-synthesizing system (Jakubowicz *et al.*, 1976). The addition of spermine to growing yeast reduced the lag time after inoculation and stimulated growth (Castelli and Rossoni, 1968).

IV. AMINE CONJUGATES

Di- and polyamines are found in the form of conjugates with cinnamic acid and its derivatives in many species of higher plants (reviewed by Smith, 1977a). The simple forms of these conjugates have been termed aromatic amides (Fig. 5). In the leaves of *Nicotiana tabacum* p-coumaroyl-, caffeoyl-, and feruloylputrescine and caffeoylspermidine have been demonstrated on floral induction, and these amides accumulate in considerable amounts (up to 2 μmol/g fr wt) in the reproductive organs of the induced tobacco plant (Cabanne *et al.*, 1976, 1977). On infection with tobacco mosaic virus, coumaroyl-, dicoumaroyl-, feruloyl-, and diferuloylputrescine accumulate in hypersensitive lesions in *N. tabacum* var. Xanthi. Coumaroyl, dicoumaroyl-, and caffeoylputrescine have been shown to inhibit multiplication of the virus. It has long been known that seeds produced by plants with a generalized virus infection are often free of virus. Furthermore, transition to the reproductive state is frequently accompanied by the development of resistance to virus multiplication. Hydroxycinnamic acid amine conjugates may therefore be natural virus inhibitors. Hypersensitivity to viruses, synthesis of aromatic amides, and floral induction are all inhibited by raising the temperature from 20° to 32°C (Cabanne *et al.*, 1976, 1977).

The anthers of the tobacco plant accumulate principally dicoumaroylputrescine, coumaroyltyramine, and dicoumaroylspermidine. The ovaries contain very large amounts of caffeoylputrescine and -spermidine. The seeds contain mainly diferuloylputrescine and feruloyltyramine (Cabanne *et al.*, 1977). After removal of the tops of tobacco plants, the aromatic amides increase in the remaining leaves from <0.1 to 10 μmol/g fr wt in 3 days, but the concentration declines as the axillary buds elongate 8 days after topping (Cabanne *et al.*, 1977). These hydroxycinnamic acid amides appear to be the main phenolic constituents in the reproductive organs of a wide range of flowering plants (Martin-Tanguy *et al.*, 1978).

Antifungal coumaroylagmatine dimers known as hordatines have been characterized in barley seedlings. These completely inhibit spore germination of *Monilinia fructicola* at 5 ppm. Many other fungi, including the barley pathogen *Helminthosporium sativum,* are also sensitive (Stoessl and Unwin, 1970). Concentrations were greatest in barley shoots less than 3 days old, but

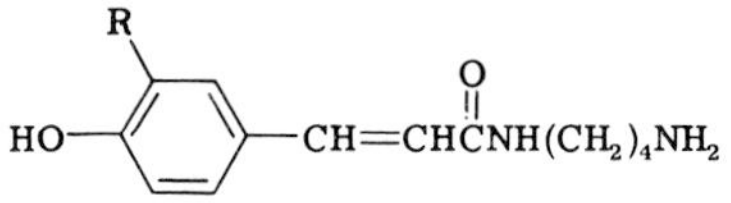

R = H   Coumaroylputrescine
R = OH   Caffeoylputrescine
R = OCH$_3$ Feruloylputrescine

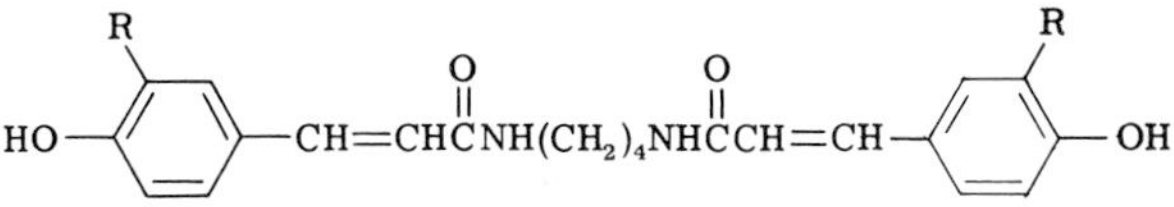

R = H   Dicoumaroylputrescine
R = OCH$_3$ Diferuloylputrescine

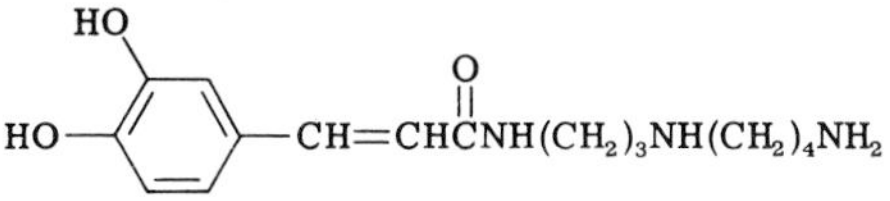

Caffeoylspermidine

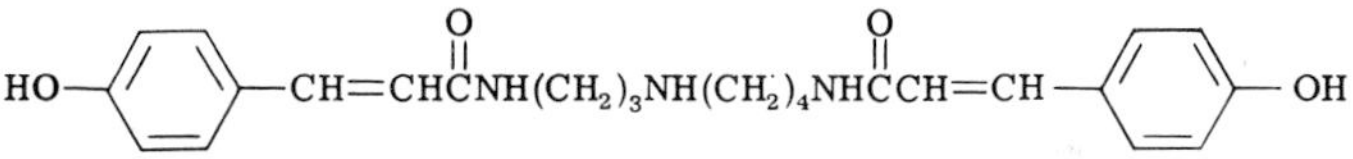

Dicoumaroylspermidine

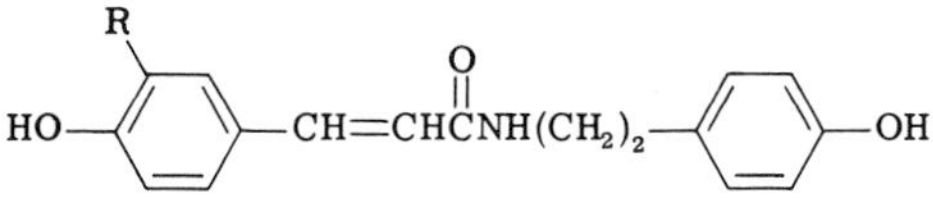

R = H   Coumaroyltyramine
R = OCH$_3$ Feruloyltyramine

Fig. 5. Aromatic amides found in various parts of *Nicotiana* (tobacco) plants.

both potassium deficiency and mildew infection increased the hordatine concentration (about sixfold each). Hordatines were found in seedlings of several species of the genus *Hordeum* but not in a variety of other cereals (Smith and Best, 1978).

The macrocyclic spermidine and spermine alkaloids that occur in a wide range of families are sometimes very complex. These are formed mainly by the condensation of moieties containing various polyamines linked to cinnamic acids by amide bonds (Hesse and Schmid, 1976). Other polyamine alkaloids (e.g., solapalmitine, Kupchan *et al.*, 1969) are derived from fatty acids in amide linkage with polyamines. Aromatic amines also form amides with organic acids (reviewed by Smith, 1977a). Amide conjugates of isobutylamine with fatty acids have been found in the Compositae, Piperaceae, and Rutaceae (reviewed by Guggenheim, 1958; Gupta *et al.*,

1977). The isobutylamide of octadecapentaenoic acid (scabrin) is a powerful insecticide (Jacobson, 1951).

V. TRYPTAMINES, PHENETHYLAMINES, AND HISTAMINE

The distribution of tryptamines and phenethylamines has been reviewed by Smith (1977b,c).

In tomato and barley plants the growth hormone IAA is derived from tryptophan with indole acetaldehyde as an intermediate. The latter may be formed either via tryptamine or via indolepyruvic acid in both species, and enzymes were demonstrated in extracts effecting all of the steps, apart from the conversion of tryptamine to indole acetaldehyde (Gibson *et al.,* 1972). Phenylacetic acid, which may also be an important plant hormone, is formed by an analogous pathway in a variety of plants. Phenylalanine is converted to phenylacetic acid either via phenylpyruvic acid or via phenethylamine (Wightman and Rauthan, 1974).

Gramine is found in the Aceraceae, Gramineae, and Leguminosae (reviewed by Smith, 1977c). Gramine and hordenine may be formed as insect-feeding inhibitors (Harley and Thorsteinson, 1967; Janzen *et al.,* 1977).

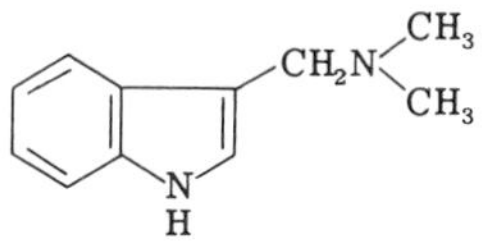

Gramine 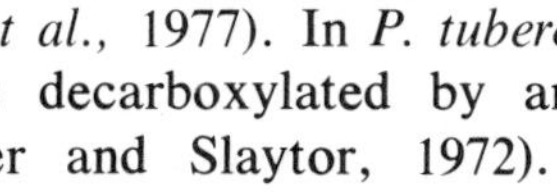   Hordenine

An increased concentration of hordenine, gramine, tryptamines, and β-carbolines probably reduces the palatability of reed canary grass (*Phalaris arundinacea*) for herbivores (Coulman *et al.,* 1977). In *P. tuberosa,* tryptophan and 5-hydroxytryptophan were decarboxylated by an enzyme that has been purified 20-fold (Baxter and Slaytor, 1972). Two *N*-methyltransferases, one specific for tryptamines and the other for *N*-methyltryptamines, using SAM as donor, were also purified (Mack and Slaytor, 1974, 1979). On feeding [3-^{14}C]tryptophan to *P. arundinacea* the route of gramine biosynthesis demonstrated in *Hordeum* and *Lupinus* appeared to operate (Leete and Minich, 1977).

Tryptamines are of common occurrence in the Gramineae and Leguminosae. Dimethyltryptamine, 5-methoxy-*N*-dimethyltryptamine, bufotenine, and mescaline and its derivatives are psychoactive (Smith, 1977b,c). Many members of the Leguminosae and Cactaceae also contain phenethylamines (Smith, 1977b). Although tryptamine and phenethylamine are physiologically important in plants as potential precursors of growth

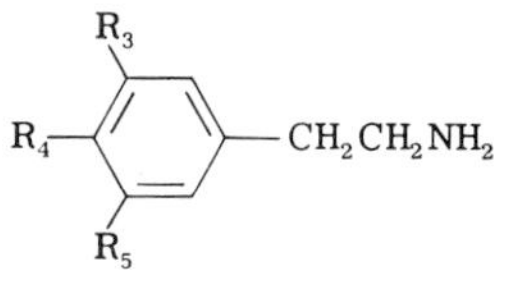

	R_3	R_4	R_5	
	H	OH	H	Tyramine
	OH	OH	H	Dopamine
	OMe	OH	H	
	OMe	OH	OH	
	OMe	OH	OMe	
	OMe	OMe	OMe	Mescaline

Fig. 6. Biosynthesis of mescaline from tyramine in the cactus *L. williamsii*. Tyrosine may also be converted to DOPA before decarboxylation to dopamine.

substances, no known metabolic function in plants has been assigned to derivatives of tryptamine or phenethylamine.

In the cactus *Lophophora williamsii* mescaline biosynthesis follows the pathway shown in Fig. 6. Tyrosine is converted to dopamine either via 3,4-dihydroxyphenylalanine (DOPA) or tyramine, and mescaline is formed by ring hydroxylation and methylation (Paul, 1973).

Ephedrine found in various species of *Ephedra* is synthesized by condensation of an uncharacterized $-C_2N$ fragment with a C_6-C_1 portion derived from cinnamic acid (Yamasaki *et al.*, 1973). Hordenine, which occurs in the roots of barley seedlings, is formed from tyrosine by decarboxylation and methylation. A specific tyrosine decarboxylase has been purified about 80-fold from barley seedlings. This enzyme is stimulated by SAM and inhibited by *p*-hydroxyphenylpyruvate, the ketoacid precursor of tyrosine (Gallon and Butt, 1971).

Histamine occurs in the stings of *Jatropha urens* (Villalobos *et al.*, 1974) and together with 5-hydroxytryptamine in the stings of the stinging nettle *Urtica dioica* (Vialli *et al.*, 1973). The distribution of histamine has been listed by Smith (1980).

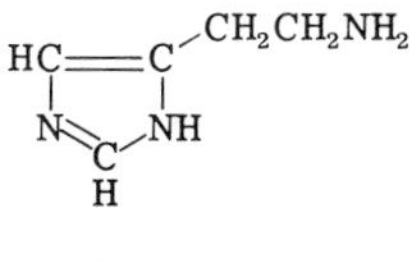

Histamine

REFERENCES

Adlakha, R. C., and Villanueva, V. R. (1980). *J. Chromatogr.* **187,** 442–446.
Aleksijevic, A., Grove, J., and Schuber, F. (1979). *Biochim. Biophys. Acta* **565,** 199–207.
Algranati, I. D., and Goldemberg, S. H. (1977). *Trends Biochem. Sci.* **2,** 272–274.
Audisio, S., Bagni, N., and Serafini Fracassini, D. (1976). *Z. Pflanzenphysiol.* **77,** 146–151.

Bagni, N., Serafini Fracassini, D., and Corsini, E. (1972). *Z. Pflanzenphysiol.* **67**, 19–23.
Baxter, C., and Slaytor, M. (1972). *Phytochemistry* **11**, 2763–2766.
Benicourt, C., and Haenni, A. L. (1976). *J. Virol.* **20**, 196–202.
Burg, S. P. (1973). *Proc. Natl. Acad. Sci. U.S.A.* **70**, 591–597.
Cabanne, F., Martin-Tanguy, J., Perdrizet, E., Vallée, J.-C., Grenet, L., Prevost, J., and Martin, C. (1976). *C.R. Hebd. Seances Acad. Sci.* **282**, 1959–1962.
Cabanne, F., Martin-Tanguy, J., and Martin, C. (1977). *Physiol. Veg.* **15**, 429–443.
Castelli, A., and Rossoni, C. (1968). *Experientia* **24**, 1119–1120.
Choih, S.-J., Ferro, A. J., and Shapiro, S. K. (1978). *J. Bacteriol.* **133**, 424–426.
Cocucci, S., and Bagni, N. (1968). *Life Sci.* **7**, 113–120.
Cohen, A. S., and Zalik, S. (1978). *Phytochemistry* **17**, 113–118.
Cohn, M. S., Tabor, C. W., and Tabor, H. (1977). *J. Biol. Chem.* **252**, 8212–8216.
Coulman, B. E., Woods, D. L., and Clark, K. W. (1977). *Can. J. Plant Sci.* **57**, 771–785.
Crocomo, O. J., and Basso, L. C. (1974). *Phytochemistry* **13**, 2659–2665.
Evans, W. C., and Somanabandhu, A. (1977). *Phytochemistry* **16**, 1859–1860.
Gallon, J. R., and Butt, V. S. (1971). *Biochem. J.* **123**, 5P.
Gibson, R. A., Schneider, E. A., and Wightman, F. (1972). *J. Exp. Bot.* **23**, 381–399.
Guggenheim, M. (1958). *In* "Handbuch der Pflanzenphysiologie" (W. Ruhland, ed.), Vol. 8, pp. 889–988. Springer-Verlag, Berlin and New York.
Gupta, O. P., Gupta, S. C., Dhar, K. L., and Atal, C. K. (1977). *Phytochemistry* **16**, 1436–1437.
Harley, K. L. S., and Thorsteinson, A. J. (1967). *Can. J. Zool.* **45**, 305–319.
Hartmann, T. (1967). *Z. Pflanzenphysiol.* **57**, 368–375.
Hartmann, T. (1972). *Biochem. Physiol. Pflanz.* **163**, 14–29.
Hartmann, T., and Aufermann, B. (1973). *Mar. Biol.* **21**, 70–74.
Hartmann, T., Dönges, D., and Steiner, M. (1972a). *Z. Pflanzenphysiol.* **67**, 404–417.
Hartmann, T., Ilert, H. I., and Steiner, M. (1972b). *Z. Pflanzenphysiol.* **68**, 11–18.
Heimer, Y. M., Misraki, Y., and Bachrach, U. (1979). *FEBS Lett.* **104**, 146–148.
Hesse, M., and Schmid, H. (1976). *Int. Rev. Sci., Org. Chem., Ser. Two* **9**, 265–307.
Hill, J. M., and Mann, P. J. G. (1964). *Biochem. J.* **91**, 171–182.
Hunter, A. R., Farrell, P. J., Jackson, R. J., and Hunt, T. (1977). *Eur. J. Biochem.* **75**, 149–157.
Jacobson, M. (1951). *J. Am. Chem. Soc.* **73**, 100–103.
Jakubowicz, T., Wolska-Mitaszko, B., Gasior, E., and Kucharzewska, T. (1976). *Acta Microbiol. Pol.* **25**, 199–204.
Jänne, J., Pösö, H., and Raina, A. (1978). *Biochim. Biophys. Acta* **473**, 241–293.
Janzen, D. H., Juster, H. B., and Bell, E. A. (1977). *Phytochemistry* **16**, 223–227.
Johne, S., Gröger, D., and Radeglia, R. (1975). *Phytochemistry* **14**, 2635–2636.
Kaur-Sawhney, R., Altman, A., and Galston, A. W. (1978). *Plant Physiol.* **62**, 158–160.
Kneifel, H. (1977). *Chem.-Z* **101**, 165–168.
Kneifel, H., Schuber, F., Aleksijevic, A., and Grove, J. (1978). *Biochem. Biophys. Res. Commun.* **85**, 42–46.
Kupchan, S. M., Davies, A. P., Barboutis, S. J., Schnoes, H. K., and Burlingame, A. L. (1969). *J. Org. Chem.* **34**, 3888–3893.
Kuttan, R., and Radhakrishnan, A. N. (1972). *Biochem. J.* **127**, 61–67.
Leete, E., and Minich, M. L. (1977). *Phytochemistry* **16**, 149–150.
Le Rudulier, D., and Goas, G. (1971). *C.R. Hebd. Seances Acad. Sci.* **273**, 1108–1111.
Le Rudulier, D., and Goas, G. (1975). *Physiol. Veg.* **13**, 125–136.
McGowan, R. E., and Muir, R. M. (1971). *Plant Physiol.* **47**, 644–648.
Mack, J. P. G., and Slaytor, M. (1974). *Proc. Aust. Biochem. Soc.* **7**, 11.
Mack, J. P. G., and Slaytor, M. (1979). *Phytochemistry* **18**, 1921–25.
Martin-Tanguy, J., Cabanne, F., Perdrizet, E., and Martin, C. (1978). *Phytochemistry* **17**, 1927–1928.

Mizusaki, S., Tanabe, Y., Noguchi, M., and Tamaki, E. (1973). *Plant Cell Physiol.* **14**, 103–110.
Montague, M. J., Koppenbrink, J. W., and Jaworski, E. G. (1978). *Plant Physiol.* **62**, 430–433.
Montague, M. J., Armstrong, T. A., and Jaworski, E. G. (1979). *Plant Physiol.* **63**, 341–345.
Naik, B. I., and Srivastava, S. K. (1978). *Phytochemistry* **17**, 1885–1887.
Nickerson, K. W., and Lane, L. C. (1977). *Virology* **81**, 455–459.
Nylén, U., and Szybek, P. (1974). *Acta Chem. Scand.* **28**, 1153–1160.
Paul, A. G. (1973). *Lloydia* **36**, 36–45.
Pösö, H., Sinervirta, R., and Jänne, J. (1975). *Biochem. J.* **151**, 67–73.
Ramakrishna, S., and Adiga, P. R. (1974). *Phytochemistry* **13**, 2161–2166.
Ramakrishna, S., and Adiga, P. R. (1975). *Eur. J. Biochem.* **59**, 377–386.
Ramakrishna, S., and Adiga, P. R. (1976). *Phytochemistry* **15**, 83–86.
Saunders, J. W., and Bush, L. P. (1979). *Plant Physiol.* **64**, 236–240.
Schütte, H. R. (1969). *In* "Biosynthese der Alkaloide" (K. Mothes and H. R. Schütte, eds.), pp. 168–182. VEB Dtsch. Verlag Wiss., Berlin.
Silver, S., and Kralovic, M. L. (1969). *Mol. Pharmacol.* **5**, 300–302.
Sindhu, R. K., and Desai, H. V. (1979). *Phytochemistry* **18**, 1937–1938.
Smith, T. A. (1969). *Phytochemistry* **8**, 2111–2117.
Smith, T. A. (1970). *Biochem. Biophys. Res. Comm.* **41**, 1452–1456.
Smith, T. A. (1971). *Biol. Rev. Cambridge Philos. Soc.* **46**, 201–241.
Smith, T. A. (1974). *Phytochemistry* **13**, 1075–1081.
Smith, T. A. (1976). *Phytochemistry* **15**, 633–636.
Smith, T. A. (1977a). *Prog. Phytochem.* **4**, 27–81.
Smith, T. A. (1977b). *Phytochemistry* **16**, 9–18.
Smith, T. A. (1977c). *Phytochemistry* **16**, 171–175.
Smith, T. A. (1977d). *Phytochemistry* **16**, 278–279.
Smith, T. A. (1977e). *Phytochemistry* **16**, 1647–1649.
Smith, T. A. (1979a). *Anal. Biochem.* **92**, 331–337.
Smith, T. A. (1979b). *Phytochemistry* **18**, 1447–1452.
Smith, T. A. (1980). *Encycl. Plant Physiol. New Ser.* **8**, 433–460.
Smith, T. A., and Best, G. R. (1978). *Phytochemistry* **17**, 1093–1098.
Smith, T. A., and Ng, W. Y. (1979). *Trans. Biochem. Soc.* **7**, 95–96.
Smith, T. A., and Wilshire, G. (1975). *Phytochemistry* **14**, 2341–2346.
Smith, T. A., Bagni, N., and Serafini Fracassini, D. (1978a). *In* "Nitrogen Assimilation of Plants" (E. J. Hewitt and C. V. Cutting, eds.), pp. 557–570. Academic Press, New York.
Smith, T. A., Best, G. R., Abbott, A. J., and Clements, E. D. (1978b). *Planta* **144**, 63–68.
Srivastava, S. K., and Prakash, V. (1977). *Phytochemistry* **16**, 189–190.
Srivastava, S. K., Prakash, V., and Naik, B. I. (1977a). *Phytochemistry* **16**, 185–187.
Srivastava, S. K., Prakash, V., and Naik, B. I. (1977b). *Phytochemistry* **16**, 1297–1298.
Stevens, L. (1970). *Biol. Rev. Cambridge Philos. Soc.* **45**, 1–27.
Stevens, L. (1975). *FEBS Lett.* **59**, 80–82.
Stoessl, A., and Unwin, C. H. (1970). *Can. J. Bot.* **48**, 465–470.
Suresh, M. R., and Adiga, P. R. (1977). *Eur. J. Biochem.* **79**, 511–518.
Suresh, M. R., and Adiga, P. R. (1978). *Biochem. J.* **172**, 185–188.
Suresh, M. R., and Adiga, P. R. (1979). *J. Biosci.* **1**, 109–124.
Suresh, M. R., Ramakrishna, S., and Adiga, P. R. (1976). *Phytochemistry* **15**, 483–485.
Suresh, M. R., Ramakrishna, S., and Adiga, P. R. (1978). *Phytochemistry* **17**, 57–63.
Tabor, C. W., and Tabor, H. (1976). *Annu. Rev. Biochem.* **45**, 285–306.
Takeo, T. (1978). *Phytochemistry* **17**, 313–314.
Terano, S., and Suzuki, Y. (1978a). *Phytochemistry* **17**, 148–149.
Terano, S., and Suzuki, Y. (1978b). *Phytochemistry* **17**, 550–551.
Unger, W., and Hartmann, T. (1976). *Z. Pflanzenphysiol.* **77**, 255–267.

Vialli, D. M., Barbetta, F., Zanotti, L., and Mihályi, K. (1973). *Acta Histochem.* **45,** 270–282.
Villalobos, J., Ramirez, F., and Moussatché, H. (1974). *Cienc. Cult. (Sao Paulo)* **26,** 690–693.
von Preusser, E. (1975). *Biol. Zentralbl.* **94,** 75–86.
Werle, E. (1955). *In* "Moderne Methoden der Pflanzenanalyse" (K. Paech and M. V. Tracey, eds.), Vol. 4, pp. 517–623. Springer-Verlag, Berlin and New York.
Wightman, F., and Rauthan, B. S. (1974). *Plant Growth Subst., Proc. Int. Conf., 8th, 1973* pp. 15–27.
Winther, M., and Stevens, L. (1978). *FEBS Lett.* **85,** 229–232.
Wolska-Mitaszko, B., Jakubowicz, T., and Gasior, E. (1976). *Acta Microbiol. Pol.* **25,** 187–197.
Yamasaki, E. F., Swindell, R., and Reed, D. J. (1970). *Biochemistry* **9,** 1206–1210.
Yamasaki, K., Tamaki, T., Uzawa, S., Sankawa, U., and Shibata, S. (1973). *Phytochemistry* **12,** 2877–2882.

Coumarins | *10*

STEWART A. BROWN

I. INTRODUCTION

Coumarins are derivatives of $2H$-1-benzopyran-2-one or benzopyrone. Well over 500 are known to occur in nature (Murray, 1978), and many hundreds more have been synthesized in the laboratory. Although they have applications ranging from use as anticoagulant drugs to use as laser dyes and fluorescent brighteners, the biochemistry of most coumarins is a quite unexplored field. To the plant biochemist interest in them is primarily twofold—their effects on the growth and development of plants and, for the naturally occurring members of the group, their biosynthesis within the plant. A discussion of these two aspects of coumarins will occupy most of this chapter.

The Biochemistry of Plants, Vol. 7

II. BIOSYNTHESIS OF COUMARINS

A. Site of Formation in the Plant*

Reppel and Wagenbreth (1958), using grafts of *Melilotus alba* on *Trigonella foenum-graecum*, found little coumarin (I) in the *M. alba* shoots and deduced that roots were required for coumarin synthesis, probably because they provided an essential precursor. However, the observations made by Gorz and Haskins (1962) during grafting experiments with coumarin-containing

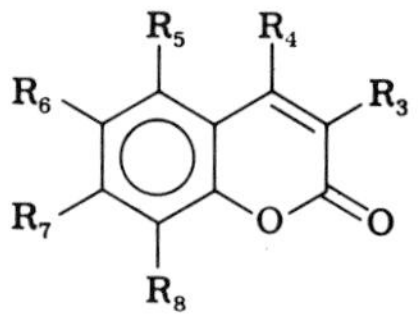

Ia. R_7 = OMe

Ib. R_6 = OMe, R_7 = OH
Ic. R_7 = OH
Id. R_7 = OH, R_8 = CH_2—CH=CMe$_2$
Ie. R_7 = O-glucosyl
If. R_6 = OMe, R_7 = O-glucosyl
Ig. R_6 = CH_2—CH=CMe$_2$, R_7 = OH
Ih. R_5 = R_7 = OMe,
R_6 = H_2C—HC——CMe$_2$

Ii. R_7 = OMe, R_8 = H_2C—HC——CMe$_2$
Ij. R_6 = R_7 = OH
Ik. R_4 = OH
Il. R_4 = Me, R_7 = OH
Im. R_6 = Me
In. R_4 = R_7 = OH
Io. R_4 = OCH_2—CH=CMe$_2$, R_7 = OH
Ip. R_3 = CMe_2—CH=CH$_2$, R_4 = R_7 = OH

and coumarin-free strains of *M. alba* have led them to conclude that the primary site of coumarin synthesis is the young, actively growing leaves and that stems and roots have a comparatively minor role. Blaim (1960a) has shown, too, that coumarin is formed from $^{14}CO_2$ equally well by detached *M. alba* shoots and intact plants, whereas herniarin (Ia) in *Lavandula officinalis* (Brown, 1963a) and scopoletin (Ib) in several species (Reznik and Urban, 1957) have been found to be formed from labeled phenylpropanoid precursors by organs other than roots. Nevertheless, root tissue cultures of *M. officinalis* synthesize coumarin (Weygand and Wendt, 1959), and scopoletin and umbelliferone (Ic) are formed from a carbohydrate source by root tissue cultures of *Atropa belladonna* (Mothes and Kala, 1955). More recently Beyrich (1967) has shown by reciprocal grafting experiments involving parsnip (*Pastinaca sativa*) that furanocoumarins in this species are formed in the fruits where they accumulate, and he has found no evidence for translocation. Steck and Bailey (1969) believe that several furanocoumarins formed during a time study of their development in *Angelica archangelica* are formed in the leaves, whereas the simple coumarin osthenol (Id) is probably formed in the roots. Cut shoots of *Thamnosma montana* synthesize a number of coumarins (Kutney *et al.*, 1973a), and grafting experiments have re-

* In all structures, where a substituent is not otherwise identified, R = H.

vealed the participation of several plant sites in the elaboration of coumarin and herniarin by *Prunus mahaleb* (Favre-Bonvin *et al.,* 1968).

It is evident that no simple, general statement can adequately cover the question of synthesis sites. There seems to be no doubt that both roots and aerial organs of *M. alba* can form coumarin, but the weight of evidence indicates that, under normal growing conditions, synthesis occurs predominantly in the leaves. The situation with other coumarins and species may vary greatly, even for different compounds formed by the same plant, and each case must be examined individually.

B. Biosynthetic Pathways

There are two distinct experimentally demonstrated pathways by which natural products incorporating the benzopyrone nucleus are formed. In a minor class, the 3- and 4-phenylcoumarins, the aromatic portion of this nucleus is polyketide-derived, with the three aliphatic carbons and the phenyl substituent originating from shikimate via a phenylpropanoid intermediate. Their formation will be treated in Section II,B,5.

In all other plant coumarins whose biosynthesis has been studied, the benzene ring and the attached three-carbon side chain of the lactone ring arise as a unit via a shikimate-derived phenylpropanoid. It was established very early in biosynthesis investigations that coumarin itself is derived from shikimic acid via phenylalanine and cinnamic acid (Kosuge and Conn, 1959; Weygand and Wendt, 1959; Brown *et al.,* 1960; Stoker and Bellis, 1962). The experiments left no doubt that coumarin originates via the shikimate–chorismate pathway leading to phenylpyruvic acid, from which arise L-phenylalanine by transamination and *trans*-cinnamic acid in turn by the action of phenylalanine ammonia-lyase. These early reactions up to the cinnamic acid stage are common to numerous phenylpropanoid-derived secondary plant products and are covered in detail in Chapter 13, Volume 5 and Chapter 20, Volume 7 of this treatise. They will not be further treated in this chapter.

1. Conversion of Cinnamic Acid to Simple Coumarins

Apart from phenylcoumarins, the coumarins to be discussed in this chapter are of two biosynthetically fundamental types: those bearing oxygenation at C-7, para to the propanoid side chain, and those lacking this feature. There are fewer than fifty examples of the latter type known, including of course coumarin itself. Thus umbelliferone (Ic) (7-hydroxycoumarin), the simplest member of the former category, can be logically regarded as the parent compound of coumarins generally.

It has been proposed by Brown (1960), on the basis of tracer investigations, that *trans*-cinnamic acid (IIa) functions in the appropriate plants as a common precursor of all coumarins and that its ortho- or para-hydroxylation

IIa. $R_1 = CH=CH-COOH$
IIb. $R_1 = CH=CH-COOH$, $R_2 = OH$
IIc. $R_1 = CH=CH-COOH$,
$\quad R_2 = O\text{-glucosyl}$
IId. $R_1 = CH=CH-COOH$, $R_4 = OH$
IIe. $R_1 = CH=CH-COOH$, $R_4 = OH$,
$\quad R_5 = OMe$

IIf. $R_1 = CH=CH-COOH$,
$\quad R_2 = O\text{-glucosyl}$, $R_4 = OMe$
IIg. $R_1 = CH=CH-COOH$, $R_4 = R_5 = OH$
IIh. $R_1 = (CH_2)_2-COOH$, $R_2 = OH$
IIi. $R_1 = (CH_2)_2-COOH$,
$\quad R_2 = O\text{-glucosyl}$

leads subsequently to the elaboration of coumarin or 7-hydroxycoumarins, respectively. Later, using lavender, one of the few species that synthesize both coumarin and a 7-oxygenated coumarin (herniarin, Ia), he showed (Brown, 1962b, 1963a) that trans-[2-^{14}C]2′-hydroxycinnamic acid (IIb) (as well as its glucoside, IIc) was selectively incorporated into coumarin, and trans-[2-^{14}C]4′-hydroxycinnamic acid (IId) into herniarin. Thus, in the elaboration of 7-oxygenated coumarins, para-hydroxylation must precede the ortho-hydroxylation that is the committed step in lactone ring formation. Confirmation for other coumarins in different species (Floss and Mothes, 1964; Austin and Meyers, 1965a) indicates this principle to be a general one. We must therefore examine the two pathways separately.

a. Cinnamic Acid to Coumarin. A "bound" coumarin long known to occur in plants (Roberts and Link, 1937; Kosuge, 1961)—coumarinyl glucoside or cis-2-glucosyloxycinnamic acid (IIc)—can be hydrolyzed in vitro to the unstable coumarinic (cis-2′-hydroxycinnamic) acid (IIb), and this product lactonizes spontaneously to "free" coumarin. It was initially assumed in biosynthetic studies that much coumarin also occurred free in the cells. That this assumption was in error, at least in Melilotus, was demonstrated by Haskins and Gorz (1961), who showed that only traces of free coumarin could be recovered from M. alba leaves that had been autoclaved before work-up. The recovery of free coumarin from plants worked up under less rigorous conditions is explained by the discovery of a β-glucosidase in Melilotus, which specifically hydrolyzes the cis-glucoside, releasing coumarin, when the cells are disrupted (Kosuge and Conn, 1961; Schön, 1966). Hence arises the odor of new-mown hay. Presumably some form of compartmentation denies this enzyme access to its substrate in the intact cell. References to coumarin in vivo must therefore be generally interpreted to mean coumarinyl glucoside, although there is some evidence for free coumarin in Hierochloë odorata, a grass (Brown, 1962a).

The conversion of trans-cinnamic acid to coumarin thus must involve ortho-hydroxylation, glucoside formation at the 2′-hydroxyl, and isomerization to the cis isomer, in that order. Kosuge and Conn (1959) and later Stoker and Bellis (1962) demonstrated the formation of *trans*-2'-hydroxycinnamic acid (IIb) or its glucoside from *trans*-[ring,3-^{14}C]cinnamic acid (IIa) by *M. alba in vivo*. More recently Gestetner and Conn (1974) identified a chloroplast enzyme in the same species mediating the ortho-hydroxylation. This enzyme, which was largely membrane-bound, had an activity peak at pH 7.0 and was over fourfold more active in the presence of glucose-6-phosphate, which apparently formed part of a generating system for reduced NADP. Light could replace the sugar phosphate, presumably as a source of reducing power for the hydroxylation system. Ellis and Amrhein (1971) have shown by feeding experiments with *o*-[^{3}H]cinnamic acid that this ortho-hydroxylation proceeds in *M. alba* with an efficient migration and retention of the ortho proton—an NIH shift.

Kinetic studies after administration of $^{14}CO_2$ to *H. odorata* (Brown, 1962a) have provided clear evidence that *trans*-2'-glucosyloxycinnamic acid (IIc) is an intermediate in the formation of coumarin *in vivo*. The fact that free coumarin isolated in these experiments had a consistently higher specific radioactivity than the bound suggested the existence of a pool of free coumarin in this plant and appears to rule out a bound coumarin–free coumarin interconversion unless one invokes separate metabolic pools. Brown suggested alternatively that free coumarin and coumarinyl glucoside could arise from a common precursor by independent pathways. A pathway from *trans*-cinnamic acid (IIa) to free coumarin via *cis*-cinnamic acid is suggested by the observation of Stoker and Bellis (1962) that the cis acid is an effective coumarin precursor. Kleinhofs *et al.* (1967) have demonstrated the glucosylation of *trans*-2'-hydroxycinnamic acid (IIb) in cell-free extracts of *M. alba*, a reaction requiring UDP-glucose and a sulfhydryl compound. A role for this glucosylating enzyme in coumarin biosynthesis is suggested by its stereospecificity—it apparently fails to act on the sodium salt of *cis*-2'-hydroxycinnamic acid. The enzyme proved too labile for purification.

It is now agreed that the trans–cis isomerization constituting the final step in the formation of bound coumarin is light- rather than enzyme-catalyzed (Haskins *et al.*, 1964; Edwards and Stoker, 1967). Haskins *et al.* (1964) concluded that, as neither steaming of *M. alba* leaflets nor maintenance at low temperatures greatly affected their ability to bring about the trans–cis conversion, the reaction could not be enzyme-mediated, and they presented evidence that it was in fact a photochemical reaction effected by wavelengths below 360 nm.

Despite considerable investigation the biochemical basis of the genetic control of coumarin biosynthesis in *Melilotus* remains, in some respects, obscure. It has long been known (Goplen *et al.*, 1957; Gorz and Haskins, 1960) that the process is influenced by two alleles, designated *Cu/cu* and *B/b*, but hypotheses advanced as to the site of their action (Schaeffer *et al.*, 1960;

Akeson *et al.*, 1963; Brown, 1963b; Kleinhofs *et al.*, 1966) have not always been supported by subsequent research. The possibility that phenylalanine ammonia-lyase is primarily controlled by the *Cu/cu* gene pair was not borne out by experiments on this enzyme from *M. alba* (Kleinhofs *et al.*, 1966). In tracer studies with several phenylpropanoid precursors of coumarin Haskins and Kosuge (1965) found that *o*-coumaric acid was extensively glucosylated in both *Cu Cu* and *cu cu* strains of *M. alba*, but that phenylalanine and *trans*-cinnamic acid were converted to *o*-coumaryl glucoside only in the *Cu Cu* strain, and inferred that the *cu*-controlled block was at the stage of ortho-hydroxylation of *trans*-cinnamic acid. However, when the *o*-hydroxylase mediating this step was later characterized (Gestetner and Conn, 1974), no significant differences in its activity among the different genotypes were observed, and it was concluded that the site of action of the *Cu/cu* allele in *Melilotus* still awaited elucidation.

The biochemical function of the *B/b* allele is clearer. Haskins and Gorz (1965) have shown that β-glucosidase activity in the *BB*, *Bb*, and *bb* genotypes is in the ratio $2.5:1:0$, the obvious inference being that it is this enzyme that is under the control of the *B/b* allele. Thus free coumarin can be formed only in the presence of the dominant *B* gene, and then only, as explained above, if cell disruption has first occurred. This same group of investigators (Gilchrist *et al.*, 1970) has identified in extracts of the *M. alba* *bb* genotype a protein that is serologically and electrophoretically similar to the β-glucosidase recovered from the *BB* plants but which lacks β-glucosidase activity. It is tempting to speculate, on this basis, that the difference in the genotypes may relate only to relatively minor modifications of tertiary protein structure influencing formation of the enzyme active site.

b. Cinnamic Acid to 7-Oxygenated Coumarins. The first reaction in this pathway is the 4'-hydroxylation of cinnamic acid. This reaction is common to the biosynthesis of a number of secondary plant products, notably lignin, and it is considered in detail by Butt and Lamb (this volume, Chapter 21).

An *o*-hydroxylase acting on 4'-hydroxycinnamic acid has also been identified, in a chloroplast fraction, by Kindl (1971) using *Hydrangea macrophylla*, which elaborates 7-oxygenated coumarins. However, its specificity differs from that of the Gestetner-Conn (1974) hydroxylase; unlike the latter, it hydroxylated cinnamic acid only at a low rate but readily converted 4'-hydroxycinnamic acid to umbelliferone (Ic), and ferulic acid (IIe) to scopoletin (Ib), presumably via the respective 2',4'-dihydroxyacids which must have rapidly undergone trans–cis isomerization and lactonization. Again unlike the other *o*-hydroxylase, it appears to have a tetrahydropteridine cofactor requirement. As the existence in plants of cinnamic acid 4'-hydroxylases is assumed to be almost universal, owing to the necessity for lignin monomer synthesis, it is the presence or absence of one or both of these

2'-hydroxylases that evidently controls whether a species contains 7-oxygenated or nonoxygenated coumarins or, in rare cases, both.

As in the case of coumarin, glucosides are significant in the formation of 7-oxygenated simple coumarins. Umbelliferone in *H. macrophylla* (Austin and Meyers, 1965b) and *Pimpinella magna* (Floss and Paikert, 1969), and herniarin in *Lavandula officinalis* (Brown, 1965), occur almost entirely in the bound form, the former predominantly as 7-β-D-glucosyloxycoumarin (skimmin, Ie) and the latter probably as *cis*-2'-glucosyloxy-4'-methoxycinnamic acid (IIf). As in the case of coumarinyl glucoside, the formation of the *cis*-glucoside of herniarin from its trans precursor is light-mediated (Edwards and Stoker, 1968). 7,8-Dihydroxycoumarin (daphnetin) exists in *Daphne* as both the 7- and 8-glucosides. Satô and Hasegawa (1969, 1971) have found a transglucosylase in *Daphne odora* flowers, which appears both to hydrolyze the 7-glucoside and to transfer a glucosyl residue from this glucoside to the 8-hydroxyl of the aglycone. In *Cichorium intybus* these authors (Satô and Hasegawa, 1972b) have described two enzymes that act in an analogous way on esculetin (Ij) 7-glucoside, converting it via hydrolysis and transglucosylation to the 6-glucoside.

The pathways to several of these coumarins have been investigated in some detail *in vivo* by tracer techniques. The formation of umbelliferone from cinnamic acid via 4'-hydroxycinnamic acid was established in this way (Brown *et al.*, 1964; Austin and Meyers, 1965a). In this case there were conflicting reports about the possible function of *trans*-[^{14}C]2'-hydroxycinnamic acid as a precursor, and it appears unlikely that it is a natural intermediate (cf. also Floss and Paikert, 1969). The synthetic route via 4'-hydroxycinnamic acid was also demonstrated for herniarin (Ia) (Brown, 1962b, 1963a). *trans*-2'-Glucosyloxy-4'-methoxycinnamic acid (IIf) appears to be an intermediate in this pathway, but the stage at which O-methylation occurs has not been established, although a hypothesis has been advanced (Brown, 1965) to account for the known facts.

The biosynthesis of scopoletin (Ib) and its glucoside scopolin (If) has been extensively studied, and the findings are in accord with a route from phenylalanine and cinnamic acid via 4'-hydroxycinnamic, caffeic (IIg), and ferulic (IIe) acids to scopoletin (Reid, 1958; Runeckles, 1963; Steck, 1967a,b; Fritig *et al.*, 1966, 1970; Loewenberg, 1970). The weight of evidence indicates that scopolin normally originates from glucosylation of scopoletin (Fritig *et al.*, 1970; Innerarity *et al.*, 1972).

Coumarin, umbelliferone, herniarin, and scopoletin are clearly derived from cinnamic acids of the corresponding oxygenation pattern. However, in the ortho-dihydroxy coumarins, esculetin (6,7-) and daphnetin (7,8-), attempts to demonstrate biosynthesis from caffeic acid (IIg) have yielded negative results (Satô and Hasegawa, 1972a). These findings, and the fact that umbelliferone is a precursor of the 7,8-dioxygenated hydrangetin (Bohm *et al.,* 1961; Kindl and Billek, 1964), suggest that additional oxygenation of the nucleus may sometimes occur in simple coumarins. As will be noted shortly, this occurs routinely in furanocoumarins.

2. Formation of Furanocoumarins

Two types of furanocoumarins are recognized: the linear (III), in which the furan ring is fused at the 6- and 7-positions of the benzene nucleus (psoralens), and the less widely distributed angular (IV), where fusion is 7, 8.

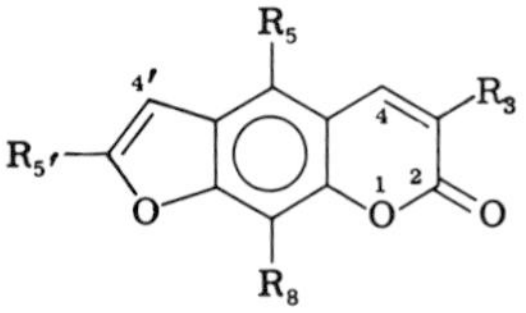

IIIa. $R_5 = R_8 = OMe$ IIId. $R_8 = OCH_2-CH=CMe_2$
IIIb. $R_8 = OMe$ IIIe. $R_4 = R_{5'} = R_8 = Me$
IIIc. $R_5 = OMe$

IVa. $R_6 = OMe$

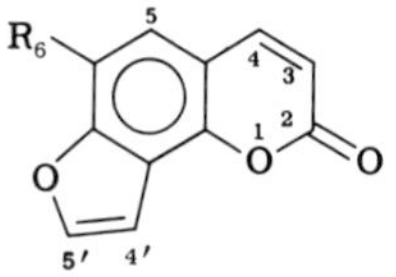

With minor exceptions, these coumarins are confined to the Umbelliferae and Rutaceae (Nielsen, 1970; Gray and Waterman, 1978; Murray, 1978). There are common features in the structure and biosynthesis of these coumarins and those of furanoquinoline alkaloids, and for a recent discussion in this context a review by Grundon (1978) is commended to the reader's attention.

a. Origin from Umbelliferone and Mevalonic Acid. Floss and Mothes (1964) showed that furanocoumarins of *Pimpinella magna* became radioactive after administration of [1-^{14}C]cinnamic acid and [2-^{14}C]umbelliferone, the latter being utilized more efficiently by a factor of 2.5. Coumarin was a poor precursor. Linear furanocoumarins in other species also arise from umbelliferone (Brown *et al.*, 1970; Caporale *et al.*, 1970), but certain derivatives of benzofuran are inactive as precursors (Caporale *et al.*, 1974). Umbelliferone can thus be regarded as the parent compound of furanocoumarins in the plant, and the base for construction of the furan ring. The coexistence of simple furanocoumarins, isopropyl dihydrofuranocoumarins such as marmesin (Va) and columbianetin (VI), and prenylated coumarins suggested to the German workers the possibility that the furan ring originates from C-1 and C-2 of an isoprenoid precursor, deriving from C-5 and C-4, respectively, of mevalonic acid (VII). Using [4-^{14}C]mevalonic acid they showed (Floss and Mothes, 1966) low incorporation into the *P. magna* furanocoumarins and in the face of very low specific activities demonstrated by ozonization that the labeled carbon of the angular furanocoumarin sphondin (IVa) was localized in the furan ring carbon not linked to the benzene ring (C-5').

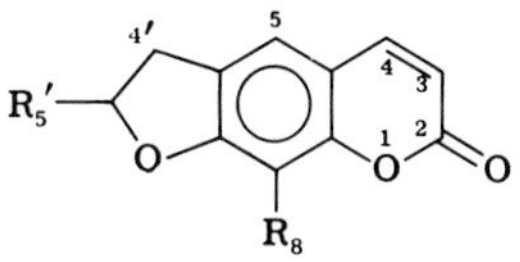

V

Va. $R_{5'}$ = CMe$_2$OH
Vb. $R_{5'}$ = CMe$_2$OH, R_8 = OH
Vc. R_3 = CMe$_2$—CH=CH$_2$
     $R_{5'}$ = CMe$_2$OAc

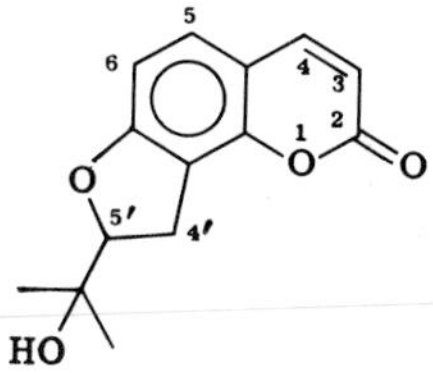

VI

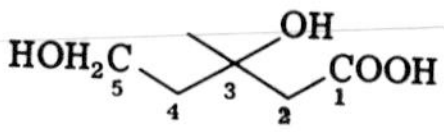

VII

Although the participation of mevalonic acid in a major biosynthetic route to furanocoumarins was subsequently questioned (Caporale *et al.*, 1964; Brown, 1970), a later intensive study by Kutney's group has established its role on a firm footing. Using shoots of *T. montana* Kutney *et al.* (1973a) showed that D,L-[2-³H]mevalonic acid lactone was not a precursor of two furanocoumarins examined, whereas feeding of the 4-³H-labeled lactone led to a low specific incorporation of activity into C-5' of isopimpinellin (IIIa), in agreement with the results of Floss and Mothes (1966). [2-¹⁴C]Acetate was more efficiently utilized than mevalonate (Kutney *et al.*, 1973b), as had been previously observed in parsnip plants (Brown, 1970). However, degradation of the isopimpinellin revealed that C-2 of the acetate was still being specifically incorporated into C-5', as would have been predicted if mevalonate were the intermediate, and not into C-4'. Most of the remaining ¹⁴C was located in the *O*-methyl carbons, which perhaps could have been derived from C-2 of acetate via glyoxylic acid, a donor of "active formate." The C-2 of glycine was also shown to be a precursor of the *O*-methyl carbons of isopimpinellin. In an extension of these studies to *T. montana* cell cultures, which yielded higher incorporations than shoots, it was shown that tritium at mevalonate positions 4 and 5 was specifically incorporated into furanocoumarin positions 5' and 4', respectively (Kutney *et al.*, 1973c).

b. Prenylation of Umbelliferone. Confirmation of the role of mevalonate was soon forthcoming through a different approach when Ellis and Brown (1974) reported the existence in *Ruta graveolens* of a particulate transferase able to mediate the prenylation of position 6 of umbelliferone by dimethylallyl pyrophosphate (VIII), an intermediate well known to be mevalonate-derived, to form the naturally occurring demethylsuberosin (6-dimethylallylumbelliferone, Ig). This prenylase requires a divalent cation and is highly specific as to length of the isoprenoid chain (cf. also Dhillon and Brown, 1976) and position of attack, there being no prenylation at position 8

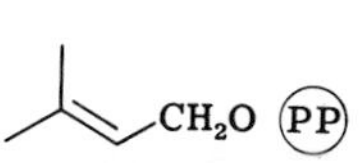

VIII

to form an angular furanocoumarin precursor. O-Methylation of the umbelliferone hydroxyl produced an inactive substrate. Dhillon and Brown (1976) adduced evidence that the transferase was a chloroplast enzyme and succeeded in purifying it over 300-fold.

c. Formation and Metabolism of Marmesin. The participation of demethylsuberosin (Ig) as an intermediate in the biosynthesis of linear furanocoumarins had already been demonstrated in plants by tracer experiments (Games and James, 1972; Austin and Brown, 1973; Brown and Steck, 1973) and has since been shown in additional species, although the observed isotope dilutions have sometimes been quite high (Innocenti *et al.*, 1977) and translocation problems have been troublesome (Brown and Steck, 1973). Epoxidation of the double bond of the prenyl side chain as a prerequisite to cyclization to a dihydrofuranocoumarin derivative had been proposed earlier, as in Scheme 1 (Whalley, 1961; Brown *et al.*, 1970). The implications of the

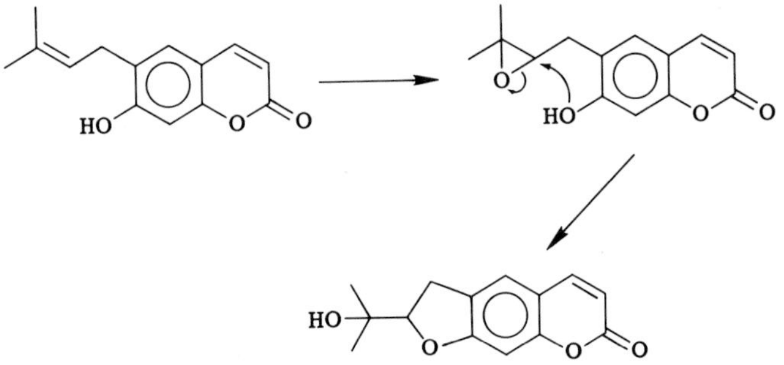

Scheme 1

stereochemistry of this reaction have since been examined (Grundon and McColl, 1975; Grundon, 1978). Although aculeatin (Ih), formed by *Toddalia aculeata*, and meranzin (Ii) of *Citrus aurantium* are precedents for such an epoxide, there has been no *in vivo* confirmation of this epoxidation reaction or the cyclization itself, which *in vitro* experiments indicate would probably occur spontaneously (Murray *et al.*, 1971). However, the role of the cyclization product, marmesin (Va), in linear furanocoumarin biosynthesis, first proposed by Abu-Mustafa *et al.* (1969), has been well established by the demonstration of its synthesis from umbelliferone and conversion to psoralens *in vivo* (Brown *et al.*, 1970; Caporale *et al.*, 1972a,b; Austin and Brown, 1973) and by trapping experiments to prove its function as a natural intermediate in species of two families (Brown *et al.*, 1970; Dall'Acqua *et al.*, 1972). Marmesin exists as enantiomers, and the *S* form (dextrorotatory in chloroform) is the metabolically active one in *R. graveolens, A. archangelica,* and *Heracleum lanatum* (Steck and Brown, 1971).

Ruta graveolens also contains 8-hydroxymarmesin (rutaretin, Vb), which is formed from marmesin (Dall'Acqua *et al.*, 1972) and is an intermediate in the biosynthesis of 8-methoxypsoralen (xanthotoxin, IIIb) by this species (Caporale *et al.,* 1971). The general significance of this compound is, however, open to question, as attempts to demonstrate its function in three species of Umbelliferae and in *Ficus carica* have given negative results (Caporale *et al.,* 1972a; Dall'Acqua *et al.,* 1975).

The mechanism by which the isopropyl residue of marmesin (Va) is eliminated and the ring double bond introduced during conversion of marmesin to psoralen (III) has not been established, although hypotheses exist (Floss, 1972; Grundon, 1978). A concerted reaction appears probable; in any event loss of the side chain does not precede double-bond formation (Brown, 1973; Marciani *et al.,* 1974). A plausible mechanism is that of Birch *et al.* (1969), who envisaged the generation of a carbocation at C-4′ of marmesin, followed by a 1,3-elimination to yield acetone and psoralen (Scheme 2). The

Scheme 2

experiments of Kutney *et al.* (1973c) with specifically tritiated mevalonate have proved that furan ring formation cannot proceed by any mechanism involving loss of the tritium originating in the 4*R*- and 5-positions of mevalonate(VII), and work on the related furanoquinoline alkaloid system (Grundon *et al.,* 1975) indicates by analogy that a 4′-oxo intermediate does not participate.

d. Elaboration of Angular Furanocoumarins. The reaction sequence from umbelliferone to psoralen in the linear furanocoumarin series appears to be paralleled by a sequence from umbelliferone to angelicin (IV) in the angular form (Steck and Brown, 1970), with columbianetin (VI) playing the role of marmesin (Va) and osthenol (8-dimethylallylumbelliferone, Id) that of demethylsuberosin (Ig) (Brown and Steck, 1973). Other reactions are assumed to be analogous, although no angular analogue of rutaretin (Va) has been reported and nothing is known about the enzymology.

e. Hydroxylation and O-Methylation. Although, as mentioned above, *R. graveolens* can 8-hydroxylate the benzene ring of a dihydrofuranocoumarin, it seems probable that most hydroxylation occurs after the psoralen or angelicin structure has been fully elaborated. Evidence for the incorporation into furanocoumarins of cinnamic acids and simple coumarins oxygenated in positions other than para to the three-carbon side chain has been negative (Floss and Paikert, 1969; Brown *et al.,* 1970; Caporale *et al.,* 1971; Dall'Acqua *et al.,* 1972), although such routes cannot yet be ruled out for all furanocoumarins. [G-³H]Psoralen (III) fed to *R. graveolens* plants yielded labeled bergapten (IIIc) and xanthotoxin (IIIb) (Brown *et al.,* 1970; Caporale *et al.*, 1971), whereas in *Ammi majus* and *Angelica archangelica* imperatorin (IIId) was also labeled (Brown *et al.*, 1970). Analogous results were obtained with two other umbellifers (Dall'Acqua *et al.*, 1975) and with *R. graveolens* cell cultures (Austin and Brown, 1973). It is presumed that specific hydroxylases exist to mediate the hydroxylation of both psoralen (III) and angelicin (IV), but no reports of such enzymes have yet appeared.

The R groups of alkoxylated furanocoumarins are of two basic types, prenyl and methyl. Little is known about the process of prenyl ether formation, but the biosynthesis of methyl ethers has been widely investigated in many organisms. Xanthotoxin (IIIb), bergapten (IIIc), and isopimpinellin (IIIa) are the most common examples of psoralen methyl ethers, although others, also incorporating dimethylallyloxy substituents, have a more limited distribution. Most of our knowledge of the O-methylation of hydroxypsoralens is derived from studies on cell-free systems, which will be discussed in detail by Poulton (this volume, Chapter 22). However, Brown and Sampathkumar (1977) have conducted *in vivo* investigations on the origin of isopimpinellin in leaves of *H. lanatum* and cell cultures of *R. graveolens*. This dimethoxypsoralen could be envisaged to arise via several possible routes from psoralen, and tracer experiments in fact showed that two of these, via the hydroxylation of either bergapten or xanthotoxin para to the methoxyl group, could function in these species, but the pathway via xanthotoxin was consistently predominant and is evidently the major synthetic route in the plant. This preference may be due to a more rapid hydroxylation of xanthotoxin than bergapten.

3. Formation of 3-(1,1-Dimethylallyl)Coumarins

Knowledge of the elaboration of these C-prenylated coumarins is sketchy, and confined to rutamarin (Vc) of *R. graveolens*. Tracer experiments have failed to show formation of this compound from demethylsuberosin (Id) (Donnelly *et al.*, 1977) or marmesin (Va) (Steck and Brown, 1971), but Donnelly *et al.* (1977) have demonstrated its biosynthesis from umbelliferone, 4,7-dihydroxycoumarin (In) and, to a much lesser extent (although specific incorporation was proved) 4-prenyloxy-7-hydroxycoumarin (Io). There is a clear implication that introduction of the C-3 side chain precedes prenylation at C-6, and a pathway was suggested (Donnelly *et al.*, 1977) involving rearrangement of Io to form Ip, and subsequent removal of the 4-oxygen function, before 6-prenylation. Although In and Io must be recognized as rutamarin precursors, there is no evidence for their occurrence in *R. graveolens*, and they cannot as yet be considered true intermediates in the pathway.

4. Formation of Pyranocoumarins

About the route to these compounds, isomeric with furanocoumarins, little has been determined. The only available direct experimental evidence is that demethylsuberosin (Ig) is a very efficient precursor of dihydroxanthyletin in *Coronilla glauca* (Brown and Steck, 1973), suggesting that the path-

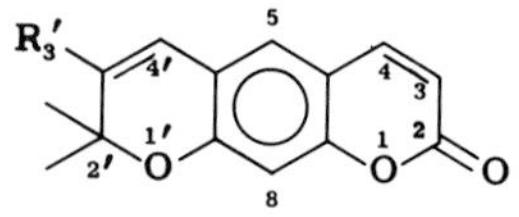

IX

IXa. $R_{3'}$ = OOC—CH=CMe$_2$
IXb. $R_{3'}$ = OH

way up to the cyclization stage coincides with that to furanocoumarins. In the light of *in vitro* experiments (Murray *et al.*, 1971; Gray and Waterman, 1978) have speculated that the mode of cyclization may depend upon the pH at the synthetic site.

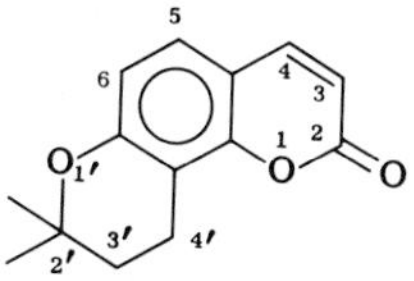

X

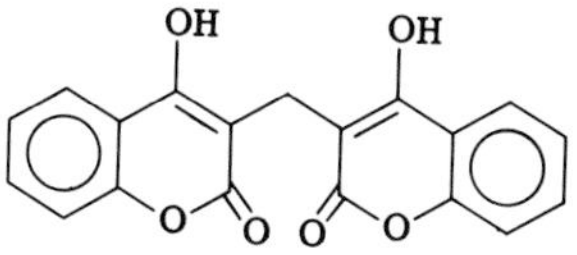

XI

5. Formation of Phenylcoumarins

As mentioned earlier, this class originates by a route very different from that of all other coumarins studied thus far. The first biosynthetic research on phenylcoumarins was conducted on coumestrol (XII), a 3-phenyl-

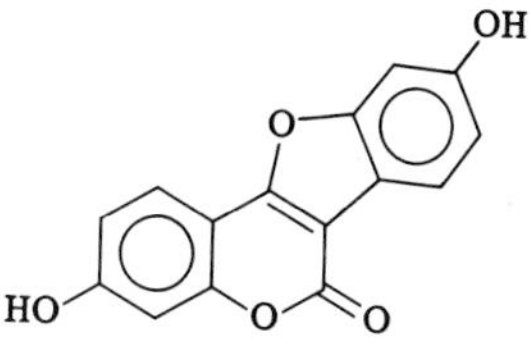

XII

coumarin of *Medicago sativa* with estrogenic properties, by Grisebach and Barz (1963). These workers, who had earlier examined the biosynthesis of isoflavonoids, found that both [1-^{14}C]acetate and [3-^{14}C]cinnamate were incorporated into coumestrol with the same labeling pattern as they had been into isoflavones: C-2 of the benzopyrone nucleus was derived from the 3-carbon of cinnamate, whereas the activity from the acetate appeared almost exclusively in the benzene ring of the nucleus. This of course argued a parallel biosynthetic origin. They later showed (Grisebach and Barz, 1964) that [β-^{14}C]4,2',4'-trihydroxychalcone 4'-glucoside, a known precursor of isoflavones, was also a precursor of coumestrol.

After further investigations on the later stages of this pathway in *Phaseolus aureus* and *Soja hispida*, Grisebach's group (Berlin *et al.*, 1972) proposed the reaction sequence shown in Scheme 3 as the major route to coumestrol.

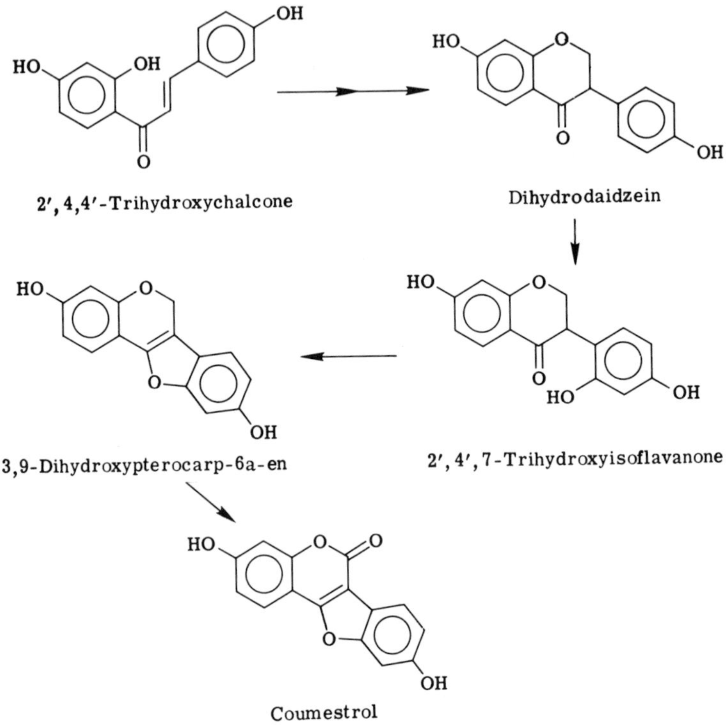

2′,4,4′-Trihydroxychalcone

Dihydrodaidzein

3,9-Dihydroxypterocarp-6a-en

2′,4′,7-Trihydroxyisoflavanone

Coumestrol

Scheme 3

4′,7-Dihydroxyisoflavone (daidzein) was shown by direct feeding experiments with labeled substrate to be an effective coumestrol precursor (Barz and Grisebach, 1966; Dewick *et al.*, 1970) but, as it is more effective than the product of its 2′-hydroxylation, 2′,4′,7-trihydroxyisoflavone (Berlin *et al.*, 1972), any route involving the latter cannot be considered to be of more than minor significance. The nonplanar nature of the dihydrodaidzein molecule is conducive on steric grounds to more ready 2′-hydroxylation at that oxidation level, and in fact its precursor efficiency is comparable to that of daidzein (Berlin *et al.*, 1972). The participation of 2′,4′,7-trihydroxyisoflavanone is supported by isotope dilution experiments (Dewick *et al.*, 1970) implying a prominent role for it (or the corresponding isoflavone). Other hypothetical sequences have been ruled out by these and other (Zilg and Grisebach, 1968) investigations. The inclusion of 3,9-dihydroxypterocarp-6a-en in the scheme rests on a chemical analogy, as pterocarp-6a-ens are oxidized with great ease to coumestans *in vitro* (Ferreira *et al.*, 1971). The results of all these studies make it quite clear that, from the biosynthetic standpoint, these compounds

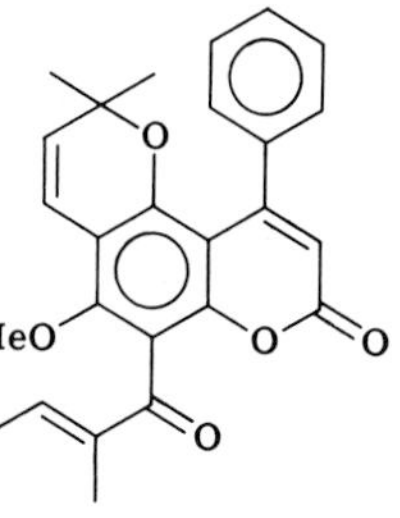

XIII 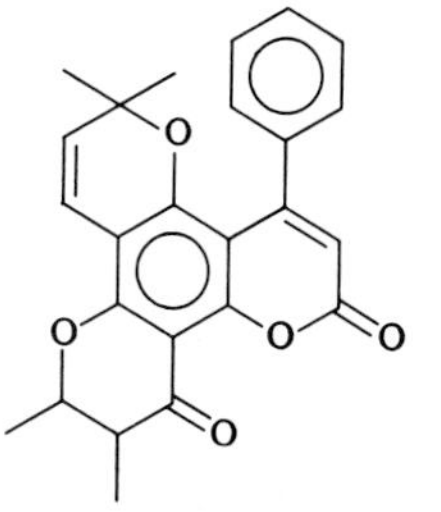   XIV 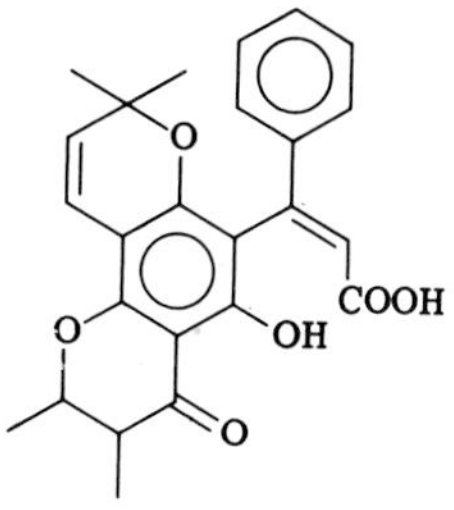   XV

are isoflavonoids, notwithstanding the fortuitous presence of the benzopyrone nucleus.

If 3-phenylcoumarins are biosynthetically isoflavonoids, it has subsequently become equally apparent from the work of Polonsky and her associates that 4-phenylcoumarins are biosynthetically neoflavonoids, a term
applied (Eyton *et al.*, 1965) to compounds incorporating a 4-arylchroman
nucleus. Kunesch and Polonsky (1967) studied the incorporation of [3-^{14}C]-
phenylalanine into the calophyllolide (XIII) elaborated by shoots of
Calophyllum inophyllum (Guttiferae) and located essentially all the ^{14}C of the
isolated coumarin in C-4, eliminating the possibility of rearrangement in the
phenylpropanoid moiety. These workers (Kunesch and Polonsky, 1969) later
demonstrated incorporation of acetate into the phloroglucinol nucleus. In
investigations with the structurally related 4-phenylcoumarin inophyllolide
(XIV), this group (Gautier *et al.*, 1972) established that calophyllic acid (XV)
had eight times the specific activity of inophyllolide after the administration
of [3-^{14}C]phenylalanine, a finding consistent with an intermediate function
for calophyllic acid in the elaboration of 4-phenylcoumarins. Their proposed
biosynthetic pathway is depicted in Scheme 4. They also showed that the
2-methylbutenoyl side chain of calophyllolide derived from isoleucine, presumably via tiglic acid (Kunesch and Polonsky, 1967), and that the isovaleryl

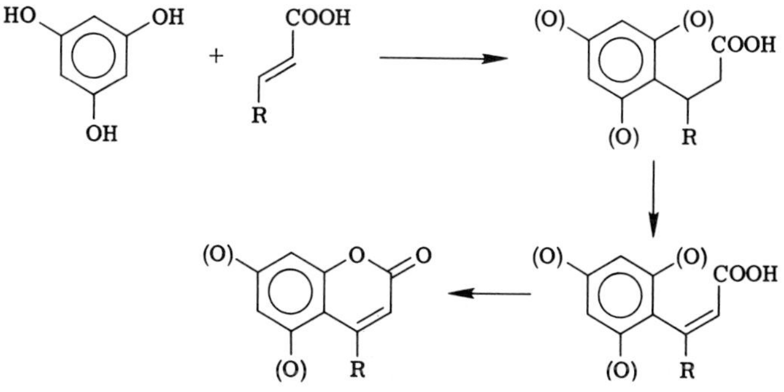

(R = C$_6$H$_5$ or *n*-alkyl)

Scheme 4

substituent in the phloroglucinol ring of the related 4-phenylcoumarin, mammeisin (XVI), originated in leucine (Kunesch *et al.,* 1969).

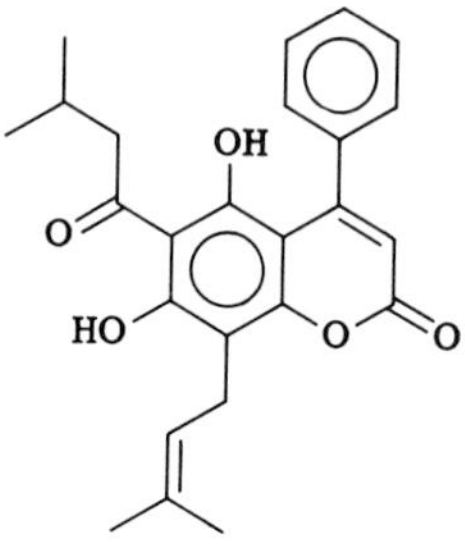

XVI

C. Subcellular Sites for the Formation of Coumarins

Evidence accumulated to date lends little support to the possibility of any highly organized system, such as a multienzyme complex located in a specific organelle, mediating the elaboration of coumarins. There is good evidence that the *o*-hydroxylases of cinnamic acids (Kindl, 1971; Gestetner and Conn, 1974; Ranjeva *et al.,* 1977) and the dimethyllalyl transferase ortho-prenylating umbelliferone (Dhillon and Brown, 1976) are chloroplast enzymes. However, *O*-methyltransferases acting on hydroxyfuranocoumarins seem not to be predominantly associated with the chloroplast (H. J. Thompson and S. A. Brown, unpublished), and phenylalanine ammonia-lyase is also apparently not a chloroplast enzyme (Camm and Towers, 1973). In other cases evidence for subcellular localization is lacking, and of course some hypothetical enzymes have not been identified at all. Speculation on this question thus seems premature at this time.

III. REACTIONS OF COUMARINS IN PLANTS

The further metabolism of coumarins in plants has not been extensively investigated, but there is no doubt that it does occur. Kosuge and Conn (1959) described the conversion of [3-^{14}C]coumarin to melilotic acid (IIh) and its glucoside (IIi) by *M. alba* shoots involving reduction and opening of the lactone ring, as well as conversion to at least two unidentified metabolites. Kuzovkina and Smirnov (1972) later noted an analogous conversion in the roots. Coumarin is also metabolized by germinating lettuce seeds to at least three unknown products (Sivan *et al.,* 1965). In contrast, umbelliferone seems resistant to further metabolism in *Hydrangea* (Austin and Meyers, 1965a).

Austin and Brown (1973) found that, when several labeled furanocoumarin precursors were fed to *R. graveolens* cells over a 7-day period, a marked increase in isotope dilution in the furanocoumarins, indicative of continuing synthesis, was accompanied by an equally marked decrease in the percentage of incorporation, explicable only on the basis of further metabolism. Thus a net turnover was indicated in the tissue, and it was suggested that the concentration of coumarins was probably held roughly constant by a balance of synthetic and degradative reactions. Double-labeling experiments with ^{14}C and tritium (Brown and Sampathkumar, 1977) have provided evidence that O-demethylation, a well-known reaction of phenolic acids in plants (Ellis, 1974), is a feature of the metabolism of bergapten (IIIc), and to a lesser extent xanthotoxin (IIIb), in *R. graveolens*. Less direct evidence for such demethylation reactions had been reported earlier (Caporale *et al.,* 1971).

IV. EFFECTS OF COUMARINS ON PLANTS

A. Toxicity and Effects on Growth

The half-century preceding the end of World War II produced only a scattering of reports on the toxicity of coumarins to plants and their effects on growth. Toxicity of coumarin to the alga *Conferva minor* was first noted in 1896 by Klebs (quoted by Schreiner and Reed, 1908). Later investigators described growth inhibition (Schreiner and Reed, 1908; Schreiner and Skinner, 1912; Sigmund, 1914; Kuhn *et al.,* 1943), antagonism to the action of a growth promoter (Veldstra and Havinga, 1943), and growth stimulation (Grace, 1938) by coumarins in various higher plant species. It was not until after the war that sustained interest in the effects of coumarins on plant growth and metabolism became evident, sparked perhaps by Nutile's (1945) work on a light-reversible dormancy of lettuce seeds induced by coumarin (for a discussion of effects on germination, see Van Sumere *et al.,* 1972; Mayer and Poljakoff-Mayber, 1975). The effects of coumarin on growth and development have now been investigated in nearly 100 plant genera, whereas the effects of its natural (and to a lesser extent synthetic) derivatives on numerous other genera are known.

An exhaustive treatment of this work would in my opinion be inappropriate in a treatise on plant biochemistry, because the basis of these physiological actions of coumarins is imperfectly understood at the molecular level. I have, however, tabulated what I regard as a representative sample of the types of effects recorded in the literature at the organ, tissue, and cellular levels for coumarin itself, which has been by far the most intensively studied, and for nine of its derivatives (Table I). Quite recent reviews by Feuer (1974) and Wolf (1974) include further discussion of this topic.

TABLE I

Selected Effects of Coumarins on Plants at the Organ, Tissue, and Cellular Levels

Coumarin and genus affected	Effect[a]	Concentration of coumarin (M)	References[b]
Aesculetin (Ij)			
Allium	+ Chromosomal changes		1
Brassica	− Root and stalk development	5×10^{-3}	2
Lepidium	− Germination		1
Aflatoxins			
Allium	+ Mitoclasic and chromatoclasic effects, chromosomic aberrations	8×10^{-5}	3
Nasturtium	− Hypocotyl and root growth		4
Angelicin (IV)			
Lactuca	− Germination; sprout and root growth	3.3×10^{-4}	5
Triticum	− Root cell mitosis	5.4×10^{-4}	6
Coumarin (I)			
Aesculus	− Budding	6.8×10^{-4}	7
Allium	+ Chromosomal changes		1
	− Mitosis	3.4×10^{-2}	8
	+ Cell polynuclearity	3.4×10^{-2}	8
Antirrhinium	− Pollen tube formation	6.8×10^{-5}	9
	− Coleoptile growth (r BAL)	3×10^{-3}	10
Brassica	− Germination (s CCC and Phosphon D; r kinetin)	5.0×10^{-4}	11
Coleus	− Petiole abscission	6.8×10^{-4}	12
Helianthus	+ Hypocotyl growth (s IAA)	3.4×10^{-4}	13
	a + Hypocotyl growth by IAA	3.4×10^{-3}	13
Hordeum	+ Flowering	10^{-5}	14
	− Flowering	10^{-4}	14
	− Root development (r GA, kinetin)	3.4×10^{-4}	15
Hydrilla	− Chloroplast movement	1.7×10^{-3}	16
Ipomoea	+ Mesophyll cell swelling	4×10^{-4}	17
Lactuca	− Germination (r light)	1.7×10^{-4}	18
	− Germination (r thiourea)	6.8×10^{-5}	19
Lagenaria	− Guard cell size	3.4×10^{-4}	20
	− Stomatal frequency	3.4×10^{-4}	20
Marchantia	−Gemma growth (s light)	10^{-4}?	21
	+ Rhizoid formation	6.8×10^{-7}	22
	− Rhizoid formation	6.8×10^{-5}	22
Melilotus	− Germination	6.8×10^{-5}	23
Musa	− Parthenocarpy	6.8×10^{-3}	24
Saccharum	− Tiller formation	10^{-3}	25
Solanum	+ Tuber initiation	1.7×10^{-4}	26
Triticum	− Leaf elongation	1.4×10^{-4}	27
4-Hydroxycoumarin (Ik)			
Cichorium	− Tuber growth	10^{-4}	28
Striga	+ Germination	6.2×10^{-6}	29

(continued)

TABLE I (Continued)

Coumarin and genus affected	Effect[a]	Concentration of coumarin (M)	References[b]
7-Hydroxy-4-methylcoumarin (Il)			
Gossypium	+ Petiole abscission	10^{-6}	30
Lycopersicon	− Rooting of cuttings	5.7×10^{-4}	31
Raphanus	− Germination	10^{-3}	32
6-Methylcoumarin (Im)			
Allium	+ Chromatid and subchromatid breaks in root meristem cells	7.5×10^{-4}	33
Psoralen (III)			
Lactuca	− Germination	3.6×10^{-4}	5
	− Sprout growth	3.6×10^{-4}	6
Raphanus	− Root growth	8.3×10^{-6}	34
	− Bud growth	1.1×10^{-5}	34
Scopoletin (Ib)			
Helianthus	+ Stomatal closure	5×10^{-4}	35
Linum	− Root growth (r IAA)	5.2×10^{-7}	36
Pisum	− Bud growth (s IAA)	10^{-4}	37
Striga	+ Germination	6.3×10^{-6}	29
Seselin (X)			
Cucumis	− Root growth (r light)	$\leq 4.4 \times 10^{-4}$	38
Lycopersicon	− Germination	$\leq 4.4 \times 10^{-4}$	38
Triticum	a + Coleoptile growth by IAA	$\leq 3.7 \times 10^{-5}$	38

[a] The following abbreviations are used: +, stimulation, induction, increase, according to context; −, inhibition, reduction, repression, retardation, according to context; a, antagonizes the effect of the agent indicated; r, effect reversed by agent indicated; s, synergistic effect with agent indicated; BAL, dimercaptopropanol; CCC, (2-chloroethyl)trimethylammonium chloride; IAA, 3-indolylacetic acid; GA, gibberellic acid.

[b] Key to references:
1. Steinegger and Leupi (1955).
2. Sigmund (1914).
3. Jacquet et al. (1971).
4. Crisan (1973).
5. Rodighiero (1954).
6. Zhamba et al. (1970).
7. Moewus (1951).
8. Vinkler et al. (1968).
9. Kuhn et al. (1943).
10. Thimann and Bonner (1949).
11. Knypl et al. (1969).
12. Gupta (1970).
13. Neumann (1960).
14. Leopold and Thimann (1949).
15. Khan (1969).
16. Abidin (1966).
17. Harada et al. (1971).
18. Nutile (1945).
19. Poljakoff-Mayber et al. (1958).
20. Inamdar and Gangadhara (1975).
21. Rousseau (1954).
22. Moewus and Schader (1952).
23. Knapp and Furthmann (1954).
24. Simmonds (1953).
25. Hrishi and Marimuthammal (1968).
26. Stallknecht (1972).
27. Isaia (1971).
28. Bagni and Serafini-Fracassini (1971).
29. Worsham et al. (1962).
30. Schwertner and Morgan (1966).
31. Lange de Moretes and Guimarães (1954).
32. Bernhard (1959).
33. Nuti Ronchi and Arcara (1967).
34. Hatsuda et al. (1960).
35. Einhellig and Kuan (1971).
36. Libbert and Lübke (1958a).
37. Libbert and Lübke (1958b).
38. Goren and Tomer (1971).

B. Other Physical Effects

Coumarin has been reported to inhibit the uptake of water at 10^{-3} M (Blaim, 1960b), potassium ions at $3 \times 10^{-4} M$ (Swenson and Burström, 1960), sucrose at $7 \times 10^{-4} M$ (Ochs and Pohl, 1959), and phenylalanine (Van Sumere et al., 1972). It seems clear that these negative effects on uptake are due to the action of coumarin in reducing the permeability of the protoplasm to water (Guttenberg and Beythien, 1951; Guttenberg and Meinl, 1954) when used in the range 7×10^{-4} to $7 \times 10^{-3} M$. However, it should be noted that exosmosis of pigments from discs of Beta vulgaris var. Rubra is enhanced by coumarin at a slightly lower concentration (10^{-4} M) (Andraud et al., 1966) and that stimulation of water uptake in mung bean cells by coumarin (7×10^{-5} to $7 \times 10^{-4} M$) has also been observed (Hara et al., 1973). Effects of coumarin on the transport of phenylalanine in pea (Van Sumere et al., 1972) and 2,4,5-trichlorophenoxyacetic acid in Phaseolus vulgaris (reversible by gibberellin) (Basler and McBride, 1977) have been recorded.

C. Effects on Metabolism

1. Enzymes

As listed in Table II, over a dozen different classes of enzymes are known to be affected by various coumarins. As in the case of the effects on growth, one can note a tendency toward a transition from a stimulatory to an inhibitory effect with increasing concentration (e.g., the effect of coumarin on α-amylase formation, scopoletin on indolylacetic acid (IAA) oxidase activity, and aflatoxin on lipase formation). Where the reverse effect is noted (scopoletin on peroxidase activity), it must be borne in mind that different sources of enzyme are involved and that the validity of a direct comparison is therefore doubtful because of species differences. Thus it appears that both activity and de novo synthesis of some enzymes can be affected both positively and negatively by coumarins.

2. Pigment Formation

Morgan and Powell (1970) have found that 10^{-3} M coumarin markedly inhibits anthocyanin synthesis in etiolated hypocotyls of P. vulgaris following exposure to room light. Considerable effort has also been expended on studies on the nature of the effects that coumarins exert on the formation and degradation of chlorophyll. Aflatoxins induced albinism in leaves of Lepidium sativum (Schoental and White, 1965), and a similar effect was observed in cotton with $1.5 \times 10^{-4} M$ aflatoxin (Jones et al., 1967). Chlorophyll formation is completely inhibited by ca. 2×10^{-4} M aflatoxins in Vigna sinensis (Adekunle and Bassir, 1973), Lemna minor, and Nasturtium officinale

(Jacquet *et al.*, 1971). Inhibition of chlorophyll formation in *Cucumis sativa* cotyledons by 3×10^{-4} *M* aflatoxin B_1 was nullified by gibberellic acid according to Singh and Sharma (1973) who believe that the toxin probably exerts its effect by restricting the hormone-induced synthesis of leaf nucleic acids and proteins. Slowatizky *et al.* (1969) found an inhibition of grana formation in maize leaves by aflatoxin, which was partially reversed in the presence of sucrose or δ-aminolevulinic acid, an early intermediate in chlorophyll formation. They concluded that aflatoxin inhibited an early step in the chlorophyll biosynthetic pathway. The effect is a potent one, since as little as 0.25 μg of aflatoxin prevents greening of the spot on a leaf where it is applied.

Coumarin, like aflatoxins, has well-defined effects on chlorophyll, as shown primarily by the extensive work of Knypl and an associate who demonstrated two principal effects. In plant leaf tissue kept in the dark, coumarin at an optimal concentration of 3×10^{-3} prevented loss of chlorophyll and subsequent yellowing (Knypl, 1967a,b; Knypl and Kulaeva, 1970a). In the light, however, it accelerated loss of chlorophyll and protein (Knypl and Kulaeva, 1970a). Both effects were correlated with inhibition of protein synthesis in *Brassica* and *Hordeum* (Knypl, 1969; Knypl and Kulaeva, 1970a). Nevertheless, the delay in the degradation of chlorophyll in the dark proceeds concurrently with inhibited protein synthesis, as shown by feeding experiments with [^{14}C]leucine (Knypl and Kulaeva, 1970c), which showed that 3×10^{-3} *M* coumarin almost totally suppressed leucine incorporation into protein. It is believed that inhibition of protease synthesis may well account for the inhibition of chlorophyll loss in these experiments (Knypl and Kulaeva, 1970b,c). As the depression of protein synthesis is not always correlated with a similar effect on RNA synthesis, the former may be due to action of the growth retardant at the translational level (Knypl and Kulaeva, 1970c). A point of further interest in this context is that coumarin may have opposite effects on different chlorophylls in the same experiment (Babaev, 1975).

3. Respiration and Oxidative Phosphorylation

Coumarin is reported to stimulate respiration in various plant tissues (Albu *et al.*, 1969; Knypl, 1961; Oshio *et al.*, 1971; Poljakoff-Mayber, 1955; Ron and Mayer, 1959) but to cause a decrease in maize seedlings (Yakushkina and Starikova, 1977), and uptake of oxygen by young rice plants is inhibited by 4-hydroxycoumarin (Ik) (Oshio *et al.*, 1971). Oxidative phosphorylation is inhibited by coumarin in lettuce (Ulitzur and Poljakoff-Mayber, 1963) and cucumber (Stenlid and Saddik, 1962) mitochondria, the phosphorus/oxygen ratio being reduced. The property of inhibiting ATP formation in mitochondria is shared by dicoumarol (XI) and coumestrol (XII) in cucumber hypocotyls (Stenlid and Saddik, 1962; Stenlid, 1970). Coumarin decreases

TABLE II

Effects of Coumarins on Enzyme Formation (F) and Activity (A)

Enzyme affected and coumarin	Effect[a]	Concentration (M)	Reference[b]
Allantoinase			
Aflatoxins	−A		1
Amylase			
Coumarin (I)	+A	2.7×10^{-6}	2
	−A		3
	−F	1.4×10^{-3}	4
Decursin (IXa)	−A		5
Decursinol (IXb)	−A		5
α-Amylase			
Coumarin (I)	−A	1.25×10^{-3}	6
	+F	$<3.4 \times 10^{-6}$	7
	−F	$>3.4 \times 10^{-6}$	7
	−F**†	3.4×10^{-4}	8
	−F	$>6.8 \times 10^{-5}$	9
	+F	$<6.8 \times 10^{-5}$	9
	+A	2.8×10^{-3}	3
Herniarin (Ia)	−A	3.1×10^{-4}	6
β-Amylase			
Coumarin			
Glucose-6-phosphate dehydrogenase			
Coumarin	−A‡	10^{-3}	10
Scopoletin (Ib)	−A	5×10^{-5}	11
IAA oxidase			
Coumarin	+A	5.7×10^{-5}	12
	−F	1.7×10^{-3}	13
Decursin	+A	3.1×10^{-5}	5
Decursinol	+A	3.3×10^{-5}	5
4-Methylumbelliferone (Ib)	+A	2.5×10^{-5}	14
Lipases			
Aflatoxin	−F†	1.5×10^{-4}	22
	+F	6.5×10^{-5}	22
Coumarin	−A	5.3×10^{-4}	4
	−F	7.0×10^{-4}	23
	−A	1.8×10^{-4}	23
Malate dehydrogenase			
Coumarin	−F	7.0×10^{-4}	24
Scopoletin	−A	5×10^{-5}	11
Menadione oxidoreductase			
Dicoumarol	−A	10^{-4}	25
Nitrate reductase			
Coumarin	−F	5×10^{-3}	26
Peroxidase			
Coumarin	+A	6.9×10^{-3}	27
	+A	5.7×10^{-5}	12
	−F	10^{-3}	28
Scopoletin	−A	1.25×10^{-6}	29
	+A	2×10^{-5}	30
Seselin	−A	1.6×10^{-6}	31, 32
	+A	7.3×10^{-5}	12
6-Phosphogluconate dehydrogenase			
Scopoletin	−A	5×10^{-5}	11
Scopolin (If)	−A	10^{-4}	11
Phenylalanine ammonia-lyase			
Scopoletin	−A	6×10^{-4}	33

Seselin (X)	+A	3.6×10^{-5}	12
Scopoletin	−A	1.3×10^{-5}	15
	+A	2.5×10^{-7}	15
	−A	5.2×10^{-4}	16
	−A	2.1×10^{-8}	17
	−A	5×10^{-4}	18
	−A	3.6×10^{-4}	19
IAA synthetase			
Coumarin	−A	10^{-4}	20
Isocitrate dehydrogenase			
Scopoletin	−A	10^{-4}	11
Lactate dehydrogenase			
Coumarin	−A	10^{-4}	21

4,5′,8-trimethylpsoralen (IIIe)*	+A	4×10^{-4}	34
Xanthotoxin (IIIb)*	+A	5×10^{-4}	34
Polyphenol oxidase			
Coumarin	−A	1.4×10^{-3}	35
Proteases			
Coumarin	−A	7.0×10^{-4}	4
Herniarin	+A	2.8×10^{-3}	3

[a] +, Stimulation; −, inhibition; *, in conjunction with radiation at 366 nm; **, reversed by kinetin; †, reversed by gibberellic acid; ‡, reversed by 3-indolylacetic acid (IAA).

[b] Key to references:

1. Verma et al. (1976).
2. Blaszkow (1969).
3. Blaszkow (1971).
4. Poljakoff-Mayber (1953).
5. Lee et al. (1976).
6. Broda (1966).
7. Dumitru et al. (1964).
8. Khan (1969).
9. Verbeek and Dumitru (1964).
10. Marrè (1953).
11. Kajinami et al. (1971).
12. Goren and Tomer (1971).
13. Knypl (1967c).
14. Andreae (1952).
15. Imbert and Wilson (1970).
16. Libbert et al. (1969).
17. Sequiera (1964).
18. Waygood et al. (1956).
19. Witham and Gentile (1961).
20. Libbert (1960).
21. Rothe (1976).
22. Jones et al. (1967).
23. Rimon (1957).
24. Mayer and Poljakoff-Mayber (1957).
25. Shichi and Hackett (1962).
26. Schrader and Hageman (1967).
27. Dumitru and Serban (1970).
28. Rychter and Lewak (1971).
29. Schaeffer et al. (1967).
30. Schafer et al. (1971).
31. Miller et al. (1975).
32. Sirois and Miller (1972).
33. Innerarity et al. (1972).
34. Hadwiger (1972).
35. Mayer (1954).

the formation of plastoquinones and ubiquinones in winter wheat (Kraszner-Berndorfer and Telegdy Kovats, 1974). Effects of coumarin on the uptake and metabolism of phosphorus compounds have also been reported (Gesundheit and Poljakoff-Mayber, 1962; Knypl and Antoszewski, 1960; Knypl, 1964).

4. Photosynthesis and Photophosphorylation

Several coumarins have been found to affect photosynthesis and photophosphorylation. Einhellig et al. (1970) concluded from carbon dioxide exchange analyses that a reduced net rate of photosynthesis was a factor contributing to reduced growth rate caused by treatment with $10^{-3} M$ scopoletin. In different species coumarin has been found both to inhibit (Yakushkina et al., 1973; Pushkina and Starikova, 1976; Yakushkina and Starikova, 1977) and to stimulate (Yakushkina and Glinina, 1973; Prokhorchik and Volynets, 1974) photophosphorylation without any clear pattern with regard to concentration, and aesculetin (Ij) is reported to have a greater stimulatory effect than coumarin (Prokhorchik and Volynets, 1974). Although the site or sites of action do not appear to have been clearly established, Berndorfer-Kraszner and Telegdy Kovats (1974) have found that the intensity of biosynthesis of plastoquinones and ubiquinones in germinating wheat is decreased by coumarin.

5. Carbohydrate Metabolism

Effects on carbohydrate metabolism *in vivo* have been described. Coumarin ($3.3 \times 10^{-3} M$) markedly reduced starch synthesis from glucose 1-phosphate (Profumo, 1953). Several groups of investigators (Ochs and Pohl, 1959; Hara et al., 1973; Hogetsu et al., 1974; Hopp et al., 1978) have reported what appears to be specific inhibition of cellulose synthesis by coumarin in several species. In the alga *Prototheca zopfii*, Hopp et al. (1978) have established that coumarin depresses transfer of the oligosaccharide chain from a glucolipid intermediate to a protein acceptor in the biosynthetic pathway.

Singh et al. (1974) demonstrated a decrease in the rate of formation of sucrose and free reducing sugars in germinating peanuts with increasing concentrations of aflatoxins. The rate of fat hydrolysis likewise decreased in this experiment, a phenomenon also noted for coumarins in lettuce seeds (Poljakoff-Mayber and Mayer, 1955).

6. Nucleic Acid Metabolism

According to Basler (1963) loss of RNA in the soluble protein and microsome fraction of cotton cotyledons is diminished by $10^{-3} M$ coumarin, and deGreef (1964) established that pea roots treated with up to $6.8 \times 10^{-4} M$ coumarin showed a general increase in DNA content, an effect reversible with time after removal of the coumarin. Knypl (1965) studied the effect of specific inhibitors of nucleic acid and protein synthesis on coumarin-induced growth of sunflower hypocotyl sections. The failure of mitomycin to affect such growth indicated that growth was independent of DNA replication, but reduced elongation under the influence of actinomycin showed a dependence on DNA-directed RNA synthesis. Although IAA-induced growth was more sensitive to these antibiotics than that induced by coumarin, the opposite held for the uracil analogues diazouracil and thiouracil, an effect reducible by uracil. Knypl concluded that coumarin, like IAA, acted through induction, acceleration, or both, of specific proteins catalyzing growth processes, believing that they may coordinate the synthesis of mRNAs.

Hadwiger (1972) has observed a decreased rate of [^{14}C]orotic acid incorporation into the RNA of excised pea pods after irradiation in the presence of 4,5′,8-trimethylpsoralen (IIIe), a potent photosensitizing furanocoumarin. Baishev (1973) found that scopoletin (Ib) and caffeic acid (IIg) inhibited synthesis of tRNA, rRNA, mRNA, and especially DNA in potato tuber buds *in vitro* and reversed stimulation of nucleic acid formation by gibberellic acid. Korableva *et al.* (1972) noted similar effects of scopoletin. Ladyzhenskaya *et al.* (1976) have reported selective inhibition by $10^{-4}\,M$ scopoletin of the incorporation of [2-^{14}C]uracil into the tRNA and DNA–RNA fractions of potato tuber meristems, with a complete suppression of high-MW RNA formation. Caffeic acid on the other hand, affected a different RNA fraction.

7. *Protein Metabolism*

In addition to work already mentioned in other contexts, Truelove *et al.* (1970) found that aflatoxin B$_1$ ($3 \times 10^{-4}\,M$ or more) inhibited the incorporation into protein of [^{14}C]leucine absorbed by *C. sativa* cotyledon discs in a time-dependent manner.

D. Summing Up

It seems probable at present that no simple explanation is likely to be forthcoming for the broad array of effects exerted by coumarins on plant growth and development. It is worth remembering that many, probably most, of these effects are not specific to coumarins, being frequently exerted more strongly by other substances. As intimated by Kefeli and Kadyrov (1971), most inhibitory effects attributed to such natural substances as coumarins may be due entirely to nonspecific influences on metabolism. Indeed, coumarins affect the activity of so many classes of enzymes (Table II) that it is tempting to accept that action at the enzyme level could account for most of the observed phenomena, along the lines suggested by Kefeli and Kutacek (1977). However, the numerous interactions reported between coumarins and other growth regulators such as IAA, cytokinins, and gibberellins make it probable that effects are also exerted at this level of control. Kefeli and Kadyrov (1971) have speculated that high concentrations of natural growth inhibitors in plants could exert an unspecified blocking effect and so maintain the biosynthesis of phytohormones at low levels, but other modes of interaction are clearly also possible.

Although the past few years have witnessed significant progress in determination of the cellular binding sites for auxins such as IAA and naphthaleneacetic acid (Batt and Venis, 1976; Ihl, 1976; Poovaiah and Leopold, 1976; Ray, 1977; Wardrop and Polya, 1977; Jacobs and Hertel, 1978; Venis and Watson, 1978), nothing appears to be known about hypothetical receptor sites for coumarins. However, the effects noted by Hadwiger (1972) and Ladyzhenskaya and her associates (1976) imply the intervention of two structurally diverse coumarins at the level of RNA synthesis, as already indicated for IAA and gibberellins (Jacobsen, 1977). An extension of this approach appears to offer considerable promise for further elucidation of the role of coumarins as growth regulators.

REFERENCES

Abidin, Z. (1966). *J. Sci. Res. (Lahore)* **1**, 33–37; *Chem. Abstr.* **69**, 84190 (1968).
Abu-Mustafa, E. A., El-Bay, F. K. A., and Fayez, M. B. E. (1969). *Planta Med.* **18**, 90–97.
Adekunle, A. A., and Bassir, O. (1973). *Mycopathol. Mycol. Appl.* **51**, 299–305.
Akeson, W. R., Gorz, H. J., and Haskins, F. A. (1963). *Crop Sci.* **3**, 167–171.
Albu, E., Spirchez, C., and Dabala, I. (1969). *Stud. Univ. Babes-Bolyai [Ser.] Biol.* **14**, No. 1, 83–90; *Chem. Abstr.* **71**, 69520 (1969).
Andraud, G., Couquelet, J., Dorel, M., and Tronche, P. (1966). *C. R. Seances Soc. Biol. Ses Fil.* **160**, 325–327.
Andreae, W. A. (1952). *Nature (London)* **170**, 83–84.
Austin, D. J., and Brown, S. A. (1973). *Phytochemistry* **12**, 1657–1667.
Austin, D. J., and Meyers, M. B. (1965a). *Phytochemistry* **4**, 245–254.
Austin, D. J., and Meyers, M. B. (1965b). *Phytochemistry* **4**, 255–262.
Babaev, D. (1975). *Izv. Akad. Nauk Turkm. SSR, Ser. Biol. Nauk* No. 4, pp. 9–16.
Bagni, N., and Serafini Fracassini, D. (1971). *Experientia* **27**, 1239–1241.
Baishev, K. S. (1973). *Fenol'nye Soedin. Ikh Fiziol. Svoistva, Mater. Vses. Simp. Fenol'nym Soedin., 2nd, 1971* pp. 73–75; *Chem. Abstr.* **81**, 86706 (1974).
Barz, W., and Grisebach, H. (1966). *Z. Naturforsch., B: Anorg. Chem., Org. Chem., Biochem., Biophys., Biol.* **21**, 1113–1114.
Basler, E. (1963). *Proc. Okla. Acad. Sci.* **43**, 35–40.
Basler, E., and McBride, R. (1977). *Plant Cell Physiol.* **18**, 939–947.
Batt, S., and Venis, M. A. (1976). *Planta* **130**, 15–21.
Berlin, J., Dewick, P. M., Barz, W., and Grisebach, H. (1972). *Phytochemistry* **11**, 1689–1693.
Berndorfer-Kraszner, E., and Telegdy Kovats, L. (1974). *Elelmez. Ip.* **28**, 65–69; *Chem. Abstr.* **81**, 88076 (1974).
Bernhard, R. A. (1959). *Bot. Gaz. (Chicago)* **121**, 17–21.
Beyrich, T. (1967). *Planta Med.* **15**, 306–310.
Birch, A. J., Maung, M., and Pelter, A. (1969). *Aust. J. Chem.* **22**, 1923–1932.

Blaim, K. (1960a). *Naturwissenschaften* **47**, 206–207.

Blaim, K. (1960b). *J. Exp. Bot.* **11**, 377–380.

Blaszkow, W. (1969). *Przem. Ferment. Rolny* **13**, 6–9; *Chem. Abstr.* **71**, 122643 (1969).

Blaszkow, W. (1971). *Przem. Ferment. Rolny* **15**, 1–4; *Chem. Abstr.* **76**, 57695 (1972).

Bohm, B. A., Ibrahim, R. K., and Towers, G. H. N. (1961). *Can. J. Biochem. Physiol.* **39**, 1389–1395.

Broda, B. (1966). *Acta Pol. Pharm. (Engl. Transl.)* **23**, 577–580.

Brown, S. A. (1960). *Z. Naturforsch., B: Anorg. Chem., Org. Chem., Biochem., Biophys., Biol.* **15**, 768–769.

Brown, S. A. (1962a). *Can. J. Biochem. Physiol.* **40**, 607–618.

Brown, S. A. (1962b). *Science* **137**, 977–978.

Brown, S. A. (1963a). *Phytochemistry* **2**, 137–144.

Brown, S. A. (1963b). *Lloydia* **26**, 211–222.

Brown, S. A. (1965). *Can. J. Biochem.* **43**, 199–207.

Brown, S. A. (1970). *Phytochemistry* **9**, 2471–2475.

Brown, S. A. (1973). *Can. J. Biochem.* **51**, 965–968.

Brown, S. A., and Sampathkumar, S. (1977). *Can. J. Biochem.* **55**, 686–692.

Brown, S. A., and Steck, W. (1973). *Phytochemistry* **12**, 1315–1324.

Brown, S. A., Towers, G. H. N., and Wright, D. (1960). *Can. J. Biochem. Physiol.* **38**, 143–156.

Brown, S. A., Towers, G. H. N., and Chen, D. (1964). *Phytochemistry* **3**, 469–476.

Brown, S. A., El-Dakhakhny, M., and Steck, W. (1970). *Can. J. Biochem.* **48**, 863–871.

Camm, E. L., and Towers, G. H. N. (1973). *Phytochemistry* **12**, 961–973.

Caporale, G., Breccia, A., and Rodighiero, G. (1964). *Prep. Bio-Med. Appl. Labeled Mol., Proc. Symp., 1964* pp. 103–107.

Caporale, G., Dall'Acqua, F., Marciani, S., and Capozzi, A. (1970). *Z. Naturforsch., B: Anorg. Chem., Org. Chem., Biochem., Biophys., Biol.* **25**, 700–703.

Caporale, G., Dall'Acqua, F., Capozzi, A., Marciani, S., and Crocco, R. (1971). *Z. Naturforsch., B: Anorg. Chem., Org. Chem., Biochem., Biophys., Biol.* **26**, 1256–1259.

Caporale, G., Dall'Acqua, F., Capozzi, A., and Marciani, S. (1972a). *Ann. Chim. (Rome)* **62**, 536–545.

Caporale, G., Dall'Acqua, F., and Marciani, S. (1972b). *Z. Naturforsch., B: Anorg. Chem., Org. Chem., Biochem., Biophys., Biol.* **27**, 871–872.

Caporale, G., Dall'Acqua, F., and Marciani, S. (1974). *Atti Ist. Veneto Sci., Lett. Arti, Cl. Sci. Mat. Nat.* **132**, 457–466.

Crisan, E. V. (1973). *Appl. Microbiol.* **26**, 991–1000.

Dall'Acqua, F., Capozzi, A., Marciani, S., and Caporale, G. (1972). *Z. Naturforsch., B: Anorg. Chem., Org. Chem., Biochem., Biophys., Biol.* **27**, 813–817.

Dall'Acqua, F., Innocenti, G., and Caporale, G. (1975). *Planta Med.* **27**, 343–348.

deGreef, J. A. (1964). *Enzymologia* **27**, 311–326.

Dewick, P. M., Barz, W., and Grisebach, H. (1970). *Phytochemistry* **9**, 775–783.

Dhillon, D. S., and Brown, S. A. (1976). *Arch. Biochem. Biophys.* **177**, 74–83.

Donnelly, W. J., Grundon, M. F., and Ramachandran, V. N. (1977). *Proc. R. Ir. Acad., Sect. B* **77**, 443–447.

Dumitru, I. F., and Serban, G. (1970). *Rom. Biochem. Lett.* pp. 69–74; *Chem. Abstr.* **75**, 605 (1971).

Dumitru, I. F., Verbeek, R., and Massart, L. (1964). *Naturwissenschaften* **51**, 490–491.

Edwards, K. G., and Stoker, J. R. (1967). *Phytochemistry* **6**, 655–661.

Edwards, K. G., and Stoker, J. R. (1968). *Phytochemistry* **7**, 73–77.

Einhellig, F. A., and Kuan, L.-Y. (1971). *Bull. Torrey Bot. Club* **98**, 155–162.

Einhellig, F. A., Rice, E. L., Risser, P. G., and Wender, S. H. (1970). *Bull. Torrey Bot. Club* **97**, 22–23.

Ellis, B. E. (1974). *Lloydia* **37**, 168–184.

Ellis, B. E., and Amrhein, N. (1971). *Phytochemistry* **10**, 3069–3072.

Ellis, B. E., and Brown, S. A. (1974). *Can. J. Biochem.* **52**, 734–738.

Eyton, W. B., Ollis, W. D., Sutherland, I. O., Gottlieb, O. R., Taveira Magalhães, M., and Jackman, L. M. (1965). *Tetrahedron* **21**, 2683–2696.

Favre-Bonvin, J., Massias, M., Mentzer, C., and Massicot, J. (1968). *Phytochemistry* **7**, 1555–1560.

Ferreira, D., Brink, C. v. d. M., and Roux, D. G. (1971). *Phytochemistry* **10**, 1141–1144.

Feuer, G. (1974). *Prog. Med. Chem.* **10**, 85–158.

Floss, H. G. (1972). *Recent Adv. Phytochem.* **4**, 143–164.

Floss, H. G., and Mothes, U. (1964). *Z. Naturforsch., B: Anorg. Chem., Org. Chem., Biochem., Biophys., Biol.* **19**, 770–771.

Floss, H. G., and Mothes, U. (1966). *Phytochemistry* **5**, 161–169.

Floss, H. G., and Paikert, H. (1969). *Phytochemistry* **8**, 589–596.

Fritig, B., Hirth, L., and Ourisson, G. (1966). *C. R. Hebd. Seances Acad. Sci., Ser. D* **263**, 860–863.

Fritig, B., Hirth, L., and Ourisson, G. (1970). *Phytochemistry* **9**, 1963–1975.

Games, D. E., and James, D. H. (1972). *Phytochemistry* **11**, 868–869.

Gautier, J., Cave, A., Kunesch, G., and Polonsky, J. (1972). *Experientia* **28**, 759–761.

Gestetner, B., and Conn, E. E. (1974). *Arch. Biochem. Biophys.* **163**, 617–624.

Gesundheit, Z., and Poljakoff-Mayber, A. (1962). *Bull. Res. Counc. Isr., Sect. D* **11**, 25–30.

Gilchrist, D. G., Haskins, F. A., and Gorz, H. J. (1970). *Genetics* **66**, 339–347.

Goplen, B. P., Greenshields, J. E. R., and Baenziger, H. (1957). *Can. J. Bot.* **35**, 583–593.

Goren, R., and Tomer, E. (1971). *Plant Physiol.* **47**, 312–316.

Gorz, H. J., and Haskins, F. A. (1960). *J. Hered.* **51**, 74–76.

Gorz, H. J., and Haskins, F. A. (1962). *Crop Sci.* **2**, 255–257.

Grace, N. H. (1938). *Can. J. Res., Sect. C* **16**, 143–144.

Gray, A. I., and Waterman, P. G. (1978). *Phytochemistry* **17**, 845–864.

Grisebach, H., and Barz, W. (1963). *Z. Naturforsch., B: Anorg. Chem., Org. Chem., Biochem., Biophys., Biol.* **18**, 466–470.

Grisebach, H., and Barz, W. (1964). *Z. Naturforsch., B: Anorg. Chem., Org. Chem., Biochem., Biophys., Biol.* **19**, 569–571.

Grundon, M. F. (1978). *Tetrahedron* **34**, 143–161.

Grundon, M. F., and McColl, I. S. (1975). *Phytochemistry* **14**, 143–150.

Grundon, M. F., Harrison, D. M., and Spyropoulos, C. G. (1975). *J. Chem. Soc., Perkin Trans. 1* pp. 302–304.

Gupta, S. K. (1970). *Indian J. Exp. Biol.* **8**, 155–156.

Guttenberg, H., and Beythien, A. (1951). *Planta* **40**, 36–69.

Guttenberg, H., and Meinl, G. (1954). *Planta* **43**, 571–5.

Hadwiger, L. A. (1972). *Plant Physiol.* **49**, 779–782.

Hara, M., Umetsu, N., Miyamoto, C., and Tamari, K. (1973). *Plant Cell Physiol.* **14**, 11–28.

Harada, H., Ohyama, K., and Cheruel, J. (1971). *Z. Pflanzenphysiol.* **66**, 307–324.

Haskins, F. A., and Gorz, H. J. (1961). *Crop Sci.* **1**, 320–323.

Haskins, F. A., and Gorz, H. J. (1965). *Genetics* **51**, 733–738.

Haskins, F. A., and Kosuge, T. (1965). *Genetics* **52**, 1059–1068.

Haskins, F. A., Williams, L. G., and Gorz, H. J. (1964). *Plant Physiol.* **39**, 777–781.

Hatsuda, Y., Murao, S., Terashima, N., and Yokota, T. (1960). *Nippon Nogeikagaku Kaishi* **34**, 484–486; *Chem. Abstr.* **58**, 11689f (1963).

Hogetsu, T., Shibaoka, H., and Shimokoriyama, M. (1974). *Plant Cell Physiol.* **15**, 265–272.

Hopp, H. E., Romero, P. A., and Pont Lezica, R. (1978). *FEBS Lett.* **86**, 259–262.

Hrishi, N., and Marimuthammal, S. (1968). *Proc. Indian Acad. Sci., Sect. B* **68B**, 131–142.
Ihl, M. (1976). *Planta* **131**, 223–228.
Imbert, M. P., and Wilson, L. A. (1970). *Phytochemistry* **9**, 1787–1794.
Inamdar, J. A., and Gangadhara, M. (1975). *Aust. J. Bot.* **23**, 13–25.
Innerarity, L. T., Smith, E. C., and Wender, S. H. (1972). *Phytochemistry* **11**, 1389–1395.
Innocenti, G., Dall'Acqua, F., Guiotto, A., and Caporale, G. (1977). *Atti Ist. Veneto Sci., Lett. Arti, Cl. Sci. Mat. Nat.* **135**, 37–47.
Isaia, A. (1971). *Planta* **96**, 175–182.
Jacobs, M., and Hertel, R. (1978). *Planta* **142**, 1–10.
Jacobsen, J. V. (1977). *Annu. Rev. Plant Physiol.* **28**, 537–564.
Jacquet, J., Boutibonnes, P., and Saint, S. (1971). *Rev. Immunol.* **35**, 159–186.
Jones, H. C., Black, H. S., and Altschul, A. M. (1967). *Nature (London)* **214**, 171–172.
Kajinami, S., Wender, S. H., and Smith, E. C. (1971). *Phytochemistry* **10**, 1501–1503.
Kefeli, V. I., and Kadyrov, C. S. (1971). *Annu. Rev. Plant Physiol.* **22**, 185–196.
Kefeli, V. I., and Kutacek, M. (1977). *Plant Growth Regul., Proc. Int. Conf., 9th, 1976* pp. 181–188.
Khan, A. A. (1969). *Physiol. Plant.* **22**, 94–103.
Kindl, H. (1971). *Hoppe-Seyler's Z. Physiol. Chem.* **352**, 78–84.
Kindl, H., and Billek, G. (1964). *Monatsh. Chem.* **95**, 1044–1052.
Kleinhofs, A., Haskins, F. A., and Gorz, H. J. (1966). *Plant Physiol.* **41**, 1276–1279.
Kleinhofs, A., Haskins, F. A., and Gorz, H. J. (1967). *Phytochemistry* **6**, 1313–1318.
Knapp, R., and Furthmann, S. (1954). *Ber. Dtsch. Bot. Ges.* **67**, 252–269.
Knypl, J. S. (1961). *Naturwissenschaften* **48**, 530–531.
Knypl, J. S. (1964). *Physiol. Plant.* **17**, 771–778.
Knypl, J. S. (1965). *Nature (London)* **206**, 844–846.
Knypl, J. S. (1967a). *Naturwissenschaften* **54**, 146.
Knypl, J. S. (1967b). *Flora (Jena), Abt. A* **158**, 230–240.
Knypl, J. S. (1967c). *Naturwissenschaften* **54**, 544.
Knypl, J. S. (1969). *Flora (Jena), Abt. A* **160**, 217–233.
Knypl, J. S., and Antoszewski, R. (1960). *Naturwissenschaften* **47**, 91.
Knypl, J. S., and Kulaeva, O. N. (1970a). *Sov. Plant Physiol. (Engl. Transl.)* **17**, 9–15.
Knypl, J. S., and Kulaeva, O. N. (1970b). *Sov. Plant Physiol. (Engl. Transl.)* **17**, 451–457.
Knypl, J. S., and Kulaeva, O. N. (1970c). *Biochemistry (Engl. Transl.)* **35**, 1053–1061.
Knypl, J. S., Bilecka, A., and Slupek, T. (1969). *Rocz. Nauk Roln., Ser. A* **95**, 167–175; *Chem. Abstr.* **72**, 89099 (1970).
Korableva, N. P., Ladyzhenskaya, E. P., Morozova, E. V., and Melitskii, L. V. (1972). *Immunitet Pokoi Rast.* pp. 190–200; *Chem. Abstr.* **79**, 62486 (1973).
Kosuge, T. (1961). *Arch. Biochem. Biophys.* **95**, 211–218.
Kosuge, T., and Conn, E. E. (1959). *J. Biol. Chem.* **234**, 2133–2137.
Kosuge, T., and Conn, E. E. (1961). *J. Biol. Chem.* **236**, 1617–1621.
Kraszner-Berndorfer, E., and Telegdy Kovats, L. (1974). *Contrib. Chem. Food Supplies, Invited Sel. Contrib. Pap. Symp., 1973* pp. 399–409; *Chem. Abstr.* **83**, 23332 (1975).
Kuhn, R., Jerchel, D., Moewus, F., Möller, E. F., and Lettré, H. (1943). *Naturwissenschaften* **31**, 468.
Kunesch, G., and Polonsky, J. (1967). *Chem. Commun.* pp. 317–318.
Kunesch, G., and Polonsky, J. (1969). *Phytochemistry* **8**, 1221–1226.
Kunesch, G., Hildesheim, R., and Polonsky, J. (1969). *C. R. Hebd. Seances Acad. Sci., Ser. D* **268**, 2143–2145.
Kutney, J. P., Verma, A. K., and Young, R. N. (1973a). *Tetrahedron* **29**, 2645–2660.
Kutney, J. P., Verma, A. K., and Young, R. N. (1973b). *Tetrahedron* **29**, 2661–2671.

Kutney, J. P., Salisbury, P. J., and Verma, A. K. (1973c). *Tetrahedron* **29**, 2673–2681.

Kuzovkina, I. N., and Smirnov, A. M. (1972). *Izv. Akad. Nauk. SSSR, Ser. Biol.* **7**, 393–397.

Ladyzhenskaya, É. P., Korableva, N. P., and Metlitskii, L. V. (1976). *Sov. Plant Physiol. (Engl. Transl.)* **23**, 644–650.

Lange de Moretes, B., and Guimarães, M. (1954). *Rev. Bras. Biol.* **14**, 333–339; *Chem. Abstr.* **49**, 4796e (1955).

Lee, C. B., Lee, M. J., Chi, H. J., and Kwon, Y. M. (1976). *Sikmul Hakhoe Chi* **19**, 7–13; *Chem. Abstr.* **85**, 73362 (1976).

Leopold, A. C., and Thimann, K. V. (1949). *Am. J. Bot.* **36**, 342–347.

Libbert, E. (1960). *Z. Bot.* **48**, 365–380.

Libbert, E., and Lübke, H. (1958a). *Flora (Jena)* **146**, 228–239.

Libbert, E., and Lübke, H. (1958b), *Flora (Jena)* **146**, 579–585.

Libbert, E., Drawert, A., and Schröder, R. (1969). *Physiol. Plant.* **22**, 1217–1225.

Loewenberg, J. R. (1970). *Phytochemistry* **9**, 361–366.

Marciani, S., Dall'Acqua, F., Innocenti, G., and Caporale, G. (1974). *Atti Ist. Veneto Sci., Lett. Arti, Cl. Sci. Mat. Nat.* **132**, 275–287.

Marrè, E. (1953). *Atti Accad. Naz. Lincei, Cl. Sci. Fis., Mat. Nat., Rend.* **15**, 433–439.

Mayer, A. M. (1954). *Enzymologia* **16**, 277–284.

Mayer, A. M., and Poljakoff-Mayber, A. (1957). *Physiol. Plant* **10**, 1–13.

Mayer, A. M., and Poljakoff-Mayber, A. (1975). "The Germination of Seeds," Chapter 6. Pergamon, Oxford.

Miller, R. W., Sirois, J. C., and Morita, H. (1975). *Plant Physiol.* **55**, 35–41.

Moewus, F., and Schader, E. (1952). *Beitr. Biol. Pflanz.* **29**, 171–184; *Chem. Abstr.* **49**, 5589i (1955).

Moewus, L. (1951). *Beitr. Biol. Pflanz.* **28**, 244–253; *Chem. Abstr.* **49**, 2572b (1955).

Morgan, P. W., and Powell, R. D. (1970). *Plant Physiol.* **45**, 553–557.

Mothes, K., and Kala, H. (1955). *Naturwissenschaften* **42**, 159.

Murray, R. D. H. (1978). *Prog. Chem. Org. Nat. Prod.* **35**, 199–429.

Murray, R. D. H., Sutcliffe, M., and McCabe, P. H. (1971). *Tetrahedron* **27**, 4901–4906.

Neumann, J. (1960). *Physiol. Plant.* **13**, 328–341.

Nielsen, B. E. (1970). *Dan. Tidsskr. Farm.* **44**, 111–286.

Nutile, G. E. (1945). *Plant Physiol.* **20**, 433–442.

Nuti Ronchi, V., and Arcara, P. G. (1967). *Mutat. Res.* **4**, 791–796.

Ochs, G., and Pohl, R. (1959). *Phyton (Buenos Aires)* **13**, 77–87.

Oshio, H., Ishizuka, K., and Mitsui, S. (1971). *Nippon Dojo Hiryogaku Zasshi* **42**, 345–348; *Chem. Abstr.* **76**, 95639 (1972).

Poljakoff-Mayber, A. (1953). *Palest. J. Bot., Jerusalem Ser.* **6**, 101–106.

Poljakoff-Mayber, A. (1955). *J. Exp. Bot.* **6**, 313–320.

Poljakoff-Mayber, A., and Mayer, A. M. (1955). *J. Exp. Bot.* **6**, 287–292.

Poljakoff-Mayber, A., Mayer, A. M., and Zacks, S. (1958). *Bull. Res. Counc. Isr., Sect. D* **6**, 118–124.

Poovaiah, B. W., and Leopold, A. C. (1976). *Plant Physiol.* **58**, 783–785.

Profumo, P. (1953). *Atti Accad. Naz. Lincei, Cl. Sci. Fis., Mat. Nat., Rend.* **15**, 135–38.

Prokhorchik, R. A., and Volynets, A. P. (1974). *Fiziol.-Biokhim. Osn. Povysh. Prod. Rast.* pp. 78–87; *Chem. Abstr.* **82**, 68777 (1975).

Pushkina, G. P., and Starikova, V. T. (1976). *Sov. Plant Physiol. (Engl. Transl.)* **23**, 171–172.

Ranjeva, R., Alibert, G., and Boudet, A. M. (1977). *Plant Sci. Lett.* **10**, 225–234.

Ray, P. M. (1977). *Plant Physiol.* **59**, 594–599.

Reid, W. W. (1958). *Chem. Ind. (London)* pp. 1439–1440.

Reppel, L., and Wagenbreth, D. (1958). *Flora (Jena)* **146**, 212–227.

Reznik, H., and Urban, R. (1957). *Naturwissenschaften* **44**, 13.

Rimon, D. (1957). *Bull. Res. Counc. Isr., Sect. D* **6**, 53–55.
Roberts, W. L., and Link, K. P. (1937). *J. Biol. Chem.* **119**, 269–281.
Rodighiero, G. (1954). *G. Biochim.* **3**, 138–146.
Ron, A., and Mayer, A. M. (1959). *Bull. Res. Counc. Isr., Sect. D* **7**, 94–96.
Rothe, G. M. (1976). *Z. Pflanzenphysiol.* **79**, 384–391.
Rousseau, J. (1954). *C. R. Hebd. Seances Acad. Sci.* **239**, 1420–1422.
Runeckles, V. C. (1963). *Can. J. Biochem. Physiol.* **41**, 2259–2267.
Rychter, A., and Lewak, S. (1971). *Phytochemistry* **10**, 2609–2613.
Satô, M., and Hasegawa, M. (1969). *Phytochemistry* **8**, 1211–1214.
Satô, M., and Hasegawa, M. (1971). *Phytochemistry* **10**, 2367–2372.
Satô, M., and Hasegawa, M. (1972a). *Phytochemistry* **11**, 657–662.
Satô, M., and Hasegawa, M. (1972b). *Phytochemistry* **11**, 3149–3156.
Schaeffer, G. W., Haskins, F. A., and Gorz, H. J. (1960). *Biochem. Biophys. Res. Commun.* **3**, 268–271.
Schaeffer, G. W., Buta, J. G., and Sharpe, F. (1967). *Physiol. Plant.* **20**, 342–347.
Schafer, P., Wender, S. H., and Smith, E. C. (1971). *Plant Physiol.* **48**, 232–233.
Schoental, R., and White, A. F. (1965). *Nature (London)* **205**, 57–58.
Schön, W. J. (1966). *Angew. Bot.* **40**, 38–54.
Schrader, L. E., and Hageman, R. H. (1967). *Plant Physiol.* **42**, 1750–1756.
Schreiner, O., and Reed, H. S. (1908). *Bot. Gaz. (Chicago)* **45**, 73–102.
Schreiner, O., and Skinner, J. J. (1912). *Bot. Gaz. (Chicago)* **54**, 31–48.
Schwertner, H. A., and Morgan, P. W. (1966). *Plant Physiol.* **41**, 1513–1519.
Sequiera, L. (1964). *Phytopathology* **54**, 1078–1083.
Shichi, H., and Hackett, D. P. (1962). *Phytochemistry* **1**, 131–136.
Sigmund, W. (1914). *Biochem. Z.* **62**, 339–386.
Simmonds, N. W. (1953). *J. Exp. Bot.* **4**, 87–105.
Singh, O. S., and Sharma, V. K. (1973). *Indian J. Exp. Biol.* **11**, 471–473; *Chem. Abstr.* **81**, 46171 (1974).
Singh, R., Singh, A., Vadhera, S., and Bhatia, I. S. (1974). *Physiol. Plant.* **32**, 359–364.
Sirois, J. C., and Miller, R. W. (1972). *Plant Physiol.* **49**, 1012–1018.
Sivan, A., Mayer, A. M., and Poljakoff-Mayber, A. (1965). *Isr. J. Bot.* **14**, 69–73.
Slowatizky, I., Mayer, A. M., and Poljakoff-Mayber, A. (1969). *Isr. J. Bot.* **18**, 31–36.
Stallknecht, G. F. (1972). *Plant Physiol.* **50**, 412–413.
Steck, W. (1967a). *Can. J. Biochem.* **45**, 889–896.
Steck, W. (1967b). *Can. J. Biochem.* **45**, 1995–2003.
Steck, W., and Bailey, B. K. (1969). *Can. J. Chem.* **47**, 2425–2430.
Steck, W., and Brown, S. A. (1970). *Can. J. Biochem.* **48**, 872–880.
Steck, W., and Brown, S. A. (1971). *Can. J. Biochem.* **49**, 1213–1216.
Steinegger, E., and Leupi, H. (1955). *Pharm. Acta Helv.* **30**, 452–461.
Stenlid, G. (1970). *Phytochemistry* **9**, 2251–2256.
Stenlid, D., and Saddik, K. (1962). *Physiol. Plant.* **15**, 369–379.
Stoker, J. R., and Bellis, D. M. (1962). *J. Biol. Chem.* **237**, 2303–2305.
Swenson, G., and Burström, H. (1960). *Physiol. Plant.* **13**, 846–854.
Thimann, K. V., and Bonner, W. D., Jr. (1949). *Proc. Natl. Acad. Sci. U.S.A.* **35**, 272–276.
Truelove, B., Davis, D. E., and Thompson, O. C. (1970). *Can. J. Bot.* **48**, 485–491.
Ulitzur, S., and Poljakoff-Mayber, A. (1963). *J. Exp. Bot.* **14**, 95–100.
Van Sumere, C. F., Cottenie, J., de Greef, J. A., and Kint, J. (1972). *Recent Adv. Phytochem.* **4**, 165–221.
Veldstra, H., and Havinga, E. (1943). *Recl. Trav. Chim. Pays-Bas* **62**, 841–852.
Venis, M. A., and Watson, P. J. (1978). *Planta* **142**, 103–107.
Verbeek, R., and Dumitru, I. F. (1964). *Arch. Int. Physiol. Biochim.* **72**, 799–818.

Verma, K., Sareen, V. K., and Singh, R. (1976). *J. Res.* (*Punjab Agric. Univ.*) **13**, 99–104; *Chem. Abstr.* **85**, 187476 (1976).

Vinkler, G. N., Kolomiets, A. F., and Shcherbakov, V. K. (1968). *Fenol'nye Soedin. Ikh. Biol. Funkts., Mater. Vses. Simp. Fenol'nym Soedin., 1st, 1966* pp. 393–399; *Chem. Abstr.* **70**, 112471 (1969).

Wardrop, A. J., and Polya, G. M. (1977). *Plant Sci. Lett.* **8**, 155–163.

Waygood, E. R., Oaks, A., and Maclachlan, G. A. (1956). *Can. J. Bot.* **34**, 905–926.

Weygand, F., and Wendt, H. (1959). *Z. Naturforsch., B: Anorg. Chem., Org. Chem., Biochem., Biophys., Biol.* **14**, 421–427.

Whalley, W. B. (1961). *In* "Recent Developments in the Chemistry of Natural Phenolic Compounds" (W. D. Ollis, ed.), pp. 20–58. Pergamon, Oxford.

Witham, F. H., and Gentile, A. C. (1961). *J. Exp. Bot.* **12**, 188–198.

Wolf, F. T. (1974). *J. Tenn. Acad. Sci.* **49**, 27–32.

Worsham, A. D., Klingman, G. C., and Moreland, D. E. (1962). *Nature* (*London*) **195**, 199–201.

Yakushkina, N. I., and Glinina, N. A. (1973). *Osob. Gormonal'n. Regulir. Rosta Rast.* pp. 65–72; *Ref. Zh., Biol. Khim.* 12F1279 (1975).

Yakushkina, N. I., and Starikova, V. T. (1977). *Sov. Plant Physiol.* (*Engl. Transl.*) **24**, 975–979.

Yakushkina, N. I., Starikova, V. T., and Churlina, V. V. (1973). *Osob. Gormonal'n. Regulir. Rosta Rast.* pp. 80–85; *Ref. Zh., Biol. Khim.* 12F1278 (1975).

Zhamba, G. E., Bukolova, T. P., Komissarenko, N. F., and Garshtya, L. Y. (1970). *Ukr. Bot. Zh.* **27**, 119–121.

Zilg, H., and Grisebach, H. (1968). *Phytochemistry* **7**, 1765–1772.

Phenolic Acids 11

G. G. GROSS

I. INTRODUCTION

Sensu strictu, the term "phenolic acid" is applicable to a large variety of different organic compounds bearing at least one phenolic hydroxyl group and a carboxyl function. It is common practice, however, to use this designation in connection with only a limited number of natural products, namely, cinnamic acids (phenylacrylic acids) and benzoic acids (phenylcarboxylic acids). On the other hand, several typical members of these two classes do not strictly fit the definition since, from obvious biogenetic relations, both the unsubstituted parent acids and compounds with substituted phenolic hydroxyl groups fall into this category.

In this chapter, the pathways leading to cinnamic and benzoic acids will be discussed. Moreover, particular emphasis will be put on the biochemical events involved in the formation of a variety of natural products derived from these phenolic acids. A final section briefly deals with the biosynthesis and metabolic fate of the closely related phenylacetic acids.

---

* This chapter covers the literature through 1978. For reasons beyond the author's responsibility an unexpectedly long period of time was required for publication. Consequently this chapter does not necessarily cover the most recent views within this area of research in every case.

The Biochemistry of Plants, Vol. 7

II. CINNAMIC ACIDS (C_6–C_3 COMPOUNDS)

A. Biosynthesis of Cinnamic Acids

It is well documented that the carbon skeleton characteristic of cinnamic acids originates from the aromatic amino acid L-phenylalanine. Feeding experiments with *Salvia splendens* (McCalla and Neish, 1959) have further demonstrated that the parent compound cinnamic acid undergoes ring substitution in a series of hydroxylation and methylation steps yielding various *p*-hydroxylated cinnamic acids (Fig. 1).

A deeper insight into this metabolic sequence has been gained from enzymatic studies. It was found that the deamination of phenylalanine was catalyzed by phenylalanine ammonia-lyase (E.C. 4.3.1.5) (Koukol and Conn, 1961); an analogous tyrosine ammonia-lyase appears to be confined mainly to grasses (Neish, 1961). These enzymes catalyzed the elimination of ammonia in an antiperiplanar reaction leading directly to *trans*-cinnamic (Hanson *et al.*, 1971; Ife and Haslam, 1971) or *trans-p*-coumaric acid (Ellis *et al.*, 1973). These studies ruled out previous theories on the intermediacy of phenylpyruvate and phenyllactate (Brown *et al.*, 1959) in this conversion.

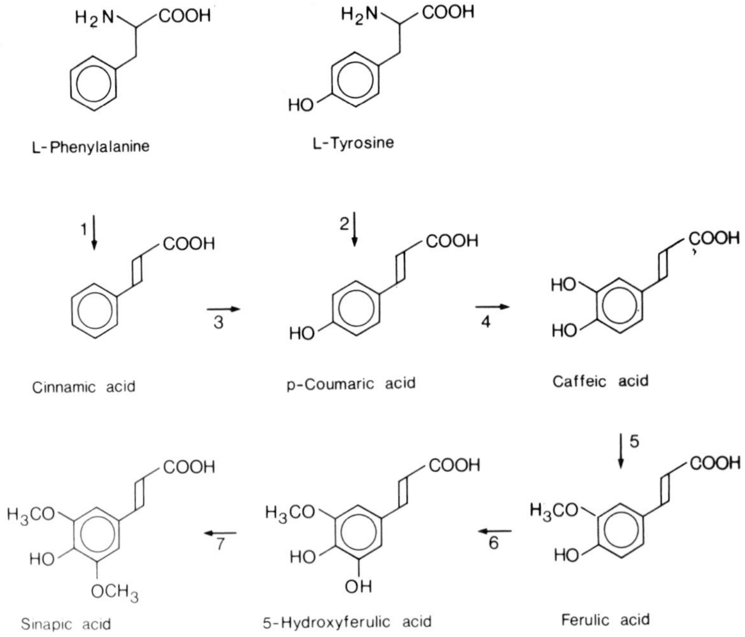

Fig. 1. The phenylalanine–cinnamate pathway. (1) Phenylalanine ammonia-lyase; (2) tyrosine ammonia-lyase; (3) cinnamate 4-hydroxylase; (4) *p*-coumarate 3-hydroxylase; (5 and 7) catechol *O*-methyltransferase; (6) ferulate 5-hydroxylase (hypothetical).

For a detailed discussion of ammonia lyases, see Hanson and Havir, this volume, Chapter 20, and the review of Camm and Towers (1973).

Among the hydroxylases involved in the cinnamate pathway, enzymes catalyzing the sequence cinnamic acid → p-coumaric acid → caffeic acid have been isolated and characterized as membrane-associated mixed-function oxygenases (see Butt and Lamb, this volume, Chapter 21). In contrast, the analogous ferulate 5-hydroxylase is still hypothetical.

Another hydroxylase, cinnamate 2-hydroxylase, which produces o-coumaric acid, has been isolated from chloroplasts of *Melilotus alba* (Gestetner and Conn, 1974). This enzyme presumably is involved in coumarin biosynthesis (see Brown, this volume, Chapter 10).

The methoxyl groups of ferulic and sinapic acids are formed by methylation of the *meta* hydroxyls of their respective precursors, caffeic and 5-hydroxyferulic acids. These reactions are known to be catalyzed by *meta*-specific catechol O-methyltransferases which utilize S-adenosyl-L-methionine as methyl donor (for details see Poulton, this volume, Chapter 22).

Until now, no conclusive data on the *para*-methylation of hydroxycinnamic acids have been available; this reaction presumably is required in synthesis of the naturally occurring isoferulic acid (3-hydroxy-4-methoxycinnamic acid) and 3,4-dimethoxycinnamic acid. Such an enzyme acting on methyl hydroxycinnamates has been isolated from the fungus *Lentinus lepideus* (Wat and Towers, 1975), and in this connection it must be recognized that secondary catabolic sequences including demethoxylation and/or demethylation and dehydroxylation reactions may be encountered (Ellis, 1974).

Also, little is known about the synthesis of 3,4-methylenedioxycinnamic acid with its typical substitution pattern (cf. piperic acid in Fig. 6). This compound might be derived from ferulic acid by a one-stage oxidation (Birch, 1963).

B. Natural Products Derived from Cinnamic Acids

Simple inspection of the structural characteristics of natural phenolic products suggests that these compounds originate from cinnamic acids. This is obvious with compounds such as cinnamyl alcohols, cinnamoyl esters and amides, and coumarins. However, cinnamic acids contribute not only to more complex molecules (e.g., flavonoids, xanthones) but also to smaller molecules (e.g., benzoic acids, simple phenols) which apparently are formed by degradation of the side chain. This conclusion is supported by the frequently observed similarity of their substitution patterns. These ideas have been substantiated for many classes of phenolic plant constituents as depicted in Fig. 2. The principles of the pathways leading to these natural products are discussed in subsequent sections.

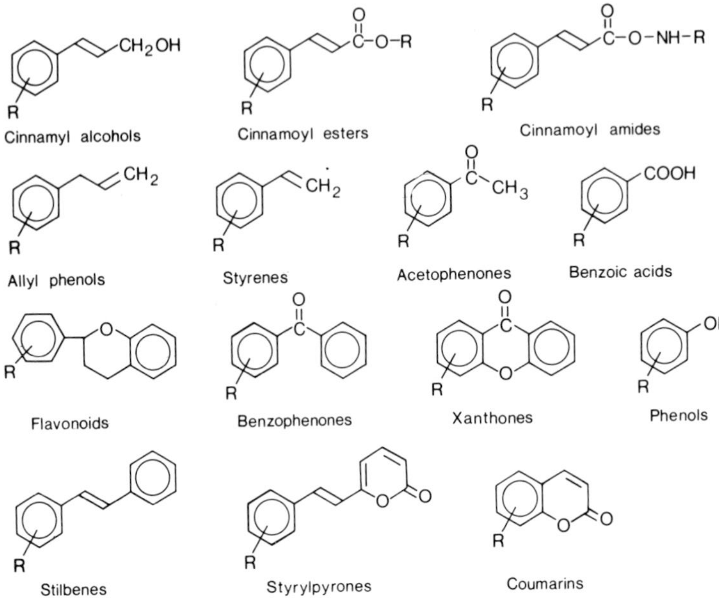

Fig. 2. Structures of natural products derived from cinnamic acids.

1. Cinnamoyl-Coenzyme A Thioesters

For thermodynamic reasons, it was apparent that the formation of many cinnamic acid derivatives would require the participation of carboxyl-activated intermediates. In particular, cinnamoyl-coenzyme A (CoA) thioesters were frequently proposed as the most likely candidates in the biosynthesis of flavonoids (Grisebach, 1962), cinnamyl alcohols (Brown et al., 1959; Neish, 1964), various esters or amides (Gross and Zenk, 1966), and benzoic acids (Zenk, 1966). Consequently, it was postulated that cinnamoyl-CoA esters occupy a central role in the metabolism of plant phenolic compounds (Zenk, 1971; Zenk and Gross, 1972). Enzymatic studies during the past few years have demonstrated unequivocally the validity of this hypothesis in several cases (Zenk, 1979).

Initially, the most serious problem in studies related to this field was the difficulty of synthesizing CoA esters of cinnamic acids because of their labile phenolic hydroxyl groups and the risk of addition reactions on the acrylic side chain. These problems were solved by using an acyl-CoA ligase from beef liver mitochondria (Gross and Zenk, 1966), which activates cinnamic acids according to the general Eqs. (1–2). More recently, similar enzymes from other sources have been used, the most efficient one being a hydroxycinnamate:CoA ligase from a pseudomonad (Zenk, 1979).

$$Ar\text{—}CH\text{=}CH\text{—}COOH + ATP \rightarrow Ar\text{—}CH\text{=}CH\text{—}CO \cdot AMP + PP_i \qquad (1)$$

$$Ar\text{—}CH\text{=}CH\text{—}CO \cdot AMP + CoA\text{—}SH \rightarrow Ar\text{—}CH\text{=}CH\text{—}CO \cdot SCoA + AMP \qquad (2)$$

(Ar = differently substituted phenyl residue)

Although these enzymatic procedures allow the preparation of considerable quantities of CoA esters, they are especially useful in the synthesis of radioactively labeled derivatives with high specific activities. One disadvantage in the use of enzymes is their limited affinity for different substrates. Consequently, chemical syntheses have been developed more recently. These methods are based on ester exchange reactions of CoA with thiophenyl or N-hydroxysuccinimide esters of cinnamic acids or on the direct reaction of the coenzyme with 4-β-D-glucocinnamic acids in the presence of dicyclohexyl carbodiimide, with subsequent removal of the protecting group by β-glucosidase (Stöckigt and Zenk, 1975).

The most prominent property of cinnamoyl-CoA esters is their characteristic uv spectrum. They exhibit a maximum near 260 nm due to the adenine moiety of CoA, and a second peak in the long-wave uv is due to absorption of the thioester linkage (Fig. 3). The maximum of this peak depends on the substitution pattern of the respective cinnamic acid, ranging from 311 nm for cinnamoyl-CoA to 352 nm for sinapoyl-CoA (Gross and Zenk, 1966; Stöckigt and Zenk, 1975). Together with the high extinction coefficient ϵ at these wavelengths (an average of 20×10^6 cm²/mol), these spectral characteristics provide the basis for a sensitive and convenient spectrophotometric assay that permits the direct measurement of reactions forming or consuming cinnamoyl-CoA thioesters (Gross and Zenk, 1966; Zenk, 1979).

That enzymes catalyzing the formation of cinnamoyl-CoA esters in fact exist in higher plants was first demonstrated only a few years ago in studies on the biosynthesis of flavonoids and lignins (cf. Hahlbrock, this volume, Chapter 14, and Grisebach, this volume, Chapter 15). Subsequently, such enzymes were recognized as common constituents of plants (for references, see, e.g., Gross 1977, 1979), and one was found in a fungus (Vance et al., 1975). Since these enzymes possess a pronounced specificity toward cinnamic acids bearing a free phenolic hydroxyl group, the systematic name hydroxycinnamate:CoA ligase has been proposed (Gross and Zenk, 1974). Examples of their proven or postulated participation in the biosynthesis of various phenolic products will be given in subsequent sections.

2. Cinnamic Acid Esters

Cinnamic acid esters are of widespread occurrence in the plant kingdom. The few examples depicted in Fig. 4 show that a great variety of hydroxylated compounds can serve as the alcoholic moiety, including sugars, alipha-

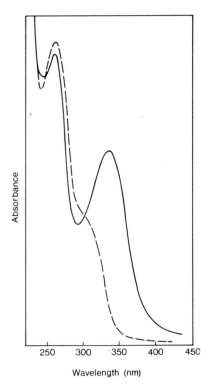

Fig. 3. Absorption spectrum of p-coumaroyl-CoA before (——) and after (---) hydrolysis. Both spectra recorded in 0.1 M phosphate buffer, pH 7.0.

tic and aromatic alcohols, and hydroxyacids. It has been postulated that the biosynthesis of these compounds involves the participation of activated cinnamoyl-CoA esters. The correctness of this assumption was demonstrated by Stöckigt and Zenk (1974) who showed that cell-free extracts from cell suspension cultures of *Nicotiana alata* catalyzed the esterification of caffeoyl–CoA and quinic acid, yielding chlorogenic acid. This result was confirmed with enzyme preparations from tomato fruits (Rhodes and Wooltorton, 1976) and cell suspension cultures from *Stevia rebaudiana* and a number of other plants (Ulbrich and Zenk, 1979).

Experiments have demonstrated the reversibility of this reaction. Moreover, substrate specificity studies revealed that the enzymes from *S. rebaudiana* and tomato were about twice as active with p-coumaroyl-CoA as substrate, yielding p-coumaroyl quinate, as compared to caffeoyl-CoA, whereas the reverse situation was found with *N. alata*. These enzymes were thus named hydroxycinnamoyl-CoA:quinate hydroxycinnamoyl transferase

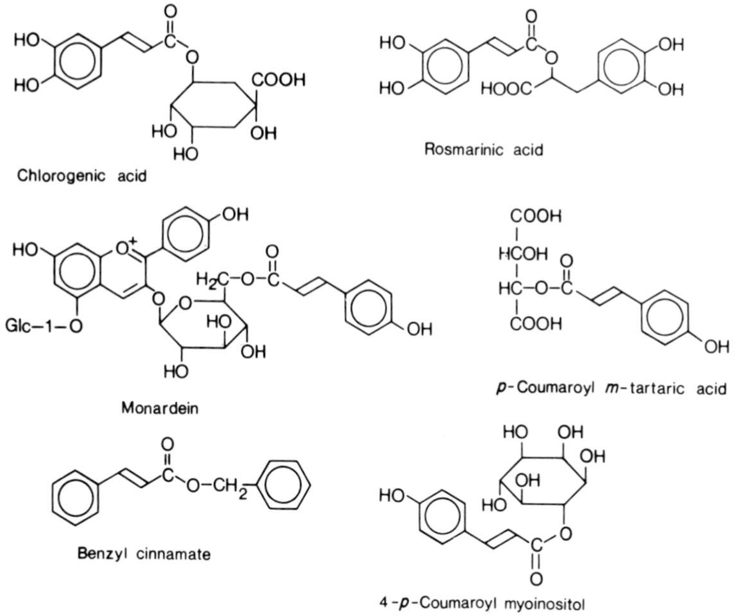

Fig. 4.   Examples of naturally occurring cinnamic acid esters.

(Fig. 5). These results are also of interest with respect to the conflicting conclusions on the biosynthesis of chlorogenic acid that were drawn from tracer experiments, i.e., whether this compound is formed directly from caffeic acid or via p-coumaroyl quinate. As pointed out by Zenk (1979), this discussion appears futile, since it must be assumed now that both possibilities occur, the actual pool sizes probably being the determining factor.

Esters other than chlorogenic acid, one of the most common phenolic products, have also recently been found to be produced via activated cinnamic acids. A hydroxycinnamoyl-CoA:shikimate hydroxycinnamoyl transferase has been isolated from N. alata (Zenk, 1979) and Cichorium endiva suspension cultures (Ulbrich and Zenk, 1979). An enzyme from pea seedlings acylates the sugar moiety of the flavonol kaempferol 3-triglucoside, with p-coumaroyl-CoA (Saylor and Mansell, 1977). Finally the formation of an O-p-coumaroyl derivative of the alkaloid lupinin catalyzed by cell-free extracts from Lupinus was stimulated by the addition of CoA, ATP, and Mg^{2+}, the cofactors of acyl-CoA ligases (Murakoshi et al., 1977).

3. Cinnamoyl Amides

Although a variety of N-cinnamoyl derivatives are known to occur in plants (examples given by Zenk, 1979), no details are known about their

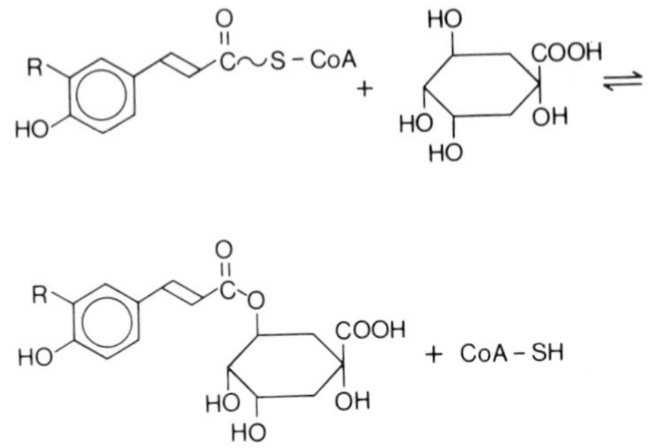

Fig. 5. Esterification of cinnamoyl-CoA esters with quinic acid by hydroxycinnamoyl-CoA:quinate hydroxycinnamoyl transferase. R = H: *p*-coumaroyl-CoA, *p*-coumaroyl quinate; R = OH: caffeoyl-CoA, chlorogenic acid.

biosynthesis. It is not unlikely, however, that cinnamoyl-CoA esters are involved. (See Smith, this volume, Chapter 9, for further information.)

4. Reduction of Cinnamic Acids

Reduction of the carboxyl group of cinnamic acids leads, via aldehydes, to the corresponding cinnamyl alcohols, and cinnamoyl-CoA esters have been proposed as activated intermediates in this conversion. This was proven in 1973 when cell-free extracts from *Forsythia* were shown to catalyze not only the formation of feruloyl-CoA according to Eqs. (1) and (2) but also the subsequent reduction of this thioester to coniferyl alcohol in a two-step reaction (Gross *et al.,* 1973):

$$Ar—CH{=}CH—CO \cdot SCoA + NADPH + H^+ \rightarrow Ar—CH{=}CH—CHO + NADP^+ \quad (3)$$

$$Ar—CH{=}CH—CHO + NADPH + H^+ \rightarrow Ar—CH{=}CH—CH_2OH + NADP^+ \quad (4)$$

Confirmation of this result was achieved subsequently in several laboratories (references in Gross, 1977, 1979). For a detailed discussion of these aspects, see Grisebach (this volume, Chapter 15).

Cinnamyl alcohols formed by the above reaction sequence are intermediates in the biosynthesis of various natural products. Undoubtedly, the most prominent among these is lignin, a macromolecule formed in higher plant cell walls by a free radical-mediated dehydropolymerization of *p*-hydroxycinnamyl alcohols. Coniferyl alcohol has also been proposed as the direct precursor of several lignans (Stöckigt and Klischies, 1977), i.e., dimeric phenylpropanes that are stereospecifically linked together at the β position. The latter suggests that their biosynthesis, in contrast to that of lignin, proceeds under enzymatic control.

Formally, further reduction of cinnamyl alcohols would lead to propenylphenols or, after an additional isomerization step, to allylphenols. Tracer studies using doubly labeled coniferyl alcohol have provided evidence for the existence of such a conversion in *Ocimum basilicum,* yielding eugenol and methyleugenol with unchanged $^3H/^{14}C$ ratios (Klischies *et al.,* 1975). It should be mentioned, however, that a different pathway involving loss of the terminal side-chain carbon at the acid level has also been proposed (Manitto *et al.,* 1974; Senanayake *et al.,* 1977).

In addition to reduction of the carboxyl group of cinnamic acids, hydrogenation of the unsaturated side chain must occur also in nature, as indicated by the existence of dihydrocinnamic acids and alcohols. Such compounds have recently been proposed as intermediates in the biosynthesis of mesembrine alkaloids (Jeffs *et al.,* 1978) and rhododendrin (Klischies and Zenk, 1978). Nothing is known, however, concerning the enzymology of these conversions.

5. Side-Chain Elongation

Grisebach postulated in 1962 that flavonoids were synthesized by the successive condensation of a cinnamoyl-CoA ester with three molecules of malonyl-CoA, with the simultaneous liberation of 3 mol of CO_2. This view was extended for a variety of related compounds thought to be formed by analogous mechanisms (Fig. 6). These predictions were confirmed several years ago by the successful isolation of flavanone synthase, an enzyme complex catalyzing the *in vitro* synthesis of naringenin in the expected manner (see Hahlbrock, this volume, Chapter 14). A similar stilbene synthase has been detected in *Rheum* (Rupprich and Kindl, 1978). Extracts of this plant are able to form resveratrol in a reaction proceeding by a different cyclization mode and involving an additional decarboxylation step.

In further studies on flavanone synthase, it was found that the α-styrylpyrone bisnoryangonin was formed under certain conditions (Kreuzaler and Hahlbrock, 1975). This compound apparently represents the condensation product of *p*-coumaroyl-CoA and only two acetate units. Most likely, analogous condensations occur in fungi in the synthesis of hispidin, a styrylpyrone with the substitution pattern of caffeic acid (Perrin and Towers, 1973; Hatfield and Brady, 1973).

With regard to the biosynthesis of xanthones, it has been suggested that they are formed in *Gentiana* from a C_6-C_1 precursor and three acetate units, yielding benzophenones which in turn undergo cyclization to xanthones (Floss and Rettig, 1964; Gupta and Lewis, 1971). However, recent experiments with *Anemarrhena* (Liliaceae) have provided strong evidence that again a C_6-C_3 compound (*p*-coumaric acid) reacts with two acetate units in the initial steps in this pathway (Fujita and Inoue, 1977).

Theoretically, the condensation of a single malonyl-CoA unit with cinnamic acid should lead to phenolic acids such as piperic acid with a five-

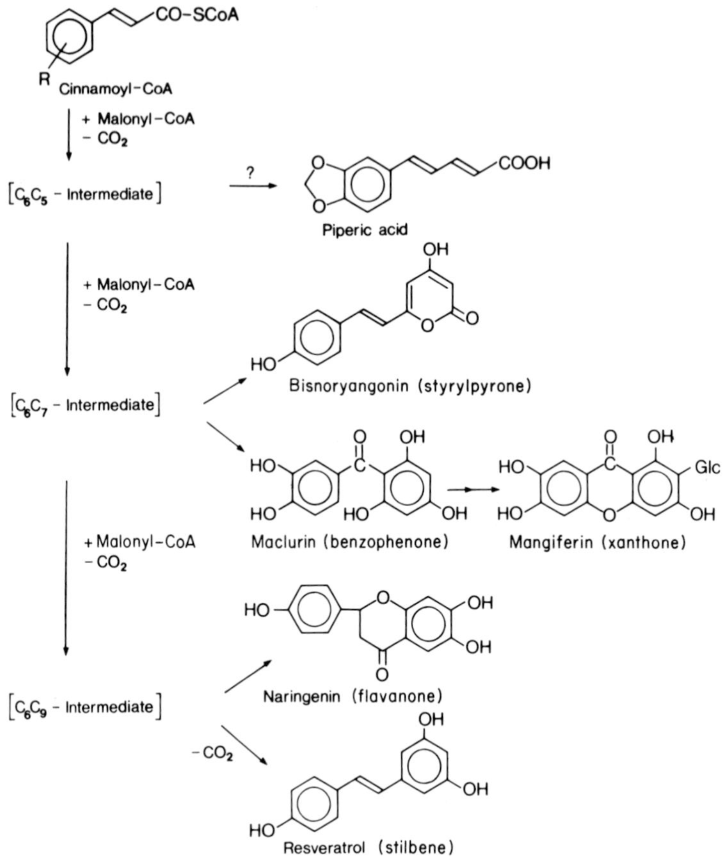

Fig. 6.   Natural products formed by condensation of cinnamoyl-CoA esters and malonyl-CoA.

carbon side chain (Fig. 6). Nothing is known, however, regarding the biosynthesis of such compounds.

6. Side-Chain Degradation

The most prominent pathway belonging to this category is the one leading to benzoic acids through the removal of an acetate unit. This conversion will be discussed in Section III,A. Decarboxylation of cinnamic acids to the corresponding styrenes is known from studies on bacteria and fungi. Retention of the stereospecific configuration of the side chain in this reaction has been demonstrated with yeast (Manitto *et al.*, 1975). No data are available about this conversion in higher plants. This applies also to the biogenesis of acetophenones which eventually might be formed in a side reaction during the β-oxidation of cinnamic acids (cf. Zenk, 1979).

III. BENZOIC ACIDS (C_6–C_1 COMPOUNDS)

A. Biosynthesis of Benzoic Acids

The most important mechanism for formation of benzoic acids in plants is the side-chain degradation of cinnamic acids. The removal of an acetate unit obviously yields the corresponding benzoic acids, with the consequence that the substitution pattern of the individual benzoic acids is determined by their respective C_6–C_3 precursors. However, hydroxylations and methylations of benzoic acids similar to those reported for cinnamic acids are also known to occur (for detailed literature, see Billek and Schmook, 1966).

A reaction sequence analogous to the β-oxidation of fatty acids was proposed by Zenk (1966; 1971; Fig. 7) to explain the formation of benzoic acids. This proposal was supported by Alibert and Ranjeva (1971) in studies on cell-free extracts from *Quercus* where the conversion of cinnamic to benzoic acid was notably stimulated upon the addition of ATP and CoA. Moreover, a *Pseudomonas* strain growing on ferulic acid as sole carbon source was found

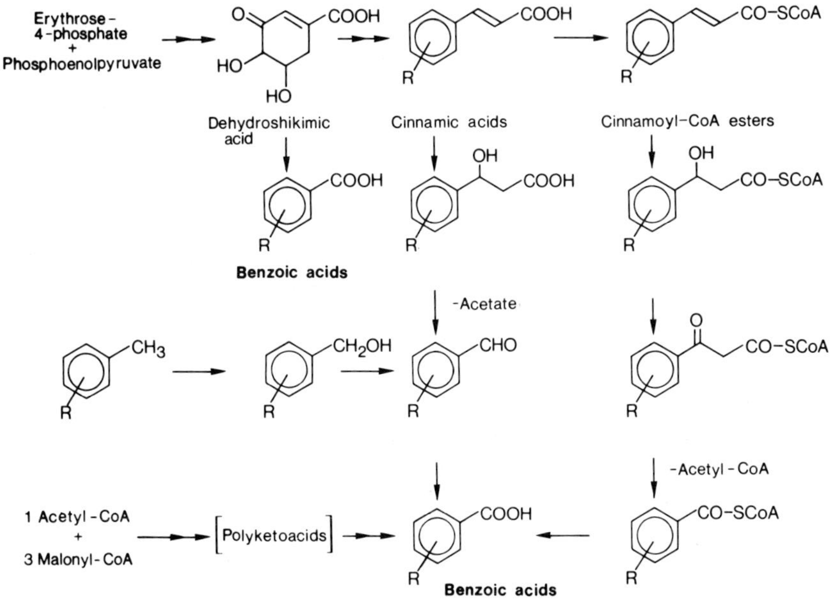

Fig. 7.   Biosynthetic pathways to benzoic acids in plants and microorganisms.

to possess an extremely active cinnamate:CoA ligase and several enzymes required for the β-oxidation sequence (Zenk, 1979). However, definitive proof of this pathway in plants is still lacking.

In addition to these observations, there are several reports on the formation of benzoic acids from C_6-C_3 precursors by membrane-associated enzyme complexes that are not stimulated by added CoA (Hagel and Kindl, 1975; Löffelhardt and Kindl, 1975; 1976). Further evidence for the existence of a different nonoxidative pathway (Fig. 7) is found in studies on *Pseudomonas acidovorans* (Toms and Wood, 1970), *Polyporus hispidus,* and potato (French *et al.,* 1976). These conflicting aspects have recently been discussed in detail by Zenk (1979).

This rather confusing picture is further complicated by the occurrence of other quite different pathways (Fig. 7). C_6-C_1 acids can also be derived directly from the skikimate pathway by aromatization, especially of dehydroshikimic acid, with the consequence that the entire alicyclic C_6-C_1 carbon skeleton is retained (the pathway via phenylpropanes involves loss of the carboxyl group). This reaction probably represents the main route to the widely occurring plant constituent gallic acid (Billek and Schmook, 1966; Zenk, 1971).

The biosynthetic sequences outlined above occur in both higher plants and microorganisms. Additional pathways have been found in the latter group. An important example is the formation of benzoic acids from acetate units via polyketoacids as demonstrated for 6-methylsalicylic acid which is synthesized by a multienzyme complex (Dimroth *et al.,* 1970). Another interesting route involves the hydroxylation and subsequent oxidation of methylphenols, e.g., *m*-cresol yielding *m*-hydroxybenzoic acid as shown in Fig. 7 (cf. Murphy and Lynen, 1975).

B. Compounds Derived from Benzoic Acids

By analogy to cinnamic acids, benzoic acids are also utilized as the precursors of a variety of natural products (Fig. 8). In contrast to the C_6-C_3 derivatives, our knowledge of the biosynthetic mechanisms involved in these conversions is rather limited. This applies especially to enzymatic studies in this field.

In some of the pathways proposed here, CoA esters were also proposed as intermediates. These considerations arose from assuming the β-oxidation of cinnamic acids, which should lead to the formation of benzoyl-CoA esters. These in turn could well be the carboxyl-activated intermediates required in the formation of benzoic acid esters (examples given in Fig. 9) or of benzaldehydes and alcohols (Zenk, 1966, 1971, 1979). Although the latter compounds are common constituents of plants, no data on their synthesis in cell-free systems are available. Thus far the *in vitro* reduction of benzoic

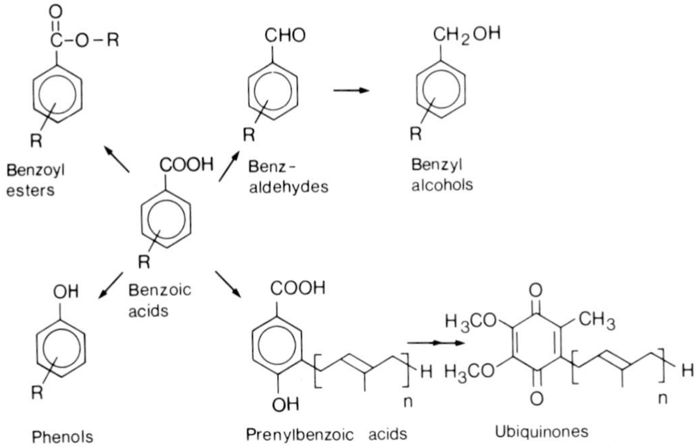

Fig. 8. Natural products derived from benzoic acids.

acids has been demonstrated only with enzymes from the ascomycete *Neurospora* (Gross and Zenk, 1969) and the basidiomycete *Polystictus* (Zenk and Gross, 1965). It was further shown that this reaction was independent of added CoA and proceeded via an intermediate acyladenylate (Gross, 1972).

Another pathway arising from benzoic acids is the route leading to ubiquinones, a sequence in which *p*-hydroxybenzoic or protocatechuic acid is prenylated with isopentenyl pyrophosphate (see, e.g., Thomas and Threlfall, 1973; Casey and Threlfall, 1978). Finally, benzoic acids can also be decarboxylated oxidatively to yield phenols (cf. Ellis, 1974) in a reaction ascribed to peroxidases (Berlin and Barz, 1975).

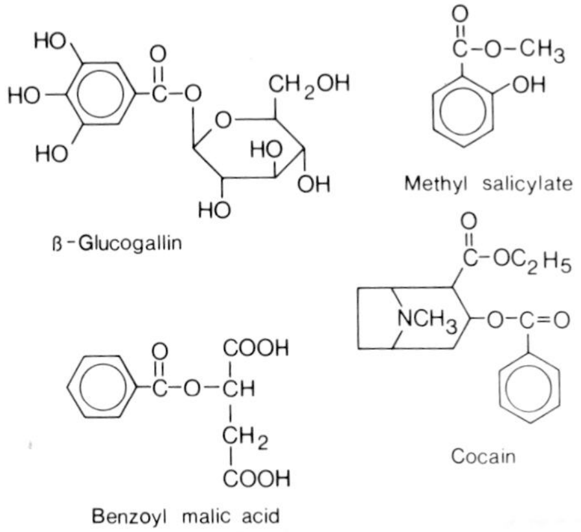

Fig. 9. Examples of naturally occurring benzoic acid esters.

IV. PHENYLACETIC ACIDS (C_6–C_2 COMPOUNDS)

Phenylacetic acids occur less frequently than the structurally related C_6–C_3 and C_6–C_1 acids. Their biosynthesis appears to involve catabolic processes (cf. Barz and Köster, this volume, Chapter 3). These C_6–C_2 acids are formed by oxidative decarboxylation of the α-ketoacids formed from phenylalanine and tyrosine (Fig. 10). *p*-Hydroxyphenylacetic acid is thus formed from *p*-hydroxyphenylpyruvic acid. An interesting alternative reaction catalyzed by the enzyme *p*-hydroxyphenylpyruvate oxidase involves oxidative decarboxylation and migration of the side chain and ring oxygenation to yield homogentisic acid. A similar reaction produces *o*-hydroxyphenylacetic acid from phenylpyruvic acid. After introduction of a second hydroxyl group into the monohydroxylated acids, all these compounds can be degraded further by cleavage of their aromatic ring (for references, see the reviews of Towers and Subba Rao, 1972, and Ellis, 1974). That these catabolic pathways in fact occur in plants and are not the result of microbial contamination has been shown by Durand and Zenk (1974a,b) with sterile plants and cell cultures from which enzymes catalyzing the above sequence could be isolated. With cell cultures a different route to phenylacetic acids was found by which tyramine and dopamine were deaminated to the corresponding C_6–C_2 acids (Meyer and Barz, 1978).

Anabolic as well as catabolic processes can be observed. For instance, homogentisic acid has been recognized as a precursor of plastoquinones (Whistance and Threlfall, 1970) or is, after decarboxylation to toluhydroquinone, incorporated into the naphthoquinone chimaphilin (Bolkart and Zenk, 1969).

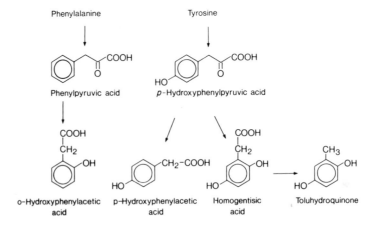

Fig. 10. Biosynthetic pathways to phenylacetic acids in plants.

V. CONCLUDING REMARKS

The results summarized above clearly demonstrate that our knowledge of the biosynthesis and metabolism of phenolic acids in plants has been significantly advanced during the past few years. This advance must certainly be attributed in large part to the increasing number of successful enzymatic studies in this area and to the wide use of plant cell cultures. Thus, starting from the solid base of knowledge provided by early tracer studies, a variety of questions have been answered in detail. These results have also stimulated research on more physiological problems in this field involving regulation and compartmentation, important aspects which, however, cannot be discussed here.

REFERENCES

Alibert, G., and Ranjeva, R. (1971). *FEBS Lett.* **19,** 11–14.
Berlin, J., and Barz, W. (1975). *Z. Naturforsch., C: Biosci.* **30C,** 650–658.
Billek, G., and Schmook, F. P. (1966). *Oesterr. Chem. Ztg.* **67,** 401–409.
Birch, A. J. (1963). *In* "Chemical Plant Taxonomy" (T. W. Swain, ed), pp. 141–166. Academic Press, New York.
Bolkart, K. H., and Zenk, M. H. (1969). *Z. Pflanzenphysiol.* **61,** 356–359.
Brown, S. A., Whright, D., and Neish, A. C. (1959). *Can. J. Biochem. Physiol.* **37,** 25–34.
Camm, E. L., and Towers, G. H. N. (1973). *Phytochemistry* **12,** 961–973.
Casey, J., and Threlfall, D. R. (1978). *FEBS Lett.* **85,** 249–253.
Dimroth, P., Walter, H., and Lynen, F. (1970). *Eur. J. Biochem.* **13,** 98–110.
Durand, R., and Zenk, M. H. (1974a). *Phytochemistry* **13,** 1483–1492.
Durand, R., and Zenk, M. H. (1974b). *FEBS Lett.* **39,** 218–220.
Ellis, B. E. (1974). *Lloydia* **37,** 168–183.
Ellis, B. E., Zenk, M. H., Kirby, G. W., Michael, J., and Floss, H. G. (1973). *Phytochemistry* **12,** 1057–1058.
Floss, H. G., and Rettig, A. (1964). *Z. Naturforsch., B: Anorg. Chem., Org. Chem., Biochem., Biophys., Biol.* **19B,** 1103–1105.
French, C. J., Vance, C. P., and Towers, G. H. N. (1976). *Phytochemistry* **15,** 564–566.
Fujita, M., and Inoue, T. (1977). *Tetrahedron Lett.* **51,** 4503–4506.
Gestetner, B., and Conn, E. E. (1974). *Arch. Biochem. Biophys.* **163,** 617–624.
Grisebach, H. (1962). *Planta Med.* **10,** 385–397.
Gross, G. G. (1972). *Eur. J. Biochem.* **31,** 585–592.
Gross, G. G. (1977). *Recent Adv. Phytochem.* **11,** 141–184.
Gross, G. G. (1979). *Recent Ad. Phytochem.* **12,** 177–220.
Gross, G. G., and Zenk, M. H. (1966). *Z. Naturforsch., B: Anorg. Chem., Org. Chem., Biochem., Biophys., Biol.* **21B,** 683–690.
Gross, G. G., and Zenk, M. H. (1969). *Eur. J. Biochem.* **8,** 413–419.
Gross, G. G., and Zenk, M. H. (1974). *Eur. J. Biochem.* **41,** 453–459.
Gross, G. G., Stöckigt, J., Mansell, R. L., and Zenk, M. H. (1973). *FEBS Lett.* **31,** 283–286.
Gupta, P., and Lewis, J. R. (1971). *J. Chem. Soc. C* pp. 629–631.
Hagel, P., and Kindl, H. (1975). *FEBS Lett.* **59,** 120–124.

Hanson, H. R., Wightman, R. H., Staunton, J., and Battersby, A. R. (1971). *Chem. Commun.* pp. 185–186.

Hatfield, G. M., and Brady, L. R. (1973). *Lloydia* **36**, 59–65.

Ife, R., and Haslam, E. (1971). *J. Chem. Soc. C* pp. 2818–2821.

Jeffs, P. W., Karle, J. M., and Martin, N. H. (1978). *Phytochemistry* **17**, 719–728.

Klischies, M., and Zenk, M. H. (1978). *Phytochemistry* **17**, 1281–1284.

Klischies, M., Stöckigt, J., and Zenk, M. H. (1975). *J. Chem. Soc., Chem. Commun.* pp. 879–880.

Koukol, J., and Conn, E. E. (1961). *J. Biol. Chem.* **236**, 2692–2698.

Kreuzaler, F., and Hahlbrock, K. (1975). *Arch. Biochem. Biophys.* **169**, 84–90.

Löffelhardt, W., and Kindl, H. (1975). *Hoppe-Seyler's Z. Physiol. Chem.* **356**, 487–493.

Löffelhardt, W., and Kindl, H. (1976). *Z. Naturforsch., C: Biosci.* **31C**, 693–699.

McCalla, D. R., and Neish, A. C. (1959). *Can. J. Biochem. Physiol.* **37**, 537–547.

Manitto, P., Monti, D., and Gramatica, P. (1974). *J. Chem. Soc., Perkin Trans. 1* pp. 1727–1731.

Manitto, P., Gramatica, P., and Ranzi, B. M. (1975). *J. Chem. Soc., Chem. Commun.* pp. 442–443.

Meyer, E., and Barz, W. (1978). *Planta Med.* **33**, 336–344.

Murakoshi, I., Ogawa, M., Toriizuka, K., Haginiwa, J., Ohmiya, S., and Otomasu, H. (1977). *Chem. Pharm. Bull.* **25**, 527–529.

Murphy, G., and Lynen, F. (1975). *Eur. J. Biochem.* **58**, 467–475.

Neish, A. C. (1961). *Phytochemistry* **1**, 1–24.

Neish, A. C. (1964). *In* "The Formation of Wood in Forest Trees" (M. H. Zimmerman, ed.), pp. 219–239. Academic Press, New York.

Perrin, P. W., and Towers, G. H. N. (1973). *Phytochemistry* **12**, 589, 592.

Rhodes, M. J. C., and Wooltorton, L. S. C. (1976). *Phytochemistry* **15**, 947–951.

Rupprich, N., and Kindl, H. (1978). *Hoppe-Seyler's Z. Physiol. Chem.* **359**, 165–172.

Saylor, M. H., and Mansell, R. L. (1977). *Z. Naturforsch., C: Biosci.* **32C**, 765–768.

Senanayake, U. M., Wills, R. B. H., and Lee, T. H. (1977). *Phytochemistry* **16**, 2032–2033.

Stöckigt, J., and Klischies, M. (1977). *Holzforschung* **31**, 41–44.

Stöckigt, J., and Zenk, M. H. (1974). *FEBS Lett.* **41**, 131–134.

Stöckigt, J., and Zenk, M. H. (1975). *Z. Naturforsch., C: Biosci.* **30C**, 352–358.

Thomas, G., and Threlfall, D. R. (1973). *Biochem. J.* **134**, 811–814.

Toms, A., and Wood, J. M. (1970). *Biochemistry* **9**, 337–343.

Towers, G. H. N., and Subba Rao, P. V. (1972). *Recent Adv. Phytochem.* **4**, 1–43.

Ulbrich, B., and Zenk, M. H. (1979). *Phytochemistry* **18**, 929–933.

Vance, C. P., Nambudiri, A. M. D., Wat, C. K., and Towers, G. H. N. (1975). *Phytochemistry* **14**, 967–969.

Wat, C. K., and Towers, G. H. N. (1975). *Phytochemistry* **14**, 663–666.

Whistance, G. R., and Threlfall, D. R. (1970). *Biochem. J.* **117**, 593–600.

Zenk, M. H. (1966). *In* "Biosynthesis of Aromatic Compounds" (G. Billek, ed.), pp. 45–60. Pergamon, Oxford.

Zenk, M. H. (1971). *In* "Pharmacognosy and Phytochemistry" (H. Wagner and L. Hörhammer, eds.), pp. 314–346. Springer-Verlag, Berlin and New York.

Zenk, M. H. (1979). *Recent Adv. Phytochem.* **12**, 139–176.

Zenk, M. H., and Gross, G. G. (1965). *Z. Pflanzenphysiol.* **53**, 356–362.

Zenk, M. H., and Gross, G. G. (1972). *Recent Adv. Phytochem.* **4**, 87–106.

Enzymology of Alkaloid Metabolism in Plants and Microorganisms

12

GEORGE R. WALLER
OTIS C. DERMER

I. INTRODUCTION

Alkaloids are metabolically active in plants and microorganisms; their biosynthesis and catabolism play an important role in the ecology and physiology of the organisms in which they occur. Yet alkaloids are more easily described than precisely defined. They are nitrogen-containing compounds having at least some basicity (hence the name), are usually heterocyclic, occur primarily in higher plants and some microorganisms, and usually ex-

The Biochemistry of Plants, Vol. 7

hibit significant physiological activity in humans and animals. They represent one of the largest and most diverse groups of secondary natural products, and they exhibit some of the most complicated molecular structures.

A few animals contain alkaloids, some derived from plants used as foods, others true products of animal metabolism. The presence of alkaloids in animals can usually be traced to a pathological condition or to a drug that was ingested; e.g., humans accumulate tetrahydroisoquinoline in phenylketonuria (Lassla and Coscia, 1979) and in Parkinson's disease treated with 3,4-dihydroxyphenylalanine (L-DOPA) (Coscia *et al.*, 1977).

There have been several major incentives for studying alkaloids. They are the oldest drugs and still have significant use in modern medicine, whether extracted from plants or produced synthetically. Organic chemists have long been fascinated by the variety and complexity of the structures represented, beginning with the isolation of morphine in 1805. Plant physiologists seek to learn the function of secondary metabolites in plants and to use them as a taxonomic aid. Why plants produce alkaloids is by no means clear, but the earlier idea that they are merely waste products of nitrogen metabolism is certainly inadequate.

In the pharmaceutical industry, research on alkaloids is devoted mainly to improving methods of production, to cataloging their chemotherapeutic effects, and to improving these compounds by chemical modifications of structure. Several hundred new alkaloids are isolated and described each year. Now, far more than formerly, physical chemical methods are directly applied in establishing the structures of these new compounds. Much additional effort of chemists goes into developing synthetic approaches to alkaloids, including the stereochemical problems the synthesis presents. The goal of phytochemists is to identify the biosynthetic pathway in the organism. They have relied for some time on the use of isotopically labeled compounds: Such compounds are fed to the plant or microorganism, and the distribution of the tracer, usually radioactive, is followed. There is also a genetic method, though much less used, of establishing metabolic pathways: Intraspecific crosses are made and, if only one set of alleles is present, the segregation in F_2 can show the interrelationships between alkaloids and other compounds present (Waller and Nowacki, 1978).

A. Metabolic Pathways

A metabolic pathway consists of a series of compounds, reactions, and enzyme catalysts. Davis (1955) advanced a terminology that has been widely accepted:

1. A *precursor* is any compound, either endogenous or exogenous, that is converted by an organism into a product.

2. An *intermediate* is a compound that is both formed and then altered by the organism under identical conditions.

3. An *obligatory intermediate* is a component of the only path by which the organism can synthesize a given substance from given precursors, or degrade a substance to given products.

A *metabolic grid* (Bu'Lock, 1965) represents the metabolism of any type of compound in which a series of parallel reactions and interconversions occur—a polydimensional array of metabolic pathways (Fig. 1). Thus a compound may be converted to an ultimate product by several different pathways, but at different rates. Exclusive or predominant use of one pathway is called *channeling,* as indicated by the heavy line in Fig. 1.

If the sequence A → B → C is suspected, and labeled B gives better yields of labeled C than labeled A, this is evidence for the sequence as written. Adelberg (1953) has pointed out, however, that it is difficult to decide whether a certain compound Y occupies position C or X in the metabolic pathway

$$A \rightarrow B \rightarrow \ C \rightarrow D \rightarrow \text{end product}$$
$$\Updownarrow$$
$$X$$

The only way to solve this problem is to show that an enzyme catalyzes the conversion B → Y, and another enzyme, the conversion Y → D; this would show that Y has position C and not position X.

"The Logic of Working with Enzymes" (Cornforth, 1973) should be required reading for investigators in this field. Not only are problems of the type described above discussed, but Cornforth also reviews temptations to draw conclusions about intermediates from inadequate evidence. One technique not cited there that seems to have merit is the following: In the metabolic pathway above, labeled A is incubated with appropriate reagents and,

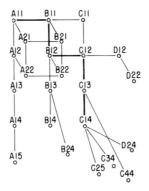

Fig. 1. A representative metabolic grid (Nowacki and Waller, 1975).

in separate experiments, with each of a series of compounds suspected of being B; if one greatly reduces the incorporation of label into C, it is likely to be B. In general, however, it is far easier by these methods to rule out pathways than to demonstrate that they are obligatory.

The value of enzymatic studies in establishing metabolic pathways has been reviewed briefly (Brown and Wetter, 1972). General reviews of the biosynthesis of alkaloids are available (Herbert, 1978; Leete, 1969, 1977; Mothes and Schütte, 1969; Waller and Nowacki, 1978). These make some reference to enzymology but reflect the predominance of feeding experiments in studies on metabolic pathways.

B. Enzymes of Metabolic Pathways

Statements about the functioning of enzymes *in vivo* usually say little or nothing about the enzymes themselves, since these are defined functionally; e.g., to note that the hydrolysis of an ester is catalyzed by a hydrolase conveys little new information. There have been repeated pleas that more attention be devoted to *in vitro* studies on enzymes of alkaloid metabolic pathways (Bu'Lock, 1965; Leete, 1967, 1969; Mothes, 1972; Robinson, 1968; Spenser, 1968; Waller and Nowacki, 1978). Indeed, publication of such work has been increasing in quantity and quality in the last decade, though the field is still dominated by feeding experiments. This chapter is intended to cover such enzymatic research. However, by no means will all work on *in vitro* synthesis or alteration of alkaloids be reviewed. Possibly to the dismay of some readers, we will not discuss alkaloid metabolism involving exogenous enzymes, i.e., those not yet demonstrated to be present in the plant or microorganism where the alkaloid occurs. Although such processes have some relevance to our subject, they are poor evidence of what actually happens *in vivo*. Thus excluded are (1) enzymes from other plants, e.g., diamine oxidase of peas (*Pisum*), (2) enzymes from microorganisms such as *Arthrobacter oxydans,* which contain nicotine oxidase, and (3) animal enzymes, such as oxidases from mammalian liver or kidney. On the other hand, cell-free extracts capable of catalyzing metabolic reactions have been considered as containing enzymes, even though the latter were not purified or even concentrated. Incidentally, this policy has made search of the literature more difficult, since such preparations are usually not indexed as enzymes; we can only hope that not many references have been missed.

C. Techniques

It is a prerequisite for studying the enzymology of alkaloid metabolism in plants and microorganisms that at least *some* compounds of the biosynthetic or degradative metabolic pathway that is being investigated have been identified. In biosynthesis the precursors are amino acids (aspartic acid, ornithine, lysine, phenylalanine, tyrosine, tryptophan) or their metabolites and a few nonnitrogenous compounds such as mevalonate. In catabolism the starting point is of course the alkaloid.

The most basic experimental evidence for participation of an enzyme system in a step or steps in the metabolic pathway is the concurrent presence of an enzyme and a precursor or product in the plant or microorganism *in vivo*. This evidence is made stronger (but is still far from conclusive) by showing parallel variation in the levels of the enzyme and the metabolic participant in different parts, or at different ages, of the organism. The first application of this appears to go back over 60 years, when True and Stockberger (1916) observed similarity in oxidase level and morphine content in *Papaver somniferum*. Indeed, this is the main method for studying the enzymology of alkaloid metabolism in plants and microorganisms *in vivo;* experiments involving the feeding of substrates do not qualify, since they deal with reactants and products but not catalysts.

The other principal way of investigating the enzymology of alkaloid metabolism in plants and microorganisms is *in vitro*. As a first step, cell-free extracts are prepared and shown to catalyze at least one step of the metabolic pathway. More detailed enzymology involves the concentration of enzyme(s) from such cell-free extracts by methods usual in such work: precipitation, centrifugation, chromatography, electrophoresis, etc. This has been well reviewed by Rhodes (1977). Progress in such purification procedures is of course followed by measurement of either the disappearance of substrate or the appearance of product, or preferably both. Since it is impossible to generalize about procedures for individual enzymes much more than this, such methods are outlined as each class of alkaloids is discussed. The investigator may find it advantageous to use plant sterile tissue culture made in nutrient medium (Barz *et al.*, 1977; Scott *et al.*, 1978a; Staba, 1969; Waller *et al.*, 1977), and mycelia from such cultures of fungi are the standard source of material. Failure to demonstrate enzyme activity in a cell-free extract does not prove absence of the enzyme *in vivo*, since the grinding of cells and blending of their contents, especially plant cells, can trigger the reaction of secondary metabolites (e.g., phenolics, quinones, and terpenes) with proteins. This may inactivate enzymes, especially in the presence of air (Brown and Wetter, 1972; Loomis, 1969, 1974; Loomis and Battaile, 1966; Mothes, 1972). It is possible to reduce such damage during mechanical disruption by adding competitive binders for these compounds: poly(vinylpyrrolidinone) (Loomis, 1969, 1974; Loomis and Battaile, 1966) or preferably this compound plus an adsorbent polystyrene (Loomis *et al.*, 1979). It is highly desirable to test such additives if attempts at isolation of enzymes or enzyme complexes fail without them. However, the investigator should be aware of the possible effects of polymers in altering individual enzyme solubilities

(Meikka and Ingham, 1978), macromolecular associations (Herzog and Weber, 1978), and aqueous–aqueous phase separations (Albertsson, 1971).

II. ALKALOID ENZYMOLOGY

A. Pyrrolidine, Piperidine, and Pyridine Alkaloids

1. Pyrrolidine Alkaloids

The only pyrrolidine alkaloid for which the enzymology of biosynthesis is known (aside from nicotine and its relatives, which are considered later) has the fused-ring indolizidine skeleton. This alkaloid, slaframine, is produced by a fungus, *Rhizoctonia leguminicola,* which grows on legumes such as red clover. Clover hay so infected when fed to animals causes excessive salivation (Guengerich *et al.,* 1973). The suggested biosynthetic pathway, based on the results of feeding labeled precursors to mold cultures, is shown in Fig. 2. The reduction of the ketone to the alcohol and the final acetylation to form slaframine (Guengerich *et al.,* 1973), and the formation of pipecolic acid (Guengerich and Broquist, 1973), have been demonstrated *in vitro* with an endogenous enzyme preparation and a commercial one, respectively.

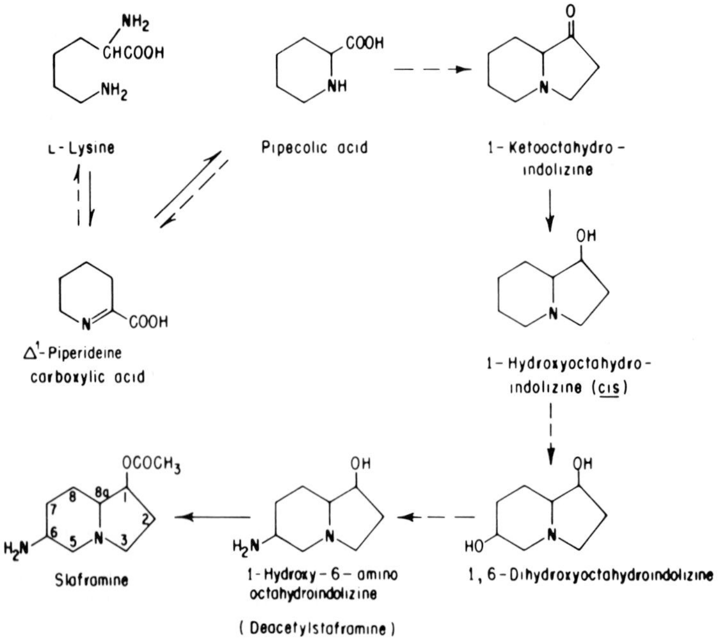

Fig. 2.   Biosynthesis of slaframine (Guengerich *et al.,* 1973).

Washed mycelia of *R. leguminicola* were ground and sonicated in phosphate buffer (pH 7.4) containing 2-mercaptoethanol, the slurry centrifuged at 37,000 *g*, and the supernatant desalted with Sephadex G-25. This crude enzyme mixture was incubated with octahydroindolizin-1-one, NADP⁺, glucose-6-phosphate, magnesium sulfate, and the buffer used for enzyme preparation. The products, separated by extraction and gas–liquid chromatography (glc), proved to be *cis*- and *trans*-octahydroindolizin-1-ols in about a 9 : 1 ratio. The same enzyme preparation incubated with [³H]deacetylslaframine, acetyl-coenzyme A (CoA), and phosphate buffer (pH 7.8) gave slaframine which was identified by thin-layer chromatography (tlc). Similarly the enzyme preparation, Δ¹-piperideine-2-carboxylic acid, NADH, NADP⁺, glucose-6-phosphate, magnesium sulfate, glucose-6-phosphate dehydrogenase, and buffer were incubated 3 h and pipecolic acid determined with ninhydrin. These reactions were proved to be enzymatic, since there was no conversion when boiled enzyme preparations were used.

2. Piperidine Alkaloids

a. γ-Coniceine. The *in vitro* synthesis of the poisonous hemlock alkaloids with isolated enzymes has been studied by Roberts (1971b, 1974b, 1977, 1978) (Fig. 3). One of her cell-free investigations (1974b) was on the formation of *N*-methylconiine from coniine and *S*-adenosyl-L-methionine (SAM), which is discussed by Poulton (this volume, Chapter 22). Earlier (1971b) she had isolated from hemlock leaves (*Conium maculatum*) an aminotransferase that catalyzed the reaction between 5-oxooctanal and L-alanine to produce

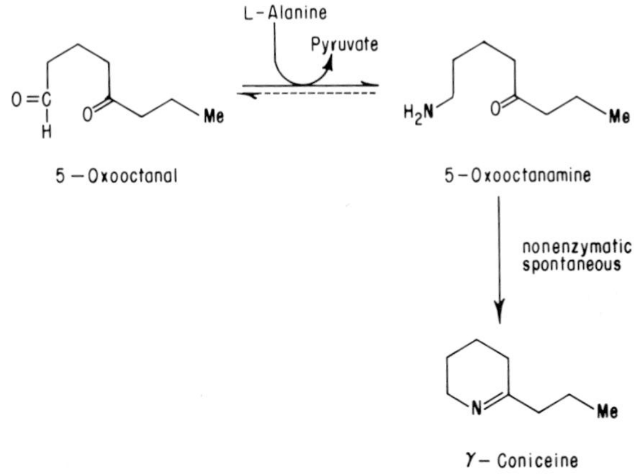

Fig. 3. Biosynthesis of γ-coniceine from 5-oxooctanal (Roberts, 1975b). Note: Since in theory all enzymatically catalyzed reactions are reversible, the ⇌ symbol has been used in representing many steps; the dashed line for the reverse reaction shows that little or nothing is known about it.

γ-coniceine. The enzyme was prepared from either ground fresh leaves or a derived acetone powder by extraction with phosphate buffer (pH 7.6), clarification by centrifuging, precipitation of protein with ammonium sulfate, and passing the protein through a Sephadex G-25 column. A fuller account of the work (Roberts, 1977) reports extraction only from the acetone powder with Tris buffer (pH 7.5) containing EDTA and dithiothreitol. Successive ensuing purification steps and their effectiveness are shown in Table I. The reaction was followed by assay for α-keto acids (as dinitrophenylhydrazones) and/or γ-coniceine (by color reaction with nitroprusside).

The purified enzyme was highly labile but could be kept longer at low temperatures and by the addition of KCl and bovine serum albumin. It had a MW of 56,200 and a pH optimum of 8.5. The K_m and V_{max} for 5-oxooctanal were 0.14 mM and 3.3 nkat/mg protein, respectively, and for L-alanine, 27 mM and 3.3 nkat/mg protein. These results indicate the operation of a binary mechanism like that known for other plant aminotransferases. The aminotransferase was relatively specific for L-alanine. It was somewhat inhibited by α-ketoglutarate, glyoxalate, and pyruvate, and moderately stimulated by pyridoxal phosphate but not by some divalent metal ions. Inhibitors that block carbonyl groups (semicarbazide, hydroxylamine, cyanide, isoniazid) and mercapto groups (iodoacetic acid, p-chloromercuribenzoate) were variably effective.

No reverse reaction (γ-coniceine to L-alanine) could be detected, probably because the presumed intermediate 5-aminooctane is much more stable in the cyclized form, γ-coniceine, and spontaneously changes to this alkaloid. In accord with this, added γ-coniceine and related alkaloids had no effect on the reaction.

The most recent study (Roberts, 1978) of γ-coniceine-forming aminotransferases reported separation by DEAE-cellulose chromatography into two such enzyme activities (A and B) and also into glutamate-oxalate aminotransferase and glutamate-pyruvate aminotransferase. A and B may have been contaminated with an amino acid:aldehyde aminotransferase responsible for aliphatic amine biosynthesis (for a discussion of such amines see Smith, this volume, Chapter 9). The procedure was generally similar to that used before, but diethyldithiocarbamate was added when acetone powders were extracted, the protamine step was omitted, and Sephacryl S-200 was substituted for calcium phosphate. Severe loss of activity occurred during the separation. Kinetic data showed significant differences between aminotransferases A and B (Table II).

Transaminases A and B both had a MW of 56,200; enzyme A had a broad pH optimum at 7.5–8.6, and enzyme B a sharp one at 8.5. They had high but not identical specificities for L-alanine as amino group donor. They were both inhibited uncompetitively by glyoxylate, A more than B; pyruvate competitively inhibited A, whereas in contrast B was unaffected or slightly stimu-

TABLE I

Purification of L-Alanine: 5-oxooctanal Aminotransferase[a, b]

Step	Fraction	Total protein (ng)	Total activity (nkat)	Specific activity (nkat/mg protein)	Yield (%)	Purification factor
1	Acetone powder extract	8160	53.8	0.007	85	1
2	Protamine sulfate supernatant	2990	46.1	0.016	83	2.1
3	45–65% $(NH_4)_2SO_4$ precipitate	1100	37.8	0.033	83	5.2
4	Calcium phosphate supernatant	348	30.5	0.087	79	12
5	DEAE-cellulose column I	39	18.3	0.476	52	66
6	DEAE-cellulose column II	15	14.3	0.953	78	132
7	Sephadex G200 gel eluate	3.5	9.3	2.666	65	369

[a] From Roberts, 1977.
[b] The γ-coniceine assay was used.

TABLE II

Kinetic Data for γ-Coniceine Formation by Aminotransferases A and B[a]

	Aminotransferase A	Aminotransferase B
K_m, L-alanine (mM)	27	55
K_m, 5-oxooctanal (mM)	1.6	0.14
V_{max} (nkat/mg protein)	1.26	3.3
Activation energy (kJ/mol)	30	0.33

[a] From Roberts, 1978.

lated. These isozymes may act in concert to regulate γ-coniceine formation via a feedback mechanism like that controlling dimethylallyltransferase through elymoclavine and agroclavine (see Floss, this volume, Chapter 7). Variations in the concentration of aminotransferases A and B in different strains of *Conium*, at different subcellular locations or at different times of day, could account for observed differences in alkaloid content (Fairbairn and Challen, 1959; Fairbairn and Suwal, 1961).

b. Interconversion of (+)-Coniine and γ-Coniceine: γ-Coniceine Reductase. Roberts (1975a,b) isolated an extract containing γ-coniceine reductase from *C. maculatum* when the plants contained coniine and methylconiine as the major alkaloids (Fig. 4). Fresh actively growing leaves or fruits were ground in Tris buffer (pH 7.5) containing 2-mercaptoethanol, EDTA, and glutathione, and the suspension was filtered and centrifuged at 39,000 g. Treatment of the supernatant with ammonium sulfate precipitated an active protein fraction at 40–60% saturation. The precipitate was dissolved in the same buffer containing EDTA and desalted with a Sephadex G-25 column; this produced the active crude enzyme studied. It was assayed by incubation with [U-^{14}C]-γ-coniceine and NADPH or NADH (or a system generating one of these) for 3.5 h at 30°C. Protein was precipitated with sulfosalicylic acid, the alkaloids were extracted and separated by paper chromatography or tlc, and the resulting coniine was recrystallized to constant specific activity and

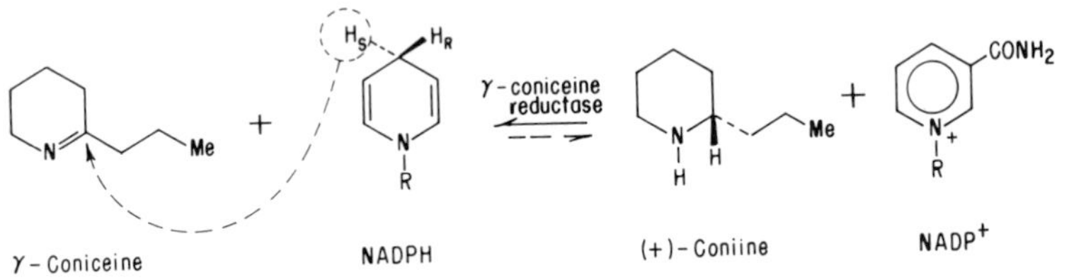

Fig. 4.   Reduction of γ-coniceine to coniine.

counted. By use of A-[4-³H]NADPH and B-[4-³H]NADPH, the hydrogen transfer to the Δ^1 double bond of γ-coniceine to form (+)-coniine was shown to be stereospecifically from the B (*pro-S*) side of NADPH. NADH can serve in the reduction of γ-coniceine only by prior conversion to NADPH.

c. Lobeline. Lobeline (Fig. 5) is used as an antidote for an overdose of narcotics. Jindra *et al.* (1966a, 1967) and Smogrovicová *et al.* (1968) prepared cell-free extracts of phenylalanine : α-ketoglutarate aminotransferase (PKA) and lysine : α-ketoglutarate aminotransferase (LKA) from *Lobelia inflata* and assayed the activities at different stages of plant development. In fresh plant material homogenized in phosphate buffer (pH 8.0) PKA was more abundant than LKA.

In 1972 Smogrovicová *et al.* used the aerial parts of *L. inflata* to produce a 105-fold purified solution of PKA. Activity was determined by measuring the resulting phenylpyruvic acid spectrophotometrically. The pH optimum was 8.0–9.0, with an apparent K_m for phenylalanine of 55 mM and for α-ketoglutarate of 0.046 mM. The enzyme preparation was preincubated with pyridoxal-5-phosphate before use. It was most active with L-aspartate (relative activity 8.8), asparagine (0.86), and phenylalanine (0.63) but also was active on other aromatic and aliphatic amino acids except lysine and serine.

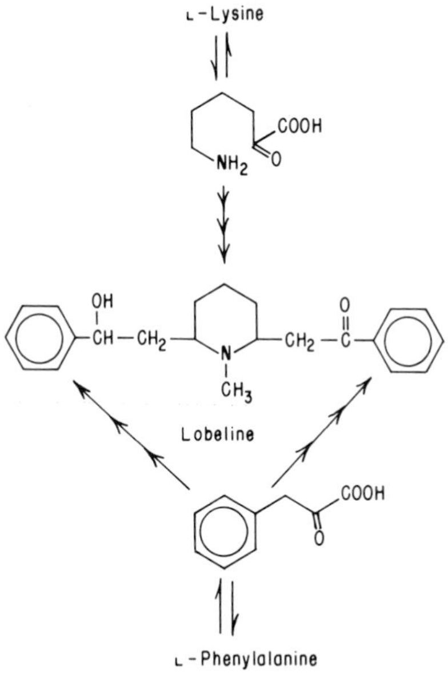

Fig. 5.   Biosynthesis of lobeline.

3. Pyridine Alkaloids

a. The Pyridine Nucleotide Cycle and Its Relationship to Precursors of Pyridine Alkaloids. In plants, condensation of a C_3 and a C_4 compound by a series of reactions yields quinolinic acid (QA), the first known intermediate in the *de novo* synthesis of pyridine nucleotides and pyridine alkaloids. It has been established conclusively that the nitrogen atom of aspartic acid becomes the nitrogen of ricinine and of course the QA nitrogen (Yang and Waller, 1965; Waller *et al.*, 1966a). The next step, at least for nucleotides, is the decarboxylative conversion of QA to nicotinic acid mononucleotide (NAMN), catalyzed by QA phosphoribosyltransferase (QAPRT) (Fig. 6). This enzyme has been obtained from various sources and even purified to crystallinity, but discussion here is limited to its preparation and properties when extracted from castor bean endosperm.

The first such preparation was made from 6-day-old seedling cotyledons by grinding, extraction with phosphate buffer at pH 7.4, straining, and centrifuging at 27,000 g; the supernatant contained the enzyme, which was shown to require 5-phosphoribosyl-1-pyrophosphate (Hadwiger *et al.*, 1963).

In more sophisticated work, etiolated seedlings of *Ricinus communis* were ground in buffer, and the suspension clarified by centrifugation, treated with QA, heated 1 min at 60°C, and recentrifuged; the supernatant was purified by dialysis, chromatography with DEAE-Sephadex A-50 and then hydroxylapatite, and isoelectric focusing. This gave 500-fold purification of QAPRT, as assayed by the production of $^{14}CO_2$ from labeled QA. Analytical disc gel electrophoresis at pH 7.5–8.3 showed the preparation to consist of three very similar proteins, all of the same MW (68,000–72,000) but differing slightly in charge. Further electrophoresis (sodium dodecyl sulfate (SDS)/polyacrylamide gel) showed that the enzyme was made up of two subunits, each of MW about 35,000 (Mann and Byerrum, 1974a).

The enzyme was stable in the presence of added sucrose and dithiothreitol, and at elevated temperatures upon addition of QA, which presumably is bound at active sites. The QAPRT was not inhibited by ATP, ricinine, or nicotinic acid, but it was inhibited by some analogues of QA, suggesting that both carboxyl groups of QA are required for substrate binding to the enzyme. The optimum pH was broad, 6.5–7.7, and K_m values were 0.012 mM (QA) and 0.045 mM (5-phosphoribosyl-1-pyrophosphate). The kinetics suggested that a ternary complex is formed by QAPRT, QA, and the pyrophosphate (Mann and Byerrum, 1974a).

b. Ricinine. The pyridine nucleotide cycle, common to nearly all organisms, is shown in Fig. 6 (see review by Waller and Nowacki, 1978). An interdependency between this cycle and ricinine biosynthesis has been shown, both by use of inhibitors *in vivo* (Johnson and Waller, 1974) and by comparison of levels of enzymes and of ricinine in *R. communis* (Mann and

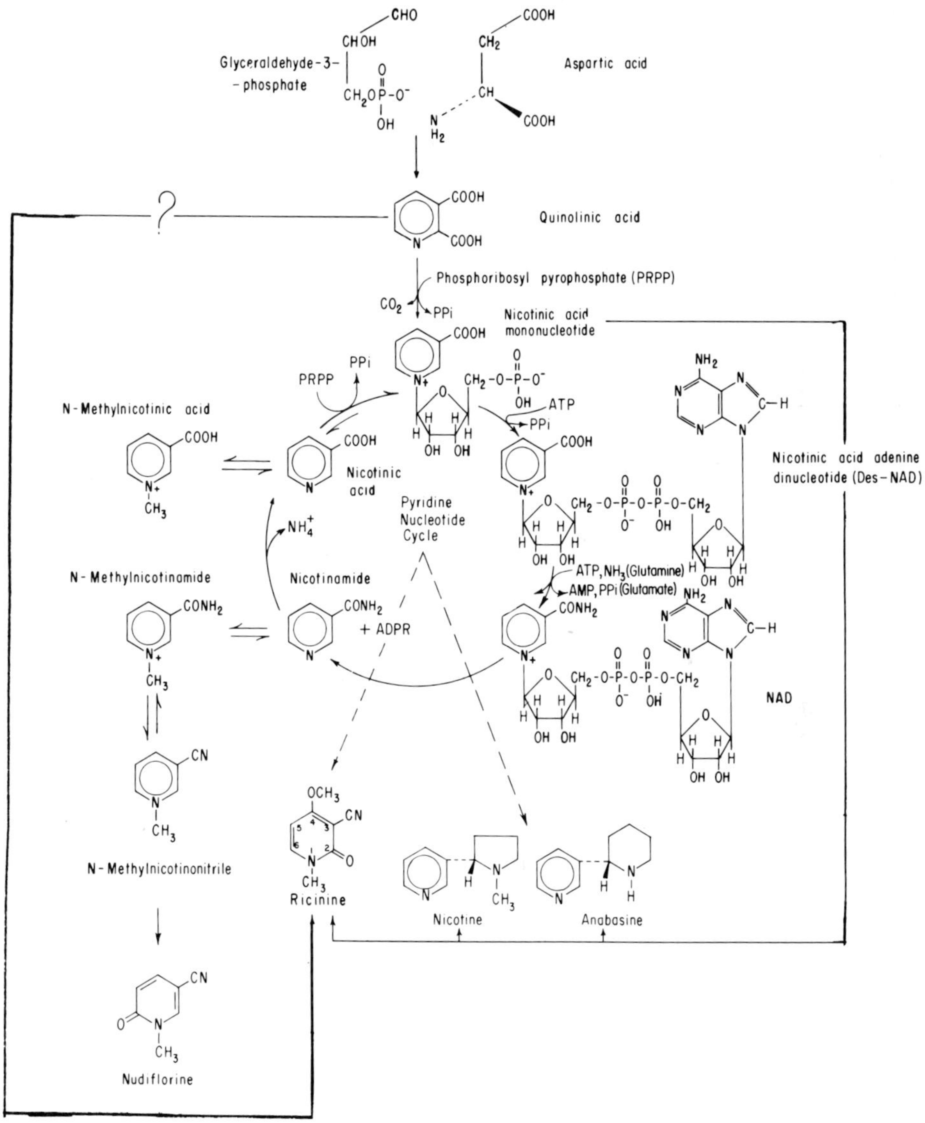

Fig. 6. Pyridine nucleotide cycle and its relationship to pyridine alkaloids. Enzymes and their functions are: quinolinic acid phosphoribosyltransferase (decarboxylating) (QAPRT)—quinolinic acid (QA) → nicotinic acid mononucleotide (NAMN); nicotinic acid mononucleotide adenyltransferase—NAMN → desamidonicotinamide adenine dinucleotide (des-NAD); NAD⁺ synthetase—des-NAD → nicotinamide adenine dinucleotide (NAD⁺); NAD⁺ glycohydrolase—NAD⁺ → nicotinamide (NAm) and adenosine diphosphoriboside (ADPR); nicotinamidase—NAm → nicotinic acid (NA); NA phosphoribosyltransferase—NA → NAMN; N-methyltransferases yield N-methylnicotinic acid or N-methylnicotinamide (Waller et al., 1966b).

Bjerrum, 1974b). Etiolated seedlings showed a sixfold increase in QAPRT in 4 days, and this preceded by only 1 day an even larger increase in ricinine (Fig. 7A and D); thereafter both soon decreased to the low levels characteristic of mature plants. However, the level of an enzyme of the pyridine nucleotide cycle, nicotinamide deamidase, was unusually low, perhaps shunting metabolites out of the cycle into ricinine biosynthesis (Fig. 7B and D). The specific activity of another enzyme of the cycle, nicotinic acid phosphoribosyltransferase, soon decreased to that of mature plants (Fig. 7C and D). This suggests that, although the seedling can produce relatively much more NAMN than the mature plant, it does not convert it to dinucleotides and thus recycle nicotinic acid. This difference in the two transferase levels, also observed by others (Waller and Yang, 1965), explains why QA when fed was

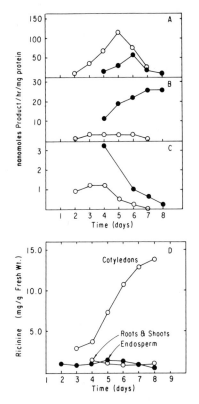

Fig. 7. Changes in specific activity of QAPRT (A), nicotinamide deamidase (B), and nicotinic acid phosphoribosyltransferase (C) in the cotyledons (●) and endosperm (○) of etiolated castor bean seedlings, and changes of ricinine content (D) of organs of such plants, with age (Mann and Byerrum, 1974b).

better incorporated into ricinine than nicotinic acid (Mann and Byerrum, 1974b; Waller et al., 1965, 1966a).

Calculations based on QAPRT levels in young cotyledons and endosperm indicate (Mann and Byerrum, 1974b) that the ricinine accumulated in cotyledons is formed there from pyridine nucleotide cycle components (such as nicotinic acid) produced in endosperm. This supports the concept of Waller and Nowacki (1978) "that alkaloids are actually side products of intermediary metabolism reactions and that the rate of their biosynthesis is directly proportional to the synthetic activities of the tissue." By feeding labeled ricinine to excised senescent leaves of R. communis and considering limiting rates of demethylation, Skursky et al. (1969) calculated that the rate of disappearance of ricinine was about 4 μg/g fr wt per day. It is hard to compare this with the rate of biosynthesis because of differences in the age of the plants, but a plant apparently accumulates ricinine throughout its life (Waller et al., 1965). The rates of anabolism and catabolism doubtless vary with the physiological stage of development, the different parts of the plant, the time of day, etc.

The intermediates between nicotinic acid or nicotinamide (Fig. 8) and ricinine are partly known. Robinson (1965) obtained a crude enzyme preparation from R. communis that catalyzed the oxidation of N-methylnicotinonitrile to the corresponding 4- and 6-pyridones; however, Waller et al. (1966b) reported that N-methylnicotinonitrile was not incorporated into ricinine in vivo. Since ricinine formation requires some pyridinium oxidase activity, we suggest that enzyme compartmentalization might have occurred or that N-methylnicotinonitrile could not be transported through the membrane defining the subcellular location of the pyridinium oxidase. It

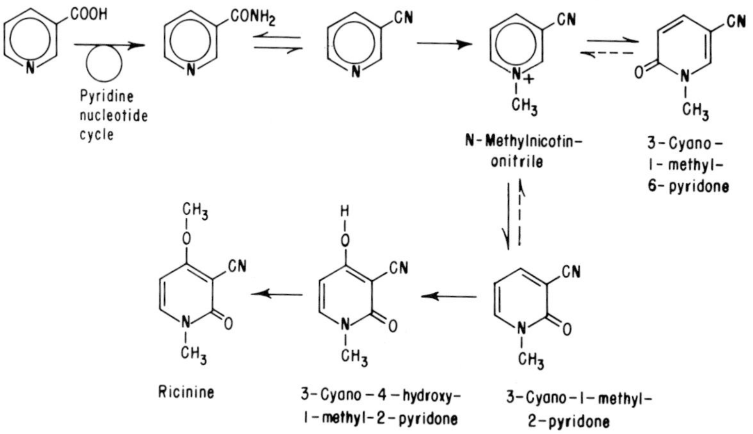

Fig. 8. Biosynthesis of ricinine from nicotinic acid or nicotinamide.

may also be that N-methylnicotinonitrile is off the main pathway and is introduced into the pathway only in altered form.

In Robinson's (1965) preparation of the enzyme solution, plant material was homogenized in potassium phosphate buffer (pH 8.5), filtered through cheesecloth, and centrifuged at 24,000 g. A fat layer and sediment were removed and the supernatant dialyzed against potassium dihydrogen phosphate to produce enzyme whose activity was assayed by the reduction of triphenyltetrazolium chloride. The pyridone products were separated and purified by column chromatography and identified by comparing with authentic standards. The enzyme was specific for the oxidation of N-methylnicotinonitrile and N-alkyl analogues.

Fu and Robinson (1970) found that 7 species (*R. communis, Trewia nudiflora, Chrogophora plicata, Jatropha gossypifolia, Tragia involucrata, Acolypha hispida*, and *Synadenium grantii*) out of 11 members of the family Euphorbiaceae tested exhibited pyridinium oxidase activity on salts of N-methylnicotinonitrile, with 4- and 6-pyridones being produced *in vitro* for both *R. communis* and *T. nudiflora*. Plants of other families, including tobacco, showed no such oxidase activity, suggesting that the enzyme system is phylogenetically specific.

The same authors (Fu *et al.*, 1972) obtained three pyridinium oxidases, probably isozymes, from *R. communis*. After the work-up used by Robinson (1965), the dialyzed crude enzyme solution was briefly heat-treated (55°C) and centrifuged, and the supernatant adjusted to pH 6.3 and again centrifuged to remove inactive proteins. The yellow supernatant (flavoproteins?) was then adjusted to pH 6.7, redialyzed, and fractionated with a DEAE-cellulose column, and the enzyme was precipitated with ammonium sulfate and purified by gel filtration (Sephadex, Sepharose, or Biogel). This gave enzymes labeled A, B, and C purified 500- to 1000-fold, but all produced 4- and 6-pyridones and otherwise resembled each other so much that they are discussed here as a single unit. They had maximum activity at pH 9.5–10.5 and fair stability from 10° to 60°C; they were specific for negatively substituted pyridinium salts and were strongly inhibited by sulfhydryl-blocking reagents, quinacrine, methanol, and cyanide (Fu *et al.*, 1972).

Fu and Robinson (1972) further purified the pyridinium oxidase B, the most stable and most abundant of the three. Elution of oxidase B from Biogel 5A and rechromatography using ECTEOLA-cellulose produced enzyme purified up to 3000-fold, but in small yield and still showing traces of impurity in polyacrylamide gel electrophoresis. The apparent MW was 250,000. The enzyme produced the 6- and 4-pyridones in a ratio of 3.5–4.1 : 1. The K_m for N-methylnicotinonitrile perchlorate was 0.54 mM and decreased with increasing pH. The value of V_{max} increased as the dipole moment of the substituent group at position 3 (the electron-withdrawing power) of the pyridine ring increased (Table III). Thus the positively charged quaternary nitrogen

TABLE III

Variation in V_m with Dipole Moment for Analogous Substrates of Pyridinium Oxidase B[a]

Substrate	V_m	Dipole moment (D)
1-Methyl-3-cyanopyridinium perchlorate	20	4.39
1-Methyl-3-nitropyridinium iodide	10	4.21
1-Methyl-3-acetylpyridinium iodide	6.6	3.00
1-Methyl-3-formylpyridinium iodide	5.7	2.76

[a] From Fu and Robinson, 1972.

atom is considered necessary for the enzyme–substrate binding site, and a strongly electron-withdrawing group enhances the oxidation reaction.

c. Nudiflorine. Mukherjee and Chatterjee (1964) reported the occurrence in *T. nudiflora* of an alkaloid, nudiflorine (1-methyl-3-cyano-6-pyridone), which was soon obtained enzymatically *in vitro* by Robinson (1965) and Fu and Robinson (1970) (Fig. 6). The enzyme extracts used were the same as those discussed for ricinine (Section II,A,3,b).

d. Nicotine. Since the two rings of nicotine are biosynthesized separately and then joined, we consider their formation separately, the pyridine portion first (Fig. 9).

The relationship of the pyridine nucleotide cycle, and of the QAPRT reaction that supplies it with metabolites, to pyridine alkaloid biosynthesis has already been mentioned (Section II,A,3,a). A comparison of the level of some enzymes relevant to the cycle in various plants is shown in Table IV (Mann and Byerrum, 1974b). The exceptionally high value for QAPRT in tobacco roots (where nicotine, anabasine, etc., are produced) compared to low values for all other enzymes suggests that the *de novo* pathway in the tobacco plant is activated for alkaloid synthesis and concurrent maintenance of adequate NAD and NADP levels.

The role of another enzyme found in *N. rustica* roots, nicotinic acid decarboxylase, in nicotine alkaloid biosynthesis is uncertain. The pyridine ring of the acid, but not the carboxyl carbon atom, is known to be incorporated; the position of the latter group, i.e., the 2-position, is the point of coupling with the pyrrolidine ring (Yang *et al.*, 1965). The decarboxylase probably is part of an enzyme system affecting this condensation. At any rate, a crude preparation enzyme with nicotinic acid decarboxylase activity was obtained from tobacco roots but not leaves or *Ricinus* roots (Gholson *et al.*, 1964). Later Chandler and Gholson (1972) found the activity localized in a particulate

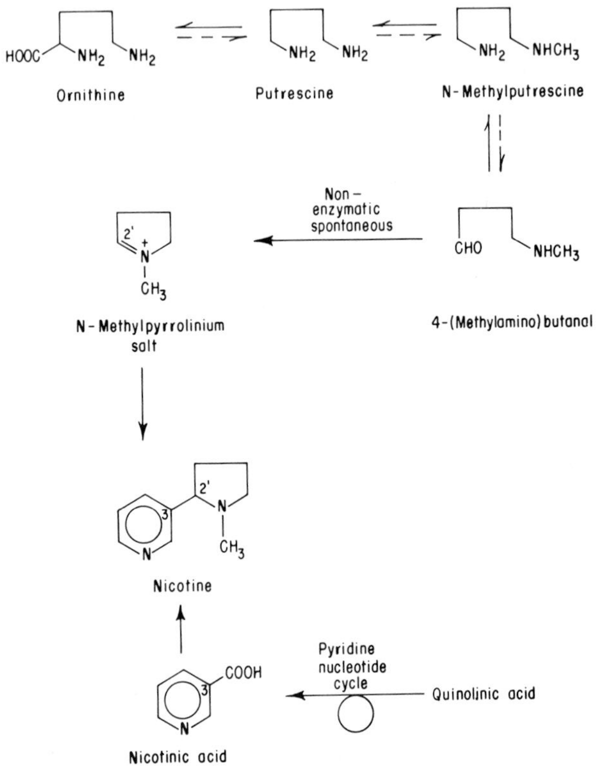

Fig. 9. Biosynthesis of the pyrrolidine ring.

fraction obtained by centrifugation at 20,000 g; this fraction also contained mitochondria, but association of the decarboxylase with mitochondria was not established. The enzyme required no cofactor but needed oxygen for the release of carbon dioxide from nicotinic acid. No free pyridine was formed. In another study, nicotinic acid decarboxylase levels in young tobacco plants, like those of most other enzymes, were depressed by a lack of nutrients (Yoshida, 1973).

The pyrrolidine ring moiety of nicotine is derived from ornithine as shown in Fig. 9.

The first enzyme work on this process involved the methyltransferase in tobacco (Mizusaki *et al.*, 1971). While *N*-methylputrescine has not been identified in the tobacco plant, it can serve as a precursor of nicotine, and the characterization of the enzyme that produces it is further evidence of its participation in nicotine synthesis. The enzyme, which is discussed further by Poulton (this volume, Chapter 22), was extracted by conventional methods from roots (the only site where it occurs) and concentrated 30-fold com-

TABLE IV

Specific Activity of Some Enzymes in Tobacco (*Nicotiana rustica*) and Other Plants[a]

| Plant source[b] | Enzyme activity (nmol product/h/mg protein) | | |
	QAPRT	Nicotinic acid phosphoribosyltransferase	Nicotinamide deamidase
Tobacco leaves (8 weeks)	0.6	0.4	2.7
Tobacco roots (8 weeks)	23.9	0.6	13.5
Tobacco leaves and roots	7.5	0.5	13.3
Soybean (4 weeks)	0.3	3.1	5.2
Sunflower (1 week)	0.3	0.2	1.5
Pea (4 weeks)	0.2	3.3	14.7
Spinach (4 weeks)	0.8	7.9	12.4

[a] From Mann and Byerrum, 1974b.

[b] Crude homogenates were prepared from the leaves unless stated otherwise.

pared to the extract. The activity was enhanced 10-fold when the plants were decapitated. It was highly specific for putrescine and had a MW of about 600,000.

N-Methylputrescine oxidase catalyzes the oxidative deamination of N-methylputrescine to 4-(methylamino)butanal which spontaneously cyclizes to the N-methyl-Δ^1-pyrrolinium ion (Fig. 9) (Mizusaki *et al.*, 1972). Tobacco roots were ground in Tris buffer (pH 7.4) containing 2-mercaptoethanol, EDTA, sodium ascorbate, and poly(ethylene glycol) and the suspension clarified by straining and centrifuging. A summary of further purification steps and yields is shown in Table V. The enzyme was measured by incubation with [N-methyl-^{14}C]methylputrescine in buffer containing catalase and determining the radioactivity of the 2-cyano-1-methylpyrrolidine produced by adding cyanide ion.

The enzyme had an optimum pH of 8.0 and a K_m for N-methylputrescine of 0.45 mM; it also had some activity toward putrescine and cadaverine but not toward the other amines tested. It was not inhibited by nicotine and related alkaloids. The use of inhibitors indicated the presence of copper and of carbonyl and mercapto functions; the essentiality of copper was proved by inactivating the enzyme by dialysis against diethyldithiocarbamate and then reactivating by the addition of copper sulfate. Although aminoguanidine caused only slight inhibition, the enzyme is regarded as a diamine oxidase.

The enzyme required for the first step of pyrrolidine ring synthesis, ornithine decarboxylase, is more widely distributed and has in fact been found

TABLE V

Purification of N-Methylputrescine Oxidase from Tobacco Roots[a]

Purification step	Volume (ml)	Total protein (mg)	Total activity (units)	Specific activity (units/mg protein)	Purification (fold)	Yield (%)
Crude extract	1,640	1,281	23,944	18.7	1	100
(NH$_4$)$_2$SO$_4$ fraction	205	564	31,945	56.7	3	133.2
DEAE-cellulose chromatography	172	116	27,110	233.7	12.5	113.2
Sephadex G-200 gel filtration	75	32	18,225	569.5	30.5	76.1
DEAE-cellulose chromatography	15	3.2	9,099	2,843.4	152	38

One unit of activity equals 1 nmol of N-methylpyrrolinium salt formed in 30 min.

[a] From Mizusaki et al., 1972.

in various plants besides tobacco (Mizusaki et al., 1973). A partially purified enzyme was obtained from Nicotiana tabacum roots by procedures including the addition of Polyclar AT and ammonium sulfate fractionation (Fischer and Leete, 1969). It required pyridoxal phosphate as cofactor and was partly inhibited by δ-N-methylornithine but not by the α-N-methyl isomer. Mizusaki et al. (1973) did not study ornithine decarboxylase alone, but in a mixture with putrescine N-methyltransferase and N-methylputrescine oxidase. Assay for the decarboxylase involved measuring the ^{14}CO$_2$ released from [1-^{14}C]ornithine. To evaluate the effects of various treatments of tobacco plants and thus to understand better the regulation of biosynthesis of the pyrrolidine moiety of nicotine, a crude enzyme solution was made from root tissue by grinding and clarifying as before (Mizusaki et al., 1971, 1972) and using a Sephadex G-25 column.

The effect of decapitating 4-week-old N. tabacum plants on increasing the enzyme activity is shown in Fig. 10; nicotine accumulation in the root followed a similar pattern (Mizusaki et al., 1973). Moreover, nicotine administered to the plants at 0.5 mM caused 50% inhibition of synthesis of the enzymes under discussion and almost total inhibition at 5 mM. Since normal plants contain enough nicotine to cause 50% inhibition, it is possible that nicotine accumulation in the roots represses the enzyme synthesis induced by shoot decapitation.

Indoleacetic acid (IAA) administered to seedlings at low concentrations at the time of decapitation stimulated enzyme production, whereas higher concentrations inhibited such production and ultimately stopped it entirely (Fig.

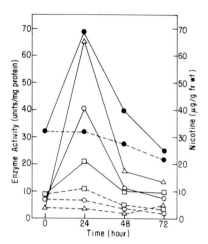

Fig. 10. Time course of changes in levels of some enzymes and of nicotine content in roots of tobacco plants. ○, Ornithine decarboxylase; △, putrescine *N*-methyltransferase; □, *N*-methylputrescine oxidase; ●, nicotine; ——, decapitated plants, ---, intact plants (Mizusaki *et al.*, 1973).

11) (Mizusaki *et al.*, 1973). Thus IAA may be important in regulating alkaloid biosynthesis.

The absence of putrescine *N*-methyltransferase and *N*-methylputrescine oxidase in leaves of *N. tabacum,* in tissue culture preparations thereof, and

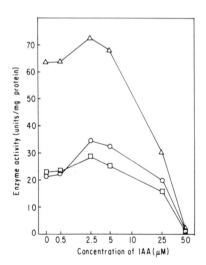

Fig. 11. Effect of IAA on levels of some enzymes in roots of decapitated tobacco plants. ○, Ornithine decarboxylase; △, putrescine *N*-methyltransferase; □, *N*-methylputrescine oxidase (Mizusaki *et al.*, 1973).

in other plants not biosynthesizing N-methylpyrrolidine alkaloids is good evidence that these enzymes indeed participate in the synthesis of nicotine alkaloids in tobacco roots.

A different kind of stress, inorganic ion deficiency, was tested for its effect on tobacco plants. Various enzyme levels were determined in roots of young plants grown in deficient media (Yoshida, 1973); the results are shown in Table VI.

The most obvious changes were those produced by nitrogen deficiency. There was a rough parallelism between nicotine levels and some of the enzyme levels.

e. Nicotine Synthetase Complex. A few laboratories have tried without success to produce nicotine *in vitro* using cell-free extracts of various parts of tobacco plants. To couple N-methylpyrrolinium ion with nicotinic acid probably requires a complicated mechanism involving several enzymes acting stereospecifically. One of the enzymes might decarboxylate nicotinic acid (nicotinic acid decarboxylase), another enzyme could be involved in breaking the double bond of the N-methylpyrrolinium ion, and a third enzyme might bind the decarboxylated nicotinic acid and the modified pyrrolidine ring and join them through a new bond between C-2' and C-3 to form nicotine (Fig. 9). This biosynthesis is somewhat similar to the joining of rings by the berberine bridge (Section II,F) and that involved in vindoline formation (Section II,H,4,h).

f. Nicotine–Nornicotine Interconversion. Bose *et al.* (1956) obtained acetone precipitates from cell-free extracts of *Nicotiana glauca* and *N. tabacum*, which catalyzed the demethylation of nicotine to nornicotine. The eliminated methyl group was transferred to ethanolamine; guanidoacetic acid was not an acceptor. However, nornicotine could not be methylated with methionine. In later work (Schröter, 1966) cell-free extracts of *Nicotiana alata* leaves, themselves alkaloid-free, catalyzed the nornicotine–nicotine interconversion. The homogenate prepared from deveined leaves with phosphate buffer was pressed through muslin and centrifuged at 20,000 g and then incubated with nicotine or nornicotine. Enzyme from the oldest (lowest) leaves promoted demethylation of nicotine; since added methyl group acceptors were not methylated by [^{14}C]nicotine, the demethylation must have been oxidative. Nornicotine was not methylated, even by SAM. Enzyme from young leaves, on the other hand, methylated nornicotine in the presence of SAM but did not demethylate nicotine significantly. These results agree with those of feeding experiments with leaves.

g. Mimosine and Related Alkaloids. (*i.*) *Mimosine.* Murakoshi *et al.* (1972a) observed that extracts of *Leucaena leucocephala* (Leguminosae) seedlings

TABLE VI

Effect of Nutrient Deficiency on Enzyme Levels in Tobacco[a,b]

Enzymes	No deficiency	$-N$	$-P$	$-K$	$-Ca$	$-Mg$	$-S$	$-B$
Ornithine decarboxylase	5.4	60.6	13.8	3.0	7.4	6.8	6.8	7.0
Putrescine N-methyltransferase	27.6	3.0	3.0	8.6	4.4	9.4	4.8	11.8
N-Methylputrescine oxidase	26.8	8.1	3.5	12.3	4.52	6.0	2.9	10.5
Nicotinic acid decarboxylase	0.25	0.092	0.015	0.015	0.010	0.022	0.051	0.079
NAD pyrophosphatase	75×10^2	32×10^2	72×10^2	96×10^2	64×10^2	80×10^2	79×10^2	76×10^2

[a] From Yoshida, 1973.

[b] Values are in nanograms per milligram of protein per hour.

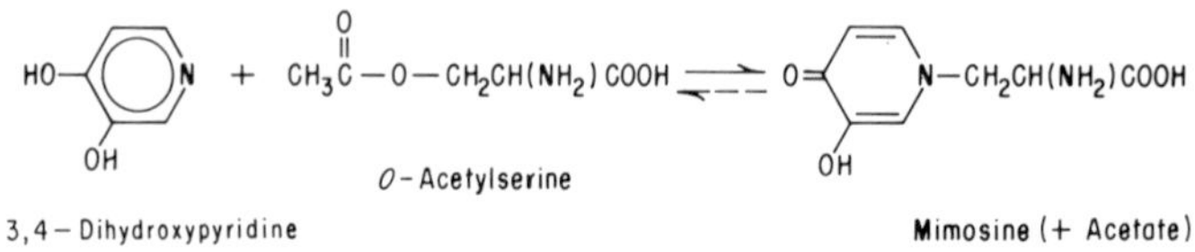

Fig. 12.   Enzymatic biosynthesis of mimosine (Murakoshi *et al.*, 1972a).

catalyzed the *in vitro* enzymatic synthesis of mimosine from 3,4-pyridinediol and *O*-acetyl-L-serine (Fig. 12). The plant tissue was macerated in phosphate buffer at pH 8.0, the suspension filtered through muslin and centrifuged at 25,000 g, and the supernatant passed through Sephadex G-25. After incubation of this preparation with substrates at pH 7.7–8.0, mimosine was separated and identified by comparison with authentic material both by paper chromatography and with an amino acid analyzer. For quantitative work [3-^{14}C]*O*-acetylserine was used and the separation of labeled mimosine by the analyzer was followed. Biosynthesis of mimosine was not dependent on pyridoxal phosphate; *O*-phosphoserine or serine could not substitute for *O*-acetylserine.

(*ii.*) *Mimoside*. Mimoside (mimosine β-D-glucoside) occurs in *Mimosa pudica* and *L. leucocephala* along with mimosine, but in smaller amounts. The *in vitro* interconversion of mimosine and mimoside was studied by Murakoshi *et al.* (1972b) (Fig. 13). An enzyme preparation was made from *L. leucocephala* seedlings grown in the dark by a procedure similar to that used for mimosine (see preceding paragraphs) except that 2-mercaptoethanol was added and a pH of 8.3 used. Incubation of the crude enzyme solution with mimosine and uridine diphosphate glucose (UDP-glucose) at pH 8.3 gave mimoside. L-Tyrosine under these conditions was glycosylated only very slowly, and DOPA not at all; similarly only UDP-glucose could serve as glucose donor. The reverse reaction, the hydrolysis of mimoside to mimosine, was catalyzed by the same enzyme preparation but in acetate buffer at pH 5.6; since this is so far from the optimum pH (8.2–8.4) for the glucosylation reaction, some nonenzymatic reaction may have intervened.

(*iii.*) *Biodegradation*. Enzyme preparations from *L. leucocephala* have been shown to separate the pyridine and amino acid moieties of mimosine. Thus mimosine was degraded in aqueous medium to 3,4-pyridinediol, pyruvate, and ammonia (Smith and Fowden, 1967). In the presence of methanethiol,

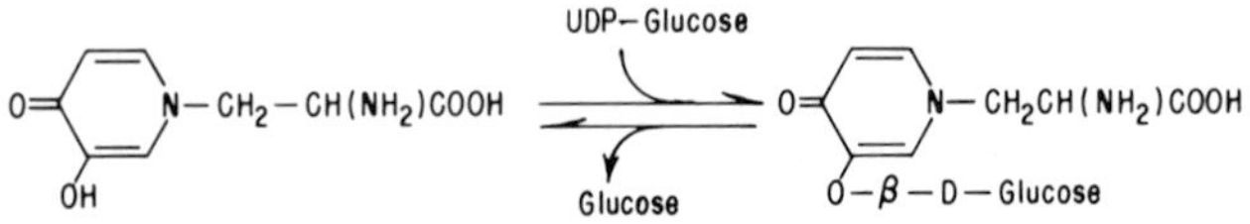

Fig. 13.   Enzymatic interconversion of mimosine and mimoside (Murakoshi *et al.*, 1972b).

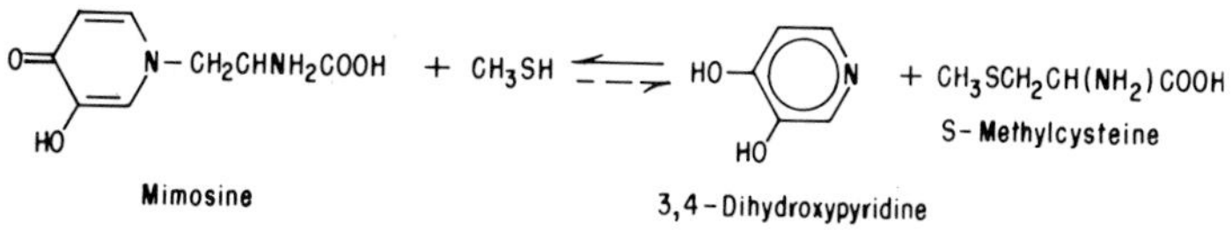

O=⟨⟩N—CH₂CHNH₂COOH + CH₃SH ⇌ HO—⟨⟩N + CH₃SCH₂CH(NH₂)COOH

S—Methylcysteine

Mimosine   3,4—Dihydroxypyridine

Fig. 14. Biodegradation of mimosine with methanethiol (Murakoshi *et al.*, 1970).

the side chain was trapped as *S*-methylcysteine (Murakoshi *et al.*, 1970) (Fig. 14). The amino acid was identified by paper chromatography and with an amino acid analyzer. Some other thiols were believed to react similarly, giving S-alkylated cysteines.

B. Tropane Alkaloids

Several alternative pathways, all starting with ornithine, for the biosynthesis of tropane alkaloids are shown in Fig. 15. These have been reviewed by Clark (1977).

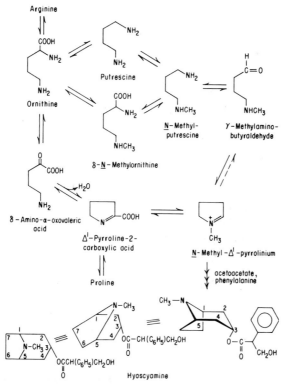

Fig. 15. Biosynthesis of tropane alkaloids from arginine.

The enzyme required for the production of ornithine (arginase) was shown to be present in various parts of *Datura tatula* of various ages and in levels paralleling alkaloid levels. This is evidence that the arginine–ornithine conversion is indeed involved (Fuller and Gibson, 1952). Repetition of this work confirmed the parallel relationship of alkaloid levels of the plant to arginase levels in roots when low concentrations of manganese dichloride were fed via the roots (Jindra, 1966). A similar study of various *Datura stramonium* callus tissues showed arginase levels high enough to maintain synthesis of tropane alkaloids (Bereznegovskaya, 1973; Jindra and Staba, 1968).

Oxidative deamination of putrescine to aldehydic products by a cell-free extract of *Atropa belladonna* was demonstrated early by Cromwell (1943). Juice from roots or etiolated shoots, with 5% added glycerol, was centrifuged, dialyzed in the cold, adjusted to pH 6.8 or 7.2, and incubated with putrescine; both ammonia and unidentified carbonyl compounds were produced.

Datura stramonium and *A. belladonna* contained relatively high levels of ornithine decarboxylase, putrescine *N*-methyltransferase, and *N*-methylputrescine oxidase. Tomato roots, barley roots, and sterile callus of the tobacco plant had a slight ornithine decarboxylase activity (Mizusaki *et al.*, 1973). *Datura stramonium* and *A. belladonna* are known to incorporate ornithine and putrescine into the *N*-methylpyrrolidine moiety of the tropane alkaloid, probably much as in nicotine biosynthesis (see Section II,A,3,d).

The tropic acid moiety of hyoscyamine–atropine probably arises from phenylalanine. In this conversion transamination to phenylpyruvic acid appears to be a reasonable step, and Jindra and co-workers (Jindra, 1966; Jindra *et al.*, 1967) have shown aminotransferase activity to be present in homogenates of various parts of *D. stramonium*, especially the roots, at flowering time. Another preparation from *D. stramonium*, tissue culture, had similar activity (Jindra and Staba, 1968), but it also catalyzed the transamination of ornithine (Jindra *et al.*, 1960; Jindra and Staba, 1968). An enzyme distinct from polyphenol oxidase but capable of catalyzing the conversion of ethyl formylphenylacetate to ethyl benzoylformate has been isolated in crude form from *Datura innoxia* roots and studied (Kalyanaraman, 1970); it is not clear whether this reaction is indeed involved in the phenylalanine–tropic acid transformation.

An esterase catalyzing both synthesis and hydrolysis of atropine *in vitro*

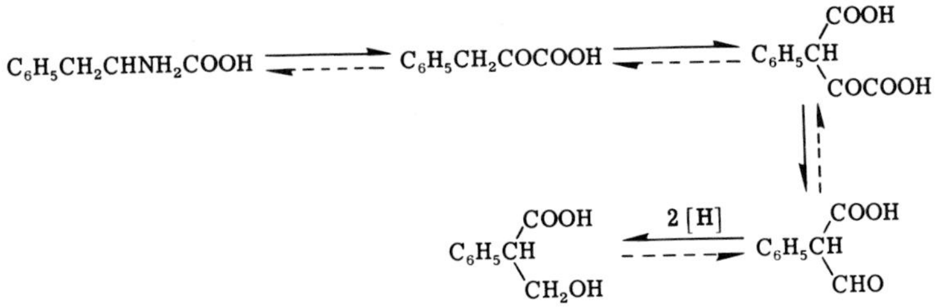

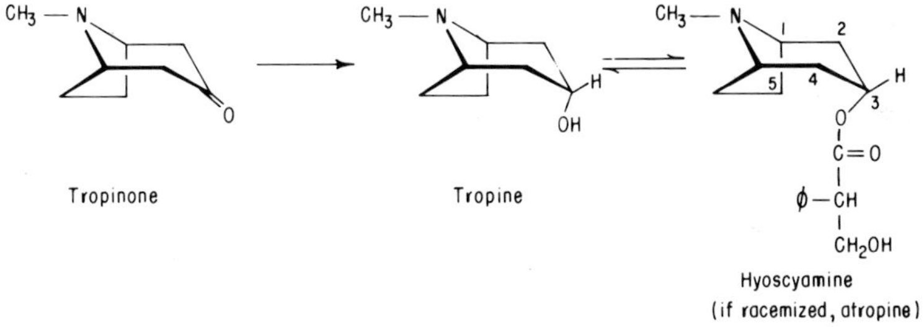

Fig. 16.   Conversion of tropinone to hyoscyamine.

was found in cell-free extracts of *D. stramonium* roots in later stages of growth; the enzyme had little specificity, acting also on homotropine and scopolamine, though not cocaine (Jindra and Cíhák, 1963; Jindra *et al.*, 1959, 1960). The esterase could be concentrated either by ammonium sulfate or acetone precipitation or by freeze-drying the extract, and its activity measured by estimation of either the tropic acid or the hyoscyamine formed. However, the esterification reaction could not be demonstrated even with the addition of ATP and CoA (Jindra *et al.*, 1962). The enzyme had maximum activity in a phosphate buffer of pH 5.3 at 30°C and was inhibited by excess tropine or physostigmine and not activated by manganese ion (Jindra, 1966; Jindra and Cíhák, 1963; Jindra *et al.*, 1964) (Fig. 16).

Similar esterases have been obtained from *Datura metel* (Kaçzkowski, 1964, 1966) and *D. tatula* (Cosson, 1968); in the latter case young leaves, as well as roots, contained some enzyme. Enzyme preparations from *D. stramonium* tissue cultures not only had esterase activity but also could cause the esterification of tropine with tropic acid provided cofactors (ATP and CoA) were present (Jindra and Staba, 1968); but later work, reported only briefly, could not effect such esterification (Romeike, 1974). Atropinase levels in *Datura* callus tissues were found to vary little with age or other variables (Doshchinshaya *et al.*, 1973).

The accumulation of alkaloids in henbane and belladonna has been found to be accompanied by diminished levels of esterase (Bereznegovskaya *et al.*, 1971, 1972).

The interconversion of hyoscyamine (I) to scopolamine (II):

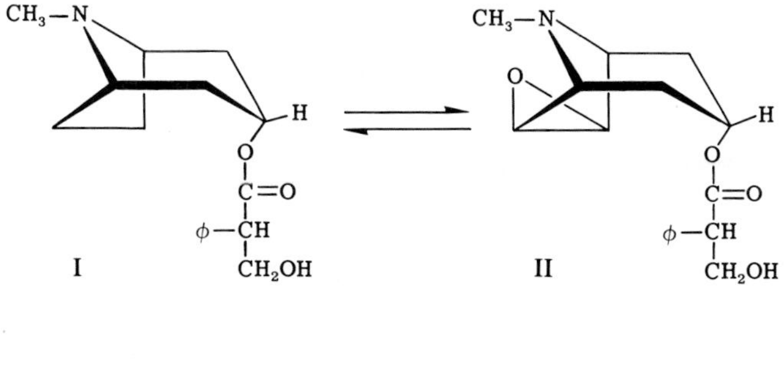

can be catalyzed by homogenates of *D. innoxia* root tissue cultures; the enzyme was heat-labile and had other expected properties (Babcock and Plotkin, 1967).

C. Amaryllidaceae Alkaloids

Very little is known about the enzymology of these compounds. An acetone powder derived from extracts of *Colchicum autumnale* had phenylalanine ammonia-lyase activity (as many other plant extracts do) and converted phenylalanine to *trans*-cinnamic acid (Schütte and Orban, 1967) (Fig. 17). A similar preparation from floral tissue of *Narcissus pseudonarcissus* was also active (Suhadolnik *et al.*, 1963). A similar extract obtained by treatment with ammonium sulfate instead of acetone gave a crude enzyme hydroxylating cinnamic acid to *p*-coumaric acid, with NADPH and tetrahydrofolate as cofactors (Suhadolnik, 1966).

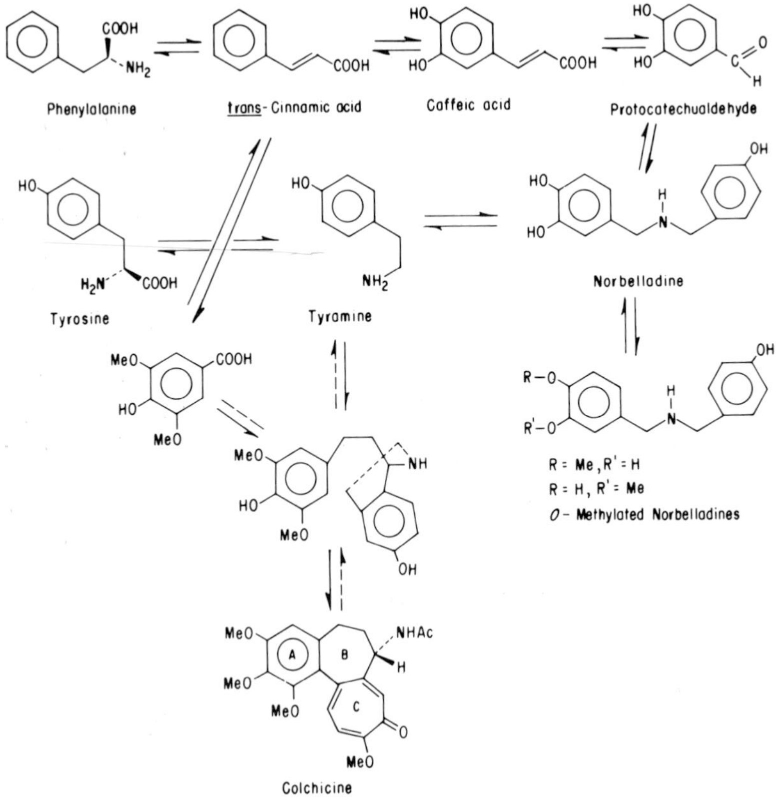

Fig. 17. Biosynthesis of Amaryllidaceae alkaloids.

Enzyme preparations have been made from *Nerine bowdenii* and *Narcissus* sp. that catalyze the conversion of norbelladine to N-isovanillyltyramine (an O-methylnorbelladine) (Fales *et al.,* 1963; Mann *et al.,* 1963; Suhadolnik, 1963); this methyltransferase is further discussed by Poulton (this volume, Chapter 22).

D. Quinolizidine (Lupine) Alkaloids

Whereas it is well-accepted from feeding experiments that lupine alkaloids are derived from lysine through cadaverine as an intermediate, no lysine decarboxylase has been found in *Lupinus luteus,* and much of the biosynthetic pathway to these alkaloids remains hypothetical. As Fig. 18 shows, lupinine is not considered an intermediate between lysine and the sparteine-lupanine alkaloids. $^{14}CO_2$ fed to *Thermopsis caroliniana* or *Lupinus angustifolius* produced labeled lupanine but not lupinine therein (Cho and Martin, 1977). *Lupinus* species do contain aminotransferases thought to be capable of causing cadaverine and α-ketoglutarate to yield 5-aminopentanal (readily cyclized to Δ^1-piperideine) and glutamic acid (Hasse and Schmid, 1963). When an insoluble crude enzyme preparation from *Lupinus polyphyllus* cell suspension cultures was used and pyruvate served as acceptor, however, no Δ^1-piperideine or other postulated intermediate could be detected, the reaction proceeding all the way to lupanine, sparteine, and related alkaloids (Wink and Hartmann, 1979). The reaction was promoted by exclusion of oxygen and inhibited by diethyldithiocarbamate, both indicating that diamine oxidase activity is not involved. Optimum pH was 7.5, and apparent K_m for cadaverine, 0.3–0.8 mM. In other work, lysine (or ϵ-N-acetyllysine) and ornithine treated with a crude enzyme preparation [$(NH_4)_2S_4$-precipitated] from *Lupinus angustifolius* were transaminated at the α-amino group to produce the expected piperideinecarboxylic or pyrrolinecarboxylic acid, respectively (Hasse *et al.,* 1967); the aminotransferase had low substrate specificity but remarkable temperature stability, retaining full activity for hours at 60°C. However, there is no evidence that Δ^1-piperideine-2-carboxylic acid is an intermediate to lupine alkaloids. Also a rather nonspecific amine oxidase was obtained by Schütte *et al.* (1966) from *L. luteus* seedlings. This oxidized putrescine to Δ^1-pyrroline via 4-aminobutanol, and cadaverine similarly to Δ^1-piperideine; whether it does so *in vivo* en route to quinolizidine alkaloids is unknown.

A condensation of lupinine with another diamino C_5 unit, presumably Δ^1-piperideine, is considered to yield sparteine, but the enzymology is unknown (Fig. 18). Cell-free extracts of *L. angustifolius,* purified by dialysis, converted sparteine to an unknown alkaloid, probably a dehydro derivative, but not to lupanine (Piechowski and Nowacki, 1958, 1959). A similar preparation from *L. albus* caused dehydrogenation of lupanine, sparteine, angus-

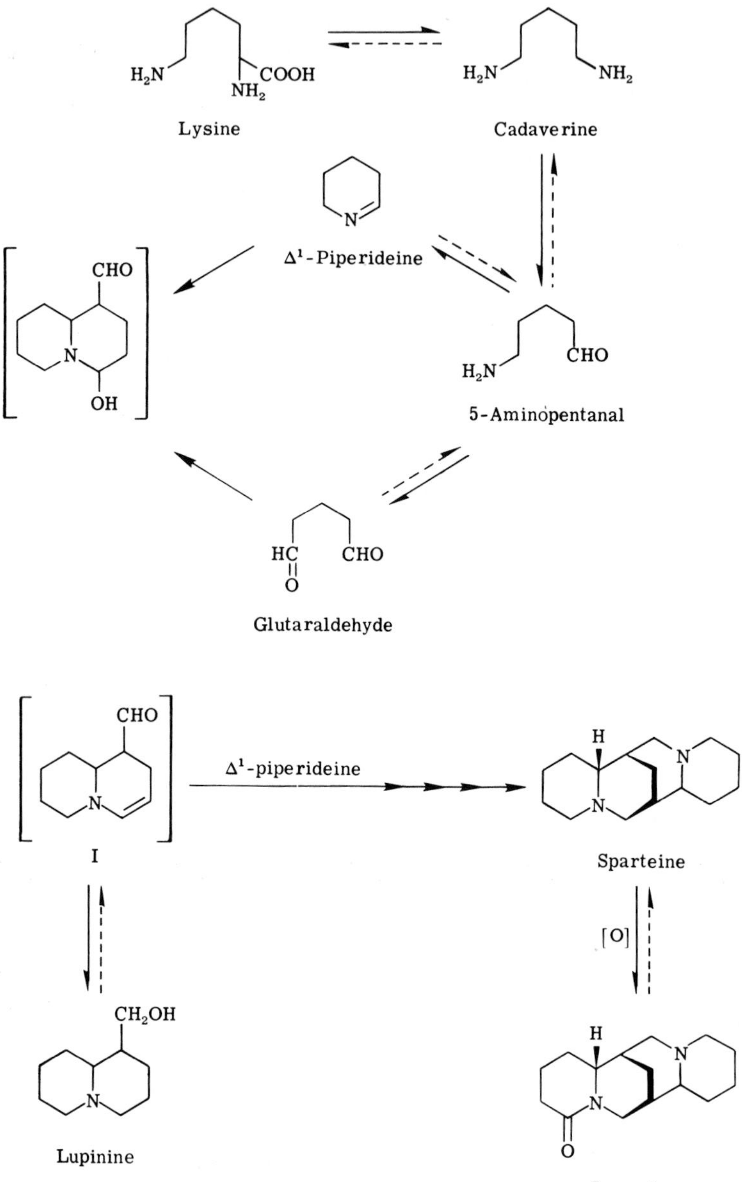

Fig. 18. Biosynthesis of sparteine, lupinine, and lupanine (in part as proposed by Golebiewski and Spenser, 1976).

tifoline, and hydroxylupanine; it could be partially purified by absorption on hydroxylapatite and elution with phosphate buffer (Nalborczyk, 1964; Nowacki, 1964). The dehydrogenation was promoted by NAD$^+$ (Nalborczyk, 1961).

The enzymatic conversion of (−)-lupinine and *trans*-4-hydroxycinnamic acid to a new alkaloid, (−)-(*trans*-4-hydroxycinnamoyl)lupinine, was studied by Murakoshi *et al.* (1977a). The reaction was catalyzed by enzymes from young seedlings of *L. luteus* but was lacking in the later stages of plant growth and in mature and immature seeds. The enzymes were prepared from the hypocotyls of seedlings of *L. luteus* by homogenization with potassium phosphate buffer (pH 7.5) containing 2-mercaptoethanol, EDTA, and sucrose. Clear supernatant solutions obtained by centrifugation at 25,000 *g* were partially purified by ammonium sulfate fractionation and Dowex-1 treatment, followed by desalting on a Sephadex G-25 column, and used as the source of enzyme. Figure 19 shows the route of biosynthesis of (−)-(*trans*-4-hydroxycinnamoyl)lupinine by such enzymes. This reaction product was not formed in reaction mixtures lacking ATP or CoA, or when the enzyme preparation had been preheated at 100°C for 15 min. Results suggest the CoA thioester as an intermediate in the esterification of (−)-(*trans*-4-hydroxycinnamoyl)lupinine, and Murakoshi *et al.* (1977a) consider that these reactions involve at least two enzymes, one a ligase catalyzing the formation of the CoA derivative of *trans*-4-hydroxycinnamic acid and the other a transferase promoting the coupling of (−)-lupinine to *trans*-4-hydroxycinnamic acid and releasing CoA. The conversion of *trans*-cinnamic

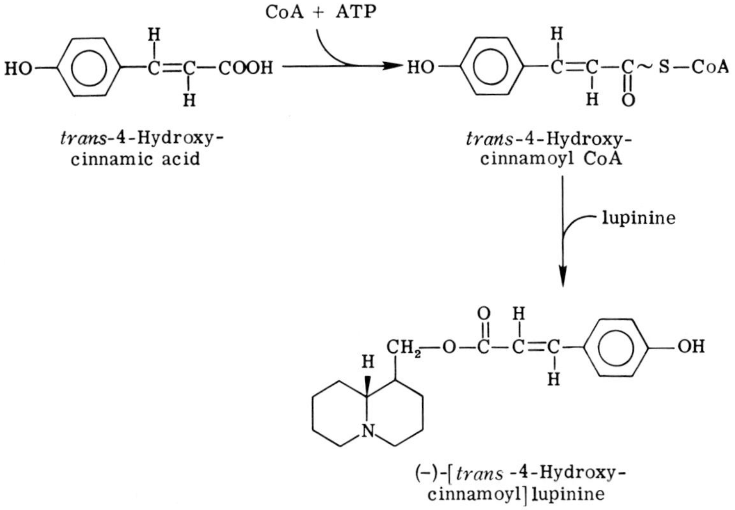

Fig. 19.   Biosynthesis of (−)-(*trans*-4-hydroxycinnamoyl)lupinine.

acid to (−)-(*trans*-cinnamoyl)lupinine by crude enzyme preparations was negligible, although the same preparation contained cinnamoyl-CoA ligase activity. Thus the intermediate role of (−)-(*trans*-cinnamoyl)lupinine has yet to be demonstrated.

Among the alkaloids occurring in *Thermopsis chinensis* (Leguminosae) are cytisine (III) and *N*-methylcytosine (IV):

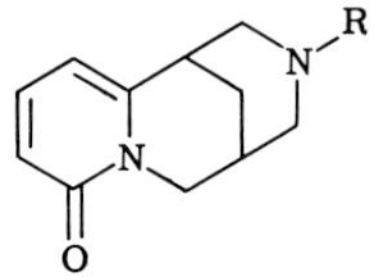

III. Cytisine, R = H
IV. *N*-Methylcytosine, R = CH₃
V. *N*-Acetylcytosine, R = COCH₃

By conventional techniques *T. chinensis* seedlings were processed to yield a crude methyltransferase catalyzing the methylation of cytisine by SAM (Murakoshi *et al.*, 1977c) (see Poulton, this volume, Chapter 22). Similar work with *Sophora flavescens* and *S. tomentosa* yielded an acetyltransferase which catalyzed the synthesis of *N*-acetylcytosine (V) from cytosine and acetyl-CoA. Seedlings were ground in phosphate buffer (pH 7.8), containing 2-mercaptoethanol. Polyclar AT was added, and the solution was clarified by straining and centrifuging. Then the supernatant was treated with ammonium sulfate (10–75% saturation) and the precipitate was redissolved and purified with Sephadex G-25. Finally, to exploit the remarkable heat stability of the enzyme, the solution was heated at 90°C for 15 min and the precipitated protein removed by centrifugation to give a supernatant in which the cytisine acetyltransferase activity of the protein was increased 12-fold. The enzyme was much inhibited by hydroxylamine but only a little by *N*-ethylmaleimide (Murakoshi *et al.*, 1978).

When fed the phytohormone zeatin, *Lupinus* seedlings metabolize it largely to lupinic acid, β-[6-(4-hydroxy-3-methylbut-*trans*-2-enylamino)purin-9-yl]alanine (Fig. 20). An enzyme preparation catalyzing this reaction (Murakoshi *et al.*, 1977b) was extracted from immature seeds of *L. luteus* (Murakoshi *et al.*, 1977a) and purified by ammonium sulfate precipitation, heat treatment, and desalting on a Sephadex G-25 column. Product formation was shown by paper chromatography and measured by the use of ninhydrin. The enzyme preparation lost about 40% of its activity in one day at 0°C; its optimum pH was sharp at 7.3–7.4. The reaction was specific for *O*-acetyl-L-serine and inhibited by hydroxylamine and cyanide and by pyridoxal phosphate at high concentrations. Some other plants were examined for the enzyme and some was detected in watermelon (*Citrullus vulgaris*) seedlings.

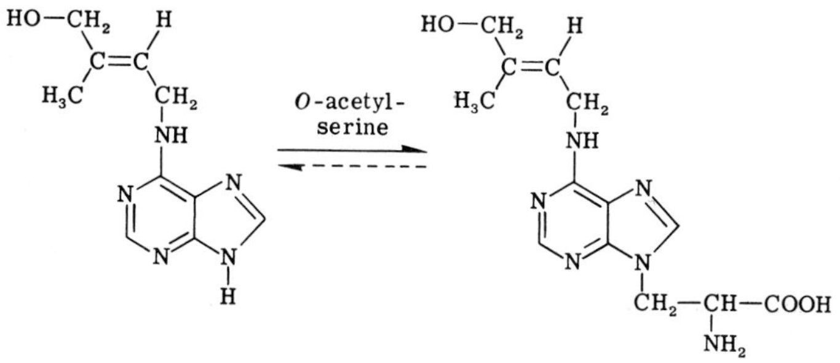

Fig. 20. Biosynthesis of lupinic acid from zeatin (Murakoshi *et al.*, 1977b).

E. Quinoline Alkaloids

It is known that acetic acid and anthranilic acid are precursors of 2-*n*-alkyl-4-quinolinols (pseudane alkaloids, VI) in *Pseudomonas aeruginosa* (Ritter and Luckner, 1971), and an enzyme preparation from this organism proved able to activate these acids, e.g., to acetyl-CoA and anthraniloyl-CoA (Geiger, 1974). However, nothing is known of the enzymology of further steps in alkaloid biosynthesis.

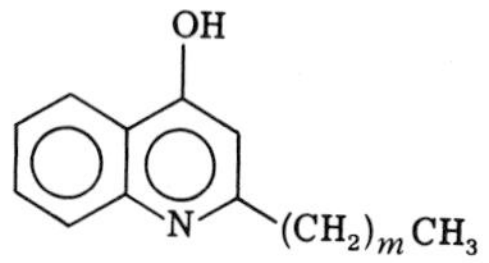

F. Isoquinoline Alkaloids

The biosynthesis of reticuline from DOPA is important because it leads to a wide range of benzylisoquinoline alkaloids (Kametani *et al.*, 1978; Sántávy, 1970, 1979; Staunton, 1979). Although a reasonable pathway for this biosynthesis is shown in Fig. 21, only two enzymes catalyzing interconversions of these compounds have been studied. One, called the berberine bridge enzyme or reticuline-converting enzyme, promotes incorporation of the *N*-methyl group of reticuline into the bridge of tetrahydroprotoberberine (scoulerine) (Barton *et al.*, 1965; Battersby *et al.*, 1965). The enzyme was isolated from cultured cells of *Macleaya microcarpa* (Papaveraceae) by Böhm and co-workers (Hofmann and Böhm, 1974; Rink and Böhm, 1975). When the cell homogenate was centrifuged, the supernatant contained the crude enzyme; it could be concentrated sevenfold by precipitation with ammonium sulfate, Sephadex G-25 chromatography, reprecipitation with ammonium sulfate, and Sephadex G-100 chromatography. It was assayed by measuring the amount of N-[$^{14}CH_3$]reticuline radioactivity appearing in scoulerine (which in *M. microcarpa* is subsequently converted to protopine and allocryptopine). The enzyme required oxygen, but none of the cofactors tested, and showed a pH optimum of 7.5–8.2. This berberine bridge enzyme, as Böhm and Rink (1975) prefer to call it, is found in other plants producing tetrahydroberberine alkaloids, such as *Chelidonium majus,* and perhaps *only* in these. It can also utilize (−)-laudanidine as substrate, but not other benzyl-tetrahydroisoquinoline alkaloids; a hydroxyl group at C-3′ appears to be essential for cyclization.

The other enzyme is one observed in cell-free extracts of *P. somniferum* that catalyzes the conversion of dopamine and 3,4-dihydroxyphenylpyruvic acid to norlaudanosoline, norlaudanosoline-1-carboxylic acid, and dehydro-norlaudanosoline, but not reticuline or later morphine alkaloids (Scott *et al.,* 1978a,b) as shown in Fig. 21. Preparation of the enzyme solution, from capsules or stems, was conventional: Plant material was homogenized in phosphate buffer (pH 6.4) containing diethyldithiocarbamate or Polyclar AT. The suspension was filtered and centrifuged at 37,000 g; the pellet contained most but not all of the activity, as assayed with [1-^{14}C]dopamine, separation by tlc, etc. The enzyme systems from capsules and stems were more active than those from seedlings or callus. Triton X-100 solubilized the pellet and quintupled its activity.

G. Morphine Alkaloids

The biosynthesis of morphine alkaloids begins with the same steps as that of isoquinoline alkaloids as far as reticuline, and nearly all the studies on enzymes effecting these processes have been concerned with the early stages and, as might be predicted, with *P. somniferum.* A metabolic grid showing hydroxylations, deaminations, and decarboxylations centered on tyrosine as an example is shown in Fig. 22. Subsequent changes involve oxidative ring closure and further transformations to the dihydrophenanthrene-type alkaloids shown in Fig. 23.

Kovács, who has been very active in this field, has obtained enzyme preparations from *P. somniferum* that may well be involved in the biosynthesis of aromatic amino acids needed as precursors of the morphine alkaloids (Benésová *et al.,* 1974; Jindra *et al.,* 1976/1977; Kovács, 1973; Kovács and Bénesová, 1976; Kovács *et al.,* 1974). A discussion of these papers, however, is beyond the scope of this chapter.

1. Aminotransferases

Roots, leaves, and seed capsules of *P. somniferum* contain amino-transferases that have been studied in tissue homogenates (Jindra *et al.,*

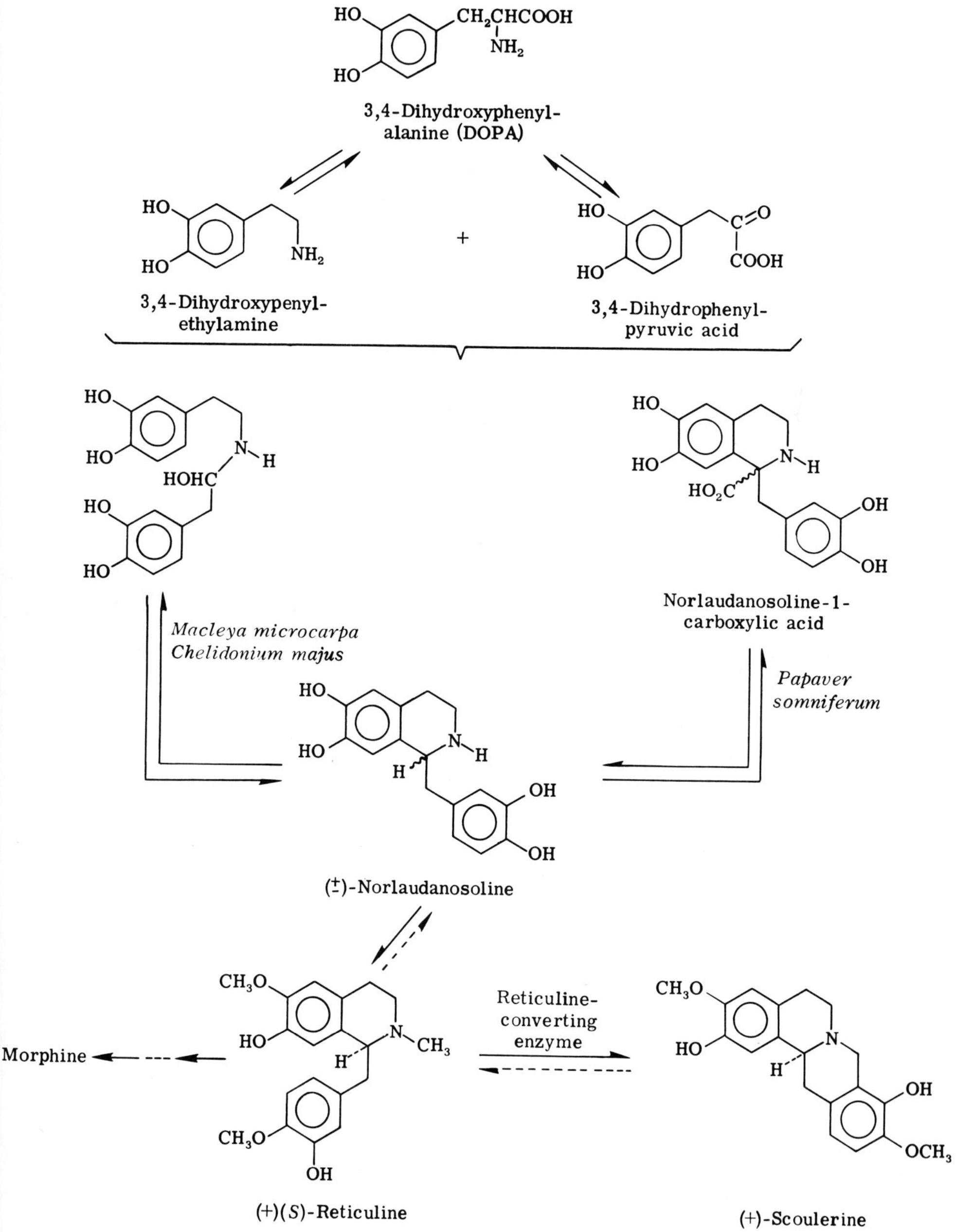

Fig. 21. Biosynthesis of reticuline from DOPA.

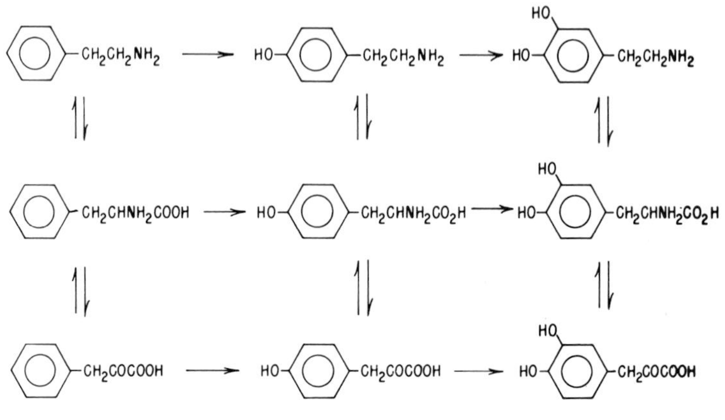

Fig. 22.   Metabolic grid of L-tyrosine enzymatic transformations.

1966b,c, 1967). α-Ketoglutaric acid was used as the amino acceptor; occurrence of the reaction was established by the identification of new keto acids,
and its extent measured at intervals by glutamic acid production. Tyrosine
aminotransferase was the most abundant, but aminotransferases for ornithine, DOPA, aspartic acid, and phenylalanine were also present (Jindra *et
al.,* 1976/1977; Kovács, 1973; Kovács and Jindra, 1970), and the levels
of the latter enzyme were roughly correlated with thebaine levels in
poppy seedlings. Partial separation of aminotransferases for phenylalanine,
aspartic acid, and ornithine could be effected on Sephadex G-100 columns;
activities were then determined by spectrophotometry of the deaminated

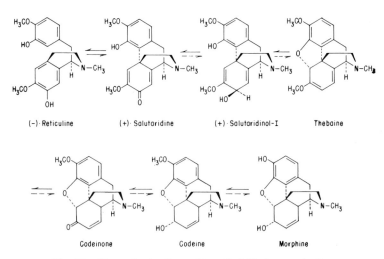

Fig. 23.   Biosynthesis of morphine alkaloids from reticuline.

acids or their derivatives (Kovács, 1970a). The greatest purification (26-fold) was achieved for phenylalanine aminotransferase. The separation permitted evaluation of temperature stability and Michaelis–Menten constants for the three enzymes (Table VII).

The phenylalanine aminotransferase had its maximum activity at pH 7.6 with α-ketoglutarate as acceptor; other ketoacids were poor or inactive as amino group acceptors. *Papaver somniferum* latex was found to contain aspartate:α-ketoglutarate and alanine:α-ketoglutarate aminotransferases, but no mention was made of those for phenylalanine and DOPA (Antoun and Roberts, 1975). Evidence for the presence of a tyrosine aminotransferase was also reported (Antoun and Roberts, 1976).

2. Decarboxylases

Homogenates of *P. somniferum* seedlings catalyzed the decarboxylation of phenylalanine, tyrosine, and DOPA, but not as well as that of glutamic acid; the reaction was followed by carbon dioxide evolved and verified by identification of the amines produced (Jindra, 1966; Jindra *et al.*, 1966b,c). Later workers, however, could not find DOPA decarboxylase (Asghar and Siddiqi, 1970). In a supernatant prepared from the latex, the only decarboxylase found in initial studies was that for glutamate (not DOPA, phenylalanine, or tyrosine) (Antoun and Roberts, 1975), but using a more sensitive assay the same authors soon reported the presence of DOPA decarboxylase also (Antoun and Roberts, 1976; Roberts and Antoun, 1978). Because of the high background radioactivity due to nonenzymatic degradation of labeled DOPA to carbon dioxide, the dopamine formed was isolated and its radioactivity measured. Ascorbate and diethyldithiocarbamate were also added to suppress oxidative degradation which also occurred. The enzyme appeared absent in the 1,000-g fraction (organelles).

DOPA decarboxylase showed a pH optimum of 7.2. Its activity was increased by pyridoxal phosphate at low concentrations but decreased by this

TABLE VII

Properties of Aminotransferases[a]

Amino acid aminotransferase	Residual activity (%) after 15 min at:			K_m (mM)
	0°C	45°C	65°C	
Aspartic acid	100	78	7	5.8
Phenylalanine	100	66	10	6.1
Ornithine	100	19	0	3.3

[a] From Kovács, 1970b.

cofactor at higher concentrations and by the substrate, L-DOPA. Among common inhibitors, p-chloromercuribenzoate and iodoacetate indicated the presence of sulfhydryl groups, and semicarbazide, hydroxylamine, and isoniazid indicated a carbonyl group, probably in pyridoxal phosphate. The enzyme had measurable but very low activity for decarboxylating L-phenylalanine, L-tyrosine, and L-histidine.

3. Oxidases

Although it has not been possible to find in *P. somniferum* a hydroxylase that catalyzes the conversion of phenylalanine to tyrosine (Jindra, 1966), the further hydroxylation of tyrosine to DOPA was catalyzed by a phenol oxidase system from this plant. The activity was contained in an acetone powder (Jindra *et al.*, 1966b; Kovács and Jindra, 1965a).

A number of investigators have studied enzyme preparations catalyzing the oxidation of monophenols or *o*-diphenols to quinones by molecular oxygen; although these are distinct activities, they are grouped together as monophenol monooxygenase, E.C. 1.14.18.1, also referred to as catechol oxidase (Fig. 24) (Mayer and Harel, 1979).

a. Preparation. Seven different enzyme preparations from seedling tissue of *P. somniferum* have been utilized in studies on the phenolase activity; several workers have used a homogenate prepared in phosphate buffer without further treatment (Jindra, 1966; Jindra *et al.*, 1966b,c; Kovács *et al.*, 1963, 1964, 1966). In later work (Jindra *et al.*, 1976/1977) the homogenate was centrifuged before use, a procedure also used by Hsu and Bills (1979). Asghar and Siddiqi (1970) converted a cell-free extract to an acetone powder which was subsequently extracted, precipitated with ammonium sulfate, redissolved and purified by dialysis and passage through carboxymethylcellulose, and finally fractionated on a DEAE-cellulose column. Purification amounted to 32-fold.

In *P. somniferum* latex the phenol-oxidizing enzymes are present only in the 1000-*g* organelle fraction and can be released by mechanical damage or the action of a surfactant (Roberts, 1971a). The latex contains two enzymes, one readily solubilized from organelle membranes and the other membrane-bound. These were purified by removing phenolics and alkaloids with Sephadex A-25 and ammonium sulfate precipitation. They were fractionated on a Sephadex A-25 DEAE column (Roberts, 1973, 1974a) with about a 20-fold concentration of the enzyme complex.

b. Assay. Although the most convenient method of measuring the effects of polyphenol oxidase is to determine the quinones formed spectrophotometrically, the oxygen uptake is probably more reliable (Mayer and Harel, 1979).

c. Substrates. The several preparations from seedlings have varying abilities to convert phenolic substrates. Some (Kovács, 1973) acted upon catechol, p-cresol, tyrosine, and DOPA; another (Asghar and Siddiqi, 1970) had neither tyrosinase nor laccase activity but catalyzed oxidation of catechol, p-cresol, DOPA, dopamine, and tyramine. The V_{max} (μl O$_2$/h) for dopamine was 1250 and 625 for DOPA; the corresponding K_m values were 0.0013 and 0.0015 mM with a pH optimum at 7–8. Still another crude preparation oxidized o-diphenols but not L-tyrosine and p-cresol (Hsu and Bills, 1974); the K_m values for dopamine and DOPA did not agree well with those reported earlier (Asghar and Siddiqi, 1970).

The latex-derived enzyme oxidized various phenolic substrates (p-cresol, catechol, p-coumaric acid, hydroquinone, tyrosine, DOPA, etc.), p-cresol and catechol being the best substrates. ($\pm$)-Reticuline and salutaridinol, phenolic intermediates in the pathway to opium alkaloids, were not oxidized; apparently both tyrosinase and laccase activities were present (Roberts, 1971a). In later work only one of the two fractions of the latex enzyme oxidized tyrosine (Roberts, 1973, 1974a).

The properties of these phenolases have not been extensively investigated. A temperature optimum of 27°C (Hsu and Bills, 1979) and pH optima of 7.0 (Hsu and Bills, 1979), 7–8 (Asghar and Siddiqi, 1970), and 8–10 (Roberts, 1971a) have been reported. The most common inhibitor recorded is diethyldithiocarbamate, but various others are known (Asghar and Siddiqi, 1970; Hsu and Bills, 1979; Jindra, 1966; Roberts, 1971a).

In the presence of a suitable substrate such as catechol the $P.$ *somniferum* phenolase complex can also oxidatively deaminate amino acids (Fig. 24). Glycine, alanine, glutamic acid, phenylalanine, tyrosine, and DOPA can all

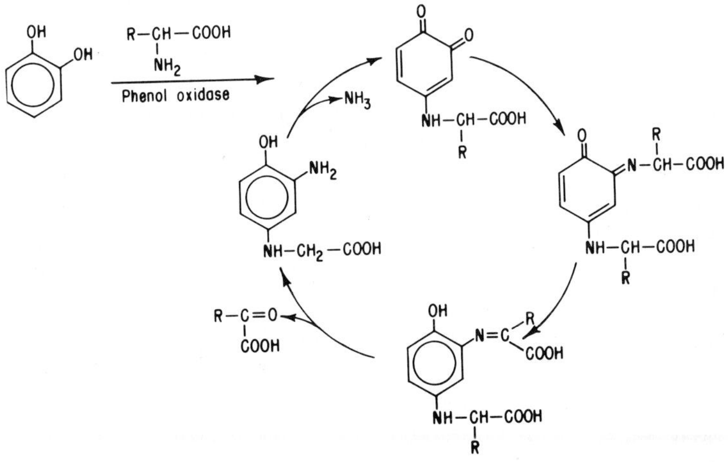

Fig. 24. Oxidative deamination of amino acids (Jindra, 1966).

be converted to keto acids by an enzyme in an acetone powder; this provides an alternative to transamination in producing certain intermediates needed in the synthesis of morphine alkaloids (Jindra, 1966; Kovács and Jindra, 1965b). This system might be significant in the synthesis of certain other secondary plant products. Jindra *et al.* (1976/1977) showed a close relationship between DOPA oxidase activity and thebaine accumulation.

d. Functions. The observation, made long ago, that phenol oxidase levels and morphine alkaloids levels are roughly correlated suggests, as just noted, that such oxidases participate in biosynthesis (Jindra, 1966; Jindra *et al.*, 1966b,c, 1976/1977; Kovács *et al.*, 1964; True and Stockberger, 1916), a finding extended to peroxidases (Farkas-Riedel, 1969). Quinones such as dopachrome are likely intermediates in the coupling reaction involved in forming morphine alkaloids, and such quinones can be isolated (as derivatives) (Asghar and Siddiqi, 1970). The reverse reaction, hydrogenation of such quinones, by enzyme preparations has also been demonstrated (Jindra *et al.,* 1966b,c; Kovács *et al.*, 1963, 1964; Roberts, 1971b). However, the results so far show merely "the possible participation of these enzymes in the synthesis of the 1-benzylisoquinoline skeleton" (Jindra, 1966).

In contrast to these early steps in morphine alkaloid biosynthesis, very little has been published on the enzymology of later steps. A cell-free extract of *P. somniferum* seedlings converted (−)-reticuline to thebaine in the presence of ATP and NADP as cofactors, but in low yield (Rapoport, 1966). More recently, homogenates of seed capsule tissue cultures were shown to cause hydrogenation of (−)-codeinone to (−)-codeine (Furuya *et al.*, 1978).

H. Indole Alkaloids

The number of known indole alkaloids exceeds 1000 (Hesse, 1964, 1968), and their complexity is impressive. Because of the importance of many members of the group as drugs, they have been extensively studied, with some attention to enzymology. Biosynthetic pathways have been established in the genera *Rauwolfia, Catharanthus,* and *Aspidosperma* of the Apocynaceae (Leete, 1969, 1977; Robinson, 1968; Spenser, 1968) and in *Claviceps* fungi (Réhácek, 1979; Thomas and Bassett, 1972). Reviews of indole alkaloids appear regularly in the Specialist Periodical Reports (Saxton, 1978).

Nearly all indole alkaloids are metabolites of tryptophan. Of the variety of types known, discussion here will be limited to those for which *in vitro* enzymatic synthesis or modification has been studied.

1. 3-(ω-Aminoalkyl)tryptophans

a. Gramine. The simplest indole alkaloid considered here, gramine, 3-(dimethylaminomethyl)tryptophan, is found in barley, *Hordeum vulgare.* A cell-free extract of young shoots of this plant with appropriate additions caused the conversion of labeled tryptophan to gramine (Breccia *et al.*, 1966). One step of this conversion can be isolated: 3-(Aminomethyl)indole and 3-(methylaminomethyl)indole are methylated to gramine in the presence of SAM and a cell-free seedling extract (Mudd, 1961).

b. *Phalaris tuberosa* (Gramineae). *Phalaris tuberosa* contains the simple alkaloids *N,N*-dimethyltryptamine and its 5-methoxy analogue. In a study of their biosynthesis, seedlings of the plant were ground in phosphate buffer (pH 7.6) containing 2-mercaptoethanol; the resulting suspension was centrifuged, protein was precipitated at 40–50% ammonium sulfate saturation, and this protein purified on DEAE-cellulose. The preparation decarboxylated and methylated L-tryptophan and its 5-hydroxy derivative to form the alkaloids mentioned but did not act on the D enantiomer. The enzyme complex required pyridoxal phosphate, was equally active between 20° and 40°C, and had maximum activity at pH 7.6. *N,N*-Dimethyltryptamine and IAA were competitive inhibitors (Baxter and Slaytor, 1972).

2. *Ergot Alkaloids*

a. History and Structure. Ergot alkaloids are produced primarily by the fungus genus *Claviceps* which grows parasitically on the pistils of many grasses (Graminae), particularly rye. The ability to synthesize these alkaloids is also possessed by other fungi such as *Aspergillus* and *Penicillium* species and by some higher plants of the Convolvulaceae family (Réháček, 1979; Snieckus, 1968). Ergot was used in childbirth and in controlling uterine hemorrhage long before it became familiar to physicians and midwives. Ergot* also has a long, notorious history of causing human suffering and death by poisoning. Fortunately, in recent times the interest in ergot alkaloids is almost wholly due to the pharmacological activity of the lysergic acid group: uterine muscle stimulation, vasodilation, anti-Parkinson's disease effect, and psychedelic effects.

Most *Claviceps* species and strains, whether living as parasites or grown in culture as saprophytes, show a growth phase, a transition period, and finally an alkaloid-producing phase (idiophase). The submerged fermentation processes for producing ergot alkaloids are well reviewed by Réháček (1974, 1979).

The ergot alkaloids are characterized by their ergoline skeletal structure (VII). They are divided into two main groups: clavine alkaloids, which have a CH_2OH or a CH_3 group at C-8, and lysergic acid alkaloids, which have a carboxyl or a derived carboxamide group at C-8. Their biosynthesis has been

---

* Ergotism is caused by bread made from rye flour contaminated with the sclerotia of *Claviceps*. The baking of bread does not destroy the ergot alkaloids, and its ingestion for several weeks results in chronic poisoning (Lewis and Elvin-Lewis, 1978).

VII

reviewed by several authors (Floss, 1976; Stadler and Stütz, 1975; Thomas and Bassett, 1972).

Like other indole alkaloids, ergot alkaloids are derived in part from tryptophan and the methyl group of methionine. However, the initial step in the biosynthesis pathway involves alkylation of tryptophan with mevalonic acid at C-4 to give 4-(γ,γ-dimethylallyl)tryptophan (DMAT) (Floss, 1976) (Fig. 25) instead of a decarboxylation. This alkylation step is believed to be rate-

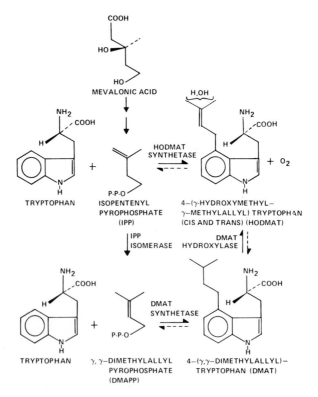

Fig. 25. Enzymatic alkylation of tryptophan.

limiting (Bu'Lock and Barr, 1968; Vining, 1970), DMAT being present in a very low steady-state concentration during alkaloid biosynthesis.

b. Clavine group. (*i*) *Complete pathway.* The first cell-free synthesis of a clavine alkaloid from tryptophan was reported in 1969 (Abe, 1972; Ohashi and Abe, 1970). Mycelia of *Claviceps* sp., *Elymus*-type strain HA-6, and *Agropyron*-type KK-2 were grown in the presence of added tryptophan, washed, and disrupted in phosphate buffer and glycerol (pH 7.2) with a French press, and the suspension centrifuged. The supernatant was incubated with DL-tryptophan, DL-mevalonolactone, methionine, 2-mercaptoethanol, and MgSO$_4$ in the phosphate buffer. When [3-^{14}C]tryptophan was used, it was incorporated extensively into setoclavine and slightly into isosetoclavine and lysergene, whereas [2-^{14}C]mevalonolactone was not incorporated at all (Fig. 26). [^{14}C]4-Dimethylallyltryptophan also appeared in setoclavine. The incorporation of tryptophan was diminished only by lysergene (of five clavine alkaloids added), which was taken to mean that lysergene is a precursor of setoclavine and other ergot alkaloids (Abe, 1972; Ohashi *et al.*, 1972a).

Another cell-free synthesis of clavine alkaloids, from tryptophan, isopentenyl pyrophosphate (IPP), and methionine, was reported by Cavender and

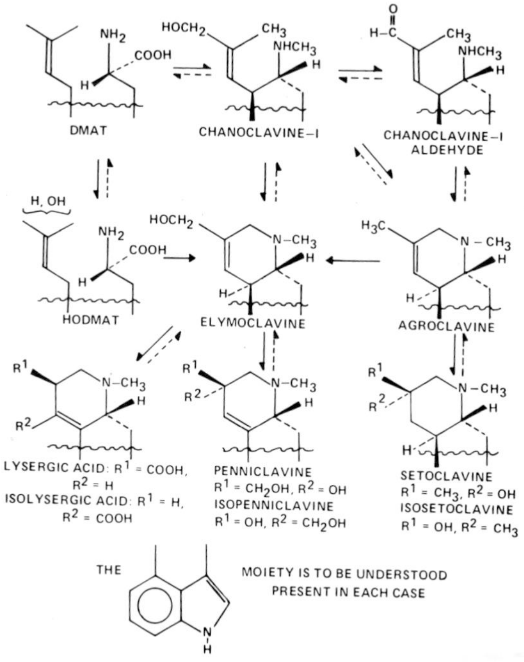

Fig. 26. Biosynthesis of clavine and lysergic acid alkaloids from DMAT via chanoclavine-I.

Anderson (1970). *Claviceps purpurea* PRL 1980 cells were grown under alkaloid-producing conditions, washed, and homogenized in phosphate buffer (pH 6.5), and the suspension centrifuged repeatedly. The 105,000 g supernatant was fractionated by ammonium sulfate precipitation, and only the 60–80% saturation fraction contained the active enzyme system. This preparation, DL-[3-^{14}C]tryptophan, and other reagents (Table VIII) were incubated at 24°C for 12 h, and the mixture of alkaloids produced was liberated, extracted, and analyzed by tlc. The conversion of labeled tryptophan to chanoclavines was only 0.15%.

(ii) *Early steps in the biosynthetic pathway.* The first intermediate, DMAT, is formed from tryptophan with the aid of dimethylallyl pyrophosphatase: L-tryptophan dimethylallyltransferase, called dimethylallyltryptophan synthetase, or DMAT synthetase (Fig. 25). The enzyme has been isolated from mycelia of *Claviceps* sp. strain S8 (probably *Claviceps fusiformis*) (Heinstein *et al.*, 1971; Lee *et al.*, 1972) and purified to apparent homogeneity (Lee *et al.*, 1976). A purification of 63-fold was achieved, with about 7% of the original activity remaining. A crude enzyme extract was made from dry mycelia by rehydration in Tris buffer (pH 8.0) containing diethyldithiocarbamate, thioglycolate, 2-mercaptoethanol, calcium chloride, and glycerol. This suspension was homogenized and centrifuged at 14,000 g, and the supernatant filtered through glass wool to remove lipids. As Table IX shows, the crude extract was further centrifuged and successfully purified by ammonium sulfate precipitation and chromatography with Sephadex G-150, DEAE-cellulose, and finally Sephadex 200. This gave an enzyme with a

TABLE VIII

Components of Incubation Mixture and Their Effect on
Chanoclavine Synthesis[a]

Incubation mixture	Chanoclavines-I and -II[b] (counts/min)
Complete[c]	3920
Less IPP	270
Less methionine	320
Less ATP	350
Less liver concentrate	200
Complete, boiled 15 min	150

[a] From Cavender and Anderson, 1970.

[b] Isolated by LC using methyl acetate–2-propanol–ammonia (45:35:20) and estimated by liquid scintillation counting.

[c] Other constituents not mentioned: enzyme preparation, 3-[^{14}C]tryptophan, $MgCl_2$, streptomycin sulfate, and potassium phosphate buffer (pH 6.5).

TABLE IX

Purification of DMAT Synthetase from *Claviceps* Strain SD 58[a]

Step	Volume (ml)	Protein (mg/ml)	Specific activity (mU/mg)[b]	Re-covery (%)	Purifi-cation (−fold)
1. Crude extract	2400	2.7	4.9	100	1
2. 10.5×10^3 g supernatant	2200	0.97	11.3	75.5	2.3
3. $(NH_4)_2SO_4$ fractionation	71.5	3.2	28.1	20.1	5.7
4. Sephadex G-150 chromatography	36.5	1.86	198.4	42.2	40.2
5. DEAE-cellulose chromatography	13.2	1.7	309.4	21.7	62.8
6. Sephadex G-200 chromatography	19.0	0.4	308.5	7.34	62.6

[a] From Lee *et al.*, 1976.

[b] One unit is defined as the amount of enzyme catalyzing the formation of 1 μmol/min of DMAT from [1-^{14}C]DMAPP under specified conditions.

specific activity of 309 mU/mg protein, which was stable for 1 week at −20°C.

In a variation of this work Maier and Gröger (1976) used mevalonate and L-tryptophan as substrates and made enzyme preparations from several strains of *Claviceps* species (including *Claviceps paspali* and *purpurea*) that produce alkaloids saprophytically. Fresh mycelia were ground either in an X-Press or with Dry Ice in the presence of a Tris buffer like that used by Lee *et al.* (1976). The supernatant from centrifuging at 15,000 g constituted the crude enzyme. This required Mg^{2+} and ATP for activity, was considerably inhibited by added tryptophan, and had a temperature optimum of 25°–30°C and a pH optimum of 7.2–7.6. The activity was concentrated by ammonium sulfate precipitation (to 30–50% saturation) but could not be purified further by Sephadex G-200 chromatography, perhaps because the mevalonate-phosphorylating part of the enzyme mixture had become separated from the DMAT synthetase fraction. For all the *Claviceps* strains studied, enzyme levels peaked at the beginning of the idiophase stage of growth and then declined, although alkaloid levels were still increasing. In general the results confirmed those of Heinstein *et al.* (1971) and Lee *et al.* (1976).

DMAT synthetase as purified by Lee *et al.* (1976) was a single protein with a MW of $7.0–7.3 \times 10^4$ and an isoelectric point at pH 5.8. The enzyme appeared to be homogeneous upon disc gel electrophoresis with and without SDS and upon isoelectric focusing. However, evidence was obtained (Lee *et al.*, 1976) that the active enzyme on storage could aggregate to an inactive trimer not readily dissociated; this is also suggested by the fact that, although disc gel electrophoresis after DEAE-cellulose chromatography removed 20% of the impurities from the enzyme, it also removed 55% of the protein.

The enzyme had a fairly broad pH optimum of 7–8. It was activated by

Fe^{2+}, Mg^{2+}, and particularly Ca^{2+}, but the function of the divalent metal ion is not known. Values of K_m for L-tryptophan and dimethylallyl pyrophosphate (DMAPP) were 0.067 and 0.2 mM, respectively, and the turnover number was about 7 sec^{-1} (Lee *et al.,* 1976; Floss *et al.,* 1979). The same data indicate but do not establish that the DMAT synthetase reaction proceeds by a sequential or ordered BiBi mechanism (Cleland, 1963a,b), with DMAPP binding to the enzyme before L-tryptophan and PP_i being released before DMAT. The specificity for L-tryptophan is great; the D enantiomer reacts only after isomerization to the L form (Heinstein *et al.,* 1971), and thiotryptophan does not serve as substrate (Krupinski *et al.,* 1976). There is a significant increase in the enzyme activity of DMAT synthetase with methyl analogues of tryptophan substituted at the 4-, 5-, 6-, and 7-positions on the indole ring. The regiospecificity of alkylation at the 4-position is guided by the stereochemistry of the enzyme surface. Substitution at a position more remote from the alkylation site is possible, but as the methyl group approaches the alkylation site, it can block the attack of DMAPP (Lee *et al.,* 1976). Further work may explain why the normally less reactive 4-position of the indole nucleus of tryptophan is alkylated in the presence of DMAT synthetase. Mevalonate and IPP gave DMAT only poorly, and this in the presence of less pure enzyme (Heinstein *et al.,* 1971; Petroski and Kelleher, 1978).

(*iii*) *Regulation of DMAT synthetase.* DMAT synthetase is inhibited by agroclavine and elymoclavine, the end products of the biosynthesis of ergot alkaloids in *Claviceps* sp. strain SD 58 (Cheng *et al.,* 1979; Floss *et al.,* 1974, 1979; Heinstein *et al.,* 1971; Heinstein and Floss, 1974). The inhibition is of a mixed type with both K_m and V_{max} being affected. It was originally suggested that ergot alkaloid synthesis was regulated by the synthesis of DMAT synthetase, so that the amounts of agroclavine and elymoclavine synthesized were controlled, but later considered that the activity of the enzyme rather than its synthesis was repressed (Floss *et al.,* 1979). For more details see Floss (this volume, Chapter 7). On the other hand, a less pure enzyme preparation from *C. purpurea* was *not* inhibited by chanoclavine-I, agroclavine, or elymoclavine (Maier and Gröger, 1976). Analogues of tryptophan increase the activity of DMAT synthetase (Floss, 1976; Krupinski *et al.,* 1976; Schmauder *et al.,* 1976). In particular, supplementation of culture media with thiotryptophan [β-(1-benzothien-3-yl)alanine], as with tryptophan itself, markedly raises both DMAT synthetase levels and alkaloid production, even in high-phosphate cultures in which alkaloid synthesis is normally impeded (Krupinski *et al.,* 1976; Robbers and Floss, 1976). The induction effect is attributed to enhanced synthesis of the enzyme, which has in fact been demonstrated with labeled amino acids and agar gel isolation of the enzyme (Hyslop *et al.,* 1979).

The enzymatic synthesis *in vitro* of precursors of clavine alkaloids, also starting with tryptophan, IPP, and methionine, was reported by Petroski (1978) and Petroski and Kelleher (1977, 1978). It produced DMAT, 4-(γ-hydroxymethyl-γ-methylallyl)tryptophan (HODMAT), and an unidentified compound, all convertible to elymoclavine and lysergic amide by feeding to cultures of *C. paspali*. The crude enzyme was prepared from mycelia of this fungus by grinding in Tris buffer containing diethyldithiocarbamate, thioglycolate, and 2-mercaptoethanol and centrifuging to give a pale-yellow supernatant. This crude enzyme solution was incubated for 1 h at 30°C with L-tryptophan, IPP, ATP, magnesium sulfate, L-methionine, and a liver concentrate in the same buffer. (The liver concentrate, boiled or unboiled, was shown not to catalyze the formation of any of the products in the absence of fungal enzyme preparation). After the reaction was stopped, the incubation mixture, less precipitated protein, was subjected to paper chromatography. In each trial three products were found: DMAT, HODMAT, and the unknown. Separate experiments showed that DMAT and HODMAT could each be converted to the unknown, and DMAT to HODMAT; but the order of efficiency of conversion to lysergic amide by feeding was DMAT > HODMAT > unknown. Thus these products can serve as precursors of clavine alkaloids, but the sequence in which they function is not clear.

The single-step hydroxylation of DMAT to HODMAT was observed by Saini and Anderson (1978) with the same enzyme preparation used for hydroxylation of agroclavine [Hsu and Anderson, 1971; see Section II,H,2,b,(V)]. The incubation mixture again contained liver concentrate, and hydroxylation was stimulated by a NADPH-generating system, indicating that a mixed-function oxygenase was operative. Conversion of labeled DMAT or HODMAT was only 0.2%, as measured by chromatography and counting. Two other unidentified products were formed, but feeding them to *C. purpurea* PRL 1980 gave only insignificant incorporation into elymoclavine.

(*iv*) *Chanoclavine-I Cyclase Complex.* All known ergot alkaloids have a closed D ring except chanoclavines, of which chanoclavine-I is a known precursor of others in the biosynthetic pathway (Fig. 26). Accordingly several research groups have obtained enzyme preparations capable of cyclizing chanoclavine-I to agroclavine and/or elymoclavine.

The properties of chanoclavine-I cyclase were studied chiefly by Erge *et al.* (1973; Erge and Maier, 1974), whose enzyme was active only for the chanoclavine-I–agroclavine conversion. It was similarly specific as to substrate; dihydrochanoclavine-I and isochanoclavine-I were unaffected, although chanoclavine-I aldehyde was converted to agroclavine well. As shown in Table X, the enzyme required ATP, a pyridine nucleotide, and Mg^{2+} for activity. Temperatures ranging from 28° to 35°C were satisfactory.

The enzyme was stable for a week at -40°C but was inactivated after 12 h at 0°C unless glycerine was added. Elymoclavine and lysergic acid caused

TABLE X

Chanoclavine-I Cyclase Preparations

Enzyme preparation	Ogunlana et al., 1970a,b	Gröger and Sajdl, 1972	Erge et al., 1973	Sajdl and Réháček, 1975
Source—*Claviceps*				
C. fusiformis	SD 58	SD 58	SD 58	
C. paspali			Li 342	MG-6
C. purpurea		Pepty 695	Pepty 695	Pla-4
?	Strain 231			
Extract preparation				
Cell breakage in phosphate buffer, pH 7.4	Freeze, grind	X-press, HOCH$_2$CH$_2$SH, EDTA	Freeze or grind	X-press, HOCH$_2$CH$_2$SH, EDTA
Centrifugation (*g*)	12,100 (after dialysis)	15,000	50,000	15,000[a]
Incubation—chanoclavine-I, enzyme solution, ATP cofactor, MgCl$_2$, NADPH, specified buffer of pH 7.4	PO$_4^{3-}$, 26–28°C, 9 h	PO$_4^{3-}$, 32°C, 4 h	Tris,[b] 32°C, 4 h	PO$_4^{3-}$, 32°C, 5 h
Analysis				
Liberation and extraction of alkaloids, tlc on SiO$_2$	3 solvent systems	3 solvent systems	1 solvent system	3 solvent systems

Identification	Quantification by ^{14}C (counting)	Cocrystallization or radiography with standard	Mass spectrometry	van Urk's reagent	Mass spectrometry
Products					
Elymoclavine[c]	+	+	+	+ and spectrometry	+
Agroclavine[c]		−	+	−	+[d]
Setoclavine				+	+
Unknown (trace)		+		−	
Other reactions					
Elymoclavine→chanoclavine-I		[e]			+[d]
Chanoclavine-I aldehyde→agroclavine				+	
Agroclavine ⇌ elymoclavine					−[f]

[a] Precipitation with $(NH_4)_2SO_4$ or use of Sephadex-25 caused great losses of activity, partly reversible by the addition of crude extracts in which enzymes had been deactivated, e.g. by heating. This suggested the need for further cofactor(s), but none could be identified.

[b] Incubation in nitrogen instead of oxygen atmosphere gave equally good production of agroclavine, hence the use of the term "chanoclavine oxidase" is inappropriate.

[c] Also produced by a cell-free extract of Claviceps not further described except that the activity was not strictly dependent on ATP (Heinstein et al., 1976).

[d] By C. paspali enzyme but not by C. purpurea enzyme.

[e] Implied by Ogunlana et al. (1970a).

[f] By C. paspali enzyme.

some slight feedback inhibition, however, experiments with tryptophan and some of its methyl derivatives indicated that chanoclavine-I cyclase was inducible (Gröger, 1976).

It must be noted, however, that there are probably several chanoclavine-I cyclases, one producing agroclavine and another elymoclavine. This accounts for some of the differences in behavior of different *Claviceps* strains and in results of different investigators.

(*v*) *Other clavine alkaloid interconversions.* At first, attempts to produce cell-free enzyme extracts from *Claviceps* species that would convert agroclavine to elymoclavine were fruitless (Agurell, 1966; Jindra and Staba, 1968; Tyler *et al.*, 1965). However, Hsu and Anderson (1970, 1971) found that the enzyme preparation that produced clavine alkaloids from tryptophan, IPP, and methionine (Cavender and Anderson, 1970) [see Section II,H,2,b,(i)] also catalyzed the aerobic hydroxylation of agroclavine to elymoclavine. Radioactively labeled agroclavine, enzyme, streptomycin sulfate, and either liver concentrate or a NADPH-generating system were incubated together at 25°C for 18 h. The reaction was followed both by spectrophotometric measurement of NADPH disappearance and by separation (tlc) and radioautographic determination of the elymoclavine produced. This accounted for about 15% of the initial radioactivity. A little setoclavine and an unknown compound were also formed.

Both an agroclavine hydroxylase and an NADPH oxidase were evidently present. Added cyanide or EDTA inhibited both enzymes, the EDTA effect suggesting that a metal ion was part of the complex; aminopterin stimulated both. Ascorbate apparently inhibited the hydroxylase but not the oxidase. The general behavior was that of a mixed-function oxidase.

The enzyme complex had a sharp pH optimum of 7.0–7.2. It caused some unknown alteration of elymoclavine as substrate but did not affect chanoclavine. It is significant that the best enzyme preparation caused consumption of 8.1 μmol NADPH/h per liter of culture medium, compared to 10 μmol/h per liter of elymoclavine produced *in vivo*.

Abe and co-workers (Abe, 1972; Ohashi *et al.*, 1970a,b, 1972a) extensively examined less purified enzyme extracts from various strains of *Claviceps* (HA-6, *Elymus* type; KK-2, *Agropyron* type; SR-134, *Secale* type; 47-A, *Pennisetum* type; and ID-2, *Paspalum* type) made by disruption of washed mycelia in buffer and clarification by centrifugation. Incubation of agroclavine or elymoclavine with such extracts and buffer caused interconversion of these two alkaloids except by type ID-2. Other conversions observed involved chanoclavine-I (by SR-134 and ID-2), festuclavine (VIII) (by KK-2), and peptide alkaloids (by HA-6) formed from agroclavine and lysergol (IX) (by all except ID-2) and peptide alkaloids (by HA-6 and SR-134) formed from lymoclavine (Abe, 1972; Ohashi and Abe, 1970). By noting the effect of added alkaloids on the production of other labeled alkaloids, it has

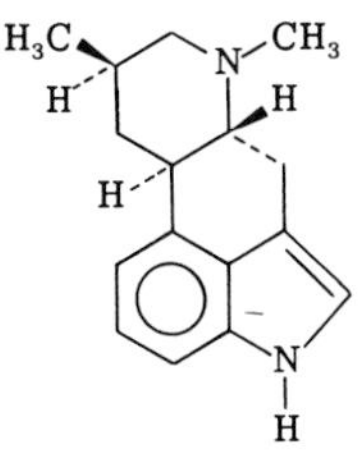

VIII 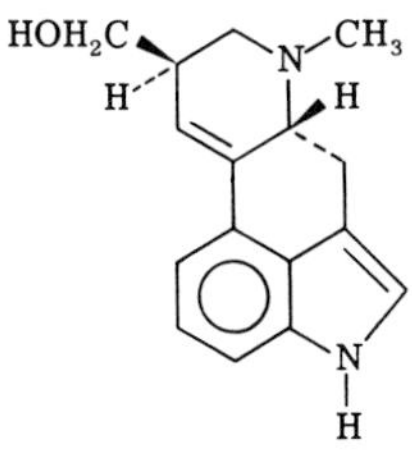   IX 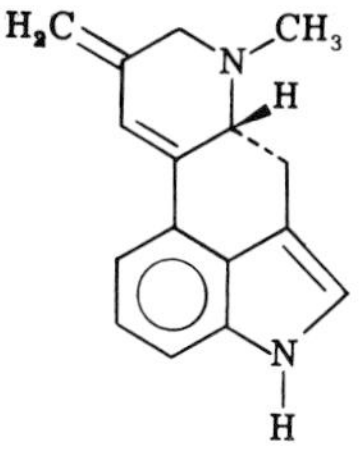   X 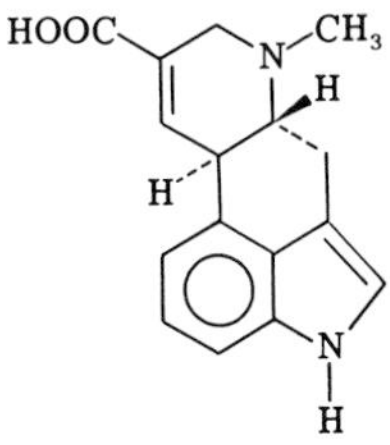   XI

been concluded that agroclavine–elymoclavine interconversions proceed via lysergol (IX) and lysergene (X) intermediates and that peptide alkaloids can be formed by a route involving lysergene (X) and not chanoclavine-I (Ohashi and Abe, 1970; Ohashi *et al.*, 1970a,b; Abe, 1972). The pathway is also considered to involve Δ^8-lysergic acid (XI).

Agroclavine was enzymatically converted to setoclavine and isosetoclavine by crude extracts from *C. fusiformis* SD 58 (Bajwa and Anderson, 1975). Mycelia from alkaloid-producing cultures were washed with phosphate buffer and homogenized, and the suspension centrifuged at 10,000 *g*. A mixture of the supernatant, agroclavine, and streptomycin sulfate was incubated at 25°C, and the products were isolated and separated by tlc. However, boiled crude extracts *also* produced 60–75% setoclavine and isosetoclavine, as well as extracts not so treated, which suggests that this species of *Claviceps* forms these alkaloids chiefly by a nonenzymatic route. This finding casts some doubt on the significance of the *in vitro* enzymatic synthesis of setoclavine and isosetoclavine reported earlier (Abe, 1972; Ohashi and Abe, 1970; Ohashi *et al.*, 1970b).

(*vi*) *Oxidations catalyzed by exogenous peroxidases.* Peroxidase (H_2O_2: donor oxidoreductase) is known to be present in ergot (Jindra *et al.*, 1968; Johansen, 1964) and is indicated in morning glory (*Ipomoea purpurea*) seedlings (Taylor *et al.*, 1966). For this reason we review research in which a peroxidase, even though not endogenous, has been used in clavine alkaloid conversions.

Filtered homogenates of tomato fruits, morning glory seedlings, or potato sprouts, all containing peroxidases, were incubated with agroclavine and/or elymoclavine and hydrogen peroxide at pH 4.6. The tomato homogenate, most studied, converted agroclavine to setoclavine, isosetoclavine, and an unstable alkaloid thought to be 10-hydroxyagroclavine, which was changed by a stronger acid to setoclavine and isosetoclavine (Taylor *et al.*, 1966). Elymoclavine similarly gave penniclavine, isopenniclavine, and another acid-unstable alkaloid. The enzymatic nature of the reaction was shown by its inhibition by catalase or cyanide ion and by boiling the homogenate before use. The morning glory and potato extracts behaved like the tomato extract (Taylor *et al.*, 1966). The same three products from agroclavine were generated when commercial horseradish peroxidase was used (Shough *et al.,* 1967; Taylor and Shough, 1967). This peroxidase also converted elymoclavine to penniclavine, isopenniclavine, 10-hydroxyelymoclavine, and some unidentified alkaloids (Lin *et al.,* 1967a; Pong, 1968). The identity of 10-hydroxyagroclavine, its rearrangement to setoclavine and isosetoclavine, and the formation of three new alkaloids (6,8-dimethyl-8,9-epoxy-10-hydroxyergoline, an isomer thereof, and norsetoclavine) were demonstrated in similar work (Lin *et al.,* 1967b; Shough, 1968; Shough *et al.,* 1967; Shough and Taylor, 1969). In more recent work, the oxidation of chanoclavine-I by horseradish peroxidase gave only two alkaloids (both unstable), one tentatively identified as 10-hydroxychanoclavine-I. Isochanoclavine-I, chanoclavine-I aldehyde, and chanoclavine-I acid were also tested but gave no identifiable products (Choong and Shough, 1979; Shough and Choong, 1978).

c. Peptide Ergot Alkaloids. The amide–peptide group of ergot alkaloids are amides of lysergic or isolysergic acid and include some dipeptide and tripeptide structures which may be cyclic (XII). The peptide amino acids always include proline; the others present are phenylalanine, leucine, or valine, but one residue is always an unusual δ-hydroxy-δ-amino acid, as shown.

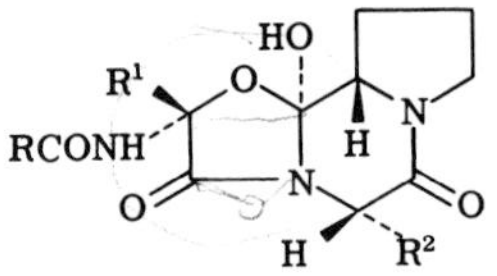

Cyclol ergot alkaloid (XII)
RCO = lysergyl or isolysergyl
e.g., in ergotamine,
$R^1 = CH_3$, $R^2 = C_6H_5CH_2$

There has naturally been interest in the enzymatic hydrolysis of such structures in connection with ergot alkaloid metabolism. The enzyme preparations used were crude homogenates of mycelia of *C. purpurea,* which were centrifuged to remove cell debris. Amici *et al.* (1966) obtained an extract that hydrolyzed amides of lysergic, isolysergic, and dihydrolysergic acid-I, but not lysergyl-NHCHOHCH$_3$ [lysergic acid methylcarbinolamide, *N*-(1-hydroxyethyl)lysergamide] or the cyclol structure. This resistance of the cyclol structure has been confirmed (Gröger *et al.,* 1976). Japanese workers have observed the enzymatic formation of ergocryptine and ergocryptinine from both elymoclavine plus labeled LLeu-DPro lactam (Abe, 1972) and lysergyl-LVal-OCH$_3$ (Ohashi *et al.,* 1972b), and of ergotamine and ergotaminine from labeled LPhe-DPro lactam (Abe, 1972). However, the evidence that peptides as such participate in these peptide alkaloid biosyntheses is not persuasive; Maier *et al.* (1974) found that both lysergyl-Ala-OCH$_3$ and lysergyl-Val-OCH$_3$ were hydrolyzed at both ester and amide linkages by a similar enzyme preparation and concluded that lysergyl amino acids were built into peptide alkaloids only after hydrolysis. A double-labeling experiment would be useful to clarify this point.

d. Clavicipitic Acid. Clavicipitic acid has been identified as a major product of cultures of some *Claviceps* species. Its production from DMAT, which involves oxidative removal of two hydrogen atoms and cyclization (Fig. 27), was demonstrated *in vitro* by an enzyme complex extracted from *C. purpurea* (Bajwa and Anderson, 1975; Saini *et al.*, 1976).

Crude enzyme preparations were obtained from two *Claviceps* strains by conventional means involving grinding in phosphate buffer, centrifugation, protamine sulfate purification, fractional precipitation with ammonium sulfate, and dialysis. More activity was found in the microsomal fraction than in the supernatant and the mitochondrial fractions (Bajwa *et al.*, 1975). However, solubilization with Triton X-100 gave maximum specific activity in the supernatant, which was further purified by DEAE-Sephadex, CM-Sephadex, and again DEAE-Sephadex column chromatography with a fivefold concentration of activity (Saini *et al.*, 1976). The clavicipitic acid produced by incubation with DMAT was isolated by tlc and estimated by measuring radioactivity.

The enzyme required oxygen for functioning but was not inhibited by carbon monoxide; no cofactor or other cosubstrate was needed. The addition of catalase or hydrogen peroxide had little effect. The enzyme was highly specific for DMAT, the L form being preferred. It was significantly inhibited by sulfhydryl reagents (*N*-ethylmaleimide, *p*-hydroxymercuribenzoate), phenanthroline (but diethyldithiocarbamate stimulated), isonicotinic hydrazide, and hydroxylamine, and somewhat by pyridoxal-5'-phosphate and by clavicipitic acid itself. No conversion of this acid to clavine alkaloids could be observed; thus these alkaloids are formed by alternative pathways utilizing DMAT. Moreover, since agroclavine did not inhibit DMAT oxidase appreciably, there is apparently no feedback control of clavicipitic acid biosyn-

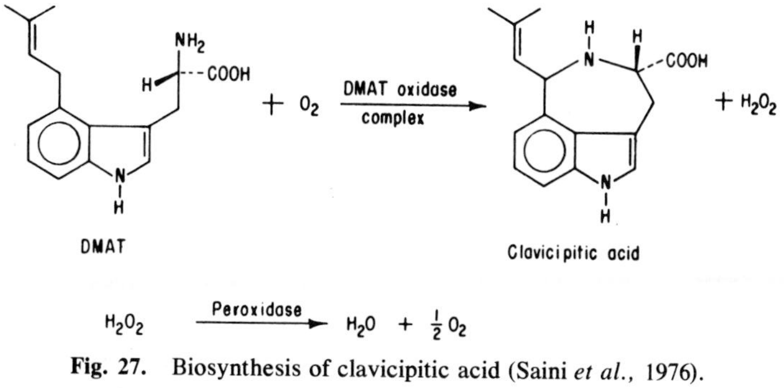

Fig. 27. Biosynthesis of clavicipitic acid (Saini *et al.*, 1976).

thesis. The pH optimum for the enzyme was remarkably high, 10.5, and the K_m value for DMAT was 0.35 mM.

3. Other Fungal Alkaloids

A mold, *Penicillium cyclopium*, contains alkaloids of the 4-isopentenyl-indole type, including β-cyclopiazonic acid (XIII). A cell-free extract of the mycelium has been shown to contain a dimethylallyl transferase that catalyzes formation of this acid from DMAPP and cycloacetoacetyltryptophanyl. This transferase does not alkylate tryptophan itself, and DMAT does not serve as a precursor of β-cyclopiazonic acid (McGrath *et al.*, 1976).

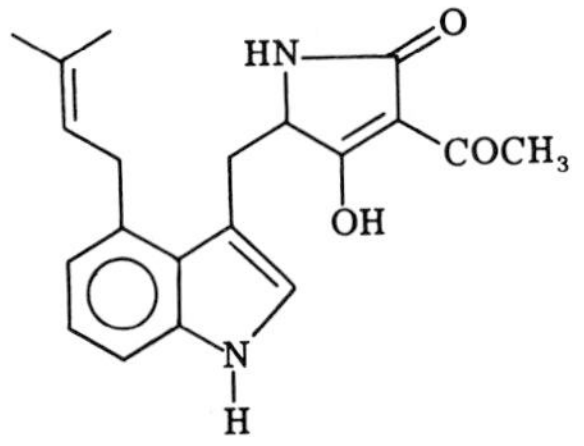

XIII

Indolmycin, an antibiotic produced by *Streptomyces griseus*, is also an indole alkaloid, though of an unusual type. Its biosynthesis from tryptophan has been shown to proceed as shown in Fig. 28. A cell-free extract of *S. griseus* was shown to catalyze the conversion of tryptophan to indolmycin

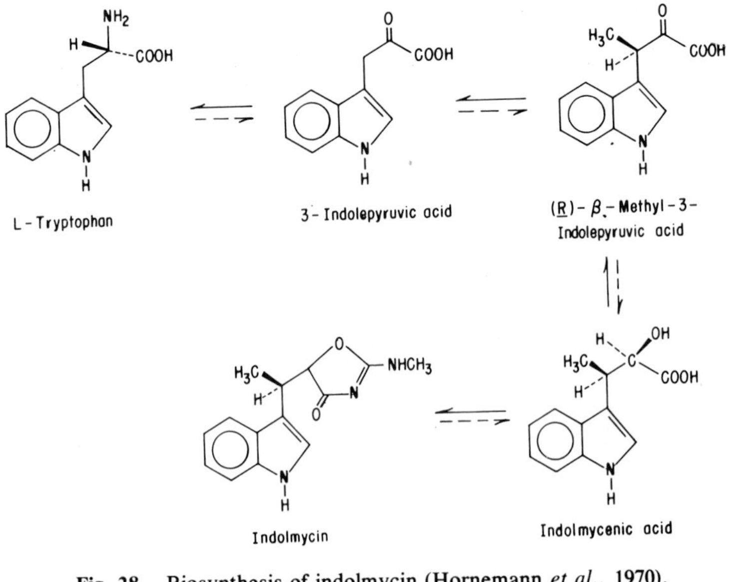

Fig. 28.   Biosynthesis of indolmycin (Hornemann *et al.*, 1970).

(Hornemann *et al.*, 1970, 1971). In a continuation and extension of this work, an enzyme preparation was obtained by cell disruption in a French press in a buffer of pH 7.0, centrifugation, and dialysis. It transmethylated 3-indolepyruvic acid to the (S)-β-methyl derivative but did not alkylate tryptophan. This transmethylase was concentrated by ammonium sulfate precipitation and purified by chromatography with Sephadex G-150, DEAE-Sephadex, and Bio-Gel A-5m (Speedie *et al.*, 1975). It is of special interest because it causes C-methylation (cf. Poulton, this volume, Chapter 22).

The cell-free extract also contained a transaminase converting tryptophan to 3-indolepyruvic acid (Hornemann *et al.*, 1971). This enzyme was partially purified (2.7-fold) by precipitation with ammonium sulfate between 45 and 60% saturation. It also transaminated L-phenylalanine, L-tyrosine, and 3-methyltryptophan, but not D-tryptophan, all provided pyridoxal phosphate was present as an amino group acceptor. The conversion of tryptophan was impeded by the addition of competing 3-methyltryptophan, but the enzyme was not inhibited by indolmycin or *p*-chloromercuribenzoate (Speedie *et al.*, 1975).

Echinulin (XIV) occurs in *Aspergillus amstelodami;* as implied by its structure, it can be formed *in vivo* from cyclo-L-alanyl-L-tryptophanyl and

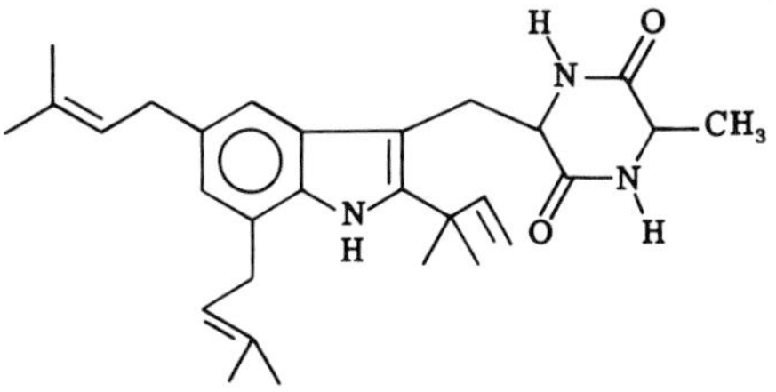

XIV

DMAPP. In an *in vitro* study Allen (1972) disintegrated washed mold mycelia by sonication in cold Tris buffer (pH 7.0) containing magnesium chloride and 2-mercaptoethanol and clarified the suspension by centrifugation. Treatment of the supernatant with ammonium sulfate precipitated enzymatically active protein, most effectively at 40–70% saturation. Incubation of the crude transferase with cyclo-L-alanyl-L-tryptophanyl and DMAPP gave a product that was not echinulin but was a good precursor thereof (Allen, 1973). Its structure was believed to be that of echinulin lacking the two 3,3-dimethylallyl groups on the benzene ring.

4. *Catharanthus Alkaloids*

Catharanthus roseus (*Vinca rosea*) or periwinkle produces more than 100 indole alkaloids (Scott, 1970), some pharmacologically active, e.g., the antitumor drugs vinblastine and vincristine (Taylor and Farnsworth, 1975). Figure 29 shows the biosynthetic pathway for these alkaloids as far as stric-

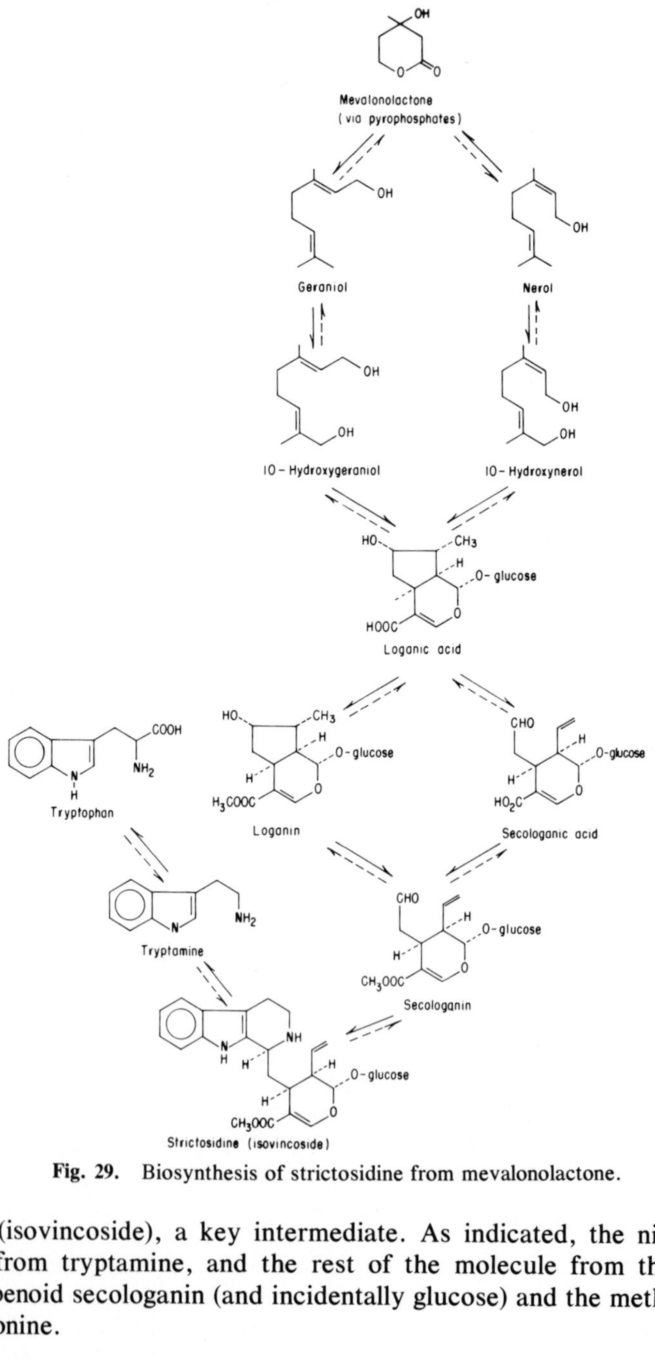

Fig. 29.   Biosynthesis of strictosidine from mevalonolactone.

tosidine (isovincoside), a key intermediate. As indicated, the nitrogen is derived from tryptamine, and the rest of the molecule from the iridoid monoterpenoid secologanin (and incidentally glucose) and the methyl group of methionine.

a. Geraniol-Nerol Hydroxylase. The enzyme that hydroxylates geraniol and nerol to the 10-hydroxy derivatives has been studied (Madyastha *et al.*, 1976; Madyastha and Coscia, 1979a) (Table XI). It is an oxygenase of the cytochrome P-450 group, as shown in part by light-reversible carbon monoxide inhibition. Maximum activity is derivable from 5-day-old *C. roseus* seedlings, but the enzyme also occurs in older plants, in *C. roseus* tissue cultures, and in *Vinca minor* and *Lonicera morrowi* (Table XI).

Catharanthus roseus seedlings were ground in Tris buffer (pH 7.6) and mixed with Polyclar AT; the strained suspension was centrifuged at 3000 *g* and the supernatant at 20,000 *g*. The pellet from the latter step was ultrasonically disintegrated in sodium cholate solution and the suspension centrifuged at 100,000 *g* to give a supernatant containing 65–70% of the activity. The reductase component was concentrated and purified by successive use of DEAE-cellulose, ultrafiltration, calcium phosphate gel, DEAE-Sephadex A, and Sephadex G-200. Alternatively, after the DEAE-cellulose step, affinity chromatography was used: a 2′,5′-ADP-Sepharose 4B column with Renex 690 detergent and elution with 2′-AMP. This was followed by DEAE-cellulose and ultrafiltration to remove residual 2′-AMP. The enzyme prepara-

TABLE XI

Geraniol-Nerol Hydroxylase and NADPH-Cytochrome c Reductase Activity of Various Plant Tissues[a,b]

Source	Hydroxylase activity (nmol/min/mg protein)	Reductase activity (nmol/min/mg protein)
Catharanthus roseus (5-day-old seedlings)	0.50	18.4
Catharanthus roseus (mature plants)	0.16	23.0
Catharanthus roseus (callus tissues from seedlings)	0.06	14.5
Catharanthus roseus (suspension cultures from callus)[c]	0.06	25.0
Vinca minor (mature plants)	0.03	
Lonicera morrowi (mature plants)	0.02	
Pisum sativum (seedlings)	0	
Persea americana (pericarp)	0.05	
Tulipa gesnerana (leaf part of bulbs)	0.01	
Solanum tuberosum (tuber)	0.006	
Strobilanthes dyeriana (callus)	0.005	

[a] From Madyastha and Coscia, 1979a; C. J. Coscia, personal communication, 1980.

[b] A 2 × 10⁴ *g* pellet was prepared from tissue extracted with 0.1 *M* Tris · HCl, pH 7.6, containing sucrose, dithiothreitol, KCl, $Na_2S_2O_5$, $MgCl_2$, and EDTA. Values represent the highest specific activities detected.

[c] Cultures were started from intact seedlings, excised roots, or cotyledons.

tions so obtained were stable for months at $-70°C$. The gel filtration procedure gave 120-fold purification, but the monoterpene hydroxylase activity could not be reconstituted well with partially purified cytochrome P-450, whereas the affinity chromatography produced 745-fold enrichment and a product capable of reconstitution (McFarlane et al., 1975; Madyastha and Coscia, 1979b).

Disc gel electrophoresis of the best reductase fractions showed two polypeptide bands with MWs of 63,000 and 78,000. The 63,000 material may be derived from the larger (78,000) by proteolytic degradation, and indeed the affinity-chromatographed preparation was richer both in the larger polypeptide and in activity. It also contained more flavins, but unless the enzyme was purified in the presence of FMN and FAD, its activity was stimulated by exogenous FMN, suggesting that it was a flavoprotein. Such stimulation was particularly observed with reductase purified by thin-layer isoelectric focusing on Sephadex G-75 carrier gels, which gave a product of pI 5.3. The visible spectrum confirmed the flavoprotein character of the reductase (Madyastha and Coscia, 1979b).

This reductase transferred electrons to ferricyanide and 2,6-dichlorophenolindophenol, as well as cytochrome c, and showed menadione-mediated NADPH oxidase activity as expected. Its K_m for NADPH was 5.7 μM and for cytochrome c, 7.8 μM. It was not inhibited by antimycin A, dicoumarol, superoxide dismutase, or catharanthine, but $NADP^+$, p-chloromercuribenzoate and cetyltrimethylammonium bromide were inhibitors.

The C. roseus oxygenase was highly specific for hydroxylation of the (E)-C-10 methyl group of geraniol and nerol (Meehan and Coscia, 1973) and did not hydroxylate the pyrophosphates (Licht, 1979; Licht et al., 1980).

Various sucrose gradient density fractionations concentrated the hydroxylase in a band at about 1.09–1.10 g/ml. Electron micrographs of particles and of fixed tissue suggest that the enzyme is localized in vacuoles or provacuoles (Madyastha et al., 1977; Meehan and Coscia, 1973).

Whereas catharanthine does not affect the cytochrome c reductase, it (but no other alkaloid tested) is a reversible linear inhibitor of the monoterpene hydroxylase, being noncompetitive with geraniol and NADPH (McFarlane et al., 1975); this may constitute feedback inhibition.

b. Geraniol-Nerol Oxidoreductase. An enzyme in the 100,000-g supernatant of C. roseus extracts catalyzes the conversion of geraniol to geranial, and of nerol to neral, in the presence of NAD^+ or $NADP^+$, and the reverse reaction when NADH or NADPH is used (Guarnaccia, 1979; Meehan, 1979). This oxidation may be involved in the biosynthesis, perhaps as an intermediate between 10-hydroxygeraniol (or -nerol) and loganic acid (Madyastha and Coscia, 1979a), but the matter has not been clarified.

c. S-Adenosyl-L-methionine: Loganic Acid Methyltransferase. This enzyme, that converts loganic acid and secologanic acid to loganin or secologanin (the methyl esters), has been partially purified (Baxter et al., 1972; Coscia et al., 1971; Madyastha et al., 1971, 1972, 1973). Details are given by Poulton (this volume, Chapter 22).

d. Tryptophan Decarboxylase. This decarboxylase catalyzing the formation of tryptamine has been extracted from C. roseus and partially purified by ammonium sulfate precipitation and gel filtration (Scott and Lee, 1975).

e. Ajmalicine Synthetase Complex. The enzyme complex converting secologanin and tryptamine to ajmalicine and related alkaloids (see Fig. 30 for the biosynthetic pathway) was first studied by Scott and Lee (1975). Callus tissue of C. roseus homogenized in Tris buffer (pH 7.0) containing 2-mercaptoethanol and Polyclar AT was centrifuged at 37,000 g and the supernatant incubated with labeled precursors and cofactors to demonstrate synthesis of ajmalicine and geissoschizine; a similar preparation from seedlings was much less active (Scott and Lee, 1975). These results were confirmed with an enzyme system from cell suspension cultures, the cells being broken up in an X-Press.

A similar extract purified merely by treating with dextran-treated charcoal was shown to produce ajmalicine, 19-epiajmalicine, and tetrahydroalstonine (Stöckigt et al., 1976), alkaloids also obtained by Scott et al. (1977a, 1978a). In the absence of reduced pyridine nucleotides, a precursor of these alkaloids accumulated; it was identified as $3\alpha(S)$-20,21-didehydroajmalicine and named cathenamine (Stöckigt, 1979; Stöckigt et al., 1977). A radioimmunoassay based on antibodies against ajmalicine proved very useful in measuring the extent of reactions produced by this enzyme system and gave the following results: the pH optimum was 6.5; but broad, substituted tryptamines were acceptable as substrates; inhibitors were transition metals and δ-D-gluconolactone. The system appears to consist (at least) of strictosidine synthetase, β-glucosidase, and a pyridine nucleotide-dependent reductase (Treimer and Zenk, 1978).

The ajmalicine-synthesizing complex was used to show that only $3\alpha(S)$-strictosidine (isovincoside), and not its $3\beta(R)$ epimer vincoside, could serve as a precursor of the Catharanthus alkaloids; the previous assumption of inversion at C-3 thus proved unnecessary (Scott et al., 1977a; Stöckigt and Zenk, 1977a,b). The system has also been used in two laboratories to evaluate the intermediacy of geissochizine in the biosynthesis of ajmalicine-type alkaloids, but with contradictory results; geissoschizine was indeed found to be convertible to the other alkaloids, but so poorly that it was considered only indirectly linked to the main biosynthetic pathway (Stöckigt, 1978). Conversely, geissochizine was found to be directly convertible to ajmalicine and thus involved, perhaps instead of cathenamine (Scott *et al.*, 1978a; Lee *et al.*, 1979). Most recently, treatment of geissochizine with a cell-free extract of *C. roseus* containing added NADPH was shown to produce (16*R*)-isositsirikine, and sodium borohydride reduction of geissochizine gave this compound but mostly the 16*S* epimer; the absolute configurations were established by correlation with sitsirikine (Hirata *et al.*, 1980).

The ajmalicine synthetase produced vallesiachotamine (XV) besides the alkaloids already mentioned as products. Among these, 19-epiajmalicine poses a problem because it has not been found in *C. roseus* (Scott *et al.*, 1978a).

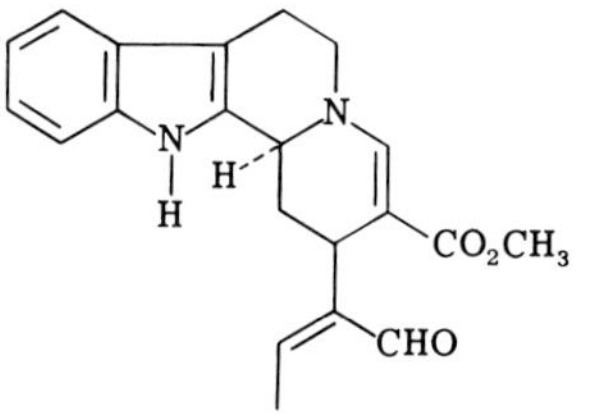

XV

In the presence of potassium borohydride used to trap unstable intermediates, a *C. roseus* enzyme preparation converted strictosidine to sitsirikine and 16-episitsirikine, presumably via 4,21-dehydrocorynantheine aldehyde (Stöckigt *et al.*, 1978). Similarly, Hirata *et al.* (1980) obtained these products by a similar reduction of corynantheine.

A good review of ajmalicine biosynthesis (Zenk, 1980) appeared too late to permit incorporating its contributions here.

f. Strictosidine (Isovincoside) Synthetase. A synthetase has been obtained from a cell-free preparation of *C. roseus* by precipitation with ammonium sulfate, redissolving in buffer (pH 7.6), and treatment with dextran-coated charcoal (Stöckigt, 1979). Various techniques were investigated to maximize strictosidine production with this preparation. The alkaloid so produced was derivatized and identified as strictosidine. However, a much more elaborate procedure was required to obtain homogeneous strictosidine synthetase (Mizukami *et al.*, 1979). Cultured *C. roseus* cells were homogenized in Tris buffer (pH 7.5), the suspension centrifuged (25,000 *g*), and the supernatant treated with ammonium sulfate to 70% saturation. The precipitate was dissolved and dialyzed, the dialysate passed through DEAE-cellulose, and the enzyme eluted with potassium chloride. Next hydroxylapatite chromatography was applied with potassium phosphate elution, and then gel filtration through Sephadex G-75. Isoelectric focusing completed the purification, which produced 740-fold enrichment and a 10% yield (Table XII). The stric-

TABLE XII

Summary of Purification of Strictosidine Synthetase[a,b]

Step	Total protein (mg)	Enzyme activity (nkat/mg protein)	Total (nkat)	Yield (%)
1. Crude extract	654	0.008	5.2	100
2. (NH$_4$)$_2$SO$_4$ precipitation	466	0.011	5.1	98
3. DEAE-cellulose	38.7	0.10	3.9	75
4. Hydroxylapatite	5.9	0.64	3.8	73
5. Sephadex G-75	0.72	2.8	2.0	38
6. Isoelectric focusing	0.08	5.9	0.5	10

[a] From Mizukami et al., 1979.
[b] 116 g of wet cells was used.

tosidine synthetase was shown to be homogeneous by both polyacrylamide gel electrophoresis and renewed isoelectric focusing.

The pure enzyme was less stable, losing half its activity in a week at 4°C. It had a broad pH optimum between 5.0 and 7.5, and K_m values of 0.46 mM for secologanin and 0.83 mM for tryptamine. The MW by gel filtration was 38,000. Its activity was not reduced by sulfhydryl-binding reagents, nor by some alkaloids derived from strictosidine; hence no feedback regulation was indicated (Mizukami et al., 1979).

g. β-D-Glucosidase. Gel filtration of the 37,000 g supernatant from seedlings or mature C. roseus plants showed the presence of four β-D-glucosidase isozymes when assayed with p-nitrophenyl β-glucoside (Scott et al., 1977b). Two such hydrolases were similarly obtained from callus cultures. One of these from whole plants and one from callus, both of 50,000 MW, could synthesize ajmalicine, but this ability was soon lost. Thus these preparations must be regarded as enzyme complexes, not merely β-glucosidases. Inhibition of this deglucosylation by the addition of δ-D-gluconolactone, a β-glucosidase inhibitor, confirmed the participation of a β-glucosidase in the sequence (Treimer and Zenk, 1978); but no isolation of such an enzyme from C. roseus has been reported.

h. Enzymes Producing Dimeric Catharanthus Alkaloids. Supernatants from homogenates of C. roseus plants catalyzed the synthesis of vindoline (Stuart et al., 1979) and the conversion of catharanthine and vindoline to 3′,4′-dehydrovinblastine (anhydrovinblastine) and leurosine, respectively (Stuart et al., 1978c) (Fig. 30). Likewise, 3′,4′-dehydrovinblastine was itself transformed into vinblastine (Baxter et al., 1979; Stuart et al., 1978b) and leurosine and catharanthine (Stuart et al., 1978a,b) by cell-free extracts,

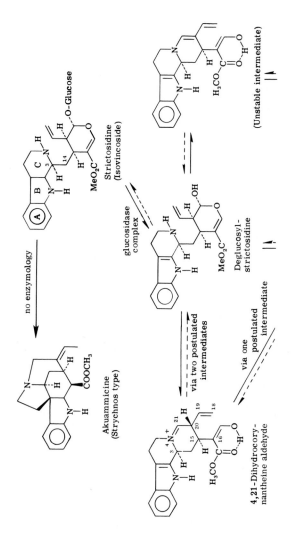

no enzymology

Strictosidine
(Isovincoside)

Akuammicine
(Strychnos type)

glucosidase
complex

Deglucosyl-
strictosidine

(Unstable intermediate)

via two postulated
intermediates

via one
postulated
intermediate

4,21-Dihydrocory-
nantheine aldehyde

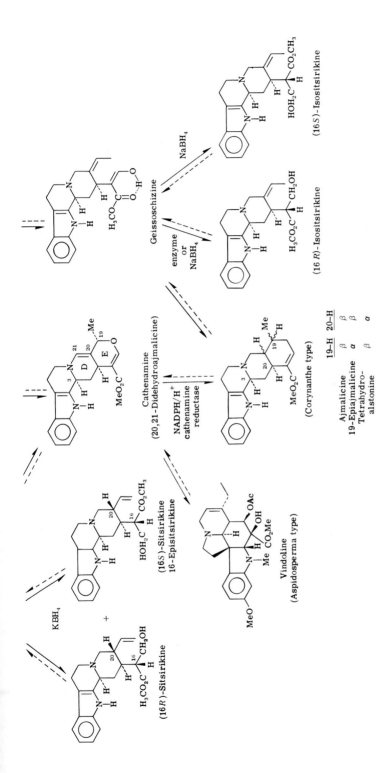

Fig. 30. Caption appears on page 380.

(*Continued*)

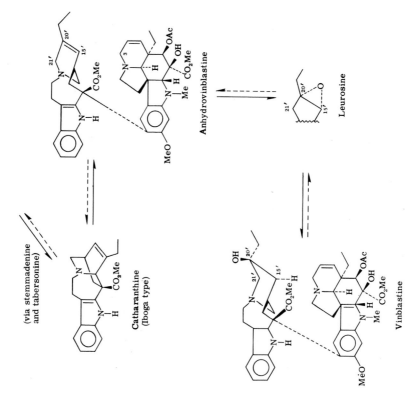

Fig. 30. Biosynthesis of other *Catharanthus* alkaloids from strictosidine.

although intact seedlings of *C. roseus* did not produce vinblastine (Langlois and Potier, 1978). It is considered very likely that 3′,4′-dehydrovinblastine is a key intermediate of vinblastine-type alkaloids.

I. Steroidal Alkaloids

Since the biochemical pathway by which steroid alkaloids are formed has not been established, our discussion is limited to glycosylation and deglycosylation reactions involving the parent alkaloids solanidine and solasodine.

1. Biosynthesis

At present, seven alkaloids of the solanine group are known to occur in the tubers and sprouts of potato, *Solanum tuberosum: α-, β-,* and *γ-*solanine and *α-, β₁-, β₂-,* and *γ-*chaconine. They are derived from the parent structure solanidine (XVI) by various kinds and degrees of glycosylation (Prelog and Joyce, 1960; Tomko and Votický, 1973).

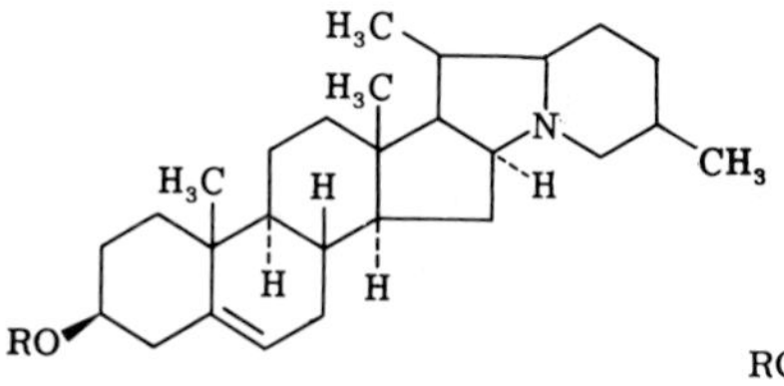

XVI 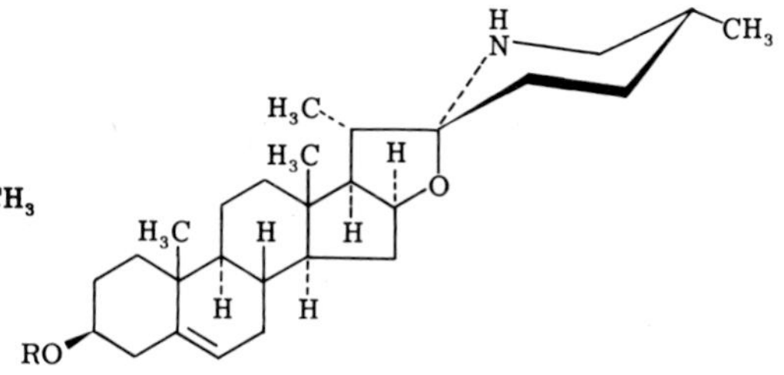   XVII

Solanidine (XVI), R = H
α-Solanine,   R = O-Rha(1 → 2)-O-Glc-(1 → 3)-Gal
β-Solanine,   R = O-Glc-(1 → 3)-Gal
γ-Solanine,   R = O-Gal
α-Chaconine,   R = O-Rha(1 → 2)-O-Rha(1 → 4)-Glc
β₁-Chaconine,   R = O-Rha(1 → 2)-Glc
β₂-Chaconine,   R = O-Rha(1 → 4)-Glc
γ-Chaconine,   R = O-Glc

Solasodine (XVII), R = H
α-Solamargine,   R = O-Rha(1-2)-O-Rha(1-4)-Glc
α-Solasonine,   R = O-Rha(1-2)-O-Glc(1-3)-Gal
Rha = α-L-rhamnopyranosyl
Glc = β-D-glucopyranosyl
Gal = β-D-galactopyranosyl

Solasodine occurs in various *Solanum* species. An enzyme preparation catalyzing its glycosylation was obtained from *Solanum laciniatum* leaves by homogenizing with poly(vinylpyrrolidone) in phosphate buffer, removing suspended matter by filtration and centrifugation, precipitating with ammonium sulfate, and dialyzing the redissolved material (Liljegren, 1971). This crude enzyme catalyzed the conversion of solasodine, UDP-glucose, and ATP to the alkaloid β-glucoside; either the solasodine or the sugar was labeled with carbon-14 in order to follow the reaction.

In very similar work potato sprouts yielded a crude enzyme catalyzing the conversion of solanidine to labeled γ-chaconine by UDP-[U-^{14}C]glucose (Jadhav and Salunkhe, 1974).

This work on transglycosylation was confirmed and extended by Lavintman et al. (1977). Starch-free supernatants from potato tubers or sprouts were centrifuged at 25,000 g, and the pellet was suspended in Tris buffer (pH 7.4), reprecipitated by recentrifuging at the same speed, and dispersed in buffer. After ammonium sulfate precipitation the original 25,000-g supernatant gave a protein mixture which was fractionally reprecipitated with the same salt; the active fraction was dissolved, and the solution dialyzed. These two preparations both caused increased incorporation of radioactivity from UDP-[^{14}C]glucose into 1-butanol-extractable products when solanidine was present. β-Sitosterol could also be glycosylated, forming both sterol glucoside and sterol glucoside esters in the presence of β-solanidine and the 25,000-g fraction, but only sterol glucoside with the ammonium sulfate fraction. β-Solanidine as substrate was glycosylated only in the presence of the ammonium sulfate fraction and then yielded two derivatives (by tlc), or six derivatives if NADPH was also added. According to their chromatographic behavior and the sugars produced on hydrolysis three of these compounds were tentatively identified as α-chaconine and α- and β-solanine; the other three might be ketodeoxyhexose analogues of β- and α-chaconine and γ-solanine. All these results suggest a connection between biosynthesis of starch and lipids such as steroid glycosides, at least in the potato, but any such relationship remains hypothetical.

In an investigation of solanidine synthesis in isolated chloroplasts of greening potatoes, Ramaswamy et al. (1976) demonstrated the presence of serine hydroxymethyltransferase, serine dehydrase, and probably pyruvic dehydrogenase activities in such chloroplasts, but no isolation of the enzymes was undertaken.

2. Biodegradation

It was shown by Petrochenko (1953) that juice from potato sprouts contained hydrolases that could remove sugar residues from steroidal alkaloids present in the juice. Work with pure alkaloids and enzyme preparations fractionally precipitated with ammonium sulfate showed that α-solanine was deglycosylated stepwise, the rhamnose residue being removed first to give β-solanine, then glucose to give γ-solanine, and finally galactose to produce solanidine (Guseva and Paseshnichenko, 1957; Swain *et al.*, 1978). α-Chaconine, on the contrary, was hydrolyzed via β_2-chaconine (loss of rhamnose) but then went directly to solanidine, not via γ-chaconine (solanidine glucoside) (Guseva and Paseshnichenko, 1957; Swain *et al.*, 1978). The enzyme is not highly specific for potato alkaloids and also cleaves tomatine and demissine (Guseva and Paseshnichenko, 1957; Prokoshev *et al.*, 1956) and α-solasonine and α-solamargine from *Solanum aviculare*. Again the two compounds differ, the α-solasonine hydrolyzing in three steps like solanine, and α-solamargine in two, like α-chaconine (Guseva and Paseshnichenko, 1959). Remarkably, however, a similar enzyme preparation from dormant tubers causes a four-step (β_1-chaconine, β_2-chaconine, γ-chaconine, and solanidine) degradation of α-chaconine, whereas for α-solanine it causes omission of γ-solanine as an intermediate product, i.e., omission of deglycosylation as a separate step (Swain *et al.*, 1978).

J. Benzodiazepine Alkaloids: Cyclopenine Group–Quinoline Group

1. Chronology of Formation of the Alkaloids and Enzymes

The fungus *P. cyclopium* produces benzodiazepine alkaloids that have been the focus of interest of the research group of Luckner in East Germany for several years (Luckner and Nover, 1977). Like other microorganisms, emerged cultures of *P. cyclopium* develop in three phases: germination occurs within 12 h after inoculation; the second phase (trophophase) is the period between 12 and 72 h; idiophase occurs 60–168 h after inoculation (Schmidt and Nover, 1974). These phases are shown in Fig. 31. In idiophase, conidia form, ripen, and detach, while hyphae start the formation of secondary products. The most prominent of these are cyclopenine and cyclopenol, which are excreted into the medium.

In the ripening of conidia several enzymatic processes affect alkaloid formation (Fig. 32) (Nover and Luckner, 1974); cyclopeptine dehydrogenase and dehydrocyclopeptine epoxidase activities appear, and cyclopenase activity develops, but only 12–24 h after detachment of conidia.

The beginning of alkaloid formation is indicated by release of the intermediates and end products, i.e., cyclopeptine, dehydrocyclopeptine, cyclopenine, and cyclopenol (Fig. 33), into the culture medium. This indicates coordinated formation of the enzymes responsible. However, the amounts of cyclopeptine dehydrogenase (CD) and dehydropeptine epoxidase (DE) measured *in vitro* do not correspond to those of the alkaloids released (Voigt *et al.*, 1978).

The last enzyme of the alkaloid-forming chain, cyclopenase, is independent of the regulatory action of the precursors, benzodiazepine alkaloids.

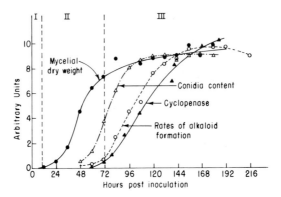

Fig. 31. Kinetics of growth processes of emerged cultures of *P. cyclopium*. Cultures were grown on nutrient solution containing glucose, NH_4^+, and phosphate. Beginning at 48 h after infection culture broth was replaced every 12 h with new broth containing only 20% of the original carbon and nitrogen levels and 2% of the phosphate content. (I) Germination phase; (II) tropophase; (III) idiophase. Ordinate: Mycelial dry weight; $10 = 5.4$ mg/cm² culture area; conidia content: $10 = 6.5$ mg conidia/cm² culture area; rate of cyclopenol excretion by hyphae: $10 = 12$ μh/h per cm² culture area (from Nover and Luckner, 1976, redrawn).

This is indicated by the occurrence of cyclopenase only in the conidiospores, whereas the enzymes of cyclopenine–cyclopenol biosynthesis are found in both hyphae and conidia (Luckner and Nover, 1971). A remarkable feature of cyclopenase is that *in vivo* it does not convert cyclopeptine and cyclopenol to the corresponding quinoline alkaloids in more than traces (Roos *et al.*, 1976). This is attributed to (1) the localization of cyclopenase on the inner

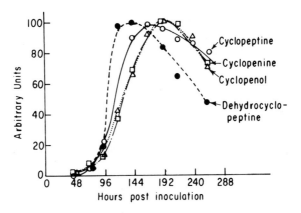

Fig. 32. Rate of benzodiazepine alkaloid excretion by cultures of *P. cyclopium*. Ordinate: Cyclopeptine: $100 = 21$ ng/h per cm² culture area; dehydrocyclopeptide: $100 = 4.4$ ng/h per cm² culture area; cyclopenine: $100 = 10.4$ μg/h cm² culture area; cyclopenol: $10 = 4.4$ μg/h per cm² culture area. From Framm *et al.*, 1973, redrawn.

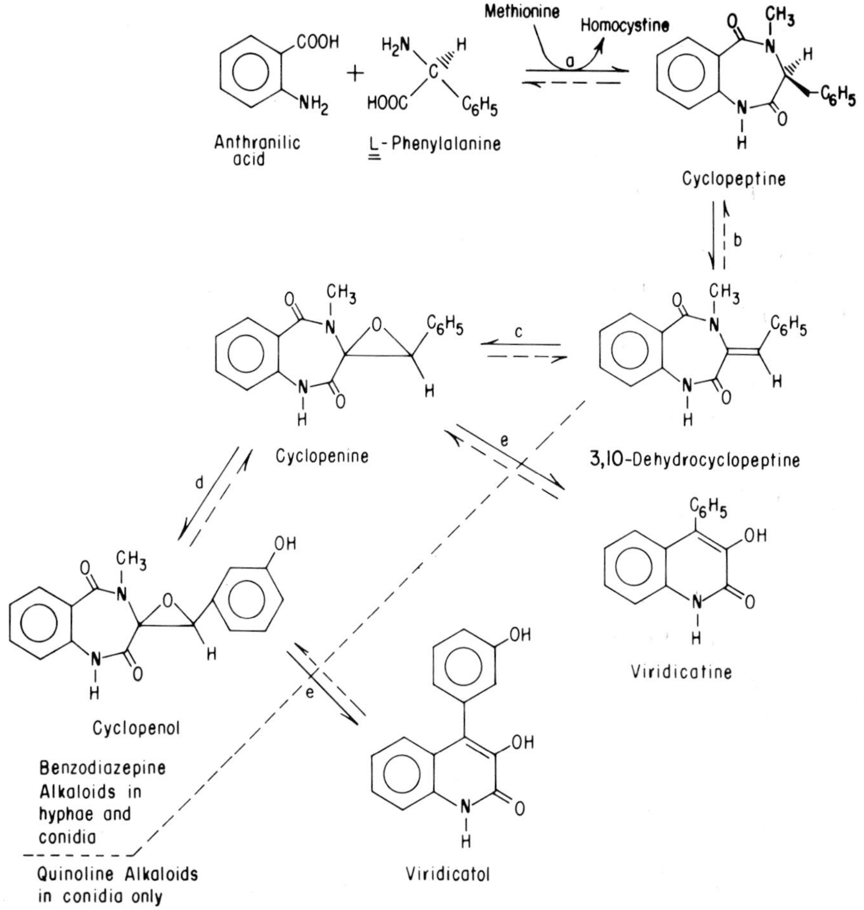

Fig. 33. Biosynthesis of benzodiazepine and quinoline alkaloids in *P. cyclopium* (Framm *et al.*, 1973; Luckner and Nover, 1977). Enzymes and other reactants involved: (a) Cyclopeptine synthetase system (hypothetical); (b) cyclopeptine dehydrogenase (CDH), hydrogen acceptor (NAD(P)$^+$); (c) dehydrocyclopeptine epoxidase (DE), oxygen, hydrogen donor; (d) cyclopenine *m*-hydroxylase (CMH), oxygen, hydrogen donor; (e) cyclopenase complex.

side of the conidiospore membrane, and (2) the very low permeability of this membrane to external benzodiazepine alkaloids (Wilson and Luckner, 1975); these two factors together prevent nearly all enzyme–substrate interactions. However, in some spores this resistance to alkaloid diffusion is abolished by treatment with azide or 2,4-dinitrophenol, whereupon quinoline alkaloids are formed but remain stored in the mycelium (Roos, 1974; Roos *et al.*, 1976). Cyclopenase can thus serve as an indicator of membrane permeability (Roos and Luckner, 1977).

It is notable that the change in alkaloid metabolism induced in *P. cyclopium* by treatment with azide or 2,4-dinitrophenol occurs normally during the idiophase of the related species *Penicillium viridicatum*. In *P. viridicatum* benzodiazepine alkaloids are excreted into the medium at the beginning of the idiophase and later taken up again and converted to quinoline alkaloids.

Nover and Luckner (1976) have summarized the sequence of events occurring during development of the enzymes and alkaloids in emerged cultures of *P. cyclopium,* particularly during idiophase (Fig. 34). Alkaloid production (as shown by the enzymes of alkaloid metabolism) in idiophase is depressed by inhibitors of gene expression, e.g., 5-fluorouracil, ethanol, and cycloheximide (Nover and Luckner, 1976; Schmidt and Nover, 1974). Mutant *dev*-63, obtained by treatment of the wild-type strain with *N*-methyl-*N'*-nitro-*N*-nitrosoguanidine, shows no cyclopenase. This mutant could be used to produce a relatively simple mixture of benzodiazepine alkaloids. The scheme shown in Fig. 34 is based primarily on the time sequence of events rather than on regulatory interdependence. More detailed study will be required to reveal the intricate mechanisms of the development of *P. cyclopium*.

2. Cyclopeptine Synthetase System

This little-known system is presumed to catalyze the conversion of anthranilic acid and phenylalanine to form cyclopeptine in the presence of methionine as a source of methyl groups. The only part of it characterized is anthranilate adenylyltransferase, which has been observed in cell-free

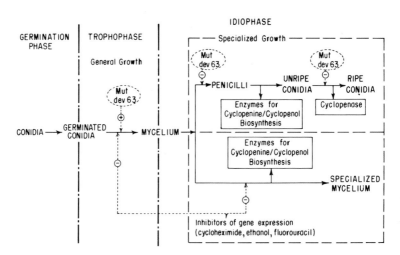

Fig. 34. Development of emerged cultures of *P. cyclopium*. Events during development are shown by solid arrows; factors promoting (+) or inhibiting (−) further events at indicated sites are shown by labeled dashed arrows.

extracts and determined spectrophotometrically (Luckner, 1979). Levels roughly parallel those of cyclopeptine dehydrogenase and cyclopenine m-hydroxylase. Phenylalanine adenylyltransferase is also present.

3. Cyclopeptine Dehydrogenase

Cyclopeptine dehydrogenase (CDH) catalyzes the interconversion of cyclopeptine and 3,10-dehydrocyclopeptine (Fig. 33). For analytical purposes, it is best extracted from *P. cyclopium* mycelium by mechanical disruption under pressure, freeze–thaw cycles, and acetone precipitation, but for CDH preparation, cell break-up by simply grinding with sand and Tris buffer (pH 7.4) at 0°C proved best (Aboutabl, 1974; Aboutabl and Luckner, 1975). Centrifugation to remove cell debris and mitochondria, precipitation with ammonium sulfate, reprecipitation with ethanol, and DEAE chromatography at pH 7.2 gave an enzyme preparation 98-fold increased in activity. Most of the CDH, however, remained in the cell wall and protoplasmic membrane function (isolated at 14,000-g centrifugation), the proportions in mitochondria and in solution being much less. The CDH was determined by incubating with cyclopeptine, NAD$^+$, and a buffer and measuring the NADH formed spectrophotometrically.

Since CDH levels and cyclopenine–cyclopenol levels are parallel during idiophase and, since inhibitors of gene expression (such as cycloheximide and 5-fluorouracil) block further formation of these alkaloids (El-Kousy *et al.*, 1975), CDH is considered to be involved in alkaloid metabolism.

The enzyme is probably a flavoprotein, since its dissociation at pH 6 released FAD and the residual enzyme so produced could be reactivated by the addition of FAD. Its activity was maximum at pH 9.1 in the presence of 10 mM Mg^{2+} at 30°C. The values of K_m with cyclopeptine and NAD$^+$ were 1.6 and 2.8 mM, respectively. The inhibition by p-chloromercuribenzoate and iodoacetamide, and its partial reversal by SH-containing compounds, suggest that mercapto groups are necessary for the function of CDH.

Comparison of the rates of dehydrogenation of various cyclopeptine analogues showed that CDH had high specificity for the naturally occurring (3S, 10R)-cyclopeptine. The stereospecificity of dehydrogenation at C-10 was established by synthesizing the 10S-^{3}H$_1$ and 10R-^{3}H$_1$ enantiomers and showing that only the latter could serve as the substrate for CDH (Aboutabl *et al.*, 1976). It was also shown that CDH removed the *pro-S* hydrogen from C-10, so that the dehydrogenation must be a cis-elimination. The acceptor, NAD$^+$, is also stereospecifically hydrogenated—at the 4-*pro-S* position; thus CDH is an A-specific dehydrogenase.

4. Dehydrocyclopeptine Epoxidase

Dehydrocyclopeptine epoxidase (DE) is a mixed-function oxidase that transforms dehydrocyclopeptine into cyclopenine (Voigt and Luckner, 1977)

(Fig. 30). It uses molecular oxygen, and NAD(P)H, ascorbate, or DL-6-methyl-5,6,7,8-tetrahydropteridine can serve as a hydrogen donor cosubstrate. It was concentrated 268-fold by a 20,000-*g* centrifugation of a sand homogenate of *P. cyclopium,* ammonium sulfate precipitation, and Sephadex chromatography. It has a MW of approximately 480,000. It is interesting that DE lost some specific activity upon precipitation with ammonium sulfate; the inhibition proved to be due to coprecipitated alkaloids (cyclopeptine, cyclopenol, viridicatin, and viridicatol) in the pellet produced by centrifugation.

The enzyme showed the following properties: maximum activity at pH 7.5; linearity of cyclopenine formation for 30 min on incubation of 35°C; K_m (dehydrocyclopeptine), 0.17 mM; K_m (NADH), 0.14 mM; K_m (NADPH), 0.12 mM; K_m (DL-6-methyl-5,6,7,8-tetrahydropteridine), 3.14 mM. The ascorbic acid optimum concentration was 7.5 mM. The enzyme had no detectable NAD(P)-transhydrogenase activity, so that the parallel use of NADH and NADPH must be a property of DE.

5. *Cyclopenine m-Hydroxylase*

Cyclopenine *m*-hydroxylase (CMH), another mixed-function oxidase, transforms cyclopenine into cyclopenol (Richter and Luckner, 1976) (Figs. 33 and 35). It is more difficult to demonstrate in strains of *P. cyclopium* that contain much cyclopenase, since cyclopenine, the substrate for both enzymes, is very rapidly converted to viridicatine. Nevertheless, isolation of CMH from both cyclopenase-rich and poor mutants is possible (Luft, 1974; Richter and Luckner, 1976), especially if the mycelia are ground with sand and sodium deoxycholate briefly so as to disrupt the hyphae but not the cyclopenase-enriched conidiospores (Richter and Luckner, 1976). Centrifugation, ammonium sulfate fractional precipitation and the use of calcium phosphate gel or Sephadex G-25 (Luft, 1974) permitted about 20-fold concentration; but rapid work was necessary, since CMH at this degree of purification has a half-life of only about 16 h at 0°C. The enzyme was assayed by its ability to hydroxylate cyclopenine to the phenol, which is determined colorimetrically (Nover and Luckner, 1974).

Molecular oxygen and a hydrogen donor [NAD(P)H, ascorbic acid, tetrahydropteridine] as a cosubstrate are required by CMH. It was inhibited by CN^- and SCN^- but not by carbon monoxide, which indicates that it is not among the mixed-function oxygenases containing a cytochrome P-450 moiety (Thompson and Siiteri, 1974), and by dicoumarol, which suggests that it is a flavoprotein. Cyclopenine was hydroxylated best, but other compounds closely related in structure to cyclopenin were also hydroxylated. Activity was maximum in the cultures at the end of idiophase, which also corresponded to maximum excretion of cyclopenol.

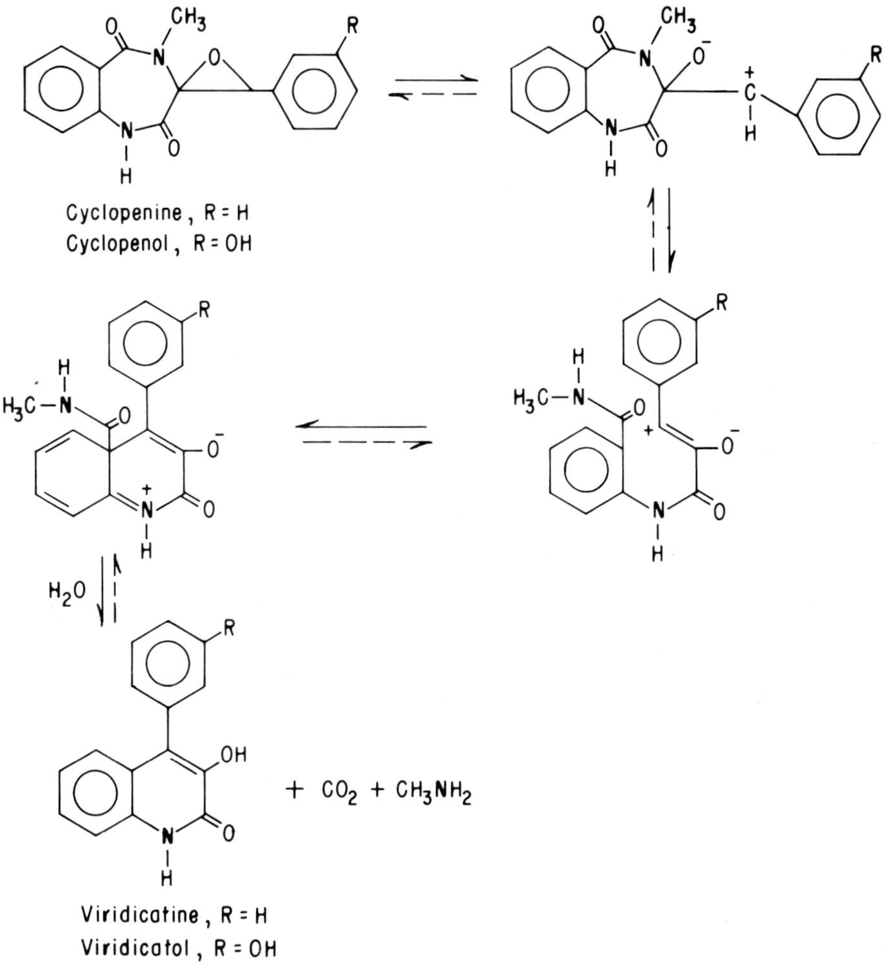

Cyclopenine, R = H
Cyclopenol, R = OH

H_2O

Viridicatine, R = H
Viridicatol, R = OH

$+ CO_2 + CH_3NH_2$

Fig. 35. Suggested mechanism for conversion of cyclopenine or cyclopenol to viridicatol by the cyclopenase complex (Luckner *et al.*, 1969).

6. Cyclopenase

As already noted (Figs. 33 and 35), cyclopenase converts benzodiazepine alkaloids (cyclopenine and cyclopenol) to quinoline alkaloids (viridicatine and viridicatol). The enzyme was recognized, obtained in cell-free extracts, and named by Luckner (1964, 1966, 1967); its production of quinoline alkaloids was traced by their color reaction with ferric ion. The alkaloid conversion, which is also effected by acids, involves the elimination of methylamide and carbon dioxide (Luckner *et al.*, 1969).

TABLE XIII

Kinetic data on cyclopenase[a]

Source	Substrate	K_m (mM)	V_{max} (mmol/min)	pH optimum
Cell wall–protoplasm	Cyclopenine	0.33	0.050	5.6
membrane fraction	Cyclopenol	0.47	0.078	5.0
Phospholipid com-	Cyclopenine	0.66	0.064	5.2
plex by $(NH_4)_2SO_4$	Cyclopenol	1.1	0.095	4.8
precipitation				
Lipid-free enzyme	Cyclopenine	0.66	0.054	5.2
Calcium-PO_4 gel-	Cyclopenol	1.15	0.099	4.8
treated				

[a] From Wilson and Luckner, 1975.

The extraction of cyclopenase from pores is best accomplished with either an X-Press or acetone–water (9 : 1), loss of activity being minimal by these methods (Wilson *et al.,* 1974). The use of a surfactant (Triton X-100) and a 12-h grinding with sand solubilized about half the enzymes of a cell wall or protoplasm membrane as a protein–phospholipid complex; this could be further dissociated with 1-butanol to yield the enzyme protein, although these separations caused some alteration and loss of activity (Wilson and Luckner, 1975). The Michaelis–Menten constants are shown in Table XIII.

It will be of interest to examine purified cyclopenase, for it is unlikely that both viridicatine and viridicatol are formed by the action of the same enzyme. The enzymes may be separable, since they split off methylamine and carbon dioxide, which is at least a two-step reaction.

K. Purine Alkaloids

There are only three papers on the enzymology of purine alkaloids, which include methylated xanthines (Fig. 36), on the biosynthesis of caffeine (XVIII) (Fig. 37) and deal mainly with transmethylases but differ in the source of plant material: Suzuki and Takahishi (1975) used young *Camellia sinensis* (*Thea sinensis*) (tea) plants 75–85 days old, Roberts and Waller (1979)

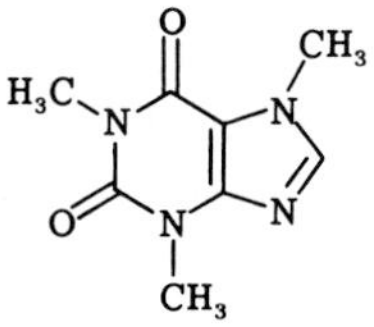

XVIII

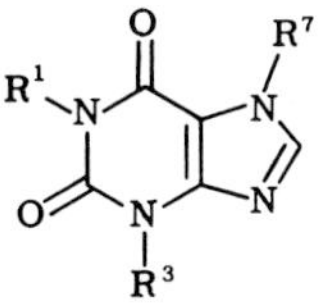

	Trivial name	R^1	R^3	R^7
Xanthine		H	H	H
1,3-Dimethylxanthine	Theophylline	CH_3	CH_3	H
3,7-Dimethylxanthine	Theobromine	H	CH_3	CH_3
1,7-Dimethylxanthine	Paraxanthine	CH_3	H	CH_3
1,3,7-Trimethylxanthine	Caffeine	CH_3	CH_3	CH_3

Fig. 36.　Some naturally occurring methylated xanthines.

used *Coffea arabica* as young seedlings and partially ripe and unripe fruits (coffee), while recently Waller *et al.* (1980) have taken steps to simplify the procedure of isolating the enzymes by using sterile tissue cultures of *C. arabica* capable of producing up to 275 times as much caffeine per unit weight of tissue as did the starting explant system (Waller and Cumberland, 1980). Each group obtained the cell-free enzyme system by essentially the same procedure: grinding, addition of Polyclar AT, centrifuging, and passing the crude preparation through Sephadex G-25 to remove caffeine and related xanthines. The coffee homogenates were used without further preparation, while the tea leaf homogenate was concentrated by precipitation with 60%

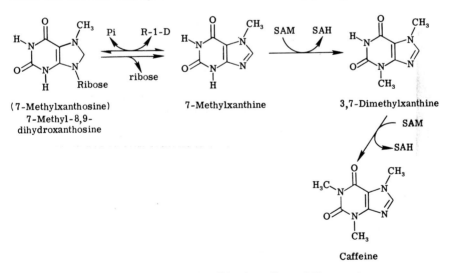

Fig. 37.　Biosynthesis of caffeine in *Coffea* and *Thea* species.

ammonium sulfate and centrifuging before the Sephadex column was used. Suzuki and Takahishi showed the presence of either two methyl transferases or one methyltransferase with differing specific activities (see Chapter 22) that catalyzed the transfer of methyl groups from S-adenosylmethionine (SAM) to 7-methylxanthine, producing theobromine and thence caffeine (Fig. 37); however, they found paraxanthine (previously not isolated from tea and coffee) to be the best acceptor of a methyl group. Similar results were obtained by Roberts and Waller, but their concentration requirement for the methylated xanthine substrates was lower by a factor of 10. However, as with the tea methyltransferase, the apparent K_m values were the same for theobromine and caffeine. Since the two reactions show similar pH optima (8.5) and behaviors with inhibitors, it remains uncertain whether one or two separate enzymes are involved in caffeine biosynthesis. Caffeine was formed rapidly by extracts of green coffee cherries, very little by those of red or yellow partially ripe cherries, and not at all by enzymes of young coffee seedlings. Furthermore, the fact that theophylline in coffee cell-free extracts is formed from 1-methylxanthine rather than 3-methylxanthine corresponds with the findings with tea cell-free extracts and thus substantiates the view in leaf discs (Looser et al., 1974) that the biosynthesis of theophylline from xanthine via 3-methylxanthine is unlikely to occur in significant amounts (Ogutuga and Northcote, 1970).

Among xanthine, hypoxanthine, xanthosine and 7-methylxanthosine, only 7-methylxanthosine could act as substrate for theobromine formation with coffee fruit cell-free extracts, thus indicating the presence of an active 7-methyl-N^9-purine nucleoside phosphorylase or 7-methyl-N^9-nucleoside hydrolase (Roberts and Waller, 1979).

Plant cell cultures are suitable systems in which to study the biosynthesis, biotransformation and biodegradation of secondary metabolites such as caffeine. Biosynthesis proceeds through 7-methylxanthosine to caffeine in the presence of an active purine nucleoside phosphorylase (7-methylxanthosine phosphorylase) or 7-methyl-N^9-nucleoside hydrolase and the action of methyl transferases with S-adenosyl-[$^{14}CH_3$]methionine on 7-methylxanthine and theobromine to give caffeine (see Fig. 37)(Waller et al., 1980). Early experiments using a cell-free extract of sterile callus tissue cultures to obtain methyltransferase(s) of C. arabica showed that the rates of incorporation of methyl groups of [$^{14}CH_3$]methionine to form caffeine were theophylline < theobromine < paraxanthine, the same as was found with the berries and the tea leaves (Waller et al., 1977).

A proposed metabolic pathway (Ogutuga and Northcote, 1970; Looser et al., 1974; Suzuki and Takahishi, 1975, 1976a,b, 1977; Waller et al., 1977, 1980; Roberts and Waller, 1978) of purines in coffee and tea plants and its relationship to caffeine is shown in Fig. 38. Looser et al. (1974) suggested

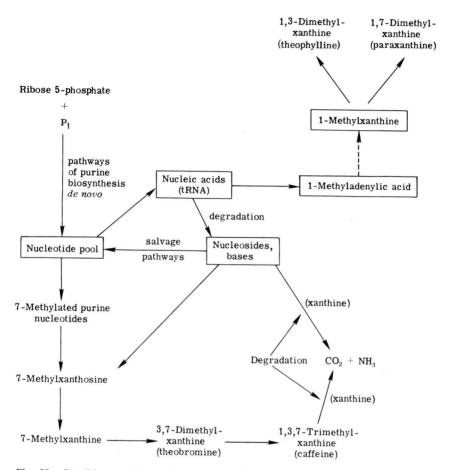

Fig. 38. Possible metabolism of purines in coffee and tea plants and its relationship to caffeine.

that free nucleotides rather than free nucleic acids are most likely to be the first compounds to be methylated in the biosynthesis of caffeine. The source of the purine ring of caffeine must be demonstrated *in vitro* and the presence of methylated purine nucleotides needs to be established in coffee and tea plants. Paraxanthine is known to be the most effective precursor of caffeine while theophylline is less active; an attractive hypothesis is that they may be derived from nucleic acid breakdown, which depends upon the observation that methylation of nucleic acids occurs in tea leaves (*in vitro* and *in vivo*) of N-1 of adenine in tRNA (Suzuki and Takahishi, 1976a). Paraxanthine has recently been found in young coffee seedlings (Chou and Waller, 1980).

Biodegradation of caffeine to CO_2 was shown to occur *in vivo* (Kalberer, 1964, 1965; Baumann and Wanner, 1972). They showed that coffee plants degrade caffeine to xanthine which is further metabolized yielding allantoin, allantoic acid, urea and finally CO_2. Furthermore it is now known that the coffee plants in tissue culture are similar to the tea plants in the metabolism of xanthine to yield urea using an *in vitro* system (Suzuki and Takahishi, 1975, 1976b; Waller *et al.,* 1980).

III. CONCLUSIONS*

The important characteristics of an enzyme include (1) its properties as a catalyst, (2) regulation of its synthesis and destruction, (3) intracellular location, and (4) tissue distribution. Of these, (1) is a chemical problem, attacked by isolating the pure enzyme protein, determining its structure (subunits, amino acid sequence, etc.), and correlating its structure with its properties. Most of the studies reviewed in this chapter have taken this approach, but few have achieved homogeneity to permit structure elucidation. Some work has shown a lack of sophistication, apparently due in part to unfamiliarity with published improvements in sample handling and in the isolation and determination of enzymes. Immunochemical approaches and affinity chromatography, for example, will doubtless be used more in the future in purifying plant enzymes.

Regulation of enzyme synthesis, activity, and catabolism in plants is of much biochemical interest but little studied in comparison to these processes in animals. The nature and function of enzyme precursors (zymogens, proenzymes) and the succession or synchrony of enzyme actions at different physiological ages have received little attention; a notable exception is the enzymology of benzodiazepine alkaloid production in *P. cyclopium* (Nover and Luckner, 1976). Similarly, controls by genetic factors, light, nutritional status, hormones, and naturally occurring inhibitors mostly remain to be investigated and interpreted. The localization of enzymes in tissues and in cells will be understood only through the joint effort of plant cytologists, plant physiologists, and analytical biochemists.

ACKNOWLEDGMENTS

We thank Drs. Robert K. Gholson, Margaret F. Roberts, Ian A. Scott, and John D. Morrison for reviewing the manuscript and making helpful comments. This is published as Article No. B-1 of the Agricultural Experiment Station, Oklahoma State University, Stillwater, Oklahoma.

* Coverage of the literature in this chapter is complete through September, 1979. We hope that any errors of commission or omission will be called to our attention.

REFERENCES

Abe, M. (1972). *Abh. Dsch. Akad. Wiss. Berlin, Kl. Chem., Geol. Biol.* **4**, 411–422.

Aboutabl, E. S. A. (1974). *Pharmazie* **29**, 73.

Aboutabl, E. S. A., and Luckner, M. (1975). *Phytochemistry* **14**, 2573–2577.

Aboutabl, E. S. A., Azzouni, A. E., Winter, K., and Luckner, M. (1976). *Phytochemistry* **15**, 1925–1928.

Adelberg, E. A. (1953). *Bacteriol. Rev.* **17**, 253–267.

Agurell, S. (1966). *Acta Pharm. Suec.* **3**, 71–100; *Chem. Abstr.* **65**, 5693 (1966).

Albertsson, P.-A. (1971). "Partition of Cell Particles and Macromolecules," 2nd ed. Wiley, New York.

Allen, C. M., Jr. (1972). *Biochemistry* **11**, 2154–2160.

Allen, C. M., Jr. (1973). *J. Am. Chem. Soc.* **95**, 2386–2387.

Amici, A. M., Minghetti, A., and Spalla, C. (1966). *Biochim. Appl.* **12**, 50–62.

Antoun, M. D., and Roberts, M. F. (1975). *Phytochemistry* **14**, 909–914.

Antoun, M. D., and Roberts, M. F. (1976). *Lloydia* **39**, 481.

Ashgar, A. S., and Siddiqi, M. (1970). *Enzymologia* **39**, 289–306.

Babcock, P. A., and Plotkin, G. R. (1967). *Lloydia* **30**, 285P.

Bajwa, R. S., and Anderson, J. A. (1975). *J. Pharm. Sci.* **64**, 343–343.

Bajwa, R. S., Kohler, R. D., Saini, M. S., Cheng, M., and Anderson, J. A. (1975). *Phytochemistry* **14**, 735–737.

Barton, D. H. R., Kirby, G. W., Steglich, W., Thomas, G. M., Battersby, A. R., Dobson, T. A., and Ranney, H. (1965). *J. Chem. Soc.* pp. 2423–2438.

Barz, W., Reinhard, E., and Zenk, M. H., eds. (1977). "Plant Tissue Culture and Its Biotechnological Application." Springer-Verlag, Berlin and New York.

Battersby, A. R., Foulkes, D. M., and Binks, A. R. (1965). *J. Chem. Soc.* pp. 3323–3332.

Baumann, T. W., and Wanner, H. (1972). *Planta Med.* **108**, 11–19.

Baxter, C., and Slaytor, M. (1972). *Phytochemistry* **11**, 2763–2766.

Baxter, C., Madyastha, K. M., Guarnaccia, R., and Coscia, C. J. (1972). *Fed. Proc., Fed. Am. Soc. Exp. Biol.* **31**, 478Abs.

Baxter, R. L., Dorschel, C. A., Lee, S.-L., and Scott, A. I. (1979). *J. Chem. Soc., Chem. Commun.* pp. 257–259.

Benésová, M., Kovács, P., and Senkpiel, K. (1974). *Biochem. Physiol. Pflanz.* **166**, 173–179.

Bereznegovskaya, L. H. (1973). *Usp. Izuch. Lek. Rast. Sib., Mater. Mezhvuz. Nauchn. 1973 Konf.*, pp. 45–46; *Chem. Abstr.* **81**, 166467 (1974).

Bereznegovskaya, L. N., Doshchinskaya, N. V., Startseva, N. A., and Fedoseeva, G. M. (1971). *Biol. Nauki (Moscow)* **14** (7), 85–91; *Chem. Abstr.* **75**, 115888 (1971).

Bereznegovskaya, L. N., Doshchinskaya, N. V., Trofimova, N., Startseva, N. A., and Fedoseeva, G. M. (1972). *Postep Dziedzinie Leku Rosl., Pr. Ref. Dosw. Wygloszone Symp., 1970* pp. 235–240; *Chem. Abstr.* **78**, 69185 (1973).

Böhm, H., and Rink, E. (1975). *Biochem. Physiol. Pflanz.* **168**, 69–77.

Bose, B. C., De, H. N., and Mohammad, S. (1956). *Indian J. Med. Res.* **44**, 91–97.

Breccia, A., Crespi, A. M., and Rampi, M. A. (1966). *Z. Naturforsch., B: Anorg. Chem., Org. Chem., Biochem., Biophys., Biol.* **21B**, 1243–1245.

Brown, S. A., and Wetter, L. R. (1972). *Prog. Phytochem.* **3**, 1–45.

Bu'Lock, J. D. (1965). "The Biosynthesis of Natural Products." McGraw-Hill, New York.

Bu'Lock, J. D., and Barr, J. G. (1968). *Lloydia* **31**, 342–355.

Cavender, F. L., and Anderson, J. A. (1970). *Biochim. Biophys. Acta* **208**, 345–348.

Chandler, J. L. R., and Gholson, R. K. (1972). *Phytochemistry* **11**, 239–242.

Cheng, L.-J., Robbers, J. E., and Floss, H. G. (1980). *J. Nat. Prod.* **43**, 329–339.

Cho, Y. D., and Martin, R. V. (1977). *Hanguk Saenghwa Hakhoe Chi* **10**, 147–152; *Chem. Abstr.* **90**, 51492c (1979).

Choong, T.-C., and Shough, H. R. (1979). Manuscript prepared for publication.

Chou, C.-H., and Waller, G. R. (1980). *Bot. Bull. Acad. Sin.* **21**, 25–33.

Clark, R. L. (1977). *Alkaloids (N.Y.)* **16**, 83–180.

Cleland, W. W. (1963a). *Biochim. Biophys. Acta* **65**, 173–187.

Cleland, W. W. (1963b). *Biochim. Biophys. Acta* **67**, 188–196.

Cornforth, J. W. (1973). *Chem. Soc. Rev.* **2**, 1–20.

Coscia, C. J., Madyastha, K. M., and Guarnaccia, R. (1971). *Fed. Proc., Fed. Am. Soc. Exp. Biol.* **30**, 1472.

Coscia, C. J., Burke, W., Jamroz, G., Lasala, J. M., McFarlane, J., Mitchell, J., O'Toole, M. M., and Wilson, M. L. (1977). *Nature (London)* **269**, 617–619.

Cosson, L. (1968). *Plant Med. Phytother.* **2**, 269–271.

Cromwell, B. T. (1943). *Biochem. J.* **37**, 722–726.

Davis, B. D. (1955). *Adv. Enzymol.* **16**, 247–312.

Doshchinskaya, N. V., Trofimova, N. A., Gachik, V. A., and Gusev, I. F. (1973). *Mater. Obl. Nauchn. Konf. Vses. Khim. O-va., Posvyashch. 75-Letiyu Khim.-Tekhnol. Fak. Tomsk. Politekh. Inst., 3rd, 1972* pp. 56–57; *Chem. Abstr.* **84**, 176698 (1976).

El-Kousy, S., Pfeiffer, E., Ininger, G., Roos, W., Nover, L., and Luckner, M. (1975). *Biochem. Physiol. Pflanz.* **168**, 79–85.

Erge, D., and Maier, W. (1974). *Pharmazie* **29**, 72.

Erge, D., Maier, W., and Gröger, D. (1973). *Biochem. Physiol. Pflanz.* **164**, 234–247.

Fairbairn, J. W., and Challen, S. B. (1959). *Biochem. J.* **72**, 556–561.

Fairbairn, J. W., and Suwal, P. N. (1961). *Phytochemistry* **1**, 38–46.

Fales, H. M., Mann, J., and Mudd, S. H. (1963). *J. Am. Chem. Soc.* **85**, 2025–2026.

Farkas-Riedel, L. (1969). *Acta Agron. Acad. Sci. Hung.* **18**, 317–323; *Chem. Abstr.* **72**, 51762 (1970).

Fischer, A. G., and Leete, E. (1969). Unpublished work, cited by Leete (1969).

Floss, H. G. (1976), *Tetrahedron* **32**, 873–912.

Floss, H. G., Robbers, J. E., and Heinstein, P. (1974). *Recent Adv. Phytochem.* **8**, 141–78.

Floss, H. G., Robbers, J. E., and Heinstein, P. F. (1979). *Proc. FEBS Meet.* **55**, 121–31.

Framm, J., Nover, L., Azzouni, A. E., Richter, H., Winter, K., Werner, S., and Luckner, M. (1973). *Eur. J. Biochem.* **37**, 78–85.

Fu, P., and Robinson, T. (1970). *Phytochemistry* **9**, 2443–2446.

Fu, P., and Robinson, T. (1972). *Phytochemistry* **11**, 95–103.

Fu, P., Kobus, J., and Robinson, T. (1972). *Phytochemistry* **11**, 105–112.

Fuller, W. C., and Gibson, M. R. (1952). *J. Am. Pharm. Assoc., Sci. Ed.* **41**, 263–266.

Furuya, T., Nakano, M., and Yoshikawa, T. (1978). *Phytochemistry* **17**, 891–893.

Geiger, U. (1974). *Pharmazie* **29**, 73.

Gholson, R. K., Chandler, J. L. R., Yang, K. S., and Waller, G. R. (1964). *Fed. Proc., Fed. Am. Soc. Exp. Biol.* **23**, 528.

Golebiewski, W. M., and Spenser, I. D. (1976). *J. Am. Chem. Soc.* **98**, 6726–6731.

Gröger, D. (1976). Unpublished work, cited by Luckner *et al.* (1976).

Gröger, D., and Sajdl, P. (1972). *Pharmazie* **27**, 188.

Gröger, D., Härtling, S., Johne, S., and Syring, U. (1976). *Biochem. Physiol. Pflanz.* **170**, 405–416.

Guarnaccia, R. (1979). Unpublished observations, cited by Madyastha and Coscia (1979a).

Guengerich, F. P., and Broquist, H. P. (1973). *Biochemistry* **12**, 4270–4274.

Guengerich, F. P., Snyder, J. J., and Broquist, H. P. (1973). *Biochemistry* **12**, 4264–4269.

Guseva, A. R., and Paseshnichenko, V. A. (1957). *Biochemistry* **22**, 792–799.

Guseva, A. R., and Paseshnichenko, V. A. (1959). *Biochemistry* **24**, 525–527.

Hadwiger, L. A., Badiei, S. E., Waller, G. R., and Gholson, R. K. (1963). *Biochem. Biophys. Res. Commun.* 13, 466–471.

Hasse, K., and Schmid, G. (1963). *Biochem. Z.* 337, 69–79.

Hasse, K., Ratych, O. T., and Salnikow, J. (1967). *Hoppe-Seyler's Z. Physiol. Chem.* 348, 843–851.

Heinstein, P., and Floss, H. G. (1974). *Nova Acta Leopold., Suppl.* 7, 299–310.

Heinstein, P., Ledesma, and Floss, H. G. (1976). Unpublished work, cited by Floss (1976).

Heinstein, P. F., Lee, S.-L., and Floss, H. G. (1971). *Biochem. Biophys. Res. Commun.* 44, 1244–1251.

Herbert, R. B. (1978). *Alkaloids (London)* 9, 1–34, and earlier similar reviews.

Herzog, W., and Weber, K. (1978). *Eur. J. Biochem.* 91, 249–254.

Hesse, M. (1964). "Indolalkaloide in Tabellen." Springer-Verlag, Berlin and New York.

Hesse, M. (1968). "Indolalkaloide in Tabellen. Ergänzungswerk." Springer-Verlag, Berlin and New York.

Hirata, T., Lee, S.-L., and Scott, A. I. (1979). *J. Chem. Soc., Chem. Commun.* pp. 1081–1083.

Hofmann, E., and Böhm, H. (1974). *Pharmazie* 29, 71–72.

Hornemann, U., Speedie, M. K., Hurley, L. H., and Floss, H. G. (1970). *Biochem. Biophys. Res. Commun.* 39, 594–99.

Hornemann, U., Hurley, L. H., Speedie, M. K., and Floss, H. G. (1971). *J. Am. Chem. Soc.* 93, 3028.

Hsu, A.-F., and Bills, D. D. (1979). *Pap., Int. Conf. Biochem., 11th, 1979.*

Hsu, J. C., and Anderson, J. A. (1970). *J. Chem. Soc., D* p. 1318.

Hsu, J. C., and Anderson, J. A. (1971). *Biochim. Biophys. Acta* 230, 518–525.

Hyslop, R., Heinstein, P. F., and Floss, H. G. (1979). Unpublished work, cited by Floss *et al.* (1979).

Jadhav, S. J., and Salunkhe, D. K. (1974). *J. Food Sci.* 38, 1099–1100.

Jindra, A. (1966). *Acta Fac. Pharm. Bohemoslov.* 13, 7–93.

Jindra, A., and Cihák, A. (1963). *Abh. Dtsch. Akad. Wiss. Berlin, Kl. Chem., Biol. Geol.* 2, 201–206.

Jindra, A., and Staba, E. J. (1968). *Phytochemistry* 7, 79–82.

Jindra, A., Zadrazil, S., and Cerná, S. (1959). *Collect. Czech. Chem. Commun.* 24, 2761–2767.

Jindra, A., Léblová, S., Sipal, Z., and Cihák, A. (1960). *Plant Med.* 8, 44–48.

Jindra, A., Sofrová, D., and Léblová, S. (1962). *Collect. Czech. Chem. Commun.* 27, 2467–2470.

Jindra, A., Cihák, A., and Kovács, P. (1964). *Collect. Czech. Chem. Commun.* 29, 1059–1064.

Jindra, A., Kovács, P., Pittnerová, Z., Pšenák, M., Sovová, M., and Smogrovicová, H. (1966a). *Herba Hung.* 5, (2-3), 30–39.

Jindra, A., Kovács, P., Pittnerová, Z., and Pšenák, M. (1966b). *Phytochemistry* 5, 1303–1315.

Jindra, A., Kovács, P., and Pittnerová, Z. (1966c). *Abh. Dtsch. Akad. Wiss. Berlin, Kl. Chem., Biol. Geol.* 3, 329–333.

Jindra, A., Kovács, P., Smogrovicová, H., and Sovová, M. (1967). *Lloydia* 30, 158–163.

Jindra, A., Ramstad, E., and Floss, H. G. (1968). *Lloydia* 31, 190–196.

Jindra, A., Kovács, P., Pšenák, M., Michels-Nyomárkay, K., and Sárkány, S. (1976/1977). *Ann. Univ. Sci. Budap. Rolando Eotvos Nominatae, Sect. Biol.* 18/19, 91–118.

Johansen, M. (1964). *Physiol. Plant.* 17, 547–559.

Johnson, R. D., and Waller, G. R. (1974). *Phytochemistry* 13, 1493–1500.

Kaçzkowski, J. (1964). *Bull. Acad. Pol. Sci.* 12, 375–378.

Kaçzkowski, J. (1966). *Abh. Dsch. Akad. Wiss. Berlin, Kl. Chem., Biol. Geol.* 3, 521–523.

Kalberer, P. (1964). *Ber. Schweiz. Bot. Ges.* 74, 62–107.

Kalberer, P. (1965). *Nature (London)* 205, 597–598.

Kalyanaraman, V. S. (1970). *Proc. Abstr. Soc. Biol. Chem. (India)* 29, 32.

Kametani, T., Fukumoto, K., and Ihara, M. (1978). *Bioorg. Chem.* **2**, 153–157.

Kovács, P. (1970a). *Biol. Plant.* **12**, 6–10.

Kovács, P. (1970b). *Biologia (Bratislava)* **25**, 359–364.

Kovács, P. (1973). *Wiss. Beitr.—Martin-Luther-Univ. Halle-Wittenberg* **16**, 17–38.

Kovács, P., and Benésová, M. (1976). *Biologia (Bratislava)* **31**, 423–430.

Kovács, P., and Jindra, A. (1965a). *Experientia* **21**, 18.

Kovács, P., and Jindra, A. (1965b). *Naturwissenschaften* **13**, 395–396.

Kovács, P., and Jindra, A. (1970). *Lloydia* **33**, 498P.

Kovács, P., Pšenák, M., and Jindra, A. (1963). *Herba Hung.* **2**, 145–154.

Kovács, P., Pšenák, M., and Jindra, A. (1964). *Cesk. Farm.* **13**, 179–181; *Chem. Abstr.* **61**, 9783 (1964).

Kovács, P., Jindra, A., and Pšenák, M. (1966). *Abh. Dtsch. Akad. Wiss. Berlin, Kl. Chem., Biol., Geol.* **3**, 335–341.

Kovács, P., Jindra, A., Nemec, P., and Benésová, M. (1974). *Pharmazie* **29**, 74.

Krupinski, V. M., Robbers, J. E., and Floss, H. G. (1976). *J. Bacteriol.* **125**, 158–165.

Langlois, N., and Potier, P. (1978). *J. Chem. Soc., Chem. Commun.* pp. 102–103.

Lassla, J. M., and Coscia, C. J. (1979). *Science* **203**, 283–284.

Lavintman, N., Tandecarz, J., and Cardini, C. E. (1977). *Plant Sci.* **8** (1), 65–70.

Lee, S.-L., Heinstein, P. F., and Floss, H. G. (1972). *Lloydia* **35**, 471.

Lee, S.-L., Floss, H. G., and Heinstein, P. F. (1976). *Arch. Biochem. Biophys.* **177**, 84–94.

Lee, S.-L., Hirata, T., and Scott, A. I. (1979). *Tetrahedron Lett.* pp. 691–694.

Leete, E. (1967). *In* "Biogenesis of Natural Compounds" (F. Bernfeld, ed.), 2nd ed., pp. 953–1023. Pergamon, Oxford.

Leete, E. (1969). *Adv. Enzymol.* **32**, 373–412.

Leete, E. (1977). *Biosynthesis* **5**, 136–239.

Lewis, W. H., and Elvin-Lewis, M. P. F. (1978). "Medical Botany," p. 24. Wiley (Interscience), New York.

Licht, H. J. (1979). Unpublished observations, cited by Madyastha.

Licht, H. J., Madyastha, K. M., Coscia, C. J., and Krueger, R. J. (1980). *In* "Microsomes, Drug Oxidations, and Chemical Carcinogenesis" (M. J. Coon, A. H. Conney and R. W. Estabrook, eds.), pp. 211–215. Academic Press, New York.

Liljegren, D. R. (1971). *Phytochemistry* **10**, 3061–3964.

Lin, W.-N. C., Ramstad, E., and Taylor, E. H. (1967). *Lloydia* **30**, 202–208.

Lin, W.-N. C., Ramstad, E., Shough, H. R., and Taylor, E. H. (1967). *Lloydia* **30**, 284.

Loomis, W. D. (1969). *In* "Methods in Enzymology" (J. M. Lowenstein, ed.), Vol. 13, pp. 555–563. Academic Press, New York.

Loomis, W. D. (1974). *In* "Methods in Enzymology" (S. Fleischer and L. Packer, eds.), Vol. 31, pp. 528–544. Academic Press, New York.

Loomis, W. D., and Battaile, J. (1966). *Phytochemistry* **5**, 423–438.

Loomis, W. D., Lile, J. D., Sandstrom, R. P., and Burbott, A. J. (1979). *Phytochemistry* **18**, 1049–1054.

Looser, E., Baumann, T. W., and Wanner, H. (1974). *Phytochemistry* **13**, 2515–2518.

Luckner, M. (1966). *Abh. Dtsch. Akad. Wiss. Berlin, Kl. Chem., Biol. Geol.* **3**, 445–453.

Luckner, M. (1967). *Eur. J. Biochem.* **2**, 74–78.

Luckner, M. (1979). *Proc. FEBS Meet.* **55**, 209–220.

Luckner, M., and Nover, L. (1971). *Abh. Dtsch. Akad. Wiss. Berlin, Kl. Chem., Biol. Geol.* **4**, 525–533.

Luckner, M., and Nover, L. (1977). *Mol. Biol., Biochem. Biophys.* **23**, 39–59.

Luckner, M., Winter, K., and Reisch, J. (1969). *Eur. J. Biochem.* **1**, 380–384.

Luckner, M., Nover, L., and Böhm, H. (1976). *Nova Acta Leopold., Suppl.* **7**, 21.

Luft, J. (1974). *Pharmazie* **29**, 73–74.

McFarlane, J., Madyastha, K. M., and Coscia, C. J. (1975). *Biochem. Biophys. Res. Commun.* **66**, 1263–1269.

McGrath, R. M., Steyn, P. S., Ferriera, N. P., and Neethling, D. C. (1976). *Bioorg. Chem.* **5**, 11–23.

Madyastha, K. M., and Coscia, C. J. (1979a). *Recent Adv. Phytochem.* **13**, 85–129.

Madyastha, K. M., and Coscia, C. J. (1979b). *J. Biol. Chem.* **254**, 2419–2427.

Madyastha, K. M., Guarnaccia, R., and Coscia, C. J. (1971). *FEBS Lett.* **14**, 175.

Madyastha, K. M., Guarnaccia, R., and Coscia, C. J. (1972). *Biochem. J.* **128**, 34P.

Madyastha, K. M., Guarnaccia, R., Baxter, C., and Coscia, C. J. (1973). *J. Biol. Chem.* **248**, 2497–2501.

Madyastha, K. M., Meehan, T. D., and Coscia, C. J. (1976). *Biochemistry* **15**, 1097–1102.

Madyastha, K. M., Ridgway, J. E., Dwyer, J. G., and Coscia, C. J. (1977). *J. Cell Biol.* **72**, 302–313.

Maier, W., and Gröger, D. (1976). *Biochem. Physiol. Pflanz.* **170**, 9–15.

Maier, W., Erge, D., and Gröger, D. (1974). *Biochem. Physiol. Pflanz.* **165**, 479–485.

Mann, D. F., and Byerrum, R. U. (1974a). *J. Biol. Chem.* **249**, 6817–6823.

Mann, D. F., and Byerrum, R. U. (1974b). *Plant Physiol.* **53**, 603–609.

Mann, J. D., Fales, H. M., and Mudd, S. H. (1963). *J. Biol. Chem.* **238**, 3820–3823.

Mayer, A. M., and Harel, E. (1979). *Phytochemistry* **10**, 3021–3027.

Meehan, T. D. (1979). Unpublished observations, cited by Madyastha and Coscia (1979a).

Meehan, T. D., and Coscia, C. J. (1973). *Biochem. Biophys. Res. Commun.* **53**, 1043–1048.

Meikka, S. I., and Ingham, K. C. (1978). *Arch. Biochem. Biophys.* **191**, 525–536.

Mizukami, H., Nordloev, H., Lee, S.-L., and Scott, A. I. (1979). *Biochemistry* **18**, 3760–3763.

Mizusaki, S., Tanabe, Y., Noguchi, M., and Tamaki, E. (1971). *Plant Cell Physiol.* **12**, 633–640.

Mizusaki, S., Tanabe, Y., Noguchi, M., and Tamaki, E. (1972). *Phytochemistry* **11**, 2757–2772.

Mizusaki, S., Tanabe, Y., Noguchi, M., and Tamaki, E. (1973). *Plant Cell Physiol.* **14**, 103–110.

Mothes, K. (1972). *Abh. Dtsch. Akad. Wiss. Berlin, Kl. Chem., Biol. Geol.* **4**, 23–24.

Mothes, K., and Schütte, H. R. (1969). "Biosynthese der Alkaloide." VEB Dtsch. Verlag Wiss., Berlin.

Mudd, S. H. (1961). *Nature (London)* **189**, 489.

Mukherjee, R., and Chatterjee, A. (1964). *Chem. Ind. (London)* pp. 1524–1525.

Murakoshi, I., Kuramoto, T., Haginiwa, J., and Fowden, L. (1970). *Biochem. Biophys. Res. Commun.* **41**, 1009–1012.

Murakoshi, I., Kuramoto, H., and Haginiwa, J. (1972a). *Phytochemistry* **11**, 177–282.

Murakoshi, I., Kuramoto, H., Ohmiya, S., and Haginiwa, J. (1972b). *Chem. Pharm. Bull.* **20**, 855–857.

Murakoshi, I., Ogawa, M., Toriizuka, K., Haginiwa, J., Ohmiya, S., and Otomasu, H. (1977a). *Chem. Pharm. Bull.* **25**, 527–528.

Murakoshi, I., Ikegami, F., Ookawa, N., Haginiwa, J., and Letham, D. S. (1977b). *Chem. Pharm. Bull.* **25**, 520–522.

Murakoshi, I., Sanda, A., Haginiwa, J., Suzuki, N., Ohmiya, S., and Otomasu, H. (1977c). *Chem. Pharm. Bull.* **25**, 1970–1973.

Murakoshi, I., Sanda, A., Haginiwa, J., Otomasu, H., and Ohmiya, S. (1978). *Chem. Pharm. Bull.* **26**, 809–812.

Nalborczyk, E. (1961). *Bull. Acad. Pol. Sci., Ser. Sci. Biol.* **9**, 409–415.

Nalborczyk, E. (1964). *Acta Soc. Bot. Pol.* **33**, 371–392.

Nover, L., and Luckner, M. (1974). *Biochem. Physiol. Pflanz.* **166**, 293–305.

Nover, L., and Luckner, M. (1976). *Nova Acta Leopold., Suppl.* **7**, 229–241.

Nowacki, E. (1964). *Genet. Pol.* **4**, 161–202.

Nowacki, E. K., and Waller, G. R. (1975). *Phytochemistry* **14**, 165–171.

Ogunlana, E. O., Tyler, V. E., Jr., and Ramstad, E. (1970a). *Lloydia* **33**, 497.
Ogunlana, E. O., Wilson, B. J., Tyler, V. E., Jr., and Ramstad, E. (1970b). *J. Chem. Soc. D* pp. 775–776.
Ogutuga, D. B. A., and Northcote, D. H. (1970). *Biochem. J.* **117**, 715–720.
Ohashi, T., and Abe, M. (1970). *Nippon Nogei Kagaku Kaishi* **44**, 519–526; *Chem. Abstr.* **74**, 136815 (1971).
Ohashi, T., Aoki, S., and Abe, M. (1970a). *Nippon Nogei Kagaku Kaishi* **44**, 527–531; *Chem. Abstr.* **74**, 136816 (1971).
Ohashi, T., Iimura, Y., and Abe, M. (1970b). *Nippon Nogei Kagaku Kaishi* **44**, 567–572; *Chem. Abstr.* **74**, 136817 (1971).
Ohashi, T., Shibuya, N., and Abe, M. (1972a). *Nippon Nogei Kagaku Kaishi* **46**, 207–213; *Chem. Abstr.* **77**, 137188 (1972).
Ohashi, T., Takahashi, H., and Abe, M. (1972b). *Nippon Nogei Kagaku Kaishi* **46**, 535–540; *Chem. Abstr.* **78**, 39411 (1973).
Petrochenko, E. I. (1953). *Dokl. Akad. Nauk SSSR* **90**, 1091–1093; *Chem. Abstr.* **47**, 10635 (1953).
Petroski, R. J. (1978). *Diss. Abstr. Int. B* **B38**, 5396.
Petroski, R. J., and Kelleher, W. J. (1977). *FEBS Lett.* **82**, 55–57.
Petroski, R. J., and Kelleher, W. J. (1978). *Lloydia* **41**, 332–341.
Piechowski, M., and Nowacki, E. (1958). *Zesz. Probl. Postepow Nauk Roln.* **20**, 219–221.
Piechowski, M., and Nowacki, E. (1959). *Bull. Acad. Pol. Sci., Ser. Sci. Biol.* **7**, 165–168.
Pong, S. F. (1968). M.S. Thesis, University of Tennessee, Memphis.
Prelog, V., and Joyce, O. (1960). *Alkaloids (N.Y.)* **7**, 343–361.
Prokoshev, S. M., Petrochenko, E. I., and Paseshnichenko, V. A. (1956). *Dokl. Akad. Nauk SSSR* **106**, 313–316; *Chem. Abstr.* **50**, 2811 (1956).
Ramaswamy, N. K., Beheri, A. G., and Nair, P. M. (1976). *Eur. J. Biochem.* **67**, 275–282.
Rapoport, H. (1966). *Abh. Dtsch. Akad. Wiss. Berlin, Kl. Chem., Biol. Geol.* **3**, 313.
Réháček, Z. (1974). *Zentralbl. Bakteriol., Parasitenkd. Infektionskr. Hyg., Abt. 2, Naturwiss.: Allg., Landwirtsch. Tech. Mikrobiol.* **129**, 20–49.
Réháček, Z. (1980). *Adv. Biochem. Eng.* **14**, 33–60.
Rhodes, M. J. C. (1977). *In* "Regulation of Enzyme Synthesis and Activity in Higher Plants" (H. Smith, ed.), pp. 245–267. Academic Press, New York.
Richter, I., and Luckner, M. (1976). *Phytochemistry* **15**, 67–70.
Rink, E., and Böhm, H. (1975). *FEBS Lett.* **49**, 396–399.
Ritter, C., and Luckner, M. (1971). *Eur. J. Biochem.* **18**, 391–400.
Robbers, J. E., and Floss, H. G. (1976). *Nova Acta Leopold., Suppl.* **7**, 243–269.
Roberts, M. F. (1971a). *Phytochemistry* **10**, 3021–3027.
Roberts, M. F. (1971b). *Phytochemistry* **10**, 3057–3060.
Roberts, M. F. (1973). *J. Pharm. Pharmacol.* **25**, Suppl., 115P.
Roberts, M. F. (1974a). *Phytochemistry* **13**, 119–123.
Roberts, M. F. (1974b). *Phytochemistry* **13**, 1847–1851.
Roberts, M. F. (1975a). *J. Pharm. Pharmacol.* **27**, Suppl., 86P.
Roberts, M. F. (1975b). *Phytochemistry* **14**, 2393–2397.
Roberts, M. F. (1977). *Phytochemistry* **16**, 1381–1386.
Roberts, M. F. (1978). *Phytochemistry* **17**, 107–112.
Roberts, M. F., and Antoun, M. D. (1978). *Phytochemistry* **17**, 1083–1087.
Roberts, M. F., and Waller, G. R. (1979). *Phytochemistry* **18**, 451–455.
Robinson, T. (1965). *Phytochemistry* **4**, 67–74.
Robinson, T. (1968). "The Biochemistry of Alkaloids." Springer-Verlag, Berlin and New York.
Romeike, A. (1974). *Pharmazie* **29**, 78.
Roos, W. (1974). *Pharmazie* **29**, 78.

Roos, W., and Luckner, M. (1977). *Biochem. Physiol. Pflanz.* **171**, 127–138.

Roos, W., Fürst, W., and Luckner, M. (1976). *Nova Acta Leopold., Suppl.* **7**, 175–182.

Saini, M. S., and Anderson, J. A. (1978). *Phytochemistry* **17**, 799–800.

Saini, M. S., Cheng, M., and Anderson, J. A. (1976). *Phytochemistry* **15**, 497–500.

Sajdl, P., and Réhácék, Z. (1975). *Folia Microbiol. (Prague)* **20**, 365–367.

Santavý, F. (1970). *Alkaloids (N.Y.)* **12**, 333–454.

Santavý, F. (1979). *Alkaloids (N.Y.)* **17**, 385–544.

Saxton, J. E. (1978). *Alkaloids (London)* **8**, 149–215, and earlier similar reviews.

Schmauder, H. P., Gerullis, C., and Gröger, D. (1976). *Biochem. Physiol. Pflanz.* **170**, 201–210.

Schmidt, I., and Nover, L. (1974). *Pharmazie* **29**, 77.

Schröter, H. B. (1966). *Abh. Dtsch. Akad. Wiss. Berlin, Kl. Chem., Geol. Biol.* **3**, 157–160.

Schütte, H. R., and Orban, U. (1967). *Naturwissenschaften* **54**, 565.

Schütte, H. R., Knöfel, D., and Heyer, O. (1966). *Z. Pflanzenphysiol.* **55**, 110–118.

Scott, A. I. (1970). *Acc. Chem. Res.* **3**, 151–157.

Scott, A. I., and Lee, S.-L. (1975). *J. Am. Chem. Soc.* **97**, 6906–6908.

Scott, A. I., Lee, S.-L., de Capite, P., Culver, M. G., and Hutchinson, C. R. (1977a). *Heterocycles* **7**, 979–984.

Scott, A. I., Lee, S.-L., and Wan, W. (1977b). *Biochem. Biophys. Res. Commun.* **75**, 1004–1009.

Scott, A. I., Lee, S.-L., Hirata, T., and Culver, M. G. (1978a). *Rev. Latinoam. Quim.* **9**, 131–138.

Scott, A. I., Lee, S.-L., and Hirata, T. (1978b). *Heterocycles* **11**, 159–163.

Shough, H. R. (1968). Ph.D. Dissertation, University of Tennessee, Memphis; *Diss. Abstr. B* **30**, 1220–1221.

Shough, H. R., and Choong, T.-C. (1978). *Lloydia* **41**, 655P.

Shough, H. R., and Taylor, E. H. (1969). *Lloydia* **32**, 315–326.

Shough, H. R., Pong, S. F., Taylor, E. H., Ramstad, E., and Lin, W. N. (1967). *Lloydia* **30**, 284.

Skursky, L., Burleson, D., and Waller, G. R. (1969). *J. Biol. Chem.* **244**, 3238–3242.

Smith, I. K., and Fowden, L. (1967). *J. Exp. Bot.* **17**, 750–761.

Smogrovicová, H., Jindra, H., and Kovács, P. (1968). *Collect. Czech. Chem. Commun.* **33**, 1967–1970.

Smogrovicová, H., Jindra, A., and Kovács, P. (1972). *Chem. Zvesti* **26**, 360–366.

Snieckus, V. (1968). *Alkaloids (N.Y.)* **11**, 1–40.

Speedie, M. K., Hornemann, U., and Floss, H. G. (1975). *J. Biol. Chem.* **250**, 7819.

Spenser, I. D. (1968). *Compr. Biochem.* **20**, 3300–3413.

Staba, E. J. (1969). *Recent Adv. Phytochem.* **2**, 75–106.

Stadler, P. A., and Stütz, P. (1975). *Alkaloids (N.Y.)* **15**, 1–40.

Staunton, J. (1979). *Planta Med.* **36**, 1–20.

Stöckigt, J. (1978). *J. Chem. Soc., Chem. Commun.* pp. 1097–1099.

Stöckigt, J. (1979). *Phytochemistry* **18**, 965–971.

Stöckigt, J., and Zenk, M. H. (1977a). *FEBS Lett.* **79**, 233–237.

Stöckigt, J., and Zenk, M. H. (1977b). *J. Chem. Soc., Chem. Commun.* pp. 646–648.

Stöckigt, J., Treimer, J., and Zenk, M. H. (1976). *FEBS Lett.* **70**, 267–270.

Stöckigt, J., Husson, H. P., Kan-Fan, C., and Zenk, M. H. (1977). *J. Chem. Soc., Chem. Commun.* pp. 164–166.

Stöckigt, J., Ruffer, M., Zenk, M. H., and Hoyer, G. A. (1978). *Planta Med.* **33**, 188–192.

Stuart, K. L., Kutney, J. P., and Worth, B. R. (1978a). *Heterocycles* **9**, 1015–1022.

Stuart, K. L., Kutney, J. P., Honda, T., and Worth, B. R. (1978b). *Heterocycles* **9**, 1391–1395.

Stuart, K. L., Kutney, J. P., Honda, T., and Worth, B. R. (1978c). *Heterocycles* **9**, 1419–1427.

Stuart, K. L., Kutney, J. P., Honda, T., Lewis, N. G., and Worth, B. R. (1979). Work cited by Stuart *et al.* (1978c).

Suhadolnik, R. (1963). Unpublished work, cited by Mothes and Schütte (1969, p. 429).

Suhadolnik, R. J. (1966). *Abh. Dtsch. Akad. Wiss. Berlin, Kl. Chem., Geol. Biol.* **3**, 369–372.

Suhadolnik, R. J., Fischer, A. G., and Zulalian, J. (1963). *Biochem. Biophys. Res. Commun.* **11**, 208–212.

Suzuki, T., and Takahishi, E. (1975). *Biochem. J.* **146**, 87–96.

Suzuki, T., and Takahishi, E. (1976a). *Phytochemistry* **15**, 1235–1239.

Suzuki, T., and Takahishi, E. (1976b). *Biochem. J.* **160**, 171–180, 181–184.

Suzuki, T., and Takahishi, E. (1977). *Drug Metab. Rev.* **6**, 213–242.

Swain, A. P., Fitzpatrick, T. J., Talley, E. A., Herb, S. F., and Osman, S. F. (1978). *Phytochemistry* **17**, 800–801.

Taylor, E. H., and Shough, H. R. (1967). *Lloydia* **30**, 197–201.

Taylor, E. H., Goldner, D. J., Pong, S. F., and Shough, H. R. (1966). *Lloydia* **29**, 239–244.

Taylor, W. I., and Farnsworth, N. R., eds. (1975). "The Catharanthus Alkaloids." Dekker, New York.

Thomas, R., and Bassett, R. A. (1972). *Prog. Phytochem.* **3**, 47–111.

Thompson, E. A., and Siiteri, P. K. (1974). *J. Biol. Chem.* **249**, 5373–5378.

Tomko, J., and Votický, Z. (1973). *Alkaloids (N.Y.)* **14**, 1–82.

Treimer, J. F., and Zenk, M. H. (1978). *Phytochemistry* **17**, 227–232.

True, R. H., and Stockberger, W. W. (1916). *Am. J. Bot.* **3**, 1–11.

Tyler, V. E., Jr., Erge, D., and Gröger, D. (1965). *Planta Med.* **13**, 315–325.

Vining, L. C. (1970). *Can. J. Microbiol.* **16**, 473–480.

Voigt, S., and Luckner, M. (1977). *Phytochemistry* **16**, 1651–1655.

Voigt, S., El-Kousy, S., Schwelle, N., Nover, L., and Luckner, M. (1978). *Phytochemistry* **17**, 1705–1709.

Waller, G. R., and Cumberland, C. (1980). *Proc. Int. Colloq. Sci. Tech. Coffee, 9th* (in press).

Waller, G. R., and Nowacki, E. K. (1978). "Alkaloid Biology and Metabolism in Plants." Plenum, New York.

Waller, G. R., and Yang, K. S. (1965). *Phytochemistry* **4**, 881–889.

Waller, G. R., Tang, M. S.-I., Scott, M. R., Goldberg, F. J., Mayes, J. S., and Auda, H. (1965). *Plant Physiol.* **40**, 803–807.

Waller, G. R., Ryhage, R., and Mayerson, S. (1966a). *Anal. Biochem.* **16**, 277–286.

Waller, G. R., Yang, K. S., Gholson, R. K., Hadwiger, A. L., and Chaykin, S. (1966b). *J. Biol. Chem.* **241**, 4411–4418.

Waller, G. R., Baumann, T. W., and Wanner, H. (1977). *Pap., Southwest Cent. States Biochem. Conf., 1977.*

Waller, G. R., Suzuki, T., and Roberts, M. F. (1980). *Proc. Int. Colloq. Sci. Tech. Coffee, 9th* (in press).

Wilson, S., and Luckner, M. (1975). *Z. Allg. Mikrobiol.* **15**, 45–51.

Wilson, S., Schmidt, I., Roos, W., Fürst, W., and Luckner, M. (1974). *Z. Allg. Mikrobiol.* **14**, 515–523.

Wink, M., and Hartmann, T. (1979). *FEBS Lett.* **101**, 343–346.

Yang, K. S., and Waller, G. R. (1965). *Phytochemistry* **4**, 881–889.

Yang, K. S., Gholson, R. K., and Waller, G. R. (1965). *J. Am. Chem. Soc.* **87**, 4184–4188.

Yoshida, D. (1973). *Bull. Hatano Tobacco Exp. Stn.* **73**, 239–244.

Biosynthesis of Plant Quinones

13

E. LEISTNER

I. STRUCTURE OF NATURALLY OCCURRING QUINONES

Chemically, quinones are compounds with either a 1,4-diketocyclohexa-2,5-dienoid or a 1,2-diketocyclohexa-3,5-dienoid moiety (Bentley and Campbell, 1974). In the former case they are named p-quinones, in the latter o-quinones. Most naturally occurring quinones are p-quinones (I, III, V) (Fig. 1); o-quinones (II, IV) (Fig. 1) are less common. In both cases, however, the quinonoid moiety consists of an alternating system of single and double bonds. This system does not occur in m-quinones; they are unstable and to the author's knowledge have not been found in nature.

The structure of many naturally occurring quinones is based on the benzoquinone (I, II), naphthoquinone (III, IV), or anthraquinone (V) ring system. These structures and their numbering system are depicted in Fig. 1.

The Biochemistry of Plants, Vol. 7

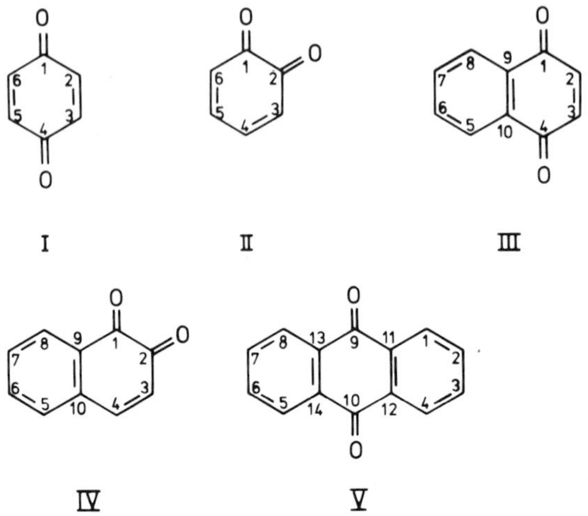

Fig. 1. Basic skeletons of naturally occurring quinones and their numbering systems.

Thomson (1971) divides quinones that do not belong to any of the above-mentioned types into anthracyclinones which occur in microorganisms, miscellaneous quinones which are partly or totally terpenoids, and extended quinones which include some of the most highly condensed aromatic ring systems found in nature. The structure, chemistry, and occurrence of these compounds have been discussed by Lindsey (1974). He draws attention to the fact that polymeric phenols such as tannins, lignins, and humic acids are easily oxidized and thus may be converted to quinones. Although it is generally accepted that no one specific formula can adequately represent humic acids, there is considerable evidence indicating the presence of quinone groups in this polymer.

Quinones may occur in a reduced state as hydroquinones (Müller and Leistner, 1978a), and on careful extraction from anthraquinone-containing plants one of the reduced forms of anthraquinones, namely, anthrones, may be isolated (Labadie, 1972). Two quinones have been described recently, which should be mentioned in this context because they represent the first quinones found to contain covalently bound sulfate. Emodin 1-(or 8)-glucoside sulfate and emodin dianthron diglucoside sulfate have been isolated from *Rumex pulcher* (Harborne and Mokhtari, 1977).

The structural elucidation of quinones by chemical and spectroscopic methods was discussed by different authors (Thomson, 1971; Zeller, 1974; Berger and Rieker, 1974), and since then the ^{13}C-nmr spectroscopy of quinones has been considerably developed (McDonald *et al.*, 1977; Höfle, 1977). The detection by esr spectroscopy of quinones in crude extracts of different plants has also been described recently (Pedersen, 1978).

II. DISTRIBUTION OF QUINONES IN PLANTS AND IN THE PLANT KINGDOM

Lipoquinones (i.e., menaquinones, phylloquinones, plastoquinones, ubiquinones, and tocopherolquinones) are likely to occur in the mitochondria and chloroplasts of every plant, where they are very probably involved in reactions of primary metabolism (Trebst, 1978; Brodie *et al.*, 1970). For this reason they are classed as primary plant products and thus will not be considered here.

Quinonoid secondary plant products are rather unevenly distributed in the plant kingdom. They occur in bacteria, fungi, lichens, gymnosperms, and angiosperms. To the author's knowledge they have not yet been detected in extracts of mosses and ferns. The distribution of quinones in angiosperms was shown in a scheme published by Zenk and Leistner (1968). Since then quinones have also been found in plant families such as the Solanaceae (Knapp *et al.*, 1972), Xyridaceae (Fournier *et al.*, 1975), and Rutaceae (Chakraborty *et al.*, 1978).

Quinones have been encountered in almost all parts and organs of plants. Thus anthraquinones have been isolated from leaves and stems (Mulchandani and Hassarajani, 1977), pods (Agrawal *et al.*, 1972), seed coats, and embryos, but not from the endosperm (Koshioka and Takino, 1978), of *Cassia* plants. Anthraquinones have been encountered in leaves of *Digitalis purpurea* (Brew and Thomson, 1971b) and in leaves and roots of *D. schischkinii* (Imre and Öztunc, 1976). Quinones have also been detected in such organs as fruits (Müller and Leistner, 1978a), flowers (Rizvi *et al.*, 1971), tubers (Ghaleb *et al.*, 1972), and leaf glands (Miyase *et al.*, 1978; Grob *et al.*, 1978). One quinone, namely, dunnione, which occurs as either (−)-dunnione or as a racemic mixture, deserves special attention because it has been isolated from three different plant species belonging to two different families (Gesneriaceae, Scrophulariaceae). In every case this quinone is excreted through the epidermal surface of the leaf (Price and Robinson, 1940; Harborne, 1966; Rüedi and Eugster, 1977).

The occurrence of quinones in the root bark (Rao and Verra Reddy, 1977; Brew and Thomson, 1971a; Zenk *et al.*, 1975) and heartwood of Rubiaceae (Balakrishna *et al.*, 1961) and Bignoniaceae (Burnett and Thomson, 1968a) has been reported. In the heartwood of *Mansonia altissima* (Sterculiaceae) and *Tectona grandis* (Verbenaceae) the concentration of quinones is highest in the center of the stem and declines almost to zero in the sapwood. Anthraquinones are localized in the ray cells (Sandermann, 1966). In *Rheum palmatum* L. *sensu lato,* which belongs to a different family (namely, Polygonaceae), the concentration of anthraquinone pigments is highest in the peripheral layers of the beet (compare Zenk *et al.*, 1975). In tissue of *Rheum* with a high anthraquinone content it has been observed that the number of cells containing these pigments is increased rather than the concentration of pigment in each cell (Schratz, 1957).

III. BIOSYNTHESIS OF PLANT QUINONES

One of the remarkable features of quinone biosynthesis in higher plants is that they are derived from a variety of different precursors and by different pathways (Zenk and Leistner, 1968; Bentley and Campbell, 1974; Bentley, 1975; Leistner, 1980).

A. The Acetate–Polymalonate Pathway

One of the most common biosynthetic pathways leading to quinones is the polyacetate or acetate–polymalonate pathway leading to compounds called acetogenins. These types of compounds often exhibit a characteristic substitution pattern reflecting their biosynthesis from acetyl-coenzyme A (CoA) and malonyl-CoA (see Vol. 4, Chapter 18 of this treatise).

A hypothetical polyketomethylene compound is postulated as an intermediate between the CoA esters and the phenols or quinones. Such a compound, which would be unstable, has never been isolated from a plant source. In an attempt to obtain clearer insight into the nature of such an intermediate Franck *et al.* (1974) synthesized a tetraketone which cyclized under mild conditions. A biomimetic synthesis of emodin (1,6,8-trihydroxy-3-methyl-9,10-anthraquinone) from a polyketomethylene precursor has also been accomplished (Harris *et al.*, 1976). A different folding mechanism for the polyketomethylene chain may occur in the biosynthesis of naphthalene derivatives (Tokoroyama and Kubota, 1971; Bauch *et al.*, 1975).

Further evidence concerning the intermediate steps in the biosynthesis of acetogenic quinones is available from the isolation and identification of the prearomatic naphthalene derivatives (+)-scytalone (VI) (Fig. 2) (Bell *et al.*, 1976) and shinanolone (VII) (Fig. 2) (Tezuka *et al.*, 1973). These compounds are associated with acetogenic naphthoquinones. A biosynthetic pathway leading from acetate to naphthoquinones in higher plants has been elucidated

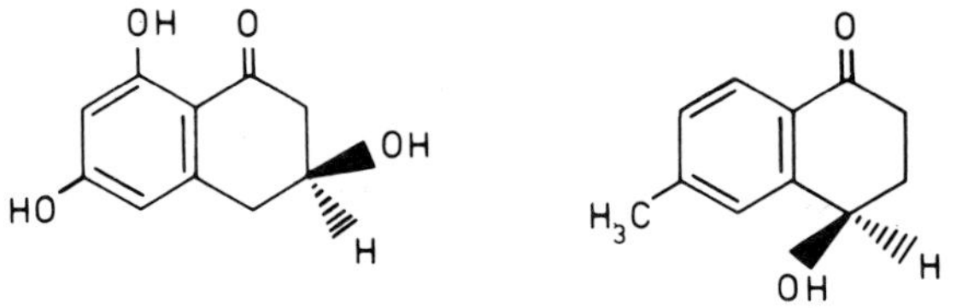

Fig. 2.   Prearomatic naphthalene derivatives, scytalone (VI) and shinanolone (VII).

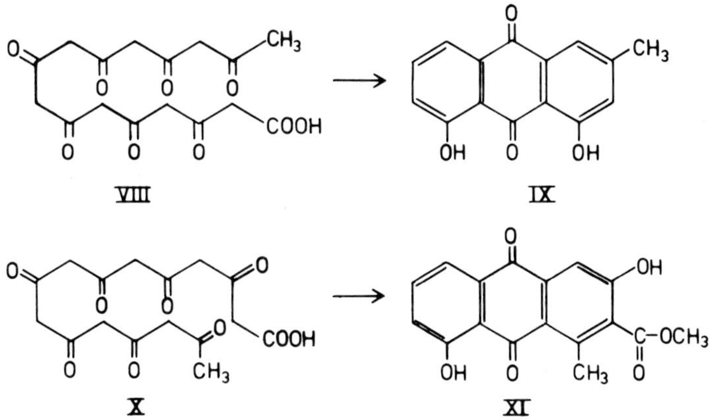

Fig. 3.   Different folding mechanisms of the polyketomethylene chain (VIII, X) involved in the biosynthesis of acetogenic anthraquinones (IX, XI).

by Durand and Zenk (1974). These investigations are of particular interest because they also led to the discovery of an aromatic ring cleavage reaction in plants. Both the ring cleavage and the biosynthesis of the naphthoquinone have been assumed to be of ecological significance.

The acetate–polymalonate route has also been shown to participate in anthraquinone biosynthesis in *Rumex* (Leistner and Zenk, 1969; Leistner, 1971; Fairbain and Muhtadi, 1972) and *Rhamnus* species (Leistner, 1971). The folding mechanism of the hypothetical polyketomethylene compound (VIII) leading to chrysophanol (IX) is depicted in Fig. 3. It is noteworthy that anthraquinones have been isolated from *Aloe saponaria* (Yagi et al., 1978) that represent an exception to the rule that a carbon is attached to the anthraquinone moiety in a β position (i.e., C-2, C-3, C-6, or C-7). Aloesaponarin (XI) carries a methyl group in one of the α positions (i.e., C-1, C-4, C-5, or C-8) of the anthraquinone moiety. This and the overall substitution pattern suggest that the folding mechanisms of the hypothetical polyketomethylene intermediates (VIII and X) are different in chrysophanol (IX) and aloesaponarin (XI) biosynthesis (Fig. 3). Both compounds IX and XI occur in the roots of *A. saponaria,* which means that both folding mechanisms have been elaborated by the same plant. Prearomatic anthracene derivatives have also been detected in *A. saponaria* (Yagi et al., 1978).

B. Biosynthesis of Quinones Derived from Aromatic Amino Acids

1. Phenylalanine as a Precursor

In many cases radiolabeled phenylalanine has served as a tool in investigating the biosynthesis of benzoquinones and naphthoquinones. This amino acid may be metabolized to substituted cinnamic and benzoic acids as outlined in chapters 11, 14 and 15 of this volume.

The resulting p-hydroxybenzoic acid (XII) (Fig. 4) or its substituted derivatives (XIII) (Fig. 4) are key intermediates in the biosynthesis of simple benzoquinones (Bolkart and Zenk, 1968a). p-Hydroxybenzoic acid (XII) has been shown to be decarboxylated oxidatively (XII → XIV) by cell-free preparations obtained from cell suspension cultures of *Glycine max*. This reaction is catalyzed by a peroxidase yielding monomeric (XV) and oligomeric benzoquinones (Berlin and Barz, 1975). On the other hand, p-hydroxybenzoic acid (XII) may serve as an acceptor for a geranyl-pyrophosphate unit. Oxidative decarboxylation, cyclization of the aliphatic side chain, aromatization, hydroxylation, and oxidation steps (XIV → XVII, XVIII) (Fig. 4) would then lead to two enantiomeric naphthoquinones (Schmid and Zenk, 1971; Inouye *et al.* 1979) known as shikonin (XVII) and alkannin (XVIII). The cyclization reaction of an isoprenoid unit attached to an aromatic ring is a reaction that has attracted the attention of Burnett and Thomson (1967) who found that under oxidative conditions the side chain chemically cyclized *in vitro*. This observation is interesting because shikonin (XVII) is a napthoquinone that accumulates in plant tissue cultures of *Lithospermum erythrorhizon* Sieb. et Zucc. when the culture is kept in the dark. In light-exposed tissue cultures, shikonin does not accumulate unless FMN is added to the culture. This suggests that FMN is a cofactor of an enzyme involved in the biosynthetic pathway leading to shikonin and that FMN is destroyed during illumination (Tabata, 1977). One may speculate that FMN plays a role in the cyclization process.

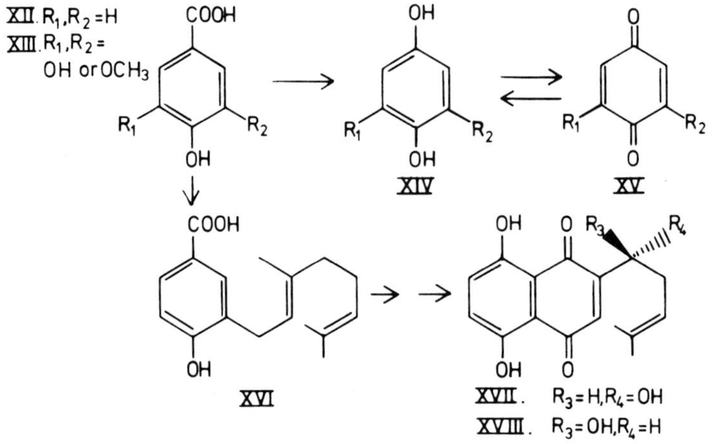

Fig. 4. The role of p-hydroxybenzoic acid(s) (XII, XIII) in benzoquinone (XV), shikonin (XVII), and alkannin (XVIII) biosynthesis.

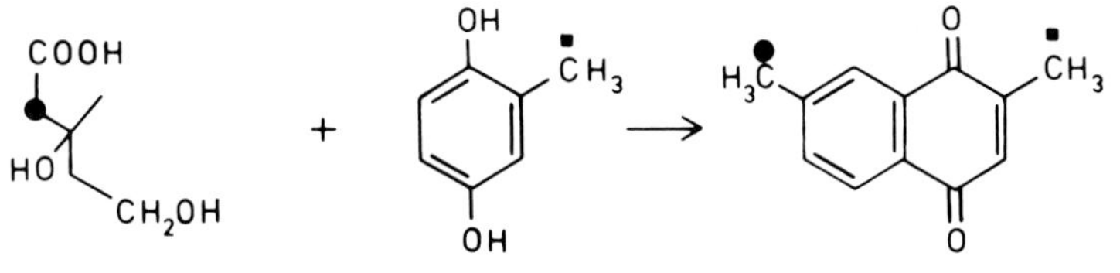

Fig. 5. Biosynthesis of chimaphilin (XXI) from mevalonic acid (XIX) and toluhydroquinone (XX).

2. Tyrosine as a Precursor

The biosynthesis of alkannin (XVIII) and chimaphilin (XXI) (Fig. 5) have one thing in common: In both cases a prenyl side chain is attached to an aromatic nucleus. Toluhydroquinone (XX) and dimethylallyl pyrophosphate give rise to an intermediate which on cyclization and aromatization yields 2,7-dimethyl-1,4-naphthoquinone (i.e., chimaphilin, XXI). The methyl group attached to C-7 of chimaphilin arises from C-2 of mevalonic acid (XIX) (Bolkart *et al.*, 1968), whereas the methyl group at position 2 of chimaphilin is derived from the β carbon of tyrosine (Bolkart and Zenk, 1968b) via homogentisic acid and toluhydroquinone (Bolkart and Zenk, 1969). The postulated intermediates in this pathway have been isolated from *Pyrola media* Sw. (Burnett and Thomson, 1968c).

C. The Shikimic Acid–*o*-Succinoylbenzoic Acid Pathway

1. Biosynthesis of Naphthoquinones in Juglans, Impatiens, and Catalpa Plants

Investigations of the biosynthesis of bacterial menaquinones (Young, 1975; Shineberg and Young, 1976; McGovern and Bentley, 1978; Meganathan *et al.*, 1980) have greatly stimulated progress in elucidating the pathways leading to plant quinones, and vice versa. Studies on the biosynthesis of plant naphthoquinones derived from the shikimic acid–*o*-succinoylbenzoic acid pathway have been carried out with plants of three species, namely, *Juglans regia* L. (Juglandaceae), *Impatiens balsamina* L. (Balsaminaceae), and *Catalpa ovata* G. Don (Bignoniaceae). These plants contain the compounds depicted in Fig. 6. Whereas compounds XXII, XXIII, and XXV were known constituents of *Juglans* and *Impatiens* plants before biosynthetic studies on plant quinones commenced, XXIV and XXVI through XXX were isolated during the investigations because they were suggested as intermediates between shikimic acid and plant quinones or their glycosides. Compounds

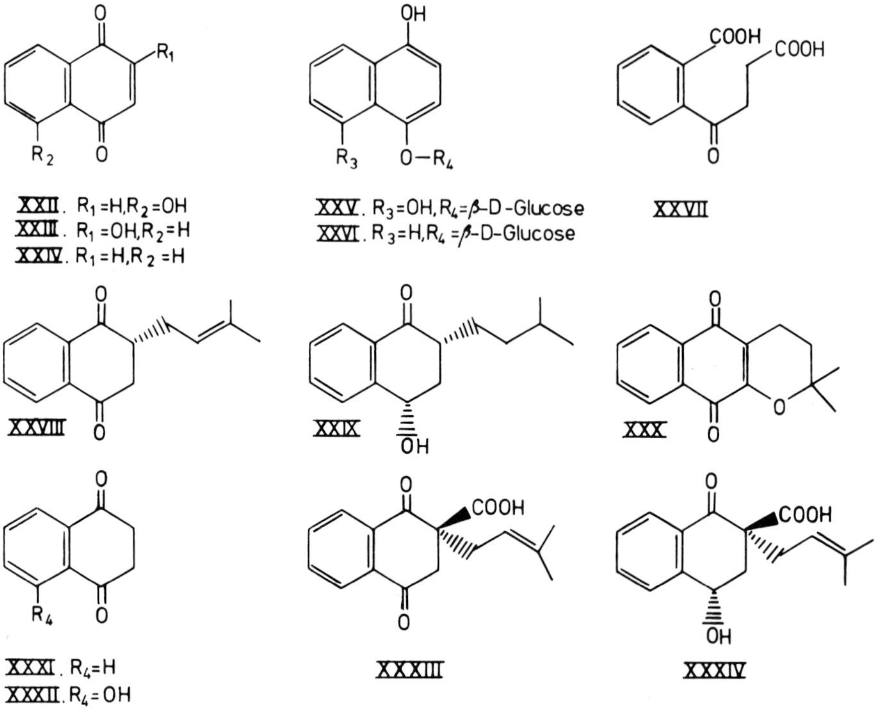

Fig. 6. Metabolites occurring in *J. regia* (XXII, XXIV, XXV, XXVI, XXXI, XXXII), *I. balsamina* (XXIII, XXVII), and *C. ovata* (XXVIII, XXIX, XXX, XXXIII, XXXIV).

XXI and XXXII are likely to occur in *Juglans* (Müller and Leistner, 1978a), whereas XXXIII and XXXIV are likely to occur in *Catalpa* (Inouye *et al.*, 1978).

Shikimic acid (XXXV) was the first compound to be incorporated specifically into juglone (XXII), and much information has been gained from incorporation studies using differently labeled samples of shikimic acid (Fig. 7). Early work (Leistner and Zenk, 1968a) had shown that radioactivity from D-[1,6-^{14}C]shikimic acid rather than L-[1,6-^{14}C]shikimic acid was predominantly incorporated into juglone in *J. regia*. Moreover it had been demonstrated that radioactivity located at C-1 and C-6 of shikimic acid (XXXV) was equally distributed among C-5, C-8, C-9, and C-10 of juglone (XXII) (Fig. 7A) and that radioactivity from the carboxyl group of shikimic acid (XXXV) contributed equally to C-1 and C-4 of juglone (Fig. 7B).

The results obtained in these experiments were interpreted in the following way: Shikimic acid (XXXV) or one of its products (chorismic or prephenic acid, but not phenylalanine or tyrosine) is a direct precursor of juglone (XXII) because the carboxyl group of shikimic acid (XXXV) was

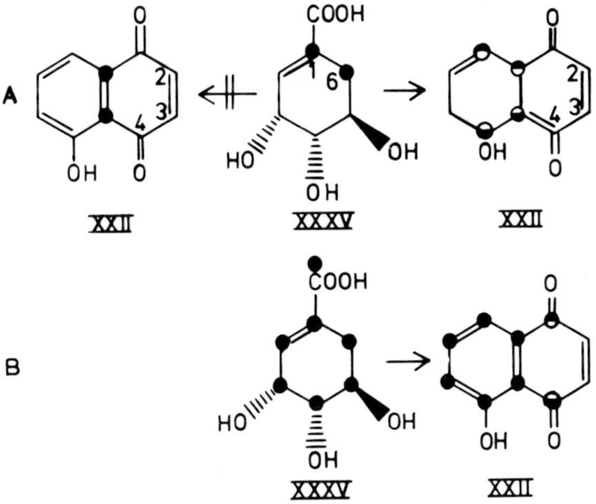

Fig. 7. Mode of incorporation of [1,6-¹⁴C]- and [U-¹⁴C]shikimic acid (XXXV) into juglone (XXII).

retained during its conversion to juglone (XXII). This is consistent with the incorporation of radioactivity from [7-¹⁴C]shikimic acid (XXXV) into juglone (XXII) (Fig. 8C and E) (Scharf *et al.*, 1971). It also can be deduced from the labeling pattern of juglone (XXII) that a C_3 unit corresponding to C-2, -3, and -4 of juglone (XXII) is attached to a hypothetical intermediate at the carbon derived from C-2 of shikimic acid (Fig. 7A). Results with tritium-labeled samples of shikimic acid as depicted in Fig. 8C and E (Scharf *et al.*, 1971), and Fig. 8D (Leduc *et al.*, 1970), can be explained only if this assumption is correct. Finally it has been concluded that a symmetrical intermediate is involved in the biosynthesis of juglone (XXII), because radioactivity from 1,6-¹⁴C (Fig. 7A) and from the carboxyl group of shikimic acid (Fig. 7B) was symmetrically distributed in juglone (XXII). The fact that tritium is located in position 8 of juglone (XXII) after feeding either [3-³H]- (Leduc *et al.*, 1970) or [6-³H]shikimic acid (XXXV) (Scharf *et al.*, 1971) (Fig. 8D and E) supports this view.

The symmetrical intermediate has been suggested to be 1,4-naphthoquinone (XXIV), and radioactive 1,4-naphthoquinone is incorporated into juglone (XXII) (Leistner and Zenk, 1968a). 1,4-Naphthoquinone (XXIV) and the glycoside of its hydroquinone (XXVI) have been isolated from *Juglans* plants, and the origin of these metabolites from precursors also involved in juglone (XXII) biosynthesis has been established (Müller and Leistner, 1976, 1978a,b).

In an attempt to distinguish among shikimic (XXXV), chorismic, and prephenic acids as the direct precursor of juglone (XXII), chirally labeled shikimic acid [(6S)- and (6R)-[7-¹⁴C, 6-³H]], shown in Fig. 8E was em-

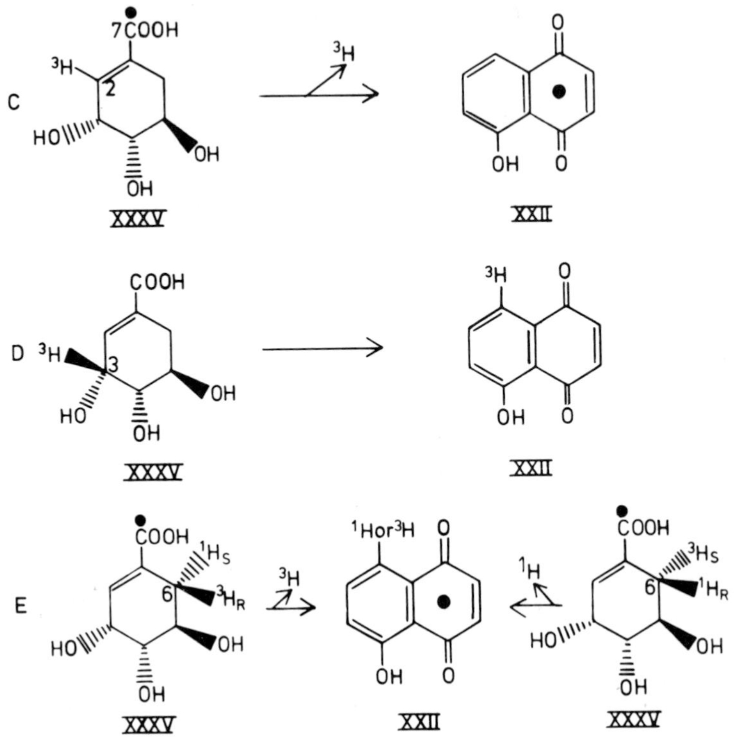

Fig. 8. Mode of incorporation of tritiated and doubly labeled shikimic acid (XXXV) into juglone (XXII).

ployed (Scharf *et al.*, 1971). It was known (Onderka and Floss, 1969; Hill and Newkome, 1969) that label in the 6*S* position would be retained during conversion of shikimic acid to chorismic acid, whereas label in the 6*R* position would be removed. The opposite steric course in the conversion of shikimic acid (XXXV) to juglone (XXII) would therefore exclude chorismic and prephenic acids as direct precursors of juglone (XXII). Where it was found that 6*S* ³H was retained and that 6*R* ³H was removed (Fig. 8E), chorismic and prephenic acids could not be excluded.

Work with bacterial mutants (Dansette and Azerad, 1970; Young, 1975) has shown that chorismic acid is likely to be the branching point leading to the quinones under discussion. However, incorporation of labeled chorismic acid into juglone (XXII) is very low (Leistner and Zenk, 1968a).

Further development in this field was stimulated by the finding that metabolites of the tricarboxylic acid (TCA) cycle (Leistner and Zenk, 1968a), particularly α-ketoglutaric acid (Grotzinger and Campbell, 1972a,b), were specifically incorporated into naphthoquinones. This indicated that

C-2, C-3, and C-4 of juglone (XXII) and lawsone (XXIII) might be derived from this source.

Hypotheses as to how α-ketoglutaric acid and shikimic acid (XXXV) or chorismic acid might join to give the next compound in the pathway have been put forward by Campbell, Bentley, Azerad, and Dansette (Bentley, 1975). Considerations of these authors have led to the proposal by Dansette and Azerad (1970) that the product in question might be *o*-succinoylbenzoic acid (XXVII).

This proposal proved to be correct, and as a result the discovery of this compound was unique: Although *o*-succinoylbenzoic acid is a stable compound and likely to be present in almost all plants, it had never been detected as a natural product until it was postulated on theoretical grounds to be an intermediate in the pathway between shikimic acid and plant quinones. *o*-Succinoylbenzoic acid (XXVII) is now an established precursor of menaquinones (Dansette and Azerad, 1970) and pheylloquinones (Thomas and Threlfall, 1974), as well as quinonoid secondary plant products (Dansette and Azerad, 1970; Leistner, 1973a; Stöckigt *et al.*, 1973; Müller and Leistner, 1978a; Inoue *et al.*, 1979). Eventually *o*-succinoylbenzoic acid (XXVII) was shown to occur in *I. balsamina* plants (Grotzinger and Campbell, 1974).

Based on early suggestions (Robins *et al.*, 1970), *o*-succinoylbenzoic acid was converted in bacterial systems (Shineberg and Young, 1976; McGovern and Bentley, 1978; Meganathan *et al.*, 1980) to 1,4-dihydroxy-2-naphthoic acid (XXXVI) (Fig. 9).

A radioactive sample of XXXVI was prepared by carboxylation of [1,4-^{14}C]1,4-naphthohydroquinone. Application of the acid to *J. regia* and *I. balsamina* plants resulted in a specific and rather high incorporation into juglone

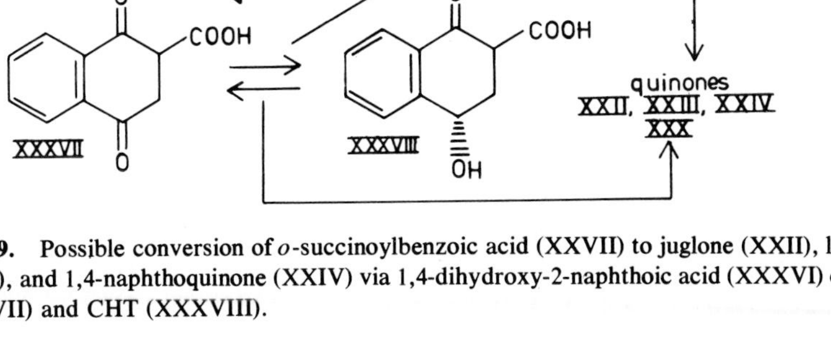

Fig. 9. Possible conversion of *o*-succinoylbenzoic acid (XXVII) to juglone (XXII), lawsone (XXIII), and 1,4-naphthoquinone (XXIV) via 1,4-dihydroxy-2-naphthoic acid (XXXVI) or COT (XXXVII) and CHT (XXXVIII).

(XXII) (6%) and lawsone (XXIII) (17%). Incorporation of the acid (XXXVI) into juglone (XXII) proceeded via 1,4-naphthoquinone (XXIV), the required symmetrical intermediate (see above). Incorporation into lawsone (XXIII), however, took place directly, possibly by oxidative decarboxylation. No 1,4-naphthoquinone (XXIV) was detectable (Müller and Leistner, 1976). This latter observation is in agreement with the fact that, unlike the situation in juglone (XXII) biosynthesis, no symmetrical intermediates are involved in the pathway leading to lawsone (XXIII) in *I. balsamina* (Grotzinger and Campbell, 1972a).

A problem remains regarding the nature of the intermediates involved in the sequence between *o*-succinoylbenzoic acid (XXVII) and quinones. 1,4-Dihydroxy-2-naphthoic acid (XXXVI) has been shown to be a possible product of *o*-succinoylbenzoic acid (XXVII) (*vide supra*). However, Inouye *et al.* (1971) have isolated, from *C. ovata* G. Don, catalponone (XXVIII) and catalponol (XXIX), the stereochemistry of which has been revised recently by Inouye *et al.* (1978). These authors assume that *o*-succinoylbenzoic acid (XXVII) is incorporated into catalponol (XXIX) and catalponone (XXVIII) without the intermediacy of an aromatic compound. During the conversion of 1-[^{14}C-carboxy]-[$2'$-^{3}H$_2$]*o*-succinoylbenzoic acid (XXVII) to catalponol (XXIX) (Fig. 10), 70% of the activity was retained in catalponol (XXIX), although tritium was located in a position adjacent to a keto group (Inouye *et al.*, 1975) (Fig. 10). The authors therefore suggested that 2-carboxy-4-oxotetralone (COT) (XXXVII) or 2-carboxy-4-hydroxytetralone (CHT) (XXXVIII) rather than 1,4-dihydroxy-2-naphthoic acid (XXXVI) was involved in the biosynthesis of quinones in *C. ovata* (Inouye *et al.*, 1978) (Fig. 9).

Although the incorporation of 1,4-dihydroxy-2-naphthoic acid (XXXVI) into juglone (XXII) and lawsone (XXIII) proved to be rather high (Müller and Leistner, 1976), incorporation following reduction was not excluded. One may argue that during the incorporation of 1,4-dihydroxy-2-naphthoic acid (XXXVI) into naphthoquinones conversion of the acid to COT (XXXVII) preceded incorporation. Aromatic (Stipanovic and Bell, 1977) and quinonoid

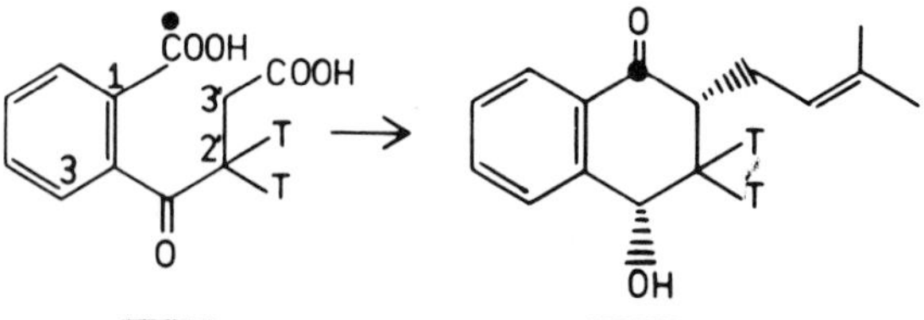

Fig. 10. Retention (70%) of tritium label relative to ^{14}C label during the conversion of 1-[^{14}C-carboxy]-[$2'$-^{3}H$_2$] succinoylbenzoic acid (XXVII) to catalponol (XXIX) in *C. ovata*. (For the numbering of XXVII the system of Inouye *et al.* (1975) was used.)

(Bell *et al.*, 1976, Inoue *et al.*, 1977a) compounds administered to plants have repeatedly been found to be reduced to tetralones, and one of these processes (Inoue *et al.*, 1977a) is assumed not to be a normal reaction *in vivo*. The assumption that nonaromatic naphthalene derivatives are involved in the pathway between *o*-succinoylbenzoic acid (XXVII) and quinones is in agreement with the observation that α-oxotetralone (XXXI) and β-hydrojuglone (XXXII) are likely to be present in *J. regia* plants and, since they are derived from *o*-succinoylbenzoic acid (XXVII) (Müller and Leistner, 1978a), they may intervene between XXXVII or XXXVIII and juglone (XXII) (Fig. 9).

For lawsone (XXIII) biosynthesis a similar proposal has been put forward (Inoue *et al.*, 1977b), and one of the possible reaction sequences favored by Inouye *et al.* (1978) for *Catalpa* has been published recently.

At present it is not possible to distinguish among COT (XXXVII), CHT (XXXVIII), and 1,4-dihydroxy-2-naphthoic acid (XXXVI), and we have to take into account that each of these compounds may play a role in the biosynthesis of plant naphthoquinones.

2. Biosynthesis of Anthraquinones in the Rubiaceae and Gesneriaceae

Naphthalene derivatives and anthraquinones have been repeatedly shown to occur in the same plant (e.g., Burnett and Thomson, 1968b). In most cases an isoprene unit is attached to these naphthalene derivatives. This observation led Sandermann and Dietrichs (1959) to suggest that a pathway other than the acetate pathway (*vide supra*) may exist in the formation of anthraquinones. Their reasoning was based on the assumption that ring closure of a prenylated naphthoquinol (or naphthoquinone) (e.g., XL) (Fig. 11) would lead to an anthraquinone ring system. If anthraquinones were indeed derived from a naphthoquinone and an isoprene unit, the question arises as to which of the precursors of naphthoquinones would yield the naphthalene-derived moiety (rings A and B of XLII and XLIII) (Fig. 11) of anthraquinones. Four alternatives would have to be considered (see above). To begin with, acetate was incorporated only into rings B and C of alizarin (XLII) (Leistner and Zenk, 1967), which was in agreement with Sandermann's proposal because acetate would have been incorporated into these anthraquinones (XLII, XLIII) via mevalonic acid (XLI) and dimethylallyl pyrophosphate. Incorporation of acetate into the whole anthraquinone molecule was not observed.

Although phenylalanine was not a precursor, shikimic acid (XXXV) was incorporated *in toto* into anthraquinones occurring in *Rubia tinctorum*. Investigations of the biosynthesis of these anthraquinones therefore paralleled those of the naphthoquinones derived from shikimic (XXXV) (or chorismic) and α-ketoglutaric acids (XXXIX) via *o*-succinoylbenzoic acid (XXVII).

In order to test the intermediacy of 1,4-naphthoquinone in alizarin biosynthesis radioactively labeled 1,4-naphthoquinone was synthesized and applied

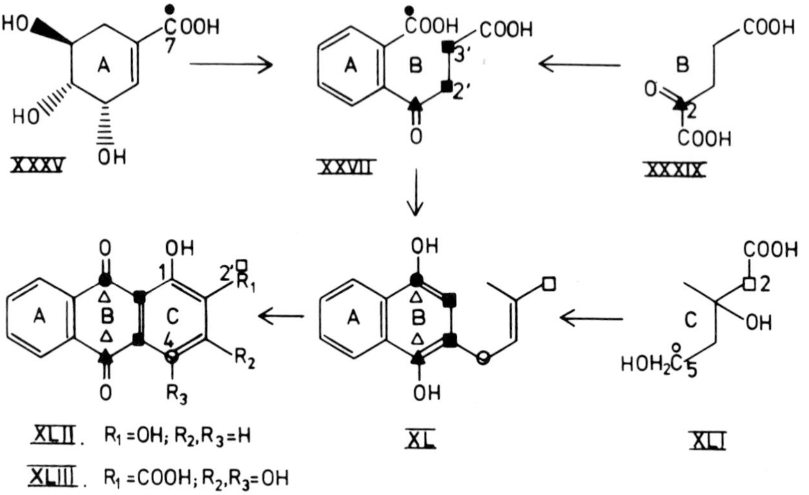

Fig. 11. Localization of radioactive carbon atoms in alizarin (XLII) or purpurincarboxylic acid (XLIII) after feeding [7-^{14}C]shikimic acid ($\bullet$, XXXV), [2-^{14}C]α-ketoglutaric acid ($\blacktriangle$, XXXIX), [2',3'-^{14}C]o-succinoylbenzoic acid ($\blacksquare$, XXVII), [2-^{14}C]-($\square$) or [5-^{14}C]-($\bigcirc$) mevalonic acid, and [1,4-^{14}C]desoxylapachol hydroquinone ($\triangle$, XL) to *R. tinctorum*.

to *Rubia* plants containing alizarin (XLII). Although 1,4-naphthoquinone was incorporated specifically, the conversion of this compound to alizarin (XLII) was later shown to be an aberrant process because [7-^{14}C]shikimic acid (XXXV) was incorporated nonsymmetrically (Leistner and Zenk, 1971) (Fig. 11) and thus could not be incorporated by way of a symmetrical intermediate such as 1,4-naphthoquinone (XXIV). Ring C of *Rubia* anthraquinones (Leistner and Zenk, 1968b) and of 1-hydroxy-2-hydroxymethyl-9,10-anthraquinone in *Streptocarpus dunnii* (Stöckigt *et al.*, 1973) was shown to be derived from mevalonic acid, possibly via dimethylallyl pyrophosphate. The C-2 atom of mevalonic acid (XLI) (Fig. 11), which gives rise to the trans methyl group of dimethylallyl pyrophosphate, contributes almost exclusively to the β carbon of the anthraquinones. After a 15-day feeding period, however, scrambling of radioactivity from [2-^{14}C]mevalonic acid between positions 1 and 2' of purpurincarboxylic acid (XLIII) was observed (Burnett and Thomson, 1968d).

   The β carbon of anthraquinones, including those that are acetate-derived, may be oxidized stepwise because all levels of oxidation ranging from a methyl, via an alcohol and an aldehyde, to a carboxyl group have been observed. Eventually the carboxyl group may be split off by decarboxylation. This is also the case with alizarin (XLII), hence it was not possible to establish the origin of ring C of alizarin using [2-^{14}C]mevalonic acid. As

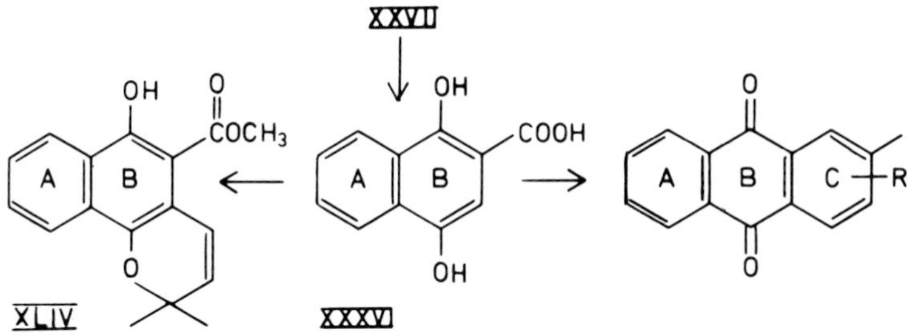

Fig. 12. The possible role of 1,4-dihydroxy-2-naphthoic acid (XXXVI) in mollugin (XLIV) and anthraquinone biosynthesis.

expected, application of [5-^{14}C]mevalonic acid (XLI) to *R. tinctorum* resulted in incorporation. Localization of radioactivity in the alizarin molecule at C-4 gave a clue to the mechanism of prenylation: the dimethylallyl pyrophosphate would be attached to the hypothetical naphthalene derivative in a position meta to the carbon derived from the carboxyl group of shikimic acid (Fig. 11). Thus, in contrast to the situation in prenylated naphthoquinones in *Catalpa* (Inouye *et al.*, 1975) and menaquinones (Baldwin *et al.*, 1974), attachment of the dimethylallyl pyrophosphate occurs on a hypothetical naphthalene derivative at a carbon derived from C-2' of *o*-succinoylbenzoic acid (Leistner, 1973a). This mechanism has been confirmed by Dansette (1972) and Inoue *et al.* (1979) using ^{13}C-labeled *o*-succinoylbenzoic acid which was administered to a cell suspension culture in a continuous culture system. As in naphthoquinone biosynthesis, 1,4-dihydroxy-2-naphthoic acid (XXXV) was assumed to be the hypothetical intermediate (Leistner and Zenk, 1971; Inoue *et al.*, 1979). Prenylation of 1,4-dihydroxy-2-naphthoic acid is a reaction that is likely to occur in the Rubiaceae, since mollugin (XLIV) (Fig. 12) has been isolated from *Galium mollugo* (Schildknecht *et al.*, 1976). This compound is associated with anthraquinones derived from *o*-succinoylbenzoic acid (Bauch and Leistner, 1978).

Decarboxylation of 2-carboxy-1,4-dihydroxy-3-(γ,γ-dimethylallyl)-1,4-naphthoquinol may lead to the hydroquinone of desoxylapachol (XL). Ring closure would complete the biosynthetic process giving rise to the ring system of anthraquinones. A radioactive sample of desoxylapachol hydroquinone (XL) has been shown to be incorporated into alizarin (XLII) (Leistner, 1973a).

Some anthraquinones have an unusual substitution pattern, and no prediction can be made as to their biosynthesis. One of these anthraquinones is benzylxanthopurpurin (Brew and Thomson, 1971a). Morindone (XLV) and soranjidiol (XLVI) have an unusual structure also, because they are substituted in both rings A and C. This pattern is typical of acetogenic an-

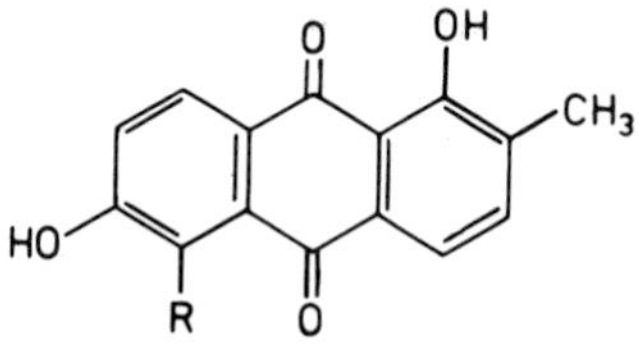

XLV. R =OH
XLVI. R =H

thraquinones. Yet both XLV and XLVI co-occur with o-succinoylbenzoic acid-derived anthraquinones (e.g., alizarin, XLII) in *Morinda citrifolia* L. Thus the question was raised whether *M. citrifolia* has developed two pathways for anthraquinone biosynthesis or whether enzymes are present in this plant that specifically hydroxylate anthraquinones in ring A. In studies on intact plants as well as cell suspension cultures the latter possibility was shown to be correct (Leistner, 1973b, 1975).

Finally, it can be concluded that four different pathways exist for the conversion of o-succinoylbenzoic acid (XXVII) to quinones. o-Succinoylbenzoic acid (XXVII) is either metabolized via at least one symmetrical intermediate (juglone, XXII) or only nonsymmetrical intermediates [lawsone (XXIII), alizarin (XLII)], and prenylation may take place either on a hypothetical naphthalene derivative at the carbon derived from C-3' of o-succinoylbenzoic acid [(catalponon (XXVIII), catalponol (XXIX)] or on a hypothetical naphthalene derivative at the carbon derived from C-2' of o-succinoylbenzoic acid [alizarin (XLII)].

D. Quinones Derived from Mevalonic Acid

There is one type of quinone that is not derived by one of the pathways mentioned in Sections III,A–C. Although the biosynthesis of this type of quinone has not been investigated in plants, their origin from mevalonic acid seems to be obvious. An example of one of these quinones is given below. The compound is named coleon A and has been isolated from *Coleus ignarius* by Eugster (1980).

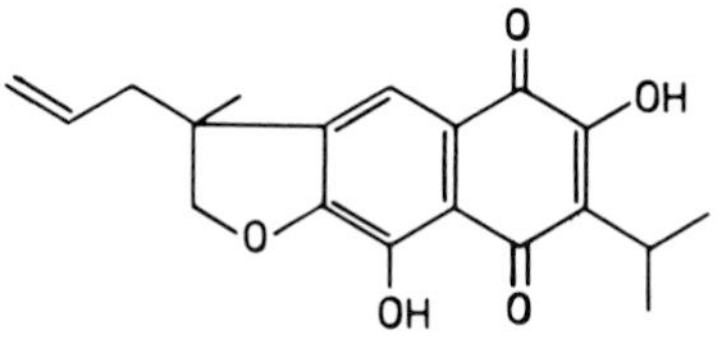

XLVII

IV. RELATION BETWEEN BIOSYNTHESIS AND DISTRIBUTION OF QUINONES

Quinones have been repeatedly used in the classification of plant taxa, and some examples and pitfalls will be discussed below. Plumbaginaceae are divided into two subfamilies, Plumbagineae on the one hand and Staticeae on the other. Naphthoquinones are restricted to the tribe Plumbagineae and are uniformly absent in Staticeae (Harborne, 1967). This is an example of the usefulness of quinones as chemical characters within a family.

7-Methyljuglone and plumbagin are naphthoquinones that are typical of the Droseraceae. Naphthoquinones do not occur in Biblidaceae and Roridulaceae, and this has been taken as an indication that these two families are not related to Droseraceae (Zenk et al., 1969).

The fact, however, that quinones are derived by different pathways may be particularly misleading if quinones are used to deduce relationships among plant taxa. Only quinones derived by the same pathway can be considered equivalent. It is therefore mandatory to compare pathways leading to quinones rather than the final quinonoid formulas (Mentzer, 1966; Merxmüller, 1967; Hegnauer, 1971).

In an early publication Mathis (1966) stated: "The presence of plumbagin which is a naphthoquinone connects the Plumbaginaceae on the one hand with Ericales (since chimaphilin and plumbagin have very similar structures) and on the other with Ebenales in which four naphthoquinones of the juglone type have been isolated." Today we know that naphthoquinones in Plumbaginaceae are acetate-derived (Durand and Zenk, 1971) and that quinones in Ebenales, including those of the juglone type, are very likely to originate from acetate (Bentley, 1975) although juglone itself stems from o-succinoylbenzoic acid (vide supra). Thus when Mathis (1966) suggested that there was a connection between Plumbaginaceae and Ebenales, he drew the right conclusion by coincidence. The statement, however, that there is a relation between Plumbaginaceae and Ericales because of the occurrence of naphthoquinones of a similar type is wrong, since we now know that quinones (chimaphilin) in Ericales (Pyrolaceae) are derived from toluhydroquinone and mevalonic acid. The relevance of considering biosynthetic pathways is also exemplified in the classification of taxa within the Rubiaceae (Hegnauer, 1971; Leistner, 1975b).

Today, in most cases, predictions regarding the biosynthesis of quinones can be made with reasonable certainty, especially if the congeners of the constituents under discussion are also inspected.

With one exception (see below) the biosynthetic pathways mentioned in Section III are restricted to certain plant taxa. Thus according to our present knowledge the pathway leading to alkannin occurs in the Boraginaceae and

Euphorbiaceae, the pathway leading to chimaphilin seems to be restricted to the Pyrolaceae only, whereas the o-succinoylbenzoic acid–mevalonic acid pathway giving rise to anthraquinones occurs in the Rubiaceae, Bignoniaceae, Gesneriaceae (vide supra), Scrophulariaceae (Dansette, 1972), and most likely (Verbenaceae (Thomson, 1971) as well.

Microorganisms have also developed biosynthetic pathways leading to quinones. These pathways are with one exception (see below) different from those in higher plants. Rifamycins are antibiotics with a naphthoquinonoid moiety derived from propionic and 3-amino-5-hydroxybenzoic acid (Gishalba et al., 1978a,b; Kibby et al., 1980), whereas the precursors of granaticin are acetate and an intact glucose unit (Floss et al., 1978). In both cases production of the quinones is restricted to species of the genus Nocardia or Streptomyces.

There is one pathway, however, leading to quinones, which occurs in bacteria, fungi, and lichens, as well as mono- and dicotyledonous plants. This is the polyacetate route. It is possible that the polyacetate route has been repeatedly acquired by different taxa during the evolution of the plant kingdom, because the starting material for the biosynthesis of this type of quinones is acetyl-CoA alone.

It may be that non-acetate-derived quinones are restricted to certain taxa because, as the result of a mixed origin, their biosynthesis requires different precursors and a series of different steps. Repeated "invention" of such a more complicated pathway seems to be less likely. This interpretation is based on Luckner's (1971) and Paech's (1950) ideas.

ACKNOWLEDGMENTS

The author's work reported herein was supported by the Deutsche Forschungsgemeinschaft.

REFERENCES

Agrawal, G. D., Rizvi, S. A. I., Gupta, P. C., and Tewari, J. D. (1972). *Planta Med.* **21**, 150–155.

Balakrishna, S., Seshadri, T. R., and Venkataramani, B. (1961). *J. Sci. Ind. Res., Sect. B* **20**, 331–333.

Baldwin, R. M., Snyder, C. D., and Rapoport, H. (1974). *Biochemistry* **13**, 1523–1530.

Bauch, H.-J., and Leistner, E. (1978). *Planta Med.* **33**, 124–127.

Bauch, H.-J., Labadie, R. P., and Leistner, E. (1975). *J. Chem. Soc., Perkin Trans. 1* pp. 689–692.

Bell, A. A., Stipanovic, R. D., and Puhalla, J. E. (1976). *Tetrahedron* **32**, 1353–1356.

Bentley, R. (1975). *Biosynthesis* **3**, 181–246.

Bentley, R., and Campbell, I. M. (1974). *In* "The Chemistry of Quinonoid Compounds" (S. Patai, ed.), pp. 683–736. Wiley, New York.

Berger, S., and Rieker, A. (1974). *In* "The Chemistry of Quinonoid Compounds (S. Patai, ed.), pp. 163–229. Wiley, New York.

Berlin, J., and Barz, W. (1975). *Z. Naturforsch., Teil C* **30**, 650–658.

Bolkart, K. H., and Zenk, M. H. (1968a). *Z. Pflanzenphysiol.* **59**, 439–444.

Bolkart, K. H., and Zenk, M. H. (1968b). *Naturwissenschaften* **55**, 444–445.

Bolkart, K. H., and Zenk, M. H. (1969). *Z. Pflanzenphysiol.* **61**, 356–359.

Bolkart, K. H., Knobloch, M., and Zenk, M. H. (1968). *Naturwissenschaften* **55**, 455.

Brew, E. J. C., and Thomson, R. H. (1971a). *J. Chem. Soc. C* pp. 2001–2007.

Brew, E. J. C., and Thomson, R. H. (1971b). *J. Chem. Soc. C* pp. 2007–2010.

Brodie, A. F., Revsin, B., Kalva, V., Phillips, P., Bogin, E., Higashi, T., Krishna Marti, C. R., Cabari, B. Z., and Marquez, E. (1970). *In* "Biological Function of Terpenoid Quinones" (T. W. Goodwin, ed.), pp. 119–143. Academic Press, New York.

Burnett, A. R., and Thomson, R. H. (1967). *J. Chem. Soc.* pp. 2100–2104.

Burnett, A. R., and Thomson, R. H. (1968a). *J. Chem. Soc. C* pp. 850–853.

Burnett, A. R., and Thomson, R. H. (1968b). *J. Chem. Soc. C* pp. 854–857.

Burnett, A. R., and Thomson, R. H. (1968c). *J. Chem. Soc. C* pp. 857–860.

Burnett, A. R., and Thomson, R. H. (1968d). *J. Chem. Soc. C* pp. 2437–2441.

Chakraborty, D. P., Islan, A., and Roy, S. (1978). *Phytochemistry* **17**, 2043.

Dansette, P. (1972). Etude sur la Biosynthèse des Naphthoquinones végétales et bactériennes. Ph.D. Thesis, Université de Paris-Sud.

Dansette, P., and Azerad, R. (1970). *Biochem. Biophys. Res. Commun.* **40**, 1090–1095.

Durand, R., and Zenk, M. H. (1971). *Tetrahedron Lett.* pp. 3009–3012.

Durand, R., and Zenk, M. H. (1974). *Phytochemistry* **13**, 1483–1492.

Eugster, C. H. (1980). *In* "Pigments in Plants" (F.-C. Czygan, ed.), pp. 149–186. G. Fischer-Verlag, Stuttgart and New York.

Fairbairn, J. W., and Muhtadi, F. J. (1972). *Phytochemistry* **11**, 215–219.

Floss, H. G., Chang, C.-J., Mascaretti, O., and Shimada, K. (1978). *Planta Med.* **34**, 345–380.

Fournier, G., Bercht, C. A. L., Paris, R. R., and Paris, M. R. (1975). *Phytochemistry* **14**, 2099.

Franck, B., Scharf, V., and Schrameyer, M. (1974). *Angew. Chem.* **86**, 160–161.

Ghaleb, H., Rizk, A. M., Hammouda, F. M., and Abdel-Gawed, M. M. (1972). *Qual. Plant. Mater. Veg.* **21**, 237–251.

Gishalba, O., and Nüesch, J. (1978a). *J. Antibiot.* **31**, 202–214.

Gishalba, O., and Nüesch, J. (1978b). *J. Antibiot.* **31**, 215–225.

Grob, K., Rüedi, P., and Eugster, C. H. (1978). *Helv. Chim. Acta* **61**, 871–884.

Grotzinger, E., and Campbell, I. M. (1972a). *Phytochemistry* **11**, 675–679.

Grotzinger, E., and Campbell, I. M. (1972b). *Tetrahedron Lett.* pp. 4685–4686.

Grotzinger, E., and Campbell, I. M. (1974). *Phytochemistry* **13**, 923–926.

Harborne, J. B. (1966). *Phytochemistry* **5**, 589–600.

Harborne, J. B. (1967). *Phytochemistry* **6**, 1415–1428.

Harborne, J. B., and Mokhtari, N. (1977). *Phytochemistry* **16**, 1314–1315.

Harris, T. M., Webb, A. D., Harris, C. M., Wittek, P. J., and Murray, T. P. (1976). *J. Am. Chem. Soc.* **98**, 6065–6067.

Hegnauer, R. (1971). *Naturwissenschaften* **58**, 585–598.

Hill, R. K., and Newkome, G. R. (1969). *J. Am. Chem. Soc.* **91**, 5893–5894.

Höfle, G. (1977). *Tetrahedron* **33**, 1963–1970.

Imre, S., and Öztunc, A. (1976). *Z. Naturforsch., Teil C* **31**, 403–407.

Inoue, K., Ueda, S., Shiobara, Y., and Inouye, H. (1977a). *Phytochemistry* **16**, 1689–1694.

Inoue, K., Shiobara, Y., and Inouye, H. (1977b). *Chem. Pharm. Bull.* **25**, 1462–1468.

Inoue, K., Shiobara, Y., Nayeshiro, H., Inouye, H., Wilson, G., and Zenk, M. H. (1979). *J. Chem. Soc., Chem. Commun.* pp. 957–959.

Inouye, H., Ueda, S., Inoue, K., Hayashi, T., and Hibi, T. (1975). *Chem. Pharm. Bull.* **23**, 2523–2533.

Inouye, H., Ueda, S., and Inoue, K. (1978). *Tetrahedron Lett.* pp. 4551–4554.

Inouye, H., Ueda, S., Inoue, K., and Matsumara, H. (1979). *Phytochemistry* **18**, 1301–1308.

Inouye, J., Okuda, T., and Hayashi, T. (1971). *Tetrahedron Lett.* pp. 3615–3618.

Kibby, J. J., McDonald, I. A., and Rickards, R. W. (1980). *J. Chem. Soc. Chem. Commun.* pp. 768–769.

Knapp, J. E., Farnsworth, N. R., Theiner, M., and Schiff, P. L. (1972). *Phytochemistry* **11**, 3091–3092.

Koshioka, M., and Takino, Y. (1978). *Chem. Pharm. Bull.* **26**, 1343–1347.

Labadie, R. P. (1972). *Pharm. Weekbl.* **107**, 541–547.

Leduc, M. M., Dansette, P. M., and Azerad, R. G. (1970). *Eur. J. Biochem.* **15**, 428–435.

Leistner, E. (1971). *Phytochemistry* **10**, 3015–3020.

Leistner, E. (1973a). *Phytochemistry* **12**, 337–345.

Leistner, E. (1973b). *Phytochemistry* **12**, 1669–1674.

Leistner, E. (1975). *Planta Med., Suppl.* pp. 214–224.

Leistner, E. (1980). *In* "Pigments in Plants" (I.-C. Czygan, ed.), pp. 352–369. G. Fischer-Verlag, Stuttgart and New York.

Leistner, E., and Zenk, M. H. (1967). *Z. Naturforsch., Teil B* **22**, 865–868.

Leistner, E., and Zenk, M. H. (1968a). *Z. Naturforsch., Teil B* **23**, 259–268.

Leistner, E., and Zenk, M. H. (1968b). *Tetrahedron Lett.* pp. 1395–1396.

Leistner, E., and Zenk, M. H. (1969). *Chem. Commun.* pp. 210–211.

Leistner, E., and Zenk, M. H. (1971). *Tetrahedron Lett.* pp. 1677–1681.

Lindsey, A. S. (1974). *In* "The Chemistry of Quinonoid Compounds" (S. Patai, ed.), pp. 793–855. Wiley, New York.

Luckner, M. (1971). *Pharmazie* **26**, 717–724.

McDonald, I. A., Simson, T. J., and Sieradowski, A. F. (1977). *Aust. J. Chem.* **30**, 1727–1734.

McGovern, E. P., and Bentley, R. (1978). *Arch. Biochem. Biophys.* **188**, 56–63.

Mathis, C. (1966). *In* "Comparative Phytochemistry" (T. Swain, ed.), pp. 245–270. Academic Press, New York.

Meganathan, R., Folger, T., and Bentley, R. (1980). *Biochemistry* **19**, 785–789.

Mentzer, C. (1966). *In* "Comparative Phytochemistry" (T. Swain, ed.), pp. 21–31. Academic Press, New York.

Merxmüller, H. (1967). *Ber. Dtsch. Bot. Ges.* **80**, 608–620.

Miyase, T., Rüedi, P., and Eugster, C. H. (1978). *Helv. Chim. Acta* **60**, 2770–2779.

Mulchandani, N. B., and Hassarajani, S. A. (1977). *Planta Med.* **32**, 357–361.

Müller, W.-U., and Leistner, E. (1976). *Phytochemistry* **15**, 407–410.

Müller, W.-U., and Leistner, E. (1978a). *Phytochemistry* **17**, 1735–1738.

Müller, W.-U., and Leistner, E. (1978b). *Phytochemistry* **17**, 1739–1742.

Onderka, D. K., and Floss, H. G. (1969). *J. Am. Chem. Soc.* **91**, 5894–5896.

Paech, K. (1950). "Biochemie und Physiologie der sekundären Pflanzenstoffe." Springer-Verlag, Berlin and New York.

Pedersen, J. A. (1978). *Phytochemistry* **17**, 775–778.

Price, J. R., and Robinson, R. (1940). *J. Chem. Soc.* pp. 1493–1499.

Rao, R. S., and Verra Reddy, G. C. (1977). *Indian J. Chem., Sect. B* **15**, 497–498.

Rizvi, S. A. J., Gupta, P. C., and Kaul, R. K. (1971). *Planta Med.* **19**, 222–233.

Robins, D. J., Campbell, I. M., and Bentley, R. (1970). *Biochem. Biophys. Res. Commun.* **39**, 1081–1086.

Rüedi, P., and Eugster, C. H. (1977). *Helv. Chim. Acta* **60**, 945–947.

Sandermann, W. (1966). *Naturwissenschaften* **53**, 513–525.

Sandermann, W., and Dietrichs, H.-H. (1959). *Holzforschung* **13**, 137–148.

Scharf, K.-H., Zenk, M. H., Onderka, D. K., Carroll, M., and Floss, H. G. (1971). *Tetrahedron Lett.* pp. 576–577.

Schildknecht, H., Straub, F., and Scheidel, V. (1976). *Justus Liebigs Ann. Chem.* pp. 1295–1306.

Schmid, H. V., and Zenk, M. H. (1971). *Tetrahedron Lett.* pp. 4151–4155.

Schratz, E. (1957). *In* "Forschungsberichte des Wirtschafts und Verkehrsministeriums Nordrhein-Westfalen" (L. Brandt, ed.), pp. 5–62. Westdeutscher Verlag, Köln, Opladen.

Shineberg, B., and Young, I. G. (1976). *Biochemistry* **15,** 2754–2758.

Stipanovic, R. D., and Bell, A. A. (1977). *Mycologia* **69,** 164–172.

Stöckigt, J., Srocka, U., and Zenk, M. H. (1973). *Phytochemistry* **12,** 2389–2391.

Tabata, M. (1977). *In* "Plant Tissue Culture and its Biotechnological Application" (W. Barz, E. Reinhard, and M. H. Zenk, eds.), pp. 3–16. Springer-Verlag, Berlin and New York.

Tezuka, M., Takahashi, C., Kuroyanagi, M., Satake, M., Yoshihiva, K., and Natori, S. (1973). *Phytochemistry* **12,** 175–183.

Thomas, G., and Threlfall, D. R. (1974). *Phytochemistry* **13,** 807–813.

Thomson, R. H. (1971). "Naturally Occurring Quinones." Academic Press, New York.

Tokoroyama, T., and Kubota, T. (1971). *J. Chem. Soc. C* pp. 2703–2708.

Trebst, A. (1978). *Philos. Trans. R. Soc. London, Ser. B* **284,** 591–599.

Yagi, A., Yamanouchi, M., and Nishioka, I. (1978). *Phytochemistry* **17,** 895–897.

Young, I. G. (1975). *Biochemistry* **14,** 399–406.

Zeller, K. P. (1974). *In* "The Chemistry of Quinonoid Compounds" (S. Patai, ed.), pp. 231–256. Wiley, New York.

Zenk, M. H., and Leistner, E. (1968). *Lloydia* **31,** 275–292.

Zenk, M. H., Fürbringer, M., and Steglich, W. (1969). *Phytochemistry* **8,** 2199–2200.

Zenk, M. H., El-Shagi, H., and Schulte, U. (1975). *Planta Med., Suppl.* pp. 79–101.

Flavonoids | *14*

KLAUS HAHLBROCK

I. INTRODUCTION

Flavonoids constitute one of the most characteristic classes of compounds in higher plants. Many flavonoids are easily recognized as flower pigments in most angiosperm families. However, their occurrence is not restricted to flowers but includes all parts of the plant. The chemical structures of flavonoids are based on a C_{15} skeleton with a chromane ring bearing a second aromatic ring B in position 2, 3, or 4 (Fig. 1). In a few cases, the six-membered heterocyclic ring C occurs in an isomeric open form or is replaced by a five-membered ring (Section III).

Various subgroups of flavonoids are classified according to the substitution patterns of ring C. Both the oxidation state of the heterocyclic ring and the position of ring B are important in the classification. Examples of the six

The Biochemistry of Plants, Vol. 7

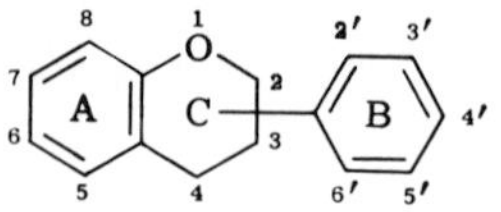

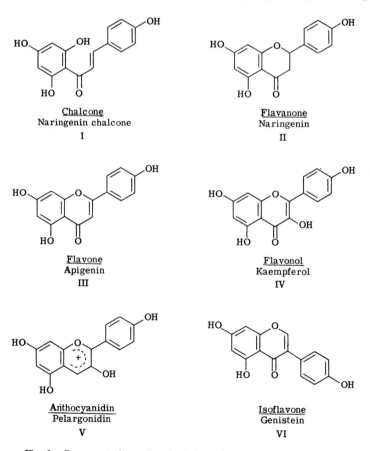

Fig. 1. Basic structure of most flavonoids. Exceptions from the numbering are used for several subgroups (e.g., Fig. 13).

major subgroups (chalcones and the isomeric flavanones, flavones, flavonols, anthocyanins, and isoflavonoids) are given in Fig. 2. Most of these (flavanones, flavones, flavonols, and anthocyanins) bear ring B in position 2 of the heterocyclic ring. In isoflavonoids ring B occupies position 3. A group of chromane derivatives with ring B in position 4 (4-phenylcoumarins = neoflavonoids) is treated by Brown (this volume, Chapter 10) together with other coumarin derivatives. Another small group comprises oligomeric

Chalcone
Naringenin chalcone
I

Flavanone
Naringenin
II

Flavone
Apigenin
III

Flavonol
Kaempferol
IV

Anthocyanidin
Pelargonidin
V

Isoflavone
Genistein
VI

Fig. 2. Representatives of each of six major subgroups of flavonoids.

flavonoids, biflavonyls, and proanthocyanidins. Several comprehensive reviews exist where structure, occurrence, and methods of identification of the flavonoids are compiled (cf. Harborne, 1967; Harborne *et al.,* 1975, Mabry *et al.,* 1970).

Altogether there are many hundreds of differently substituted flavonoid aglycones. Most of these occur as glycosides with different combinations of sugars attached to hydroxyl groups. The sugars are often further substituted by acyl residues, such as malonate, 4-coumarate, caffeate, and ferulate. Some flavonoids occur as *C*-glycosyl derivatives in position 6 or 8.

The major flavonoids occur almost universally in higher plants, including mosses and ferns. In the Centrospermae and Cactaceae, anthocyanins are replaced by betalains, a chemically different class of colored compounds (see Piatelli, this volume, Chapter 19). Despite several positive reports, the occurrence of flavonoids in bacteria, algae, and fungi is doubtful (Harborne, 1967). The function of flavonoids in higher plants as attractants of animals involved in fertilization is obvious. Other important functions are attributed to flavonoids as protective agents against uv light (Robberecht and Caldwell, 1978) or infection by phytopathogenic organisms (Grisebach and Ebel, 1978) (see also Bell, this volume, Chapter 1). Flavonoids are often rapidly metabolized after synthesis. The mechanisms of flavonoid degradation are discussed by Barz and Köster (this volume, Chapter 3). The rates of both synthesis and degradation of flavonoids vary greatly during different stages of plant development. This important aspect is dealt with by Wiermann (this volume, Chapter 4). Furthermore, the important implications of compartmentation of the sites of synthesis, storage, and degradation of flavonoids (see Stafford, this volume, Chapter 5) should be borne in mind.

The early steps in the biosynthesis of the various subgroups of flavonoids are closely related. Earlier experiments with radioactively labeled precursors established that the carbon skeleton of all flavonoids is derived from acetate and phenylalanine. Ring A is formed from three acetate units (malonate), and phenylalanine gives rise to ring B and C-2, C-3, and C-4 of the heterocyclic ring C. A central intermediate in the formation of all flavonoids is the chalcone or the isomeric flavanone. The biosynthetic relationship of flavonoids as concluded mainly from isotope experiments is illustrated in Fig. 3. Most of the results from tracer studies have been summarized in detailed reviews (cf. Hahlbrock and Grisebach, 1975).

More recent work at the enzymatic level has largely confirmed these original hypothetical steps postulated for the incorporation of acetate and phenylalanine into flavonoids. In particular, extensive study of the enzymology and regulation of flavone and flavonol glycoside biosynthesis has revealed many further details of the individual reactions. However, relatively little beyond the results of tracer studies is known about the synthesis of other flavonoids. If the biochemistry of flavone and flavonol glycosides ap-

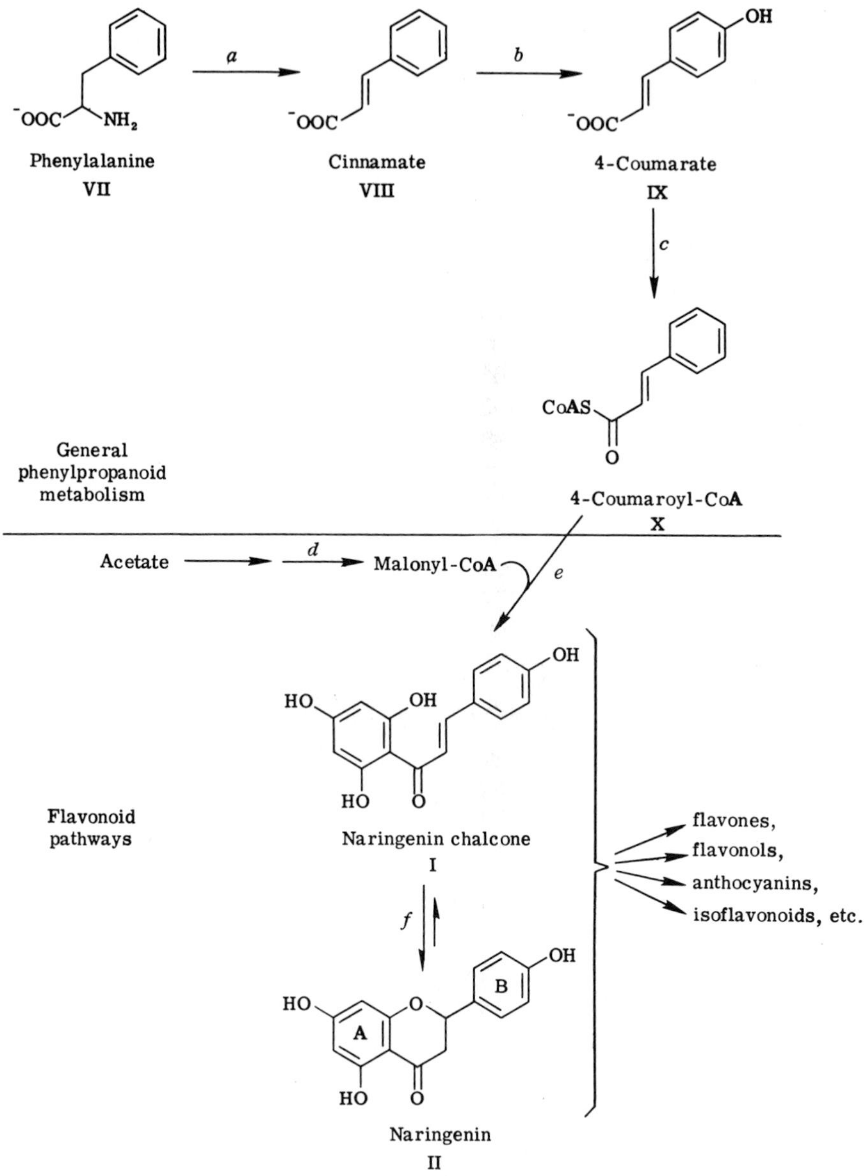

Fig. 3. Common steps in the biosynthesis of all flavonoids. For enzymes a–f, see Table I.

pears overemphasized in some of the following sections, this therefore not only reflects their relative frequency of occurrence but also the relatively advanced state of knowledge of their synthesis. Cell cultures from various plants were especially useful in studies on the biochemistry of phenylpropanoid compounds (see also Chapters 2, 3, and 15), and most of the studies on flavonoid biosynthesis and degradation in recent years have been conducted with plant cell cultures.

II. FLAVONES AND FLAVONOLS

A. Biochemical Relations

Chemically, flavonols are 3-hydroxyflavones. Except for this difference, flavonol aglycones are often similar in their substitution patterns to flavones occurring in the same tissue. A striking example of the structural similarity of co-occurring flavones and flavonols is given below for the aglycones from cell cultures of parsley. The same example of cultured parsley cells also demonstrates the great similarity of the two biosynthetic pathways. Several steps in the biosynthesis of flavones and flavonols are catalyzed by the same enzymes. However, a major difference between the two types of flavonoids is frequently related to the substitution pattern of the 3-hydroxyl group of flavonols. This position is usually more-or-less highly glycosylated with various sugars which can be further substituted with acyl residues.

B. Biosynthesis in Parsley

Leaves and seeds from parsley plants can contain several percent flavone and flavonol glycosides on a dry weight basis. A high concentration of these compounds occurs also in irradiated suspension cultures of parsley cells derived from leaf petioles, and these cell cultures have been used as a convenient system for studies on the biosynthesis of flavone and flavonol glycosides. Particular advantages of this system are the large amounts of tissue and the high levels of enzyme activity that can be obtained, as well as the defined conditions under which the regulation of biosynthesis can be investigated (Section II,B,3). Most of the enzymes of flavonoid glycoside biosynthesis known to date were isolated from parsley cell cultures.

1. Chemical Structures

The more than 20 different flavonoid glycosides occurring in irradiated parsley cells are derived from only three flavone and three flavonol aglycones, all of which have very similar substitution patterns (Kreuzaler and Hahlbrock, 1973). The chemical structures of the aglycones and their possible biosynthetic relations are shown in Fig. 4. Except for the characteristic 3-hydroxyl group of flavonols, the six aglycones differ only with respect to substitution in the 3'-position.

All flavones and flavonols in parsley occur as glycosides (Kreuzaler and

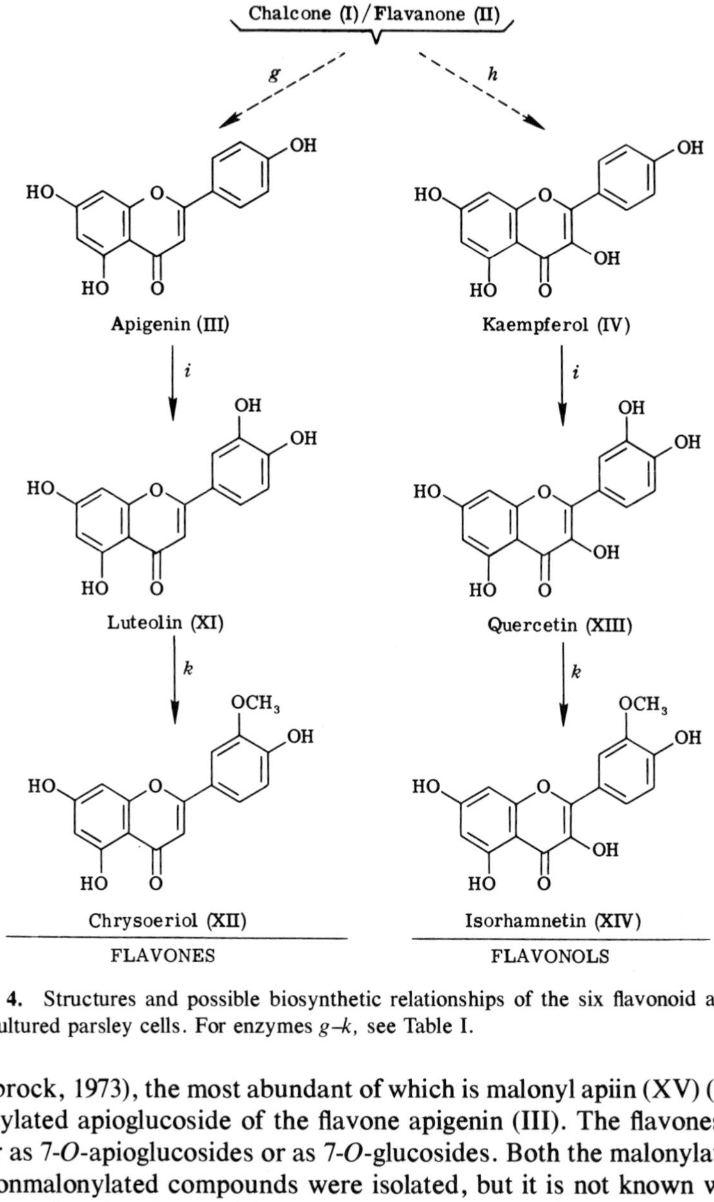

Fig. 4. Structures and possible biosynthetic relationships of the six flavonoid aglycones from cultured parsley cells. For enzymes *g–k,* see Table I.

Hahlbrock, 1973), the most abundant of which is malonyl apiin (XV) (Fig. 5), an acylated apioglucoside of the flavone apigenin (III). The flavones occur either as 7-*O*-apioglucosides or as 7-*O*-glucosides. Both the malonylated and the nonmalonylated compounds were isolated, but it is not known whether the nonmalonylated glycosides accumulate in substantial amounts *in vivo* or are merely artifacts formed during isolation. Flavonols are either 7-O-monoglucosylated or 3,7-O-bisglucosylated. Bis-, mono-, and nonmalonylated flavonol glycosides have been isolated. Figure 5 shows, as one example, the structure of the bismalonylated form of isorhamnetin 3,7-O-

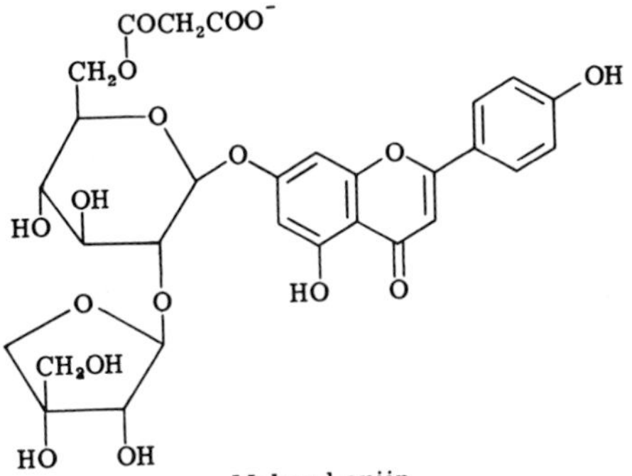

Malonyl apiin
XV

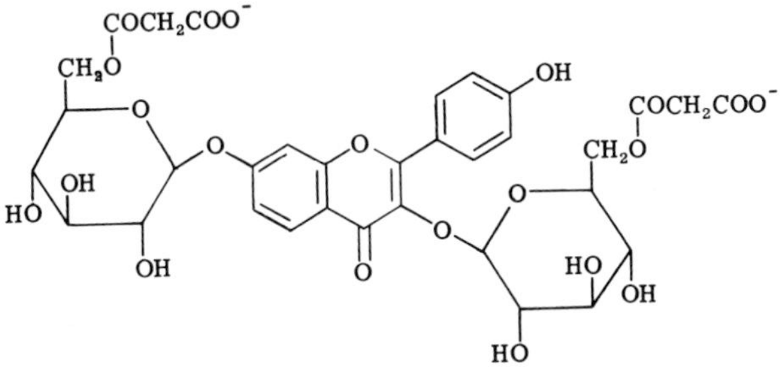

Bismalonyl isorhamnetin 3,7-O-bisglucoside
XVI

Fig. 5. Structures of two major acylated flavonoid glycosides from cultured parsley cells.

bisglucoside (XVI). The site of malonylation of flavone and flavonol glycosides is not known with certainty but is deduced from indirect evidence to be the 6-position of the glucose moiety (Hahlbrock, 1972).

2. Biosynthetic Steps

As indicated in Fig. 3, the enzymatic reactions leading to the formation of flavonoids from compounds of intermediary metabolism can be classified into two groups (Section II,B,3,b). The first group (I) comprises the reactions catalyzed by the enzymes of general phenylpropanoid metabolism. The result is the conversion of L-phenylalanine (VII) to 4-coumaroyl-coenzyme A (CoA) (X). This sequence of reactions precedes the formation not only of all flavonoids but also of most other phenylpropanoid compounds, such as lignin and cinnamate esters (see Gross, this volume, Chapter 11, and Grisebach, this volume, Chapter 15).

The second group (II) consists of about 13 enzymes and can be further subdivided into several major steps. The first step is the conversion of acetyl-CoA to malonyl-CoA which serves as substrate for three enzymes of this pathway. In the second step, the C_{15} skeleton is formed by a synthase reaction. The product is then further converted to the basic flavone and flavonol structures. The two subsequent steps are the partial hydroxylation and methylation of the aglycones and their glycosylation. In a last step, the glycosides are acylated to give the final products.

a. General Phenylpropanoid Metabolism. The first two enzymes of general phenylpropanoid metabolism, phenylalanine ammonia-lyase (*a*) and cinnamate 4-hydroxylase (*b*) (Fig. 3), are discussed in detail by Hanson and Havir (this volume, Chapter 20) and Butt and Lamb (this volume, Chapter 21). The enzymes from parsley cells (Zimmerman and Hahlbrock, 1975; Pfändler *et al.*, 1977) are remarkably similar in their properties to those from other plant tissues. They are specific for the conversion of L-phenylalanine to *trans*-4-coumarate via *trans*-cinnamate. The third enzyme of this sequence, 4-coumarate:CoA ligase (*c*) has marked specificity for *trans*-4-coumarate. The K_m value for this substrate is lower and the V/K_m value is higher than for some of the most closely related compounds, including caffeate and ferulate (Knobloch and Hahlbrock, 1977). In addition to 4-coumarate, caffeate and ferulate might be considered possible substrates for flavonoid biosynthesis in parsley, since their substitution patterns correspond to those of the 3'-hydroxylated (XI, XII) and 3'-methoxylated flavonoids (XII, XIV), respectively. However, an actual function as precursor is uncertain for caffeate and very unlikely for ferulate (Sections II,B,2,c and II,B,2,e).

The CoA esters of 4-coumarate and of other substituted cinnamic acids play a central role as intermediates at the branching point between general phenylpropanoid metabolism and the various subsequent specific pathways. The pronounced substrate specificities and interesting regulatory properties of 4-coumarate:CoA ligases from different plants indicate that this enzyme represents a sensitive point of control in the distribution of CoA esters among the different pathways (Hahlbrock, 1977; Hahlbrock and Grisebach, 1979). In parsley cells, the ligase is thought to be specifically involved in the synthesis of flavonoids, and perhaps of cinnamate esters which also occur in this material (Knobloch and Hahlbrock, 1977; T. Kühnl, unpublished results). The enzyme has been purified and studied extensively. It is strongly inhibited by AMP and exhibits sigmoidal saturation kinetics for the two cosubstrates of 4-coumarate, ATP and CoASH. This behavior was interpreted to be an indication of homotropic allosteric effects and a possible regulation of ligase activity through the ATP/AMP ratio (Knobloch and Hahlbrock, 1977).

The activities of all three enzymes of general phenylpropanoid metabolism are regulated interdependently with the subsequent enzymes of flavonoid glycoside biosynthesis in irradiated parsley cells. Under certain conditions, the activity of phenylalanine ammonia-lyase might be rate-limiting for the accumulation of flavonoids (Section II,B,3,b). Some of the characteristic properties of the enzymes discussed here are listed in Table I.

b. Acetyl-CoA Carboxylase. This carboxylase is normally regarded as a typical enzyme of fatty acid biosynthesis (see Vol. 4). However, its product, malonyl-CoA, is not only an important building block for the carbon skeleton of flavonoid aglycones but is also the substrate for the final malonylation of the glycosides in parsley cells. In fact, synthesis of acylated glycosides requires more molecules of malonyl-CoA than of any other substrate. It has been deduced from the regulation of acetyl-CoA carboxylase activity in parsley cells (Section II,B,3,b) that this enzyme is an integral part of flavonoid biosynthesis (Ebel and Hahlbrock, 1977). The reaction catalyzed by acetyl-CoA carboxylase (d) is

$$CH_3-CO-SCoA + HCO_3^- + ATP \rightleftharpoons HOOC-CH_2-CO-SCoA + ADP + P_i$$

c. Chalcone Synthase. The step catalyzed by chalcone synthase (e) is the key reaction in flavonoid biosynthesis (Fig. 3). For several years, the enzyme was thought to be a flavanone synthase (Kreuzaler and Hahlbrock, 1975a), but the immediate product of the reaction has now been identified as naringenin chalcone Heller and Hahlbrock, 1980). In all earlier publications concerning the enzyme from parsley (and probably from other sources, too) the term "flavanone synthase" should therefore be replaced by "chalcone synthase." The chalcone undergoes chemical cyclization very rapidly at the pH optimum of the synthase reaction (pH 8), and special conditions (e.g., low pH and the absence of chalcone isomerase) must be employed to allow the chalcone to be detected as an intermediate in flavanone formation.

Naringenin chalcone (I) and the isomeric flavanone (II) are not accumulated in significant amounts in parsley cells (Kreuzaler and Hahlbrock, 1973). Nevertheless, substantial quantities of a homogeneous preparation of the synthase are easily isolated from irradiated cells (Kreuzaler and Hahlbrock, 1975a; Kreuzaler et al., 1979). The enzyme catalyzes the sequential condensation of one molecule of 4-coumaroyl-CoA with three molecules of malonyl-CoA. The aliphatic sidechain of 4-coumarate is thereby elongated by three acetate units (Kreuzaler and Hahlbrock, 1975a,b; Hrazdina et al., 1976). The concept of a stepwise addition of acetate units to the starter molecule is supported by the additional formation under certain assay conditions of synthase products containing one or two acetate units less than naringenin (Kreuzaler and Hahlbrock, 1975b; Hrazdina et al., 1976). As further side reactions, chalcone synthase catalyzes both CO_2 exchange and decarboxylation of malonyl-CoA (see below).

The various results obtained with purified synthase preparations suggest the reaction mechanism depicted in Fig. 6. The proposed stepwise elongation

TABLE I

List of Enzymes Involved in the Biosynthesis of Flavonoid Glycosides from Compounds of Intermediary Metabolism in Cultured Parsley Cells[a]

Group	Reaction	Enzyme	E.C. number	Most efficient substrate(s) tested	$K_m \times 10^6$	Approximate MW
I	a	Phenylalanine ammonia-lyase	4.3.1.5	L-Phenylalanine (VII)	32	330,000
I	b	Cinnamate 4-hydroxylase	1.14.13.11	trans-Cinnamate (VIII)	70	—
I	c	4-Coumarate:CoA ligase	6.2.1.12	trans-4-Coumarate (IX)	14	60,000
II	d	Acetyl-CoA carboxylase	6.4.1.2	—	—	—
II	e	Chalcone synthase	—	Malonyl-CoA,	9	80,000
				4-Coumaroyl-CoA (X)	1.6	
II	f	Chalcone isomerase	5.5.1.6	Naringenin chalcone (I)	16[b]	50,000
II	g	"Flavonoid oxidase 1"[c]	—	—	—	—
II	h	"Flavonoid oxidase 2"[d]	—	—	—	—
II	i	"Flavonoid 3'-hydroxylase"[d]	—	—	—	—
II	k	S-Adenosylmethionine:flavonoid 3'-O-methyltransferase	2.1.1.42	Luteolin (XI) (or 7-O-glucoside)	46	50,000
II	l	UDP-glucose:flavonoid 7-O-glucosyltransferase	2.4.1.82	Luteolin (XI)	1.5	50,000
II	m	UDP-apiose/UDP-xylose synthase	—	UDP-D-glucuronate (XX)	2	86,000
II	n	UDP-Apiose:flavone 7-O-glucoside apiosyltransferase	2.4.2.25	Apigenin 7-O-glucoside	66	50,000
II	o	UDP-glucose:flavonol 3-O-glucosyltransferase	2.4.1.-	Quercetin (XIII) (or 7-O-glucoside)	—	50,000
II	p	Malonyl-CoA:flavonoid 7-O-glycoside malonyltransferase	2.3.1.-	Apiin	10	50,000
II	q	Malonyl-CoA:flavonol 3-O-glucoside malonyltransferase	2.3.1.-	Kaempterol 3-O-glucoside	4	50,000

[a] For references see text.
[b] Enzyme from parsley plants.
[c] Enzyme from parsley plants.
[d] Hypothetical reaction; see Sections II,B,2,d and 2,h.

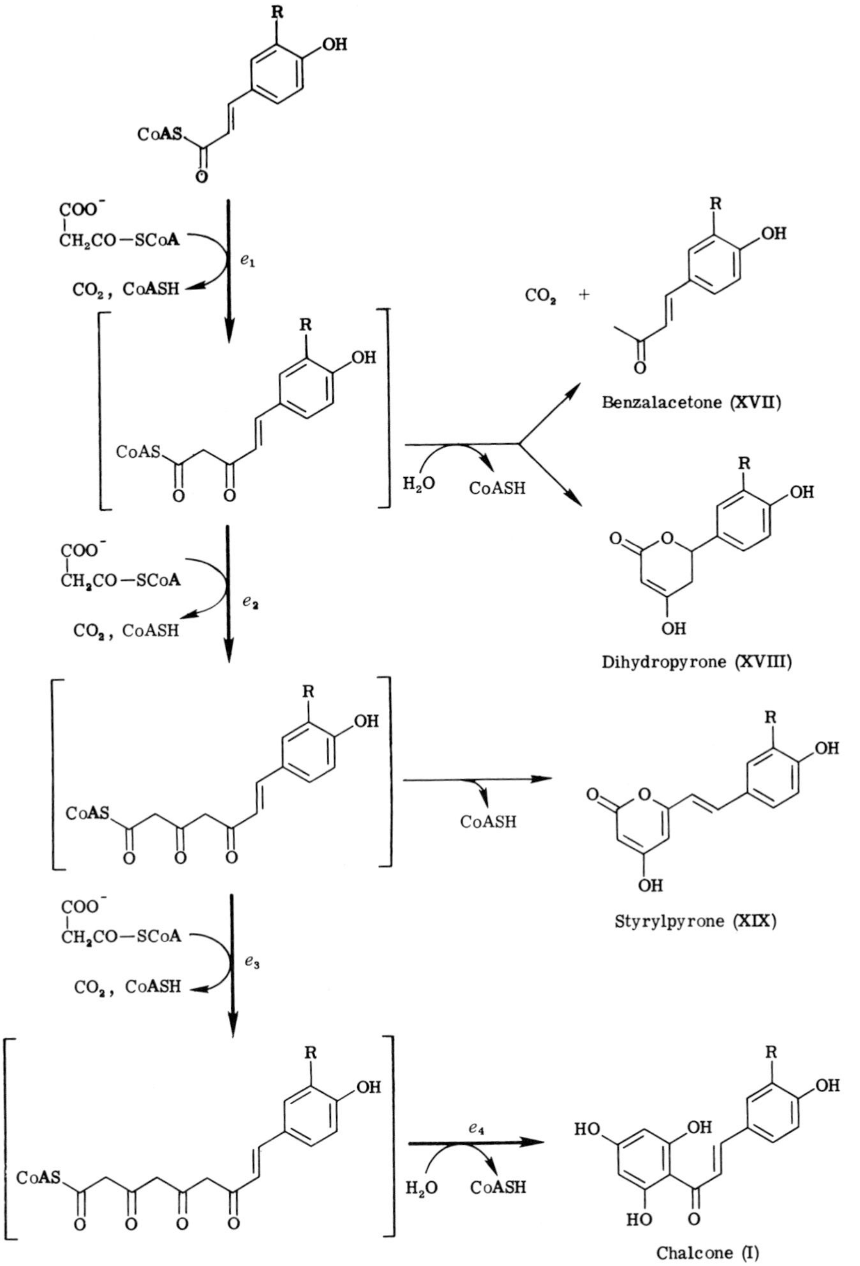

Fig. 6. Proposed mechanism of action of chalcone synthase from parsley. Broad arrows (reactions e_1–e_4), major reactions *in vivo*; other arrows, side reactions *in vitro*. R = H or OH. (Modified after Hrazdina *et al.*, 1976).

(reactions e_1-e_3) leads to the formation of a hypothetical tri-β-ketoacyl thioester as the last open-chain intermediate. This compound is probably highly unstable and cyclizes either spontaneously or in an enzyme-mediated reaction (e_4). The intermediate occurrence of the preceding mono- and di-β-ketoacyl thioesters was deduced from the formation of a benzalacetone (XVII), a dihydropyrone (XVIII), and a styrylpyrone (XIX) derivative as by-products of the synthase reaction. These compounds are not found in parsley cells *in vivo*. Their synthesis *in vitro* was attributed to a premature release of incomplete condensation products from the synthase under the artificial conditions of the enzyme assay. The putative immediate release products, the intermediate thioesters, are probably unstable and might rapidly hydrolyze and then undergo either decarboxylation or cyclization, as indicated in Fig. 6 (Kreuzaler and Hahlbrock, 1975b; Hrazdina *et al.*, 1976).

Decarboxylation and CO_2 exchange of malonyl-CoA are further side reactions that might provide some insight into the mechanism of the action of chalcone synthase. Under appropriate conditions, both reactions are catalyzed by a purified synthase preparation at considerable rates (Kreuzaler *et al.*, 1978). The occurrence of these side reactions suggests the involvement of an acetyl-CoA carbanion in the condensation reaction (Fig. 7). Under

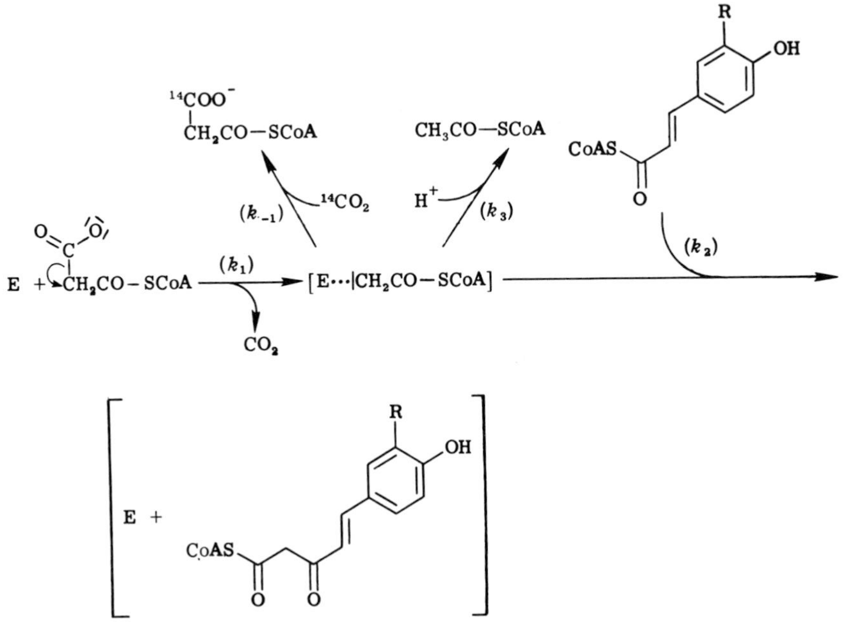

Fig. 7. Proposed mechanism of reaction e_1 (Fig. 6) of chalcone synthase (E) from parsley. Broad arrows, putative reactions *in vivo*; other arrows, additional reactions *in vitro*. R = H or OH. (Modified after Kreuzaler *et al.*, 1978).

normal conditions *in vivo*, the electrophilic condensation partner is assumed to be the acyl residue either of 4-coumaroyl-CoA, the starter molecule of the condensation reaction, or of the subsequent intermediates in chain elongation. An alternative reaction *in vitro* for stabilizing the carbanion could be either protonation or reversal of its formation, that is, carboxylation (Kreuzaler *et al.*, 1978). The respective K_m values for malonyl-CoA in the resulting decarboxylation and CO_2 exchange reactions are relatively high (approximately 0.4 and 2 nmol/liter) when compared with a value of about 9 μmol/liter in the complete synthase reaction (S. E. Gardiner and K. Hahlbrock, unpublished results). This suggests that these futile reactions do not take place at significant rates *in vivo*.

An interesting phenomenon has been reported with respect to the substrate specificity of chalcone synthase. Enzyme preparations from cell cultures of both parsley and *Haplopappus gracilis* are specific for 4-coumaroyl-CoA as substrate at a pH optimum of about pH 8 (Hrazdina *et al.*, 1976; Saleh *et al.*, 1978). However, equal rates of conversion of 4-coumaroyl-CoA and caffeoyl-CoA (Fig. 6, R = OH) to the corresponding products occur at pH 6.5. It has been speculated that both the substrate concentration and the pH at the site of action of the synthase play a role in determining the relative rates of formation of the two compounds isolated, naringenin (II) and eriodictyol 3'-hydroxynaringenin (Saleh *et al.*, 1978). Feruloyl-CoA (R = OCH_3) is not converted with significant efficiency to the corresponding chalcone or flavanone (Hrazdina *et al.*, 1976), suggesting that methylation takes place at a later stage during flavonoid biosynthesis (see also Section II,B,2,e).

The mechanisms of chain elongation are similar in the chalcone synthase reaction and in the synthesis of fatty acids or of other "polyacetate" compounds derived from malonyl-CoA (see Vol. 4). However, one peculiarity of chalcone synthase is the utilization of CoA esters as immediate substrates for the condensation reaction. Purified chalcone synthase from parsley cells does not contain an acyl carrier protein nor a pantetheinyl residue directly bound to the enzyme (Kreuzaler *et al.*, 1979). Such components have been found in other enzymes synthesizing "acetogenins" derived from malonyl-CoA (see Vol. 4). The formation of acetyl-CoA from malonyl-CoA in the decarboxylation reaction mentioned above is a strong indication that the putative acetyl carbanion intermediate reacts as a CoA ester and not in an enzyme-bound form (Kreuzaler *et al.*, 1978).

d. Formation of the Basic Flavone and Flavonol Structures. Details of these reactions are largely unknown. The only well-known reaction involved in this step is that catalyzed by chalcone isomerase (*f*). This enzyme has been isolated from a variety of plants and studied in great detail (Section III,B). Chalcone isomerase from parsley is specific for 4,2',4',6'-tetrahydroxychalcone (I) and naringenin (II) as a substrate–product pair (Fig. 3). It does not react with chalcones that lack one of the hydroxyl groups or bear an O-glucosyl residue (Hahlbrock *et al.*, 1970a).

Sutter *et al.* (1975) reported the oxidation of flavanones to the corresponding flavones with cell-free extracts from young parsley leaves. A chalcone did not serve as substrate unless it was cyclized to the flavanone prior to oxidation. Some properties of the enzyme catalyzing flavanone oxidation suggest that it is not a peroxidase. Despite considerable effort, it has not been possible to demonstrate the enzyme in cell cultures of parsley. In parsley, nothing is known about the mechanism of formation of the flavonol structure from the chalcone or flavanone precursor.

e. Hydroxylation and Methylation of Aglycones. Since very little is known about the preceding step of flavone and flavonol formation, it is not absolutely certain whether hydroxylation of these compounds in the 3'-position (possible reaction *i*) is required. As mentioned above, 3'-hydroxynaringenin (eriodictyol) can be synthesized under certain conditions *in vitro,* presumably via the corresponding chalcone (Section II,B,2,c). If this compound were synthesized in sufficient amounts *in vivo* and accepted as substrate for flavone and flavonol formation, the 3'-hydroxyl group would not need to be introduced at a later stage.

All the remaining reactions of this pathway are well known (Figs. 4, 8, and 10). Methylation (*k*) of the 3',4'-dihydroxy-substituted flavonoids, luteolin (XI) and quercetin (XIII), in the 3'-position is catalyzed by a specific S-adenosyl-L-methionine : flavonoid 3'-O-methyltransferase (Ebel *et al.*, 1972; Ebel and Hahlbrock, 1977). Under appropriate conditions (e.g., pH 9.3), the enzyme also methylates caffeate, but with a K_m that is about 35-fold higher than for luteolin. This, together with the fact that feruloyl-CoA is not a substrate of chalcone synthase (Section II,B,2,c), indicates that the methyltransferase is involved in flavonoid biosynthesis predominantly or exclusively following completion of the C_{15} skeleton. The same methyltransferase is probably involved in both the flavone and the flavonol glycoside pathways (Ebel and Hahlbrock, 1977). Further details concerning this and related enzymes are given by Poulton (this volume, Chapter 22).

f. Glycosylation. The first reaction of this step (*l*) is identical for flavones and flavonols. The enzyme UDP-glucose : flavone/flavonol 7-O-glucosyltransferase had a broad specificity for several flavone and flavonol aglycones but does not glucosylate various other phenolic substances, including isoflavones (Sutter *et al.*, 1972). A flavonol (quercetin) 3-O-glucoside is not further glucosylated in the 7-O position. This indicates that the reaction sequence is $l \rightarrow o$ for the 7-O- and 3-O-glucosylations of flavonols (Fig. 8) and

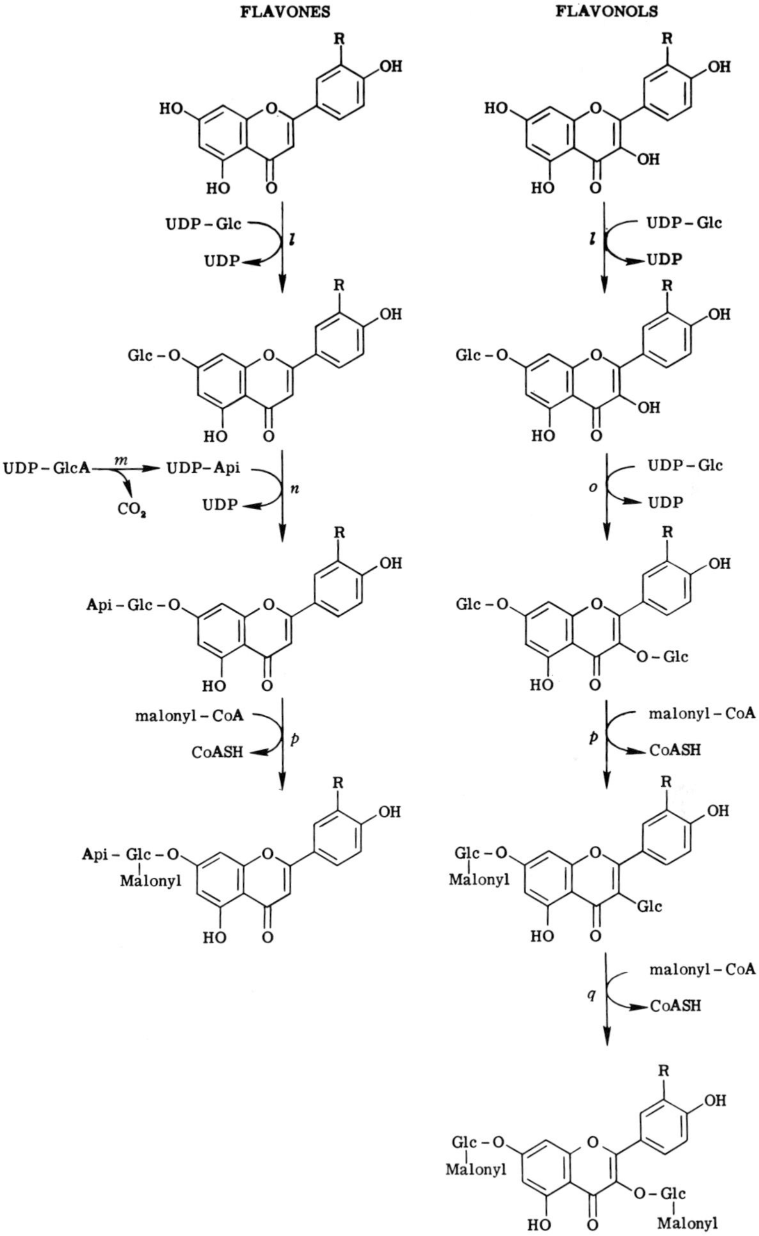

Fig. 8. Sequence of glycosylation and acylation reactions in cultured parsley cells. R = H, OH, or OCH$_3$. For enzymes l–q, see Table I.

that this sequence cannot be reversed. The 7-*O*-glucosyltransferase as well as the following glycosyltransferases are described in greater detail by Hösel (this volume, Chapter 23).

The branched-chain sugar apiose is found in parsely only in flavone glycosides (Sandermann, 1969). The synthesis (*m*) of the sugar nucleotide, UDP-D-apiose, is therefore specifically related to flavonoid biosynthesis. UDP-apiose synthase (complete name: UDP-D-apiose/UDP-D-xylose synthase) catalyzes the formation of two products *in vitro*, UDP-D-apiose (XXI) and UDP-D-xylose (XXII) from UDP-D-glucuronate (XX). The reaction is shown in Fig. 9 (see also Vol. 3). The significance of the synthesis of UDP-xylose in addition to UDP-apiose is not known. Its formation might be an artificial side reaction occurring only under the assay conditions *in vitro*. The UDP-apiose/UDP-xylose ratio of formation does not change during a 1400-fold purification of the synthase to apparent homogeneity. Results of extensive studies on the enzyme by Matern and Grisebach (1977) suggest that an L-*threo*-4-pentosulose is an intermediate in the synthesis of both products. The enzyme is composed of two proteins with MWs of 65,000 and 86,000, which occur in approximately equal molar amounts. Each protein contains two apparently identical subunits. Only the large protein exhibits synthase activity. The function of the small protein may be associated with the stabilization of enzyme activity.

Transfer of the apiosyl residue to flavone glucosides is catalyzed by

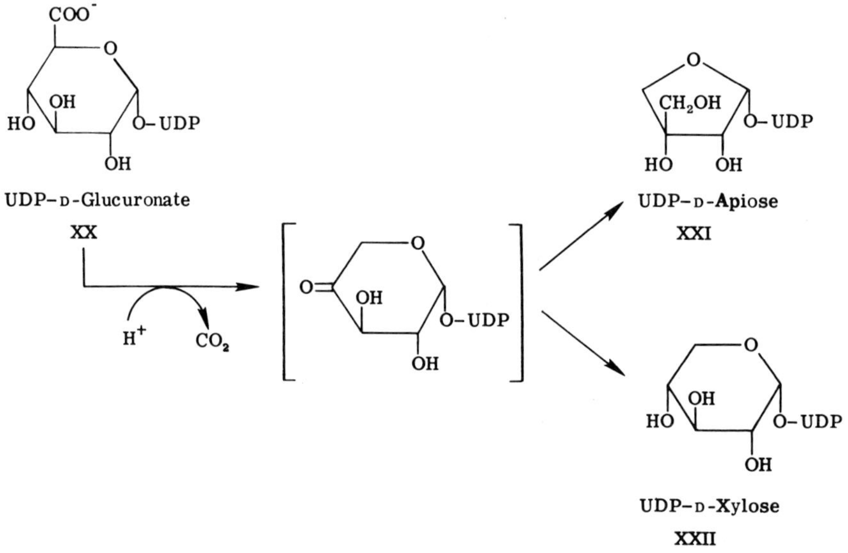

Fig. 9. Proposed mechanism of action of UDP-apiose synthase. (after Matern and Grisebach, 1977).

UDP-D-apiose : flavone 7-O-glucoside apiosyltransferase (n). The reaction product is a flavone 7-O-β-D-apiofuranosyl(1 → 2)-β-D-glucoside, e.g., apiin (cf. malonylapiin in Fig. 5). Ortmann et al. (1972) purified the enzyme partially and studied some of its catalytic properties (see also Hösel, this volume, Chapter 23). The apiosyltransferase is specific for UDP-D-apiose as glycosyl donor and acts with good efficiency on the 7-O-glucosides of all three flavones occurring in parsley cells (Section II,B,1). The enzyme does not accept flavonol 7-O- or 3-O-glucoside as a substrate. This is in agreement with the nonoccurrence of flavonol apioglucosides in parsley. Together with UDP-apiose synthase, and perhaps with the insufficiently characterized "flavone oxidase" (g), the apiosyltransferase from parsley cells is therefore one of the few enzymes of flavonoid glycoside biosynthesis exclusively involved in the formation of flavone derivatives.

A similarly small number of enzymes are specific for flavonol derivatives. One is UDP-D-glucose : flavonol 3-O-glucosyltransferase (o). This enzyme was studied by Sutter and Grisebach (1973, 1975). It has a strict positional specificity and catalyzes the 3-O-glucosylation of flavonols. Substrates in vitro can be either aglycones or their 7-O-glucosides. It is not known whether both types of compounds are substrates in vivo, but it is probable that all flavonols are 7-O-glucosylated prior to 3-O-glucosylation, since flavonol 3-O-glucosides are not accepted as substrates by the 7-O-glucosyltransferase. A particularly interesting feature of the 3-O-glucosyltransferase reaction is its free reversibility (see Hösel, this volume, Chapter 23).

g. Acylation. Two malonyltransferases are involved in the acylation of the various flavone and flavonol glycosides. One enzyme (p) is specific for the glucosyl moieties of flavone and flavonol 7-O-glycosides, and the other (q) is specific for the 3-O-glucosyl moieties of flavonol mono- and bisglucosides (Hahlbrock, 1972; U. Matern, unpublished results). The enzymes can be easily separated by conventional methods of protein fractionation. They have a similar MW of about 50,000. Both are soluble enzymes which appear to be localized in the cytoplasm. The position of malonylation is probably C-6 of the glucosyl residues (Hahlbrock, 1972).

h. Proposed Sequence of Reactions. A scheme for the sequence of all individual reactions specifically related to the biosynthesis of flavone and flavonol glycosides in parsley is depicted in Fig. 10. The scheme clearly distinguishes between the enzymes of general phenylpropanoid metabolism (group I) and of the flavone and flavonol glycoside pathways (group II). Some operational criteria for this classification are given in Section II,B,3,b. As pointed out above, details of reactions g–i are not known. In fact, it is possible that reaction i is not involved in this pathway at all or utilizes the flavanone rather than the flavone and flavonol as substrate.

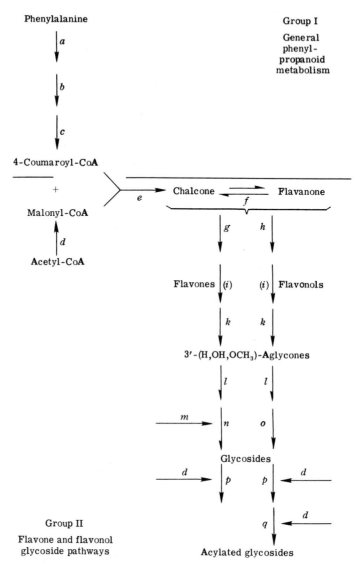

Fig. 10. Complete sequence of steps involved in the biosynthesis of flavone and flavonol glycosides in cultured parsley cells. For details, see Section II,B.

3. Regulation of Enzyme Activities

a. Regulation in Intact Plants. During the development of parsley plants, only very young cotyledons and leaves possess high activities of the enzymes involved in flavonoid biosynthesis. The enzyme activities decrease rapidly to very low levels in later stages of development. The enzymes of general phenylpropanoid metabolism and of the flavonoid glycoside pathways seem to be regulated in this plant in a coordinated manner (see, Wiermann, this volume, Chapter 4).

b. Regulation in Cell Cultures. In cell cultures of parsley, the two groups of enzymes can be distinguished more clearly than in intact plants (for the formal classification, see Fig. 10 and Section II,B,2). For example, group I enzymes can be induced by dilution of a cell culture into fresh medium in the absence of light, whereas the enzymes of group II are not induced under these conditions (Hahlbrock and Wellmann, 1973). Rapid increases and subsequent decreases in all three enzyme activities of group I occur in a highly coordinated manner (Fig. 11), but the significance of the induction is not known with certainty (Hahlbrock and Grisebach, 1979).

Another criterion for classification of the enzymes into two groups is the difference in the time courses of their activity changes in irradiated cells (Hahlbrock et al., 1976; Ebel and Hahlbrock, 1977). Both groups are induced simultaneously under these conditions, and the general shapes of the activity curves are similar for all enzymes. However, under appropriate conditions, the enzymes of group I reach maximal activities several hours earlier and then decline much more rapidly than the enzymes of group II. This difference is schematically illustrated in Fig. 11. Moreover, the length of a short but

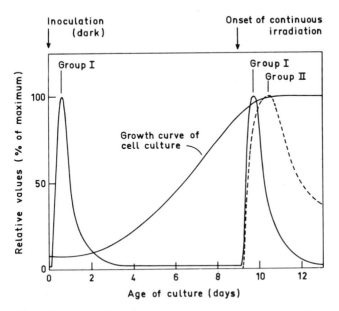

Fig. 11. Scheme illustrating the induction of enzymes of groups I and II by inoculation or irradiation of cultured parsley cells.

distinct apparent lag period preceding detectable increases in the enzyme activities differs between the two groups (see below). Finally, the enzymes of group I are maximally inducible in irradiated cells shortly before the stationary phase of the culture, and those of group II much earlier during the lag phase (Hahlbrock and Grisebach, 1979).

The light-induced activity changes are due to temporarily increased rates of *de novo* synthesis of the enzymes. Using one characteristic enzyme of each group, phenylalanine ammonia-lyase and chalcone synthase, Schröder *et al.* (1979) demonstrated a quantitative relationship between the changes in enzyme activity and corresponding changes in mRNA activity. The different peak positions for the two enzyme activities (Fig. 11) can be explained by different lengths of the periods during which the mRNAs are induced by light. The results of a typical experiment are shown in Fig. 12. The quantitative correlation between the curves for enzyme and mRNA activities obeys the equation

$$dE(t)/dt = {}^{\circ}k_s(t) - {}^{1}k_d E(t) \tag{1}$$

Equation (1) describes the relationship between enzyme activity (E), the zero-order rate constant of enzyme synthesis (or mRNA activity, ${}^{\circ}k_s$), both changing with time (t), and the first-order rate constant of enzyme degradation (${}^{1}k_d$). The latter constant can easily be calculated from the half-life of enzyme activity ($t_{1/2}$) with the equation

$${}^{1}k_d = \ln 2/t_{1/2} \tag{2}$$

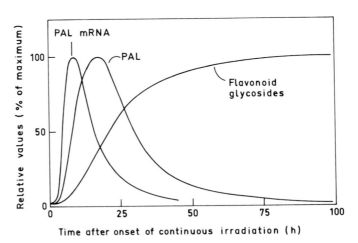

Fig. 12. Schematic representation of light-induced changes in phenylalanine ammonia-lyase (PAL) mRNA and enzyme activities, and in the concentration of total flavonoids in cultured parsley cells. Relations among the three curves are expressed in Eqs. (1)–(3).

Under the conditions used by Schröder *et al.* (1979), the half-lives of phenylalanine ammonia-lyase and chalcone synthase are on the order of about 5–10 h.

The initial lag preceding detectable increases in enzyme activity is not a true lag but rather a period during which small changes are below the limit of detectability. Although the rate of phenylalanine ammonia-lyase synthesis increases with no detectable lag, the increase is slow during the first few hours of mRNA induction (Betz *et al.*, 1978). The increase in this enzyme activity lies therefore within the limits of detectability until about 2 h after the onset of irradiation. The induction of chalcone synthase mRNA starts at a much lower level than that of phenylalanine ammonia-lyase mRNA. This causes a comparatively long lag of about 4 h preceding detectable increases in synthase activity (Schröder *et al.*, 1979).

Mathematical calculations can be applied also to the relationship between enzyme activity and the rate of product formation. In old cell cultures of parsley, where degradation of flavonoids is negligible (Barz, 1977; see also Barz and Köster, this volume, Chapter 3), the equation

$$P(t) = \int_o^t E(\tau)\, d\tau \tag{3}$$

is applicable. This equation describes the changes in the amount of products (P) accumulated with time (t, τ) and enzyme activity (E) changes in response to irradiation of the cells. Coincidence of the measured curve for flavonoid accumulation and the curve calculated from Eq. (3) is to be expected only for the rate-limiting enzyme. In parsley cell cultures from a late growth stage (cf. Fig. 11), such coincidence exists only for phenylalanine ammonia-lyase. Figure 12 illustrates the direct correlation between mRNA and catalytic activities of this enzyme and flavonoid glycoside accumulation in parsley cells, as calculated using Eqs. (1–3). It must be concluded that phenylalanine ammonia-lyase can under certain conditions be the rate-limiting enzyme in the biosynthesis of flavonoids. Furthermore, the clear-cut relationship between enzyme activity and product accumulation leaves no room for considering end product or other feedback inhibition of phenylalanine ammonia-lyase in this case.

4. Relation to Quality and Quantity of Light

The formation of flavonoids often requires light, but in contrast to the usual mechanism of action of red light via phytochrome, red/far-red control of flavonoid biosynthesis in parsley cells requires activating irradiation with uv light. Following uv irradiation, which is alone sufficient for induction, the activities of the enzymes of both groups I and II may be modified by changes in the phytochrome state. The action spectrum for the uv effect indicates a maximal response below 300 nm, but the receptor has not been identified. At low doses, the dose–response relationship for the enzyme induction with uv light is linear (Wellmann, 1974, 1975).

C. Biosynthesis in Other Plants

Information on the biosynthesis of flavones and flavonols in other plants is scarce but is in agreement with the scheme shown in Fig. 10 for parsley. One particularly useful system where the enzymology and regulation of flavonoid biosynthesis are studied in relation to organ development is tulip anthers. Among the enzymes investigated are chalcone synthase and chalcone isomerase (Herdt *et al.*, 1978). The anther system is described in detail by Wiermann (this volume, Chapter 4).

In addition to those mentioned here, flavonoid-specific methyltransferases and glycosyltransferases are treated by Poulton (this volume, Chapter 22) and Hösel (this volume, Chapter 23).

Cell suspension cultures of soybean (*Glycine max*) contain a flavonoid-specific 3'-*O*-methyltransferase (Poulton *et al.*, 1976) and a UDP-glucose: flavonol 3-*O*-glucosyltransferase (Poulton and Kauer, 1977). The high specificity of the methyltransferase for flavonoid substrates suggests that methylation of these compounds takes place in soybean, as in parsley (Section II,B,2,e), at the level of the complete flavonoid structures (see Poulton, this volume, Chapter 22, for a more detailed discussion). Furthermore, one of the two isoenzymes of 4-coumarate: CoA ligase found in soybean cells is strikingly similar in its catalytic properties to the ligase from parsley cells (Section II,B,2,a). Its function might therefore likewise be closely related to the biosynthesis of flavonoids (Knobloch and Hahlbrock, 1975; Hahlbrock and Grisebach, 1979). Both the methyltransferase and the ligase, together with the other known enzymes related to this pathway, exhibit a sharp peak in activity at a distinct growth stage of soybean cell cultures. This growth stage is characterized by a short period during which the cells accumulate apigenin (III) and various other flavonoids. Light stimulates phenylpropanoid metabolism in these cells but is not required for enzyme induction (Hahlbrock, 1977). An isoflavonoid phytoalexin, glyceollin (XXX), is produced on treatment of cultured soybean cells with a specific elicitor (Ebel *et al.*, 1976).

D. Catabolism

An important aspect of the biochemistry of flavonoids is their catabolism, both in plants and in microorganisms. The turnover rates of flavonoids in flavonoid-producing plants are closely linked to the developmental stage of the tissue concerned. The case discussed above, where parsley cell cultures from a certain stage of the growth cycle show no significant rate of flavonoid degradation, is probably the exception rather than the rule. In many cases, a given concentration of flavonoids in a plant tissue is the result of a steady-state equilibrium between synthesis and degradation. For details of flavonoid catabolism, see Barz and Köster (this volume, Chapter 3).

III. CHALCONES, FLAVANONES, AND AURONES

A. Biochemical Relations

The most abundant chalcones, flavanones, and aurones have either a phloroglucinol-type (I) or resorcinol-type (XXIII) hydroxylation pattern of the acetate-derived ring A (Harborne, 1967; Bohm, 1975a,b). The biosynthetic relations between the 7,4'-dihydroxyflavanone, liquiritigenin (XXIII), and the corresponding chalcone (XII) and aurone (XXV) are indicated in Fig. 13 (note the different numbering of the carbon atoms for the three types of compounds). The interconversion of chalcone and flavanone is catalyzed by chalcone isomerase (Section III,B). All naturally occurring flavanones so far isolated without racemization are levorotatory and have the (−)-2S configuration.

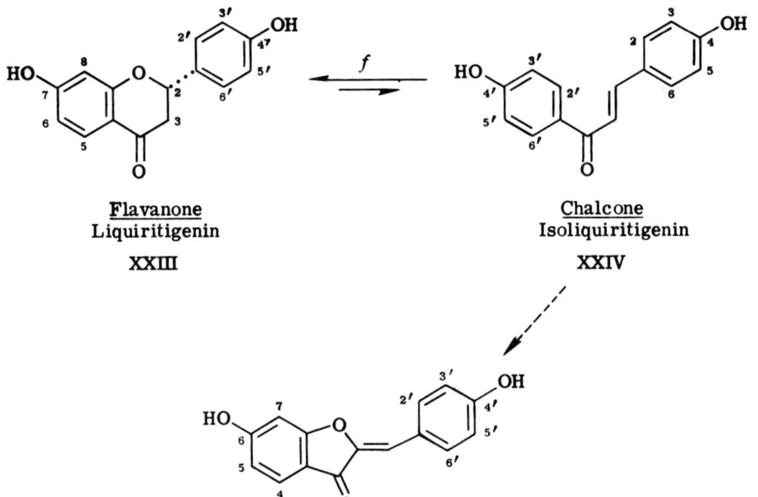

Flavanone
Liquiritigenin
XXIII

Chalcone
Isoliquiritigenin
XXIV

Aurone
Hispidol
XXV

Fig. 13. Proposed biosynthetic relations between chalcones, flavanones, and aurones. f, Chalcone isomerase.

Formation of the aurone from the chalcone probably requires at least two enzymatic steps, one of which might be a peroxidase-catalyzed reaction. The overall conversion is catalyzed by cell-free extracts from various plants, but details about the enzymes involved are not known (Wong, 1965; Rathmale and Bendall, 1972; Wong and Wilson, 1972; see also Hahlbrock and Grisebach, 1975). Wong and Wilson (1972) proposed the function of a 2-(α-hydroxybenzyl)coumaranone derivative as intermediate in the synthesis of the aurone hispidol (XXV) from the corresponding chalcone (XXIV).

The roles of naringenin chalcone (I) and the corresponding flavanone (II) in the synthesis of flavones and flavonols in parsley cells is described in Section II,B,2,d. There is little doubt that the isomeric pair chalcone–flavanone is generally the common central intermediate in the synthesis of all flavonoids (Hahlbrock and Grisebach, 1975). However, the actual mechanism through which flavanones and chalcones function as intermediates is less clear in most other plants than in parsley. Wong and Grisebach (1969) interpreted results from competition feeding experiments and kinetic studies on the rates of conversion of a labeled chalcone–flavanone pair to the corresponding flavone, dihydroflavonol, flavonol, and isoflavone as favoring the chalcone as the more immediate precursor. However, a definite answer to this question will certainly have to await further enzymatic studies. The key enzymes in this connection are chalcone synthase, chalcone isomerase, and the enzymes converting flavanones and/or chalcones to flavones, flavonols, anthocyanidins, isoflavones, aurones, etc. Beyond the work with parsley (Section II) and *H. gracilis* (Section IV) very little is known about the properties of chalcone synthase and flavanone–chalcone-oxidizing enzymes, except peroxidase. On the other hand, chalcone isomerase has been isolated and characterized from many plants.

B. Chalcone Isomerase

Chalcone isomerase was the first enzyme to be isolated that is exclusively involved in flavonoid metabolism (Moustafa and Wong, 1967). Subsequently, the enzyme was partially purified from a variety of sources and extensively studied. Most of the work was summarized by Hahlbrock and Grisebach (1975). A recent report of an extensive purification and some kinetic properties of chalcone isomerase from soybean seed (*G. max*) includes an estimated value of approximately 10/min for the catalytic center activity of the enzyme (Boland and Wong, 1975).

In essence, chalcone isomerases (with flavanones as products) have no cofactor requirements and are specific for the formation (or opening) of the six-membered heterocyclic ring of flavanones. The chemical equilibrium of this reaction lies far to the side of the flavanone. Chalcone isomerase often occurs in multiple forms which can be separated by the usual techniques of

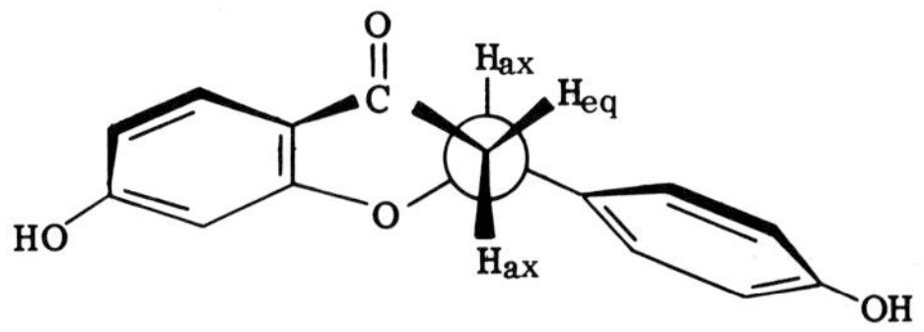

Fig. 14. Stereochemistry of the flavanone liquiritigenin (XXIII).

enzyme fractionation and differ considerably with respect to their pH optima and K_m values. It is therefore likely that the multiple forms are true isoenzymes and not isomeric forms differing only in protein conformation. Usually, the substrate specificities of the isomerases correspond to the substitution patterns of the flavonoids found in the plant tissues from which the enzymes were isolated.

Chalcone glucosides usually do not serve as substrates for chalcone isomerase. The various naturally occurring chalcone and flavanone glycosides are probably biosynthetic end products and might be reutilized only after hydrolysis. The aglycones can be either end products or intermediates in the synthesis of other flavonoids.

The complete stereochemistry of the isomerase reaction was elucidated by nmr studies with deuterated flavanones (Hahlbrock *et al.*, 1970b). The cyclization of isoliquiritigenin (XXIV) by either one of the two isoenzymes from mung bean seedlings (*Phaseolus aureus*) can be carried out in the presence of highly enriched deuterium oxide. In both cases, a deuteron is introduced into the flavanone (XXIII), specifically in the axial position at C-3 (Fig. 14). Conversely, a chalcone deuterated at this carbon atom gives rise to a flavanone labeled in the equatorial position. The configuration of the adjacent optically active center is $(-)$-(2*S*) (Moustafa and Wong, 1967; Hahlbrock *et al.*, 1970b).

Kuhn *et al.* (1978) investigated the genetic control of chalcone isomerase activity in *Callistephus chinensis*. Continuation of such studies might add significantly to our knowledge of the *in vivo* function of this enzyme in flavonoid biosynthesis.

IV. ANTHOCYANINS

Next to chlorophyll, anthocyanins are the most important group of plant pigments visible to the human eye. The aglycones (anthocyanidins) are structurally closely related to each other. There are six major anthocyanidins whose chemical structures are all based on the structure of pelargonidin (V). They differ from pelargonidin by bearing one or two additional hydroxyl or methoxyl groups in the 3'- and 5'-positions. For example, one of the most common representatives, cyanidin (XXVII), bears a 3'-hydroxyl group (Fig. 15). Several rarer anthocyanidins are further substituted, lack the 3-hydroxyl group (3-deoxyanthocyanidins), or both. The aglycones as such are chemically rather unstable. They occur *in vivo* as 3-glycosides, 3,5-bisglycosides, or 3,7-bisglycosides. The glycosides are often further substituted by acyl residues.

Altogether the anthocyanidins constitute a large family of differently colored compounds and occur in countless mixtures in practically all parts of most higher plants (cf. Section I). They are of great economic importance as fruit pigments and thus also as pigments of fruit juices and wine. Furthermore, anthocyanins have been important markers since the early days of plants genetics. An enzyme that is specific for anthocyanin biosynthesis, UDP-glucose:cyanidin 3-*O*-glucosyltransferase, was included in studies on the genetic control of this pathway (Dooner and Nelson, 1977; Kho *et al.*, 1977).

The chemistry, natural occurrence, inheritance, and taxonomy of anthocyanins have been studied in great detail. Several excellent and comprehensive reviews of these aspects are available (Harborne, 1967; Hess, 1968; Jurd, 1972; Timberlake and Bridle, 1975). On the other hand, except for the results of tracer studies, the biochemistry of anthocyanins is very poorly understood. Tracer studies with ^3H- and ^{14}C-labeled precursors have confirmed the general rule that anthocyanins, like other flavonoids, are synthesized from acetate and phenylalanine (Fig. 3). Dihydrokaempferol (XXVI) is an efficient precursor of cyanidin glycosides (Fig. 15) (cf. Hahlbrock and Grisebach, 1975; Kho *et al.*, 1977). However, virtually noth-

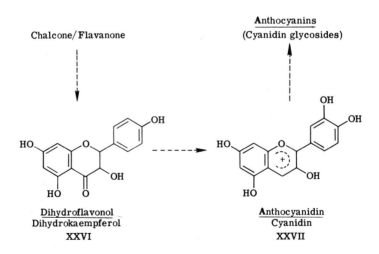

Fig. 15. Proposed sequence of reactions involved in the biosynthesis of anthocyanins.

ing is known about the mechanism by which the characteristic flavylium cation is formed from this precursor (Fig. 15).

Cell suspension cultures of *H. gracilis* are a useful system for studies on anthocyanin biosynthesis. As in parsley cells (Section II,B,4), uv light greatly stimulates flavonoid accumulation in previously dark-grown *H. gracilis* cells (Wellmann *et al.*, 1976). Within a day of irradiation, these cells develop a deep, dark-red color which is due to the production of one or more cyanidin glycosides. Labeled dihydrokaempferol, when administered to irradiated cells, is converted to the pigment at a much higher incorporation rate than the corresponding chalcone or flavanone, or phenylalanine (Fritsch *et al.*, 1971). This indicates the close biosynthetic relationship between the dihydroflavonol and cyanidin, although the conversion has not been demonstrated in cell-free extracts. The activities of phenylalanine ammonia-lyase, chalcone synthase, and chalcone isomerase increase simultaneously with anthocyanin accumulation (Wellmann *et al.*, 1976). This coincidence supports the general applicability of the scheme shown in Fig. 3 to the early steps of flavonoid biosynthesis, including anthocyanins. Furthermore, the catalytic properties of synthase preparations from *H. gracilis* (Saleh *et al.*, 1978) and parsley are strikingly similar. This similarity includes the surprising difference in the pH optima for 4-coumaroyl-CoA and caffeoyl-CoA as substrates in the formation of naringenin and eriodictyol, respectively (cf. Section II,B,2,c).

V. ISOFLAVONOIDS

In contrast to most other flavonoids, isoflavonoids have a rather limited taxonomic distribution, mainly within the Leguminosae. On the other hand, their chemical structures exhibit an unusually wide range of modifications. The large structural variability is indicated by the examples given in Fig. 16 for some of the major classes of isoflavonoids. An extensive review of the chemistry and taxonomic distribution of isoflavonoids was compiled by Wong (1975).

From a biochemical point of view the frequent absence of a 5-hydroxyl group and the frequent presence of a 2'-hydroxyl group in isoflavonoids are especially interesting. Furthermore, there are several instances where neighboring hydroxyl groups are either condensed to give a dihydrofuran or furan ring structure, as in glyceollin (XXX) and medicagol (XXXII), or connected by a methylene bridge, as in neotenone (XXVIII) and medicagol (XXXII). Another characteristic feature of many isoflavonoids is the occurrence of isoprenoid or other carbon–carbon linked side chains, as in neotenone (XXVIII), rotenone (XXIX), glyceollin (XXX), and licoricidin (XXXI).

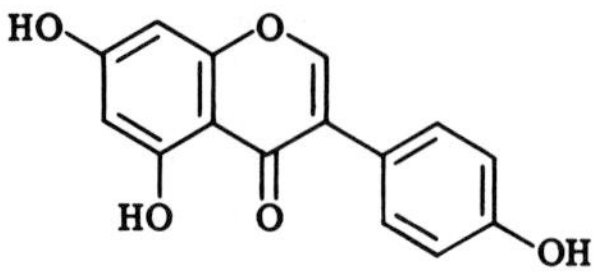

Isoflavone
Genistein

VI

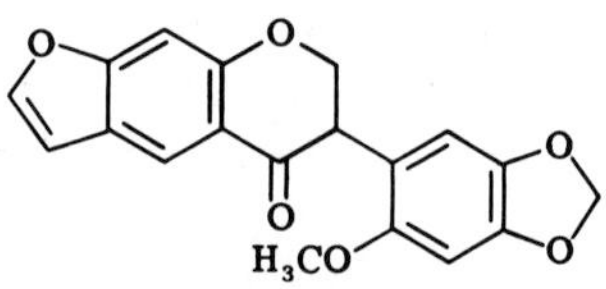

Isoflavanone
Neotenone

XXVIII

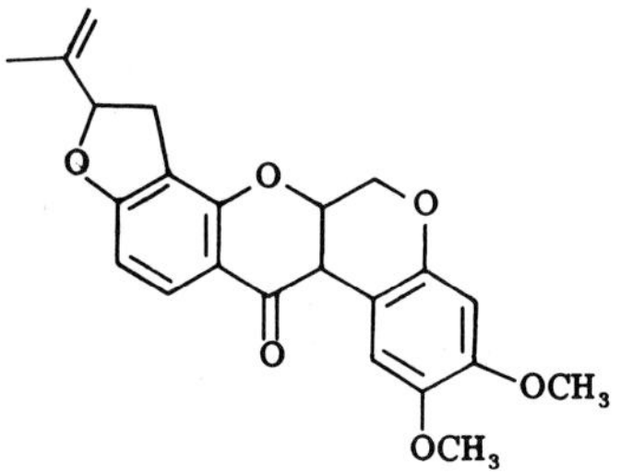

Rotenoid
Rotenone

XXIX

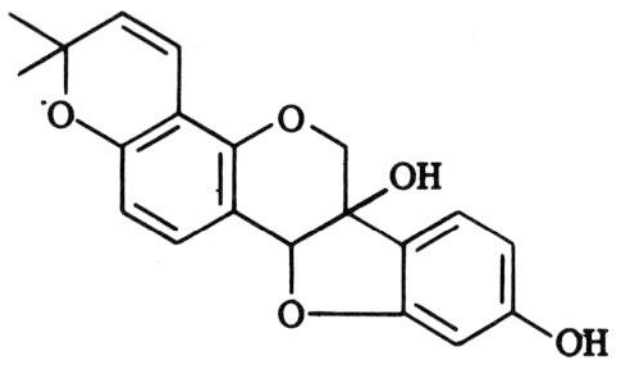

Pterocarpan
Glyceollin

XXX

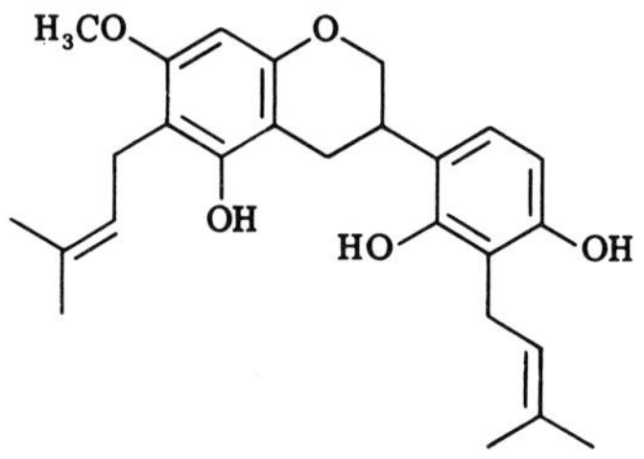

Isoflavan
Licoricidin

XXXI

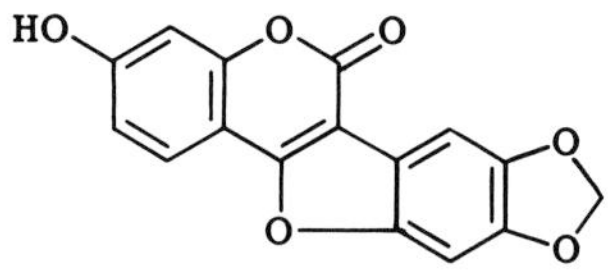

Coumestan
Medicagol

XXXII

Fig. 16. Representatives of each of six major subgroups of isoflavonoids.

As in the case of anthocyanins, most of our knowledge about the biosynthesis of isoflavonoids originates from studies with radioactive isotopes. Again, it has been established that acetate gives rise to ring A and that phenylalanine, cinnamate, and cinnamate derivatives are incorporated into ring B and C-2, -3, and -4 of the heterocyclic ring (Hahlbrock and Grisebach, 1975). The mode of incorporation of phenylpropanoid compounds into isoflavonoids is depicted in Fig. 17. Since chalcones and flavanones are efficient precursors of isoflavonoids, the required aryl migration of ring B

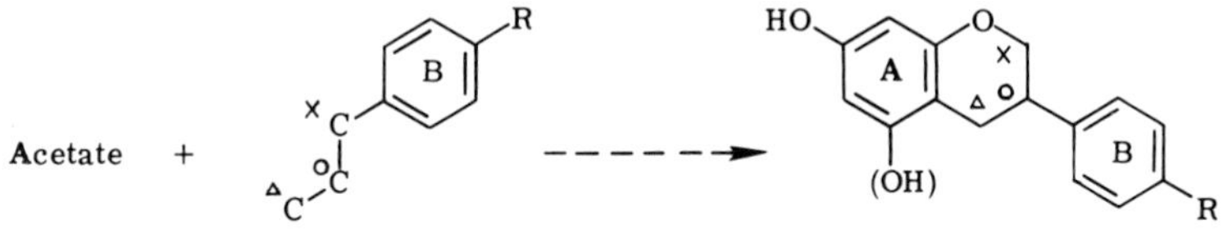

Acetate +

Fig. 17. Mode of incorporation of phenylpropanoids into isoflavonoids. Symbols indicate the labeling pattern with radioactive precursors.

from the former β carbon to the former α carbon of the phenylpropanoid precursor must take place after formation of the basic C_{15} skeleton. However, the actual mechanism of aryl migration is not known. Even the possible significance of a postulated unsubstituted 4'-hydroxyl group which might be required for aryl migration (Pelter et al., 1971) is unclear. Feeding experiments with ^{3}H- and ^{14}C-labeled 4-methoxycinnamate and 4'-methoxychalcone were inconclusive because of rapid demethylation of these compounds (Hahlbrock and Grisebach, 1975). The interconversion of isoflavonoids is well documented (e.g., Dewick and Ward, 1978).

Work at the enzymatic level has contributed indirect evidence that methylation of ring B of isoflavonoids takes place after aryl migration. Cell suspension cultures of chick-pea (Cicer arietinum) contain an S-adenosylmethionine:isoflavone 4'-O-methyltransferase that is specific for the methylation of isoflavones, e.g., genistein (VI), in the 4'-position (Wengenmayer et al., 1976; see also, Poulton, this volume, Chapter 22). The specificity of this enzyme for isoflavones might indicate that the biosynthesis of methylated isoflavones does not involve earlier methylated intermediates.

Glyceollin (XXX), like several other pterocarpans, belongs to the chemically heterogeneous class of phytoalexins. These are produced in various plants in response to attack by a pathogen (Grisebach and Ebel, 1978). The accumulation of glyceollin in challenged soybean cotyledons (G. max) coincides with large increases in the activity of phenylalanine ammonialyase, cinnamate 4-hydroxylase, 4-coumarate:CoA ligase, and chalcone synthase (Zähringer et al., 1978; J. Ebel, unpublished results). This coincidence confirms the conclusions drawn from tracer studies that the same enzymes are involved in the initial biosynthetic steps of isoflavonoids and other flavonoids (cf. Section II,B,2,a–c).

VI. OTHER FLAVONOIDS

Further groups of flavonoids not treated in the previous sections are biflavonoids, proanthocyanidins, C-glycosyl flavonoids, and neoflavonoids. One example of each group is given in Fig. 18. The four groups can be classified as condensed flavonoids (biflavonoids, except condensed proan-

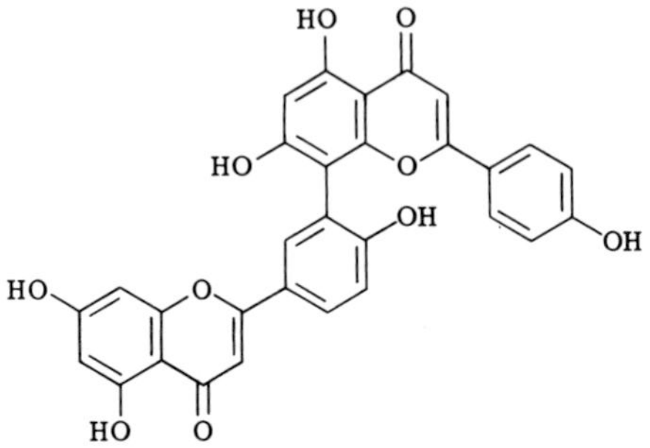

Biflavonoid
Amentoflavone
XXXIII

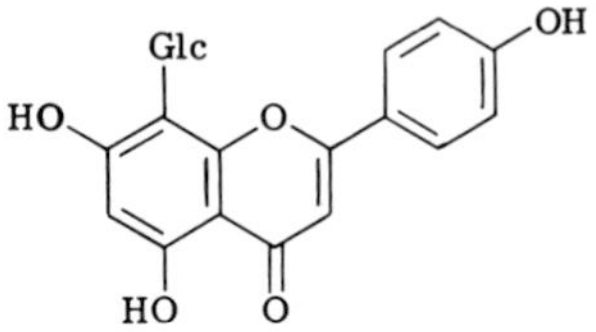

C-Glycosyl flavonoid
Vitexin
XXXIV

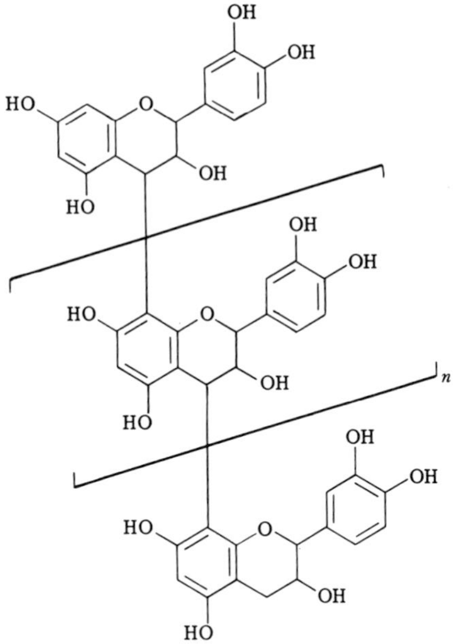

Proanthocyanidin (condensed)
Procyanidin
XXXV

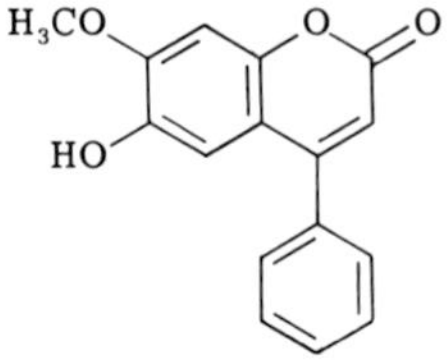

Neoflavonoid
Dalbergin
XXXVI

Fig. 18. Representatives of each of four minor subgroups of flavonoids.

thocyanidins), compounds yielding anthocyanidins on acid treatment (proanthocyanidins, monomeric or condensed), flavonoids linked to sugars through carbon–carbon bonds (C-glycosyl flavonoids), and derivatives of 4-phenylcoumarin (neoflavonoids).

A detailed discussion of these relatively minor groups of flavonoids is beyond the scope of this chapter. They are synthesized *in vivo* according to the general scheme of flavonoid biosynthesis (Fig. 3). Their chemistry, biochemistry, and taxonomic distribution was summarized by Geiger and Quinn (1975) (biflavonoids), Haslam (1975) (proanthocyanidins), Chopin and Bouillant (1975) (*C*-glycosyl flavonoids), and Donelly (1975) (neoflavonoids). For neoflavonoids, see also Brown (this volume, Chapter 10).

REFERENCES

Barz, W. (1977). *In* "Plant Tissue Culture and Its Biotechnological Application" (W. Barz, E. Reinhard, and M. H. Zenk, eds.), pp. 153–177. Springer-Verlag, Berlin and New York.

Betz, B., Schäfer, E., and Hahlbrock, K. (1978). *Arch. Biochem. Biophys.* **190**, 126–135.

Bohm, B. A. (1975a). *In* "The Flavonoids" (J. B. Harborne, T. J. Mabry, and H. Mabry, eds.), pp. 442–504. Chapman & Hall, London.

Bohm, B. A. (1975b). *In* "The Flavonoids" (J. B. Harborne, T. J. Mabry, and H. Mabry, eds.), pp. 560–631. Chapman & Hall, London.

Boland, M. J., and Wong, E. (1975). *Eur. J. Biochem.* **50**, 383–389.

Chopin, J., and Bouillant, M. L., (1975). *In* "The Flavonoids" (J. B. Harborne, T. J. Mabry, and H. Mabry, eds.), pp. 632–691. Chapman & Hall, London.

Dewick, P. M., and Ward, D. (1978). *Phytochemistry* **17**, 1751–1754.

Donelly, D. M. X. (1975). *In* "The Flavonoids" (J. B. Harborne, T. J. Mabry, and H. Mabry, eds.), pp. 801–865. Chapman & Hall, London.

Dooner, H. K., and Nelson, O. E. (1977). *Biochem. Genet.* **15**, 509–519.

Ebel, J., and Hahlbrock, K. (1977). *Eur. J. Biochem.* **75**, 201–209.

Ebel, J., Hahlbrock, K., and Grisebach, H. (1972). *Biochim. Biophys. Acta* **269**, 313–326.

Ebel, J., Ayers, A. R., and Albersheim, P. (1976). *Plant Physiol.* **57**, 775–779.

Fritsch, H., Hahlbrock, K., and Grisebach, H. (1971). *Z. Naturforsch., Teil B* **26**, 581–585.

Geiger, H., and Quinn, C. (1975). *In* "The Flavonoids" (J. B. Harborne, T. J. Mabry, and H. Mabry, eds.), pp. 692–742. Chapman & Hall, London.

Grisebach, H., and Ebel, J. (1978). *Angew. Chem.* **90**, 668–681.

Hahlbrock, K. (1972). *FEBS Lett.* **28**, 65–68.

Hahlbrock, K. (1977). *In* "Plant Tissue Culture and Its Biotechnological Application (W. Barz, E. Reinhard, and M. H. Zenk, eds.), pp. 95–111. Springer-Verlag, Berlin and New York.

Hahlbrock, K., and Grisebach, H. (1975). *In* "The Flavonoids" (J. B. Harborne, T. J. Mabry, and H. Mabry, eds.), pp. 866–915. Chapman & Hall, London.

Hahlbrock, K., and Grisebach, H. (1979). *Annu. Rev. Plant. Physiol.* **30**, 105–30.

Hahlbrock, K., and Wellmann, E. (1973). *Biochim. Biophys. Acta* **304**, 702–706.

Hahlbrock, K., Wong, E., Schill, L., and Grisebach, H. (1970a). *Phytochemistry* **9**, 949–958.

Hahlbrock, K., Zilg, H., and Grisebach, H. (1970b). *Eur. J. Biochem.* **15**, 13–18.

Hahlbrock, K., Knobloch, K.-H., Kreuzaler, F., Potts, J. R. M., and Wellmann, E. (1976). *Eur. J. Biochem.* **61**, 199–206.

Harborne, J. B. (1967). "Comparative Biochemistry of the Flavonoids". Academic Press, New York.

Harborne, J. B., Mabry, T. J., and Mabry, H., eds. (1975). "The Flavonoids." Chapman & Hall, London.

Haslam, E. (1975). *In* "The Flavonoids" (J. B. Harborne, T. J. Mabry, and H. Mabry, eds.), pp. 505–559. Chapman & Hall, London.

Heller, W., and Hahlbrock, K. (1980). *Arch. Biochem. Biophys.* **200**, 617–619.

Herdt, E., Sütfeld, R., and Wiermann, R. (1978). *Cytobiology* **17**, 433–441.

Hess, D. (1968). "Biochemische Genetik." Springer-Verlag, Berlin and New York.

Hrazdina, G., Kreuzaler, F., Hahlbrock, K., and Grisebach, H. (1976). *Arch. Biochem. Biophys.* **175**, 392–399.

Jurd, L. (1972). *Recent Adv. Phytochem.* **5**, 135–164.

Kho, K. F. F., Bolsman-Louwen, A. C., Vuik, J. C., and Bennink, G. J. H. (1977). *Planta* **135**, 109–118.

Knobloch, K.-H., and Hahlbrock, K. (1975). *Eur. J. Biochem.* **52**, 311–320.

Knobloch, K.-H., and Hahlbrock, K. (1977). *Arch. Biochem. Biophys.* **184**, 237–248.

Kreuzaler, F., and Hahlbrock, K. (1973). *Phytochemistry* **12**, 1149–1152.

Kreuzaler, F., and Hahlbrock, K. (1975a). *Eur. J. Biochem.* **56**, 205–213.

Kreuzaler, F., and Hahlbrock, K. (1975b). *Arch. Biochem. Biophys.* **169**, 84–90.

Kreuzaler, F., Light, R. J., and Hahlbrock, K. (1978). *FEBS Lett.* **94**, 175–178.

Kreuzaler, F., Ragg, H., Heller, W., Tesch, R., Witt, I., Hammer, D., and Hahlbrock, K. (1979). *Eur. J. Biochem.* **99**, 89–96.

Kuhn, B., Forkmann, G., and Seyffert, W. (1978). *Planta* **138**, 199–203.

Mabry, T. J., Markham, K. R., and Thomas, M. B. (1970). "The Systematic Identifaction of Flavonoids." Springer-Verlag, Berlin and New York.

Matern, U., and Grisebach, H. (1977). *Eur. J. Biochem.* **74**, 303–312.

Moustafa, E., and Wong, E. (1967). *Phytochemistry* **6**, 625–632.

Ortmann, R., Sutter, A., and Grisebach, H. (1972). *Biochim. Biophys. Acta* **289**, 293–302.

Pelter, A., Bradshaw, J., and Warren, R. F. (1971). *Phytochemistry* **10**, 835–850.

Pfändler, R., Scheel, D., Sandermann, H., and Grisebach, H. (1977). *Arch. Biochem. Biophys.* **178**, 315–316.

Poulton, J. E., and Kauer, M. (1977). *Planta* **136**, 53–59.

Poulton, J. E., Grisebach, H., Ebel, J., Schaller-Hekeler, B., and Hahlbrock, K. (1976). *Arch. Biochem. Biophys.* **173**, 301–305.

Rathwell, W. G., and Bendall, D. S. (1972). *Biochem. J.* **127**, 125–132.

Robberecht, R., and Caldwell, M. M. (1978). *Oecologia* **32**, 277–287.

Saleh, N. A. M., Fritsch, H., Kreuzaler, F., and Grisebach, H. (1978). *Phytochemistry* **17**, 183–186.

Sandermann, H. (1969). *Phytochemistry* **8**, 1571–1575.

Schröder, J., Kreuzaler, F., Schäfer, E., and Hahlbrock, K. (1979). *J. Biol. Chem.* **254**, 57–65.

Sutter, A., and Grisebach, H. (1973). *Biochim. Biophys. Acta* **309**, 289–295.

Sutter, A., and Grisebach, H. (1975). *Arch. Biochem. Biophys.* **167**, 444–447.

Sutter, A., Ortmann, R., and Grisebach, H. (1972). *Biochim. Biophys. Acta* **258**, 71–87.

Sutter, A., Poulton, J. E., and Grisebach, H. (1975). *Arch. Biochem. Biophys.* **170**, 547–556.

Timberlake, C. F., and Bridle, P. (1975). *In* "The Flavonoids" (J. B. Harborne, T. J. Mabry, and H. Mabry, eds.), pp. 214–266. Chapman & Hall, London.

Wellmann, E. (1974). *Ber. Dtsch. Bot. Ges.* **87**, 267–273.

Wellmann, E. (1975). *FEBS Lett.* **51**, 105–107.

Wellmann, E., Hrazdina, G., and Grisebach, H. (1976). *Phytochemistry* **15**, 913–915.

Wengenmayer, H., Ebel, J., and Grisebach, H. (1976). *Eur. J. Biochem.* **65**, 529–536.

Wong, E. (1965). *Biochim. Biophys. Acta* **111**, 358–363.

Wong, E. (1975). *In* "The Flavonoids" (J. B. Harborne, T. J. Mabry, and H. Mabry, eds.), pp. 743–800. Chapman & Hall, London.

Wong, E., and Grisebach, H. (1969). *Phytochemistry* **8**, 1419–1426.

Wong, E., and Wilson, J. M. (1972). *Phytochemistry* **11**, 875.

Zähringer, U., Ebel, J., and Grisebach, H. (1978). *Arch. Biochem. Biophys.* **188**, 450–455.

Zimmermann, A., and Hahlbrock, K. (1975). *Arch. Biochem. Biophys.* **166**, 54–62.

Lignins 15

HANS GRISEBACH

I. INTRODUCTION

Lignin (Latin: *lignum* = wood) is found as an integral cell-wall constituent of all vascular plants including herbaceous species. Lignin does not occur in algae. Recent investigations have shown that mosses do not contain true lignin, but do contain other phenolic cell-wall constituents such as sphagnum

The Biochemistry of Plants, Vol. 7

ISBN 0-12-675407-1

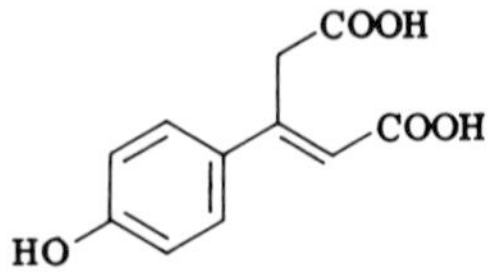

Fig. 1. 4-Hydroxy-β-(carboxymethyl)cinnamic acid (sphagnum acid) from the cell wall of mosses.

acid, the structure of which is shown in Fig. 1 (Tutschek, 1975). The earlier concept of gradual evolution from a primitive moss lignin to a more developed hardwood lignin is therefore no longer tenable.

Lignin is considered to contribute to the compressive strength of the cell wall, but it has no influence on tensile properties. The ability of plants to form lignin must have been a decisive factor in the adaptation of plants to the terrestrial habitat. Only with lignified cell walls was it possible to build the rigid stems of woody plants and trees and the conducting cell elements for water transport. In the phylogeny of plants, lignin appears either concurrently with or prior to the development of conducting tissues.

Lignin represents about 20–35% of the cell wall of conifer wood. The rest is composed of cellulose (40%), hemicelluloses (30%), and some protein. Lignin is not only in intimate physical contact with hemicellulose but seems to be bound to it by covalent bonds. Covalent linkages to cell-wall protein may also occur, and it has been suggested that an early step in lignification may be cross-linking of protein of the primary wall during polymerization of lignin monomers (Whitmore, 1978).

In 1933 Erdtman postulated that lignin was formed by enzymatic dehydrogenation of cinnamyl alcohols (e.g., 4-coumaryl alcohol), coniferyl alcohol, and sinapyl alcohol (Fig. 2). This hypothesis was later confirmed and elaborated by Freudenberg and his associates. Important contributions to the structure and biosynthesis of lignin were also made by Adler, Kratzl, Neish, and others. One of the experimental approaches of Freudenberg was to polymerize cinnamyl alcohols *in vitro* with fungal laccase or with peroxidase for various lengths of time. Structural elucidation of the dimeric or

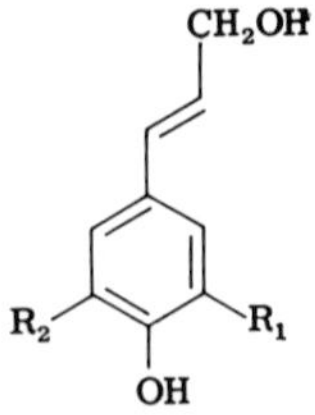

Fig. 2. Structure of lignin precursors (monolignols). 4-Coumaryl alcohol, $R_1 = R_2 = H$; coniferyl alcohol, $R_1 = OCH_3$, $R_2 = H$; sinapyl alcohol, $R_1 = R_2 = OCH_3$.

oligomeric reaction products provided clues to the nature of the various bond types existing in lignin. The high-polymer material also obtained with this method had an ir spectrum very similar to that of lignin isolated from plants.

Recent structural studies on lignin involving ^{13}C-nmr spectroscopy (Nimz, 1974) have in principle confirmed the structure that had been proposed earlier based on analytical data, model experiments, and biosynthetic investigations. Figure 3 shows the structure of beech lignin proposed on the basis of degradative studies and ^{13}C-nmr spectroscopy.

Gymosperm lignin is made up mainly of coniferyl alcohol, whereas lignin from angiosperm dicotyledons consists of both guaiacyl and syringyl units (Fig. 3) in approximately equal amounts. In monocotyledons the lignins of the Cyperaceae and Gramineae are guaicyl syringyl lignins with a relatively high percentage of 4-hydroxyphenyl groups. The latter can be attributed at least in part to 4-coumaric acid esterified with lignin (Higuchi et al., 1967).

Earlier studies on the biosynthesis of lignin were carried out in vivo with isotopically labeled precursors. These studies confirmed that cinnamyl alcohols are the primary building blocks of lignin. It was also shown that cinnamyl alcohols themselves were probably derived from L-phenylalanine via the corresponding cinnamic acids. Only in recent years has it been possible

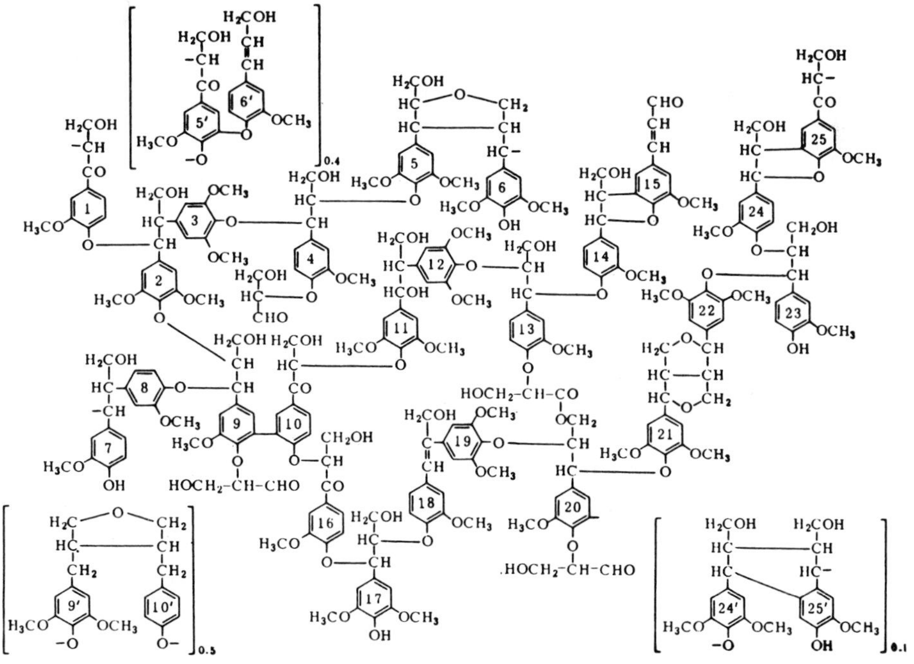

Fig. 3. Structure of beechwood lignin after Nimz (1974) (1) Guaicyl unit; (2) syringyl unit.

to investigate the biosynthesis of lignin precursors at an enzymatic level. Important progress has also been made in our knowledge of the enzymes involved in the polymerization process.

A standard text on lignins has been edited by Sarkanen and Ludwig (1971), which contains all pertinent information on occurrence, formation, structure, and reactions. Recent reviews on the chemistry and biochemistry of lignin (Gross, 1978), the biosynthesis of lignin (Grisebach, 1977; Gross, 1977a), and the enzymatic controls in the biosynthesis of lignin (Hahlbrock and Grisebach, 1979) are available. This chapter focuses on the biochemistry of lignification.

II. GENERAL PHENYLPROPANOID METABOLISM

A. General Outline

The biosynthesis of cinnamyl alcohols originates from L-phenylalanine, the aromatic amino acid ultimately derived from carbohydrate metabolism via the shikimic acid pathway (Vol. 5, Chapter 13). It is still an open question whether there exists only one or several pools of this amino acid supplying the precursors for protein synthesis, lignification, flavanoid formation, and the synthesis of other phenylpropanoids or compounds derived from them.

General phenylpropanoid metabolism is defined as the sequence of reactions involved in the conversion of L-phenylalanine to activated cinnamic acids (Hahlbrock and Grisebach, 1975). The reaction sequence of this pathway together with the corresponding enzymes is shown in Fig. 4. It is not implied by this figure that only one set of enzymes is present in a particular plant. Parallel pathways could occur in different cell compartments, and metabolic channeling may occur in the presence of multienzyme complexes or isoenzymes (see Stafford, this volume, Chapter 5). Examples of the occurrence of isoenzymes will be discussed below.

The first enzyme of this pathway, phenylalanine ammonia-lyase (PAL), catalyzing the trans-elimination of ammonia from L-phenylalanine to form *trans*-cinnamic acid, is discussed in detail by Hanson and Havir (this volume, Chapter 20). Cinnamic acid is then converted, by a sequence of hydroxylation and methylation reactions, to several substituted acids that can be activated as their corresponding esters of coenzyme A (CoA). These activated acids can then enter different biosynthetic pathways leading to lignin, flavonoids, stilbenes, benzoic acids, and other compounds. General phenylpropanoid metabolism is therefore one of the most important pathways in higher plants, and details of it are discussed by Gross (this volume, Chapter 11) and Hahlbrock (this volume, Chapter 14). A few reactions that are important to lignin formation are outlined here.

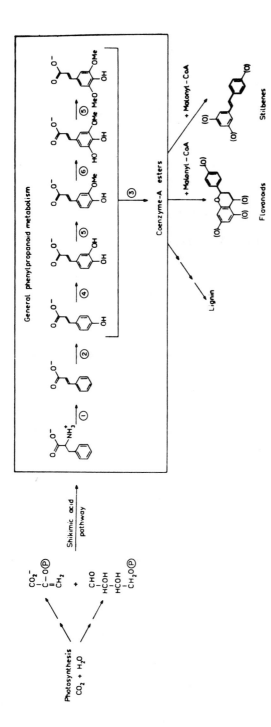

Fig. 4. General phenylpropanoid metabolism. Enzymes: (1) PAL; (2) cinnamate 4-hydroxylase; (3) 4-coumarate:CoA ligase; (4) 4-coumarate 3-monooxygenase; (5) SAM:caffeate 3-*O*-methyltransferase; (6) "ferulate 5-hydroxylase" (hypothetical).

B. Hydroxylases

Hydroxylation of cinnamic acid in the 4-position is mediated by a microsomal mixed-function oxidase which is described in detail by Butt and Lamb (this volume, Chapter 21). Cinnamate 4-hydroxylase is specific for the trans isomer of cinnamic acid (Pfändler *et al.,* 1977). The product, 4-coumaric acid, can be further hydroxylated in the 3-position by the action of phenolase to yield caffeic acid (Butt and Lamb, this volume, Chapter 21). A further hydroxylation reaction must be involved in the formation of sinapic acid, but this reaction has not yet been demonstrated in a cell-free system. Two possibilities exist for the formation of sinapic acid—hydroxylation of caffeic acid to 3,4,5-trihydroxycinnamic acid and subsequent methylation to sinapic acid, or hydroxylation of ferulic acid to 5-hydroxyferulic acid followed by methylation to yield sinapic acid. These pathways are shown in Fig. 5. From the substrate specificity of the methyltransferase (see below) the second pathway seems more probable, but the pathway via trihydroxycinnamic acid cannot be excluded.

C. Methyltransferases

Since methyltransferases are described in detail by Poulton (this volume, Chapter 22), only a few details pertinent to lignin biosynthesis are discussed

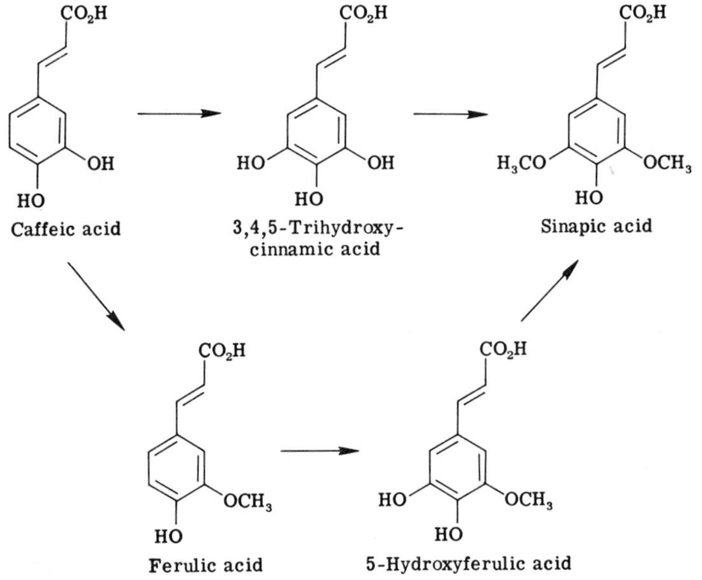

Fig. 5. Possible pathways for the biosynthesis of sinapic acid from caffeic acid. According to the substrate specificity of methyltransferase the pathway via 5-hydroxyferulic acid is more likely.

here. It is now certain that two distinct S-adenosylmethionine (SAM):3,4-dihydric phenol 3-O-methyltransferases occur in plants. One enzyme is thought to be specific for the methylation of flavonoid substrates such as quercetin and luteolin, whereas the other methylates substituted cinnamic acids such as caffeic acid and 5-hydroxyferulic acid. The latter enzyme is therefore thought to be involved in lignin formation.

The SAM:caffeic acid 3-O-methyltransferase (CMT) has been purified from a number of plant sources. The same enzyme is responsible for the methylation of caffeic acid to ferulic acid and the methylation of 5-hydroxyferulic acid or 3,4,5-trihydroxycinnamic acid to sinapic acid (Section II,B). However, there is a marked difference between the enzymes from angiosperms and gymnosperms. Whereas the angiosperm methyltransferase from bamboo, poplar (Shimada *et al.*, 1970), and soybean cell cultures (Poulton *et al.*, 1976) is meta-specific for 3,4-dihydroxy-, 3,4,5-trihydroxy-, and 3,4-dihydroxy-5-methoxycinnamic acids, sliced tissue from *Gingko biloba* (Shimada *et al.*, 1970) and a methyltransferase from Japanese black pine (*Pinus thunbergii*) failed to methylate or only very poorly methylated 3,4,5-trihydroxycinnamic acid and 3,4-dihydroxy-5-methoxycinnamic acid (Kuroda *et al.*, 1975). This difference in substrate specificity of the angiosperm and gymnosperm methyltransferases would account for the much higher proportion of syringyl units in angiosperm lignin. Another possible reason for this difference could be a lower rate of hydroxylation in the 5-position in gymnosperms. Furthermore, the substrate specificity of 4-coumarate:CoA ligase (see below) may play a role in the determination of lignin composition.

D. 4-Coumarate:CoA Ligases

The last step of general phenylpropanoid metabolism is the activation of cinnamic acids to form CoA thioesters. This reaction is catalyzed by 4-coumarate:CoA ligase [recommended name: 4-coumaroyl-CoA synthase or 4-coumarate:CoA ligase (AMP), E.C. 6.2.1.12] and proceeds in analogy to the activation of fatty acids with ATP and CoASH according to the following equation

$$\text{Cinnamic acid} + \text{CoASH} + \text{ATP} \rightleftharpoons \text{cinnamoyl-CoA} + \text{AMP} + \text{PP}_i$$

Details of these enzymes are described by Gross (this volume, Chapter 11). Most interesting in connection with lignification is the existence of isoenzymes of the ligase and the fact that according to its substrate specificity one of these isoenzymes could be specifically involved in the lignin pathway. In cell cultures of soybean (*Glycine max*), which form a guaiacyl syringyl lignin, two isoenzymes of 4-coumarate:CoA ligase are present that differ in substrate specificity. The substrates with the highest V/K_m ratios for ligase 1 are 4-coumaric, ferulic, and sinapic acids. In contrast, ligase 2 has the highest V/K_m ratios for 4-coumaric and caffeic acids but a 17-fold smaller ratio for ferulic acid than ligase 1 and cannot use sinapic acid as substrate. It has therefore been postulated that ligase 1 belongs to the lignin pathway and ligase 2 to the flavonoid pathway (Knobloch and Hahlbrock, 1975). A similar situation exists in pea seedlings where only one of the two isoenzymes detected can activate sinapic acid (Wallis and Rhodes, 1977). In *Petunia* leaves three isoenzymes of the ligase seem to be present. One isoenzyme, for which the best substrate is caffeic acid, is assumed to be involved in ester formation; the second isoenzyme (ferulate:CoA ligase) could be involved in flavonoid and lignin biosynthesis; the third isoenzyme, which activates sinapic acid, could take part in the formation of syringyl lignin in older tissues (Ranjeva *et al.*, 1976).

In lignifying tissue of *Forsythia* the presence of only one ligase, which does not activate sinapic acid, has been reported (Gross and Zenk, 1974). Since the lignin of *Forsythia suspensa,* an angiosperm, would be expected to contain syringyl residues, a second ligase might exist in this plant that has not yet been discovered.

It has been observed that the percentage of syringyl residues in lignin increases with the age of plants. The activity of ligases that activate sinapic acid may therefore increase with the age of the plant.

III. REDUCTION OF CINNAMOYL-CoA ESTERS TO CINNAMYL ALCOHOLS

A. General Remarks

The reactions that have been described lead not only to lignin but to other secondary plant products as well. However, the reduction of cinnamoyl-CoA esters to cinnamyl alcohols is specific for the pathway of lignin biosynthesis. Evidence for the reduction of ferulic acid to coniferyl alcohol via coniferyl aldehyde has come from tracer experiments (Higuchi and Brown, 1963). It follows from thermodynamic considerations that the reduction of a carboxylic acid to an aldehyde requires an activated acid. The reduction potential of the cinnamic acid–cinnamaldehyde half-reaction is not known, but an estimate can be made from the acetate–acetaldehyde half-reaction for which the standard reduction potential is -0.6 V. Since the potential of NAD^+–NADH is -0.320 V at pH 7.0, a change in free enthalpy of $+12.9$ kcal/mol for the reduction of acetate with NADH as coenzyme is obtained. Although this value could be quite different under physiological conditions, reduction of a free carboxylic acid is very unlikely. In contrast, the reduction potential for the couple acetyl-CoA–acetaldehyde is -0.41 V. This reduces the ΔG value for the reduction of acetate with NAD(P)H to $+4.1$ kcal/mol.

The possibility of the participation of CoA thioester in lignin biosynthesis

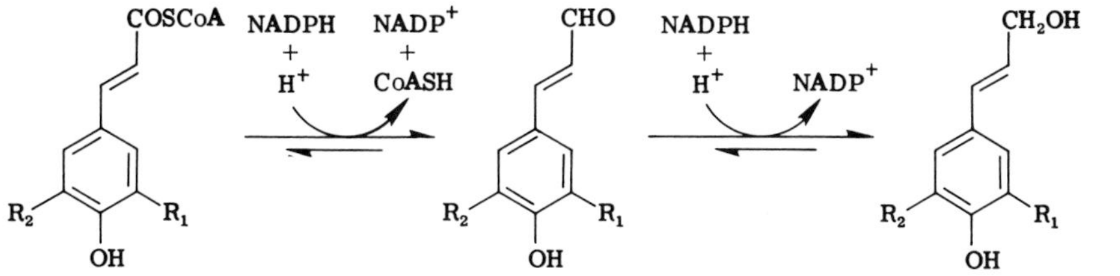

Fig. 6. Reactions catalyzed by cinnamoyl-CoA:NADPH oxidoreductase and cinnamyl alcohol dehydrogenase.

was discussed in 1959 by Stewart Brown *et al.* However, it was not until 1970 that the formation of CoA esters of cinnamic acids was described with a cell-free system from higher plants (Hahlbrock and Grisebach, 1970), and only recently has it been shown that reduction to the cinnamyl alcohols indeed takes place with these esters (Ebel and Grisebach, 1973; Mansell *et al.*, 1972). The overall reaction requires two enzymes (cinnamoyl-CoA:NADPH oxidoreductase and cinnamyl alcohol dehydrogenase) and proceeds via the cinnamaldehydes. The reaction sequence catalyzed by these two enzymes is shown in Fig. 6. Some properties of these enzymes are now described.

B. Cinnamoyl-CoA:NADPH Oxidoreductase

Enzymes catalyzing the reduction of differently substituted cinnamoyl-CoA esters to the corresponding aldehydes have been reported to be present in cell suspension cultures of soybean (Wengenmayer *et al.*, 1976), lignifying tissues of *F. suspensa* (Gross and Kreiten, 1975), swede root (*Brassica napobrassica*) tissue (Rhodes and Wooltorton, 1975), *Acer* sp. (Gross, 1978), and spruce (*Picea abies*) (Th. Lüderitz, unpublished results). The enzyme from soybean cell cultures has been studied in the greatest detail. A 1660-fold purification of the soybean oxidoreductase was achieved by conventional purification procedures and affinity chromatography using CoA–hexane–agarose or the "general ligand" 5′-AMP attached to Sepharose. With both affinity methods the enzyme could be eluted specifically with $NADP^+$.

The apparent MW of the oxidoreductase obtained from the elution volume on Sephadex G-100 was 38,000. Sodium dodecyl sulfate (SDS) gel electrophoresis gave only one band with a MW of 38,000. These results indicate that the enzyme consists of only one polypeptide chain.

The enzyme exhibits high specificity in the reduction of differently substituted cinnamoyl-CoA esters with NADPH and is one of the B-specific (H_S) dehydrogenases. The enzyme from *Forsythia* has also been shown to have B specificity. The reaction rate with NADH is only about 5% of that found with NADPH.

The reversibility of the reduction was proved by the isolation of feruloyl-CoA from an incubation mixture containing coniferaldehyde, $NADP^+$, CoASH, and the oxidoreductase. The reaction proceeds according to the following equation:

$$\text{Feruloyl-CoA} + \text{NADPH} + \text{H}^+ \rightleftharpoons \text{coniferaldehyde} + \text{NADP}^+ + \text{CoASH}$$

The pH optimum for the forward reaction is about 6.2 and for the reverse reaction about 7.2.

Double reciprocal plots for substrates and coenzyme were all linear, as were secondary plots of slopes and intercepts; these yielded the K_m and V values listed in Table I. The soybean enzyme has the highest affinity for feruloyl-CoA and 5-hydroxyferuloyl-CoA, but 4-coumaroyl-CoA and sinapoyl-CoA are also reduced with good efficiency. The substrate specificity of the reductase is consistent with the occurrence of a guaiacyl syringyl lignin in soybean cell cultures. A random sequential mechanism was postulated for the enzyme from kinetic studies (Grisebach et al., 1977).

The enzyme from Forsythia had the highest relative activity with feruloyl-CoA (100%). 5-Hydroxyferuloyl-CoA, sinapoyl-CoA, and 4-coumaroyl-CoA all showed equal relative activities of 20%. The oxidoreductase from swede root discs aged in the presence of ethylene separated into two active fractions on a DEAE-cellulose column. Both active fractions had similar K_m values for feruloyl-CoA ($\sim 5 \ \mu M$), and it is not certain that these fractions represented isoenzymes.

It would be interesting to compare oxidoreductases from angiosperms and gymnosperms with regard to their ability to reduce sinapoyl-CoA esters. Nakamura et al. (1974) reported that sliced xylem tissue from shoots of both poplar (Populus nigra) and cherry (Prunus yedonsis) reduced ferulic and sinapic acids to the corresponding aldehydes and alcohols, whereas tissue from gymnosperms such as Japanese red pine (Pinus densiflora) and ginkgo

TABLE I

Substrate Specificity of the Cinnamoyl-CoA:NADPH Oxidoreductase from Soybean Cell Cultures

Substrate	$10^5 \times K_m$ (M)	V (nkat/mg)	$10^{-5} \times V/K_m$ (nkat/mg/mol)
NADPH	2.8	230	82
Feruloyl-CoA	7.3		31.5
NADPH	12	69	5.8
Sinapoyl-CoA	40		1.7
NADPH	29	30	1.0
p-Coumaroyl-CoA	10		3.0
5-Hydroxyferuloyl-CoA	5	58	11.6

(*G. biloba*) could reduce only ferulic acid. Since the reduction was not found in the gymnosperms, it appears that either the ligase could not activate sinapic acid to sinapoyl-CoA or reduction of the latter compound was not possible.

C. Cinnamyl Alcohol Dehydrogenase

This enzyme catalyzes the last step in the series of reactions leading from L-phenylalanine to substituted cinnamyl alcohols. $NADP^+$-dependent cinnamyl alcohol dehydrogenases have been isolated from *F. suspensa* (Mansell *et al.*, 1974) and soybean cell cultures (Wyrambik and Grisebach, 1975). The enzyme is widely distributed in the plant kingdom (Mansell *et al.*, 1974, 1976), and in a few plants multiple forms have been detected by starch gel electrophoresis.

In cell cultures of soybean two $NADP^+$-dependent isoenzymes were present, which have been separated from each other and from a NAD^+-dependent aliphatic alcohol dehydrogenase on DEAE-cellulose and hydroxyapatite. Whereas isoenzyme 1 is specific for coniferyl alcohol, isoenzyme 2 can oxidize a number of substituted cinnamyl alcohols (or reduce the corresponding aldehydes) including 4-coumaryl alcohol, coniferyl alcohol, and sinapyl alcohol. The enzyme from *Forsythia* has a substrate specificity similar to that of isoenzyme 2 from soybean cells. The Michaelis constants for substituted cinnamaldehydes (1.7–9.1 μM) are 3–11 times lower for isoenzyme 2 than for the corresponding alcohols (11–57 μM). For the equilibrium constant

$$K = \frac{[\text{coniferaldehyde}]\,[\text{NADPH}]\,[H^+]}{[\text{coniferyl alcohol}]\,[\text{NADP}^+]}$$

a mean value of $2.8(\pm1.1) \times 10^{-9}$ has been obtained. These results indicate that under physiological conditions reduction of cinnamaldehydes is favored.

Isoenzyme 2 from soybean cell cultures was purified about 3760-fold to apparent homogeneity (Wyrambik and Grisebach, 1979). The purification procedure included affinity chromatography on a $NADP^+$–agarose column with biospecific elution of the enzyme with a $NADP^+$ gradient. The enzyme of MW 69,000 (Sephadex G-100) was composed of two apparently identical subunits (MW 40,000 on SDS gel electrophoresis). Inhibition of the dehydrogenase by metal-chelating agents, metal analysis, and incorporation of $^{65}Zn^{2+}$ into the enzyme proved the presence of Zn^{2+} in the dehydrogenase (Wyrambik and Grisebach, 1979).

The stereospecificity of the reaction was investigated with the dehydrogenases from *Forsythia* (Mansell *et al.*, 1974; Klischies *et al.*, 1978) and soybean (Grisebach *et al.*, 1977). With respect to NADPH the enzyme belongs to the A-specific (H_R) dehydrogenases, and the *pro-R* hydrogen of coniferyl alcohol is removed during oxidation to the aldehyde.

The cinnamyl alcohol dehydrogenase from soybean bears a close resemblance to alcohol dehydrogenases from yeast, horse liver, and rat liver with respect to its subunit MW, inhibition by metal chelators, zinc content, steady state kinetics (Grisebach *et al.*, 1977), and stereospecificity with regard to coenzyme and substrate. These similarities are consistent with the assumption that these enzymes evolved from a common ancestral gene.

IV. THE ROLE OF CINNAMYL ALCOHOL GLUCOSIDES IN LIGNIFICATION

A. General Remarks

Cinnamyl alcohol glucosides [e.g., coniferin (coniferyl alcohol β-D-glucoside) and syringin (sinapyl alcohol β-D-glucoside)] have been found in the sap of conifers and other gymnosperms but seem to occur only rarely in angiosperms (Sarkanen and Ludwig, 1971). A notable exception among the angiosperms is lilac (*Syringa vulgaris*), which contains higher concentrations of syringin in its bark (Freudenberg *et al.*, 1951). A more extended search for these glucosides in angiosperms is, however, necessary before a more definite statement can be made about their distribution.

Experiments with labeled coniferin have shown that this compound can act as a lignin precursor in a variety of species (Brown, 1966). However, the role of coniferin and of the other cinnamyl alcohol glucosides in lignification remains uncertain. Since xylem tissue seems to be autonomous for lignin biosynthesis (Gross, 1977a, 1978), the original suggestion by Freudenberg (1965) that cinnamyl alcohols were translocated in the form of their glucosides from the cambial zone into lignifying tissue has been questioned. Another suggestion was that cinnamyl glucosides acted in conifers as a reservoir to augment the supply of precursor for lignifying cells.

In recent years the turnover of coniferin, and its formation from UDP-glucose and coniferyl alcohol and the enzymatic hydrolysis of coniferin, have been investigated.

B. Turnover of Coniferin

The turnover of coniferin has been studied in spruce (*P. abies*) seedlings. Spruce seeds do not contain detectable amounts of coniferin, but it accumulates in the stems and roots of seedlings. Forty-day-old seedlings contain about 70 μg coniferin per gram fresh weight, which is about equally distributed between stems and roots. Pulse-labeling experiments with L-[^{14}C]phenylalanine or $^{14}CO_2$ showed a rapid synthesis of coniferin and a slower turnover of this compound with a half-life in the range of 60–120 h (Marcinowski and Grisebach, 1977). A quantitative correlation between coniferin turnover and lignin synthesis was not possible in the seedlings. The small pool size of coniferin of about 50–70 nmol coniferin per seedling (average weight 190–250 mg) and its relatively slow turnover suggest that only part of lignin synthesis occurs via coniferin.

C. UDP-Glucose:Coniferyl Alcohol Glucosyltransferase

An enzyme that catalyzed the transfer of glucose from UDP-glucose to coniferyl alcohol with the formation of coniferin was isolated from suspension cultures of 'Paul's Scarlet' rose (Ibrahim and Grisebach, 1976). The enzyme was purified about 120-fold by conventional procedures. It had a pH optimum of 7.5 in Tris–HCl buffer and required SH groups for activity. Mg^{2+} or Ca^{2+} had no influence on enzyme activity. As estimated from the elution volume of a Sephadex G-100 column the MW was about 52,000. The glucosyltransferase had a strong affinity for coniferyl and sinapyl alcohol and a significantly lower one for coniferyl aldehyde. The apparent K_m values for these substrates were found to be 3.3×10^{-6}, 5.6×10^{-6}, and $6.5 \times 10^{-5}M$, respectively. There was some activity with the corresponding C_6–C_3 acids and with o-dihydroxy compounds such as caffeic acid and quercetin. The enzyme is therefore quite distinct from those reported to catalyze the transfer of glucose to simple hydroxyphenols, phenolic acids, and flavonoids (see Hösel, this volume, Chapter 23).

A glucosyltransferase with properties quite similar to those of the enzyme from rose cell cultures was isolated from lignifying stem segments of For-sythia orata (Ibrahim, 1977). Ibrahim investigated the distribution of the glucosyltransferase in a number of plant species belonging to different taxonomic groups. Of all the plants examined gymnosperms exhibited the highest levels of glucosyltransferase activity. The enzyme was also detected in angiosperms, especially in woody species, and low activity seemed to be present in bryophytes, which form no lignin.

Recently, the corresponding glucosyltransferase has also been isolated from cambial sap of spruce, which contains a high amount of coniferin (Schmid and Grisebach, unpublished results). The Michaelis constants for coniferyl alcohol ($2.5 \times 10^{-4}M$) and UDP-glucose ($2.2 \times 10^{-4}M$) have been determined with a 265-fold purified enzyme and are considerably higher than the K_m values of the enzymes from rose cultures and F. orata.

D. Glucosidases for Cinnamyl Alcohol Glucosides

Freudenberg et al. (1955) had detected β-glucosidase activity in sections of Araucaria excelsa in cells adjacent to the cambium using the indican reaction as a histochemical method. In this reaction indoxyl β-D-glucoside is hydrolyzed by glucosidase activity, and the liberated indoxyl is oxidized in situ to indigo. With the same technique β-glucosidase activity was also found in many hardwood species. Until recently this glucosidase activity had never been characterized, and it was not known whether one or several glucosidases were present and what their specificity toward cinnamyl glucosides was.

β-Glucosidases with good activity toward coniferin were recently isolated from spruce seedlings (Marcinowski and Grisebach, 1978) and chick-pea (*Cicer arietinum* L.) cell cultures (Hösel *et al.,* 1978). A survey of these and other glycosidases is given by Hösel (this volume, Chapter 23).

Spruce seeds contain a soluble β-glucosidase that does not catalyze the hydrolysis of coniferin but is active with 4-nitrophenyl β-D-glucoside. In contrast, a cell-wall fraction from hypocotyls and roots of spruce seedlings has good activity toward coniferin with a pH optimum between 4.5 and 5.5. A number of other aryl β-glycosides were also hydrolyzed by these particulate fractions, and from kinetic investigations it was concluded that more than one glycosidase was present in the cell-wall fractions. It was possible to solubilize about half of the glucosidase activity from the hypocotyl particulate fraction by treatment with 0.6 M NaCl. The solubilized enzyme could be separated on Sepharose 6B into two fractions with glucosidase activity (glucosidases I and II). Glucosidase I has been purified to apparent homogeneity. The enzyme is composed of only one polypeptide chain with MW of about 58,000. It is a glycoprotein with about 8.5% carbohydrate and binds reversibly to concanavalin A–Sepharose (S. Marcinowski, unpublished results). Comparison of the homogeneous glucosidase I with that of the particulate glucosidase from cell walls showed that they were very similar with respect to pH optimum and substrate specificity toward coniferin and a number of other aryl β-glucosides.

In chick-pea cell cultures, high activity for the hydrolysis of coniferin is also present in cell wall preparations. Various β-glucosidase activities have been solubilized from these preparations by 0.5 M NaCl treatment. One of these glucosidases has a high activity for hydrolysis of coniferin.

The substrate specificities of the glucosidases from spruce and chick-pea are compared in Table II. Both glucosidases have about the same activity for coniferin, but the spruce glucosidase has a higher V/K_m ratio for syringin. Indican was a very poor substrate for both glucosidases. With the enzyme from chick-pea cell cultures, the relative specific activity with indican was only about 1% of that found with coniferin. A positive indican reaction therefore does not necessarily mean the presence of a glucosidase that is able to hydrolyze coniferin.

β-Glucosidase preparations from sweet almonds and from roots of chick-pea plants can hydrolyze coniferin only very poorly.

Specific antibodies for glucosidase 1 from spruce were raised in rabbits and used for localization of this enzyme in hypocotyls of spruce seedlings by an immunofluorescent technique (Marcinowski *et al.,* 1979). The specific fluorescence of the antibody bound to the glucosidase at the inner layer of

TABLE II

Substrate Specificities of Pure Glucosidase I from Spruce Hypocotyl and Pure Glucosidase from *Cicer arietinum* Cell Cultures

Substrate	Spruce glucosidase			*Cicer* glucosidase		
	K_m (mM)	V (mkat/kg)	V/K_m (mkat/kg/mmol)	K_m (mM)	V (mkat/kg)	V/K_m (mkat/kg/mmol)
Coniferin	1.3	144	111	0.8	100	125
Syringin	1.8	121	67	5	28	6
2-Nitrophenyl β-D-glucoside	0.6	257	428	0.3	118	393
4-Methylumbelliferyl β-D-glucoside	1.0	102	102	0.4	148	370
L-Picein[a] (4-hydroxyacetophenone β-D-glucoside)	6.3	139	22	0.6	33	55
4-Nitrophenyl β-D-glucoside	0.7	11	16	0.4	32	79

[a] Occurs in spruce needles.

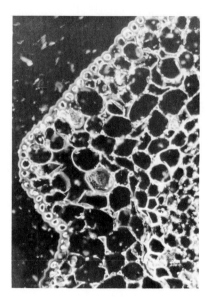 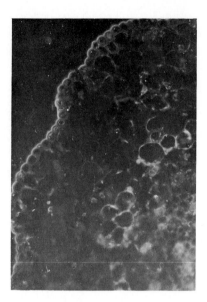

Fig. 7. (a) Immunofluorescent labeling of glucosidase in a transverse microtome section of a 19-day-old hypocotyl showing specific fluorescence at the inner layer of the secondary cell walls: × 320. (b) Control with serum of unimmunized rabbits.

the secondary wall is shown in Fig. 7 together with the control, which is completely devoid of fluorescence. Further confirmation for localization of the glucosidase in the cell wall is the finding that protoplasts prepared from the tissue of the germination hook of *P. abies* have only little residual glucosidase activity.

In contrast to the results of Freudenberg *et al.* (1955), who found glucosidase activity in sections of *A. excelsa* only in a layer adjacent to the cambial zone, the glucosidase in spruce seedlings is present in all cells of the hypocotyl. As judged from the fluorescence intensity, there appears to be higher glucosidase activity in the epidermal layer and in the vascular bundles. These are the cell populations in which lignification first appears during seedling development.

V. POLYMERIZATION OF CINNAMYL ALCOHOLS TO LIGNIN

A. Peroxidases

The polymerization of monolignols (4-coumaryl, coniferyl, and sinapyl alcohols) is initiated by oxidation of the phenolic hydroxyl group of these monomers, yielding mesomeric phenoxy radicals. Coupling of these radicals leads to dilignols which, after reoxidation and radical coupling, are converted to oligomeric intermediates which finally combine to form the lignin macromolecule. The two enzymes laccase (*o*- and *p*-diphenol oxidase, E.C. 1.14.18.1) and peroxidase (E.C. 1.11.1.7) have been implicated in the polymerization process. The participation of laccase and peroxidase in lignin formation was estimated (Harking and Obst, 1973) by the oxidation of syringaldazine to the purple tetramethoxyazo-*p*-methylene quinone (Fig. 8) by laccase and oxygen or by peroxidase and H_2O_2. Application of a solution of syringaldazine to cross sections of microtomed surfaces of numerous sample stubs from angiosperms and gymnosperms did not produce any coloration indicating the presence of laccase. In contrast, when H_2O_2 was added, an intense purple ring formed in the xylem tissue adjacent to the cambium. From these and other results not described here, it was concluded that the physiologically active enzyme in the zone of incipient lignification was exclusively peroxidase.

Peroxidase occurs in plants as isoenzymes, and cell-wall-bound isoenzymes have been found in several plants (Stafford, 1974). These cell-wall-bound peroxidases are very probably involved in the lignification process. For example, three groups of peroxidase isoenzymes were detected in tobacco (Mäder, 1976). One group was localized only in the cell wall (G_I). The two other isoenzyme groups were localized mainly in the protoplasts, but they were also bound to the walls, covalently in the case of G_{II} and ionicly in the case of G_{III}. Furthermore it was shown that *in vitro* polymerization of coumaryl and coniferyl alcohol to ligninlike substances was catalyzed at substantial rates by groups G_I and G_{II} but only at a negligible rate by group G_{III} (Mäder *et al.*, 1977). Quantitative data (K_m, V_{max}) were not provided, but

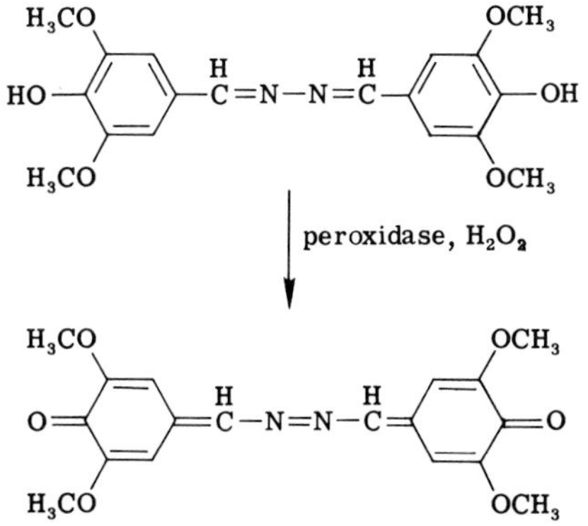

Fig. 8. Oxidation of syringaldazine to the purple tetramethoxyazo-*p*-methylene quinone by peroxidase and H_2O_2.

the results indicated that peroxidase isoenzymes specific for the initiation of lignification were present in plant cell walls.

B. Formation of Hydrogen Peroxide

H_2O_2 required by the peroxidases has been detected in high concentrations in the xylem and bark of poplar (Sagisaka, 1976). Experiments on the origin of H_2O_2 were carried out with horseradish cell walls (Elstner and Heupel, 1976) and cell walls isolated from *Forsythia* xylem (Gross and Janse, 1977). These cell walls catalyze the formation of H_2O_2 in a complex reaction requiring NAD(P)H and Mn^{2+}. Evidence was provided that a peroxidase bound to the cell wall was responsible for H_2O_2 formation. The NAD(P)H required for this reaction could be supplied by a malate dehydrogenase also bound to the cell wall (Gross, 1977b). The combined activities of the bound malate dehydrogenase and peroxidase allow the formation of H_2O_2. This coupled system can promote the polymerization *in vitro* of coniferyl alcohol to a ligninlike dehydrogenation polymer. The NAD(P)H oxidation by the bound peroxidase is a complex radical mechanism in which superoxide anion (O_2^-) could be involved.

The observation that H_2O_2 formation was stimulated by various monophenols and that the strongest stimulation was obtained with 3×10^{-5} M coniferyl alcohol (Gross *et al.*, 1977) indicates that hydroxycinnamyl alcohols could be directly involved in regulating the H_2O_2 supply needed for polymerization. The possibility has been discussed that a malate–oxalacetate shuttle across the plasmalemma would allow the transport of cytoplasmic reducing equivalents into the cell wall compartment and that H_2O_2 originates within the cell wall at the site of lignification.

VI. REGULATION OF LIGNIFICATION

Virtually nothing is known about the regulation of lignification *in vivo*. Several possible regulatory mechanisms that have been found *in vitro* can be considered, but it is difficult to judge their individual significance. Three main areas for control of lignification can be discussed: (1) supply of lignin precursors, (2) transport of lignin monomers into the cell wall, and (3) polymerization of hydroxycinnamyl alcohols in the cell wall.

The conversion of phenylalanine to cinnamic acid is the point of entry into phenylpropanoid metabolism. Since phenylalanine is also an amino acid essential for protein biosynthesis, it can be expected that the enzyme PAL is under regulatory control. Numerous internal and external factors, e.g., substrates, products, light, and hormones, can indeed affect the synthesis and (or) activity of this enzyme (Details are discussed by Hanson and Havir, this volume, Chapter 20). Activation of cinnamic acids is another potential control point. 4-Coumarate:CoA ligases from parsley and soybean cultures were shown to be competitively inhibited by AMP (Knobloch and Hahlbrock, 1975). Inhibition by AMP was also reported for the enzyme from *Forsythia* (Gross and Zenk, 1974). The activity of these ligases is therefore under control of the adenylate energy charge of the cell. For the parsley ligase, a sigmoidal dependence of enzyme activity on ATP and CoASH was found, indicating an allosteric character for this enzyme (Knobloch and Hahlbrock, 1975).

If the energy charge of the cell can be important for the activation of cinnamic acids, the reduction charge could also play a role in regulating the reduction of cinnamoyl-CoA esters to cinnamyl alcohols. The term "reduction charge" was introduced for the ratio of NADPH to NADPH plus NADP$^+$ (Wildner, 1975). In the case of cinnamoyl-CoA oxidoreductase, NADP$^+$ is a mixed noncompetitive inhibitor with a K_i of 0.5 mM with respect to NADPH as substrate. With cinnamyl alcohol dehydrogenase, NADP$^+$ is a competitive inhibitor of NADPH with a K_i of 15 μM (Grisebach et al., 1977).

The possible role of compartmentation and of multienzyme complexes in lignin biosynthesis is discussed in detail by Stafford (this volume, Chapter 5). Czichi and Kindl (1977) have expressed the opinion that the low Michaelis constant for phenylalanine in the PAL reaction makes it hard to imagine how a significant concentration of L-phenylalanine can be maintained for protein synthesis without physical separation of the two processes for protein and phenylpropane biosynthesis. The enzymological evidence for multienzyme complexes is still rather fragmentary. Microsomal fractions from *Quercus penduculata* catalyze the three steps from phenylalanine to caffeic acid (Alibert et al., 1972), and microsomal fractions from potato (Czichi and Kindl, 1975), and cucumber cotyledons (Czichi and Kindl, 1977) catalyze the conversion of phenylalanine to 4- and 2-coumaric acids.

In cell suspension cultures hormones can influence lignin synthesis, and some examples in which this effect has been studied on an enzymatic level are mentioned here. In cell cultures of tobacco, lignification was observed only in a kinetin culture and not in the presence of auxin, 2,4-dichlorophenoxyacetic acid (2,4-D), or indolbutyric acid. *S*-Adenosylmethionine:catechol *O*-methyltransferase activity in the kinetin cultures was considerably higher than in the controls, and a rise in enzyme activity coincided with the onset of lignification (Yamada and Kuboi, 1976). Lignin formation in other tissue cultures grown in the presence of kinetin was also paralleled by enhanced activity of PAL (Rubery and Fosket, 1969). Another hormone that has a stimulatory effect on lignin synthesis is ethylene.

In swede roots, large increases in the activity of PAL, cinnamate 4-hyroxylase, and cinnamate:CoA ligase were observed when swede root discs were aged in air containing 8 ppm ethylene (Rhodes *et al.*, 1976).

With respect to regulation of lignification, it is also important that the enzymes involved in the biosynthesis of lignin precursors be tissue-specific and be located predominantly or exclusively in the xylem where lignification occurs (Gross, 1977a). Xylem tissue seems therefore to be autonomous for lignin biosynthesis.

According to this finding intercellular transport of coniferin as suggested by Freudenberg (Section IV,A) does not seem necessary. This glucoside could be an intracellular storage metabolite or the form in which coniferyl alcohol is excreted into the cell wall. The presence of β-glucosidases in the cell wall, described in Section IV,C, is consistent with the latter view. Some experimental evidence suggests that the transport of lignin monomers from the cytoplasm to the cell wall is mediated by vesicles (the Golgi apparatus?) (Pickett-Heaps, 1968). Once the lignin precursors have reached the cell wall, initiation of polymerization could be controlled by the liberation of free cinnamyl alcohols from their glucosides (see above) and/or by the supply of H_2O_2 (Section V,B).

The connection between tissue differentiation and lignin formation is discussed by Wiermann (this volume, Chapter 4).

VII. LIGNIFICATION IN RELATION TO DISEASE RESISTANCE

Lignification is a common response to infection or wounding in plants (Asada *et al.*, 1975). A definite role for lignin formation in relation to disease resistance is indicated by a number of more recent experiments. When tuber discs of susceptible and resistant varieties of potato were inoculated with a sporangial suspension of *Phytophtera infestans*, there were increases in the level of PAL, chlorogenic acid, and lignin (Friend *et al.*, 1973). Lignification in the resistant strain was faster and was concentrated in the center of the infected area. Similar observations were made upon infection of radish roots (Asada *et al.*, 1975) and wheat leaves (Ride, 1975). In wheat leaves, for example, lignin was rapidly synthesized around the areas infected by non-pathogenic fungi, and subsequent fungal growth was limited to the wounds. The rate of lignification was slower in response to pathogenic fungi, and these fungi were not restricted to the wounds. Clarke (1973) has suggested that virulent pathogens inhibit the development of lignin barriers by diverting lignin precursors into other metabolic pathways. Direct experimental evidence for this suggestion is, however, not available.

Localized lignin formation in leaf epidermal cell walls has also been implicated as a general mechanism of resistance of reed canary grass (*Phalaris arundinacea*) (Vance and Sherwood, 1976). The fungi did not penetrate the modified wall. An approximately two-fold increase in the activity of PAL, hydroxycinnamate:CoA ligase, and peroxidase was found to parallel the induced lignification. When inoculated tissues were treated with the protein synthesis inhibitor cycloheximide, the increases in lignin content and enzyme activity associated with lignin biosynthesis were inhibited, and the tissue became susceptible to fungal penetration (Vance *et al.*, 1976). These observations support the idea that lignin formation is important in actively preventing complete fungal penetration of epidermal cells.

REFERENCES

Alibert, G., Ranjeva, R., and Boudet, A. (1972). *Biochim. Biophys. Acta* **279**, 282–289.
Asada, Y., Oguchi, T., and Matsumoto, I. (1975). *Rev. Plant Prot. Res.* **8**, 104.
Brown, S. A. (1966). *Annu. Rev. Plant Physiol.* **17**, 223–244.
Brown, S. A., Wright, D., and Neish, A. C. (1959). *Can. J. Biochem. Physiol.* **37**, 25–34.
Clarke, D. (1973). *Physiol. Plant Pathol.* **3**, 347–358.
Czichi, U., and Kindl, H. (1975). *Planta* **125**, 115–125.
Czichi, U., and Kindl, H. (1977). *Planta* **134**, 133–143.
Ebel, J., and Grisebach, H. (1973). *FEBS Lett.* **30**, 141–143.
Elstner, E. F., and Heupel, A. (1976). *Planta* **130**, 175–180.
Freudenberg, K. (1965). *Science* **148**, 595–600.
Freudenberg, K., Kraft, R., and Heimburger, W. (1951). *Chem. Ber.* **84**, 472–476.
Freudenberg, K., Reznik, H., Fuchs, W., and Reichert, M. (1955). *Naturwissenschaften* **42**, 29–35.
Friend, J., Reynolds, S. B., and Aveyard, M. A. (1973). *Physiol. Plant Pathol.* **3**, 495–507.
Grisebach, H. (1977). *Naturwissenschaften* **64**, 619–625.
Grisebach, H., Wengenmayer, H., and Wyrambik, D. (1977). *In* "Pyridine Nucleotide-Dependent Dehydrogenases" (H. Sund, ed.), pp. 458–471. de Gruyter, Berlin.
Gross, G. G. (1977a). *Recent Adv. Phytochem.* **11**, 141–184.
Gross, G. G. (1977b). *Phytochemistry* **16**, 319–321.
Gross, G. G. (1978). *Recent Adv. Phytochem.* **12**, 177–220.
Gross, G. G., and Janse, C. (1977). *Z. Pflanzenphysiol.* **84**, 447–452.
Gross, G. G., and Kreiten, W. (1975). *FEBS Lett.* **54**, 259–262.
Gross, G. G., and Zenk, M. H. (1974). *Eur. J. Biochem.* **42**, 453–459.
Gross, G. G., Janse, C., and Elstner, E. F. (1977). *Planta* **136**, 271–276.
Hahlbrock, K., and Grisebach, H. (1970). *FEBS Lett.* **11**, 62–64.
Hahlbrock, K., and Grisebach, H. (1975). *In* "The Flavonoids" (J. B. Harborne, T. J. Mabry, and H. Mabry, eds.), pp. 866–915. Chapman & Hall, London.
Hahlbrock, K., and Grisebach, H. (1979). *Annu. Rev. Plant Physiol.* **30**, 105–130.
Harkin, J. M., and Obst, J. R. (1973). *Science* **180**, 296–298.
Higuchi, T., and Brown, S. A. (1963). *Can. J. Biochem. Physiol.* **41**, 621–628.
Higuchi, T., Ito, Y., and Kawamura, I. (1967). *Phytochemistry* **6**, 875–881.
Hösel, W., Surholt, E., and Borgmann, E. (1978). *Eur. J. Biochem.* **84**, 487–492.
Ibrahim, R. K. (1977). *Z. Pflanzenphysiol.* **85**, 253–262.
Ibrahim, R. K., and Grisebach, H. (1976). *Arch. Biochem. Biophys.* **176**, 700–708.
Klischies, M., Stöckigt, J., and Zenk, M. H. (1978). *Phytochemistry* **17**, 1523–1525.

Knobloch, K. H., and Hahlbrock, K. (1975). *Eur. J. Biochem.* **52**, 311–320.
Kuroda, H., Shimada, M., and Higuchi, T. (1975). *Phytochemistry* **14**, 1759–1763.
Mäder, M. (1976). *Planta* **131**, 11–15.
Mäder, M., Nessel, A., and Bopp, M. (1977). *Z. Pflanzenphysiol.* **82**, 247–260.
Mansell, R. L., Stöckigt, J., and Zenk, M. H. (1972). *Z. Pflanzenphysiol.* **68**, 286–288.
Mansell, R. L., Gross, G. G., Stöckigt, J., Franke, H., and Zenk, M. H. (1974). *Phytochemistry* **13**, 2427–2435.
Mansell, R. L., Babbel, G. R., and Zenk, M. H. (1976). *Phytochemistry* **15**, 1849–1853.
Marcinowski, S., and Grisebach, H. (1977). *Phytochemistry* **16**, 1665–1667.
Marcinowski, S., and Grisebach, H. (1978). *Eur. J. Biochem.* **87**, 37–44.
Marcinowski, S., Falk, H., Hammer, D. K., Hoyer, B., and Grisebach, H. (1979). *Planta* **144**, 161–165.
Nakamura, Y., Fushiki, H., and Higuchi, T. (1974). *Phytochemistry* **13**, 1777–1784.
Nimz, H. (1974). *Angew. Chem.* **86**, 336–344.
Pfändler, R., Scheel, D., Sandermann, H., Jr., and Grisebach, H. (1977). *Arch. Biochem. Biophys.* **178**, 315–316.
Pickett-Heaps, J. D. (1968). *Protoplasma* **65**, 181–205.
Poulton, J. E., Hahlbrock, K., and Grisebach, H. (1976). *Arch. Biochem. Biophys.* **176**, 449–456.
Ranjeva, R., Boudet, A. M., and Faggion, R. (1976). *Biochimie* **58**, 1255–1262.
Rhodes, M. J. C., and Wooltorton, L. S. C. (1975). *Phytochemistry* **14**, 1235–1240.
Rhodes, M. J. C., Hill, C. R., and Wooltorton, L. S. C. (1976). *Phytochemistry* **15**, 707–710.
Ride, J. P. (1975). *Physiol. Plant Pathol.* **5**, 125–134.
Rubery, P. H., and Fosket, D. E. (1969). *Planta* **87**, 54–62.
Sagisaka, S. (1976). *Plant Physiol.* **57**, 308–309.
Sarkanen, K. V., and Ludwig, C. H. (1971). in "Lignins." Wiley (Interscience), New York.
Shimada, M., Ohashi, H., and Higuchi, T. (1970). *Phytochemistry* **9**, 2463–2470.
Stafford, H. A. (1974). *Annu. Rev. Plant Physiol.* **25**, 459–486.
Tutschek, R. (1975). *Z. Pflanzenphysiol.* **76**, 353–365.
Vance, C. P., and Sherwood, R. T. (1976). *Plant Physiol.* **57**, 915–919.
Vance, C. P., Anderson, J. O., and Sherwood, R. T. (1976). *Plant Physiol.* **57**, 920–922.
Wallis, P. J., and Rhodes, J. M. C. (1977). *Phytochemistry* **16**, 1891–1894.
Wengenmayer, H., Ebel, J., and Grisebach, H. (1976). *Eur. J. Biochem.* **65**, 529–536.
Whitmore, F. W. (1978). *Plant Sci. Lett.* **13**, 241–245.
Wildner, G. F. (1975). *Z. Naturforsch.*, Teil C **30**, 756–760.
Wyrambik, D., and Grisebach, H. (1975). *Eur. J. Biochem.* **59**, 9–15.
Wyrambik, D., and Grisebach, H. (1979). *Eur. J. Biochem.* **97**, 503–509.
Yamada, Y., and Kuboi, T. (1976). *Phytochemistry* **15**, 395–396.

Cyanogenic Glycosides | *16*

E. E. CONN

I. INTRODUCTION

The ability of living organisms to produce hydrocyanic acid (HCN) is known as *cyanogenesis*. The early and excellent work on this subject involved the phenomenon as it occurs in higher plants and has been reviewed by Robinson (1930). Renewed interest in this subject has broadened the list of cyanogenic organisms to include bacteria, fungi, millipedes, and moths.

HCN does not occur free in higher plants but is released from *cyanogenic precursors* as the result of enzymatic action. Early work established that these precursors are usually glycosides of α-hydroxynitriles (cyanohydrins); when the cellular integrity of *cyanophoric* plant tissue is disrupted, such *cyanogenic glycosides* are brought in contact with catabolic enzymes which hydrolyze the glycosides and dissociate the α-hydroxynitriles. In one family

The Biochemistry of Plants, Vol. 7

of higher plants, the Sapindaceae, the α-hydroxynitriles are stabilized by esterification with a long-chain fatty acid to form a *cyanogenic lipid*. Such lipids are hydrolyzed, presumably by specific lipases, releasing the α-hydroxynitrile which then can dissociate and form HCN. Cyanogenic lipids, together with other cyanolipids, have been recently reviewed (Mikolajczak, 1977) and will not be treated in this chapter.

The cyanogenic process in bacteria and fungi does not appear to involve a stable cyanogenic precursor such as a glycoside or ester of a fatty acid. While labile cyanohydrins have been isolated from fungi, their role, if any, as intermediates in cyanogenesis has not been established (Tapper and Mac-Donald, 1974). Cyanogenic bacteria, like fungi, produce HCN directly from glycine, and intermediates have not been identified. This field has been reviewed by Knowles (1976) and was one of the subjects treated in a recent symposium (Vennesland, 1981). The production of HCN by millipedes has received additional attention recently (Duffey and Towers, 1978), and cyanogenic glycosides have been established as precursors of HCN in moths (Davis and Nahrstedt, 1979).

Renewed interest in cyanogenesis in the past two decades has resulted in a number of reviews emphasizing the chemistry (Conn, 1969, 1980a; Eyjolffson, 1970; Seigler, 1977, 1980), biosynthesis (Conn, 1973, 1979a, 1980a), toxicology (Conn, 1979b), physiological role (Jones, 1972, 1978, 1979), and chemotaxonomic significance (Hegnauer, 1973, 1977) of cyanogenic compounds. This chapter will attempt to summarize recent developments on the nature, occurrence, and metabolism of these interesting compounds.

II. CHEMICAL STRUCTURE AND PROPERTIES

Cyanogenic glycosides have the general structure I and therefore have α-hydroxynitriles as their aglycone. In all but five of the known cyanogenic

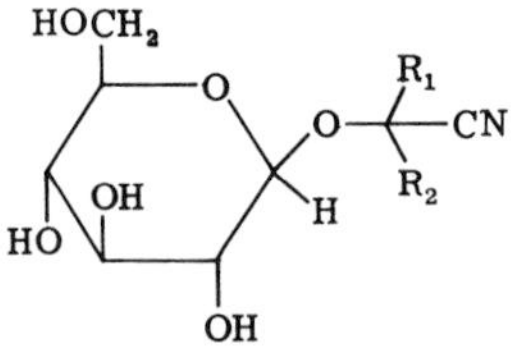

I

glycosides, the sugar moiety is D-glucose-linked by an O-β-glucosyl bond. The five exceptions are amygdalin, vicianin, lucumin, linustatin, and neolinustatin, which have disaccharides as their sugar component. R_1 and R_2 in I may be aliphatic or aromatic substituents (or hydrogen), and in one group are the elements of a cyclopentene ring. Since R_1 is usually not identical to R_2, the carbinol carbon of the aglycone is chiral, and this introduces the possibility of epimeric compounds. Several such epimeric pairs exist in nature, usually not in the same species or even in related families.

A group of compounds occurring in the family Cycadaceae have been termed *pseudo* cyanogenic glycosides (Lythgoe and Riggs, 1949). Strictly speaking these compounds are erroneously termed cyanogenic, since they are glycosides of methylazoxymethanol. While these compounds undergo enzymatic hydrolysis on cellular disruption of cycad tissues containing them, the aglycone, methylazoxymethanol, does not dissociate to form HCN. Only if the aglycone is treated with dilute base and then acidified will HCN be produced, and then only in small amounts relative to the methylazoxymethanol.

Cyanogenic glycosides can be classified according to the chemical (i.e., aliphatic, aromatic, alicyclic) nature of their aglycones (Eyjolfsson, 1970). However, the present author and others (Conn, 1980a; Seigler, 1977; Hegnauer, 1977) have preferred a classification based on the biosynthetic origin, either established or presumed, of the aglycone, and this will be used in the following brief description of these compounds. For greater detail, the reader should consult Conn (1980a) and Seigler (1981).

A. Structures

1. Glycosides Derived from Valine and Isoleucine

Linamarin (II) and (R)-lotaustralin (III) are two of the most common cyanogenic glycosides found in plants; they occur in the families

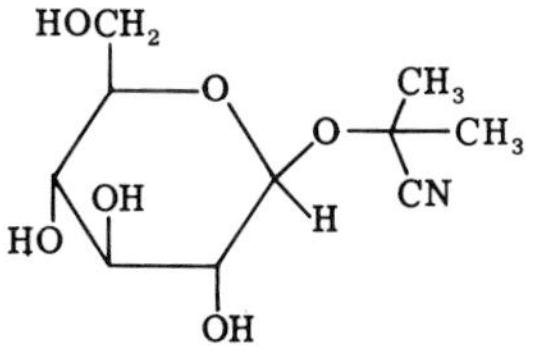

II 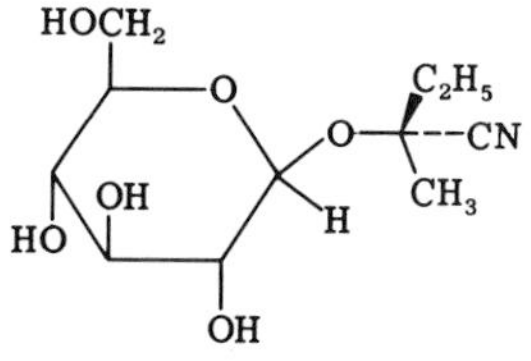   III

Leguminosae and Euphorbiaceae, which are noted for cyanogenesis (see below). The aglycones of linamarin and (R)-lotaustralin are the cyanohydrins of acetone and 2-butanone; the latter contains a chiral center, and in lotaustralin the configuration is R. These two aglycones are derived (see below) from L-valine and L-isoleucine, respectively. Without exception, linamarin and lotaustralin always occur together in the same species, presumably because of the existence of a single set of biosynthetic enzymes that can act either on valine or isoleucine and the corresponding intermediates (Conn, 1973).

Recently, two new cyanogenic glycosides, which are disaccharides having the same aglycones as linamarin and lotaustralin, have been described (Smith *et al.*, 1980). These are linustatin (IV) and neolinustatin (V), which

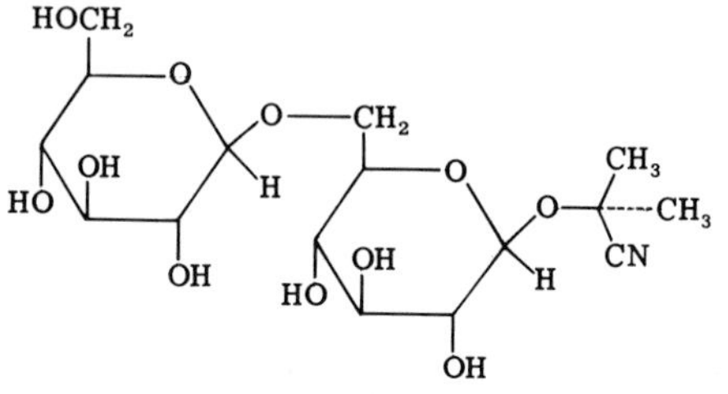

IV

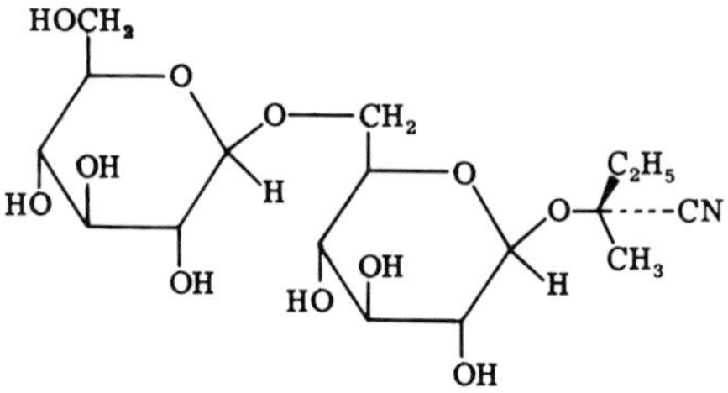

V

were isolated from seed of linen flax (*Linum usitatissimum*). In these compounds, the sugar is D-gentiobiose; the relationship between linamarin and linustatin therefore is analogous to that between prunasin and amygdalin found in the Rosaceae, even including the fact that the disaccharide is found in seed whereas the monosaccharide is more often found in vegetative tissue.

2. Glycosides Derived from L-Leucine

Five glycosides presumably derive their aglycones from L-leucine; these are (S)-proacacipetalin (VI), epiproacacipetalin (VII), (S)-heterodendrin (VIII), (S)-cardiospermin (IX), and (S)-cardiospermin p-hydroxybenzoate (X). Proacacipetalin is isomeric with acacipetalin (XI) which was isolated by

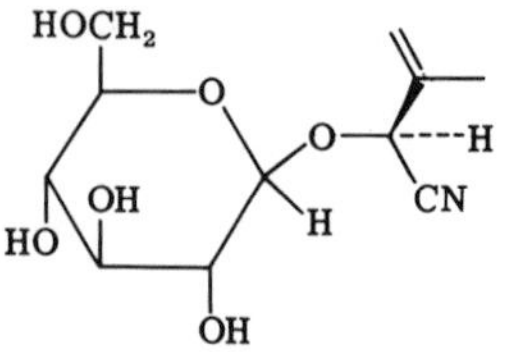

VI

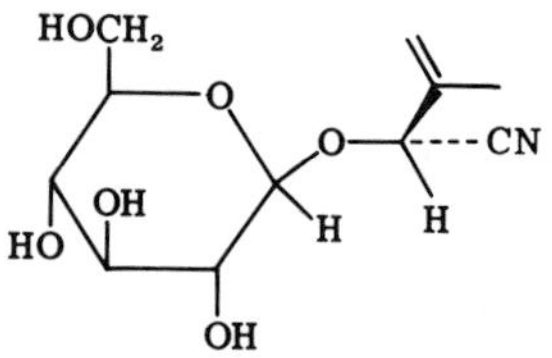

VII

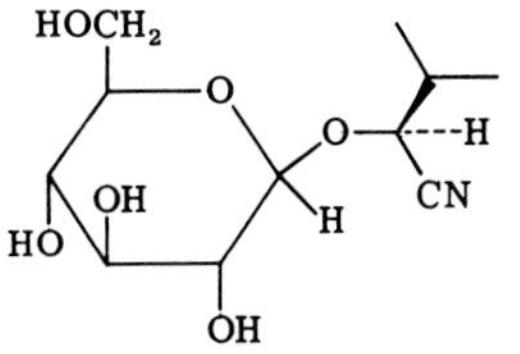

VIII 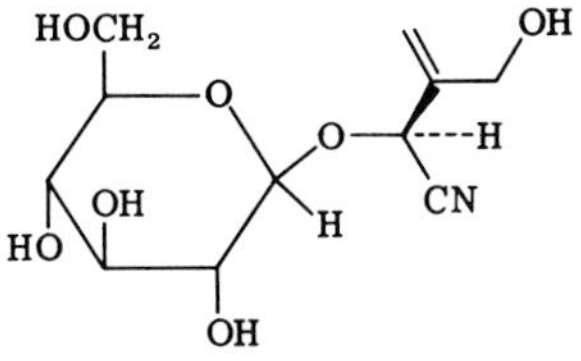   IX

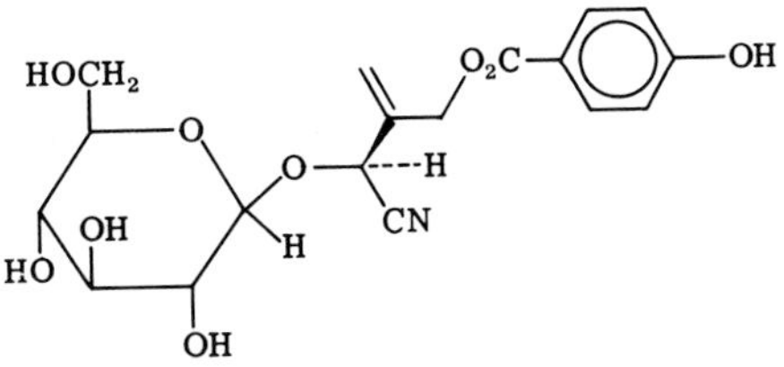

X 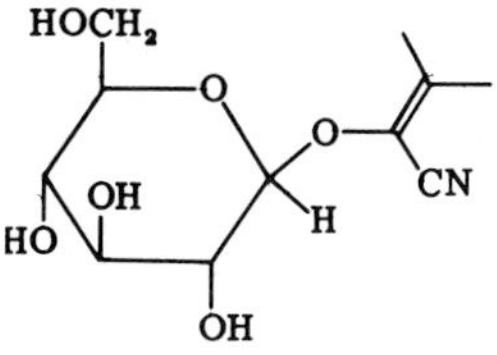   XI

Steyn and Rimington (1935; Rimington, 1935) nearly half a century ago from two African acacias. More recent work (Ettlinger *et al.,* 1977) clearly demonstrates that the naturally occurring glycoside in *Acacia sieberiana* DC var. *woodii* (Burt Davy) Keay and Brenan is (*S*)-proacacipetalin and that acacipetalin is probably an artifact of isolation. Epiproacacipetalin, with the *R* configuration, has been isolated from *Acacia globulifera* (D. S. Seigler, unpublished). The saturated compound corresponding to proacacipetalin is known as heterodendrin (VIII); it also has the *S* configuration and was isolated from *Heterodendron olaefolium* (Nahrstedt and Hübel, 1978). Cardiospermin (IX) was isolated from *Cardiospermum hirsutum* (Sapindaceae), which is known to contain one or more cyanolipids of closely related structures (Seigler *et al.,* 1974). The *p*-hydroxybenzoate ester (X) of cardiospermin occurs in a member of the Rosaceae known as *Sorbaria arborea* (Nahrstedt, 1976); this is an unusual finding in that all other cyanogens isolated from members of the Rosaceae are derived from phenylalanine.

3. Glycosides Derived from L-Phenylalanine

L-Phenylalanine is probably the precursor of the aglycone of seven cyanogenic glycosides including two epimeric pairs, (*R*)-prunasin (XII) and (*S*)-sambunigrin (XIII), (*R*)-holocalin (XIV) and (*S*)-zierin (XV), and three disaccharides known as amygdalin (XVI), vicianin (XVII), and lucumin (XVIII). (*R*)-Prunasin differs from (*S*)-sambunigrin in configuration at the carbinol carbon atom. As with other epimeric pairs, this appears to be a subtle difference. However, the real effect is profound, for the *R* epimer is the compound found in the rose family (Rosaceae) along with the related disaccharide amygdalin, whereas (*S*)-sambunigrin was first reported in *Sambucus nigra* (Caprifoliaceae). Sambunigrin was erroneously reported to

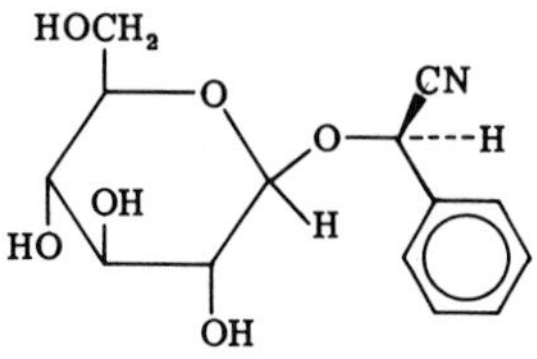

XII

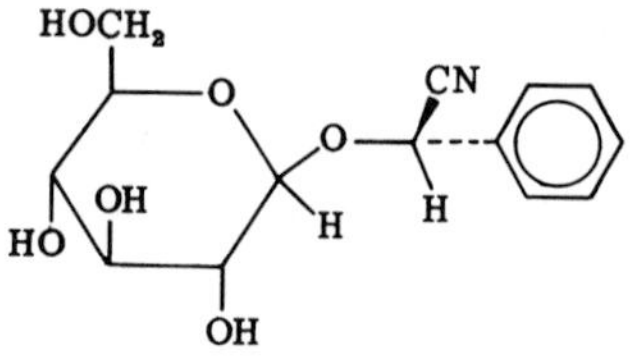

XIII

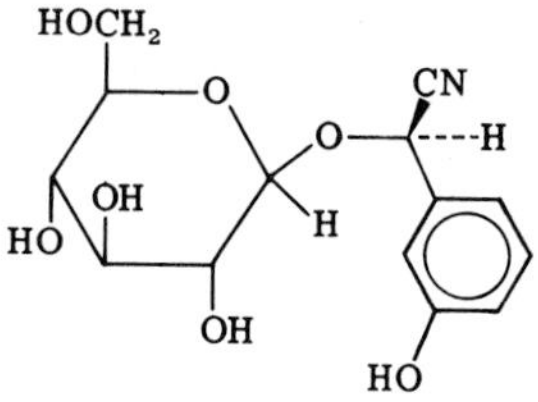

XIV

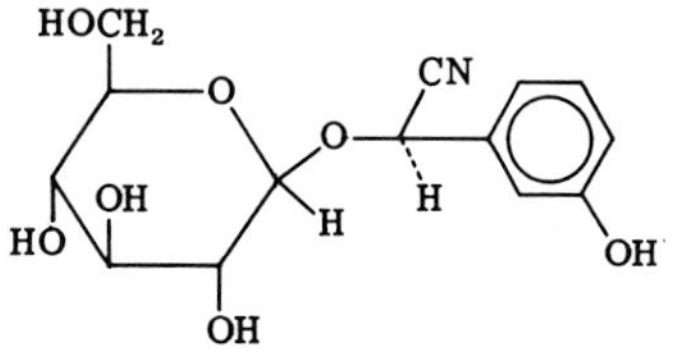

XV

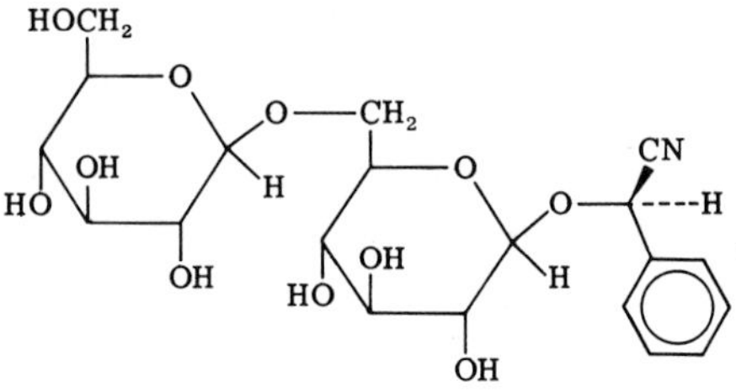

XVI

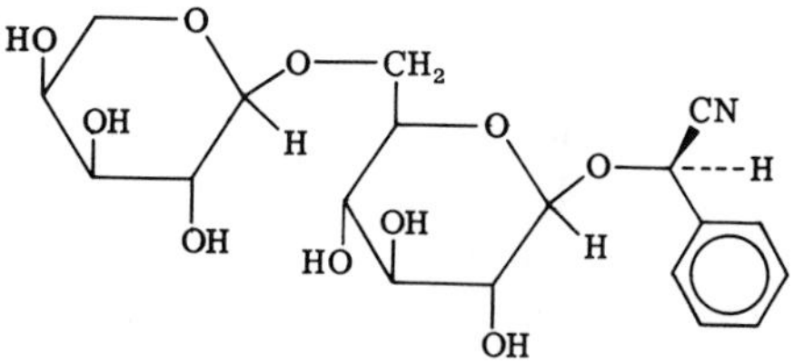

XVII

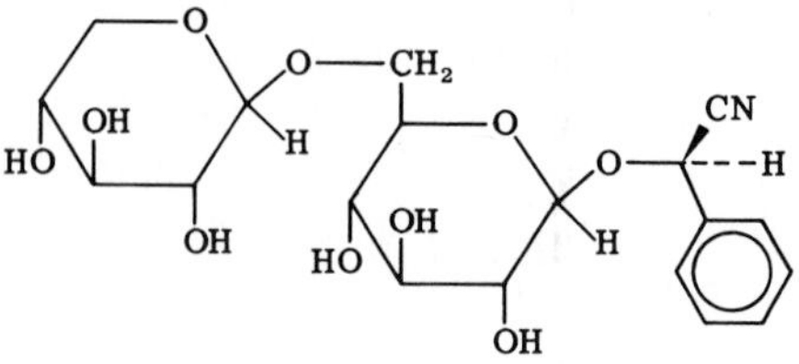

XVIII

occur naturally as a epimeric mixture with (R)-prunasin in leaves of cherry laurel (Herissey, 1906). Thirty years later, Plouvier (1935) showed the epimeric mixture to be an artifact of isolation, due to the ease of epimerization of the chiral center in mild alkali. The glycosides occur in other families (Table I) as well but never cooccur in the same species, with the exception of certain specimens of *S. nigra* collected in Denmark (Jensen and Nielsen, 1973).

4. Glycosides Derived from L-Tyrosine

Six cyanogens are known that arise from L-tyrosine. These include another epimeric pair, (S)-dhurrin (XIX) and (R)-taxiphyllin (XX), which again do not cooccur at the generic level. Also derived from tyrosine is *p*-glucosyloxymandelonitrile (XXI) which has a limited distribution (Table I) but is unique in that the α-hydroxynitrile group is not stabilized, either as a glycoside or a fatty acid ester. This lability makes the compound difficult to study, although the p-glucosyloxybenzaldehyde formed when the HCN has dissociated is well characterized. In (S)-proteacin (XXII), the labile

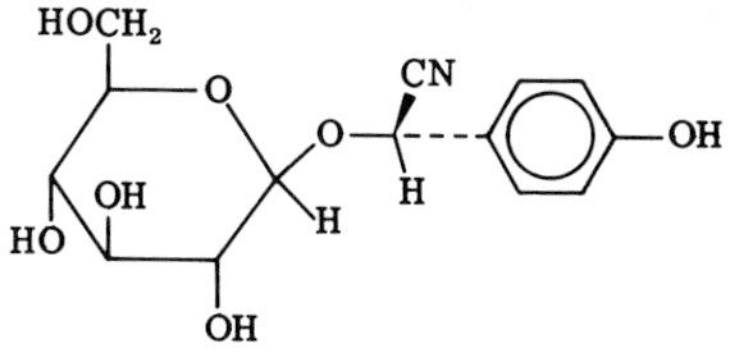

XIX

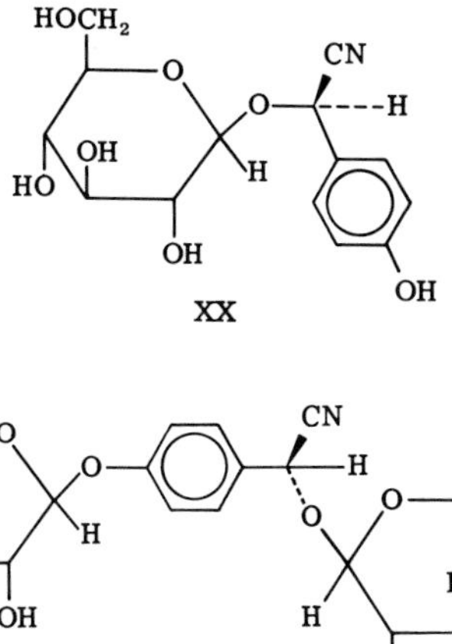

XX

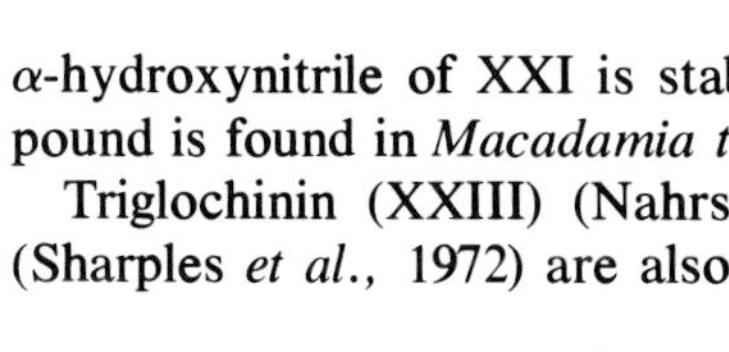

XXI

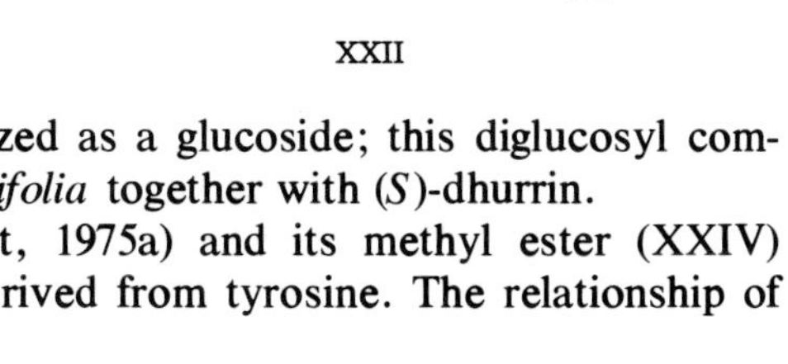

XXII

α-hydroxynitrile of XXI is stabilized as a glucoside; this diglucosyl compound is found in *Macadamia ternifolia* together with (S)-dhurrin.

Triglochinin (XXIII) (Nahrstedt, 1975a) and its methyl ester (XXIV) (Sharples *et al.*, 1972) are also derived from tyrosine. The relationship of

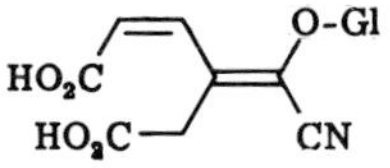

XXIII

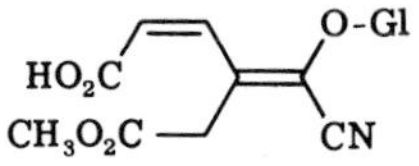

XXIV

TABLE I

Distribution of Cyanogenic Glycosides[a]

Glycoside	Family	Genera
Amygdalin	Rosaceae	*Cotoneaster* spp., *Cydonia, Eriobotrya, Malus* spp., *Photinia, Prunus* spp., *Sorbus*
Cardiospermin	Sapindaceae	*Cardiospermum, Heterodendron*
Cardiospermin, *p*-hydroxy-benzoyl ester	Rosaceae	*Sorbaria*
Deidaclin	Passifloraceae	*Deidamia*
Dhurrin	Gramineae	*Sorghum* spp.
	Proteaceae	*Macadamia, Stenocarpus*
	Trochodendraceae	*Trochodendron*
Epitetraphyllin B	Passifloraceae	*Adenia*
Gynocardin	Flacourtiaceae	*Gynocardia, Pangium*
Heterodendrin	Leguminosae	*Acacia*
	Sapindaceae	*Heterodendron*
Holocalin	Caprifoliaceae	*Sambucus*
	Leguminosae	*Holocalyx*
	Liliaceae	*Chlorophytum* spp.
p-Glucosyloxymandelo-nitrile	Berberidaceae	*Nandina*
	Leguminosae	*Goodia*
	Ranunculaceae	*Thalictrum* spp.
Linamarin	Compositae	*Dimorphotheca* spp., *Osteospermum*
	Euphorbiaceae	*Cnidoscolus, Hevea, Manihot* spp.
	Leguminosae	*Acacia, Lotus* spp., *Phaseolus, Trifolium*
	Linaceae	*Linum* spp.
	Papaveraceae	*Papaver*
Lotaustralin	Compositae	*Dimorphotheca* spp., *Osteospermum*
	Euphorbiaceae	*Manihot*
	Leguminosae	*Acacia, Lotus* spp., *Phaseolus, Trifolium*
	Linaceae	*Linum* spp.
	Papaveraceae	*Papaver*
Lucumin	Sapotaceae	*Lucuma*
Proacacipetalin	Leguminosae	*Acacia* spp. (African and American)
Proteacin	Proteaceae	*Macadamia*
	Ranunculaceae	*Thalictrum*
Prunasin	Caprifoliaceae	*Sambucus*
	Compositae (Asteraceae)	*Achillea, Centaurea, Chaptalia*
	Leguminosae	*Acacia* spp. (Australian), *Holocalyx*
	Myoporaceae	*Eremophila*
	Myrtaceae	*Eucalyptus*
	Oliniaceae	*Olinia*
	Polypodiaceae	*Cystopteris, Pteridium*

(continued)

TABLE I (Continued)

Glycoside	Family	Genera
Prunasin	Rosaceae	*Cotoneaster* spp., *Cydonia*, *Prunus* spp., *Sorbus*
	Saxifragaceae	*Jamesia*
	Scrophulariaceae	*Linaria* spp.
Sambunigrin	Caprifoliaceae	*Sambucus*
	Leguminosae	*Acacia* spp., (Australian)
	Olacaceae	*Ximenia*
Taxiphyllin	Calycanthaceae	*Calycanthus* spp., *Chimonanthus*
	Cupressaceae	*Juniperus*
	Euphorbiaceae	*Phyllanthus*
	Gramineae	*Bambusa* spp., *Bouteloua, Chloris*, spp., *Cynodon, Dendrocalamus* spp., *Eleusine* spp., *Glyceria* spp., *Melica* spp., *Molinia, Sieglingia, Stipa* spp., *Tridens*
	Magnoliaceae	*Liriodendron*
	Taxaceae	*Taxus* spp.
	Taxodiaceae	*Metasequoia*
Tetraphyllin A	Passifloraceae	*Tetrapathaea*
Tetraphyllin B	Passifloraceae	*Adenia, Barteria, Tetrapathaea*
Triglochinin	Araceae	*Alocasia, Anthurium, Arum, Dieffenbachia, Lasia, Pinellia*
	Campanulaceae	*Campanula* spp.
	Juncaginaceae	*Scheuzeria, Triglochin*
	Lilaeaceae	*Lilaea*
	Magnoliaceae	*Liriodendron*
	Platanaceae	*Platanus* spp.
Triglochin, methyl ester	Ranunculaceae	*Thalictrum*
Vicianin	Leguminosae	*Vicia* spp.
	Polypodiaceae	*Davallia* spp.
Zierin	Caprifoliaceae	*Sambucus*
	Rutaceae	*Zieria*

[a] After Seigler, 1977, 1980, with permission.

these compounds to dhurrin or taxiphyllin is not immediately obvious until one notes that hydroxylation of the aromatic ring of the latter compounds at the 3-position followed by oxidative ring opening and tautomerization yields triglochinin. The configuration of the double bonds in triglochinin has been studied (Nahrstedt, 1975a); isotriglochinin (XXV), which has been isolated in small amounts from plants containing triglochinin as the major cyanogen, may not be naturally occurring because of the ease of isomerization of triglochinin. The methyl ester may also arise during the purification of triglochinin and therefore be artifactual.

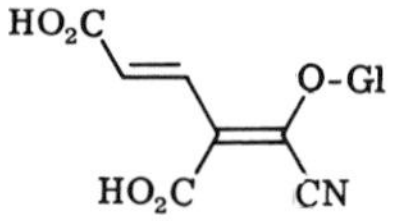

HO$_2$C
O-Gl
HO$_2$C   CN

XXV

5. Glycosides Having Cyclopentenyl Rings

Five cyanogenic glycosides have an aglycone that is cyclopentenoid in nature; of these, gynocardin (XXVI) is the only one with an established configuration (Kim *et al.*, 1970). Deidaclin (XXVII) and tetraphyllin A (also XXVII) apparently differ at the chiral carbinol atom (see Seigler, 1977, for discussion); tetraphyllin B (XXVIII) [presumably identical to barterin (Paris

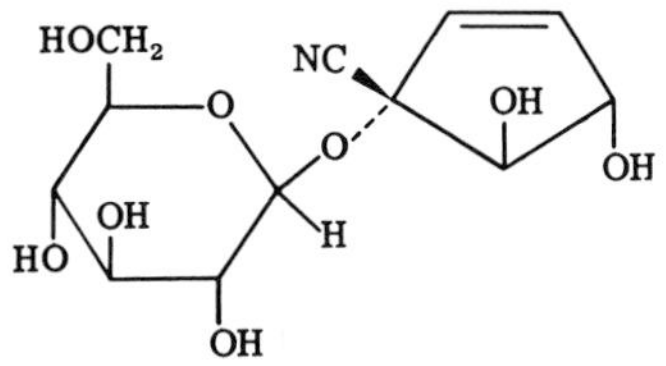

XXVI 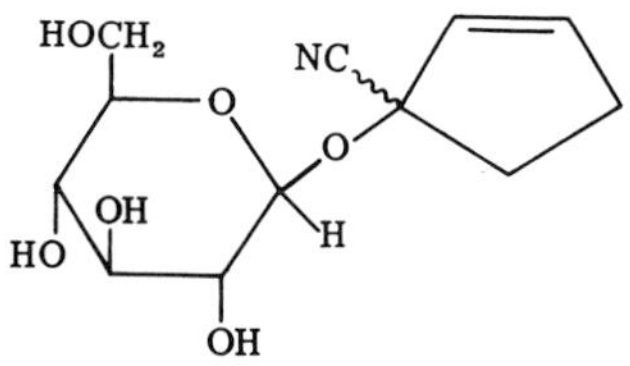   XXVII

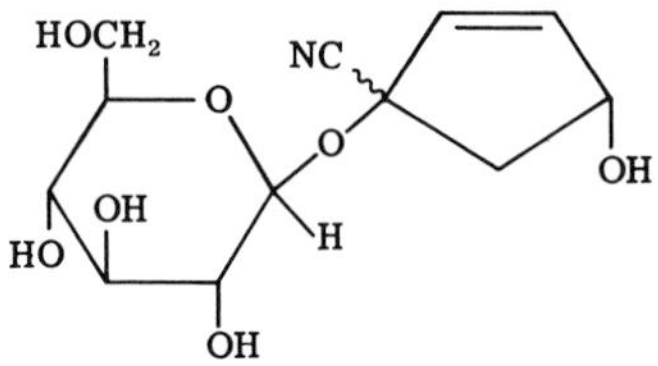

XXVIII

et al., 1969)] and epitetraphyllin B have been isolated from the same species, *Adenia volkensii* (Gondwe *et al.*, 1978). (Whereas the reported coexistence of epimers in a single species is always subject to criticism, care was taken in this work to avoid epimerization.) These five compounds clearly cannot be derived directly from a protein amino acid like the other cyanogenic glycosides. The obvious precursor of these cyclopentenoid cyanogens is cyclopentenyl glycine, a nonprotein amino acid whose biosynthesis has been studied by Cramer *et al.* (1980).

B. Chemical Properties

The reviews of Eyjolfsson (1970), Seigler (1977), and Nahrstedt (1981) contain much information on the chemical and physical properties of these compounds. Only the more prominent properties will be described here.

Cyanogenic glycosides are cleaved in dilute acid at elevated temperatures (in excess of 60°C) to yield their sugar component and the α-hydroxynitrile. The latter is unstable and can dissociate to HCN and an aldehyde or ketone in a pH-dependent reaction (The ketone produced on hydrolysis of trichlochinin reacts with water to form triglochinic acid.) Hydrolysis in concentrated acid can yield the α-hydroxyacid and NH_3 if the nitrile group in the glycoside is hydrolyzed more rapidly than the β-glycosidic bond (Uribe and Conn, 1966). Treatment with a mild base [e.g., saturated $Ba(OH)_2$] hydrolyzes the nitrile group of several cyanogenic glycosides to form the corresponding glycosidic acid (i.e., amygdalinic acid). However, dhurrin and taxiphyllin are labile in dilute alkali and decompose to HCN, glucose, and p-hydroxybenzaldehyde, the products obtained on acid hydrolysis.

Alkali facilitates epimerization at the carbinol carbon of cyanogenic glycosides having electron-withdrawing groups (i.e., an aromatic ring) adjacent to that carbon. Nahrstedt (1975b) has carried out a detailed study on this process with amygdalin and showed that epimerization can occur even at pH 7.0 at elevated temperatures. This obviously requires that care be exercised in extraction procedures involving boiling aqueous solvent if epimerization is to be avoided.

Seigler (1977) has gathered together data on the uv, ir, and nmr spectra of cyanogenic glycosides. Of these three types of spectra, nmr has been most widely used in structural studies, especially on trimethylsilyl (TMS) esters. Recent work with high-field instruments and with ^{13}C-nmr spectroscopy has been reported by Nahrstedt (1981). The uv spectral properties of cyanogens having aromatic rings have been utilized analytically; there is a large increase in light absorption due to the benzaldehyde and p-hydroxybenzaldehyde released on hydrolysis of these compounds.

It should be noted that only three cyanogenic glycosides—amygdalin, linamarin, and prunasin—are available commercially. All others would have to be isolated from their natural sources if experimental studies were contemplated. There are readily available plant sources of a few other cyanogens, namely, dhurrin, linustatin, and neolinustatin. However, the isolation and purification of these compounds from natural sources is moderately difficult for two reasons: First, enzymes (see below) within the plant source frequently decompose the glycoside during extraction. Second, sugars and other glycosides often cooccur with the cyanogenic glycoside and make purification difficult.

III. DETECTION

The presence of a cyanogenic substance in fresh plant material is indicated by the release of HCN when the plant material is crushed or otherwise destroyed. This process brings the glucoside into contact with a β-glucosidase and a hydroxynitrile lyase capable of degrading it (Section V,B); HCN is formed and can be measured qualitatively or quantitatively (see Conn, 1980a, for details). The fact that HCN is released only after tissue disruption has been explained by assuming that the cyanogenic glucoside is located in a compartment separate from the degradative enzymes. This explanation has recently been confirmed in the case of the cyanogenic glucoside dhurrin which occurs in young sorghum leaves. Kojima *et al.* (1979) showed that dhurrin was located exclusively in the epidemial cells of 6-day-old green sorghum leaves. Catabolic enzymes, on the other hand, were located in the mesophyll protoplasts isolated from such leaves. The third major tissue in the leaves, namely, the bundle sheath strands, lacked both dhurrin and its catabolic enzymes.

In addition to measuring the HCN released on hydrolysis of cyanogenic glycosides the intact glycosides can be measured by quantitative gas–liquid chromatography (glc) of their TMS ethers. This technique also can be used for qualitative identification, since the TMS derivatives of each glycoside have a characteristic elution time under standardized conditions (Nahrstedt, 1970, 1973).

Application of these methods has disclosed that cyanogenic glycosides can occur in any of several organs (i.e., leaves, stems, flowers, roots, seeds) or parts of cyanogenic plants. The cyanogen may also be present at one period in the life cycle of the plant and not at others. Thus the dry seed of sorghum contains no dhurrin or other cyanogen, but a 4-day-old dark-grown young seedling can contain 5% dhurrin (dry weight). The wild lima bean contains 3–4% linamarin (fresh weight) which is not metabolized on germination. Indeed, the linamarin appears to be quantitatively transferred from the cotyledons of the seed into the newly formed seedling tissue where, during the first 25 days or so, it may provide protection from herbivores (Clegg *et al.*, 1979). In general, young, actively developing tissue seems to contain the highest concentrations of cyanogen, but there are exceptions to this statement. This subject is discussed in more detail by Conn (1980a).

IV. DISTRIBUTION

Recent estimates of the number of cyanogenic plants exceed 2000 species (Gibbs, 1974; Hegnauer, 1977) representing about 110 families and including both gymnosperms and monocotyledonous and dicotyledonous angiosperms. Families that are especially noted for cyanogenesis are the Araceae (50 species), Compositae (50), Euphorbiaceae (50), Gramineae (100), Leguminoseae (125), Passifloraceae (30), and Rosaceae (150) (Gibbs, 1974). In the vast majority of the 2000 or so cyanogenic species only the ability to produce HCN has been observed and the source of the HCN has not been identified. Indeed, the cyanogenic precursor has been identified in only about 200, or 10%, of the known cyanogenic plants, primarily because of difficulties in isolation, purification, and characterization. These studies have given us the 29 known glycosides described in Section II. They also cause us to wonder whether the cyanogenic species that have not been examined contain other, unidentified glycosides or other precursors of HCN (see Conn, 1980b, for discussion).

Table I lists the plant families and genera in which cyanogenic glycosides have been identified. This listing shows that prunasin, taxiphyllin, triglochinin, linamarin, and lotaustralin are widely distributed in nature. Other glycosides are much more restricted in their occurrence. Thus amygdalin has been reported only in the Rosaceae, cardiospermin only in the Sapindaceae, and proacacipetalin and epiproacacipetalin only in the Leguminosae (only in the genus *Acacia*), whereas the structurally related gynocardin, tetraphyllins A and B, epitetraphyllin B, and deidaclin are known only from the closely related Passifloraceae, Flacourtiaceae, and Turneraceae. Other restricted distributions can be found in Table I. Dhurrin and taxiphyllin were once considered to be restricted in distribution, but recent work in a number of laboratories has altered this viewpoint (see Seigler, 1980, for discussion). The distribution of triglochinin, derived as it is from tyrosine, seems to be linked to that of taxiphyllin and dhurrin.

Almost from the time that cyanogenesis was described, there has been an interest in its utility as a taxonomic tool. Hegnauer's laboratory clearly leads the research in this field, as demonstrated by a treatise (Hegnauer, 1962–1973), reviews (Hegnauer, 1973, 1977), and a continuing series of papers from his laboratory (e.g., Fikenscher and Hegnauer, 1977; Tjon Sie Fat, 1979). These publications and that of Gibbs (1974) should be consulted if one wishes to determine whether a given species has ever been reported to be cyanogenic. Monographs on poisonous plants (e.g., Watt and Breyer-Brandwijk, 1962; Kingsbury, 1964; Everist, 1974) also are an excellent source to consult for such questions. However, another approach is that of Seigler who has compiled two useful lists of the cyanogenic plants found in Oklahoma and Texas (Seigler, 1976b) and in the northeastern United States (Seigler, 1976a).

V. METABOLISM

A. Biosynthesis

Research carried out in several laboratories over the past 20 yr has provided much information regarding the biosynthesis of cyanogenic glycosides. This research has been summarized in several reviews (Conn, 1973, 1979a,

1980a) and therefore will not be repeated here, except for the salient features.

All the known cyanogenic glycosides except those having a cyclopentenoid aglycone can be derived from five protein amino acids, L-phenylalanine, L-tyrosine, L-valine, L-isoleucine, and L-leucine. The precursor–product relationship, which is shown in Fig. 1 for a few compounds, dictates that the carboxyl carbon of the amino acid be lost, the amino group be oxidized to a nitrile, and the β-carbon be hydroxylated and subsequently glycosylated. While these relationships were being established with the use of ^{14}C- and ^{15}N-labeled amino acids, work on the intermediate steps was initiated, and after nearly two decades it was possible to write the biosynthetic pathway shown in Fig. 2.

Experimental evidence of three types supports the pathway shown in Fig. 2. Several cyanogenic species, notably sorghum, flax, and cherry laurel, were fed ^{14}C-labeled amino acids and related intermediates, and the incorporation of isotope into the glycoside end product was monitored and compared. A second type of study involved "trapping" experiments in which the proposed intermediates were fed simultaneously with ^{14}C-labeled amino

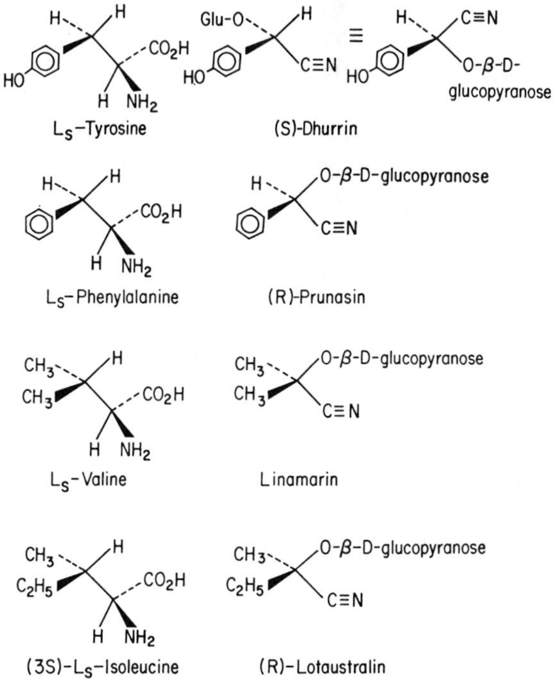

Fig. 1. The precursor–product relationship for four cyanogenic glycosides. Reproduced with permission from Conn, 1979a.

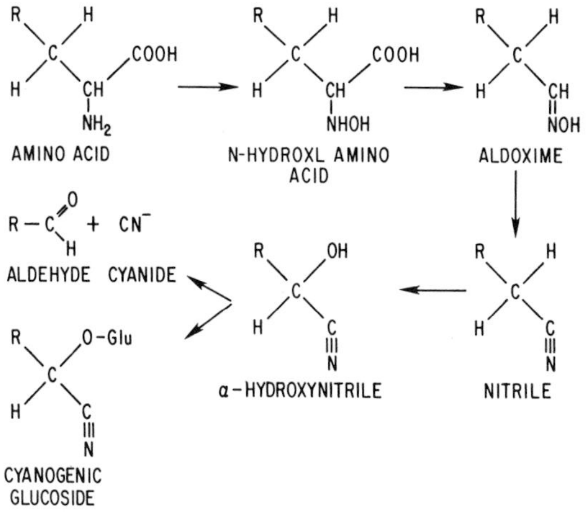

Fig. 2. The biosynthetic pathway for cyanogenic glycosides. Reproduced with permission from Conn, 1979a.

acids as primary precursors and observed to acquire radioactivity. A third type of evidence came from enzymatic studies; indeed, enzymes catalyzing the last step in linamarin biosynthesis (Hahlbrock and Conn, 1970) and dhurrin biosynthesis (Reay and Conn, 1974) were isolated and characterized several years ago.

More recent enzymatic studies have involved the use of microsomal preparations that catalyze the conversion of L-tyrosine to p-hydroxy-(S)-mandelonitrile (McFarlane et al., 1975). Such microsomal preparations have been invaluable in sorting out the details of the pathway shown in Fig. 2, especially in establishing that an N-hydroxyamino acid [N-hydroxytyrosine in the case of sorghum (Møller and Conn, 1979)] is the first intermediate in the pathway. Sorghum microsomes were also indispensable in ordering the sequence of later intermediates in the pathway (Shimada and Conn, 1977). These experiments have been extensively reviewed and evaluated (Conn, 1973, 1980a).

One of the more puzzling aspects of dhurrin biosynthesis was the repeated difficulty in detecting in the intact sorghum plant any of the intermediates shown in Fig. 2. Studies with sorghum microsomes have provided data indicating (Møller and Conn, 1980) that the biosynthetic pathway is a highly *channeled* process catalyzed by enzymes organized in such a way that they utilize their substrate when produced by the preceding enzyme in the sequence in decided preference to the same compound added externally. For example, p-hydroxyphenylacetonitrile produced from tyrosine in sorghum microsomes is used 120 times more effectively than nitrile added externally to the particles. It is this channeling phenomenon that presumably accounts for the failure of investigators to detect any significant pool of nitrile as an intermediate in sorghum seedlings. More important, however, channeling provides for the rapid and efficient flow of carbon and nitrogen atoms from L-tyrosine into dhurrin and at the same time perhaps protects labile intermediates (e.g., N-hydroxytyrosine) from wasteful side reactions (see Conn, 1980b, for discussion).

Hösel and Nahrstedt (1980) have recently obtained microsomes from dark-grown seedlings of *Triglochin maritima* (arrow grass) that catalyze the conversion of L-tyrosine to *p*-hydroxymandelonitrile. *Triglochin maritima* contains (*R*)-taxiphyllin in addition to triglochinin, and in the presence of the microsomes, NADPH, UDP-glucose, and a soluble protein fraction containing a UDP glucosyltransferase, Hösel and Nahrstedt observed the *in vitro* synthesis of (*R*)-taxiphyllin from L-tyrosine. Detailed studies on the enzymatic properties of arrow grass microsomes indicate that the oxidation of L-tyrosine in this system is also a channeled process (Cutler *et al.*, 1981).

In their work with *T. maritima* Hösel and Nahrstedt (1980) observed that it was essential to remove any seed coats that adhered to the arrow grass seedlings in order to obtain active microsomal preparations. Following this clue, A. J. Cutler and M. Sternberg (unpublished observations) have successfully prepared microsomal preparations from flax seedlings that catalyze the oxidation of L-valine and L-isoleucine in the presence of NADPH and produce cyanohydrins of acetone and 2-butanone. These particles also utilize *N*-hydroxyvaline, isobutyaldoxime, and isobutyronitrile, and produce HCN, presumably because these compounds are intermediates being metabolized according to the pathway shown in Fig. 2. Any significant amount of seed coat not removed from the flax seedlings is strongly inhibitory, since the* microsomes isolated from such seedlings have little or no enzymatic activity.

B. Catabolism

As noted in Section III, cyanogenic glycosides in plants are usually enzymatically hydrolyzed and their HCN is released when plant tissues containing these compounds are disrupted. This process is known as cyanogenesis and is represented in Fig. 3 for the cyanogenic glucoside dhurrin. Two enzymes are involved in this sequential process, the first being a β-glucosidase that catalyzes hydrolysis of the β-glucosidic bond of dhurrin. The second enzyme, a hydroxynitrile lyase (oxynitrilase, E.C. 4.1.2.11), catalyzes the stereospecific dissociation of the (*S*)-cyanohydrin produced when dhurrin is hydrolyzed.

The activities of these enzymes can be monitored in several ways. The glucose (or other sugar) released in the first step can be measured directly by chemical or enzymatic methods. If, however, the oxynitrilase is present in

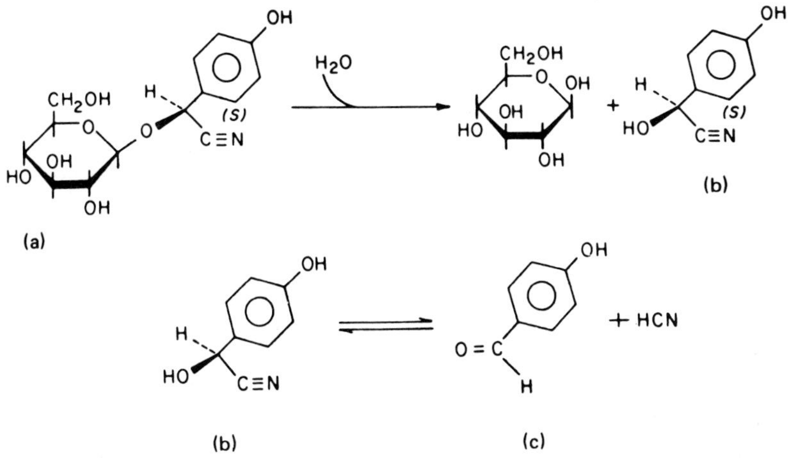

Fig. 3. The catabolism of dhurrin in disrupted seedlings of *Sorghum bicolor*. Dhurrin (a) is hydrolyzed by dhurrin β-glycosidase to yield glucose and *p*-hydroxy-(*S*)-mandelonitrile (b). The latter then dissociates enzymatically in the presence of hydroxynitrile lyase and nonenzymatically to release HCN and *p*-hydroxybenzaldehyde. Reproduced with permission from Kojima *et al.* (1979).

excess so that the second reaction is not rate-limiting, the hydrolytic step can be assayed by measuring the formation of HCN or *p*-hydroxybenzaldehyde.

Studies carried out on these enzymes involved in cyanogenesis have revealed some important facts. Perhaps most important, the β-glucosidases acting on cyanogenic glycosides show a high degree of substrate specificity for the cyanogen involved. There were indications of this specificity in earlier studies by Haisman and Knight (1967), Butler *et al.* (1965), and Mao and Anderson (1967), which showed that enzymes active on aromatic glycosides such as prunasin and dhurrin had little activity on linamarin and lotaustralin. However, the work of Hösel and Nahrstedt (1975) on a β-glucosidase from *Alocasia macrorrhiza* first clearly demonstrated the extreme specificity of this group of enzymes. The glucosidase from *A. macrorrhiza* exhibited maximum catalytic activity toward triglochinin, the cyanogen found in this species. It also could attack several chromogenic substrates (at much lower rates) and dhurrin at 1% of the rate for triglochinin. However, it was inactive on seven other cyanogenic glycosides and several other naturally occurring glycosides. These findings clearly call into question the significance of the many studies that have utilized chromogenic substrates such as *p*-nitrophenyl β-glucosides. Obviously, enzyme assays using these substrates are easy to carry out and, for certain comparisons, it is important to have a single aglycone (e.g., *p*-nitrophenol) linked with different sugars (e.g., glucose, galactose, ribose) and with a different stereochemistry (α vs β).

However, it is now clear, at least in the case of cyanogenic glycosides, that the enzymes studied with chromogenic substrates may be entirely different from the enzyme responsible for cyanogenesis.

Other natural products exist naturally as β-glycosides in intact plant tissues, and their hydrolysis is initiated by cellular disruption as in cyanogenesis. Such reactions play an important part in the turnover or further metabolism of such compounds (see Barz and Köster, this volume, Chapter 3). The subject of β-glycosidases is reviewed by Poulton (this volume, Chapter 22), and there is a section on the enzymes that hydrolyze cyanogenic glycosides.

Oxynitrilases (Fig. 3) have attracted further interest recently, based on earlier findings that the enzyme in almonds is a flavoprotein utilizing FAD, whereas that in sorghum is not (Seeley et al., 1966). The requirement of the almond enzyme for FAD is remarkable in view of the fact that the reaction catalyzed does not involve oxidation–reduction. Jorns (1979) has now examined the almond enzyme in greater detail and concluded that the role of the flavin is structural. Recently F. J. Carvalho (unpublished) has discovered and purified an oxynitrilase in cassava leaves and petioles that catalyzes dissociation of the cyanohydrins of acetone and 2-butanone. As expected from the specificity of the sorghum and almond oxynitrilase for aromatic cyanohydrins, the flax enzyme is inactive on aromatic cyanohydrins.

In view of the established lability of cyanohydrins, one could question the role of oxynitrilases. The answer may lie in the dependence of the cyanohydrin dissociation reaction on pH. F. J. Carvalho (unpublished) has shown that acetone cyanohydrin dissociates at very low rates at pH values from 1 to 5. Between pH 5 and 6 the rate is easily measured, but above pH 6.0 the reaction is difficult to follow because of its speed. Such data suggest that there is a need for an oxynitrilase in the process of cyanogenesis where the pH of the homogenized plant tissue is approximately 5.

C. Metabolic Turnover and HCN Detoxification

There is much evidence that many secondary plant products are not metabolically inert but are actively turned over (see Barz and Köster, this volume, Chapter 3). This was first realized for cyanogenic glycosides when Abrol and Conn (1966) and Abrol et al. (1966) showed that ^{14}C administered to cyanogenic plants as labeled amino acids was incorporated in significant amounts into the amide carbon of asparagine. This finding indicated that cyanogenic glycosides were being synthesized and catabolized and that the HCN produced in the intact plant was subsequently converted into asparagine by sequential action of the two enzymes β-cyanoalanine synthase (E.C. 4.4.1.9) (Hendrickson and Conn, 1969) and β-cyanoalanine hydrolase (E.C. 4.2.1.65) (Castric et al., 1972) (Fig. 4). In this process, the enzyme

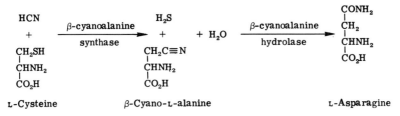

Fig. 4. Metabolism of HCN in plants.

β-cyanoalanine synthase, discovered by Blumenthal-Goldschmidt *et al.* in 1963, appears to play an important role in detoxifying HCN (see Conn, 1980a, for further discussion).

In a recent study Miller and Conn (1980) further examined the question of how HCN is metabolized in plants. They measured the amounts of three enzymes, β-cyanoalanine synthase, rhodanese (E.C. 2.8.1.1), and formamide hydrolyase (E.C. 4.2.1.66), that utilize HCN as a substrate; a total of 16 species, both cyanogenic and noncyanogenic, was examined. They found that β-cyanoalanine synthase occurred in all the plants and that the amount of the enzyme was significantly higher in strongly cyanogenic species. However, the synthase was present even in noncyanogenic plants. Rhodanese was present in four of the species studied, but only two of these were cyanogenic, namely, *Sorghum bicolor* and *Eschscholzia californica*. They could not confirm the presence of formamide hydrolyase in any plant. These findings imply an important role for β-cyanoalanine synthase in HCN detoxification in plants known to produce HCN. Why the enzyme is present in other plants remains unknown.

VI. PHYSIOLOGICAL ROLE

Cyanogenic plants (and more recently cyanogenic glycosides) have attracted the interest of scientists for nearly two centuries because the HCN they produce is a toxic substance that inhibits cytochrome oxidase and other respiratory enzymes. The literature on poisonous plants (see Everist, 1974; Kingsbury, 1964) stresses this point and cites many examples where poisoning of humans and other animals has occurred. Two recent reviews have discussed the toxicology of cyanogenic glycosides (Conn, 1979b; Montgomery, 1980), and another (Lewis, 1977) has reviewed the controversy surrounding amygdalin which has been proposed as a cure for cancer for many years.

The physiological role of cyanogenic glycosides in plants is undoubtedly related to their ability to produce toxic amounts of HCN. Presumably these glycosides, like many other secondary compounds, have provided survival value to the plant containing them by acting as toxicants or feeding deterrents to herbivores that could threaten the continued existence of the plant. In the specific example of cyanogenic glycosides there are data strongly supporting such a protective role (Jones, 1978, 1979). Two recent symposia have treated the general subject of plant–herbivore interaction and cyanogenic glycosides have been discussed (Rosenthal and Janzeu, 1979; Harborne, 1978).

VII. CONCLUSION

The cyanogenic compounds occurring in living organisms, especially plants, have interested scientists since Schrader's (1803) early report of prussic acid in bitter almonds. This interest, originally centered on the toxicity of the HCN produced, has expanded into studies dealing with their chemistry, biochemistry, chemotaxonomy, toxicology, and physiological role. A body of literature dealing with these several aspects is available for these interesting compounds; this chapter has attempted to provide an introduction to that literature.

REFERENCES

Abrol, Y. P., and Conn, E. E. (1966). *Phytochemistry* **5**, 237–242.
Abrol, Y. P., Conn, E. E., and Stoker, J. R. (1966). *Phytochemistry* **5**, 1021–1027.
Blumenthal-Goldschmidt, S., Butler, G. W., and Conn, E. E. (1963). *Nature (London)* **197**, 718–719.
Butler, G. W., Bailey, T. W., and Kennedy, L. D. (1965). *Phytochemistry* **4**, 369–381.
Castric, P. A., Farnden, K. J. F., and Conn, E. E. (1972). *Arch. Biochem. Biophys.* **152**, 62–69.
Clegg, D. O., Conn, E. E., and Janzen, D. H. (1979). *Nature (London)* **278**, 343–344.
Conn, E. E. (1969). *Agric. Food Chem.* **17**, 519–526.
Conn, E. E. (1973). *Biochem. Soc. Symp.* **38**, 277–302.
Conn, E. E. (1979a). *Naturwissenschaften* **66**, 28–34.
Conn, E. E. (1979b). *Int. Rev. Biochem.* **27**, 21–43.
Conn, E. E. (1980a). *Encycl. Plant Physiol., New Ser.* **8**, 461–492.
Conn, E. E. (1980b). *Annu. Rev. Plant Physiol.* **31**, 433–451.
Cramer, U., Rehfeldt, A. G., and Spener, F. (1980). *Biochemistry* **19**, 3074–3080.
Cutler, A. J., Hösel, W., Sternberg, M., and Conn, E. E. (1981). *J. Biol. Chem.*, **256**, 4253–4258.
Davis, R. H., and Nahrstedt, A. (1979). *Comp. Biochem. Physiol. B* **64**, 395–397.
Duffey, S. S., and Towers, G. H. N. (1978). *Can. J. Zool.* **56**, 7–16.
Ettlinger, M. G., Jaroszewski, J. W., Jensen, S. R., Nielsen, B. J., and Nartey, F. (1977). *J. Chem. Soc., Chem. Commun.* pp. 952–953.
Everist, S. L. (1974). "Poisonous Plants of Australia." Angus & Robertson, Sydney.
Eyjolfsson, R. (1970). *Fortschr. Chem. Org. Naturst.* **28**, 74–108.
Fikenscher, L. H., and Hegnauer, R. (1977). *Pharm-Weekbl.* **112**, 11–20.

Gibbs, R. D. (1974). "Chemotaxonomy of Flowering Plants," Vols. I-IV. McGill-Queens Univ. Press, Montreal.

Gondwe, A., Seigler, D. S., and Dunn, J. E. (1978). *Phytochemistry* **17**, 271–274.

Hahlbrock, K., and Conn, E. E. (1970). *J. Biol. Chem.* **245**, 917–922.

Haisman, D. R., and Knight, D. J. (1967). *Biochem. J.* **103**, 528–534.

Harborne, J. B., ed. (1978). "Biochemical Aspects of Plants and Animal Coevolution." Academic Press, New York.

Hegnauer, R. (1962–1973). "Chemotaxonomie der Pflanzen," Vols. I-VI. Birkhaueser, Basel.

Hegnauer, R. (1973). *Biochem. Syst.* **1**, 191–197.

Hegnauer, R. (1977). *Plant Syst. Evol., Suppl.* **1**, 191–209.

Hendrickson, H. R., and Conn, E. E. (1969). *J. Biol. Chem.* **244**, 2632–2640.

Herissey, H. (1906). *J. Pharm. Chim., Paris* **23**, 5–14.

Hösel, W., and Nahrstedt, A. (1975). *Hoppe-Seyler's Z. Physiol. Chem.* **356**, 1265–1275.

Hösel, W., and Nahrstedt, A. (1980). *Arch. Biochem. Biophys.* **203**, 753–757.

Jensen, S. R., and Nielsen, B. J. (1973). *Acta Chem. Scand.* **27**, 2661–2662.

Jones, D. A. (1972). *In* "Phytochemical Ecology" (J. B. Harborne, ed.), Chapter 7, pp. 103–124. Academic Press, New York.

Jones, D. A. (1978). *In* "Biochemical Aspects of Plant and Animal Coevolution" (J. B. Harborne, ed.), Chapter 2, pp. 21–34. Academic Press, New York.

Jones, D. A. (1979). *Am. Nat.* **113**, 445–451.

Jorns, M. S. (1979). *J. Biol. Chem.* **254**, 12145–12152.

Kim, H. S., Jeffrey, G. A., Panke, D., Clapp, R. C., Coburn, R. A., and Long, L., Jr. (1970). *J. Chem. Soc., Chem. Commun.* p. 381.

Kingsbury, J. M. (1964). "Poisonous Plants of the U.S. and Canada." Prentice-Hall, Englewood Cliffs, New Jersey.

Knowles, C. J. (1976). *Bacteriol. Rev.* **40**, 652–680.

Kojima, M., Poulton, J. E., Thayer, S. S., and Conn, E. E. (1979). *Plant Physiol.* **63**, 1022–1028.

Lewis, J. P. (1977). *West. J. Med.* **127**, 55–62.

Lythgoe, B., and Riggs, N. V. (1949). *J. Chem. Soc.* pp. 2716–2718.

McFarlane, I. J., Lees, E. M., and Conn, E. E. (1975). *J. Biol. Chem.* **250**, 4708–4713.

Mao, C.-H., and Anderson, L. (1967). *Phytochemistry* **6**, 473–483.

Mikolajczak, K. L. (1977). *Prog. Chem. Fats Other Lipids* **15**, 97–130.

Miller, J. M., and Conn, E. E. (1980). *Plant Physiol.* **65**, 1199–1202.

Møller, B. L., and Conn, E. E. (1979). *J. Biol. Chem.* **254**, 8575–8583.

Møller, B. L., and Conn, E. E. (1980). *J. Biol. Chem.* **255**, 3049–3056.

Montgomery, R. D. (1980). *In* "Toxic Constituents of Plant Food Stuffs" (I. E. Liener, ed.), 2nd ed., pp. 143–160. Academic Press, New York.

Nahrstedt, A. (1970). *Phytochemistry* **9**, 2085–2087.

Nahrstedt, A. (1973). *Plant Med.* **24**, 83–89.

Nahrstedt, A. (1975a). *Phytochemistry* **14**, 1339–1340.

Nahrstedt, A. (1975b). *Arch. Pharm. (Weinheim, Ger.)* **308**, 903–910.

Nahrstedt, A. (1976). *Z. Naturforsch., Teil C* **31**, 397–400.

Nahrstedt, A. (1981). *In* "Cyanide in Biology" (B. Vennesland, E. E. Conn, C. J. Knowles, J. Westley, F. Wissing, eds.) Academic Press, London (in press).

Nahrstedt, A., and Hubel, W. (1978). *Phytochemistry* **17**, 314–315.

Paris, M., Bouquet, A., and Paris, R. R. (1969). *C. R. Hebd. Seances Acad. Sci., Ser. D* **268**, 2804–2806.

Plouvier, V. (1935). *C.R. Hebd. Seances Acad. Sci.* **200**, 1985.

Reay, P. F., and Conn, E. E. (1974). *J. Biol. Chem.* **249**, 5826–5830.

Rimington, C. (1935). *Onderstepoort J. Vet. Sci. Anim. Ind.* **5**, 445.

Robinson, M. E. (1930). *Biol. Rev. Cambridge Philos. Soc.* **5**, 126–141.

Rosenthal, G. A., and Janzeu, D. H., eds. (1979), "Herbivores, their Interaction with Secondary Plant Metabolites", Academic Press, New York.

Schrader, J. C. C. (1803). *Anal. Phys. (Leipzig)* [1] **13**, 503–504.

Seeley, M. K., Criddle, R. S., and Conn, E. E. (1966). *J. Biol. Chem.* **241**, 4457–4462.

Seigler, D. S. (1976a). *Econ. Bot.* **30**, 395–407.

Seigler, D. S. (1976b). *Proc. Okla. Acad. Sci.* **56**, 95–110.

Seigler, D. S. (1977). *Prog. Phytochem.* **4**, 83–120.

Seigler, D. S. (1981). *Rev. Latinoam. Quim.* **12**, 39–48.

Seigler, D. S., Eggerding, C., and Butterfield, C. (1974). *Phytochemistry* **13**, 2330–2332.

Sharples, D., Spring, M. S., and Stoker, J. R. (1972). *Phytochemistry* **11**, 2999–3003.

Shimada, M., and Conn, E. E. (1977). *Arch. Biochem. Biophys.* **180**, 199–207.

Smith, C. R., Jr., Weisleder, D., Miller, R. W., Palmer, I. S., and Olson, O. E. (1980). *J. Org. Chem.* **45**, 507–510.

Steyn, D. C., and Rimington, C. (1935). *Onderstepoort J. Vet. Sci. Anim. Ind.* **4**, 51–■■.

Tapper, B. A., and MacDonald, M. A. (1974). *J. Microbiol.* **20**, 563–566.

Tjon Sie Fat, L. (1979). *Proc. K. Ned. Akad. Wet., Ser. C* **82**, 197–209.

Uribe, E., and Conn, E. E. (1966). *J. Biol. Chem.* **241**, 92–94.

Vennesland, B., Conn, E. E., Knowles, C. J., Westley, J., Wissing, F., eds. (1981). "Cyanide in Biology," Academic Press, London (in press).

Watt, J. M., and Breyer-Brandwijk, M. G. (1964). "The Medicinal and Poisonous Plants of Southern and East Africa," 2nd ed. Livingstone, Edinburgh and London.

Glucosinolates

17

PEDER OLESEN LARSEN

The Biochemistry of Plants, Vol. 7

I. INTRODUCTION

Glucosinolates have the general formula I, thus all possess the same functional group. In plants they can be enzymatically degraded to give D-glucose, sulfate, and isothiocyanates (II), organic cyanides (III) plus sulfur, or thiocyanates (IV) (besides minor products; see Section V). Glucosinolates occur in important crop plants and vegetables, mainly from the family Cruciferae. The enzymatic decomposition products have toxic properties. Glucosinolates are biochemically closely related to the widely distributed cyanogenic glycosides (see Conn, this volume, Chapter 16), and they are of well-established importance in the relationship between plants and insects. Therefore glucosinolates have attracted wide interest and have been the subject of numerous studies.

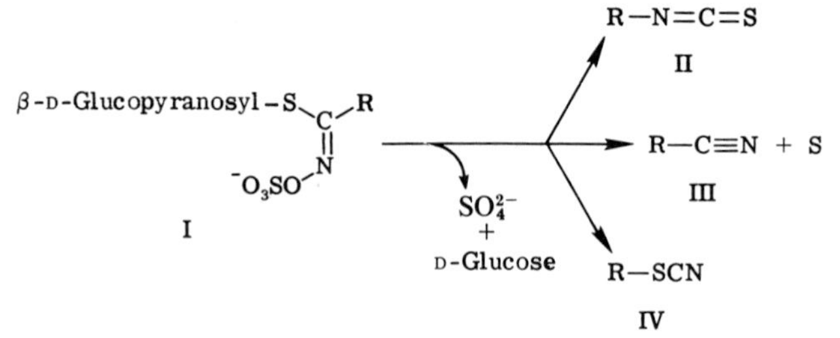

$$
\begin{array}{ccc}
 & & R\!-\!N\!=\!C\!=\!S \\
 & & II \\
\beta\text{-}D\text{-Glucopyranosyl}\!-\!S & & \\
\quad\quad\quad\quad\quad\quad \underset{N}{\overset{\parallel}{C}}\!-\!R & \longrightarrow & R\!-\!C\!\equiv\!N + S \qquad (1) \\
\quad\quad\quad {}^{-}O_3SO & & III \\
 & SO_4^{2-} & \\
I & + & \\
 & D\text{-Glucose} & R\!-\!SCN \\
 & & IV
\end{array}
$$

The main emphasis in this chapter will be placed on the general chemical, biochemical, and biological properties of glucosinolates, whereas the variations in side chains and the distribution pattern in plants will be only briefly treated. A number of reviews on glucosinolates are available in the literature, both comprehensive (Kjær, 1960, 1976; Ettlinger and Kjær, 1968; Appelqvist, 1972; VanEtten and Tookey, 1979; Underhill, 1980; Tookey et al., 1980) and covering specific aspects such as biosynthesis (Underhill et al., 1973; Kjær and Larsen, 1973, 1976, 1977, 1980), catabolism and enzymatic degradation (Ohtsuro, 1975; Björkman, 1976; Benn, 1977), distribution (Kjær, 1974), and distribution and analysis (Rodman, 1978). However, more recent original reports are of importance to many aspects of the field. In the following only selected references will be given. For additional information references are available in the reviews listed above.

II. CHEMISTRY

A. Structure and Nomenclature

Formula I was originally proposed on the basis of chemical studies on allylglucosinolate and p-hydroxybenzylglucosinolate (Ettlinger and Lundeen, 1956). The Z configuration shown around the C=N double bond was proposed on the basis of chemical evidence (Ettlinger *et al.,* 1961) and subsequently proved for allylglucosinolate by X-ray analysis (Marsh and Waser, 1970). Structure I is assumed to be valid for all glucosinolates so far identified, although in most cases this has not been strictly proven. The assumption implies that variability occurs only in the side chain R, whereas the carbohydrate moiety is always a β-D-glucopyranosyl residue and the configuration around the C=N double bond is always Z. In many cases the glucosyl moiety has been established only by enzymatic hydrolysis and chromatographic identification of the glucose liberated. The Z configuration has been established rigorously only for allylglucosinolate, as mentioned above. However, the assumption seems justified in view of the chemical and spectroscopic similarities among all natural glucosinolates and of their degradation by the same enzymes.

Because of the low pK value of the sulfonic acid group and because of the instability of glucosinolates in strong acid, they invariably occur in nature in the anionic form. According to the semisystematic nomenclature introduced in 1961 (Dateo, 1961; Ettlinger and Dateo, 1961) and generally now accepted, the parent compound glucosinolate is the anion I, where R = H. The various glucosinolates are derived by naming the side chain R as a prefix, and crystalline glucosinolate salts are therefore defined only by giving the name of the cation involved.

Before the introduction of this semisystematic nomenclature trivial names were given to glucosinolates. Some of these names refer to defined salts. For example, sinigrin is potassium allylglucosinolate, progoitrin is sodium $(2R)$-hydroxy-3-butenylglucosinolate, and sinalbin is the sinapine salt of the p-hydroxybenzylglucosinolate anion (sinapine is a rather widely distributed quarternary ammonium base composed of choline esterified with 3,5-dimethoxy-4-hydroxycinnamic acid). In other cases the counterion is not defined and the trivial name may refer to both the anion and one or more of its salts. Because of this ambiguity and because by now more than 80 glucosinolates are known, the use of trivial names should be abandoned completely.

Before introduction of the semisystematic nomenclature glucosinolates were generally called mustard oil glucosides (in German, *Senfölglucosiden*) after the mustard oils (isothiocyanates) liberated by enzymatic degradation.

A complete list of the 50 glucosinolates identified up to 1968 is available (Ettlinger and Kjær, 1968). Since then, the structures of more than 25 new glucosinolates have been established (Kjær and Larsen, 1973, 1976, 1977, 1980). Glucosinolates are presumably all derived biosynthetically from amino acids in a reaction series in which the carboxyl group is lost and the α carbon transformed into the central carbon in the glucosinolates (Section

IV). The side chain R therefore is identical with the substituent on the α carbon in the amino acid. Only seven of the glucosinolates correspond to protein amino acids (alanine, valine, leucine, isoleucine, phenylalanine, tyrosine, and tryptophan). The remaining glucosinolates are derived by modifications of the side chains, probably taking place at the glucosinolate level, and/or are derived from nonprotein amino acids produced from protein amino acids by a chain-lengthening process (cf. Section IV,C).

Table I is a list of the structural types encountered among glucosinolates. This table also gives examples of each of the structural groups involved, mainly the simplest member of the group. A notable feature is the occurrence of series of glucosinolates deviating only in the number of CH_2 groups in the side chain. Most conspicuous is the whole series of methyl-

TABLE I

Structural Types of Glucosinolates

Structural type	Examples	R in I
1. Aliphatic side chains		
1.1. Saturated, without functional groups	Methylglucosinolate	CH_3
1.2. With double bonds	Allylglucosinolate	$CH_2{=}CHCH_2$
1.3. With alcohol groups, free or acylated	1-Methyl-2-hydroxyethyl-glucosinolate	$HOCH_2CH(CH_3)$
	3-Benzoyloxypropylglu-cosinolate	$C_6H_5COO(CH_2)_3$
1.4. With keto groups	4-Oxoheptylglucosinolate	$CH_3(CH_2)_2CO(CH_2)_3$
1.5. With methylthio groups	3-Methylthiopropylglucosinolate	$CH_3S(CH_2)_3$
1.6. With methylsulfinyl groups	3-Methylsulfinylpropylglu-cosinolate	$CH_3SO(CH_2)_3$
1.7. With methylsulfonyl groups	3-Methylsulfonylpropyl-glucosinolate	$CH_3SO_2(CH_2)_3$
1.8. With esterified carboxyl groups	3-Methyloxycarbonylpropyl-glucosinolate	$CH_3OCO(CH_2)_3$
2. Aromatic side chains		
2.1. Without functional groups	Benzylglucosinolate	$C_6H_5CH_2$
	2-Phenylethylglucosinolate	$C_6H_5(CH_2)_2$
2.2. With phenol groups, free, methylated or glycosylated	p-Hydroxybenzylglucosinolate	$p\text{-}HOC_6H_4CH_2$
2.3. With alcohol groups	2-Hydroxy-2-phenylethylglu-cosinolate	$C_6H_5CHOHCH_2$
3. Heteroaromatic side chains	Indol-3-ylmethylglucosinolate	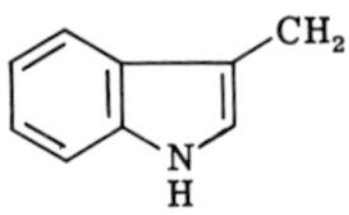

sulfinylglucosinolates in which 3–11 carbons have been found in the alkyl side chain. In some of the other groups in Table I similar series (1.2, 1.5, 1.7) can be found. The table gives no information on the configuration of chiral centers in the side chains, but in nearly all cases the configurations have been established. (2R)- and (2S)-hydroxy-3-butenylglucosinolate have both been found in species from *Cruciferae* but not in the same species (Ettlinger and Kjær, 1968).

The classification given in Table I is based solely on structure. Other classifications could, however, be devised, for example, based on distribution or on biosynthesis and distinguishing between glucosinolates directly derived from protein amino acids and from other amino acids. Classification could also be done by selecting the groups of glucosinolates containing side chains conferring special reaction possibilities on the glucosinolates or their decomposition products.

The polarity of the side chain is of importance in various analytical procedures, and a distinction between glucosinolates with polar and nonpolar side chains can therefore be useful.

B. Properties of Glucosinolates

Because of their glucose moiety and ionic forms, glucosinolates are hydrophilic, nonvolatile compounds. The salts are easily soluble in water and insoluble in nonpolar solvents. A number of glucosinolates have been obtained as crystalline salts with potassium, sodium, tetramethylammonium, or sinapine as the counterion.

Generally, however, crystalline salts are not easily obtained from glucosinolates. This is due partly to difficulties in crystallization and partly to difficulties in separating mixtures of glucosinolates into individual compounds. In many cases identification has been based solely on spectroscopy and on transformation to isothiocyanates. In a few instances peracetylation of glucosinolates introducing four acetyl groups into the glucose moiety and additional acetyl groups on free hydroxy groups in the side chain has led to the isolation of crystalline salts.

Glucosinolates are fairly stable at neutral pH values. Systematic investigations on their stability in water with or without heating have, however, not been carried out. In special instances, instability has been observed, for example, for indol-3-ylmethylglucosinolate (Gmelin and Virtanen, 1961). Glucosinolates can be decomposed by the action of both strong acids and strong bases. With acid, decomposition to give a carboxylic acid and hydroxylamine occurs. Thus allylglucosinolate (V) is transformed into 3-butenoic acid (VI) and hydroxylamine, whereas p-hydroxybenzylglucosinolate (VII) is transformed into p-hydroxyphenylacetic acid (VIII) and hydroxylamine:

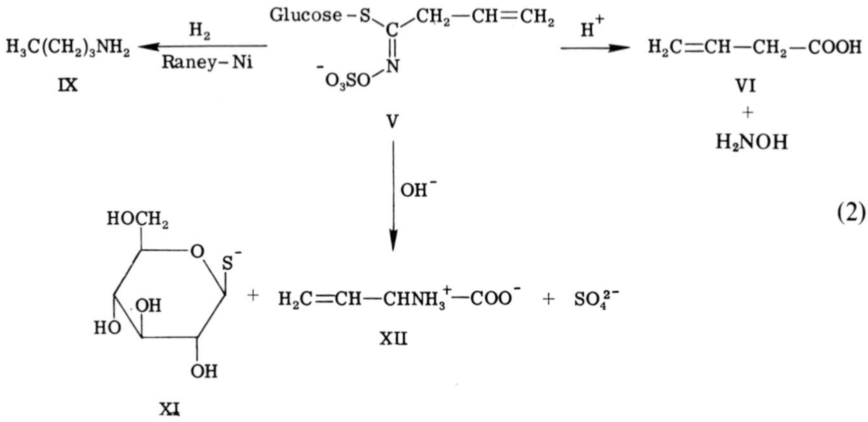

$$H_3C(CH_2)_3NH_2 \xleftarrow[\text{Raney-Ni}]{H_2} \quad \text{Glucose–S}\diagdown\underset{\underset{O_3SO}{\overset{\|}{N}}}{C}\diagup CH_2\text{—CH}=CH_2 \xrightarrow{H^+} H_2C=CH\text{—}CH_2\text{—COOH}$$

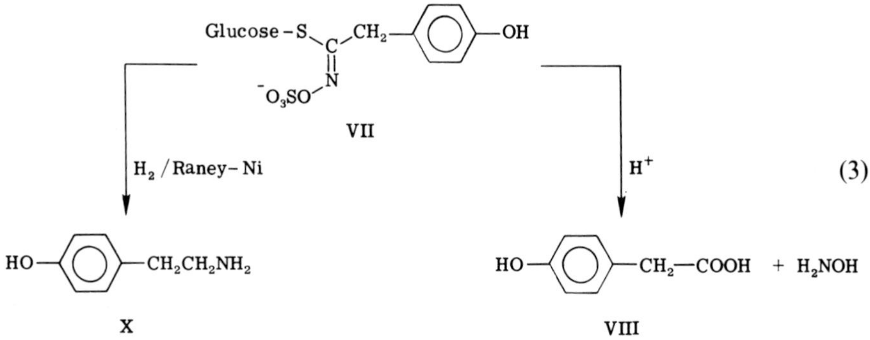

IX V VI

+

H_2NOH

(2)

OH^-

XI + $H_2C=CH\text{—}CHNH_3^+\text{—}COO^-$ + SO_4^{2-}

XII

$$\text{Glucose–S}\diagdown\underset{\underset{O_3SO}{\overset{\|}{N}}}{C}\diagup CH_2\text{—}\bigcirc\text{—OH}$$

VII

H_2/Raney–Ni   H^+   (3)

$HO\text{—}\bigcirc\text{—}CH_2CH_2NH_2$ $HO\text{—}\bigcirc\text{—}CH_2\text{—}COOH$ + H_2NOH

X   VIII

On treatment with hydrogen and Raney nickel, amines are produced from glucosinolates; thus butylamine (IX) is formed from allylglucosinolate and tyramine (X) is formed from p-hydroxybenzylglucosinolate (Ettlinger and Lundeen, 1956).

Treatment of allylglucosinolate with an aqueous base at room temperature results in production of the anion of β-D-thioglucopyranose (XI), sulfate, and 2-amino-3-butenoic acid (XII). The reaction probably involves a Neber-type rearrangement. Similarly, by treatment with a base, benzylglucosinolate gives the β-thioglucose anion, sulfate, and phenylglycine, but in this case other minor products are observed. Both allyl- and benzylglucosinolate have high kinetic acidity of the hydrogen atoms at C-2 in the aglucone, and this probably greatly facilitates the arrangement. Glucosinolates other than the allyl and benzyl compounds can also give α-amino acids with an aqueous base, but only at elevated temperatures (Friis et al., 1977). Other reactions with a base may take place, however, as shown by the degradation of p-hydroxybenzylglucosinolate to give thiocyanate (Gmelin and Virtanen, 1960), of indol-3-ylmethylglucosinolate and N-methoxyindol-3-ylmethyl-glucosinolate to a large number of indol derivatives including indol-3-ylacetic

acid and indol-3-ylmethylcyanide (indoleacetonitrile) and the corresponding
N-methoxy compounds (Gmelin and Virtanen, 1961, 1962), and of 2-hydroxy-
3-butenylglucosinolate to 5-vinyloxazolidin-2-one and 2-hydroxy-3-butene-
nitrile in an aqueous solution with borax (Gronowitz et al., 1978).

Treatment of glucosinolates with $AgNO_3$ or $Hg(OAc)_2$ results in the production of α-D-glucose and silver or mercury derivatives which again can be
decomposed to give nitriles, for example, 3-4-pentenenitrile from allylglucosinolate.

C. Properties of Isothiocyanates

Isothiocyanates (XIII), produced by enzymatic degradation of glucosinolates, are generally volatile compounds with a strong smell and taste. Compounds with strongly polar and large side chains have, however, restricted
volatility. Isothiocyanates are chemically very reactive. With ammonia and
amines they produce substituted thioureas (XIV):

$$R-N=C=S \; + \; R'-NH_2 \longrightarrow R-NH-\underset{\underset{S}{\parallel}}{C}-NH-R' \qquad (4)$$

$$\text{XIII} \hspace{4cm} \text{XIV}$$

With alcohols they produce thionocarbamates, and especially 2- and
3-hydroxy-substituted isothiocyanates [(XV) and (XVI)] spontaneously form
oxazolidine-2-thiones (XVII) and tetrahydro-1,3-oxazine-2-thiones (XVIII):

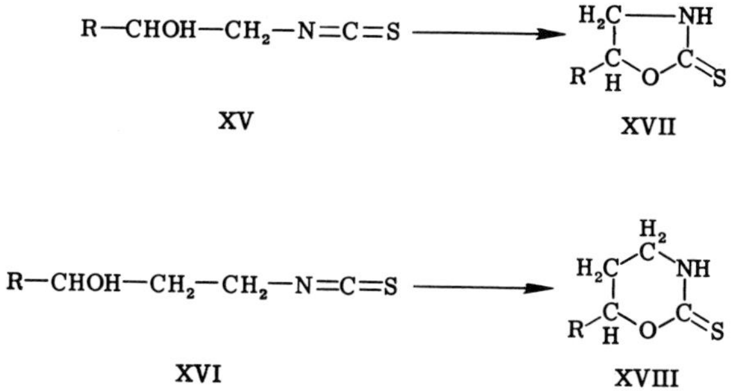

$$(5)$$

Some isothiocyanates are rather unstable, especially those that produce stable carbonium ions through loss of the thiocyanate ion. Thus
p-hydroxybenzylisothiocyanate easily gives the thiocyanate ion and
p-hydroxybenzyl alcohol in aqueous solution (Gmelin and Virtanen, 1960;
Kawakishi et al., 1967). Indol-3-ylmethylisothiocyanate and N-methoxy-
indol-3-ylmethylisothiocyanate also easily give thiocyanate ions (Gmelin
and Virtanen, 1961, 1962).

D. Isolation and Analysis

Glucosinolate-containing plants also possess enzymes catalyzing their degradation (myrosinases). Therefore, during isolation of the native glucosinolates the enzymes must be denatured, for example, by extraction of the plant material with boiling 70% methanol. Subsequent steps in isolation procedures exploit the ionic properties by use of ion-exchange resins (Kjær, 1960; Underhill and Wetter, 1969; Björkman, 1972; VanEtten and Daxenbichler, 1977), but there are possibilities for improvement in these steps. Methods for paper chromatography and thin-layer chromatography of glucosinolates have been described. Gas chromatography of properly derivatized glucosinolates has also been introduced, although again there seems to be room for improvement (Thies, 1976; Olsson et al., 1976). Most analytical procedures involve, however, enzymatic degradation to give isothiocyanates, sulfate, and glucose, followed by determination of one of the products. The great variability in the products of enzymatic decomposition (Section V) may cause erroneous results. Isothiocyanates can be determined by gas chromatography or, after transformation to thioureas, by paper or thin-layer chromatography. Oxazolidine-2-thiones can be determined by uv spectroscopy. Glucose can be determined enzymatically and sulfate gravimetrically. The thiocyanate ion produced from phenolic and indolic glucosinolates can easily be measured colorimetrically (VanEtten et al., 1974; Daxenbichler and VanEtten, 1977; Rodman, 1978).

III. OCCURRENCE

Glucosinolates have been found only in dicotyledonous plants. They occur mainly in the order Capparales in sensu Dahlgren (Dahlgren, 1975) in the families Capparidaceae, Cruciferae, Resedaceae, Tovariaceae, Moringaceae, Limnanthaceae, Salvadoraceae, Tropaeolaceae, and Gyrostemonaceae (Limnanthaceae and Tropaeolaceae are sometimes placed in the order Geraniales). In at least the first three of these families glucosinolates seem to be present in all species. The glucosinolates in the Cruciferae are of special interest, since a number of important vegetables, herbs, and agricultural crops belong to this family. In addition, glucosinolates occur in the families Caricaceae and Euphorbiaceae (only in very few species in this large family).

A great variability in side chains is found in the Capparidaceae, Cruciferae, and Resedaceae. In the remaining families only glucosinolates either derived directly from protein amino acids or with additional hydroxy groups have been found. This simple pattern is similar to that known for cyanogenic glycosides.

Glucosinolates and cyanogenic glycosides have never been encountered in the same species. The literature contains scattered reports on cyanogenesis in glucosinolate-containing plants, but the precursor of cyanide has never been identified in these cases. On the other hand, and in contrast to glucosinolates, cyanogenic glycosides are found in many species of the Euphorbiaceae. Glucosinolates may be present in additional families, as indicated in various literature reports, but clear proof is lacking in most instances (Ettlinger and Kjær, 1968). The distribution of glucosinolates in families, genera, and species has often been used in chemotaxonomic considerations (Kjær, 1974; Rodman, 1978). Many glucosinolate-containing plants, especially among the Cruciferae, contain, however, several glucosinolates together, often with very large differences in concentration between the major and minor components. As many as 15 glucosinolates have been identified in one species. Furthermore, the glucosinolate content, including the relative concentrations of the individual compounds, often varies considerably in different organs of the same plant. Population variations in glucosinolate content can also occur (Kjær, 1960; Rodman, 1978). Therefore the major problem in chemotaxonomy, that of determining the absence of a specific compound, is also of importance in work with glucosinolates.

Glucosinolates generally occur in all parts of a plant but in widely different concentrations. Only a few systematic investigations have been reported on the glucosinolate concentrations in various parts of a single species throughout the whole life cycle. One example is available—indole glucosinolates in *Isatis tinctoria* (Elliott and Stowe, 1971b). Studies on the biosynthesis of indole glucosinolates in *I. tinctoria* indicate a rapid glucosinolate turnover with a half-life of about 2 days (Mahadevan and Stowe, 1972). On the other hand, numerous analyses of concentrations in vegetables and other crops have been reported. Generally, the levels in fresh plant parts are about 0.1% or less based on fresh weight (Cole, 1976; VanEtten *et al.,* 1976), whereas levels in seeds may be up to 10% of the dry weight (Josefsson, 1972; VanEtten *et al.,* 1974; Wetter and Dyck, 1973). Only limited knowledge is available on the glucosinolate concentration in individual plant cells, but in seeds the endosperm is the site of accumulation (Kjær, 1960). Recently, the glucosinolates in horseradish root cells have been localized in the vacuoles (Grob and Matile, 1979; Matile, 1980)

In this connection it is worthwhile to mention that cations must be present to balance the negative charges on glucosinolate anions. In seeds with glucosinolate anions contributing up to 10% of the dry weight the concentrations of cations required must be considerable. The isolation of most glucosinolates as potassium salts does not necessarily mean that potassium ion is the physiological counterion but only that potassium ions are introduced during the standard ion-exchange procedure employed. However, the classical isolations of potassium allylglucosinolate from *Brassica nigra* and

Armoracia lapathifolia (horseradish) indicate that the potassium ion is the dominating counterion in these species. Similarly the isolation of *p*-hydroxybenzylglucosinolate as the sinapine salt from *Sinapis alba* also indicates that sinapine is the dominating counterion here. Basic proteins may also act as counterions (compare Finlayson, 1976).

Glucosinolates in higher plants always occur together with myrosinase (Section V,A).

IV. BIOSYNTHESIS

A. Amino Acids as Precursors

Glucosinolates are presumably all derived from amino acids in the series of steps outlined in Scheme 1. The derivation from amino acids was first demonstrated by incorporation of DL-[β-^{14}C]tryptophan into indol-3-ylmethylglucosinolate in *Brassica oleraceae* (Kutáček *et al.*, 1962) and by the incorporation of ^{14}C-labeled phenylalanine into benzylglucosinate in *Tropaeolum majus* (Underhill *et al.*, 1962; Benn, 1962). The incorporation of ^{14}C from phenylalanine was specific; the activity from [α-^{14}C]phenylalanine was located in the thiohydroximate carbon, whereas that from [β-^{14}C]-phenylalanine was found in the benzyl carbon. No activity was present in the glucosinolate when [*carboxyl*-^{14}C]phenylalanine was fed.

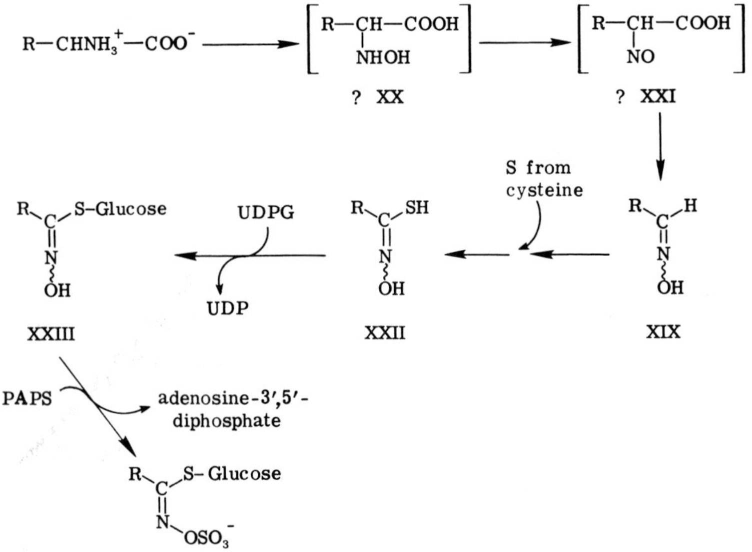

Scheme 1. The biosynthetic pathway from amino acids to glucosinolates.

An origin from amino acids has subsequently been confirmed for other glucosinolates by labeling studies, usually with labeling in specific positions and subsequent degradation to show that the isotope is present in the expected positions. Thus it has been shown that p-hydroxybenzylglucosinolate is derived from tyrosine, isopropylglucosinolate from valine, 3-methoxy-carbonylpropylglucosinolate from 2-aminoadipic acid, and phenylethyl-glucosinolate from 2-amino-4-phenylbutanoic acid. In the last case incorporation percentages of up to 40% have been obtained. Also, a large number of experiments have been performed on the incorporation of amino acids, but involving changes in the side chains (probably at the glucosinolate level; see below), which are in complete agreement with the origin of these compounds from amino acids.

In the biosynthesis of a glucosinolate from an amino acid the nitrogen is preserved. This was first shown by the incorporation of double-labeled L-[^{14}C,^{15}N]phenylalanine into benzylglucosinolate in *T. majus* with no change in the ^{14}C/^{15}N ratio (Underhill and Chisholm, 1964). The result proved that the amino nitrogen and the carbon skeleton of phenylalanine, except for the carboxyl carbon, were incorporated intact into benzylglucosinolate and that all the intermediates must contain the nitrogen atom.

B. The Intermediates between Amino Acids and Glucosinolates

The occurrence of aldoximes (XIX) as intermediates in the biosynthesis of glucosinolates was first demonstrated by the incorporation of phenyl[1-^{14}C]-acetaldoxime into benzylglucosinolate in *Lepidium sativum* (Tapper and Butler, 1967) and *T. majus* (Underhill, 1967). Phenylacetaldoxime formed from phenylalanine was also found to be a natural constituent of *T. majus*. Subsequently a large number of detailed investigations have supported aldoximes as intermediates. Double-labeling experiments have shown the incorporation of 4-methylthiobutanaldoxime into allylglucosinolate and of 5-methylthiopentanaldoxime into 2-hydroxy-3-butenylglucosinolate with no change in the ^{14}C/^{15}N ratios.

Aldoximes are intermediates not only in the biosynthesis of glucosinolates but also in that of cyanogenic glucosides (cf. Chapter 16) and other plant constituents. The general role of aldoximes has been the subject of a comprehensive review (Mahadevan, 1973).

At least one intermediate must occur between the amino acids and the aldoximes. Experiments with ^{14}C-labeled N-hydroxyphenylalanine have been interpreted to support the idea that N-hydroxyamino acids (XX) are intermediates (Kindl and Underhill, 1968). However, this conclusion is not justified, since the incorporation into benzylglucosinolate observed may be due to chemical decomposition of N-hydroxyphenylalanine to give phenylacetaldoxime. However, the occurrence of N-hydroxyamino acids as intermediates is supported by the recent definite establishment of N-hydroxytyrosine as an intermediate between tyrosine and p-hydroxyphenylacetaldoxime in the biosynthesis of the cyanogenic glucoside dhurrin in *Sorghum* seedlings. Presumably, in this biosynthesis 3-(p-hydroxyphenyl)-2-nitrosopropanoic acid is the intermediate between the N-hydroxyamino acid and the aldoxime (cf. Chapter 16). The inclusion of N-hydroxyamino acids and nitroso acids (XXI) in Scheme 1 is therefore based solely on analogy.

Aldoximes are transformed into glucosinolates via thiohydroximic acids (XXII) and desulfoglucosinolates (XXIII). The occurrence of thiohydroximic acids and desulfoglucosinolates as intermediates was first demonstrated by the incorporation of phenylacetothiohydroximic acid labeled with both ^{14}C and ^{35}S, and of desulfobenzylglucosinolate labeled with ^{14}C both in the glucose moiety and the aglucone, into benzylglucosinolate in *T. majus* (Underhill and Wetter, 1969). Subsequently, phenylacetothiohydroximic acid was identified as a natural product in *T. majus,* and trapping experiments showed that it was derived from phenylacetaldoxime. An enzyme transforming the thiohydroximic acid and UDP-glucose into desulfoglucosinolate and UDP has been identified in *T. majus.* Also, 4-methylthiobutanethiohydroximic acid has been shown to be a precursor of allylglucosinolate.

The intermediates between the aldoximes and the thiohydroximic acids have not been identified. It is known that the sulfur in cysteine is readily incorporated into the thiohydroximic acid (and 1-thioglucose has been excluded as a precursor). Nitro compounds have been proposed as intermediates, being produced from the aldoximes by oxidation and after isomerization to their aci forms being transformed into S-alkylated nitroso compounds by the addition of an appropriate thiol. Subsequent tautomerization could give S-alkylated thiohydroximic acids (Ettlinger and Kjær, 1968). In agreement with this proposal the incorporation of 1-nitro-2-[1,2-^{14}C$_2$]phenylethane into benzylglucosinolate was demonstrated in *T. majus,* although the incorporation percentages obtained were low. The nitro compound is a natural constituent in this species, and its derivation from phenylacetaldoxime has been demonstrated by trapping experiments (Matuso *et al.,* 1972). The pathway can, however, be considered only tentative until formation of the nitro compound from the aldoxime and its subsequent conversion into the thiohydroximic acid have been confirmed in cell-free systems and until the efficiency of other nitro compounds as precursors of glucosinolates is established.

Desulfoglucosinolates have also been established beyond doubt as the last intermediates in the biosynthetic pathway. As mentioned above, an enzyme catalyzing formation of the desulfobenzylglucosinolate by glucosyl transfer from UDPG to phenylacetothiohydroximic acid has been found in glucosinolate-producing plants. The enzyme (UDP-glucose : thiohydroximate glucosyltransferase) possesses a high degree of specificity for the thiohydroximate functional group and for the nucleotide donor. Little specificity is exhibited for the side chain of the thiohydroximates. Desulfo-*p*-hydroxybenzylglucosinolate is a natural constituent of *S. alba* and is labeled after feeding the corresponding ^{14}C-labeled aldoxime. A sulfotransferase has been found in various glucosinolate-producing plants that catalyzes the production of benzylglucosinolate and adenosine-3',5'-diphosphate from desulfobenzylglucosinolate and 3'-phosphoadenosine-5'-phosphosulfate (PAPS). The enzyme shows very little specificity for the side chain of the desulfoglucosinolate. Enzymes from different plants are generally capable of catalyzing the sulfation of desulfoglucosinolates not normally present in the plant from which the enzyme is isolated (Underhill *et al.*, 1973).

The intermediates in Scheme 1, aldoximes, thiohydroximic acids, and desulfoglucosinolates, all must show geometrical isomerism around the C=N bond. No definite information seems to be available on their configuration, but generally the same configuration as that found in the glucosinolates themselves is assumed.

C. Chain-Lengthening of Amino Acids

A large number of glucosinolates have carbon skeletons that do not correspond to those of protein amino acids. These glucosinolates belong to two groups, those with a straight aliphatic chain containing either a double bond or a methylthio, methylsulfinyl, or methylsulfonyl group, and those with a phenylethyl side chain (cf. Section II,A and Table I). It has been shown that these glucosinolates are derived from nonprotein amino acids which again are derived from protein amino acids by a chain-lengthening process similar to that taking place in the formation of leucine from valine, of glutamic acid from aspartic acid (via oxaloacetic acid and 2-ketoglutaric acid in the tricarboxylic acid cycle), and of 2-aminoadipic acid from glutamic acid. In some cases these chain-lengthening processes are followed by transformations in the side chains (Section IV,D). The chain-lengthening process involving the incorporation of acetic acid is outlined in Scheme 2.

The pathway has mainly been established by studies on the biosynthesis of allylglucosinolate from methionine via 2-amino-5-methylthiopentanoic acid (homomethionine) and of 2-phenylethylglucosinolate from phenylalanine via 2-amino-4-phenylbutanoic acid. In horseradish, specifically labeled methionine and acetate are incorporated into allylglucosinolate, so that the label in the thiohydroximate carbon is derived from the methyl carbon of acetate and the alkyl carbons are derived from C-2, -3, and -4 of methionine. Neither the carboxyl groups of acetate and methionine nor the methyl carbon of methionine is incorporated. Also, in *Brassica juncea* the thiohydroximate carbon is derived from the methyl carbon of acetate. 2-Amino-5-methylthiopentanoic acid is incorporated with higher efficiency

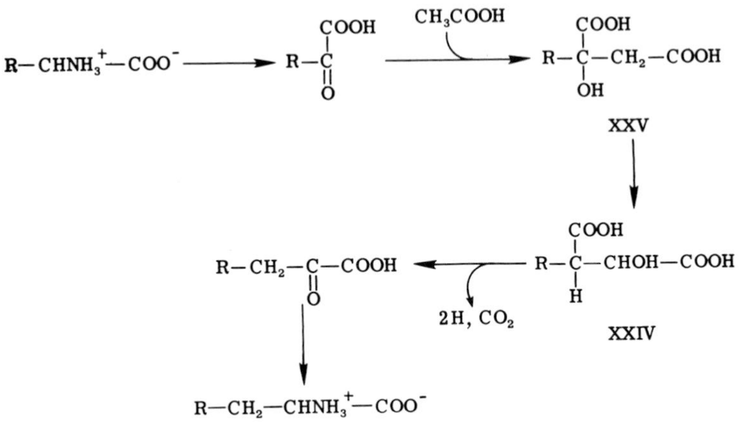

$$R-CHNH_3^+-COO^- \longrightarrow R-\underset{O}{\underset{\|}{C}}-COOH \xrightarrow{\ CH_3COOH\ } R-\underset{OH}{\underset{|}{C}}-CH_2-COOH$$

XXV

$$R-CH_2-\underset{O}{\underset{\|}{C}}-COOH \longleftarrow \underset{2H,\ CO_2}{\longleftarrow} R-\underset{H}{\underset{|}{C}}-CHOH-COOH$$

XXIV

$$R-CH_2-CHNH_3^+-COO^-$$

Scheme 2. Chain lengthening of amino acids in the biosynthesis of glucosinolates.

than methionine. $^{14}C,^{15}N$-labeled 2-amino-5-methylthiopentanoic acid is incorporated into allylglucosinolate without a change in the $^{14}C/^{15}N$ ratio. Furthermore, 2-amino-5-methylthiopentanoic acid occurs naturally in horseradish and is derived from methionine as shown by labeling experiments.

Correspondingly, it has been shown that labeled carbon atoms from acetate and phenylalanine are incorporated in the expected positions in 2-phenylethylglucosinolate; 40% incorporation of labeled 2-amino-4-phenylbutanoic acid into the glucosinolate has been obtained. 2-Amino-4-phenylbutanoic acid has been found in trace amounts in *Nasturtium officinale*. Labeled 3-benzylmalic acid (XXIV, $R = C_6H_5CH_2$) is incorporated into the glucosinolate and into 2-amino-4-phenylbutanoic acid. Labeled 2-benzylmalic acid (XXV, $R = C_6H_5CH_2$) is incorporated into 3-benzylmalic acid, 2-amino-4-phenylbutanoic acid, and the glucosinolate. Additional evidence for the chain-lengthening pathway has also been obtained by studies on the biosynthesis of 2-hydroxy-2-phenylethylglucosinolate in *Reseda luteola* (Underhill *et al.*, 1973, Kjær and Larsen, 1973, 1976, 1977).

The chain-lengthening process is presumed not only to produce amino acids and glucosinolates with one additional carbon atom but also to be responsible for production of the entire homologous series of glucosinolates (Section II,A). Thus in studies on the biosynthesis of 3-butenylglucosinolate, 2-hydroxy-3-butenylglucosinolate, and 4-pentenylglucosinolate in *Brassica campestris*, label was incorporated from methionine, acetate, and 2-amino-5-methylthiopentanoic acid in the expected positions. The results were interpreted to support formation of the two first glucosinolates from 2-amino-6-methylthiohexanoic acid and of the pentenylglucosinolate from 2-amino-7-methylthioheptanoic acid. 2-Amino-6-methylthiohexanoic acid has been shown to be the precursor of 2-hydroxy-3-butenylglucosinolate. Methionine and 2-amino-5-methylthiopentanoic acid are precursors of 3-methylthiopropylglucosinolate and 3-methylsulfinylpropylglucosinolate.

D. Side-Chain Modification in Glucosinolates

The side-chain modifications taking place in glucosinolates are transformation of methylthio groups into methylsulfinyl groups, into methylsulfonyl groups, and—by elimination—into a terminal double bond, and hydroxylations. In all cases investigated these changes took place after completion of the parent glucosinolate or at least at a very late stage in the biosynthetic sequence. Thus 2-amino-4-pentenoic acid is only poorly incorporated into allylglucosinolate in horseradish and into 3-butenylglucosinolate, 2-hydroxy-3-butenylglucosinolate, and 4-pentenylglucosinolate in *B. campestris*. Again, 2-amino-4-phenylbutanoic acid is more efficiently incorporated than 2-amino-4-hydroxy-4-phenylbutanoic acid into 2-hydroxy-2-phenylethylglucosinolate in *Reseda luteola*. Also, 2-phenylethylglucosinolate is efficiently transformed into the hydroxylated glucosinolate. 2-Hydroxy-3-butenylglucosinolate is efficiently produced from 5-methylthiopentanaldoxime, suggesting that hydroxylation and double-bond formation occur later in the biosynthetic sequence than the oxime.

The transformation of both methylthiopropylglucosinolate and methylsulfinylpropylglucosinolate into allylglucosinolate was demonstrated in horseradish. It could, however, not be established whether elimination took place at the methylthio or methylsulfinyl level. Reversible oxidation and reduction between 3-methylthiopropylglucosinolate and 3-methylsulfinylpropylglucosinolate has been demonstrated in *Cheiranthus kewensis*.

Studies on indoleglucosinolates in *I. tinctoria* indicate that both 1-methoxy- and 1-sulfoindol-3-ylmethylglucosinolate are derived from indol-3-ylmethylglucosinolate. On the other hand, studies on the biosynthesis of *p*-hydroxybenzylglucosinolate in *S. alba* indicate that tyrosine is the precursor and that the phenolic group is therefore present at the amino acid stage of the biosynthesis.

E. Specificity and Control

Very little is known about the control of glucosinolate biosynthesis. As mentioned above, a number of the later enzymes in the biosynthetic sequence have low specificities. It is therefore likely that some sort of control is exerted in the first step, the activation (probably N-hydroxylation) of the amino acid. Glucosinolates occur in well-defined patterns. However, different organs of the same plant may contain strongly varying ratios of diffferent glucosinolates. This often is the case in plants containing glucosinolates with the same carbon skeleton but either with methylthio, methylsulfinyl, or methylsulfonyl groups or with a terminal double bond. The regulation of these ratios must obviously be due to some sort of metabolic control at the glucosinolate level.

Studies on the biosynthesis of indol-3-ylmethylglucosinolate and p-hydroxybenzylglucosinolate in *S. alba* indicate that two different enzyme systems are involved in formation of the two glucosinolates (Bergmann, 1970).

Breeding to produce plants with a low glucosinolate content has been performed, mainly in *Brassica napus*. Varieties with a lowered glucosinolate content have been obtained, but it has not been possible to obtain varieties with no glucosinolates (Lööf and Appelqvist, 1972). The glucosinolate content of *B. napus* 'Bronowski' is low compared with that of the cultivar 'Regina II'. In a search to find the metabolic block that causes the low glucosinolate content, a number of labeled precursors including methylthiopentanaldoxime and desulfo-3-butenylglucosinolate were fed. It was concluded that 2-hydroxy-3-butenylglucosinolate was derived from 3-butenylglucosinolate in 'Regina II' and that metabolic blocks occurred both in a step prior to the oxime and in the hydroxylation in 'Bronowski' (Josefsson, 1971, 1973).

V. DEGRADATION

A. Myrosinases (Thioglucoside Glucohydrolases)

All plants containing glucosinolates also contain enzymes decomposing these compounds. The normal products of the enzymatic degradations are glucose and isothiocyanates (see, however, Sections V,D and E). The enzymes normally are called myrosinases. Their role is to catalyze hydrolysis of the thioglucoside bond (see below), and a more systematic designation would be thioglucoside glucohydrolases (E.C. 3.2.3.1).

In the earlier literature it had been suggested that myrosinases were multienzymes also catalyzing hydrolysis of the sulfate ester group. However, the possibility of sulfatase activity has been excluded by a number of experiments (Björkman, 1976). Thus it has been shown that allylglucosinolate can be hydrolyzed with a sulfatase from animal sources to give the desulfoallylglucosinolate (desulfoglucosinolates are also intermediates in the biosynthesis of glucosinolates; see Section IV,B). This compound could be cleaved by myrosinase, but the reaction rate was slow and no isothiocyanate was produced (Nagashima and Uchiyama, 1959b,c). Also, desulfobenzylglucosinolate is cleaved very slowly by myrosinase (Reese *et al.,* 1958; Ettlinger and Lundeen, 1957). When glucosinolates are treated with $AgNO_3$, silver derivatives of the aglucones are produced. When treated with $Na_2S_2O_3$ or NaCl, these aglucones decompose spontaneously to isothiocyanates, whereas they produce nitriles on treatment with H_2S or HCl:

$$\begin{array}{c} R\!\!-\!\!\underset{\underset{N\diagdown OSO_3^-}{\|}}{C}\!\!-\!\!S\text{- Glucose} \quad \xrightarrow[\underset{\alpha\text{-D-gluco-}}{AgNO_3}]{} \quad R\!\!-\!\!\underset{\underset{N\diagdown OSO_3^-}{\|}}{C}\!\!-\!\!SAg \end{array} \nearrow \begin{array}{l} R\!\!-\!\!NCS \\ \scriptstyle Na_2S_2O_3 \\ \scriptstyle or\ NaCl \end{array} \searrow \begin{array}{l} H_2S\ or \\ HCl \\ R\!\!-\!\!CN\ +\ S \end{array} \tag{6}$$

It has also been demonstrated with allylglucosinolate that the aglucone generated chemically or by myrosinase action decomposes by two competing nonenzymatic pathways: a proton-independent isothiocyanate-forming one and a proton-dependent nitrile-forming one (Miller, 1965; Benn, 1977) (see Section V,B).

Glucose is produced in the enzymatic reaction as β-D-glucopyranose (Ettlinger and Thompson, 1962).

Myrosinases have been extensively purified from various plants, and the presence of isoenzymes has been established in a number of cases. A characteristic feature is that some myrosinases require ascorbic acid for activation, whereas others are insensitive to ascorbic acid (Nagashima and Uchiyama, 1959a; Ettlinger et al., 1961). Both types are often present in the same plant. The action of ascorbic acid is not due to the oxidation–reduction properties of this compound. The effect is probably allosteric, and extensive kinetic studies and binding studies have been carried out. Studies on the most purified myrosinase preparations indicate MWs of about 125,000–150,000, with two or four subunits. The enzymes are glycoproteins containing SH groups which are essential for the activity as shown by complete inactivation with the SH reagents p-mercuribenzoate and 5,5'-dithio-bis-2-nitrobenzoic acid (Ellman's reagent) (Björkman, 1976).

Myrosinases from different plants have apparently the same specificity. Various glucosinolates are degraded with different velocity, but all naturally occurring glucosinolates are substrates. The enzyme also hydrolyzes other thioglucosides. Simple β-glucosides are also substrates, although they are degraded very slowly compared with glucosinolates (Reese et al., 1958; Tsuruo and Hata, 1968).

Myrosinase is separated from the glucosinolates in intact plant tissues. The enzyme is generally assumed to be present in special cells, so-called idioblasts, whereas glucosinolates are distributed widely in the parenchymal tissue (Kjær, 1960), but this interpretation of the anatomical finding has recently been questioned (Jørgensen *et al.*, 1977; Iversen *et al.*, 1979). The distribution of myrosinase in different parts of plants and its cellular localization have been the subject of extensive investigations (Björkman, 1976; Rodman, 1978; Pihakaski and Pihakaski, 1978; Pihakaski and Iversen, 1976; Jørgensen *et al.*, 1977). Recently evidence has been presented for the location of myrosinase in horseradish roots in connection with the cell walls and membrane-enclosed intracellular bodies (Matile, 1980).

Thioglucosidases capable of hydrolyzing glucosinolates have been reported from bacteria, fungi, and mammalian tissues (Reese *et al.*, 1958; Björkman, 1976) and have also been found in the cabbage aphid (*Brevicoryne brassicae* (MacGibbon and Allison, 1971) (cf. Section VI,A).

B. Formation of Isothiocyanates and Nitriles

The enzymatically produced loss of glucose to give thiohydroxamate *O*-sulfonates (XXVI) normally is followed by a Lossen-type rearrangement to give isothiocyanates.

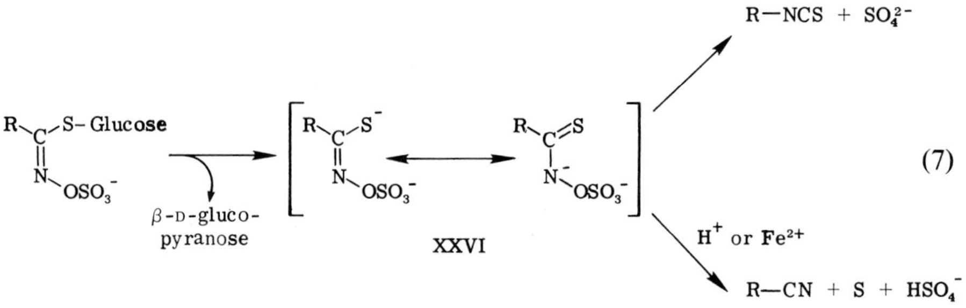

$$R{-}NCS + SO_4^{2-}$$

$$\text{(7)}$$

$$H^+ \text{ or } Fe^{2+}$$

$$R{-}CN + S + HSO_4^-$$

The stereochemistry of the rearrangement has never been established, but complete retention at C-1 of the side chain is likely. This conclusion is based on the fact that the configuration of 2-butylisothiocyanate produced by the degradation of 2-butylglucosinolate is the same as the configuration at C-3 in isoleucine (Kjær and Hansen, 1957) and the assumption that the glucosinolate is derived from isoleucine without a change in configuration at this chiral center.

Not only isothiocyanates but also the corresponding nitriles are formed in greater or lesser amounts with the concomitant liberation of elemental sulfur (Daxenbichler *et al.*, 1977). Nitrile formation is favored by low pH values. In studies with the aglucone of allylglucosinolate, generated chemically or by myrosinase action, it has been shown (Miller, 1965) that 97% of the product is the nitrile at pH 2, whereas at pH 5 97% of the product is the isothiocyanate. At pH 3.5 the two products are formed in equal amounts. Nitrile formation can also be promoted by ferrous ions and other cations. The half-life of the thiohydroximate O-sulfonate derived from allylglucosinolate is about 30 s in aqueous solution at 24°C.

C. Formation of Oxazolidine-2-thiones and Tetrahydro-1,3-oxazine-2-thiones

As described in Section B, II, isothiocyanates with a hydroxy group in the 2- or 3-position cyclize to give, respectively, oxazolidine-2-thiones and tetrahydro-1,3-oxazine-2-thiones. These transformations are of special interest in connection with the toxic properties of glucosinolates and their derivatives (cf. Section VI,B).

D. Formation of Organic Thiocyanates, Thiocyanate Ions, and Ascorbigen

In a number of cases the crushing of glucosinolate-containing plants gives rise to organic thiocyanates and inorganic thiocyanate ions. This was first demonstrated for allylglucosinolate in *Thlaspi arvense* (Gmelin and Virtanen, 1959). It has also been demonstrated that benzylglucosinolate in *Lepidium ruderale* and *L. sativum* can be transformed into benzylthiocyanate. Thiocyanate formation has also been observed from 4-methylthiobutyl-glucosinolate in *Eruca sativa* (Benn, 1977).

It has been proposed that thiocyanates are produced directly from glucosinolates by a special enzyme; it has also been suggested that thiocyanates are produced from isothiocyanates by the action of an "isomerase." Both of these possibilities have, however, been excluded (Miller, 1965), and evidence has been put forward that thiocyanates are produced enzymatically from the aglucones of the glucosinolates (thiohydroxamate O-sulfonates):

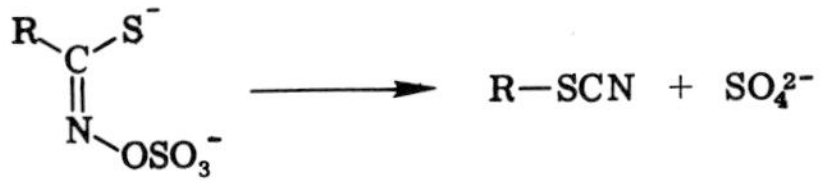

$$\underset{\substack{\text{N}\\\text{OSO}_3^-}}{\overset{\substack{\text{R}\qquad\text{S}^-}}{\text{C}}} \longrightarrow R-SCN + SO_4^{2-} \qquad (8)$$

The enzyme presumably responsible for this rearrangement has, however, never been isolated or properly characterized. It must be rather efficient in order to be able to compete with the chemical degradation of the thiohydroximate O-sulfonates described in Section V,B. [$1'$-^{14}C]Allylglucosinolate is transformed into [3-^{14}C]allylthiocyanate under the influence of *T. arvense* seed flour extracts. The mechanism of the reaction has been thoroughly discussed, but too little experimental evidence is available to permit definite conclusions (Benn, 1977; Lüthy and Benn, 1977).

Because of the great stability of the allyl and benzyl carbonium ions the organic thiocyanates can give rise to thiocyanate ions (and the corresponding alcohols), especially at higher pH values. Indol-3-ylmethylglucosinolate also gives rise to thiocyanate ions. In this case, however, it has not been proposed that SCN^- is produced via the aglucone and an organic thiocyanate. Instead the experimental evidence has been taken as support for formation from the enzymatically produced isothiocyanate by dissociation to give SCN^- and the relatively stable indolylmethylcarbonium ion subsequently transformed into 3-hydroxymethylindole. Indol-3-ylmethylglucosinolate also gives rise to SCN^- ions in a nonenzymatic reaction after heating in aqueous solution (Gmelin and Virtanen, 1961; Elliott and Stove, 1971a; see also McGregor, 1978). Also, p-hydroxybenzylglucosinolate gives rise to thiocyanate ions by the formation of p-hydroxybenzylisothiocyanate followed by dissociation (Kawakishi et al., 1967).

The indol-3-ylmethylcarbonium ion can react nonenzymatically with ascorbic acid to give (by C-alkylation) ascorbigen (in fact a mixture of two diastereoisomers) (Kiss and Neukom, 1966). Ascorbigen has long been known as a bound form of ascorbic acid isolated from cabbage. The compound is formed only on crushing and heating and may therefore be considered an artifact (Gmelin and Virtanen, 1961). The reaction is of importance in reducing the amount of ascorbic acid available in cabbage. Ascorbigen has an antiscorbutic effect which, however, is considerably smaller that that of ascorbic acid itself (Kiesvaara and Virtanen, 1962).

E. Formation of Cyanoepithioalkanes

In special cases cyanoepithioalkanes (XXVII) can also be produced from glucosinolates having double bonds in the side chain by enzymatic degradations in crushed plants:

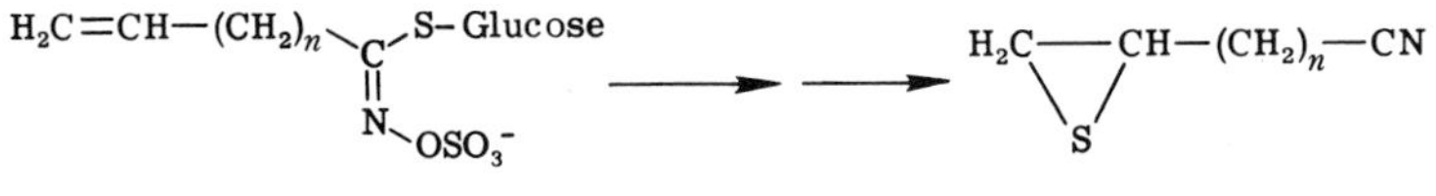

$$H_2C=CH-(CH_2)_n-\underset{\underset{OSO_3^-}{\overset{|}{N}}}{\overset{S-Glucose}{\underset{|}{C}}} \longrightarrow \longrightarrow H_2C\underset{S}{\overset{CH-(CH_2)_n-CN}{\diagdown\diagup}} \tag{9}$$

XXVII

$n = 1-3$

Thus 3-butenylglucosinolate can give the episulfide 1-cyano-3,4-epithiobutane (XXVII, $n = 2$) in *B. campestris,* whereas the corresponding 2-hydroxy-substituted glucosinolate can undergo a corresponding reaction in *Crambe abyssinica* and *B. napus.* The reaction is not restricted to glucosinolates with a C_4 side chain but can also take place with allylglucosinolate ($n = 1$) and 4-pentenylglucosinolate ($n = 3$). The reaction probably takes place by the action of myrosinase to give the aglucone of the glucosinolate and subsequent formation of the episulfides in a new enzyme-catalyzed reaction. The enzyme system is labile and has not been further characterized. Possible reaction mechanisms have again been discussed, but no definite conclusions have been made (Benn, 1977; Daxenbichler *et al.*, 1977).

F. Transformation in Intact Plants

Myrosinase action probably takes place only after some kind of mechanical injury to a plant. Isothiocyanates and the other decomposition products described above are not considered to be normally present in intact plants. There is some evidence, however, of the endogenous production of isothiocyanates in intact plants at very low levels (Tang, 1973; Mahadevan, 1973).

Indole-containing glucosinolates may be transformed into auxins, including 3-indolylacetonitrile. Conflicting evidence is found in the literature, and it is not clear whether these transformations are of importance in intact plants or occur only on disintegration of plant material (Elliott and Stowe, 1971a,b; Rodman, 1978).

A few investigations indicate a fast turnover of glucosinolates, but it is not known whether the reaction catalyzed by myrosinase is the only means of degradation of glucosinolates in plants or whether other catabolic routes exist. It may be of interest that in a number of cases glucosinolates and the corresponding amines with one carbon less have been found simultaneously in plants. The amines may arise by degradation of isothiocyanates, although precautions have been taken to prevent glucosinolate decomposition during the isolation of amines. Furthermore, amines are not produced easily from all the isothiocyanates in question. In any case the occurrence of the γ-glutamyl derivative of isopropylamine in *Lunaria annua* together with isopropylglucosinolate indicates that at least in this case amine production takes place in the intact plant. Isopropylamine is very seldom encountered in nature, and a biogenetic relationship with the glucosinolate is probable (Larsen, 1965).

VI. GLUCOSINOLATES AND ANIMALS

A. Glucosinolate–Insect Relationships

Glucosinolates and isothiocyanates provide classical examples of the importance of secondary plant products in plant–insect interrelationships. Thus it was shown in 1910 that potassium allylglucosinolate could induce larvae of *Pieris brassica* and *P. rapae* to eat plant material that they normally rejected (Verschaeffelt, 1910).

Glucosinolates are nonvolatile and can therefore be recognized by insects only on contact. Isothiocyanates, on the other hand, are generally volatile and can induce olfactory responses. Glucosinolates are toxic to insects that normally do not feed on *Crucifer* or other glucosinolate-containing plants (Erickson and Feeny, 1974; Blau *et al.*, 1978), and in a number of instances glucosinolates have been demonstrated to inhibit feeding in such insects (Klingauf *et al.*, 1972; Nault and Styer, 1972). On the other hand, many examples are available of the stimulation of feeding behavior of feeders on species of *Cruciferae* by glucosinolates (see, for example, Thorsteinson, 1953; Nayar and Thorsteinson, 1963; David and Gardiner, 1966; Klingauf *et al.*, 1972; Ma, 1972; Nault and Styer, 1972; Hicks, 1974; Nielsen, 1978). Isothiocyanates are also known to be attractants for a number of adult insects associated with species of *Cruciferae,* and some insect larvae are attracted by isothiocyanates from short distances (see, for example, Finch and Skinner, 1974; Tanton, 1977).

Glucosinolates stimulate oviposition in insects associated with species of *Cruciferae*. Also, isothiocyanates have been shown to induce oviposition in a number of cases (see, for example, Ma and Schoonhoven, 1973; Nair *et al.*, 1976).

Isothiocyanates presumably are recognized by insects with sensory organs situated on the antennae or the palpae. Glucosinolates in some instances have been shown to be recognized by tarsal receptors (Schoonhoven, 1968; Ma, 1972; Ma and Schoonhoven, 1973). There seems to be no close correlation between the amounts of particular glucosinolates and their acceptability to insects (Nielsen, 1978; Nielsen *et al.*, 1979).

No detailed information is available on the degradation of glucosinolates by insects (Benn, 1977) but, as mentioned in Section V,A, myrosinase has been found in insects feeding on species of *Cruciferae* (MacGibbon and Allison, 1971).

B. Toxicology of Glucosinolates

Glucosinolates are precursors of compounds with goitrogenic action in mammals, including humans. The active antithyroid compounds are of three types: the isothiocyanates themselves, oxazolidine-2-thiones, and the thiocyanate ion. Cabbage, containing allylglucosinolate, (2*R*)-hydroxy-3-butenylglucosinolate, and indol-3-ylmethylglucosinolate, can thus be the source of goitrogenous allylisothiocyanate, (*S*)-5-vinyl-oxazolidine-2-thione (goitrin), and SCN$^-$. Also goitrogenic substances are produced in substantial amounts by rape seed. A large number of investigations have been carried out on the occurrence of these agents in food and their liberation in animals, transfer from cattle feed to milk, and physiological significance (Greer, 1962; Podoba and Langer, 1964; Ettlinger and Kjær, 1968; Josefsson, 1972; Van-Etten and Tookey, 1979; Tookey *et al.*, 1980). The toxicity of glucosinolates is," however, due not only to the goitrogens but also to the very reactive isothiocyanates and to organic nitriles (Josefsson, 1972). Isothiocyanates also have antibacterial and antifungal properties (Ettlinger and Kjær, 1968).

VII. CONCLUSION

Although glucosinolates have been extensively investigated for many years, there are still important gaps in most areas of our knowledge of these compounds. This comment applies both to the basic chemistry and to the biosynthetic pathway where some individual steps are unknown and where very little is known about control systems. The comment applies as well to the enzymatic and nonenzymatic degradation of glucosinolates and to the turnover and the occurrence of these compounds in plants. Hopefully, these gaps will be filled in by renewed efforts in the near future, both because of the great practical and economic importance of glucosinolates and because they constitute a well-defined group of natural products with unique and intriguing chemical, biochemical, and biological properties.

REFERENCES

Appelqvist, L. (1972). *In* "Rapeseed" (L. Appelqvist and R. Ohlson, eds.), pp. 123–173. Elsevier, Amsterdam.

Benn, M. (1977). *Pure Appl. Chem.* **49,** 197–210.

Benn, M. H. (1962). *Chem. Ind. (London)* p. 1907.

Bergmann, F. (1970). *Z. Pflanzenphysiol.* **62,** 362–375.

Björkman, R. (1972). *Acta Chem. Scand.* **26,** 1111–1116.

Björkman, R. (1976). *In* "The Biology and Chemistry of the Cruciferae" (J. G. Vaughan, A. J. MacLeod, and B. M. G. Jones, eds), pp. 191–205. Academic Press, New York.

Blau, P. A., Feeny, P., Contardo, L., and Robson, D. S. (1978). *Science* **200,** 1296–1298.

Cole, R. A. (1976). *Phytochemistry* **15,** 759–762.

Dahlgren, R. (1975). *Bot. Not.* **182,** 119–147.

Dateo, G. P. (1961). Ph.D. Thesis, Rice University, Houston, Texas.

David, W. A. L., and Gardiner, B. O. C. (1966). *Entomol. Exp. Appl.* **9,** 247–255.

Daxenbichler, M. E., and VanEtten, C. H. (1977). *J. Assoc. Off. Anal. Chem.* **60,** 950–953.

Daxenbichler, M. E., VanEtten, C. H., and Spencer, G. F. (1977). *J. Agric. Food Chem.* **25,** 121–124.

Elliot, M. C., and Stowe, B. B. (1971a). *Plant Physiol.* **47,** 366–372.

Elliott, M. C., and Stowe, B. B. (1971b). *Plant Physiol.* **48,** 498–503.

Erickson, J. M., and Feeny, P. (1974). *Ecology* **55,** 103–111.

Ettlinger, M. G., and Dateo, G. P. (1961). "Studies of Mustard Oil Glucosides," Final Report Contract DA19-129-QM-1059. U.S. Army Natick Laboratories, Natick, Massachusetts.

Ettlinger, M. G., and Kjær, A. (1968). *Recent Adv. Phytochem.* **1,** 58–144.

Ettlinger, M. G., and Lundeen, A. J. (1956). *J. Am. Chem. Soc.* **78,** 4172–4173.

Ettlinger, M. G., and Lundeen, A. J. (1957). *J. Am. Chem. Soc.* **79,** 1764–1765.

Ettlinger, M. G., and Thompson, C. P. (1962). "Studies of Mustard Oil Glucosides (II)," Final

Report Contract DA-19-129-QM1689 (AD-290 747). U.S. Dept. of Commerce, Washington, D.C.

Ettlinger, M. G., Dateo, G. P., Harrison, B. W., Mabry, T. J., and Thompson, C. P. (1961). *Proc. Natl. Acad. Sci. U.S.A.* **47**, 1875–1880.

Finch, S., and Skinner, G. (1974). *Ann. Appl. Biol.* **77**, 213–226.

Finlayson, A. J. (1976). *In* "The Biology and Chemistry of the Cruciferae" (J. G. Vaughan, A. J. MacLeod, and B. M. G. Jones, eds.), pp. 279–306. Academic Press, New York.

Friis, P., Larsen, P. O., and Olsen, C. E. (1977). *J. Chem. Soc., Perkin Trans. 1* pp. 661–665.

Gmelin, R., and Virtanen, A. I. (1959). *Acta Chem. Scand.* **13**, 1474–1475.

Gmelin, R., and Virtanen, A. I. (1960). *Acta Chem. Scand.* **14**, 507–510.

Gmelin, R., and Virtanen, A. I. (1961). *Ann. Acad. Sci. Fenn., Ser. A2* **107**, 1–25.

Gmelin, R., and Virtanen, A. I. (1962). *Acta Chem. Scand.* **16**, 1378–1384.

Greer, M. A. (1962). *Recent Prog. Horm. Res.* **18**, 187–212.

Grob, K., and Matile, P. (1979). *Plant Sci. Lett.* **14**, 327–335.

Gronowitz, S., Svensson, L., and Ohlson, R. (1978). *J. Agric. Food Chem.* **26**, 887–890.

Hicks, K. L. (1974). *Ann. Entomol. Soc. Am.* **67**, 261–264.

Iversen, T. H., Baggerud, C., and Beisvaag, T. (1979). *Z. Pflanzenphysiol.* **94**, 143–154.

Jørgensen, L. B., Behnke, H., and Mabry, T. J. (1977). *Planta* **137**, 215–224.

Josefsson, E. (1971). *Physiol. Plant.* **24**, 161–175.

Josefsson, E. (1972). *In* "Rapeseed" (L. Appelqvist and R. Ohlson, eds.), pp. 354–377. Elsevier, Amsterdam.

Josefsson, E. (1973). *Physiol. Plant.* **29**, 28–32.

Kawakishi, S., Namiki, M., Watanabe, H., and Muramatsu, K. (1967). *Agric. Biol. Chem.* **31**, 823–830.

Kiesvaara, M., and Virtanen, A. I. (1962). *Acta Chem. Scand.* **16**, 510–512.

Kindl, H., and Underhill, E. W. (1968). *Phytochemistry* **7**, 745–756.

Kiss, G., and Neukom, H. (1966). *Helv. Chim. Acta* **49**, 989–992.

Kjær, A. (1960). *Fortschr. Chem. Org. Naturst.* **18**, 122–176.

Kjær, A. (1974). *In* "Chemistry in Botanical Classification" (G. Bendz and J. Santesson, eds.), pp. 229–234. Academic Press, New York.

Kjær, A. (1976). *In* "The Biology and Chemistry of the Cruciferae" (J. G. Vaughan, A. J. MacLeod, and B. M. G. Jones, eds.), pp. 207–219. Academic Press, New York.

Kjær, A., and Hansen, S. E. (1957). *Acta Chem. Scand.* **11**, 898–900.

Kjær, A., and Larsen, P. O. (1973). *Biosynthesis* **2**, 71–105.

Kjær, A., and Larsen, P. O. (1976). *Biosynthesis* **4**, 179–203.

Kjær, A., and Larsen, P. O. (1977). *Biosynthesis* **5**, 120–135.

Kjær, A., and Larsen, P. O. (1980). *Biosynthesis* **6**, p. 155–180.

Klingauf, F., Sengonca, C., and Bennewitz, H. (1972). *Oecologia* **9**, 53–57.

Kutáček, M., Procházka, Z., and Vereš, K. (1962). *Nature (London)* **194**, 393–394.

Larsen, P. O. (1965). *Acta Chem. Scand.* **19**, 1071–1078.

Lööf, B., and Appelqvist, L. (1972). *In* "Rapeseed" (L. Appelqvist and R. Ohlson, eds.), pp. 101–122. Elsevier, Amsterdam.

Lüthy, J., and Benn, M. H. (1977). *Can. J. Biochem.* **55**, 1028–1031.

Ma, W. C. (1972). *Meded. Landbouwhogesch. Wageningen* **72-11**, 1–162.

Ma, W. C., and Schoonhoven, L. M. (1973). *Entomol. Exp. Appl.* **16**, 343–357.

MacGibbon, D. B., and Allison, R. M. (1971). *N. Z. J. Sci.* **14**, 134–140.

McGregor, D. I. (1978). *Can. J. Plant Sci.* **58**, 795–800.

Mahadevan, S. (1973). *Annu. Rev. Plant Physiol.* **24**, 89–114.

Mahadevan, S., and Stowe, B. B. (1972). *Plant Physiol.* **50**, 43–50.

Marsh, R. E., and Waser, J. (1970). *Acta Crystallogr., Sect. B* **26**, 1030–1037.

Matile, P. (1980). *Biochem. Physiol. Pflanz.* **175**, 722–731.

Matsuo, M., Kirkland, D. F., and Underhill, E. W. (1972). *Phytochemistry* 11, 697–701.

Miller, H. E. (1965). M.A. Thesis, Rice University, Houston, Texas.

Nagashima, Z., and Uchiyama, M. (1959a). *Nippon Nogei Kagaku Kaishi* 33, 980–984.

Nagashima, Z., and Uchiyama, M. (1959b). *Nippon Nogei Kagaku Kaishi* 33, 1068–1071.

Nagashima, Z., and Uchiyama, M. (1959c). *Nippon Nogei Kagaku Kaishi* 33, 1144–1149.

Nair, K. S. S., McEwen, F. L., and Snieckus, V. (1976). *Can. Entomol.* 108, 1031–1036.

Nault, L. R., and Styer, W. E. (1972). *Entomol. Exp. Appl.* 15, 423–437.

Nayar, J. K., and Thorsteinson, A. J. (1963). *Can. J. Zool.* 41, 923–929.

Nielsen, J. K. (1978). *Entomol. Exp. Appl.* 24, 41–54.

Nielsen, J. K., Dalgaard, L., Larsen, L. M., and Sørensen, H. (1979). *Entomol. Exp. Appl.* 25, 227–239.

Ohtsuru, M. (1975). *Kyoto Daigaku Shokuryo Kagaku Kenkyusho Hokoku* 38, 13–32.

Olsson, K., Theander, O., and Åman, P. (1976). *Swed. J. Agric. Res.* 6, 225–229.

Pihakaski, K., and Iversen, T. H. (1976). *J. Exp. Bot.* 27, 242–258.

Pihakaski, K., and Pihakaski, S. (1978). *J. Exp. Bot.* 29, 335–345.

Podoba, J., and Langer, P., eds. (1964). "Naturally Occurring Goitrogens and Thyroid Function." Slovak Acad. Sci., Bratislava.

Reese, E. T., Clapp, R. C., and Mandels, M. (1958). *Arch. Biochem. Biophys.* 75, 228–242.

Rodman, J. E. (1978). *Phytochem. Bull.* 11, No. 1-2 (Phytochem. Sect., Bot. Soc. Am.), 6–31.

Schoonhoven, L. M. (1968). *Annu. Rev. Entomol.* 13, 115–136.

Tang, C. (1973). *Phytochemistry* 12, 769–773.

Tanton, M. T. (1977). *Entomol. Exp. Appl.* 22, 113–122.

Tapper, B. A., and Butler, G. W. (1967). *Arch. Biochem. Biophys.* 120, 719–721.

Thies, W. (1976). *Fette, Seifen, Anstrichm.* 78, 231–234.

Thorsteinson, A. J. (1953). *Can. J. Zool.* 31, 52–72.

Tookey, H. L., VanEtten, C. H., and Daxenbichler, M. E. (1980). *In* "Toxic Constituents of Plant Foodstuffs" (I. E. Liener, ed.), pp. 103–142. Academic Press, New York.

Tsuruo, I., and Hata, T. (1968). *Agric. Biol. Chem.* 32, 1425–1431.

Underhill, E. W. (1967). *Eur. J. Biochem.* 2, 61–63.

Underhill, E. W. (1980). *Encycl. Plant Physiol., New Ser.* 8, 493–511.

Underhill, E. W., and Chisholm, M. D. (1964). *Biochem. Biophys. Res. Commun.* 14, 425–430.

Underhill, E. W., and Wetter, L. R. (1969). *Plant Physiol.* 44, 584–590.

Underhill, E. W., Chisholm, M. D., and Wetter, L. R. (1962). *Can. J. Biochem. Physiol.* 40, 1505–1514.

Underhill, E. W., Wetter, L. R., and Chisholm, M. D. (1973). *Biochem. Soc. Symp.* 38, 303–326.

VanEtten, C. H., and Daxenbichler, M. E. (1977). *J. Assoc. Off. Anal. Chem.* 60, 946–949.

VanEtten, C. H., and Tookey, H. L. (1979). *In* "Herbivores: Their Interaction with Secondary Plant Metabolites" (G. A. Rosenthal and D. H. Janzen, eds.), pp. 471–500. Academic Press, New York.

VanEtten, C. H., McGrew, C. E., and Daxenbichler, M. E. (1974). *J. Agric. Food Chem.* 22, 483–487.

VanEtten, C. H., Daxenbichler, M. E., Williams, P. H., and Kwolek, W. F. (1976). *J. Agric. Food. Chem.* 24, 452–455.

Verschaeffelt, E. (1910). *Proc. K. Ned. Acad. Wet.* 13, 536–542.

Wetter, L. R., and Dyck, J. (1973). *Can. J. Anim. Sci.* 53, 625–626.

Vegetable Tannins *18*

EDWIN HASLAM

I. INTRODUCTION

The enormous effort that has been devoted to studies on the chemistry and more recently the biogenesis of secondary metabolites still leaves scientists largely ignorant of their function or biochemical significance (Thomas, 1978). Whereas ignorance prevails at the experimental level, not surprisingly, theories and speculation abound at the philosophical level, adding to the fascination that secondary metabolites hold as objects of scientific study. Opposing viewpoints have developed. One theory argues that they are waste products or "accidents of metabolism" (Muller, 1969; and, for example, the case of alkaloids, Mothes, 1969), notwithstanding the fact that many are toxic to the plant or microorganism unless dissipated into the environment (e.g., volatile monoterpenes), or that they are harmlessly sequestered in the plant itself (e.g., phenolic glycosides). A contrary opinion maintains that these substances possess (or possessed) pertinent biological functions (e.g., Fraenkel, 1959; Janzen, 1969), and the example of vegetable tannins is frequently cited in support of this view. The importance of vegetable tannins to plants lies, it is believed, in their effectiveness as repellants to predators, animal or micro-

The Biochemistry of Plants, Vol. 7

bial. According to Bate-Smith (1954) the relevant property is *astringency,* which for animals renders the tissue unpalatable by precipitation of salivary proteins and for parasitic organisms impedes invasion of the plant tissue by immobilizing extracellular enzymes. It is persuasively argued that this strong association with proteinaceous materials is a primary function that has been of considerable evolutionary significance in the plant kingdom. In complete contrast, another suggestion, which has some experimental support, indicates (Corcoran *et al.,* 1972; Corcoran and Green, 1975) that vegetable tannins may exert a growth regulatory function by inhibiting growth caused by gibberellins.

After the early encouragement emerging from Emil Fischer's outstanding contributions to the chemistry of vegetable tannins (Fischer, 1919) chemists were slow to recognize the complexity of the problems these substances posed, and over the next 40–50 yr it became one of the untidy corners of organic chemistry. Nonetheless the knowledge of structure, molecular shape, biosynthesis, and chemical reactivity appertaining to the principal vegetable tannins that has recently been obtained (Haslam, 1966, 1977; Mayer, 1973) now makes possible, for the first time, a systematic examination of their biochemical and biological properties. This must surely be the aim of future work.

It is not possible to give a concise definition of the word "tannin," and this has led to numerous misunderstandings in the literature. Tanning is a process whereby an animal skin is turned into leather. At present, most leather is produced by synthetic tanning agents (such as basic chromium salts), but in earlier times, and to a limited extent today, extracts of various plants have been used for this purpose. A few dicotyledonous plants provide the bulk of the vegetable extracts used for tanning, and some of the principal sources still used commercially are shown in Table I (Howes, 1953; White, 1957). In these plants the accumulation in particular tissues of compounds that tan proteins is particularly high and, as Table I shows, this abundance may occur in tissues as diverse as root, stem, fruit, pods, bark, wood, and leaves. The purpose of tanning is to bring about cross-linking of the collagen chains in the skin and thus to protect the protein fibers from microbial attack and give the skin greater stability to water, heat, and abrasion (Gustavson and Holm, 1952; Haslam, 1966). During this process, using vegetable extracts, the skin may adsorb up to half its weight in tannins. In this sense the implication of the word "tannin," and indeed its original use by Seguin (1796), clearly indicates a substance that produces leather from hide. In plant extracts these substances are polyphenols of varying molecular size and complexity. Invariably they constitute only a limited proportion of the total polyphenols in a plant tissue, and the failure to make this critical distinction has led one writer (White, 1957) to suggest that "much of the botanical data concerning the occurrence of tannins in plants is of doubtful validity since it is based on tests

TABLE I

Sources of Vegetable Tannins of Commercial Importance

Anacardiaceae
    Quebracho (*Schinopsis lorentzii, S. balansae*; heartwood), condensed
    Sumac (*Rhus typhina, R. coriaria*; leaves), hydrolyzable, gallotannin
    Chinese (*R. semialata*; galls), hydrolyzable, gallotannin
Rhizophoraceae
    Mangrove (bark, various species), condensed
Leguminosae
    Wattle or mimosa (*Acacia mearnsii*; wood, bark)—condensed
    Burma cutch (*A. catechu; wood*), condensed
    Divi-divi (*Caesalpinia coriaria*; fruit pods), hydrolyzable, ellagitannin
    Tara (*C. spinosa*; fruit pods), hydrolyzable, gallotannin
    Algarobilla (*C. brevifolia*; fruit pods), hydrolyzable, ellagitannin
Fagaceae
    Chestnut (*Castanea* sp.; wood, bark, leaves), hydrolyzable, ellagitannin
    Valonea (*Quercus aegilops*; acorn cups), hydrolyzable, ellagitannin
    Oak (*Quercus* sp.; wood, bark), hydrolyzable, ellagitannin
    Turkish (*Q. infectoria*; galls), hydrolyzable, gallotannin
Myrtaceae
    Myrtan (*Eucalyptus* sp.; wood, bark, leaves), condensed and hydrolyzable, ellagitannin
Rubiaceae
    Gambier (*Uncaria gambier*; leaves, twigs), condensed
Combretaceae
    Myrabolans[a] (*Terminalia chebula*; fruit), hydrolyzable, ellagitannin

[a] Various spellings of this word may be encountered such as myrobalans, myrabalans, and myrabolams.

which are insufficiently specific." Briefly, the general criteria for phenols (color tests, etc.) are quite inadequate for determining the presence of vegetable tannins.

Bate-Smith and Swain (1962) have adopted the earlier ideas of White (1957) in formulating a definition of vegetable tannins which, considering present knowledge, is the most useful one to follow. These authors define vegetable tannins as "water-soluble phenolic compounds having molecular weights between 500 and 3,000 and, besides giving the usual phenolic reactions, they have special properties such as the ability to precipitate alkaloids, gelatin and other proteins." This is the definition utilized in this chapter with the added proviso that the phenols are normal metabolic products and are not *in vitro* transformation products formed by chemical or other means. Nonetheless it is well to note that this definition derives from a consideration of the ways in which polyphenols tan protein fibers of animals skins, and as such it groups together a series of phenolic compounds that possess this common characteristic. Clearly this property may be a purely fortuitous one and, from the point of view of plant metabolism and plant biochemistry as a whole, it may in the final analysis be a quite misleading one.

This ability to complex with proteins makes vegetable tannins distinctive metabolic products, since the formation of substantial quantities of free tannin molecules within the cytoplasm would presumably cause precipitation of structural and catalytic proteins. In this respect they resemble several antibiotics that are more toxic to the producing organism than their precursors (Demain and Katz, 1977). Attempts to define the site of synthesis of these complex polyphenols within the plant cell have met with limited success. Baur and Walkinshaw (1974) studied the tannin depositions in tissue cultures of *Pinus elliotti* and concluded that synthesis occurred in the rough and smooth endoplasmic reticulum. Substantial tannin deposits were also observed (with electron microscopy) to be surrounded by at least one unit membrane within the cytoplasm or isolated from the cytoplasmic membrane by deposition within the vacuole. Tannin toxicity to the cell appeared to occur only after a large concentration of material accumulated within the cells or membrane integrity was lost.

The presence of vegetable tannins in plant tissues has several important practical consequences. Interaction of the tannins with salivary proteins and glycoproteins in the mouth renders the tissue astringent to the taste, and this characteristic may determine one's enjoyment of particular fruits (e.g., blackberry, strawberry, cranberry, and apple). Firmly established in the biochemical literature (Howes, 1953; Goldstein and Swain, 1963) is the belief that changes in the palatability of many fruits that occur on ripening are associated with concomitant changes in the concentration of tannins present in the fruit. A widely expressed view is that the astringency of green immature fruit is due to the presence of tannins but that on ripening these compounds are much reduced in quantity or are modified in some unspecified way. This long-standing and attractive hypothesis is nevertheless supported by only meager scientific evidence. The taste not only of fruit but of drinks derived from fruit, such as cider and wine, is also dependent on the tannins present. Thus a wine with insufficient tannins appears flat and insipid, whereas too great a concentration may lend a wine a harsh, rough quality.

Tannins also influence the development of soil profiles (Handley, 1961) by affecting the ways in which soil organic matter is decomposed. These effects are due to the capacity of vegetable tannins to combine with both minerals and organic matter and to the resistance of the resulting complexes to microbial attack. This property may be ecologically significant where particular plants completely dominate the vegetation of a certain habitat. Thus plants such as heather (*Calluna vulgaris*), mountain cranberry (*Vaccinium vitis-idaea*), bearberry (*Arctostaphylos uva-ursi*), and bracken (*Pteridium aquilinum*)—all of which metabolize substantial quantities of polyphenols including tannins—often appear to exclude many other forms of vegetation on heaths and moorlands. When these plants die, the polyphenols present may well help to produce a soil in which only their progeny can thrive. Alternatively the same effect may arise from the phytotoxic effect vegetable tannins exert once they leach from the plant into the surrounding soil. Tannins also have toxic effects on potential pathogens, and the resistance of various plant tissues to attack by microorganisms and viruses has been ascribed to these polyphenols, either alone or as their quinone oxidation products, associating with the protein components of viruses or similarly inactivating microbial exoenzymes. Similarly, when tannins are polymerized (probably via quinone oxidation products formed by the enzyme phenolase), they form an insoluble protective barrier which prevents microbial attack. Such a change invariably occurs at points where physical damage has been inflicted upon the plant.

II. CLASSIFICATION AND DISTRIBUTION

The most acceptable classification of vegetable tannins is based on structural types first suggested by Freudenberg (1920). This separates vegetable tannins into two groups—condensed or nonhydrolyzable and hydrolyzable. Recent work has shown condensed tannins, now more properly referred to as *proanthocyanidins* (Haslam, 1977), to possess the general type of structure depicted in I—a "polymeric" flavan-3-ol in which the interflavan bonds are most commonly C-4 to C-8 but in which some C-4 to C-6 bonds may be

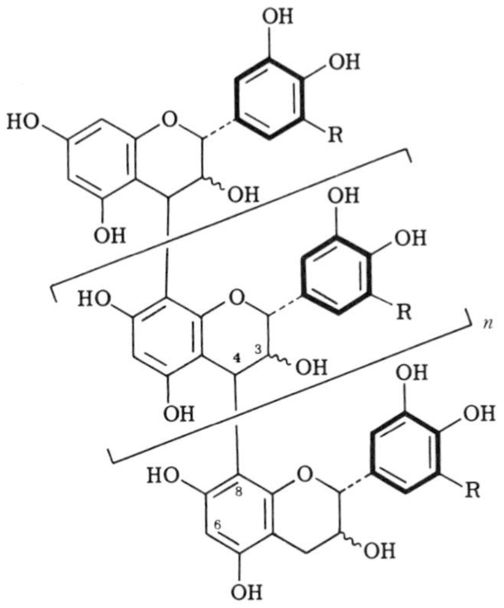

Proanthocyanidins (I. R = H, OH)
$n = 0, 1, 2, 3, 4, 5$ ----------

present. Hydrolyzable tannins are split by acids, bases, and in some cases hydrolytic enzymes (e.g., tannase; Haslam *et al.,* 1961a; Haslam and Stangroom, 1966) into sugars (usually D-glucose) or related polyols and a phenolic acid. In the case of *gallotannins* this is gallic acid (IV), and the vegetable tannins are simply polygalloyl esters (II), e.g., Chinese gallotannin, *syn*-tannic acid, from galls of *Rhus semialata* (VI). The closely related *ellagitannins* are characterized by the hexahydroxydiphenoyl ester group (III), (as in pedunculagin (VII), galls of *Quercus robur*), or in a structurally modified form (oxidation, reduction, or hydrolytic fission of one aromatic ring may have occurred), as in chebulagic acid (VIII, myrabolans or fruit of *Terminalia chebula*). When the hexahydroxydiphenoyl group (III) is cleaved from the molecule, the parent acid rapidly lactonizes to yield the insoluble dilactone ellagic acid (V). Gallotannins and ellagitannins are presumed to be closely related biogenetically (Mayer and Schmidt, 1956), although no experimental proof of this relationship has been furnished. The Freudenberg classification, based as it is on structural considerations, also represents a subdivision based on biogenetic origins; proanthocyanidins are products of the flavonoid pathway of biosynthesis (polyketide and cinnamate) and gallo- and ellagitannins are both derived from the shikimate pathway (Haslam, 1974).

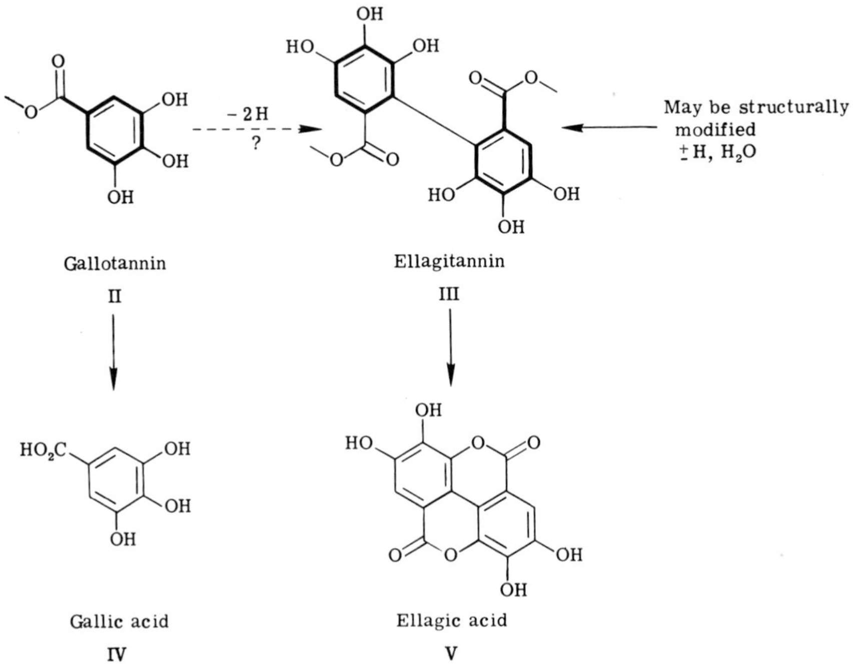

Gallotannin
II

Ellagitannin
III

May be structurally modified
$\pm$ H, H_2O

Gallic acid
IV

Ellagic acid
V

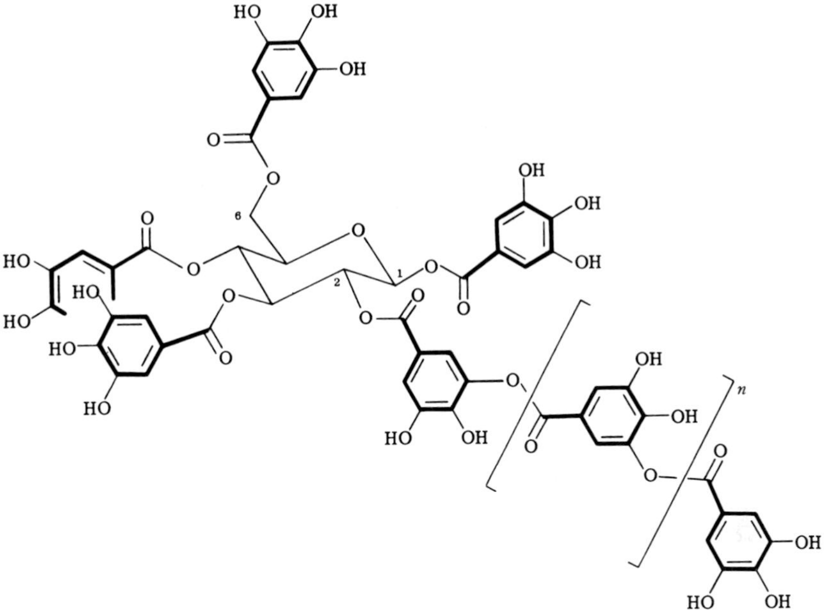

Heterogeneity: $n = 0, 1, 2$
      C–1 galloyl group may be absent

Tannic acid, Chinese gallotannin (VI)
Twig galls,   *Rhus semialata*
Leaves,   *Rhus typhina* (sumac)

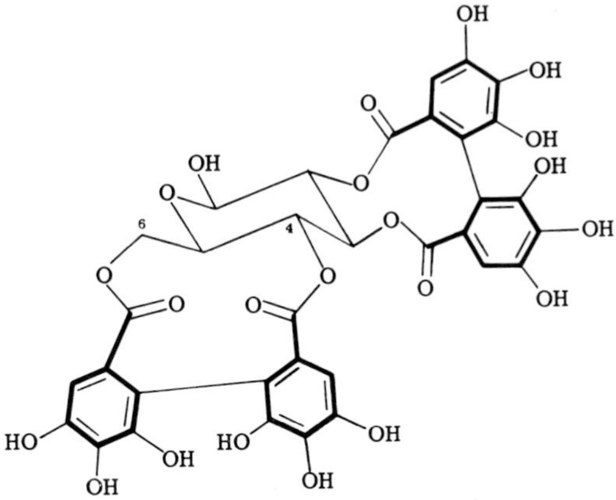

Pedunculagin (VII)
Oak galls
*hhda, 2, 3 :4, 6 (–)

*Hexahydroxydiphenic acid

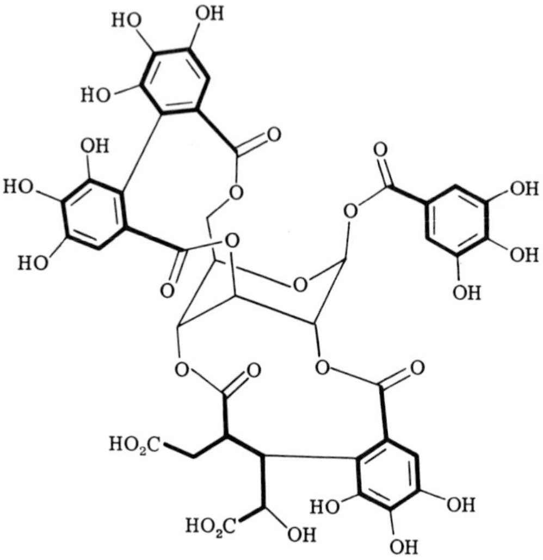

Chebulagic acid (VIII)
Myrobalans
*hhda, 3, 6 (+)

A common structural feature of both major categories of tannins (I, VI–VIII) is the accumulation within a molecule of moderate size of a substantial number of unconjugated phenolic groups, many of which are associated with an o-dihydroxy or o-trihydroxy orientation within a phenyl ring. Overwhelming circumstantial evidence unmistakeably suggests that the distinctive properties of vegetable tannins derive from this particular structural feature.

The separation of vegetable tannins into condensed and hydrolyzable is in a sense misleading, since in acidic media both types are hydrolytically degraded. Thus with acid proanthocyanidins give anthocyanidins, gallotannins give gallic acid (IV), and ellagitannins give ellagic acid (V) or a derivative thereof. Bate-Smith, in his classical pioneering work on the distribution of phenolics in plants (Bate-Smith, 1962; Bate-Smith and Lerner, 1954; Bate-Smith and Metcalfe, 1957), made use of this lability to acid as a means of mapping out the occurrence of vegetable tannins throughout the plant kingdom. Each hydrolysis product was detected by chromatography, color (butanol–hydrochloric acid, cyanidin λ_{max} 547 nm, delphinidin λ_{max} 558 nm), or fluorescence under uv light (gallic acid, violet; ellagic acid, "soft" violet). This work led to several important conclusions. First, vegetable tannins are of little importance as constituents of fungi, algae, mosses, liverworts, and grasses but are of great significance in many dicotyledons. Bate-Smith further concluded that the histological reaction for tannins in plant tissues is most commonly due to the presence of proanthocyanidins (referred to in the original papers as "leuco-anthocyanins") and that the capacity to synthesize tannins is a primitive character which tends to be lost with increasing phylogenetic specialization.

In a more recent series of papers Bate-Smith (1972, 1973, 1975, 1977) has amplified these data and provided semiquantitative information on the distribution of tannins in plants. The total amount of vegetable tannin in a tissue was determined by heme analysis and expressed as the tannic acid equivalent (TAE). In this procedure the total amount of tannin in a leaf extract is estimated by precipitation of the protein of hemolyzed blood and colorimetric determination of the residual hemoglobin. The TAE is then determined using tannic acid (VI) as a standard. The contributions to the TAE of the individual classes of tannin present are then estimated approximately using different colorimetric methods: proanthocyanidins by heating with hydrochloric acid in butanol (vide supra), ellagitannins by treatment with nitrous acid under nitrogen (blue color, λ_{max} 600 nm; cf. Schmidt, 1955), and gallotannins by reaction with potassium iodate (pink color, λ_{max} 500 nm). Bate-Smith was thus able to comment on the status of vegetable tannins as markers of evolutionary trends in plants. Proanthocyanidins are present in the most primitive of vascular plants such as ferns and gymnosperms, but hydrolyzable tannins are not. These appear for the first time in dicotyledons, where ellagitannins are much more common than gallotannins. This phylogenetic order—proanthocyanidins, ellagitannins, gallotannins—proposed by Bate-Smith (1972) parallels their order of increasing efficiency as precipitants of proteins.

Further interesting facets of vegetable tannin metabolism have been revealed by other studies (Haslam et al., 1972; Haslam, 1977; and unpublished observations). Several plant families (e.g., Aceraceae, Ericaceae, Fagaceae, and Rosaceae) retain the capacity to synthesize different types of tannin, although one particular plant may often specialize in one particular form. Thus in the Ericaceae, Calluna sp., Erica sp., and Rhododendron sp. are all rich sources of proanthocyanidins, but A. uva ursi has only a minimal capacity for proanthocyanidin synthesis and combines this with a high level of gallotannin metabolism. An interesting feature of the plant family Aceraceae (Haslam, 1965; Bate-Smith, 1977) is the low level of proanthocyanidin synthesis that is combined in individual species with that of differing patterns of hydrolyzable tannins. Some species metabolize an ellagitannin, others a gallotannin, and yet a third group produce as a major phenolic metabolite 2,6-digalloyl-1,5-anhydro-D-glucitol (Perkin and Uyeda, 1922; referred to as aceritannin, although in terms of the tannin definition given earlier it is not a member of this class of phenols). An examination of different tissues of the same plant often reveals differing patterns of vegetable tannin metabolism. Thus in Q. robur young leaves principally synthesize ellagitannins, in fresh

bark and young shoots there is a balance between ellagitannin and proanthocyanidin metabolism, whereas in young acorn cups the synthesis of
proanthocyanidins is the dominant form of vegetable tannin metabolism.

III. STRUCTURE AND BIOSYNTHESIS

Investigations of the chemistry of vegetable tannins concentrated for a
great many years on extracts that were of commercial importance (Table I).
Because of their complexity, progress was in many cases very slow. Substantial progress has since been made in the investigation of proanthocyanidins as they occur generally in plants but where their concentration does not
reach levels that make them valuable as sources of tannins. Conversely,
although considerable structural information has resulted from studies on
hydrolyzable tannin extracts (Table I), what is absent here is an appreciation
of the importance of these structures as secondary constituents in plants as a
whole. Emphasis is placed in the following discussion upon the aspects of
structure and biosynthesis of vegetable tannins that are presumed to be of
greatest significance to plant biochemistry.

A. Gallotannins and Ellagitannins

Schmidt and Mayer's elegant schemes for the biosynthesis of ellagitannins (Mayer and Schmidt, 1956; Schmidt, 1956) postulate a direct biogenetic
link between these polyphenols and gallotannins. They suggest that the
hexahydroxydiphenoyl group (III) and its various modified forms, which are
characteristic of ellagitannins, are derived initially by oxidative coupling of
two suitably disposed galloyl ester groups (II) in a gallotannin. A particular
chirality is imposed on the twisted biphenyl system of III at this point by the
chirality of the alcohol portion of the ester (Table II). The oxidative coupling
of two galloyl ester groups in this way has several other important consequences. It imposes rigidity on a molecule that previously had considerable
flexibility, and the molecule concomitantly changes shape from a flat, disc-
like structure to a compact, much more globular form. These changes produce comparable changes not only in solubility but also in the ability to
precipitate proteins. The driving force for these changes and their biological
significance is not yet clear, although it is worthy of note that gallotannins
are relatively rarely found in plants although ellagitannins are much more
common.

Although many simple mono- and digalloyl esters of sugars, polyols,
glycosides, and other phenols have been isolated from plants (e.g., Myers
and Roberts, 1960; Perkin and Uyeda, 1922; Mayer et al., 1965; Britton and
Haslam, 1965), they have little propensity to precipitate proteins. Naturally

TABLE II

Structure of Ellagitannins Based on D-Glucose[a]

Corilagin (myrabolans)	β-1-Galloyl; (+)-hexahydroxydiphenic acid linked 3,6; glucose conformation 1B (Schmidt *et al.*, 1954; Schmidt and Lademann, 1950; Schmidt and Schmidt, 1952)
Peduncalagin, (knopper nuts, oak galls)	(−)-Hexahydroxydiphenic acid linked 2,3 and 4,6; glucose conformation C1 (Harreus *et al.*, 1965)
Chebulinic acid	β-1,3,6-trigalloyl; chebulic acid (bound form, XV) linked 2,4 as in chebulagic acid (VIII); glucose conformation B-3 or 1C — 1B (Jochims *et al.*, 1968; Haslam and Uddin, 1967)
Chebulagic acid (myrabolans)	β-1-Galloyl; (+)-hexahydroxydiphenic acid linked 3,6; chebulic acid (bound form, XV) linked 2,4 as in VIII; glucose conformation 1C (Jochims *et al.*, 1968; Haslam and Uddin, 1967)
Terchebin (myrabolans)	β-1,3,6-Trigalloyl; (+) form of isohexahydroxydiphenic acid (XVI); glucose conformation B-3 or 1C $\leftrightarrows$ 1B (Schmidt *et al.*, 1967)
Brevilagin 1 (algarobilla)	(−) form of dehydrohexahydroxydiphenic acid (XVII) linked β-1,3 and 4,6; glucose conformation 2B (Eckert *et al.*, 1967)
Brevilagin 2 (algarobilla)	(−)-Hexahydroxydiphenic acid linked 4,6; residue, (−) form of dehydrohexahydroxydiphenic acid (XVII) linked β-1,3; glucose conformation 2B $\leftrightarrows$ B-3 (Groebke *et al.*, 1967)
Geraniin (*Geranium* sp. and *Mallotus japonicus*)	Corilagin structure plus dehydrohexahydroxydiphenic acid (XVII) linked 2,4 (Nayeshiro *et al.*, 1977)
Vescalin, castalin (oak and chestnut)	Based on nonahydroxytriphenic acid (XXI) linked 1,2,3,5; open-chain form of D-glucose (Kuhlmann *et al.*, 1971)
Vescalagin, castalagin (oak and chestnut)	Based on nonahydroxytriphenic acid (XXI) linked 1,2,3,5; (−)-hexahydroxydiphenic acid linked 4,6; open-chain form of D-glucose (Jochims *et al.*, 1971)
Valolaginic acid (*Valonea*)	Based on trilloic acid (XXIII) linked 1,2,3,5; (−)-hexahydroxydiphenic acid linked 4,6; open-chain form of D-glucose (Bilzer *et al.*, 1976b)
Isovalolaginic acid, (*Valonea*)	Based on trilloic acid (XXIII) linked 1,2,3,5; (1)-hexahydroxydiphenic acid linked 4,6; open-chain form of D-glucose (Mayer *et al.*, 1976; Busath *et al.*, 1976)
Castavaloninic acid (*Valonea*)	Based on valoneic acid (XXII) and nonahydroxytriphenic acid (XXI) esterified to D-glucose (Bilzer *et al.*, 1976a)
Mallotusinic acid, (*M. japonicus*)	Based on valoneic acid (XXII) and dehydrohexahydroxydiphenic acid (XVII) esterified to 1-*O*-galloyl-β-D-glucose (Okuda and Seno, 1978)
Mallotinic acid (*M. japonicus*)	Based on valoneic acid (XXII) esterified to 1-*O*-galloyl-β-D-glucose (Okuda and Seno, 1978)
Punicalagin (pomegranate)	Based on gallagic acid (XXIV) esterified 2,6 to D-glucose and (−)-hexahydroxydiphenic acid linked 3,4 (Andra *et al.*, 1977)
Punicalin (pomegranate)	Based on gallagic acid (XXIV) esterified 2,6 to D-glucose (Andra *et al.*, 1977)

[a] D-Glucose conformations as defined by Reeves (1951) and Jochims *et al.* (1968).

occurring galloyl esters having this propensity usually have MWs in the region of 1000 and often possess a chain of depsidically linked galloyl units. Relatively few gallotannins have been discovered in plants and thoroughly documented, of these Chinese, Turkish, sumac, and tara (Table I) have been studied in most detail.

Fischer and Freudenberg, in the period 1910–1919, first represented the structure of Chinese gallotannin as a core of β-penta-O-galloyl-D-glucose to which other galloyl residues (~5) were attached. Later work (Armitage *et al.*, 1961; Haslam and Haworth, 1964; Haslam, 1967) made various refinements to this structure, principally establishing the presence of a chain of depsidically linked galloyl ester groups (on average three or four) attached to the hydroxyl at C-2 of the β-penta-O-galloyl-D-glucose core. The heterogeneity of gallotannin samples as they are frequently obtained from plant materials is believed to be attributable to variations in the number of depsidically linked galloyl groups and to the absence in some molecules of the galloyl ester at C-1 (VI). Sumac (Table I) and dhava gallotannins (*Anogeissus latifolia*, leaves) have structures identical to that of Chinese gallotannin (Armitage *et al.*, 1961; Nayudamma *et al.*, 1964), and preliminary evidence suggests that gallotannins from the leaves of *Cotinus coggyria, Acer platanoides, A. campestre, A. rubrum,* and *Pelargonium* sp. possess the same structure. Gayuba tannin (*A. uva-ursi,* leaves) is a prototype of these gallotannins possessing the same basic structure but with a polygalloyl chain containing (on average) only one depsidically linked galloyl group (Britton and Haslam, 1965). The analysis of Turkish gallotannin was complicated by the presence of ellagitannins (a characteristic of *Quercus* sp.), but a structure analogous to VI was put forward (Armitage *et al.*, 1962) with the variable-length polygalloyl chain at C-6 and the C-2 hydroxyl group unesterified.

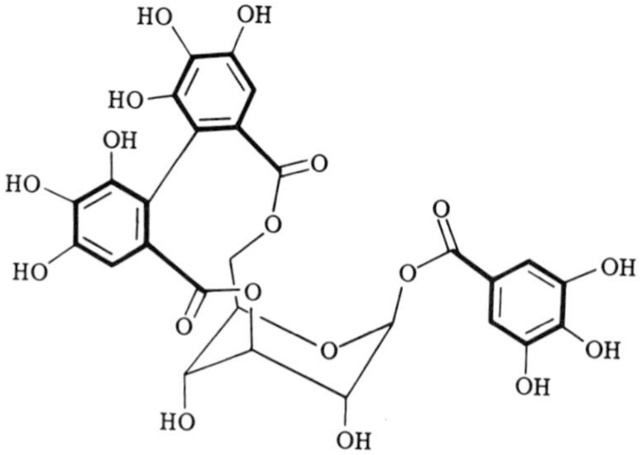

Corilagin (IX)
Myrobalans, divi-divi
*hhda, 3, 6 (+)

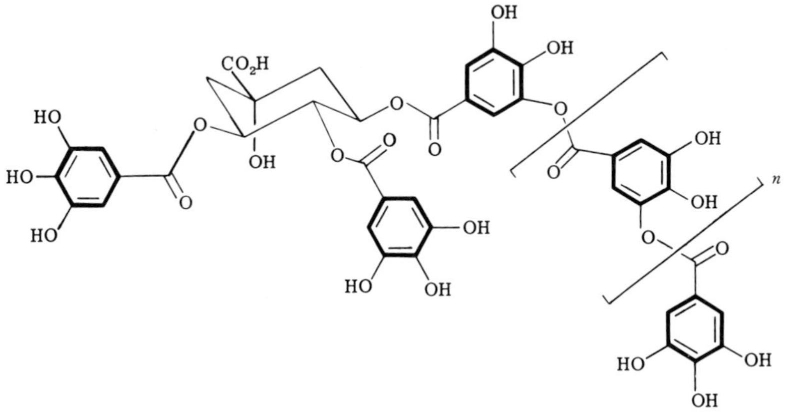

Tara galliotannin (**X**, n = 0, 1, 2)
Caesalpinia spinosa

Structure X was proposed for the gallotannin from tara (Table I), in which the galloyl ester groups are attached to a core of D-quinic acid (Haslam *et al.*, 1962).

Present evidence indicates that gallotannins are not widely distributed in plants, but the real extent of their occurrence requires further investigation and documentation. The biochemical mode of the derivation of gallic acid (IV)—the phenolic residue that forms the structural basis of gallotannins—is similarly in question. This particular problem indeed highlights some of the weaknesses and limitations of the isotopic tracer technique as it is used to delineate biosynthetic pathways in higher plants. Experimental support for at least three routes has been obtained (Fig. 1) and, although it is conceivable that a network of pathways may exist, it seems unlikely that they all constitute normal metabolic routes to gallic acid.

3,4,5-Trihydroxycinnamic acid has been found in plants only as its various methyl ethers, and several observations indicate that gallic or ellagic acid is the systematic taxonomic equivalent of this "missing" acid (Bate-Smith, 1962; Haslam, 1974). Zenk (1964) formulated a conventional pathway (*a*) from L-phenylalanine to 3,4,5-trihydroxycinnamic acid followed by β-oxidation to form gallic acid (Fig. 1). This conclusion was based on feeding studies using ^{14}C-labeled L-phenylalanine, L-tyrosine, and benzoic acid in *Rhus typhina*. Neish, Towers, and their collaborators (Chen *et al.*, 1964) favored a variation (*b*) in which β-oxidation occurred at the caffeic acid stage to give protocatechuic acid which was then further hydroxylated to give gallic acid. Finally, work with the mold *Phycomyces blakesleeanus* and later *R. typhina* and *Acer* sp., using ^{14}C-labeled D-glucose and shikimic acid, demonstrated (Haslam *et al.*, 1961b; Cornthwaite and Haslam, 1965; Dewick and

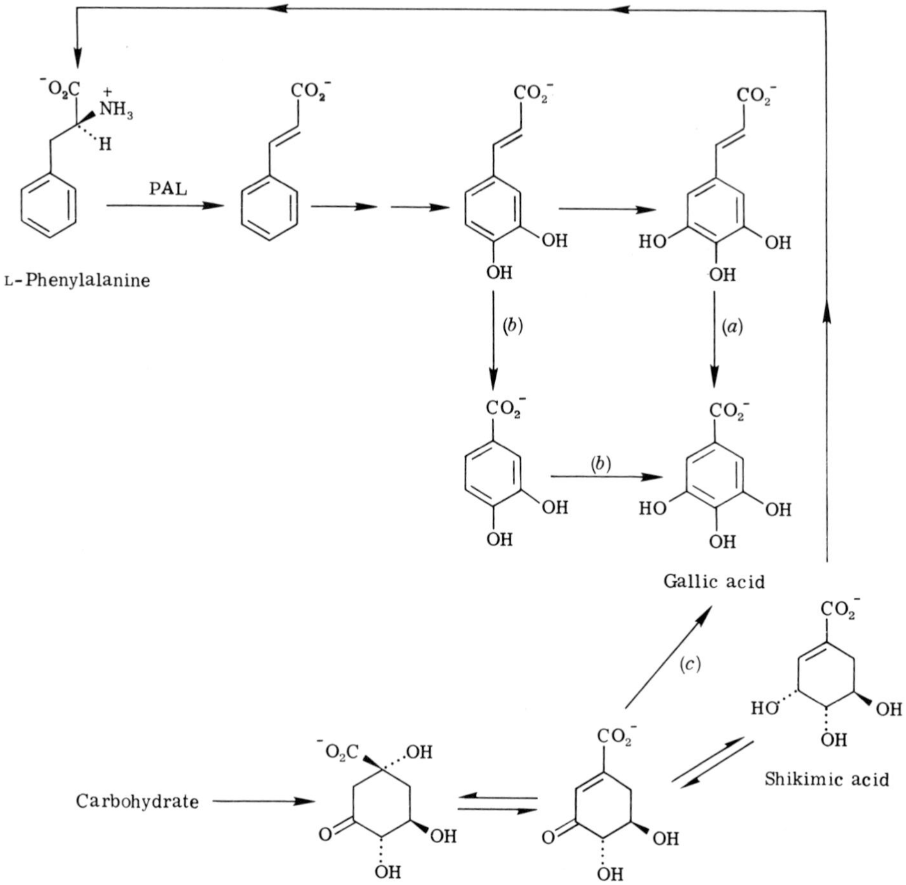

Fig. 1. Biosynthetic pathways to gallic acid. (All acids are formulated as anions.)

Haslam, 1969) that a third route (c)—the direct dehydrogenation of 3-dehydroshikimic acid—existed to gallic acid. Conn and Swain (1961) supported this latter conclusion in work with *Geranium* sp.

The major commercial sources of ellagitannins are myrabolans, divi-divi, algarobilla, valonea, and the bark of oak and Spanish chestnut (Table I). It is with the ellagitannins derived from these plant sources that the outstanding work of Schmidt and Mayer has been concerned. Bate-Smith and Metcalfe (1957) have, however, recorded the presence of ellagic acid in many plants, and a survey of these plants to detect hexahydroxydiphenyl esters (E. Haslam, unpublished observations) has provided a preliminary basis on which the work of Mayer and Schmidt can be assessed in relation to the presence of ellagitannins in plants generally. Whereas valonea, oak, and chestnut extracts have features in common with ellagitannins in the plant family Fagaceae, the extracts derived from myrabolans, divi-divi, and algarobilla are distinctive and there appears to be little overlap with the ellagitannins detected in other plants. There thus remains a considerable amount of work to be carried out in this field, but with this important proviso in mind the principal features of the Heidelberg school's elegant work forms the basis for a discussion of ellagitannin structure and biosynthesis.

Most ellagitannins give ellagic acid (V) on acid hydrolysis, and in the manufacture of leather the aesthetically pleasing bloom on many finished leathers is formed by a crystalline deposit of this substance. In addition, other phenolic compounds—notably chloroellagic acid (XI), chebulic acid (XII), dehydrodigallic acid (XIII), flavogallonic acid (XVIII), valoneic acid dilactone (XIX), trilloic acid trilactone (XX), and gallagic (XXIV) acid (Figs. 2 and 3)—have been obtained from ellagitannins or ellagitannin extracts. Each of these acids is derived by cleavage of a particular phenolic residue from an ellagitannin, and Schmidt and Mayer have proposed that these groups in turn are derived initially by oxidative coupling of two or more galloyl ester groups in a polygalloyl ester precursor (Figs. 2 and 3). Dehydrodigallic acid (XIII) may thus be visualized as being formed from a phenolic residue (XIV), itself derived by C—O as opposed to C—C oxidative coupling of two galloyl ester groups (Fig. 2). Analogously, flavogallonic acid (XVIII) and valoneic acid dilactone (XIX) are obtained from precursors formed by oxidative coupling of three galloyl ester groups (Fig. 3). Determination of the structure of chebulic acid (XII, Mayer and Schmidt, 1951; Haworth and de Silva, 1951, 1954; Haslam and Uddin, 1967) pointed the way to a putative biogenetic relationship with hexahydroxydiphenic acid, which was fully elaborated by Mayer and Schmidt (1956). The hypothesis is based on the proposal that one of the aromatic residues of a bound hexahydroxydiphenoyl group can undergo oxidation, reduction, or rearrangement to give "modified" phenolic groups (XV, XVI, XVII, XXIII). Thus chebulic acid, as it is bound to D-glucose in chebulinic acid and chebulagic acid (VIII), is in the form (XV), and trilloic acid (XXIII) is similarly presumed to be formed by the modification of one galloyl ester group in nonahydroxytriphenic acid (XXI) (Fig. 3). Support for this biogenetic relationship has been established by the further isolation of ellagitannins containing within their structures the bound form of isohexahydroxydiphenic acid (XVI), in which one aromatic ring is unusually in its triketo tautomeric form, and the bound form of dehydrohexahydroxydiphenic acid (XVII) in which one aromatic ring is oxidized to the corresponding o-quinone. In Table II details are summarized of some of the ellagitannin structures, based on D-glucose, in which the hexahydroxydiphenoyl group and the corresponding triphenoyl group and their

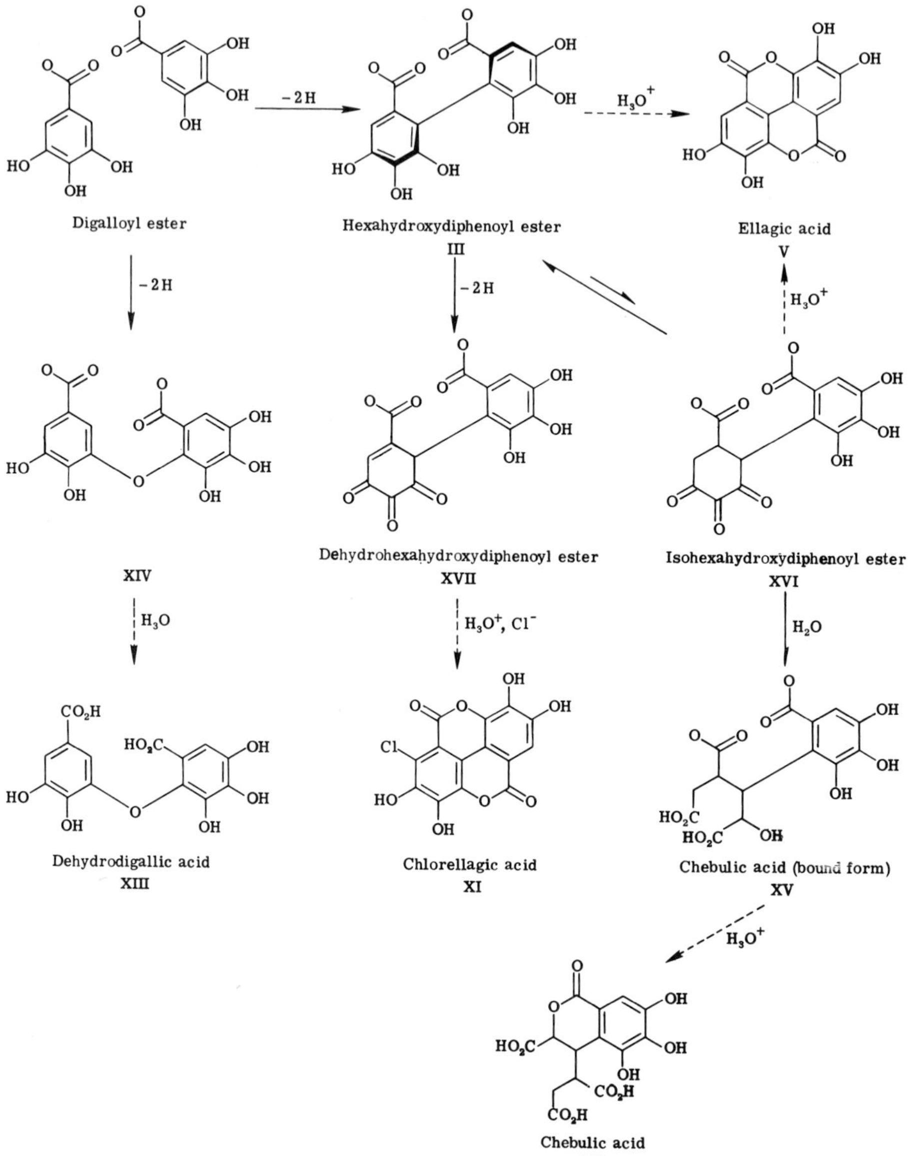

Fig. 2. Biogenetic relationships and transformations of the hexahydroxyphenoyl ester group. Solid arrows represent proposed biosynthetic reactions; dashed arrows represent hydrolysis reactions.

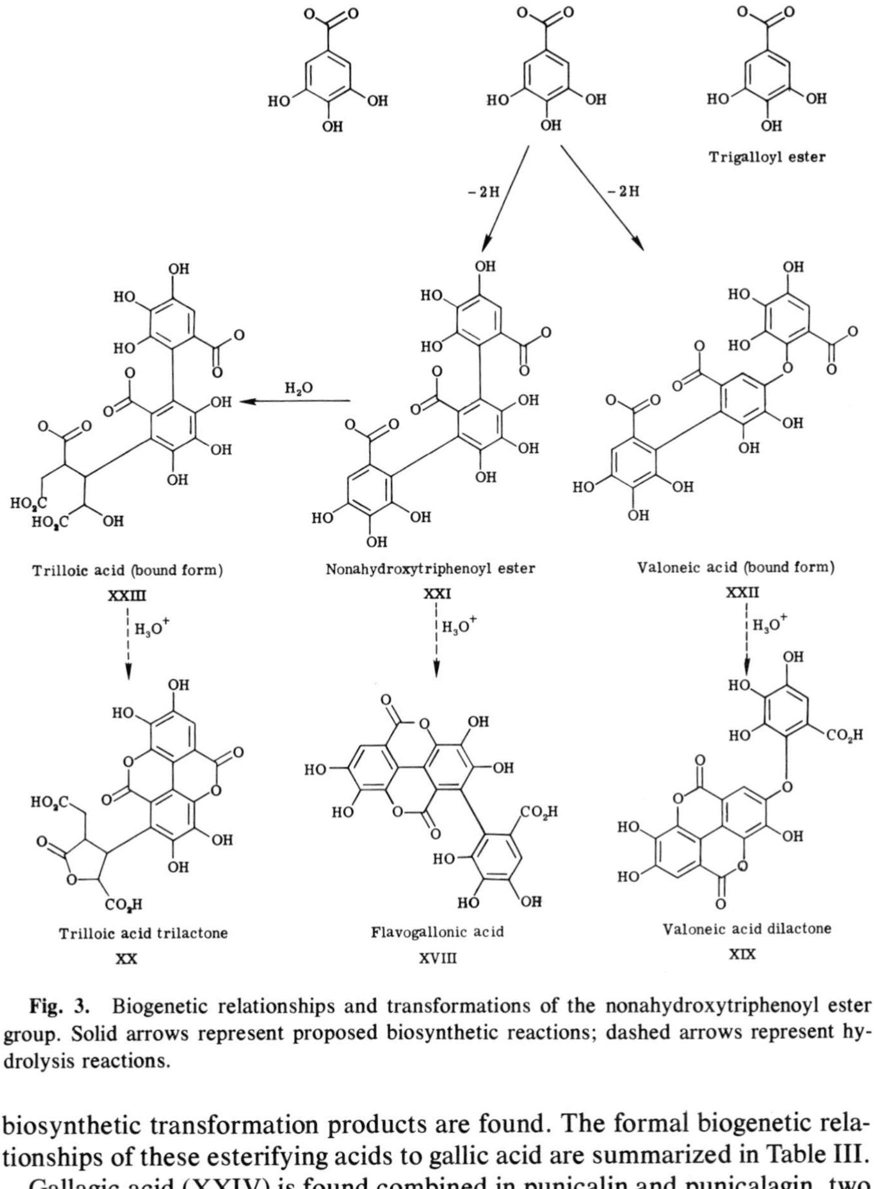

Fig. 3. Biogenetic relationships and transformations of the nonahydroxytriphenoyl ester group. Solid arrows represent proposed biosynthetic reactions; dashed arrows represent hydrolysis reactions.

biosynthetic transformation products are found. The formal biogenetic relationships of these esterifying acids to gallic acid are summarized in Table III.

Gallagic acid (XXIV) is found combined in punicalin and punicalagin, two tannins obtained from pomegranate (Mayer, Gorner, and Andra, 1977), and the two carboxyl groups esterify the hydroxyl groups at C-2 and C-4 of D-glucose. Although this acid and the tannins are clearly related to the ellagitannins discussed above, the Schmidt–Mayer biogenetic hypothesis has to be modified to accomodate the formation of this phenolic residue.

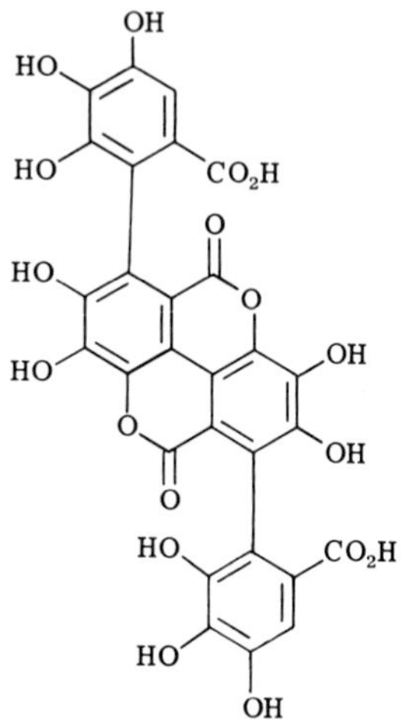

Gallagic acid

XXIV

The structural patterns present in ellagitannins metabolized by a very limited number of plants (Table II) are nevertheless highly reminiscent of those involved in secondary metabolism in other organisms, for example, polyketide biosynthesis in molds and fungi. In these cases a plausible explanation of the phenomenon of secondary metabolism is that an accumulation of intermediates is built up in some primary metabolic pathway and that enzymes are induced for the synthesis of secondary products from one or more of these primary metabolites or linkage compounds. Usually one key parent secondary compound is visualized as being produced, and this then undergoes a wide range of structural modifications ("chemical embroidery") leading to the formation of a variety of secondary metabolites only slightly

TABLE III.

Formal Biogenetic Relationship of Gallic Acid to the Esterifying Acids of Ellagitannins (Figs. 2 and 3)

Dehydrodigallic acid		$-2H$
Hexahydroxydiphenic acid		$-2H$
Isohexahydroxydiphenic acid	$2 \times$ gallic acid	$-2H$
Dehydrohexahydroxydiphenic acid		$-4H$
Chebulic acid (bound form)		$-2H, +2H_2O$
Valoneic acid		$-4H$
Nonahydroxytriphenic acid	$3 \times$ gallic acid	$-4H$
Trilloic acid		$-4H, +2H_2O$
Gallagic acid	$4 \times$ gallic acid	$-6H, -2H_2O$

different from each other. What matters from the organism's point of view is to dispose of the primary metabolic accumulation products. A typical example is the production of 6-methylsalicylic acid by microfungi. This in turn functions as the parent compound for a large family of metabolites (e.g., *m*-cresol, 3-hydroxyphthalic acid, gentisic acid, patulin) depending on the species and culture conditions (Bu'Lock, 1978).

Circumstantial evidence suggests that the biosynthesis of gallotannins is intimately linked to that of ellagitannins, and with the analogy above in mind, this feature is utilized in a hypothesis that tentatively links the biosynthesis of these polyphenols to the overall pathway of aromatic amino acid metabolism (Fig. 4). The parent compound serving as a precursor of both groups of vegetable tannins is suggested to be β-penta-*O*-galloyl-D-glucose, and from this ellagitannins are metabolized by processes of the type outlined above. Conversely, gallotannins are formed by the addition of further galloyl groups, as depsides, to the parent β-penta-*O*-galloyl-D-glucose. In many plants there appears to be a capacity to carry out principally one or the other of these processes to give either gallotannins or ellagitannins. However, in some plants (e.g., *Quercus* sp.) there appears to be a capacity to carry out both types of synthesis simultaneously.

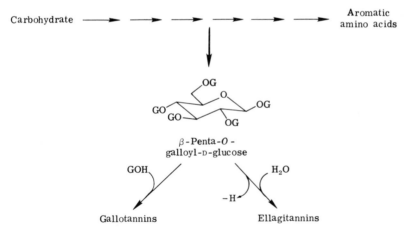

Fig. 4.   Tentative relationships in hydrolyzable tannin biosynthesis.

B. Proanthocyanidins

Structural information on the actual components of the condensed tannin extracts of commerce (Table I), apart from wattle (Roux, 1972), is still in an undeveloped state. Several extracts have long been known to be rich in catechins (flavan-3-ols), and once Freudenberg had formulated the structure for (+)-catechin (XXVII) and had proposed his "catechin hypothesis," it became customary to regard these substances as the sole basis of condensed tannin chemistry (Freudenberg, 1920; Haslam, 1966). Wide-ranging studies with flavan-3-ols and flavan-3,4-diols as models led to proposals for their self-condensation to complex polymers thought to be characteristic of the vegetable tannin extracts themselves. However, because of the almost complete absence of data on the components of the extracts, it is not possible to give a realistic appraisal of this work in its relationship to condensed tannin formation in the extracts shown in Table I. However, it is in the case of wattle (*Acacia mearnsii*) that Roux and his collaborators (Roux, 1972; Drewes *et al.*, 1967a,b) have characterized a group of proanthocyanidins from the wood and bark of this tree, accompanied by a high concentration of their apparent precursor 2,3-*trans*-3,4-*trans*-(+)-mollisacacidin (XXV). Roux has formulated schemes whereby this substance participates—via the initial simulated acid-catalyzed formation of the carbocation (XXVI)—in condensation reactions with itself and with various flavan-3-ols to give the condensed tannins of wattle (Fig. 5).

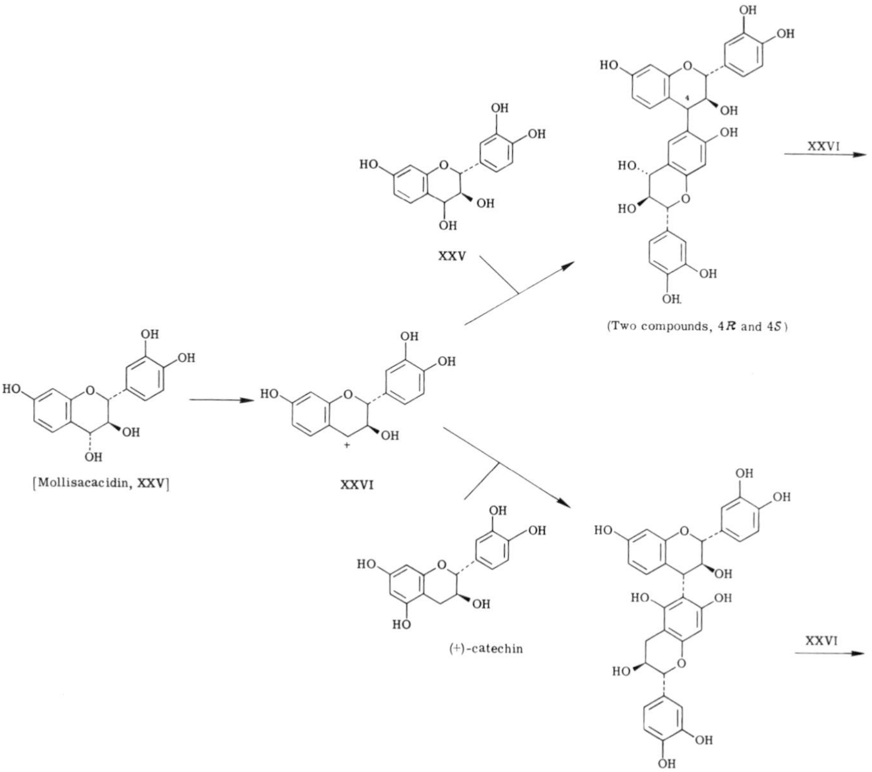

Fig. 5.   Biogenesis of proanthocyanidins of wattle (*Acacia mearnsii*).

Somewhat surprisingly, the structural problems posed by condensed tannins have come nearest to solution for the members of this class that are widely distributed in plants but which, because of their relatively low concentration, have little importance as commercial sources. Rosenheim (1920) began the first critical study of these substances during an investigation of pigmentation in the grape. In view of their distinctive reaction with acids to give anthocyanidins he called them *leucoanthocyanins,* a name that was later changed to *leucoanthocyanidins* and finally to *proanthocyanidins* (Freudenberg and Weinges, 1960). Robinson and Robinson (1933) and later Bate-Smith and Lerner (1954) examined their systematic distribution in plants. Bate-Smith noted that the majority yielded cyanidin, and a few delphinidin, with acid. In no case did he observe the presence of prodelphinidins without the concomitant occurrence of procyanidins. Bate-Smith also drew attention to the fact that proanthocyanidins were mostly confined to plants with a woody growth habit and, with Swain (Bate-Smith and Swain, 1954) he pointed out the great similarity in systematic distribution to that of substances botanists defined as tannins. Forsyth and Roberts (1960), working with the procyanidins of the cocoa bean, were the first to obtain evidence showing that these substances were probably flavan-3-ol dimers and higher oligomers, and subsequent investigations (Kaltenhauser *et al.,* 1968; Gortiz *et al.,* 1968; Bahr *et al.,* 1969; Haslam *et al.,* 1972; Fletcher *et al.,* 1977) confirmed these observations. Attention has focused on procyanidins, the most commonly found proanthocyanidins, and this work is outlined below.

Plant surveys reveal the presence of characteristic paper chromatographic "fingerprints" for procyanidins. Probably the most commonly encountered is that found, for example, in hawthorn (*Crataegus monogyna*). Co-occurring with (−)-epicatechin is one major procyanidin dimer [B-2, formally two (−)-epicatechin units linked C-4 to C-8], a minor dimer [two (−)-epicatechin units linked C-4 to C-6], a major trimer, and various higher oligomers. An alternative but highly characteristic pattern is found, for example, in the sallow willow catkin (*Salix caprea*), where (+)-catechin, a major procyanidin dimer [B-3, formally two (+)-catechin units linked C-4 to C-8], a minor dimer [two (+)-catechin units linked C-4 to C-6], a trimer, and various oligomers constitute the fingerprint. The two remaining characteristic patterns are found, for example, in *Rubus* sp. [(−)-epicatechin, procyanidin B-4, etc.] and in *Sorghum vulgare* [(+)-catechin, procyanidin B-1, etc.] and are the most interesting from the biosynthetic point of view since the two flavan-3-ol fragments formally composing the dimers (B-4 and B-1) are of opposite stereochemistry at C-3. It is possible using this form of analysis to classify procyanidin-bearing plants according as to whether they belong to one of the four "genetically homogeneous" forms or to any particular combination.

Apart from the procyanidin polymer derived from sorghum (Gupta and Haslam, 1978) comparatively little is known concerning the higher oligomeric forms, but it is assumed that their structures are essentially extensions of the simple dimeric and trimeric structures in which the molecular size is increased by the addition of further flavan units mainly by C-4 to C-8 linkage (I). This increase in MW causes a decrease in solubility, and the polymers are frequently difficult to solubilize. In terms of their ability to precipitate proteins, hence their astringency to the palate, the dimeric procyanidins have a relative astringency of about 10% compared to tannic acid. As the molecular size of the procyanidin increases, so also does its astringency, but as yet no detailed study has been carried out to determine which are the important components of the procyanidin complex that contribute the astringent taste of plant material.

Procyanidins (B-1–B-4, Fig. 6) occur free and unglycosylated and almost invariably with one or both of the diastereoisomeric pair (−)-epicatechin and (+)-catechin, although the procyanidins of the Palmae are found in association with (+)-epicatechin (Ferrari *et al.*, 1972). Procyanidins are degraded by acid to give the carbocation (e.g., XXIXa) (Fig. 7) from the upper half of the molecule and the flavan-3-ol from the lower half. The carbocation is normally rapidly converted by proton loss and oxidation to cyanidin (XXX)—this is the basis of the characteristic color reactions of proanthocyanidins, *vide supra*—but it may also be intercepted under appropriate conditions by nucleophiles such as thiols or another flavan-3-ol molecule. Chemical and spectroscopic investigations established the structure and stereochemistry

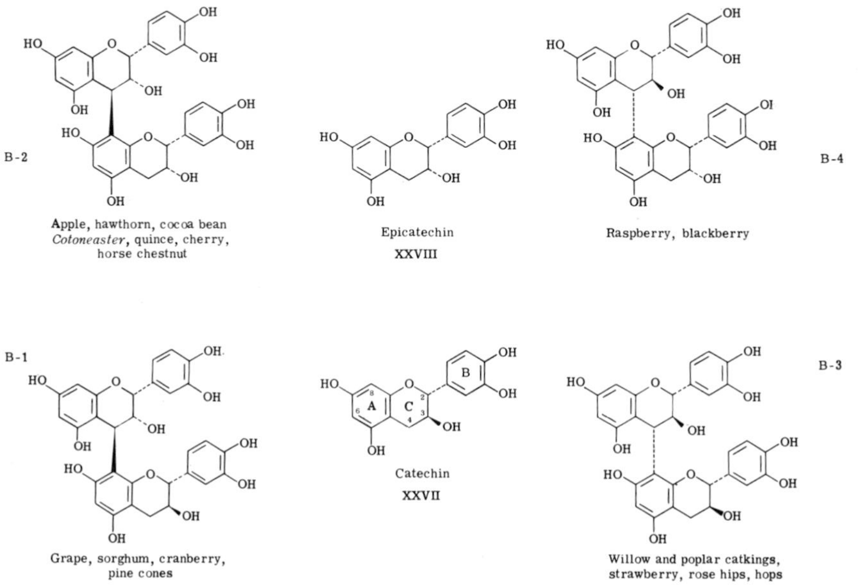

B-2

Apple, hawthorn, cocoa bean
Cotoneaster, quince, cherry,
horse chestnut

Epicatechin
XXVIII

B-4

Raspberry, blackberry

B-1

B-3

Grape, sorghum, cranberry,
pine cones

Catechin
XXVII

Willow and poplar catkings,
strawberry, rose hips, hops

Fig. 6.   Major dimeric procyanidins. Occurrence and structure.

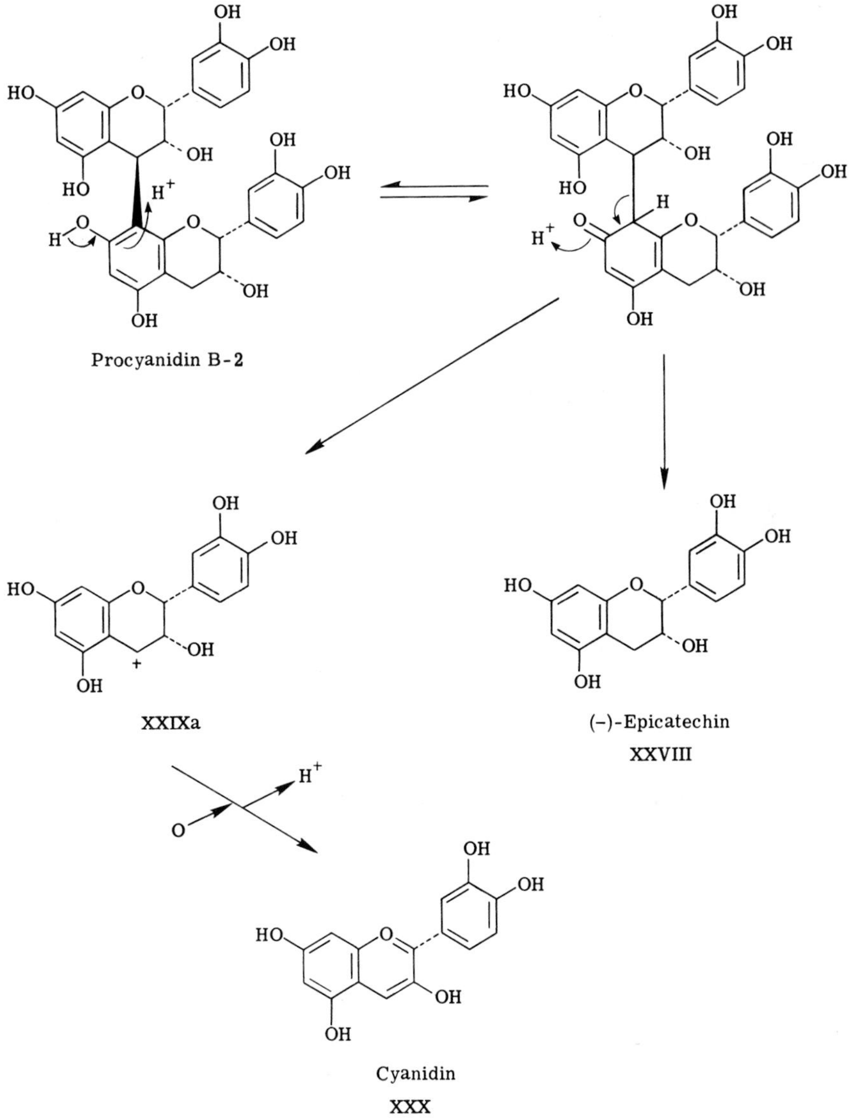

Fig. 7. Acid-catalyzed degradation of procyanidin B-2.

of the four major procyanidin dimers and showed that for each dimer there was restricted rotation about the interflavan bond. Molecular models show that as a result each procyanidin dimer has a preferred conformation; significantly perhaps, those of B-1 and B-2 bear a pseudo-mirror image relationship to those of B-3 and B-4.

Elaboration of the oligomeric forms of procyanidins by the addition of further flavan-3-ol units (e.g., I), bearing in mind the conformational restraints about the interflavan bond, leads to two helical structures. The central core of these linear polymers is composed of rings A and C of the flavan repeat unit, and ring B (the o-dihydroxyphenyl ring) projects laterally from this core. Significantly, those formed from repeat units related to $(-)$-epicatechin (e.g., extension of the procyanidin B-1 or B-2 type) are *left-hand* helices, whereas those of the types formed from $(+)$-catechin repeat units, related to procyanidins B-3 and B-4, are *right-hand* helices.

Studies on the biosynthesis of procyanidins have been conducted with a range of fruit-bearing plants—horse chestnut (*Aesculus hippocastanum, A.* x *carnea*), willow catkins (*Salix caprea, S. inorata*), raspberry, and blackberry (*Rubus idaeus, R. fructicosus*)—using a number of ^{14}C- and ^{3}H-labeled cinnamic acid precursors (Fig. 8) (Haslam *et al.*, 1977). The results showed that the C^6—C^3 carbon skeleton of the cinnamate precursor was incorporated intact into the flavan units; H_a was retained ($\sim$80–90%), H_c was lost (90–100%), and H_b was retained (50% retention, one NIH shift occurring on the first hydroxylation of the phenyl ring). A crucial observation was that the two apparently identical structural units that form the procyanidin dimer were labeled to different extents. This significant feature of the evidence was interpreted as showing that the two flavan-3-ol-type units of the procyanidin molecule were derived from different metabolic entities. Coupled with the circumstantial evidence implying a close connection with flavan-3-ol biosynthesis, this information has been used to formulate a projected scheme of biosynthesis for procyanidins as shown in Fig. 9.

The main proposal of the biosynthetic scheme is that the procyanidins are formed as by-products of flavan-3-ol and cyanidin metabolism. If the reduction of flav-3-en-3-ol to flavan-3-ol is envisaged as a two-step process in which stereospecific proton addition to give a bridged protonated species precedes specific (cis or trans to proton addition) delivery of hydride ion (or its equivalent from, say, NADPH), then the carbocations (XXIX, with the appropriate 3S or 3R absolute stereochemistry) may possibly result from a situation in which the supply of biological reductant is rate-limiting. The carbocations (XXIX) would then be formed by leakage of the bridged protonated species from the active site of the enzyme and, it is postulated, react with the final reduction product, the nucleophilic flavan-3-ol remaining in the vicinity of the enzyme, to yield procyanidins. The trimers, tetramers, and various higher oligomers also formed can then be visualized to result from reactions of the appropriate dimer and further carbocation. The full stereochemical details of the hypothesis have been outlined (Haslam, 1977).

Support for this view of procyanidin metabolism has been furnished by the discovery of *in vitro* laboratory reactions that mimic the *in vivo* situation (Fletcher *et al.*, 1977). Acid-catalyzed degradation of procyanidin dimers (Fig. 7) is also a source of carbocation intermediates (XXIX), and treatment of a mixture of procyanidin B-2 and (−)-epicatechin with acid gives, after degradation and resynthesis, a mixture of procyanidins whose paper chromatographic fingerprint exactly matches that given by the procyanidins in *Malus* sp., *Prunus* sp., and *Crataegus* sp. noted above. Analogous reactions can be similarly set up as model syntheses for the three other "genetically homogeneous" types of procyanidin biosynthesis encountered in plants. The close correspondence between the *in vitro* pattern of products (formed by the entirely promiscuous encounter between the appropriate carbocation and nucleophilic flavan-3-ol) and the procyanidin profile set up *in vivo* raises the important question whether the actual procyanidin biosynthetic reactions in plants are under enzymatic control or not.

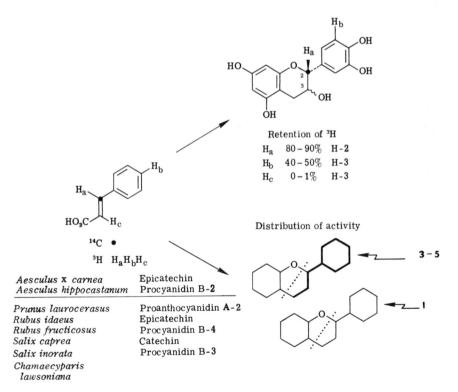

Retention of ³H

Ha   80–90%   H-2
Hb   40–50%   H-3
Hc   0–1%   H-3

Distribution of activity

Aesculus x carnea   Epicatechin
Aesculus hippocastanum   Procyanidin B-2

Prunus laurocerasus   Proanthocyanidin A-2
Rubus idaeus   Epicatechin
Rubus fructicosus   Procyanidin B-4
Salix caprea   Catechin
Salix inorata   Procyanidin B-3
Chamaecyparis
  lawsoniana

Fig. 8.   Tracer experiments in flavan-3-ol and procyanidin biosynthesis. Summary of results.

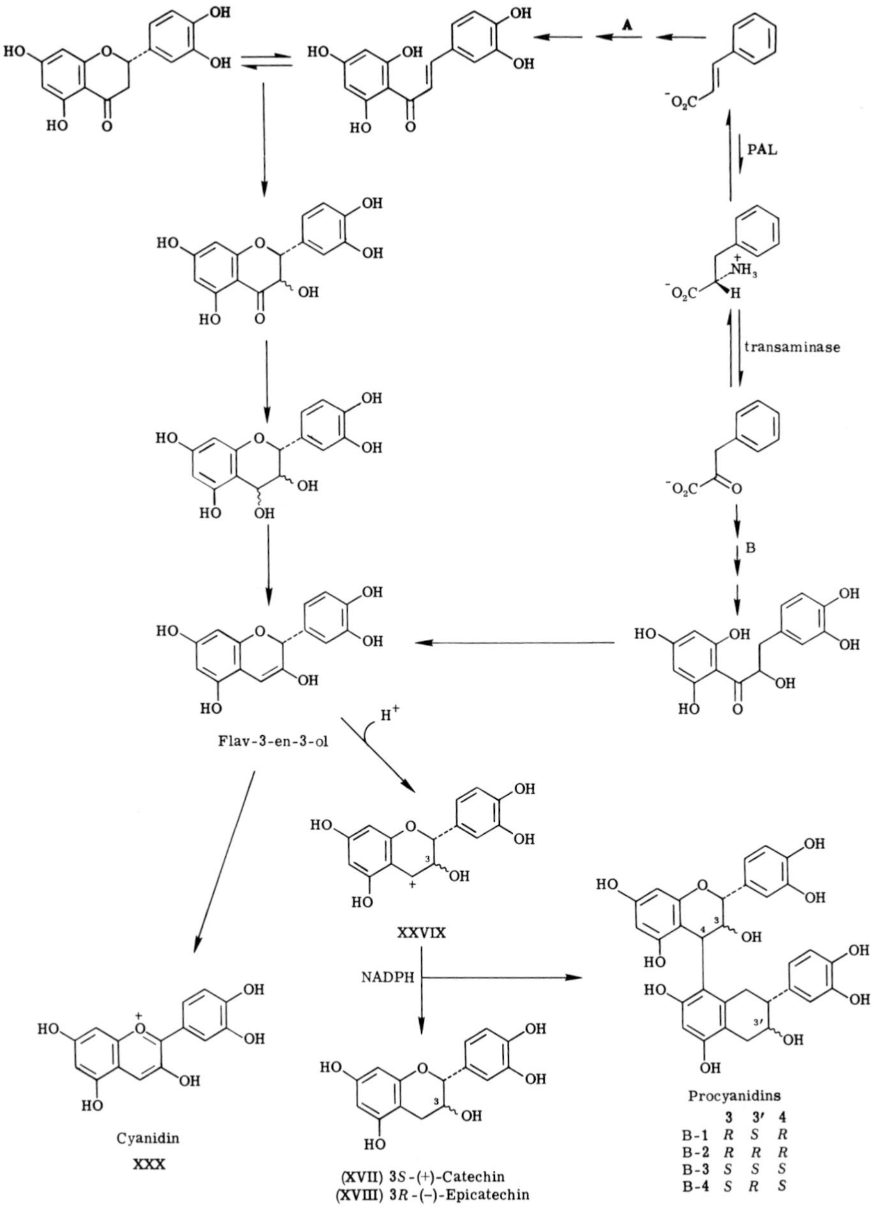

Fig. 9. Suggested pathways of biosynthesis of cyanidin, (+)-catechin, (−)-epicatechin, and procyanidins.

Comparable chemical and biosynthetic details have not yet been obtained for the prodelphinidin class of proanthocyanidins, although there is every reason to suppose that in the majority of instances their chemistry and biochemistry closely resemble that of procyanidins and are similarly associated with the metabolism of (+)-gallocatechin and (−)-epigallocatechin in plants. Finally an intriguing metabolic situation that has some relationship to both hydrolyzable tannins and proanthocyanidins should perhaps be briefly noted. The leaves of some plants (e.g., the tea plant, *Camellia sinensis* and various *Bergenia* sp.) are rich in polyphenols, particularly flavan-3-ols and their 3-*O*-galloyl esters (Myers and Roberts, 1960). It is interesting to note that in these instances the leaves contain only very small quantities of proanthocyanidins.

IV. CONCLUSION AND FUTURE PROSPECTS

Future work in this area must address the problems presented by vegetable tannins from a much more biological point of view. Although it is not the intention unnecessarily to add further speculation to that already present in the literature, it is, however, finally worth drawing attention to two features of the chemistry of vegetable tannins that structural studies have now revealed and which may have some relevance in the biological context.

Despite their apparently disparate modes of construction, condensed and hydrolyzable tannins display many structural similarities. In particular, attention should be drawn to the *o*-dihydroxy or trihydroxy phenolic nuclei (I, II, III) which are probably of prime importance in the association with proteins and which all project toward the outer surface of the molecular structures. There are, moreover, remarkable similarities in the distances separating the phenolic groups of different catechol and pyrogallol nuclei in the various classes of tannins. For many a separation of 12–15 Å may be discerned, and a notable feature concerns the relationship of these structures to DNA. The overall shape and the spatial relationship between the phenolic groups in vegetable tannins show that many are the optimum size to fit into the larger groove of the DNA double helix and at the same time permit hydrogen bonding of the catechol (or pyrogallol) nuclei to phosphate groups in the DNA chains on opposite sides of the groove. Whereas this may well be a purely coincidental relationship, it may alternatively point to some functional significance (either now or in the past) of vegetable tannins.

Second, with reference to a point made earlier, is the umbrella classification of vegetable tannins a misleading one? In this context it is worth noting that their elaboration during the biosynthesis of the difficultly soluble polymeric forms of proanthocyanidins raises the question whether these metabolites have a role that is principally structural in character. Bate-Smith has consistently drawn attention to the empirical relationship existing between the presence of proanthocyanidins in the leaves and the "woodiness" of a particular plant species or family. Proanthocyanidins first appear, from the phylogenetic point of view, as plants develop a vascular character. In plants such as ferns proanthocyanidins are frequently encountered, and it may be significant that it is the polymeric forms that constitute the major proportion of the metabolites found in these plants. Although Bate-Smith generally discounted any connection with lignification itself, an examination of proanthocyanidins to ascertain if they have some structural role may well repay further study.

ACKNOWLEDGMENTS

As a young scientist, the author was once advised to "choose a difficult problem and one that may seem at the time to interest very few others." Vegetable tannins have provided such a problem. Although once dismissed as unwanted debris of plant metabolism, they have now, 20 yr later, not only assumed a key significance in numerous interactions between plants and other organisms but apparently will play a leading role in the debate when the questions of secondary plant metabolism are finally resolved. The author wishes to acknowledge his debt to Professor R. D. Haworth, F. R. S., for his introduction to this field and for his constant encouragement, to Professor W. Mayer and the late Professor O. Th. Schmidt for their criticism and advice, and to Dr. E. C. Bate-Smith, to whom much of our present knowledge of the biochemistry of vegetable tannins is due, for his unfailing enthusiasm and for the pertinent nagging questions he has continually posed.

REFERENCES

Andra, K., Gorner, A., and Mayer, W. (1977). *Justus Liebig's Ann. Chem.* p. 1976.

Armitage, R., Bayliss, G. S., Gramshaw, J. W., Haslam, E., Haworth, R. D., Jones, K., Rogers, H. J., and Searle, T. (1961). *J. Chem. Soc.* p. 1842.

Armitage, R., Haslam, E., Haworth, R. D., and Searle, T. (1962). *J. Chem. Soc.* p. 3808.

Bahr, W., Ebert, W., Goritz, K., Marx, H.-D., and Weinges, K. (1969). *Fortschr. Chem. Org. Naturst.* **27**, 158.

Bate-Smith, E. C. (1954). *Food* **23**, 124.

Bate-Smith, E. C. (1962). *J. Linn. Soc. London, Bot.* **58**, 95.

Bate-Smith, E. C. (1972). *Phytochemistry* **11**, 1153, 1755.

Bate-Smith, E. C. (1973). *Phytochemistry* **12**, 907.

Bate-Smith, E. C. (1975). *Phytochemistry* **14**, 1107.

Bate-Smith, E. C. (1977). *Phytochemistry* **16**, 1421.

Bate-Smith, E. C., and Lerner, N. H. (1954). *Biochem. J.* **58**, 126.

Bate-Smith, E. C., and Metcalfe, C. R. (1957). *J. Linn. Soc. London, Bot.* **55**, 669.

Bate-Smith, E. C., and Swain, T. (1954). *Chem. Ind. (London)* p. 433.

Bate-Smith, E. C., and Swain, T. (1962). *In* "Comparative Biochemistry" (H. S. Mason and A. M. Florkin, eds.), Vol. 3, p. 764. Academic Press, New York.

Baur, P. S., and Walkinshaw, C. H. (1974). *Can. J. Bot.* **52**, 615.

Bilzer, W., Mayer, W., and Schilling, G. (1976a). *Justus Liebig's Ann. Chem.* p. 876.

Bilzer, W., Busath, H., Günther, A., Mayer, W., and Schilling, G. (1976b). *Justus Liebig's Ann. Chem.* p. 987.

Britton, G., and Haslam, E. (1965). *J. Chem. Soc.* p. 7312.

Bu'Lock, J. D. (1978). *In* "Comprehensive Organic Chemistry" (E. Haslam, ed.), Vol. 5, p. 946. Pergamon, Oxford.

Busath, H., Mayer, W., and Schilling, G. (1976). *Justus Liebig's Ann. Chem.* p. 2169.

Chen, D., El-Basyouni, S. Z., Ibrahim, R. K., Neish, A. C., and Towers, G. H. N. (1964). *Phytochemistry* **3**, 485.

Conn, E. E., and Swain, T. (1961). *Chem. Ind. (London)* p. 592.

Corcoran, M. R., and Green, F. B. (1975). *Plant Physiol.* **56**, 801.

Corcoran, M. R., Geissman, T. A., and Phinney, B. O. (1972). *Plant Physiol.* **47**, 323.

Cornthwaite, D., and Haslam, E. (1965). *J. Chem. Soc.* p. 3008.

Demain, A. L., and Katz, E. (1977). *Bacteriol. Rev.* **41**, 465.

Dewick, P. M., and Haslam, E. (1969). *Biochem. J.* **113**, 537.

Drewes, S. E., Eggers, S. H., Feeney, J., and Roux, D. G. (1967a). *J. Chem. Soc. C* p. 1217.

Drewes, S. E., Eggers, S. H., Feeney, J., Roux, D. G., and Saayman, H. M. (1967b). *J. Chem. Soc. C* p. 1302.

Eckert, R., Schanz, R., Schmidt, O. T., and Wurmb, R. (1967). *Justus Liebig's Ann. Chem.* **706**, 131.

Ferrari, F., Marini-Bettolo, G. B., Monache, F. D., and Poce-Tucci, A. (1972). *Phytochemistry* **11**, 2333.

Fischer, E. (1919). "Untersuchungen uber Depside und Gerbstoffe." Springer-Verlag, Berlin and New York.

Fletcher, A. C., Gupta, R. K., Haslam, E., and Porter, L. J. (1977). *J. Chem. Soc. Perkin Trans.* p. 1628.

Forsyth, W. G. C., and Roberts, J. B. (1960). *Biochem. J.* **74**, 374.

Fraenkel, G. S. (1959). *Science* **129**, 1466.

Freudenberg, K. (1920). "Die Chemie der Naturlichen Gerbstaffe." Springer-Verlag, Berlin and New York.

Freudenberg, K., and Weinges, K. (1960). *Tetrahedron* **8**, 336.

Goldstein, J. L., and Swain, T. (1963). *Phytochemistry* **2**, 271.

Goritz, K., Nader, F., and Weinges, K. (1968). *Justus Liebig's Ann. Chem.* **715**, 164.

Groebke, W., Schanz, R., Schmidt, O. T., and Wurmb., R. (1967). *Justus Liebig's Ann. Chem.* **706**, 154.

Gupta, R. K., and Haslam, E. (1978). *J. Chem. Soc., Perkin Trans.* p. 892.

Gustavson, K. H., and Holm, B. (1952). *J. Am. Leather Chem. Assoc.* **47**, 700.

Handley, W. R. C. (1961). *Plant Soil* **15**, 37.

Harreus, A., Schmidt, O. T., and Wurtell, L. (1965). *Justus Liebig's Ann. Chem.* **690**, 150.

Haslam, E. (1965). *Phytochemistry* **4**, 495.

Haslam, E. (1966). "Chemistry of Vegetable Tannins." Academic Press, New York.

Haslam, E. (1967). *J. Chem. Soc. C* p. 1734.

Haslam, E. (1974). "The Shikimate Pathway." Butterworth, London.

Haslam, E. (1977). *Phytochemistry* **16**, 1625.

Haslam, E., and Haworth, R. D. (1964). *Prog. Org. Chem.* **6**, 1.

Haslam, E., and Stangroom, J. E. (1966). *Biochem. J.* **99**, 28.

Haslam, E., and Uddin, M. (1967). *J. Chem. Soc. C* p. 2381.

Haslam, E., Haworth, R. D., Jones, K., and Rogers, H. J. (1961a). *J. Chem. Soc.* p. 1829.

Haslam, E., Haworth, R. D., and Knowles, P. F. (1961b). *J. Chem. Soc.* p. 1854.

Haslam, E., Haworth, R. D., and Keen, P. C. (1962). *J. Chem. Soc.* p. 3815.

Haslam, E., Jacques, D., Tanner, R. J. N., and Thompson, R. S. (1972). *J. Chem. Soc., Perkin Trans. 1* p. 1837.

Haslam, E., Jacques, D., Opie, C. T., and Porter, L. J. (1977). *J. Chem. Soc., Perkin Trans. 1* p. 1637.

Haworth, R. D., and de Silva, L. B. (1951). *J. Chem. Soc.* p. 3511.

Haworth, R. D., and de Silva, L. B. (1954). *J. Chem. Soc.* p. 3611.

Howes, F. N. (1953). "Vegetable Tanning Materials." Butterworth, London.

Janzen, D. H. (1969). *Science* **165**, 415.

Jochims, J. C., Schmidt, O. T., and Taigel, G. (1968). *Justus Liebig's Ann. Chem.* **717**, 169.

Jochims, J. C., Mayer, W., Schauerte, K., Schilling, G., and Seitz, H. (1971). *Justus Liebig's Ann. Chem.* **751**, 60.

Kaltenhauser, W., Marx, H.-D., Nader, E., Nader, F., Perner, J., Seiler, D., and Weinges, K. (1968). *Justus Liebig's Ann. Chem.* **711**, 184.

Kuhlmann, F., Mayer, W., and Schilling, G. (1971). *Justus Liebig's Ann. Chem.* **747**, 51.

Mayer, W. (1973). *Justus Liebig's Ann. Chem.* p. 1759.

Mayer, W., and Schmidt, O. T. (1951). *Justus Liebig's Ann. Chem.* **571**, 1, 15.

Mayer, W., and Schmidt, O. T. (1956). *Angew. Chem.* **68**, 103.

Mayer, W., Kunz, W., and Loebich, F. (1965). *Justus Liebig's Ann. Chem.* **688**, 232.

Mayer, W., Schick, H., and Schilling, G. (1976). *Justus Liebig's Ann. Chem.* p. 2178.

Mothes, K. (1969). *Experientia* **25**, 225.

Muller, C. H. (1969). *Science* **164**, 197.

Myers, M., and Roberts, E. A. H. (1960). *J. Sci. Food Agric.* **11**, 153.

Nayeshiro, H., Okuda, T., and Seno, K. (1977). *Tetrahedron Lett.* p. 4421.

Nayudamma, Y., Rajadurai, S., Reddy, K. K., and Sastry, K. N. S. (1964). *Aust. J. Chem.* **17**, 238.

Okuda, T., and Seno, K. (1978). *Tetrahedron Lett.* p. 134.

Perkin, A. G., and Uyeda, Y. (1922). *J. Chem. Soc.* p. 66.

Reeves, R. E. (1951). *Adv. Carbohydr. Chem.* **6**, 107.

Robinson, G. M., and Robinson, R. (1933). *Biochem. J.* **27**, 206.

Rosenheim, O. (1920). *Biochem. J.* **14**, 278.

Roux, D. G. (1972). *Phytochemistry* **11**, 1219.

Schmidt, O. T. (1955). *In* "Moderne Methoden der Pflanzenanalyse" (K. Paech and M. V. Tracey, eds.), Vol. 3, p. 526. Springer-Verlag, Berlin and New York.

Schmidt, O. T. (1956). *Fortschr. Chem. Org. Naturst.* **14**, 71.

Schmidt, O. T., and Lademann, R. (1950). *Justus Liebig's Ann. Chem.* **569**, 149.

Schmidt, O. T., and Schmidt, D. M. (1952). *Justus Liebig's Ann. Chem.* **578**, 25.

Schmidt, O. T., Schmidt, D. M., and Herok, J. (1954). *Justus Liebig's Ann. Chem.* **587**, 67.

Seguin, A. (1876). *Ann. Chim. (Paris)* [14]**20**, 15.

Thomas, R. (1978). *In* "Comprehensive Organic Chemistry" (E. Haslam, ed.), Vol. 5, p. 871. Pergamon, Oxford.

White, T. (1957). *J. Sci. Food Agric.* **8**, 378.

Zenk, M. H. (1964). *Z. Naturforsch., Teil B* **19**, 63.

The Betalains: Structure, Biosynthesis, and Chemical Taxonomy

19

MARIO PIATTELLI

I. INTRODUCTION

The brilliant colors of flowers and fruits, which not only elicit considerable aesthetic interest but have also invited scientific investigation, are mainly due to two types of natural pigments, namely, carotenoids and anthocyanins. While the lipid-soluble carotenoids account for the majority of yellow hues, the water-soluble anthocyanins are responsible for most orange, scarlet, crimson, purple, violet, and blue colors. In a restricted number of plants, however, a single class of water-soluble nitrogenous pigments, betalains, provides the colors of flowers and fruits (and sporadically of other plant parts) from yellow through various shades of orange and red to violet. Owing

The Biochemistry of Plants, Vol. 7

ISBN 0-12-675407-1

largely to the inherent difficulty of isolation, which was overcome only with the introduction of chromatographic and electrophoretic techniques, progress in the chemistry of these pigments was for a long time extremely slow. The first real breakthrough occurred in 1957 with the isolation, in crystalline form, of betanin, the red-violet glucoside of beet root (Wyler and Dreiding, 1957; Schmidt and Schönleben, 1957), and in the successive decade the fundamentals of betalain chemistry were firmly established.

After the structural studies, attention was directed toward total synthesis and, with less but not negligible success, toward the machinery of biosynthesis and the intricacies of the mechanism whereby formation of betalains is regulated in the living plant.

II. STRUCTURE AND OCCURRENCE OF BETALAINS

The term "betalain" was introduced (Mabry and Dreiding, 1968) to embrace all compounds whose structure is based on the general expression I and are therefore immonium derivatives of betalamic acid (II). Conjugation

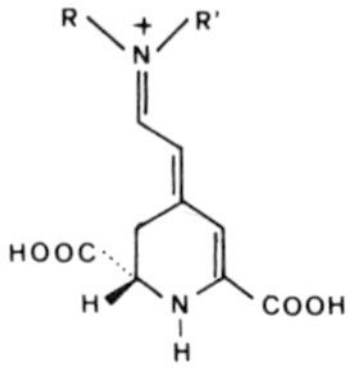

$$\text{I} \qquad\qquad \text{II}$$

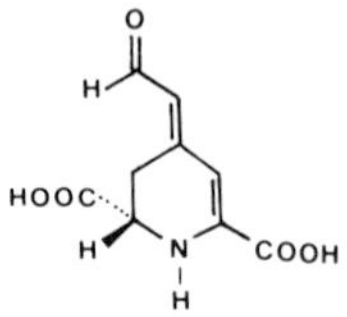

of a substituted aromatic nucleus to the 1,7-diazaheptamethinium chromophore shifts the absorption maximum from about 480 nm in yellow betaxanthins to about 540 nm in red-violet betacyanins.

A. Naturally Occurring Betaxanthins

The archetypal compound representing naturally occurring betaxanthins (Fig. 1) is indicaxanthin, isolated from cactus fruits (*Opuntia ficus-indica*) and shown to have the absolute structure III (Piattelli *et al.*, 1964c). The closely related portulaxanthin (IV) occurs in flowers of *Portulaca grandiflora* (Piattelli *et al.*, 1965a), and vulgaxanthin-I (V) and vulgaxanthin-II (VI) have been obtained from beet root (*Beta vulgaris*) (Piattelli *et al.*, 1965b). The yellow flowers of *Mirabilis jalapa* yield a complex mixture of pigments from which four miraxanthins (VII—X) have been isolated (Piattelli *et al.*, 1965c). Dopaxanthin (XI) has been found in flowers of *Glottiphyllum longum* (Impellizzeri *et al.*, 1973a). Muscaaurin-I (XII), in which the dihydropyridine moi-

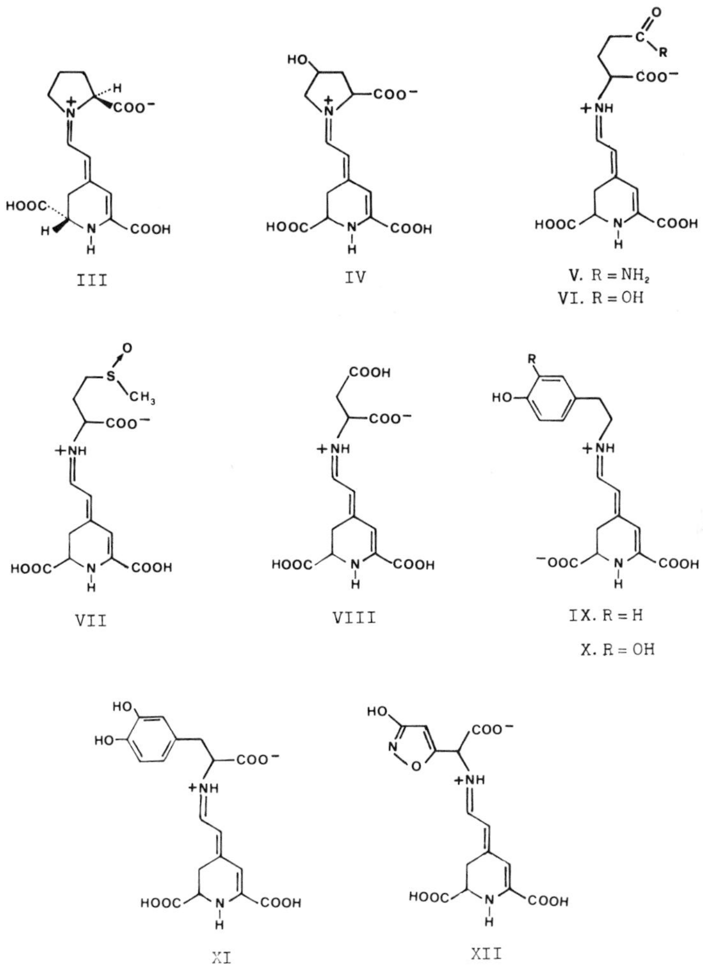

Fig. 1. Naturally occurring betaxanthins.

ety is linked to a nonprotein amino acid (ibotenic acid), has been isolated from the toadstool *Amanita muscaria* (fly agaric) (Musso, 1973). The structures of all these pigments have been substantiated by partial synthesis, but the absolute configuration of the betalamic acid moiety has not been established.

B. Naturally Occurring Betacyanins

A considerable number of different betacyanins may be derived from two basic nuclei, betanidin (XIIIa) and its C-15 epimer isobetanidin (XIIIb)

(Wyler *et al.*, 1963; Wilcox *et al.*, 1965b), by glycosidation of one of the hydroxyl groups located at positions 5 and 6 (no betacyanin is known to have both positions substituted with sugar residues). The sugars present in betacyanins include monosaccharides, disaccharides, and trisaccharides (Fig. 2), and their hydroxyl groups may be acylated by one or more of a variety of inorganic and organic, aliphatic and cinnamic, acids.

Two pairs of isomeric monosides have been recorded: betanin (XIVa) and isobetanin (XIVb), 5-O-β-D-glucopyranosides of betanidin and isobetanidin, respectively (Piattelli *et al.*, 1964a; Wilcox *et al.*, 1965a), and gomphrenin-I (XVa) and -II (XVb), obtained from globe amaranth, *Gomphrena globosa,* and shown to be 6-O-β-D-glucopyranosides of the isomeric aglycones (Minale *et al.*, 1967). Four pairs of biosides are known. Amaranthin (XVIa), the 5-O-[2-O-(β-D-glucopyranosyluronic acid) β-D-glucopyranoside] of betanidin, and its epimer, isoamaranthin (XVIb), have

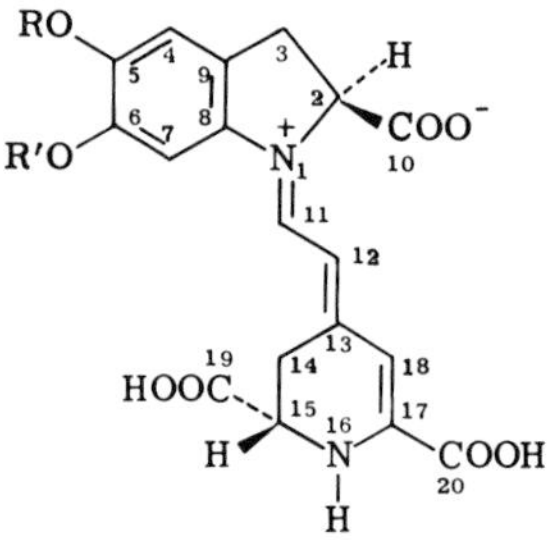

Betanidin series (a) 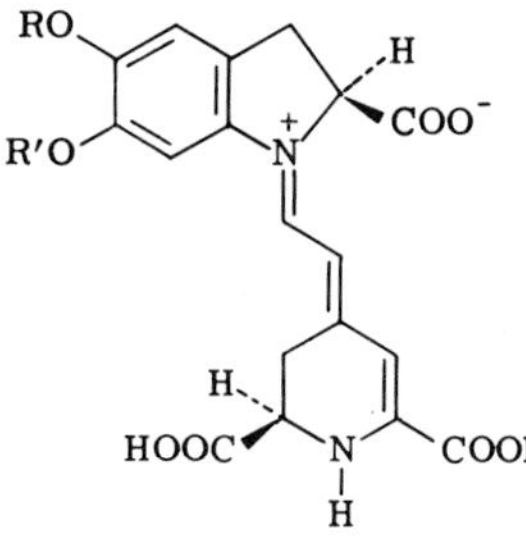   Isobetanidin series (b)

R	R'	a	b
H	H	Betanidin (XIIIa)	Isobetanidin (XIIIb)
β-D-Glucosyl	H	Betanin (XIVa)	Isobetanin (XIVb)
H	β-D-Glucosyl	Gomphrenin-I (XVa)	Gomphrenin-II (XVb)
2'-O-(β-D-Glucosyl-uronic acid)-β-D-glucosyl	H	Amaranthin (XVIa)	Isoamaranthin (XVIb)
β-Sophorosyl	H	Bougainvillein-r-I (XVIIa)	Isobougainvillein-r-I (XVIIb)
H	β-Sophorosyl	Pigment DP3 from bougainvillein-v's (XVIIIa)	Pigment DP4 (XVIIIb)
β-Cellobiosyl	H	Pigment DO1 from oleracins (XIXa)	Pigment DO2 (XIXb)
H	2^G-Glucosyl-rutinosyl	Pigment DP1 from bougainvillein-v's (XXa)	Pigment DP2 (XXb)
H	H	2-Decarboxybetanidin (XXI)	
COOH at C-2 missing			

Fig. 2.   Naturally occurring nonacylated betacyanins.

been isolated from leaves of *Amaranthus tricolor* (Piattelli *et al.*, 1964b; Piattelli and Minale, 1966). From purple bracts of a horticultural variety of *Bougainvillea* ('Mrs. Butt') two epimeric betacyanins, bougainvillein-r-I (XVIIa) and isobougainvillein-r-I (XVIIb), have been isolated and proved to be 5-*O*-β-sophorosides of betanidin and isobetanidin (Piattelli and Imperato, 1970a). The 6-*O*-β-sophoroside of betanidin (XVIIIa) and its epimer (XVIIIb) have been obtained by deacylation of the pigments, bougainvillein-v's, from bracts of *Bougainvillea glabra* var. *sanderiana* (Piattelli and Imperato, 1970b). The isomeric 5-*O*-β-cellobiosides (XIXa and XIXb) have been recovered on deacylation of the pigments, oleracin-I and -II, isolated from *Portulaca oleracea* (Imperato, 1975a). The only known triosides (XXa and XXb), 6-(2^G-glucosylrutinosides) of betanidin and isobetanidin, respectively, have been obtained, along with the 6-*O*-β-sophorosides, by deacylation of the bougainvillein-v's (Piattelli and Imperato, 1970b; Imperato, 1975b).

A large number of acylated pigments have been reported, and the structures of the following monoacyl derivatives totally elucidated: prebetanin, a minor pigment from *B. vulgaris*, identified as betanin 6'-sulfate (Wyler *et al.*, 1967); rivinianin from *Rivinia humilis*, betanin 3'-sulfate (Imperato, 1975c); phyllocactin from *Phyllocactus hybridus*, the malonic acid 6'-half-ester of betanin (Minale *et al.*, 1966); iresinin, the major pigment from *Iresine herbstii*, the 3-hydroxy-3-methylglutaric acid 6'-half-ester of amaranthin (Minale *et al.*, 1966); gomphrenin-III and -V, from *Gomphrena globosa*, 6'-*p*-coumaroyl and 6'-feruloyl esters of gomphrenin-I, respectively (Minale *et al.*, 1967). The occurrence of C-15 epimers of these pigments has also been reported. In addition, a number of acylated pigments of partially known structure have been recorded (Minale *et al.*, 1966, 1967; Piattelli and Impellizzeri, 1969; Piattelli and Imperato, 1970a,b, 1971; Impellizzeri *et al.*, 1973b; Imperato, 1975a). For most of these, the ratio of acid(s) to deacylated pigment has been determined, whereas the exact location of the acyl group(s) has not been ascertained.

Decarboxybetanidin (XXI) isolated from flowers of *Carpobrotus acinaciformis* (Piattelli and Impellizzeri, 1970) is unique among betacyanins in containing a modified aglycone moiety.

III. TOTAL SYNTHESIS OF BETALAINS

The main problem posed by the total synthesis of betalains, namely, synthesis of betalamic acid, has been recently solved (Hermann and Dreiding, 1977; Büchi *et al.*, 1977). This imino acid, originally obtained as a degradation product of betanin (Kimler *et al.*, 1971) or betanidin (Sciuto *et al.*, 1972), reacts readily with amino or imino compounds to give the related betalains.

IV. BIOSYNTHESIS OF BETALAINS

A. Formation of the Betalamic Acid Moiety

Wyler *et al.* (1963) first drew attention to the fact that the betanidin molecule can be dissected into two C_9N units, both formally derivable from 3,4-dihydroxyphenylalanine (DOPA) (Fig. 3). According to this attractive suggestion, one molecule of DOPA undergoes ring cleavage and subsequent ring closure, giving betalamic acid, and a second molecule undergoes oxidative cyclization to cyclo-DOPA. Condensation of betalamic acid with cyclo-DOPA would yield betanidin, whereas reaction with other amino or imino compounds could by analogy yield a variety of betalains. Preliminary experiments showed that labeled DOPA was incorporated into betanidin occurring in the root and hypocotyl of *B. vulgaris* (Hörhammer *et al.*, 1964) and into amaranthin of *Amaranthus* seedlings (Garay and Towers, 1966). However, no attempt was made to locate the label in the radioactive pigments, and the observed incorporation did not establish the involvement of DOPA in the biosynthesis of the betalamic acid moiety. Minale *et al.* (1965), who have investigated the biosynthesis of indicaxanthin in fruits of *O. ficus-indica*, observed that [2-^{14}C]DOPA was incorporated into the dihydropyridine moiety of betalains. However, it should be pointed out that the pigments were not isolated in crystalline form in these studies, and thus their radiopurity is open to question.

In later studies Miller *et al.* (1968) showed that DOPA was incorporated into betanin in high yield in fruits of *Opuntia decumbens* and *O. bergeriana* and that the major proportion of the radioactivity (ca. 90%) appeared in the

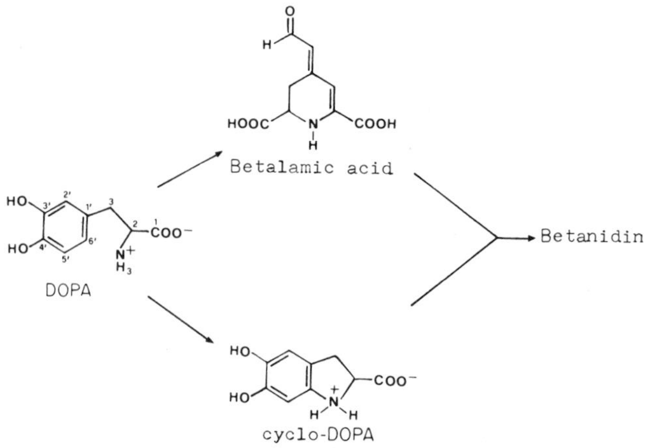

Fig. 3. Speculative scheme for the biosynthesis of betanidin.

dihydropyridine portion of the pigment. In seedlings of *B. vulgaris* considerably less overall incorporation but a higher relative percent of incorporation into the cyclo-DOPA moiety were observed. This result has been confirmed by Liebisch *et al.* (1969) by feeding experiments with beets employing [1-^{14}C, ^{15}N] and [2-^{14}C, ^{15}N] tyrosine. These studies showed incorporation of the labeled atoms into both heterocyclic units of betanin, the cyclo-DOPA group being labeled twice as much as the dihydropyridine portion.

Interest was then centered on more detailed study of the pathway leading from DOPA to the dihydropyridine system of betalains. A priori, this could originate either from an opening of the aromatic ring between the two hydroxyl groups (intradiol cleavage) followed by the bonding of nitrogen to C-5′, or by ring opening between C-4′ and C-5′ (extradiol cleavage) and subsequent closure by the bonding of nitrogen to C-3′. In an attempt to distinguish between these two possibilities, the incorporation of doubly labeled L-[1-^{14}C, 3′,5′-^{3}H]tyrosine into indicaxanthin in fruits of *O. ficus-indica* has been studied (Impellizzeri and Piattelli, 1972). The rationale of this tracer experiment is that, when [3′,5′-^{3}H]tyrosine is enzymatically hydroxylated, half the tritium is statistically eliminated (Guroff *et al.*, 1957) from a position ortho to the hydroxyl group, resulting in the formation of [5′-^{3}H]DOPA without hydroxylation-induced migration (NIH shift) of tritium. The isolation of labeled pigment with a ^{3}H/^{14}C ratio half that of the administered precursor was consistent with extradiol cleavage of the aromatic ring of DOPA distal with respect to the side chain. The alternative route involving an intradiol cleavage would have given a ^{14}C-labeled indicaxanthin free of tritium, although the occurrence of an NIH shift at the hydroxylation step would have invalidated this conclusion. Fischer and Dreiding (1972) obtained the same result in experiments incorporating L-[3′,5′-^{3}H]tyrosine into the betanin of cactus fruits (*O. decumbens*) and supplied independent proof of the absence of an NIH shift during the oxidation of tyrosine to DOPA in this experiment. Finally, Chang *et al.*, (1974) fed DL-[1-^{14}C]DOPA and DL-[2-^{14}C]DOPA to yellow flower buds of *P. grandiflora* and isolated [^{14}C]betalamic acid of the same *S* configuration as L-DOPA. From the above evidence it is possible to propose the following biosynthetic pathway to betanidin and betaxanthins (Fig. 4).

Although it is possible that closure of the dihydropyridine ring occurs after condensation with amino acids or amines, the fact that betalamic acid occurs free in a number of Centrospermae (Kimler *et al.*, 1971) lends support to the argument that this compound is a common intermediate of all betalains.

The proposed mechanism for betalamic acid biosynthesis requires an enzyme of the metapyrocatechase type analogous to those originally obtained from microorganisms (Nozaki *et al.*, 1968) and more recently found in higher plants as well (Saito and Komamine, 1976). If such an enzyme, not wholly specific with respect to the mode of ring opening, catalyzes extradiol cleav-

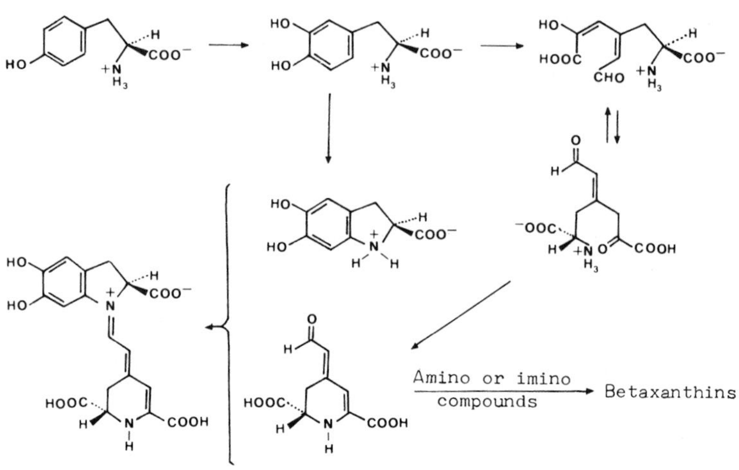

Fig. 4. Possible mechanism for betalain formation.

age of the aromatic ring *proximal* rather than *distal* to the side chain, subsequent closure of the open-chain intermediate (XXII) will lead to an isomer of betalamic acid having structure XXIII. In fact, a yellow compound isolated by Döpp and Musso (1973a,b), along with derivatives of betalamic acid, from fly agaric (*A. muscaria*) and originally assigned structure XXIV, was subsequently shown to possess the biogenetically more plausible dihydroazepine structure XXIII (von Ardenne *et al.*, 1974). This structure has been recently confirmed by a biomimetic type of synthesis (Barth *et al.*, 1979).

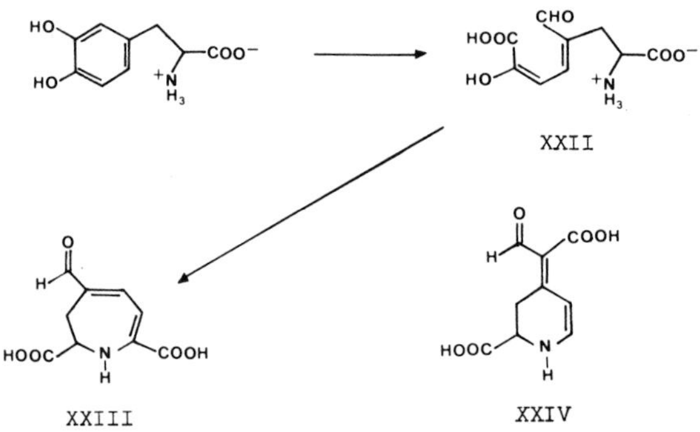

B. Glycosylation

Knowledge of the mechanism by which individual betacyanins are formed from betanidin and isobetanidin is still at a rudimentary stage. Labeled betanidin was efficiently incorporated into betanin when fed to fruits of *Opuntia dillenii* (Sciuto *et al.*, 1972), a result indicating glucosylation of betanidin. However, subsequent work (Sciuto *et al.*, 1974) on the biosynthesis of amaranthin raises the possibility that the observed incorporation of betanidin into betanin is a reaction induced by feeding betanidin. In these experiments, various nonlabeled presumptive precursors of betanin [(S)-cyclo-DOPA, (S)-cyclo-DOPA 5-O-β-D-glucoside, and betanidin] were fed to seedlings of a betaxanthin-producing yellow variety ('Golden Feather') of *Celosia plumosa*. This plant is able to produce amaranthin, the betacyanin normally synthesized by red varieties of the same species, when fed suitable precursors. The much higher rate of utilization of cyclo-DOPA and cyclo-DOPA glucoside as compared to that of betanidin and betanin suggests that the sequence cyclo-DOPA → cyclo-DOPA glucoside → cyclo-DOPA (glucuronic acid) glucoside → amaranthin constitutes the main pathway for amaranthin synthesis in this plant material. This inference is obviously valid only if all the exogenous precursors reach the site of biosynthesis with approximately the same efficiency.

C. Betanidin and Isobetanidin Derivatives

In a given plant, betacyanins are usually found as pairs of C-15 diastereoisomers. In a few plants, however, some betanidin glycosides occur unaccompanied by their isobetanidin counterparts (Piattelli and Minale, 1964), whereas the opposite is never observed. Furthermore, it has been reported (Piattelli *et al.*, 1969) that seedlings of *A. tricolor* contain only amaranthin, while in mature leaves of the same plant the pigment is accompanied by its isomer. These facts might suggest that betanidin pigments are the primary products of metabolism and that the isobetanidin derivatives are formed by spontaneous epimerization.

V. REGULATION OF BETALAIN BIOSYNTHESIS

A. Genetic Control

The most important genetic studies on betalain inheritance have been carried out with *P. grandiflora* by Japanese geneticists (Yasui, 1920; Ikeno, 1921, 1922, 1924, 1928; Enamoto, 1923, 1927), who established that white forms, not always completely devoid of pigment, are recessive to all colored forms. Within the pink and yellow series, three color shades, expressed by multiple allelomorphs of the *C* gene, have been distinguished (Yamaguti, 1935). More recently, a number of color forms and their F_1 hybrids have been surveyed for betalains by Ootani and Hagiwara (1969); purple flowers contain primarily betacyanins and trace amounts of betaxanthins, whereas the opposite applies to yellow flowers. Red, orange, and intermediate colors are due to the presence of varying proportions of betacyanins and betaxanthins. Betacyanins are, in general, dominant to betaxanthins and, in F_1 hybrids, the higher betacyanin content in purple forms is dominant to the lower content in pink forms, whereas the higher content of betaxanthin in yellow-orange forms is recessive to the lower content in pink forms. A genetic analysis of betalain formation in petals of *P. grandiflora* in sterile culture has also been carried out (Adachi and Katayama, 1970).

Little genetic information is available regarding betalains in other plants of the Centrospermae. Butterfass (1968) has established that gene *R* for hypocotyl color of *B. vulgaris* is located on chromosome II, and Shevtsov and Trukhanov (1969) have carried out preliminary studies on the genetics of betalain biogenesis in this plant.

B. Extrinsic Factors

1. Light

The effect of light on betalain biosynthesis has received a great deal of attention. Light stimulates additional betalain synthesis in species that normally form substantial amounts in the dark; in contrast, it is a compulsory requirement for pigment synthesis in species that in darkness synthesize betalins to a very limited extent or not at all (Wohlpart and Mabry, 1968).

Far-red reversibility, suggesting phytochrome involvement, of light-induced betalain synthesis has been demonstrated in seedlings of *Amaranthus tricolor* (Piattelli *et al.*, 1969; Colomas and Bulard, 1975), *A. caudatus* (Köhler, 1972b; French *et al.*, 1973), *Chenopodium rubrum* (Wagner and Cumming, 1970), *Celosia plumosa* (Giudici de Nicola *et al.*, 1973a), and *C. cristata* (Giudici de Nicola *et al.*, 1974). However, no phytochrome control could be demonstrated in *Amaranthus salicifolius* (Heath and Vince, 1962). Kendrick and Frankland (1969) have shown that the concentration of phytochrome in *A. caudatus* progressively increases from germination for 72 h and attains a plateau thereafter; the light requirement of the seedlings for betalain synthesis evolves concomitantly, the pigment formation in response to illumination being maximal between 72 and 96 h (Köhler, 1972b).

Photocontrol of betalain synthesis presumably occurs via gene activation. This presumption is based on the action of inhibitors of nucleic acid and protein biosynthesis on pigment synthesis (Birnbaum and Köhler, 1970; Köhler and Birnbaum, 1970b; Stobart *et al.*, 1970; Piattelli *et al.*, 1970a,b; Giudici de Nicola *et al.*, 1972a; Köhler, 1972a, 1975) and is in accord with Mohr's (1966a,b, 1970) extensive studies on anthocyanin synthesis in mustard seedlings. Photoreceptor(s) other than phytochrome may also be involved (Giudici de Nicola *et al.*, 1972a; Köhler, 1973). No doubt, although light clearly controls the ultimate expression of gene activation, demonstration of increased rates of enzyme synthesis is also required in order to conclude that gene activation is the primary regulatory process following photoreception, since direct activation of a preexisting enzyme could provide the same end result.

Under continuous illumination with white light (i.e., under conditions of "high-energy reaction"), photosynthesis seems to be the most important factor governing pigment synthesis. Indeed, the inhibition of chlorophyll formation following the administration of levulinic acid is paralleled by a strong decrease in pigment accumulation. Furthermore, betalain formation stimulated by continuous illumination, in contrast to pigment accumulation induced by short-term irradiation, is inhibited not only by salicylaldoxime and 2,4-dinitrophenol but also by 3-(3,4-dichlorophenyl)-1,1-dimethylurea, a specific inhibitor of photophosphorylation (Giudici de Nicola *et al.*, 1972b, 1973a, 1974).

The evidence outlined thus far indicates that the effect of light on betalain synthesis may be exercised at two different levels involving the activation of genes (direct or not) and the availability of energy-rich compounds. The effect of light on the gene system appears to be mediated by photoreceptor(s) other than phytochrome, whereas that on the availability of the energy-rich compounds seems to be mediated by phytochrome in the early hours of irradiation and successively by the photosynthetic system.

The far-red high-energy reaction associated with betalain synthesis in species able to synthesize substantial amounts of pigment in darkness (*C. plumosa, C. cristata,* and *A. caudatus*) is dependent on photosynthesis, whereas continuous far-red light is without effect on *A. tricolor,* which has an absolute light requirement for pigment production (Giudici de Nicola *et al.,* 1973b, 1974). With the assumption that the genes involved in pigment formation are inactive in etiolated seedlings of this species, and taking into account the fact that continuous far-red light is known to maintain a low but constant level of active P_{fr} form (Hartmann, 1966), one might conclude that gene activation is not the primary action of phytochrome. Moreover, the fact that, in *A. tricolor,* pigment synthesis is strongly stimulated by continuous white light, whereas far-red light is ineffective, suggests that photoactivation of the genes coding for the enzymes involved in betalain formation requires wavelengths other than 720 nm (Giudici de Nicola *et al.,* 1973b).

Rast *et al.* (1973) have observed that dibutyryl 3',5'-cyclic adenosine monophosphate (cAMP) can satisfy the light requirement for the synthesis of betacyanins in etiolated seedlings of *Amaranthus paniculatus*. They therefore have proposed that phytochrome exerts its controlling role through the agency of cAMP. However, Elliott and Murray (1975) found no significant increase in cAMP content in *Amaranthus* seedlings under conditions where induction was taking place. Their results suggest that dibutyryl cAMP acts as a cytokinin analogue rather than as a cyclic nucleotide. The additive effects of light and dibutyryl cAMP on amaranthin accumulation (Giudici de Nicola *et al.*, 1975b) provide a further argument against the involvement of cAMP in the photoactivation of betalain synthesis.

Little information is available concerning which steps of the biosynthetic pathway are subject to photocontrol, and studies on the enzymology of betalain synthesis are clearly required before assessment of the level of control by individual enzymes can be made. Experiments by French *et al.* (1973, 1974) indicated that exogenously supplied DOPA elicited a moderate stimulation of dark synthesis of amaranthin, and tyrosine a still smaller stimulation in seedlings of *A. caudatus* var. *viridis*. In illuminated seedlings the accumulation of pigment is considerably greater and of similar magnitude in the presence of either of these precursors. On the basis of this and related observations French *et al.* (1973, 1974) have suggested that photocontrol is exerted on at least two parts of the biosynthetic pathway, one between tyrosine and DOPA, and a second between DOPA and amaranthin. However, Giudici de Nicola *et al.* (1975a), who used isolated cotyledons of *A. caudatus* and *A. tricolor* to prevent oxidation of substrates by roots to melanin-like black pigments, found no evidence for a controlling step between tyrosine and DOPA. In this case DOPA had a stimulating effect greater than that of tyrosine under all the experimental conditions employed. In illuminated cotyledons fed with either precursor amaranthin accumulation was much greater than in dark controls, whereas cyclo-DOPA failed to enhance pigment production both in darkness and in the light. Such results suggest that the photocontrol of betalain synthesis is exerted at the level of formation of the dihydropyridine portion of the molecule.

2. Chemical Induction

a. Hormones and Growth Modifiers. Kinetin (6-furfurylaminopurine) is able to replace the light requirement for betalain production (Bamberger and Mayer, 1960; Köhler, 1967, 1972b; Piattelli *et al.*, 1971; Giudici de Nicola *et al.*, 1973a; Mazin *et al.*, 1976). However, experiments with kinetin are difficult to interpret, since it is known to affect an extremely wide range of developmental processes in plants. Nevertheless, it is readily apparent that the effect of kinetin on pigment accumulation is large in plants with a light requirement and moderate in species capable of synthesis in the dark. Since the effect of kinetin has been ascribed to gene activation in analogy to the effects of light and the depressing action of inhibitors of nucleic acid and protein biosynthesis, this differential response to the hormone has been interpreted as being due to the fact that the genes for betalain synthesis are already "open" in etiolated seedlings of plants lacking an absolute light requirement (Giudici de Nicola *et al.*, 1973a,b). Salicylaldoxime and 2,4-

dinitrophenol inhibit the pigment biosynthesis stimulated by kinetin, suggesting that oxidative phosphorylation is involved in the regulation of betalain production (Giudici de Nicola *et al.,* 1972b). In general, these results have suggested that kinetin, like light, acts at two different levels, namely, gene activation and control of the availability of energy-rich compounds.

Evidence has been presented (Elliott, 1977) that betacyanin synthesis in *Amaranthus* seedlings is enhanced by K^+ ions and by fusicoccin which is known to control K^+ transport across membranes. It therefore has been suggested that the primary general action of cytokinins in enzyme induction is to increase membrane permeability to K^+ ions. Since light-induced activation of the genetic system could also be a secondary manifestation of altered membrane permeability to K^+ (Kroeger and Lezzi, 1966), it is not impossible that a unique regulatory process is operative in both light- and kinetin-stimulated synthesis, i.e., a change in permeability to potassium ions.

Exogenous gibberellic acid (GA_3) substantially inhibits light-induced amaranthin production in *A. caudatus* seedlings, and growth retardants (Phosphon D, CCC, AMO 1618) enhance the pigment level (Stobart *et al.,* 1970; Kinsman *et al.,* 1975a,b; Stobart and Kinsman, 1977). Apparently GA_3 controls the availability of tyrosine, possibly by its diversion to increased protein synthesis (Laloraya *et al.,* 1976). Comparable results have been obtained (Laloraya, 1970; Srivastav and Laloraya, 1973, 1976) for dark-grown *Celosia argentea* var. *plumosa* seedlings. Kinetin-stimulated pigment synthesis in *A. tricolor* is also inhibited by GA_3 (Colomas *et al.,* 1973). It remains doubtful whether these effects of gibberellin are of real physiological importance; however, the endogenous GA_3 level, as well as applied GA_3, affects pigment production (Kinsman *et al.,* 1975b).

Abscisic acid inhibits light-stimulated betacyanin synthesis in seedlings of *A. tricolor* (Stobart *et al.,* 1970) and both light- and kinetin-induced synthesis in *A. caudatus,* possibly by decreasing membrane permeability (Biddington and Thomas, 1977).

b. Other Chemical Inductors. Numerous other chemical agents have been reported to affect betalain synthesis.

Precursors such as DOPA and tyrosine increase betalain accumulation when administered to seedlings of *Amaranthus* (Garay and Towers, 1966; Köhler, 1967; French *et al.,* 1973, 1974; Giudici de Nicola *et al.,* 1975a). However, no such increase was observed for betanin synthesis in callus cultures (Constabel and Nassif-Makki, 1971) and in excised leaf discs (Wohlpart and Black, 1973) of *B. vulgaris*. Consequently it has been suggested that the developmental stage of the tissue rather than the availability of exogenous precursors is the limiting factor in pigment formation (Constabel and Nassif-Makki, 1971).

Added sucrose stimulates betacyanin accumulation in leaf discs of red beet plants (Wohlpart and Black, 1973). Potassium nitrate enhances pigment synthesis in *A. caudatus* seedlings (Köhler and Birnbaum, 1970a).

Endress (1977) studied the effect of cAMP, theophylline, papaverine, and ammonium nitrate on betacyanin accumulation in callus of *P. grandiflora*. Theophylline ($\geqq 5 \times 10^{-5}\ M$) and ammonium nitrate enhanced pigment accumulation through stimulation of *de novo* synthesis of enzymes. Inhibition by papaverine and theophylline ($\leqq 10^{-5}\ M$) was ascribed to regulation of callus phosphodiesterase which could influence the concentration of nucleotides and subsequently affect tyrosine biosynthesis.

VI. FUNCTION OF BETALAINS

As in the case of other "secondary metabolites", it is impossible to assign a definite *raison d'être* to betalains in the economy of the organisms that produce them. When present in flowers or fruits they may, like anthocyanins, act as attractants for vectors that ensure adequate pollination or as agents of seed dispersal by animals. However, their occurrence in other plant parts, such as leaves, stems, and roots, seems to have no immediate function. Also, the transient coloration of many seedlings of the Centrospermae and the reddening of senescent leaves of several plants of this order have no obvious physiological or ecological significance. Whatever its significance, the process resembles the analogous phenomenon observed in anthocyanin-producing species. Betalain formation in wounded tissues is perhaps a defense mechanism against viral infection, since these pigments have an inhibitory effect on viral reproduction (Sosnova, 1970).

Stenlid (1976) has found that betalains (betanin and vulgaxanthin) are effective inhibitors of indoleacetic acid (IAA) oxidase and that betanin counteracts the inhibitory effect of IAA on wheat root elongation. There are no indications that betalains play any role *in vivo* as regulators of IAA activity, but they must be considered potential modifiers of auxin metabolism.

VII. CHEMICAL TAXONOMY OF BETALAINS

The potential importance of betalains to plant taxonomy was evaluated by several authors long before their structure was clarified. As early as 1876 Bischoff was aware that an unusual violet pigment occurred in a restricted group of taxonomically related plants. His list included representatives of the families Amaranthaceae, Chenopodiaceae, Phytolaccaceae, Portulacaceae and, erroneously, Polygonaceae (the color tests on which he had to rely to distinguish between what we now call betacyanins and anthocyanins were not infallible). The last family was shown to have been incorrectly included by Gertz (1906), who not only brought to seven the number of families recognized as containing *Rübenroth* (betacyanin) with the addition of examples from the Nyctaginaceae, Aizoaceae, and Basellaceae but also stated clearly that all the *Rübenroth* families belonged to a single order, the Centrospermae. Kryz (1919, 1920) perceived that the red pigment of the flowers and fruits of *Nopalxochia phyllanthoides* gave the typical color reactions of *Rübenroth*. He thus further expanded the list of betacyanin families with the addition of the Cactaceae, a family sometimes treated as a separate order (Cactales syn. Opuntiales) but considered as properly belonging to the order Centrospermae in most modern classifications. These early observations, which were essentially based on color reactions, have been confirmed by Reznik (1955, 1957) with the aid of the more refined techniques of chromatography and electrophoresis. Recent additions to the list of betalain families are the Didieraceae (Rauh and Reznik, 1961), a small family from Madagascar, and the Stegnospermaceae (Mabry *et al.*, 1963), a taxon generally considered a subfamily (Stegnospermatoideae) of the Phytolaccaceae but elevated to the family status and placed in the Pittosporales by Hutchinson (1959).

Among the classically constituted Centrospermae, only the Caryophyllaceae synthesize anthocyanins instead. This remarkable correlation between chemical and morphological characters has led Mabry (1964) to propose that the order Centrospermae (Chenopodiales), including the Cactaceae, be reserved for betalain-containing families and that the anthocyanin-containing Caryophyllaceae be separated into the related but distinct order Caryophyllales. The totally different chemical structure of betalains and anthocyanins, the fact that they are mutually exclusive (other classes of flavonoids are, however, common in betalain families), and the restricted distribution of betalains are good arguments in favor of the paramount taxonomic significance ascribed by Mabry to these pigments. Such arguments are not lessened by the finding of their occurrence in *A. muscaria* (Döpp and Musso, 1973a,b), which lacks any phylogenetic relation to Centrospermae plants. This occurrence may simply be a case of chemical convergence under evolutionary pressure, or perhaps no more than an isolated biochemical oddity.

On the basis of present-day knowledge, it is difficult to say whether there is any correlation between betalain patterns and the botanical classification of plants of the Centrospermae at lower systematic levels. However, some generalities can be extracted from the data at hand. Thus, for instance, amaranthin appears to be restricted to the Amaranthaceae and Chenopodiaceae, and phyllocactin to the Cactaceae and Chenopodiaceae. Hopefully further investigation will provide data shedding light on this point and at the same time describing more precisely the order Centrospermae on the basis of the betalain criterion. In this connection there are a number of small angiosperm families that have been considered by one or more authorities, on purely morphological grounds, as belonging to (or being closely related) the betalain-producing families and that await examination of their pigments:

1. Achatocarpaceae. Placed by Hutchinson (1959) in the Bixales near the Thymeleales, this small group has been included in the Centrospermae by Engler (1964).

2. Agdestidaceae. Often assigned to the Phytolaccaceae, this taxon from central and tropical South America has been elevated to the family rank by Hutchinson (1959) and placed in the Chenopodiales near the Basellaceae.

3. Barbeuiaceae. The genus *Barbeuia* from Madagascar, formerly in the Phytolaccaceae, has been considered a family of the Chenopodiales by Hutchinson (1959).

4. Batidaceae. Included in the Caryohyllales and the Chenopodiales by Bessey (1915) and Hutchinson (1959), respectively, this group has been recognized as a separate order, Batales, by Engler (1964) and placed near the Rosales.

5. Dysphaniaceae. The Australian genus *Dysphania,* considered by Hutchinson (1959) to belong to the Chenopodiaceae, has been elevated to family status by Engler (1964).

6. Gyrostemonaceae. A small group confined to Australia and Tasmania and usually included in the Phytolaccaceae, it was elevated to the rank of family and placed in the Centrospermae by Engler (1964).

7. Halophytaceae. This family from South America contains the single genus *Halophytum,* included by Hutchinson (1959) in the Chenopodiaceae.

REFERENCES

Adachi, T., and Katayama, Y. (1970). *Bull. Fac. Agric., Univ. Miyazaki* **16,** 137–145.

Bamberger, E., and Mayer, A. M. (1960). *Science* **131,** 1094–1095.

Barth, H., Kobayashi, M., and Musso, H. (1979). *Helv. Chim. Acta* **62,** 1231–1235.

Bessey, C. E. (1915). *Ann. Mo. Bot. Gard.* **2,** 109–164.

Biddington, N. L., and Thomas, T. H. (1977). *Physiol. Plant.* **40,** 312–314.

Birnbaum, D., and Köhler, K.-H. (1970). *Biochem. Physiol. Pflanz.* **161,** 521–531.

Bischoff, H. (1876). Inaugural Dissertation, University of Tübingen.

Büchi, G., Fliri, H., and Shapiro, R. (1977). *J. Org. Chem.* **42,** 2192–2194.

Butterfass, T. (1968). *Theor. Appl. Genet.* **38,** 348–350.

Chang, C., Kimler, L., and Mabry, T. J. (1974). *Phytochemistry* **13,** 2771–2775.

Colomas, J., and Bulard, C. (1975). *Planta* **124,** 245–254.

Colomas, J., Laytou, J.-L., and Bulard, C. (1973). *C. R. Hebd. Seances Acad. Sci.* **277,** 173–176.

Constabel, F., and Nassif-Makki, H. (1971). *Ber. Dtsch. Bot. Ges.* **84,** 629–636.

Döpp, H., and Musso, H. (1973a). *Naturwissenschaften* **60,** 477–478.

Döpp, H., and Musso, H. (1973b). *Chem. Ber.* **106**, 3473–3482.

Elliott, D. C. (1977). *In* "Regulation of Cell Membrane Activities in Plants" (E. Marrè and O. Ciferri eds.), pp. 317–323. Elsevier, Amsterdam.

Elliott, D. C., and Murray, A. W. (1975). *Biochem. J.* **146**, 333–337.

Enamoto, N. (1923). *Jpn. J. Bot.* **1**, 137–151.

Enamoto, N. (1927). *Jpn. J. Bot.* **3**, 267–288.

Endress, R. (1977). *Phytochemistry* **16**, 1549–1554.

Engler, A. (1964). *In* "Syllabus der Pflanzenfamilien" (H. Melchior, ed.), 12th ed., Vol. II. Borntraeger, Berlin.

Fischer, N., and Dreiding, A. S. (1972). *Helv. Chim. Acta* **55**, 649–658.

French, C. J., Pecket, R. C., and Smith, H. (1973). *Phytochemistry* **12**, 2887–2891.

French, C. J., Pecket, R. C., and Smith, H. (1974). *Phytochemistry* **13**, 1505–1511.

Garay, A. S., and Towers, G. H. N. (1966). *Can. J. Bot.* **44**, 231–236.

Gertz, O. (1906). Dissertation, University of Lund.

Giudici de Nicola, M., Piattelli, M., Castrogiovanni, V., and Molina, C. (1972a). *Phytochemistry* **11**, 1005–1010.

Giudici de Nicola, M., Piattelli, M., Castrogiovanni, V., and Amico, V. (1972b). *Phytochemistry* **11**, 1011–1017.

Giudici de Nicola, M., Piattelli, M., and Amico, V. (1973a). *Phytochemistry* **12**, 353–357.

Giudici de Nicola, M., Piattelli, M., and Amico, V. (1973b). *Phytochemistry* **12**, 2163–2166.

Giudici de Nicola, M., Amico, V., and Piattelli, M. (1974). *Phytochemistry* **13**, 439–442.

Giudici de Nicola, M., Amico, V., Sciuto, S., and Piattelli, M. (1975a). *Phytochemistry* **14**, 479–481.

Giudici de Nicola, M., Amico, V., and Piattelli, M. (1975b). *Phytochemistry* **14**, 989–991.

Guroff, G., Daly, J. W., Jerina, D. M., Renson, J., Witkop, B., and Udenfriend, S. (1957). *Science* **157**, 1524–1530.

Hartmann, K. M. (1966). *Photochem. Photobiol.* **5**, 349–366.

Heath, O. V. S., and Vince, D. (1962). *Symp. Soc. Exp. Biol.* **16**, 114–137.

Hermann, K., and Dreiding, A. S. (1977). *Helv. Chim. Acta* **60**, 673–683.

Hörhammer, L., Wagner, H., and Fritzsche, W. (1964). *Biochem. Z.* **339**, 398–400.

Hutchinson, J. (1959). "The Families of Flowering Plants," Vol. I. Oxford Univ. Press (Clarendon), London and New York.

Ikeno, S. (1921). *J. Coll. Agric., Tokyo Imp. Univ.* **8**, 93–133.

Ikeno, S. (1922). *Z. Indukt. Abstamm.- Vererbungsl.* **29**, 122–135.

Ikeno, S. (1924). *Jpn. J. Bot.* **2**, 45–62.

Ikeno, S. (1928). *Jpn. J. Bot.* **4**, 189–218.

Impellizzeri, G., and Piattelli, M. (1972). *Phytochemistry* **11**, 2499–2502.

Impellizzeri, G., Piattelli, M., and Sciuto, S. (1973a). *Phytochemistry* **12**, 2293–2294.

Impellizzeri, G., Piattelli, M., and Sciuto, S. (1973b). *Phytochemistry* **12**, 2295–2296.

Imperato, F. (1975a). *Phytochemistry* **14**, 2091–2092.

Imperato, F. (1975b). *Phytochemistry* **14**, 2526.

Imperato, F. (1975c). *Phytochemistry* **14**, 2526–2527.

Kendrick, R. E., and Frankland, B. (1969). *Planta* **86**, 21–32.

Kimler, L., Larson, R. A., Messenger, R., Moore, J. B., and Mabry, T. J. (1971). *Chem. Commun.* pp. 1329–1330.

Kinsman, L. T., Pinfield, N. J., and Stobart, A. K. (1975a). *Planta* **127**, 149–152.

Kinsman, L. T., Pinfield, N. J., and Stobart, A. K. (1975b). *Planta* **127**, 207–212.

Köhler, K.-H. (1967). *Ber. Dtsch. Bot. Ges.* **80**, 403–415.

Köhler, K.-H. (1972a). *Phytochemistry* **11**, 127–131.

Köhler, K.-H. (1972b). *Phytochemistry* **11**, 133–137.

Köhler, K.-H. (1973). *Biol. Zentralbl.* **92**, 307–336.

Köhler, K.-H. (1975). *Biochem. Physiol. Pflanz.* **168**, 113–122.

Köhler, K.-H., and Birnbaum, D. (1970a). *Biol. Zentralbl.* **89**, 201–211.

Köhler, K.-H., and Birnbaum, D. (1970b). *Biochem. Physiol. Pflanz.* **161**, 511–520.

Kroeger, H., and Lezzi, M. (1966). *Annu. Rev. Entomol.* **11**, 1–22.

Kryz, F. (1919). *Z. Unters. Nahr.- Genussm. Gebranchsgegenstaende* **38**, 364–365.

Kryz, F. (1920). *Oesterr. Chem.-Ztg.* **23**, 55–56.

Laloraya, M. M. (1970). *Indian J. Plant Physiol.* **13**, 1–14.

Laloraya, M. M., Srivastav, H. N., and Guruprasad, K. N. (1976). *Planta* **128**, 275–276.

Liebisch, H. W., Matschiner, B., and Schuette, H. R. (1969). *Z. Pflanzenphysiol.* **61**, 269–278.

Mabry, T. J. (1964). *In* "Taxonomic Biochemistry and Serology" (C. A. Leone, ed.), pp. 239–254. Ronald Press, New York.

Mabry, T. J., and Dreiding, A. S. (1968). *Recent Adv. Phytochem.* **1**, 145–160.

Mabry, T. J., Taylor, A., and Turner, B. R. (1963). *Phytochemistry* **2**, 61–64.

Mazin, V., Shashkova, L. S., Andreev, L. N., Komizerko, E. I., Zhloba, N. M., and Kefeli, V. I. (1976). *Dokl. Akad. Nauk SSSR* **231**, 506–509.

Miller, H. E., Rösler, H., Wohlpart, A., Wyler, H., Wilcox, M. E., Frohofer, H., Mabry, T. J., and Dreiding, A. S. (1968). *Helv. Chim. Acta* **51**, 1470–1474.

Minale, L., Piattelli, M., and Nicolaus, R. A. (1965). *Phytochemistry* **4**, 593–597.

Minale, L., Piattelli, M., De Stefano, S., and Nicolaus, R. A. (1966). *Phytochemistry* **5**, 1037–1052.

Minale, L., Piattelli, M., and De Stefano, S. (1967). *Phytochemistry* **6**, 703–709.

Mohr, H. (1966a). *Photochem. Photobiol.* **5**, 469–483.

Mohr, H. (1966b). *Z. Pflanzenphysiol.* **54**, 63–83.

Mohr, H. (1970). *Biol. Rundsch.* **23**, 187–194.

Musso, H. (1973). *Chimia* **27**, 659.

Nozaki, M., Ono, K., Nakazawa, T., Kotani, S., and Hayaishi, O. (1968). *J. Biol. Chem.* **243**, 2682–2690.

Ootani, S., and Hagiwara, T. (1969). *Jpn. J. Genet.* **44**, 65–79.

Piattelli, M., and Impellizzeri, G. (1969). *Phytochemistry* **8**, 1595–1596.

Piattelli, M., and Impellizzeri, G. (1970). *Phytochemistry* **9**, 2553–2556.

Piattelli, M., and Imperato, F. (1970a). *Phytochemistry* **9**, 455–458.

Piattelli, M., and Imperato, F. (1970b). *Phytochemistry* **9**, 2557–2560.

Piattelli, M., and Imperato, F. (1971). *Phytochemistry* **10**, 3133–3134.

Piattelli, M., and Minale, L. (1964). *Phytochemistry* **3**, 547–557.

Piattelli, M., and Minale, L. (1966). *Ann. Chim. (Rome)* **56**, 1060–1064.

Piattelli, M., Minale, L., and Prota, G. (1964a). *Ann. Chim. (Rome)* **54**, 955–962.

Piattelli, M., Minale, L., and Prota, G. (1964b). *Ann. Chim. (Rome)* **54**, 963–968.

Piattelli, M., Minale, L., and Prota, G. (1964c). *Tetrahedron* **20**, 2325–2329.

Piattelli, M., Minale, L., and Nicolaus, R. (1965a). *Rend. Accad. Sci. Fis. Mat., Naples* **32**, 55–56.

Piattelli, M., Minale, L., and Prota, G. (1965b). *Phytochemistry* **4**, 121–125.

Piattelli, M., Minale, L., and Nicolaus, R. A. (1965c). *Phytochemistry* **4**, 817–823.

Piattelli, M., Giudici de Nicola, M., and Castrogiovanni, V. (1969). *Phytochemistry* **8**, 731–736.

Piattelli, M., Giudici de Nicola, M., and Castrogiovanni, V. (1970a). *Atti Accad. Naz. Lincei, Cl. Sci. Fis., Mat. Nat., Rend.* [8] **48**, 255–260.

Piattelli, M., Giudici de Nicola, M., and Castrogiovanni, V. (1970b). *Phytochemistry* **9**, 785–789.

Piattelli, M., Giudici de Nicola, M., and Castrogiovanni, V. (1971). *Phytochemistry* **10**, 289–293.

Rast, D., Skřivanová, R., and Bachofen, R. (1973). *Phytochemistry* **12**, 2669–2672.

Rauh, W., and Reznik, H. (1961). *Bot. Jahrb.* **81**, 94–105.

Reznik, H. (1955). *Z. Bot.* **43**, 499–530.

Reznik, H. (1957). *Planta* **49**, 406–434.

Saito, K., and Komamine, A. (1976). *Eur. J. Biochem.* **68**, 237–243.

Schmidt, O. T., and Schönleben, W. (1957). *Z. Naturforsch., Teil B* **12**, 262–263.

Sciuto, S., Oriente, G., and Piattelli, M. (1972). *Phytochemistry* **11**, 2259–2262.

Sciuto, S., Oriente, G., Piattelli, M., Impellizzeri, G., and Amico, V. (1974). *Phytochemistry* **13**, 947–951.

Shevtsov, I. A., and Trukhanov, V. A. (1969). *Tsitol. Genet.* **3**, 164–167.

Sosnova, V. (1970). *Biol. Plant.* **12**, 425–427.

Srivastav, H. N., and Laloraya, M. M. (1973). *Nature (London)* **243**, 224.

Srivastav, H. N., and Laloraya, M. M. (1976). *Plant Biochem. J.* **3**, 134–136.

Stenlid, G. (1976). *Phytochemistry* **15**, 661–663.

Stobart, A. K., and Kinsman, L. T. (1977). *Phytochemistry* **16**, 1137–1142.

Stobart, A. K., Pinfield, N. J., and Kinsman, L. T. (1970). *Planta* **94**, 152–155.

von Ardenne, R., Döpp, H., Musso, H., and Steglich, W. (1974). *Z. Naturforsch., Teil C* **29**, 637–639.

Wagner, E., and Cumming, B. G. (1970). *Can. J. Bot.* **48**, 1–18.

Wilcox, M. E., Wyler, H., Mabry, T. J., and Dreiding, A. S. (1965a). *Helv. Chim. Acta* **48**, 252–258.

Wilcox, M. E., Wyler, H., and Dreiding, A. S. (1965b). *Helv. Chim. Acta* **48**, 1134–1147.

Wohlpart, A., and Black, S. M. (1973). *Phytochemistry* **12**, 1325–1329.

Wohlpart, A., and Mabry, T. J. (1968). *Plant Physiol.* **43**, 457–459.

Wyler, H., and Dreiding, A. S. (1957). *Helv. Chim. Acta* **40**, 191–192.

Wyler, H., Mabry, T. J., and Dreiding, A. S. (1963). *Helv. Chim. Acta* **46**, 1745–1748.

Wyler, H., Rösler, H., Mercier, M., and Dreiding, A. S. (1967). *Helv. Chim. Acta* **50**, 545–561.

Yamaguti, Y. (1935). *Jpn. J. Genet.* **11**, 109–112.

Yasui, K. (1920). *Bot. Mag.* **34**, 55–65.

Phenylalanine Ammonia-Lyase*

20

KENNETH R. HANSON
EVELYN A. HAVIR

I. INTRODUCTION

$$X-\text{C}_6\text{H}_4-\underset{H'\ NH_3^+}{\overset{H_S\ H_R^*}{\text{C}}}-\text{COO}^- \rightleftharpoons X-\text{C}_6\text{H}_4-\underset{H}{\overset{H^*}{\text{C}}}=\underset{H}{\text{C}}-\text{COO}^- + NH_4^+ \qquad (1)$$

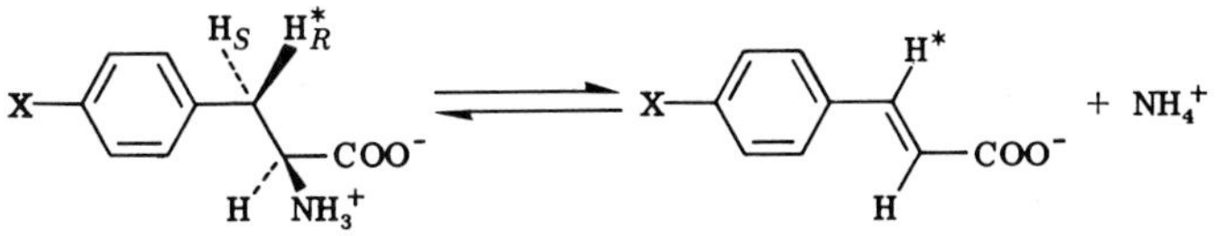

Phenylalanine ammonia-lyase (PAL) catalyzes the elimination of NH_3 from L-phenylalanine to give trans-cinnamate [Eq. (1), X = H] (Koukol and

* The preparation of this chapter was supported in part by National Science Foundation grant PCM 77-09093.

The Biochemistry of Plants, Vol. 7

577

Conn, 1961). The enzyme from many sources, notable grasses and certain fungi, also acts on L-tyrosine to yield *trans-p*-coumarate and may be said to show tyrosine ammonia-lyase (TAL) activity [Eq. (1), X = OH] (Camm and Towers, 1973; Jangaard, 1974; Neish, 1961; Young *et al.,* 1966). The enzyme derives its importance from being the first committed enzyme of phenylpropanoid metabolism in higher plants and from the novelty of the enzymatic mechanism responsible for the elimination process. In fungi the enzyme may play a catabolic role, or metabolism may lead to compounds with specific functions (Wat and Towers, 1979); e.g., methyl *cis*-ferulate is a natural germination inhibitor of rust uredospores (Faudin and Macko, 1974).

This chapter concerns itself first with the properties of isolated PAL and then with its role in plant metabolism. Such an order is followed because the metabolic questions raised go far beyond the matter of recording PAL activity. To discuss the regulation of metabolic flux in the phenylpropanoid pathway, a knowledge of the kinetic and structural properties of the enzyme is required. An understanding of regulatory processes, in turn, is required to interpret the changes in PAL activity that occur during growth and development and in response to invasion by plant pathogens or wounding. Only passing reference will be made to these larger topics.

As PAL may show TAL activity, it is pertinent to ask whether it is closely related to any other plant enzyme. There are close structural and mechanistic resemblances between PAL and histidine ammonia-lyase (HAL) from bacteria and mammals (Hanson and Havir, 1972b). It is not known whether a distinct plant HAL exists. Both HAL and PAL activity have been observed in peroxisomes from spinach and sunflower leaves and glyoxysomes from castor bean endosperm. The preparation from sunflower also showed tryptophan ammonia-lyase activity (Ruis and Kindl, 1971). A distinct enzyme tryptophan ammonia-lyase has not been described from any source. It is reported that AMP and cyanocobalamine together eliminate the lag period and enhance the activity of HAL in dialyzed extracts of *Vicia faba,* but the significance of this observation is not clear (Kamel and Maksoud, 1978). The 3-indolylacrylic acid in lentil seedlings is believed to derive from *N*-trimethyl-L-tryptophan (Hofinger *et al.,* 1975) by an enzymatic Hofmann elimination reaction analogous to the conversion of ergothionine to thiolurocanic acid (Wolff, 1962). It seems possible therefore that PAL can show a relatively broad specificity toward aromatic amino acids when subject to constraints that have not yet been duplicated *in vitro*. Aspartate and 3-methylaspartate ammonia-lyases appear to be structurally and mechanistically unrelated to PAL (Hanson and Havir, 1972b). The only discussion of ammonia lyases other than PAL will therefore be passing references to HAL. The two enzymes may be related by divergent evolution, but no direct evidence on this point exists.

Considerable progress toward defining the properties of PAL has been made in the last 8 yr. Pertinent reviews from this period include the following: Creasy and Zucker (1974), Camm and Towers (1973), Hanson and Havir (1972a,b, 1977), Hahlbrock (this volume, Chapter 14), Hahlbrock and Grisebach (1979), Huffaker and Peterson (1974), Margna (1977), Schopfer (1977), Stafford (1974a,b), and Zucker (1972).

II. ENZYMOLOGY

A. Assay

A number of techniques are available for measuring PAL activity. The formation of cinnamic acid from phenylalanine can be followed spectrophotometrically (Koukol and Conn, 1961; Zucker, 1965; Havir and Hanson, 1968a), by gas chromatography (Czichi and Kindl, 1975a), by high-pressure liquid chromatography (Murphy and Sutte, 1978), and by thin-layer chromatography when a radioactive substrate is used (Havir *et al.,* 1971). In addition, a method that measures activity by the release of tritium from specifically labeled phenylalanine has been described (Amrhein *et al.,* 1976). The choice of assay method is determined in large part by the levels of activity to be measured and the type of tissue from which the enzyme is extracted. The problems associated with spectrophotometric assays have been discussed (Erez, 1973; Camm and Towers, 1973).

The above techniques may be adapted for the assay of TAL, but certain difficulties must be emphasized. In a spectrophotometric assay (Havir *et al.,* 1971; Neish, 1961) pH and temperature must be very carefully controlled because the absorbance of both tyrosine and p-coumarate is affected by a temperature-sensitive phenolic ionization. Claims that TAL activity has been detected in tissue extracts by spectrophotometric measurements should be verified by experiments using radioactively labeled tyrosine and proper controls. In many extracts TAL activity is very low. Thus PAL/TAL ratios are liable to systematic errors, and variations in the ratio may reflect assay difficulties. Different assays may give different ratios (Emes and Vining, 1970).

B. Sources and Structure

Phenylalanine ammonia-lyase has been purified and characterized from a number of plant and fungal sources (for earlier references, see Camm and Towers, 1973). It appears that the divergence of properties between the enzyme from one plant and another are no greater than those between plants and fungi. There have been no reports of the enzyme in animals and only one study on enzyme from a prokaryotic organism, *Streptomyces verticillatus* (an actinomycete) (Emes and Vining, 1970). The enzyme is also present in certain algae, e.g., *Dunaliella marina* (Czichi and Kindl, 1975b). Only a limited search has been made for the enzyme in bacteria and blue-green algae.

Table I summarizes MW determinations for the intact enzyme and for the subunits released by sodium dodecyl sulfate (SDS)–mercaptoethanol treatment. It seems likely that the normal subunits are either identical or very similar in size. The additional smaller subunit protein reported for enzyme

TABLE I

Molecular Weights, Subunit Sizes, and Kinetic Properties[a]

Sources of PAL	Molecular weight[b]		Kinetic parameters[c]				
	Intact	Subunits	K_m^H (μM)	K_m^L (μM)	Hill coefficient, h	PAL/TAL[d]	Reference[e]
Cuscuta vine	280,000	n.r.	100[f], 140	—	n.r.	n.r.	1
Gherkin	316,000[g]	n.r.	290	43	0.65	n.r.	2
Maize	306,000	83,000	270	29[h]	n.r.	9	3,4,5,6
Mustard	300,000[g]; 240,000	n.r.; 55,000[i]	150	—	1.08[i]	un.T.	7 8
Parsley	330,000	83,000	240	32	0.60	n.r.	9
Potato	330,000[g]	83,000	260[j]	38	0.75	un.T.	4,10
Soybean	330,000[k]	80,000[k]	78[l]	6.5[l]	0.56[l]	un.T.[m]	11,12
Sweet potato	275,000–330,000	80,000	16	11	0.94	un.T.[n]	13
Wheat	320,000	85,000 +75,000	91[o]	44[o]	0.71[o]	4–20[p]	14,15
Rhizoctonia solani	330,000	90,000 +70,000	5,000	180	0.50	6	16
Rhodotorula glutinis	275,000–300,000; 278,000	83,000	250[q]	—	n.r.	1.7	4,17 8
Sporobolomyces pararoseus	275,000–300,000	n.r.	300	—	n.r.	3.5	18
Streptomyces verticillatus	226,000	n.r.	160	—	n.r.	n.r.	19

[a] n.r., Not reported; —, only a single K_m value reported (apparently no cooperativity); un.T., undetectable TAL activity, therefore PAL/TAL ratio not calculated.

[b] Intact enzyme values usually based on molecular sieve or sucrose density gradient measurements or both. Subunit sizes estimated by SDS polyacrylamide gel electrophoresis. Where two bands were observed the assumption has been made that there were 2α and 2β subunits.

[c] Some additional values: buckwheat, $K_m = 45 \mu M$ (Amrhein and Gödeke, 1977); pea buds, $K_m^H = 800 \mu M$. $K_m^L = 52 \mu M$ (Attridge *et al.*, 1971); tobacco leaves, $K_m = 220 \mu M$ (O'Neal and Keller, 1970), and tissue culture, $K_m = 30 \mu M$ (Innerarity *et al.*, 1972); *Ustilago hordei*, $K_m = 450 \mu M$, un.T. (Subba Rao *et al.*, 1967).

[d] Additional sources of PAL/TAL ratios: Bandoni *et al.* (1968), Camm and Towers (1969), Jangaard (1974), Neish (1961), Young *et al.* (1966).

from a number of sources (e.g., Nari *et al.*, 1972) could be generated from the normal subunit either by proteolytic nicking or by the spontaneous cleavage of a labile bond. The amount of the smaller protein may increase on storage (unpublished results with maize enzyme) or as a result of variant SDS–mercaptoethanol treatments (Zimmermann and Hahlbrock, 1975). Direct comparisons of electrophoretic mobility in maize and potato, maize and *Rhodotorula,* and maize and soybean enzyme suggest that the differences in subunit size are less than 4% (Havir and Hanson, 1973, 1975, Havir, 1980). The MWs estimated by comparisons with reference proteins place the normal subunit size at 83,000 and indicate that the enzyme is tetrameric (for statistical arguments, see Havir and Hanson, 1973). Recent studies on the mustard enzyme also indicate a tetrameric enzyme with identical or very similar subunits, but the subunits have a MW of 55,000 or 60,000 and the intact enzyme a MW of about 240,000 (Gupta and Acton, 1979). This estimate is similar to that for the intact enzyme from *S. verticillatus* (Emes and Vining, 1970). The results are also reminiscent of those for HAL from various

---

[e] 1) Nagaiah *et al.* (1977), 2) Iredale and Smith (1974), 3) Hanson and Havir (1977), 4) Havir and Hanson (1973), 5) Havir *et al.* (1971), 6) Jangaard (1974), 7) Schopfer (1971), 8) Gupta and Acton (1979), 9) Zimmermann and Hahlbrock (1975), 10) Havir and Hanson (1968a,b), 11) Havir (1981a), 12) Hanson (1981a), 13) Tanaka and Uritani (1976), 14) Nari *et al.* (1972), 15) Neish (1961), 16) Kalghatgi and Subba Rao (1975), 17) Hodgins (1971), 18) Parkhurst and Hodgins (1971, 1972), and 19) Emes and Vining (1970).

[f] Values determined with PAL I from tip and lignifying regions, respectively.

[g] Also forms high-MW aggregates.

[h] Cooperativity varies; a preparation showing no cooperativity with respect to L-phenylalanine has been studied (Hanson, 1981a).

[i] The subunit MW is calculated to be 60,000 if the *Rhodotorula* enzyme subunits are 83,000. The Hill coefficient probably does not differ significantly from unity (Gupta and Acton, 1979).

[j] Fitted values assuming $k = k'$ and corresponding to the symmetrical semilog plot shown in Fig. 3A are $K_S = 105$ μM, $K_S' = 262$ μM, hence $K_S'/K_S = 2.5$, $[S]_{0.5} = 166$ μM, $K_m^L = 66$ μM, and slope at $[S]_{0.5}$ for semilog plot $= 0.446$ and for Hill plot $= 0.77$.

[k] Enzyme from cell suspension cultures. Molecular weights estimated by direct comparison with the maize enzyme (Havir, 1981a).

[l] Fitted values assuming $k = k'$ are $K_S = 12$ μM, $K_S' = 78$ μM, hence $K_S'/K_S = 6.5$, $[S]_{0.5} = 30.6$ μM, K_m as listed, and slope at $[S]_{0.5}$ for semilog plot $= 0.324$ and for Hill plot as listed (Hanson, 1981a). For changes in cooperativity on purification, see Havir (1981a).

[m] No TAL activity was observed for enzyme from soybean tissue culture or from leaves (Havir, 1981a), however, the ratio PAL/TAL $= 2.8$ has been reported for young plants (Jangaard, 1974).

[n] The PAL from cut-injured sweet potato showed TAL activity (Minmikawa and Uritani, 1965).

[o] Calculated from published data of Nari *et al.* (1974); also calculated $k'/k = 1.75$, $K_S = 79$ μM, $K_S' = 640$ μM, hence $K_S'/K_S = 8.1$, $[S]_{0.5} = 621$ μM, slope at $[S]_{0.5}$ for semilog plot $= 0.41$ (see Fig. 3B).

[p] Ratio varied from 4 to 20 depending on purification step (Young and Neish, 1966).

[q] Some samples show a small amount of cooperativity; e.g., $K_m^H = 650$ μM, $K_m^L = 530$ μM (Hanson and Havir, 1977).

sources (four subunits each of MW 54,000, see Hanson and Havir, 1972b). It may be relevant that parsley cells fed L-[^{35}S]methionine yielded immuno-precipitated PAL with 83,000-MW subunits, but when protein synthesis was carried out *in vitro* with polyribosomes and PAL protein isolated by im-munoprecipitation, additional radioactive peaks of MW 65,000, 56,000, and 42,000 were observed (Schröder *et al.,* 1976).

An important possibility raised by the above mustard results is that the 83,000 subunit of PAL contains a terminal sequence which constitutes a separate domain of the folded protein. This section of the protein is not involved in catalysis or subunit association, but it may have an unknown function not required in mustard. The smaller subunits could be produced in mustard from the standard 83,000-MW subunit by cleavage before or after folding, or it could be coded for by a shorter message. In the latter case it could represent survival of a primitive form of PAL. During evolution an extra protein sequence could have been acquired to form a new domain. A distant possibility, suggested by the subunit weight data, is that the subunits have a basic domain structure of MW 27,500, which has been duplicated or triplicated in the process of evolution.

Two forms of the enzyme from oak have been separated by DEAE-Sephadex chromatography, and differences in catalytic properties observed (Alibert *et al.,* 1972; Section III,D). Recent studies mention only one form of the enzyme from sliced sweet potato roots (Tanaka and Uritani, 1977), al-though two forms were earlier reported to be separable by DEAE-cellulose chromatography (Minamikawa and Uritani, 1965). Three forms of PAL separable by DEAE-cellulose chromatography have been reported from spinach leaves: two from the chloroplast (one activated by reduced thiore-doxin; Section III,D) and one from the cytoplasm (Nishizawa *et al.,* 1979). No MWs have been reported for PAL from oak and spinach. Isoelectric focusing of PAL from mustard showed a main peak at pH 5.6 and a minor peak at pH 5.3. The PAL from the minor peak was not further characterized (Gupta and Acton, 1979). Higher-MW enzyme was observed to be present in potato extracts (Havir and Hanson, 1968a) and extracts of parsley cell cultures (Zimmermann and Hahlbrock, 1975), but the nature of the association was not established.

The subunits of PAL are dissociated with difficulty. High concentrations of urea, guanidinium chloride, or SDS–mercaptoethanol are required, and no restoration of the tetrameric form has been reported. The enzyme from some sources is very stable: That from *Rhizoctonia solani* retained full activity on heating at 60°C for 1 h (Kalghatgi and Subba Rao, 1975). On the other hand, unexpected losses of activity are encountered in some purification steps. In some cases this may be caused by the removal of bound carbohydrate (Havir, 1979).

Polysaccharide remained with the maize enzyme even through a hydroxylapatite purification step yielded PAL of very high specific activity (Havir, 1979). Partial separations of enzyme from polysaccharide were achieved by DEAE-Sephadex chromatography and by other means. These observations suggest that association with polysaccharide may complicate MW determinations. As the subunit MW for the enzyme from maize did not differ significantly from the MWs of the potato and *Rhodotorula* enzyme subunits (Havir and Hanson, 1973, 1975), it seems unlikely that covalently bound carbohydrate contributes a significantly variable amount to the subunit MW. Carbohydrate was present, however, in the trichloroacetic acid (TCA) precipitate, thus some carbohydrate may be covalently attached. The presence of an associated polysaccharide in purified PAL could complicate attempts to prepare immunospecific antiserum to the enzyme.

Amino acid compositions reported for purified enzyme from wheat (Nari *et al.*, 1972), maize and potato (Havir and Hanson, 1973), and *R. solani* (Kalghatgi and Subba Rao, 1975) are close to that for a hypothetical average protein. The MW of the subunits is too high for amino acid composition to be used as a reliable guide for calculating a revised subunit MW based on the assumption that all subunits are identical (Havir and Hanson, 1973).

An enzyme with the above characteristics might be coded for by a single structural gene. Histidine ammonia-lyase appears to contain the same prosthetic group as PAL and have a similar subunit structure (Hanson and Havir, 1972b). Studies on the *hut* operon in *Baccillus subtilis* suggest that HAL is coded for by a single structural gene (Chasin and Magasanik, 1968). When labeled at its four —SH groups, HAL from a *Pseudomonas* species gave rise to a single tryptic peptide containing these groups. The yield was too high to be derived from only two of the four subunits (Hassall and Soutar, 1974; Klee and Gladner, 1972).

Labeling experiments indicate that there are two active sites per tetramer, because the prosthetic group is formed from the terminal sequences of two subunits, because the conformation of the tetramer is such that two potential active sites are hidden within the enzyme, or because there are indeed two structural genes for the enzyme. A number of nucleophilic "carbonyl" reagents inactivate the enzyme by attacking the electrophilic prosthetic group. Incorporation of the radioactive reagent was proportional to the loss in activity (Hodgins, 1971). About 1.4 mol/mol of $^{14}CN^-$ was incorporated into the *Rhodotorula glutinis* enzyme (Hodgins, 1971) and 1.6 mol into the *Sporobolomyces pararoseus* enzyme (Parkhurst and Hodgins, 1972). More than 1 mol/mol of $^{14}CH_3NO_2$ was incorporated into the potato enzyme (Hanson and Havir, 1970). Between 0.3 and 3 mol/mol of ^3H from tritiated borohydride was incorporated into the potato enzyme (Hanson and Havir, 1970) and between 0.7 and 1.2 mol/mol into the *Rhodotorula* enzyme (Hodgins, 1971). It is also relevant that cinnamate protected about two of the four —SH groups of the *Rhodotorula* enzyme from reaction with Ellman's reagent

(Hodgins, 1972). All these experiments involve assumptions about the purity of the enzyme used. In addition Nari *et al.* (1974) suggested that the kinetics of inhibition of the wheat enzyme by benzoic acid implied a two-protomer enzyme (Section II,E). Labeling experiments also suggest that HAL is a two-protomer enzyme (see Hanson and Havir, 1972b). The interpretation of these labeling experiments should not be confused with the concept of "half-of-the-sites reactivity." This occurs when extreme negative cooperativity with respect to the substrate causes only two of four sites to be catalytically active at one time (Levitzki and Koshland, 1976; Section II,E).

The enzyme from some sources is inactivated by sulfhydryl reagents such as *p*-chloromercuribenzoate, *p*-chloromercuriphenylsulfonic acid, *N*-ethyl-maleimide, and iodoacetic acid. The sources of sensitive enzymes include barley (Koukol and Conn, 1961), *R. glutinis* (Hodgins, 1971), and *S. verticillatus* (Emes and Vining, 1970); the insensitive enzymes include potato (Havir and Hanson, 1968b), maize (Havir *et al.,* 1971), and *Ustilago hordei* (Subba Rao *et al.,* 1967). Competitive inhibitors of the *S. pararoseus* enzyme were incapable of fully protecting the enzyme against inactivation by Ellman's reagent, suggesting that the —SH groups are not at the active site (Parkhurst and Hodgins, 1972). Whereas mercaptoethanol had no effect on the activity of the barley enzyme, reduced glutathione stimulated activity by 75% and L-cysteine inhibited the enzyme (Koukol and Conn, 1961). It is possible that L-cysteine binds to the active site and thus inhibits as an analogue in the same way that glycine inhibits PAL and HAL. The stimulation by reduced glutathione, however, implies that —SH groups on the enzyme either form bonds to glutathione or can form disulfide bridges. The activation may be analogous to that reported by Buchanan and associates (Nishizawa *et al.,* 1979) for one of the two forms of PAL from spinach chloroplasts (Section III,D). The activity of this enzyme is slowly increased threefold by reduced thioredoxin. Oxidized glutathione or dehydroascorbate suffices to deactivate the enzyme. It seems likely that in the thioredoxin-regulated form of PAL two —SH groups are placed so that —SS— bridges can be formed with the assistance of conformational changes brought about by the reagent. The barley enzyme may be entirely or partly thioredoxin-regulated. A reexamination of PAL from various sources in the light of these findings is desirable.

C. Catalysis

1. Chemical Limitations by β Substituents

In most enzymatic eliminations of HX from HC $^\beta$C $^\alpha$X, a carboxylate ion is a β substituent. The substrates of PAL and HAL have aromatic β substituents instead, and this affects both the overall equilibrium for the elimination process and the acidity of the β proton. Whereas the equilibrium constant for ammonia elimination from L-aspartate at 30°C and pH 7 is $5.5 \times 10^{-3} M$, that from L-phenylalanine at 30°C and pH 8.5 is $4.1 M$, and that from L-histidine is only slightly less. The greater equilibrium potential of the aromatic amino acids for releasing ammonia may be attributed both to resonance differences and differences in hydration between the ionic species involved (Hanson and Havir, 1972b; Havir and Hanson, 1968b).

The kinetic potential of the β-phenyl group for promoting ammonia elimination appears to be less than that of the β-carboxylate ion. The turnover number per active site (k_{cat}) at the optimum pH for aspartate ammonia-lyase from *Bacterium cadaveris* is ~165/sec, whereas that for PAL is ~3/sec (see Hanson and Havir, 1972b). This value is lowered 13- to 40-fold if the substrate for PAL is 2,5-dihydro-L-phenylalanine. The lowering may be attributed to the electronic effect of replacing the benzene ring of phenylalanine by a poorer electron sink, namely, an isolated double bond (Hanson *et al.*, 1979). The kinetic disadvantages of an aromatic β substituent appear to have been offset for HAL by coordination of the imidazole ring to a metal ion (Mildvan, 1971); k_{cat} for HAL is essentially the same as for aspartate ammonia-lyase. The electrophilic prosthetic group mechanism for PAL and HAL discussed below is undoubtedly more complicated than the mechanism for aspartate ammonia-lyase. The evolution of such a mechanism may be attributed to the necessity for overcoming the electronic disadvantages of the aromatic β substituent in order to achieve an enzyme with an adequate k_{cat}.

2. pH Effects

The pH optimum for PAL at saturating substrate concentrations is about 8.7. A bell-shaped pH–activity curve has been reported in all cases; the curve for the potato enzyme may be defined by the critical dissociations $pK_\beta = 7.25$ and $pK_\alpha = 10.25$ (Havir and Hanson, 1968b). Slight inconsistencies and shifts arise on going from one buffer to another (e.g., sweet potato enzyme; Tanaka and Uritani, 1977). With L-tyrosine as the substrate for the maize enzyme the pH optimum is shifted to about 7.7 (Havir *et al.*, 1971), however, no shift is observed for PAL from *R. solani* (Kalghati and Subba Rao, 1975). These properties relate to critical "bottleneck" ionizations in the catalytic process. The high pH shoulder may reflect ionization of the enzyme-bound phenylalanine and the lower shoulder the ionization of the basic group on the enzyme that abstracts the β proton of the substrate. The $\tilde{V}_{max}/\tilde{K}_m$ curve* should reflect the same ionizations of free enzyme and substrate and also be bell-shaped, and $\tilde{K}_m$ should vary insofar as binding changes these ionizations. No study on $\tilde{V}_{max}/\tilde{K}_m$ or $\tilde{K}_m$ changes has been published. The results for HAL are complicated by ion effects (Klee *et al.*, 1975; Peterkofsky, 1962). Those for PAL, in addition, would probably be complicated by cooperativity effects (Section II,E). The neglect of pH dependence stud-

---

* The ~ superscripts are used to distinguish empirical catalytic constants studied as a function of pH, [I], etc., from constants defined as part of a detailed model.

ies is unfortunate, as *in vivo* the enzyme must operate under conditions of varying pH. Phenylalanine ammonia-lyase is present in cytosol and chloroplast stroma (Section III,B). Whereas the cytosol remains more or less neutral, the stroma may change in pH from 7 to 9 when the tissue is exposed to light. This form of PAL activation must be taken into account in any attempt to simulate mathematically the kinetic regulation of phenylpropanoid metabolism *in vivo* (Section III,D).

3. Catalytic Sequence

Several lines of evidence indicate that the catalytic sequence for PAL acting on L-phenylalanine (or L-tyrosine) is Ordered Uni-Bi with cinnamate (or p-coumarate) released before ammonia. The free enzyme binds cinnamate (or p-coumarate) to the active site in such a manner that access to the prosthetic group by nucleophilic reagents is blocked. Some inhibitors bind to the free enzyme and some to both the free enzyme and the amino-enzyme intermediate. These conclusions are summarized in Eq. (2). The binding of compounds to regulatory sites and cooperativity effects on binding are discussed below (Sections II,D and E).

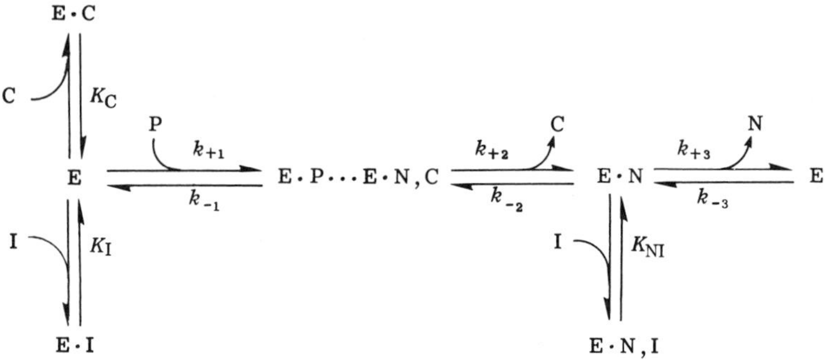

The fact that a single active site acts on both phenylalanine and tyrosine has been firmly established for the maize enzyme. The PAL/TAL ratio was the same in seedlings raised under different conditions and during purification (Reid *et al.*, 1972). Phenylalanine or cinnamate inhibited TAL activity, and tyrosine or p-coumarate inhibited PAL activity. Partial inactivation by the nucleophilic reagent $NaBH_4$ in the presence of either product as a protecting substance did not alter the PAL/TAL ratio (Havir *et al.*, 1971). When tyrosine was used as a substrate in the presence of [^{14}C]cinnamate, the rate of L-[^{14}C]phenylalanine formation was much faster than the reverse reaction in the presence of high concentrations of ammonia. This shows not only binding to the same site but formation of an amino-enzyme intermediate capable of accepting [^{14}C]cinnamate. The complementary experiment was also performed (Wightman *et al.*, 1972). The rate of incorporation of [^{14}C]-cinnamate into phenylalanine was shown to increase with increased phenylalanine concentration and therefore with the concentration of amino-enzyme intermediate (*S. pararoseus* enzyme; Parkhurst and Hodgins, 1972). ("Ferry-boat labeling" of this type would allow [^{14}C]cinnamate fed to a plant to end up in alkaloids derived from phenylethylamine.)

The specificity of enzymes such as PAL from potato or soybean that act on phenylalanine but not tyrosine is not simply attributable to lack of space to accommodate a para substituent in the binding region of the active site. The potato enzyme was found to accept substrates with groups larger than OH in the para position, yet it failed to bind L-tyrosine. Some special destabilizing effect of the para OH group appears to be involved (Hanson and Havir, 1977). No cases have been reported of preparations with TAL activity failing to act on phenylalanine. There are indications that some tissues contain distinct forms of the enzyme with differing PAL/TAL ratios. These have not been followed up, e.g., by careful protection experiments.

Protection experiments can yield some of the constants of Eq. (2). The ability of ammonia, cinnamate, or inhibitors to bind to the active site at pH 8.7 may be investigated by studying the effect of the compound on the pseudo-first-order rate of inactivation of the enzyme by nitromethane (Hanson, 1979). The method is a refinement of that used to study the protection of the potato enzyme against NaBH$_4$ attack (Havir and Hanson, 1968b) and *Rhodotorula* enzyme against inactivation by NaCN or NaHSO$_3$ (Hodgins, 1968; Hanson and Havir, 1972a). If ligand binding does not result in cooperative interactions between the enzyme subunits, protection may be described by a simple titration function defined by a constant K_p, the dissociation constant of the complex of a ligand with a single active site:

$$p/P_{max} = ([L]/K_p)/(1 + [L]/K_p) \tag{3}$$

As [L] increases, $p = 1 - k^p/k^0$ approaches P_{max}. The pseudo-first-order rate constants for inactivation, k^p and k^0, are determined in the presence and absence of the ligand. If there is complete protection on saturating with L, then $k^p = 0$ and $P_{max} = 1$. The results of a series of protection experiments with different ligands may be shown conveniently in plots of p against log [L] (Section II D).* The empirical K_p values may be equated with constants of Eq. (2). In an unpublished study on soybean enzyme that had lost its cooperativity properties and had a K_m of 93 μM, protection experiments gave for cinnamate, $K_p = K_C = 14 \mu M$; for NH$_4^+$, $K_p = k_{+3}/k_{-3} = 520$ mM; and for D-phenylalanine, $K_p = K_I = 85 \mu M$. Such determinations are more direct and use less enzyme than inhibition studies of the type discussed below. If ligand binding produces cooperative interactions, then appropriate constants may be determined (Section II,E).

---

* As Eq. (3) is analogous to the Michaelis–Menten equation, the same graphical and iterative least-squares methods may be applied to estimate P_{max} and K_p as to estimate V_{max} and K_m (see Section II,C).

TABLE II

Kinetic Relationships Implied by Equation 2 Assuming No Cooperativity Effects[a]

Inhibition[b]	$\tilde{V}_{max}/V_{max}$	$(\tilde{V}_{max}/\tilde{K}_m)/(V_{max}/K_m)$	$K/\tilde{K}_m$
I binds to E, pure competitive	1	$\dfrac{1}{1 + [I]/K_I}$	$\dfrac{1}{1 + [I]/K_I}$
I binds to E and EN, pure mixed non-competitive	$\dfrac{1}{1 + [I]/K_{ni}}$	$\dfrac{1}{1 + [I]/K_I}$	$\dfrac{1 + [I]/K_{ni}}{1 + [I]/K_I}$
C binds to E and EN	$\dfrac{1}{1 + [C]/K_{nc}}$	$\dfrac{1}{(1 + [C]/K_C)(1 + [C]/K'_{nc})}$	$\dfrac{1 + [C]/K_{nc}}{(1 + [C]/K_C)(1 + [C]/K'_{nc})}$
N binds to E, pure competitive	1	$\dfrac{1}{1 + [N]/K_n}$	$\dfrac{1}{1 + [N]/K_n}$

[a] Equations expressed to emphasize titration relationships to the dissociation constant K_I or K_C or the weighted dissociation constant K_{ni}, K_{nc}, K'_{nc}, or K_n.

[b] The terminology is that advocated by Wong (1975).

The results of inhibition studies are compatible with Eq. (2). Table II shows the implied kinetic relationships. The various dissociation constants K_I, K_{ni}, etc., define rectangular hyperbolas which can be regarded as titration functions. Their significance may be explained as follows:

1. Binding to the free enzyme leads only to "pure competitive" inhibition. At high substrate concentrations all the enzyme is present as E · S and I has no effect on $\tilde{V}_{max}$. At low substrate concentrations I titrates $\tilde{V}_{max}/\tilde{K}_m$ to zero as [I] increases relative to κ_I. In a study (Hanson, 1981a) of maize PAL that had lost its cooperativity properties D-phenylalanine was found to be a true competitive inhibitor ($K_m = 450\ \mu M$, $K_I = 2.4$ mM). Both [S] and [I] were varied over their full titration ranges. The effect of cooperativity is discussed in Section II,E.
2. Binding to both free enzyme and the amino enzyme intermediate leads to "pure mixed noncompetitive" inhibition. The intermediate E · N is in a steady state: $[E \cdot P \ldots E \cdot N, C]k_{+2} = [E \cdot N]k_{+3}$. At high substrate concentrations increasing [I] diverts all the enzyme to E · N,I and therefore $\tilde{V}_{max}$ titrates to zero as [I] increases relative to the weighted constant $K_{ni} = K_{NI}(1 + k_{+3}/k_{+2})$. At low substrate concentrations [I] titrates $\tilde{V}_{max}/\tilde{K}_m$ to zero. When $K_{ni} = K_I$, the result is "pure classic noncompetitive" inhibition and $\tilde{K}_m$ is unchanged. When $K_{ni} \gg K_I$, "pure competitive" inhibition is observed, and when $K_I \gg K_{ni}$, "pure noncompetitive" inhibition. Phenylpropiolic acid, $PhC{\equiv}CCOOH$, an

analogue of cinnamic acid, has been studied as an inhibitor of the *Rhodotorula* enzyme ($K_m = 250$ μM, $K_I = 24$ μM, $K_{ni} = 33$ μM) (Hodgins, 1971) and of maize enzyme that had lost its cooperativity properties ($K_m = 450$ μM, $K_I = 6$ μM, $K_{ni} = 4$ μM) (Hanson, 1981b). The constant K_{ni} is not directly observable, but as other results suggest the enzyme-substrate complex is not mainly present as $E \cdot N$, this implies that binding of I to $E \cdot N$ is weaker than to E. There is no discussion in the literature of the structural features that prevent binding to $E \cdot N$.

3. At high substrate concentrations cinnamate controls $\tilde{V}_{max}$ by influencing the partitioning between $E \cdot P \ldots E \cdot N,C$ and $E \cdot N$. As [C] increases relative to $K_{nc} = (k_{+3} + k_{+2})/k_{-2}$, $\tilde{V}_{max}$ titrates to zero. At low substrate concentrations $\tilde{V}_{max}/\tilde{K}_m$ is modified by both competition and partitioning. For the *Rhodotorula* enzyme, cinnamate acted as a competitive inhibitor ($K_C = 26$ μM) (Hodgins, 1971); presumably both partitioning constants K_{nc} and K'_{nc} have much higher values than 160 μM, the highest concentration of cinnamate used.

4. High concentrations of ammonia had no significant effect on $\tilde{V}_{max}$ for the potato enzyme (Havir and Hanson, 1968b). A change would be expected for a random release mechanism but not for Eq. (2).

The kinetic evidence that Eq. (2) applies is incomplete. The reverse reaction has not been studied, and the reaction between amino-enzyme and cinnamate requires further quantitation. This being the case, emphasis must be placed on the protection experiments with cinnamate. If the enzyme and amino-enzyme bind cinnamate in the same manner, then as cinnamate fully blocks access to the prosthetic group, cinnamate must be released before ammonia. There are indications, noted in the following section, that cinnamate binding also prevents release of the β proton removed in the elimination process.

4. Rate-limiting Step

The low turnover number for the enzyme suggests that there is an intrinsically slow step in the catalytic sequence and therefore that quasi-equilibrium conditions prevail. If essentially all the enzyme-substrate complex were present as $E \cdot N$, the step $E \cdot N \rightarrow E + N$ would be fully rate-limiting. Changes in substituents of the aromatic ring that did not alter the rate-limiting step would have very little effect on k_{cat}. A regression study on log k_{cat} for para-substituted phenylalanines as a function of substituent parameters showed that a significant portion of the variance in log k_{cat} could be explained in terms of substituent size (Hanson and Havir, 1977). It appears therefore that the para-substitutions did not change the rate-limiting step. Since k_{cat} changed, the step $E \cdot N \rightarrow E + N$ was not fully rate-limiting. Presumably, HAL has the same amino-enzyme hydrolysis step in its catalytic sequence.

As k_{cat} for HAL is much higher than for PAL, it is clear that the step is not intrinsically slow. It seems probable therefore that the step is not even partially rate-limiting for PAL; i.e., the rate-limiting step is cinnamate release or a step prior to this.

Because of the low k_{cat} for PAL, it might be expected that the elimination process would be rate-limiting. As the tritium isotope effect on V_{max}/K_m was low (1.3), it is possible that C^β—H_S bond cleavage is associated with an equilibrium step prior to the rate-limiting step (potato enzyme; Wightman *et al.*, 1972). An equilibrium perturbation method is now available that could be used to study any ^{13}C, ^{15}N, and ^2H isotope effects associated with the elimination process (Cleland, 1977). The above study with tritiated phenylalanine also indicated that the β proton of the substrate did not exchange with the medium, via the enzyme, at a rate that was fast compared to the release of cinnamate. This could imply that cinnamate binding completely blocks exchange with the base on the enzyme that accepts the substrate β hydrogen. The lowering of k_{cat} observed on replacing L-phenylalanine by the analogue 2,5-cyclohexadienyl-L-alanine can be attributed to an electronic effect on the elimination step (Hanson *et al.*, 1979). It is not known whether the rate-limiting step becomes C^β—H_S cleavage as a result of this change in substrate structure.

5. Stereochemistry

The stereochemistry of the elimination process has been studied in a number of laboratories using both L-phenylalanine and L-tyrosine stereospecifically labeled in the β position (review: Hanson and Havir, 1972b; also Bartl *et al.*, 1977; Müller and Crout, 1975). The *pro-3S* hydrogen is eliminated, and the product is *trans*-cinnamate [Eq. (1)]. The α hydrogen is not lost and does not exchange with the medium during the reaction (Hanson *et al.*, 1971). Figure 1 shows the elimination as a minimal motion process in which the plane containing the benzene ring and that containing the carboxylate ion collapse into a single plane. The nitrogen of the substrate is presumed to be linked to the prosethetic group of the enzyme. The balloons represent p orbitals, and the concerted process maximizes π–p and p–p orbital overlap. If the elimination is stepwise, the motions may still approximate the minimal motion process.

6. Prosthetic Group

The structure of the electrophilic prosthetic group of the enzyme is only partially understood. It is assumed that the substrate nitrogen is chemically modified by combination with the prosthetic group (forward shuttle steps). The resultant electron withdrawn from the C—N bond assists the elimination process. After the elimination step(s), cinnamate is released and the aminated prosthetic group is hydrolyzed (reverse shuttle).

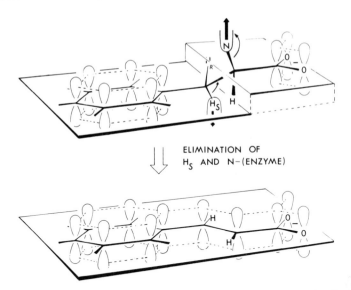

Fig. 1. Anti concerted elimination process. (Hanson and Rose, 1975; reprinted with permission, copyright American Chemical Society).

In addition to the carbonyl reagents mentioned in Section II,B, PAL is inhibited or inactivated by phenylhydrazine, O-methylhydroxylamine, 2-aminooxyacetic acid, and related compounds. Whereas $NaBH_4$, $NaBH_3CN$, and CH_3NO_2 always inactivate the enzyme, the above $-NH_2$ compounds, CN^-, and bisulfite appear to form adducts reversibly. With enzyme from some sources the initial adducts are irreversibly changed. Adducts with aminoxy compounds such as 2-aminooxyacetic acid and the super inhibitors discussed in Section II,D, hydrolyze with halftimes of several minutes (Hanson, 1981b).

The absorbancy spectrum of the enzyme is that of a normal protein with, at most, a shoulder in the 340-nm region. This rules out the possibility that tightly bound pyridoxal phosphate is present. Urocanase, which is also inactivated by carbonyl reagents that attack tightly bound NAD^+, shows a difference spectrum with a peak at 335 nm on borohydride reduction. This is attributable to the formation of 1,4- plus 1,2- and 1,6-NADH (Egan and Phillips, 1977). As the spectrum of PAL was unchanged on borohydride reduction (Havir and Hanson, 1968b), PAL does not contain NAD^+, however, the possibility that the mechanism of the enzyme is an internal oxidation–reduction process cannot be eliminated entirely. The weak 340-nm shoulder in the spectrum of PAL may imply that a firmly bound transition metal is present (Dixon et al., 1976). Coordination to the substrate nitrogen could assist the elimination process.

Inactivation by CH_3NO_2 (pK_a 10.3) takes place with a dependence on pH described by a simple titration curve. Ionization of the reagent was assisted by binding to the active site as the apparent pK_a ranged from 8.5 to 9, depending on the source of the enzyme (Havir and Hanson, 1975). The results imply that the electrophilic group being attacked does not change by ionization over the range 7.5–10.0. In agreement with this, the rate of inactivation by $NaBH_3CN$ does not change over the same range (E. A. Havir, unpublished results). The prosthetic group is thus unlikely to be a complex system in equilibrium, which reacts in one form with the substrate and in other forms with carbonyl reagents. The $NaBH_3CN$ experiments also establish that the prosthetic group is not an aldehyde or ketone, as these groups are not appreciably reduced above pH 5. On proteolysis or CNBr cleavage $^{14}CH_3NO_2$-labeled enzyme yielded a mixture of labeled peptides, thus the electrophilic group is covalently bound to the enzyme (E. A. Havir, unpublished result).

Several lines of evidence indicate that irreversible inactivation involves nucleophilic attack on C-3 of an alanine skeleton. The enzyme inactivated by a labeled reagent was hydrolyzed under vacuum, and the products examined. Tritiated $NaBH_4$ yielded DL-[3-^{3}H]alanine and THO (Hanson and Havir, 1969, 1970; Havir et al., 1971; Hodgins, 1971); $^{14}CH_3NO_2$ yielded DL-[4-^{14}C]aspartic acid, $^{14}CO_2$, and $H^{14}COOH$ (Havir and Hanson, 1975); $^{14}CN^-$ yielded DL-[4-^{14}C]aspartic acid (Hodgins, 1971; Parkhurst and Hodgins, 1972). On $^{14}CH_3NO_2$ treatment the *Rhodotorula* enzyme, but not the maize enzyme, also yielded [1,2-^{14}C]glycine which may be derived from a symmetrical intermediate generated by another part of the prosthetic group (Havir and Hanson, 1975).

It was proposed in 1969 (Hanson and Havir) that the prosthetic group contained a dehydroalanine residue activated either through the amino group or the carboxyl group. Experiments with HAL and $^{14}CH_3NO_2$ suggested that the dehydroalanine residue could be activated through Schiff base formation with the amino group (Givot et al., 1969):

$$R_1R_2C{=}N{-}\overset{\displaystyle C^\beta H_2}{\underset{}{C^\alpha}}{-\!-\!-}\overset{\displaystyle O}{\underset{}{C}}{-}X$$

β-Activated dehydroalanine derivatives are formed as intermediates in pyridoxal-catalyzed enzymatic β replacement reactions; e.g., such nucleophiles as ^-OH, ^-SH, $^-SCH_3$, and indole add to the β carbon of the intermediate in the tryptophanase-catalyzed reaction. The hypothesis explains the isolation of racemic amino acids labeled by β attack, but physical masking of the imino group must be postulated to explain why $NaBH_4$ does not reduce the system completely. The main difficulty with the proposal is that the modified amino group of the phenylalanine adduct would not obviously be a much better leaving group than $-NH_3^+$. To meet this objection it has been suggested that the initial adduct becomes the vinylogue of an amide by a prototropic rearrangement (Hanson and Havir, 1970, 1972b). There is no spectroscopic or other direct evidence to support this possibility. The anti stereochemistry rules out concerted mechanisms in which the adduct abstracts the β hydrogen of the substrate.

A labile prosthetic group of the above type could be formed by the modification of normal amino acids. Serine, cysteine, or their derivatives yield dehydroalanyl residues by β-elimination. Oxidation of an amino acid side chain to an aldehyde group in a precursor form of the enzyme could allow Schiff base formation with an N-terminal cysteine which would then readily eliminate H_2S.

The active site must contain a counter ion to the carboxyl group of the substrate. Hodgins (1972) has reported that the $R.$ $glutinis$ enzyme is inhibited by halide ions (iodine is the strongest inhibitor with $K_I = 80$ mM), and this suggests that electrostatic attraction to the counter ion, rather than hydrogen bonding, is important. On the other hand, a careful search has not been made to rule out the possibility that the counter ion is a divalent metal ion capable of forming a second sphere complex with the carboxylate ion, as in the case of mandelate racemase (Kenyon and Hageman, 1979; Maggio et $al.,$ 1975). An essential amino group may be present, as the $Rhodotorula$ enzyme was protected against inactivation by acetic anhydride or 2,4,6-trinitrobenzenesulfonic acid by inhibitors (Hodgins, 1972). The tyrosine reagent N-acetylimidazole did not inactivate the enzyme (Hodgins, 1972). The base for abstracting the β proton of the substrate is apparently not an —SH group but, on the basis of the pH profile, it could be imidazole.

D. Conformational Adjustments on Ligand Binding and during Catalysis

Many X-ray diffraction studies show that proteins can undergo major conformational adjustments on ligand binding. The changes may include interactions between protomers so that binding on one protomer influences binding at a similar site on another protomer (Section II,E), they may result in movement of domains within a protomer, or they may involve more local readjustments. Different ligands may move the protein to binding states that are the same or distinct.

If different ligands on binding cause the protein to be in the $same$ state, then linear free energy arguments may be applied to relate structural differences to differences in the free energy of dissociation of the enzyme–ligand complex as defined by $\delta \Delta G = RT \ln (K_2/K_1)$. Structural differences among ligands may be expressed in terms of electronic, hydrophobic, and steric factors. This approach, which has been widely used in discussing drug design and aspects of enzymology, has been applied to the action of PAL on a series of para-substituted L-phenylalanines (OH, F, Cl, Br, I, NO_2) (Hanson and Havir, 1977).

If two ligands on binding cause the protein to be in *distinct* states, then intrinsically different contributions to the binding equilibrium are made by the differences in protein structure and one will not in general find that free energies of dissociation can be treated as additive functions of structural changes. It is pertinent therefore to examine data on the free energy of dissociation of ligand complexes in order to see if there is evidence for lack of additivity. It is sufficient here to consider ratios K_2/K_1 for inhibition or protection constants, bearing in mind that free energy values refer to log K values. Where there is evidence that binding at one site influences binding at a second site (Section II,E), the relevant average is the concentration of ligand producing 50% dissociation. One may also assume that K_m values are dissociation constants. A limited amount of data is available from inhibition studies on the enzyme from *R. glutinis* (Brand and Harper, 1976; Hodgins, 1968), *S. pararoseus* (Parkhurst and Hodgins, 1972), and buckwheat (Amrhein *et al.*, 1976), and from protection and inhibition studies on the maize enzyme (Hanson, 1981b). The inhibitors D- and L-2-aminooxy-3-phenylpropionic acids (D- and L-AOPP) are of particular interest in that they are effective at the nanomole level. (The inhibitory properties of D- and L-AOPP and of 2-aminooxyacetic acid were first reported by Amrhein *et al.* (1976) for the buckwheat enzyme and have been studied also for the maize enzyme (Hanson, 1981b). Inhibition is normally reversible but L-AOPP irreversibly inactivates PAL from soybean (Havir, 1981a). The compounds have proved to be important tools for studying phenylpropanoid metabolism (Amrhein and Gödeke, 1977; Amrhein and Hollander, 1979).)

The main evidence that PAL is induced to adopt different states when different ligands bind is as follows:

1. Glycine binds to PAL with a K_I (*Rhodotorula* enzyme) or K_p (maize enzyme) slightly higher than the K_m for phenylalanine, although alanine, isoleucine, leucine, and valine do not bind significantly. The benzene ring must make an important contribution to binding phenylalanine. A corresponding amount of free energy, at least 5 kcal (Hanson, 1981b), must be released on going from the protein state that binds phenylalanine to the state that binds glycine, since the glycine molecule forms part of the phenylalanine molecule. Attention has been drawn to the binding of glycine to PAL (Parkhurst and Hodgins, 1972) and to HAL (Givot *et al.*, 1969), and the conclusion drawn that an induced fit of enzyme to substrate takes place. The point emphasized here is that phenylalanine binds to a less stable protein state than glycine.

2. The dissociation constant for *O*-methylhydroxylamine is slightly lower

than that for NH_3 (K_p data, maize enzyme), the factor being about 2.7 (Hanson, 1981b). This probably indicates a small difference in the chemical equilibrium for interaction with the prosthetic group. The dissociation constant for 2-aminoxyacetic acid is lower than that for glycine by a factor of about 100. On going from the protein state that binds glycine to that which binds aminoxyacetic acid, free energy is therefore released corresponding to the somewhat larger factor $100/2.7 = 37$. The same factor for L-AOPP and L-phenylalanine is, however, at least 10,000. This is far too large to be explained in terms of stronger covalent bonding. As the binding interactions with the protein for L-phenylalanine and L-AOPP must be the same apart from their $-NH_2$ and $-ONH_2$ groups, the analogue must bind to a much more stable state of the protein.

The conclusion that there is a large release of free energy in going from the state binding L-phenylalanine to that binding the super inhibitor L-AOPP has special implications. It indicates that the compound is a transition state analogue for the elimination process (Lienhard, 1973; Wolfenden, 1972). If phenylalanine binds to the enzyme in a relatively *unfavorable* free energy state, free energy can be released from the protein at the same time that the bound phenylalanine passes to the transition state with adsorption of free energy. Conformational changes in the enzyme thus contribute to lowering the net free energy of activation for elimination. The transition state alignment must be a matter of very fine adjustment, since the corresponding hydrazine analogues are only slightly better inhibitors than L-phenylalanine (Brand and Harper, 1976), and the difference could easily be attributed to the relative affinities of NH_3 and hydrazine for the prosthetic group.

The surprising finding that D- and L-AOPP are almost equally effective superinhibitors implies that both bring the protein into the same transition state-related conformation. Although the K_m for L-phenylalanine varies according to the source of the enzyme, K_p or K_I for D-phenylalanine is always close to the K_m for its enantiomer. As the compounds have the same hydrophobic and electronic characteristics, this similarity must imply interactions with the protein that ignore their steric difference. If L-phenylalanine fits the active site as shown in Fig. 1, then its enantiomer may also fit with the amino nitrogen, benzene ring, and $-COO^-$ in essentially the same position but with a mirror image arrangement of C-2 and C-3. The *pro*-3R hydrogen of D-phenylalanine, when so fitted to the active site, is not well-positioned for catalysis of proton elimination. To achieve such mirror image packing of D- and L-AOPP into the active site something has to be altered. If the nitrogens of the aminooxy groups are in the same positions postulated for D- and L-phenylalanine and the benzene rings are in the plane as before, then the $-CO_2^-$ groups are lowered. This corresponds to the transition state conformation for the substrate implied by the collapse of two planes into one plane as represented in Fig. 1. The hypothesis that AOPP is a transition state analogue is thus compatible with the stereochemical evidence concerning the elimination process (Hanson, 1981b).

E. Subunit Cooperativity on Ligand Binding

1. The Need for Models

The changes in enzyme conformation brought about by ligand binding and, in principle, those that occur during the catalytic sequence (Section II,D) may include the whole enzyme and not just a single subunit. Binding to one site may thus assist binding of the same or another ligand to a second site (positive cooperativity) or hinder such binding (negative cooperativity). Binding at one site may also affect k_{cat} at another site. The PAL from many plant and fungal sources appears to show negative cooperativity with respect to substrate binding: The apparent K_m decreases with increasing [S]. When this was first observed with the potato enzyme (Havir and Hanson, 1968b), it was evident that the allosteric model of Monod *et al.* (1965) could not explain the phenomenon. The more general induced-fit models described by Koshland *et al.* (1966) have been extended by Ricard *et al.* (1974) and applied specifically to PAL (Nari *et al.*, 1974). In this section a simple two-protomer model that suffices to explain the available results will be outlined.*

A large number of regulatory proteins are now known that show negative cooperativity (Levitzki and Koshland, 1976). There is thus a reasonable presumption that the cooperative interactions shown by PAL play a role in the short-term regulation of phenylpropanoid metabolism. A descriptive model is therefore needed, so that empirical constants can be used in the mathematical simulation of regulatory changes. Negative cooperativity is not always observed, and further study will be needed to determine whether this is the result of loss of cooperativity during purification or whether PAL *in vivo* can behave as if each active site is completely separate.

The two-protomer model will first be presented for substrate binding and then extended to include competitive inhibition. The problem of ligand binding to regulatory sites will not be considered in detail.

2. Substrate Cooperativity

If, as indicated in Section II,A, PAL has only two active sites, then it may be said to consist of two protomers each having two subunits. The model

---

* The model is mathematically equivalent to that presented by Nari *et al.* (1974). In developing their "structural" models they treated the rate and equilibrium constants discussed here as the products of interaction coefficients and intrinsic rate and equilibrium constants. The conceptual value of this approach is discussed by Ricard *et al.* (1972, 1974).

shown in Fig. 2 is said to be partially concerted, as the states E($\_\_$) and E(SS) have twofold axes of symmetry, whereas the protein of the state E(S$\_$) or E($\_$S) does not. If quasi-equilibrium conditions are assumed, then dissociation constants may be written for each protomer: K_S when one substrate molecule is bound and K'_S when both sites are occupied. The derivation of the initial velocity relationship Eq. (4) resembles the derivation of a titration function for a dibasic acid. Equation (4) passes through a maximum if $k > 2k'$ as the formation E(SS) becomes inhibitory. As [S] increases, $v \rightarrow V_{Lt} = 2k'[E^0]$. It is convenient therefore to reexpress Eq. (4) as a velocity relative to V_{Lt} as in Eq. (5).

$$\frac{v}{[E^0]} = \frac{2k[S]/K_S + 2k'[S]^2/K_SK'_S}{1 + 2[S]/K_S + [S]^2/K_SK'_S} \tag{4}$$

$$v_{rel} = \frac{v}{V_{Lt}} = \frac{(k/k')[S]/K_S + [S]^2/K_SK'_S}{1 + 2[S]/K_S + [S]^2/K_SK'_S} \tag{5}$$

The model expresses the fact that cooperativity effects may be on binding (K'_S/K_S) or on catalysis (k'/k). Such cooperativity may be positive, zero, or negative, hence there are nine combinations hereafter indicated as (binding, catalysis). If both are zero (0, 0), then the equations reduce to the Michaelis–Menten form: $v_{rel} = ([S]/K_S)/(1 + [S]/K_S)$.

If there is only binding cooperativity, then the possibilities are (+, 0) or (−, 0). As $k' = k$, the curve does not pass through a maximum and $V_{max} = V_{Lt}$. This has been termed the "dimeric simple sequential" model (Ricard *et al.*, 1974). Plots of v_{rel} against log [S] are symmetrical about the point for [S] = $(K_SK'_S)^{1/2}$; they correspond to the titration curves for dibasic acids. Figure 3A presents our original data for the potato enzyme reexpressed as a semilog plot and superposed on the (−, 0) curve; $K_S = 105$ mM and $K'_S = 262$ mM [hence $K'_S/K_S = a^2 = 2.6$ and $(K_SK'_S)^{1/2} = [S]_{0.5} = 166$ mM]. The dotted curve marked rev. K_S,K'_S is the case $K'_S = 105$ mM, $K_S = 262$ mM.

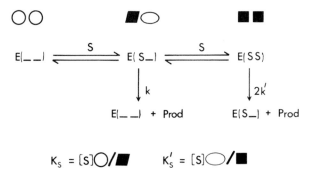

Fig. 2. Two-promoter partially concerted model; substrate only. (This is a partial diagram. For simplicity, the corresponding equilibria for E($\_$S) formation and the conversion of this to E($\_$ $\_$) and products have been omitted.)

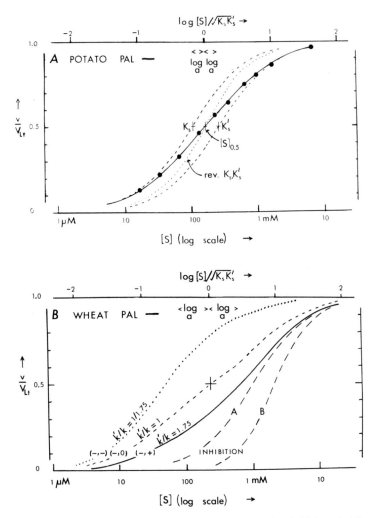

Fig. 3. Semilog plots showing the relationship of relative initial velocities to log phenylalanine concentrations. The continuous curves show experimental data. The envelope curves K_S and K'_S in (A) are defined by the Michaelis–Menten relationship. The curves in (A) and curve $k'/k = 1$ in (B) are symmetrical about the points marked + and illustrate the "binding cooperativity only" restriction. Curves A and B in (B) illustrate the shift from negative binding cooperativity (continuous line) through zero to positive apparent cooperativity as inhibitor concentrations change K_A without changing K'_A.

The former exemplifies negative cooperativity on binding with K'_S/K_S greater than unity, and the latter positive cooperativity. In the semilog plot negative cooperativity has a slope greater than 0.57 at the symmetry point and positive cooperativity less. This slope corresponds to that at the midpoint of the Michaelis–Menten curve as exemplified by the envelope curves for K_S and K'_S (dashed). If K'_S/K_S were ≥ 100, the negative cooperativity curve would be seen as the fusion of two **S**-shaped curves.

If there is only catalytic cooperativity, the possibilities are $(0, +)$ or $(0, -)$. In most circumstances they would not be differentiated from the normal Michaelis–Menten equation. The observed K_m, however, would be greater than K_S for a $(0, +)$ curve and less for a $(0, -)$ curve.

The general case may be exemplified by the wheat enzyme (Fig. 3B). The continuous curve corresponds to the parameters estimated by Nari *et al.* (1974). Their values may be converted to the constants of Eq. (5); $K_S = 0.079$ mM, $K'_S = 0.64$ mM (hence $(K_S K'_S)^{1/2} = 0.224$ mM, $K'_S/K_S = a^2 = 8.1$), and $k'/k = 1.75$. The cooperativity is negative therefore with respect to binding, but positive with respect to catalysis. This $(-, +)$ curve may be compared with curves having the same K and K' values: the symmetrical $(-, 0)$ dashed curve and the $(-, -)$ dotted curve where the values of k' and k are reversed. $[S]_{0.5} = (K_S K'_S)^{1/2}$ only for the symmetrical $(-, 0)$ case. A curve $(+, +)$ will be exemplified in discussing inhibition of the wheat enzyme.

Given the two-protomer partially concerted model it is necessary to decide how best to handle experimental data. First, one must determine whether there is a departure from Michaelis–Menten kinetics. Departure from linearity may be observed readily in v-against-$v/[S]$ plots. However, there are advantages to the semilog plot shown here, especially when used in conjunction with an iterative least-squares fit.* The graphed data may be overlaid on the theoretical simple titration curve using a light table, and any skewness in the distribution of data points noted. The same skewness should be apparent in a table of residuals for the fitted curve.

If the results cannot be defined by the Michaelis–Menten equation, one may determine whether only binding cooperativity is important by testing the data points for symmetry of distribution on the semilog plot. Two copies of the semilog plot, one rotated 180°, may be superposed on a light table. The symmetry point gives log $[S]_{0.5} = 0.5$ log $K_S K'_S$. Appropriate constants may

---

* Programs for iterative least-squares fits to the Michaelis–Menten equation (Cleland, 1967; Hanson *et al.*, 1967) and to Eqs. (5) and (6) (Cleland, 1967) have been described. We find that these FORTRAN programs rewritten in BASIC are convenient for use with laboratory computers such as the Wang 2200.

be estimated by overlaying the data points on curves calculated for different K'_S/K_S values. As a check, the tangential value of the Michaelis constant at high substrate concentrations may be estimated from v-against-$v/[S]$ plots. In general $K^T_m = K^H_m = K'_S(2 - k/k')$ and, if it is assumed that $k = k'$, then $K^H_m = K'_S$. The values of K^H_m and $[S]_{0.5}$ suffice to calculate K_S. The asymptotic value of the Michaelis constant at a low substrate concentration is less useful, since in general $K^A_m = K^L_m = K_S/(2 - k' K_S/kK'_S)$ (Nari et al., 1974). A more objective alternative is to obtain the constants by an iterative least-squares fit and to examine the residuals for skewness.

The Hill plot of log $[v/(V - v)]$ against log $[S]$ is related to the semilog plot. The Hill coefficient h is the slope of the plot at log $[S]_{0.5}$. When only binding cooperativity is important, $h < 1$ indicates negative binding cooperativity and $h > 1$ positive binding cooperativity. The value $h = 1$ is thus analogous to the slope of 0.57 at log $[S]_{0.5}$ in the semilog plot. It should be noted that, for values of K'_S/K_S between 1 and 100, the semilog plot is approximately linear for $v_{rel} = 0.2$–0.8. The semilog plot is more useful, as it does not transform the velocity values and can be employed for a number of purposes.

If the results obviously depart from the above $k' = k$ case, then the only method available is iterative least squares fitting to the full equation. Even if the $k' = k$ case appears to apply, it is of interest to compare the variances for the fits to the full equation and the restricted equation to see if there is any justification for preferring the former. A large number of data points covering the full range in v_{rel} are required, and the standard errors of the constants calculated may be large. Even uncertain constants, however, are of value when the purpose is to describe the empirical curve. The Hill coefficient and the slope of the semilog plot at $[S]_{0.5}$ are also useful as descriptive constants.

Table I gives published K^H_m, K^L_m, and h values. In general K^L_m is subject to more uncertainty than K^H_m. It would be useful in the future to record at least $[S]_{0.5}$ and the slope of the semilog plot at $[S]_{0.5}$. Symmetry should be tested for and, if shown, K_S and K'_S should be calculated, since this can be accomplished readily by graphical methods.

3. Protection and Cooperativity

Ligand binding to the active sites of a two-protomer enzyme may be described in terms of a partially concerted model. A model for complete protection could resemble Fig. 2, except that the pseudo-first-order rate constants for inactivation by nitromethane k and k' appear as $\text{E}(\text{--}) \xrightarrow{2k}$ inactive enzyme and $\text{E}(\text{L-}) \xrightarrow{k'}$ inactive enzyme. The model leads to Eq. (6) where $p = 1 - k^p/k^0$, $w = 2 - k'/k$, and $P_{Lt} = 1$, cf. Eq. (3). Binding cooperativity ($w = 2$) sufficed to account for the observed cinnamate protection of the maize enzyme; $K_p = 20$, $K'_p = 88$ (Hanson, 1981b).

$$\frac{p}{P_{Lt}} = \frac{w[L]/K + [L]^2/K_p K'_p}{1 + 2[L]/K_p + [L]^2/K_p K'_p} \tag{6}$$

4. Inhibition and Cooperativity

Inhibitors that bind only to the free enzyme and not the amino-enzyme intermediate give rise to competitive inhibition kinetics if there is no cooperativity between subunits (Section II,C). If cooperativity occurs both for inhibitor binding and substrate, the two-protomer partially concerted model may be extended as in Fig. 4. This leads to Eq. (7). Because the inhibition is competitive, V_{Lt} is unchanged by [I]:

$$\frac{v}{V_{Lt}} = \frac{(Q'/Q)[S]/K_A + [S]^2/K_A K'_A}{1 + 2[S]/K_A + [S]^2/K_A K'_A} \tag{7}$$

where K_A and K'_A are apparent dissociation constants which reduce to K_S and K'_S in the absence of inhibitor as

$$K_A = (P/Q)K_S \qquad\qquad K'_A = QK'_S$$
$$P = 1 + 2[I]/K_I + [I]^2/K_I K'_I \qquad Q = 1 + [I]/K''_I$$

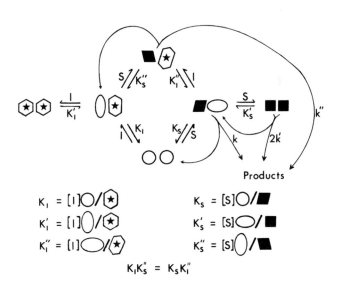

$$K_I = [I]\bigcirc/\varhexstar \qquad\qquad K_S = [S]\bigcirc/\blacksquare$$
$$K'_I = [I]\bigcirc/\varhexstar \qquad\qquad K'_S = [S]\bigcirc/\blacksquare$$
$$K''_I = [I]\bigcirc/\varhexstar \qquad\qquad K''_S = [S]\bigcirc/\blacksquare$$
$$K_I K''_S = K_S K''_I$$

Fig. 4. Two-protomer partially concerted model; inhibitor binds competitively to free enzyme but not to amino-enzyme intermediate; no restrictions on E(IS) formation. (This is a partial diagram. The corresponding equilibria and reactions for E(__S), E(I__), and E(SI) have been omitted.)

Cooperativity with respect to catalysis is expressed by the term $Q' = k/k' + (k''/k')[I]/K_I''$. It follows

$$(K_A K_A')^{1/2} = p^{1/2} \times (K_S K_S')^{1/2} \text{ and } ((K_A K_A')^{1/2}/[I])_{Lt\,I\to\infty} = (K_S K_S')^{1/2}/(K_I K_I')^{1/2}$$

$$K_A'/K_A = (Q^2/P)(K_S'/K_S) \quad \text{and} \quad (K_A'/K_A)_{Lt\,I\to\infty} = (K_I K_I'/K''^2)(K_S'/K_S)$$

The properties of the model are most readily understood for the case in which there is no cooperativity with respect to catalysis ($Q' = Q$). In the semilog plot all curves at constant [I] are symmetric, with $[S]_{0.5} = (K_A K_A')^{1/2}$. Both K_S and K_S' are determined in the absence of inhibitor. In the special case where $K_I = K_I'$, a plot of $[S]_{0.5}$ against [I] yields a straight line. In general $(K_I K_I')^{1/2}$ and K_I can be estimated by trial-and-error fitting. The dependence of K_A' on [I] yields K_I'' and the limiting value of K_A'/K_A gives a check on the assigned values. In the special case where L-phenylalanine has a very large effect on D-phenylalanine binding, so that K_I'' is very much larger than the inhibitor concentrations studied, K_A' does not change with [I], as E(SI) is not formed. The variation in the shape of the curves with increasing [I] is a noteworthy aspect of the model. When there is negative cooperativity in the absence of inhibitor, then the slope may decrease or increase according to the relative values of K_I'', K_I, and K_I'. When $K_I < (K_I K_I')^{1/2}$, the limiting K_A'/K_A exceeds K_S'/K_S and the negative cooperativity is increased. When $K_I = (K_I K_I')^{1/2}$, the limit for K_A'/K_A is K_S'/K_S. Computations show that the variation in K_A'/K_A with [I] is very small for moderate differences in K_I and K_I'. When $K_I > (K_I K_I'')^{1/2}$, the negative cooperativity decreases. Negative cooperativity becomes positive cooperativity as [I] increases when the limiting K_A'/K_A value is greater than unity. A change from negative to essentially zero cooperativity was observed with D-phenylalanine in our original studies on the potato enzyme (Havir and Hanson, 1981b) and has recently been found for the enzyme from soybean (Hanson, 1981a). In the latter case the estimated constants are K_S, K_S', and $K_S'' = 12, 78,$ and $57\ \mu\text{mol}$; K_I, K_I', and $K_I'' = 34, 136,$ and $160\ \mu\text{mol}$. A change from negative to positive cooperativity with increasing [I] has been observed for the wheat enzyme (Nari et al., 1974; see next paragraph) and that from R. solani (Kalghatgi and Subba Rao, 1975).

In the general case in which Q' differs from Q, values of K_A and K_A' obtained by curve fitting could, in principle, be examined as above. In studying D-phenylalanine inhibition for the wheat enzyme, however, Nari et al. (1974) considered the special case in which E[SI] is not formed, so that $Q' = k/k'$ and $Q = 1$. Figure 3B shows computed curves based on the data, i.e., for $k'/k = 1.75$ and $K_S'/K_S = 8.1$. The solid line is for [I] = 0, curve A for $P = 8.1$, hence $K_A'/K_A = 1$, and curve B for $P = 65.6$, hence $K_A'/K_A = 1/8.1$.

When Q' is a constant, Eq. (7) gives Eq. (8), where v_I and v_S are initial velocities in the presence and absence of inhibitor at constant [S]:

$$\frac{v_S}{v_I} - 1 = \frac{2[I]/K_I + [I]^2/K_I K_I'}{1 + 2[S]/K_S + [S]^2/K_S K_S'} \tag{8}$$

This equation applies also when $k' = k$. It was found for the wheat enzyme that at constant [S] the expression $(v_S/v_I) - 1$ was proportional to [I] and parabolically related for benzoate. These workers concluded therefore that for the former neither E(SI) nor E(II) was formed; i.e., D-phenylalanine binding to one site prevented binding of D- or L-phenylalanine to the other site. For benzoate E(SI) was not formed, but E(II) was. Unfortunately Nari et al. (1974) did not derive K_S or K_I values from their studies and use them to simulate the entire data set. Linear plots for $(v_S/v_I) - 1$ against [I] were also reported for the Rhizoctonia enzyme inhibited by D-phenylalanine (Kalghatgi and Subba Rao, 1975). It is not known whether k' differs significantly from k for this enzyme, as only v-against-$v/[S]$ plots were reported. It seems evident that there is considerable variation from source to source in the negative cooperativity associated with D-phenylalanine binding.

The model could be extended to include binding to the amino-enzyme intermediate [Eq. (2)] but the equations would be more complicated. The mixed noncompetitive kinetics for the noncooperative case (Table II) imply that V_{Lt} will titrate to zero as [I] increases. No studies on such inhibition with enzyme showing substrate cooperativity have been reported.

5. Effector Binding

There are obvious difficulties in distinguishing the results of the above type of inhibition from the consequences of effector site binding. Two possible examples of effector site binding have been reported. Gallic acid stimulates V_{max} for PAL from oak leaves and roots (Boudet et al., 1971). The stimulation passes through a maximum in gallic acid concentration at about $10^{-3} M$, suggesting that there are separate effector sites on each protomer. The effector shows negative cooperativity with respect to its own binding. Binding to the first site increases V_{max}, but a second binding restores symmetry and decreases V_{max}. The second example is the inhibition of PAL from pea seedlings by quercetin and related flavonoids (Attridge et al., 1971). The inhibition appeared to reach a plateau value at about 1 mM quercetin. Treatments such as freezing and thawing decreased the response. Both these examples are discussed further below (Section III,D).

The physical separation of the active sites or of effector sites and active sites may be succeptible to examination using cinnamic acids linked by chains of varying length (Mack, 1979).

In summary, considerable progress has been made in understanding the cooperativity of PAL. The mathematical tools are available for expressing results in terms of a simple two-protomer model, but a physical picture of the enzyme and its distribution of sites is lacking. It is not yet clear which phenomena are of metabolic importance.

III. METABOLISM

A. Pathways

Phenylpropanoid compounds have been estimated to accumulate in certain plant tissues at a rate of 22 μmol carbon/g fr wt per hour (Creasy, 1968; Creasy and Zucker, 1974). In photosynthetic tissues a net CO_2 fixation of 100–150 μmol carbon/g fr wt per hour is typical. The drain on the pools of phenylalanine and tyrosine may thus be large, and integration of phenylpropanoid metabolism with the shikimate pathway of aromatic acid biosynthesis must be presumed. Such integration will be particularly important if developmental changes or responses to stimuli or infection cause sudden increases in the synthesis of certain phenylpropanoid compounds. Three factors involved in integration will be considered in the next three sections: (1) The relevant enzymes are likely to be localized within a particular region of the cell; (2) the activation, synthesis, and degradation of the enzymes must be coordinated to provide a coarse regulation of metabolic flux; (3) there must be short-term interactions between metabolites and enzymes to control the pool sizes of intermediates—this kinetic regulation provides the fine tuning of the metabolic pathway.

General flowcharts for phenylpropanoid metabolism occur elsewhere in this volume. It is convenient to refer to cinnamic acid and the derived 4-, 3,4-, and 3,4,5-substituted acids together with their coenzyme A (CoA) esters as the core sequence. Cinnamate 4-hydroxylase, a cytochrome-P-450-containing monooxygenase complex, bears a special relationship to PAL because p-coumarate may be formed in two ways in many tissues: from phenylalanine via cinnamate, but also from L-tyrosine. Approximately six metabolic options may be distinguished: Substituted cinnamates lead via cinnamoyl-CoA esters (1) to conjugates such as chlorogenic acid, (2) to lignin, or (3) to a flavonoid compound; (4) the ortho-hydroxylation of cinnamic acids by a cytochrome P-450 enzyme system leads to coumarins; (5) a coupling reaction may lead to 4-phenylcoumarins; and (6) chain shortening of the cinnamic acids yields benzoate and substituted benzoates. The CoA esters may be intermediates in the shortening process, but the details of the benzoate synthase complex have not yet been worked out.

The more the different metabolic options are confined to specialized cells or localized within the cell, the fewer the complications likely to be involved in the kinetic regulation of PAL. With completely separated metabolic sequences, only activation or inhibition along the direct line of the pathway is important in determining pool size. If there is no separation, or a leaky separation, the pathways may compete and there must be cross-regulation between pathways to ensure that the metabolites of one do not totally inhibit the other. Because plant metabolism is extensively packaged into chloroplasts, mitochondria, peroxisomes, etc., regulation by transport, redox, and pH may replace types of kinetic regulation observed in bacteria (Sanwal, 1970).

B. Subcellular Distribution

Evidence has accumulated that PAL occurs not only in the cytoplasm but also in plastids, mitochondria, and microbodies (Table III; Alibert et al., 1977; McClure, 1979; Ranjeva et al., 1979). Most of the enzyme activity is released by disrupting the cell or cell organelles, but a portion remains associated with miscellaneous membrane fragments (microsomal fractions) and with thylakoid preparations from chloroplasts. The nature of this association is unknown. It is possible that prior to disruption much of the "soluble" enzyme is membrane-bound. On the other hand, random occlusions and binding of soluble enzyme to carbohydrate associated with organelles could explain some of the distribution data. In two cases where different forms of the enzyme were observed, the differences in properties could be correlated with differences in localization. In spinach leaves one form, accounting for 75% of the PAL activity, is present in the extrachloroplast cytoplasm (the amount appears to be regulated by phytochrome control). Two forms, one of which was activated threefold by reduced thioredoxin, were located in the chloroplast. Whereas the leaf extracts contained some of the chloroplast enzymes, the chloroplast extracts contained little of the cytoplasmic form (Nishizawa et al., 1979). In leaves and roots of oak the form known as PAL I was associated with a mitochondria–microbody fraction and with benzoate synthase activity, whereas PAL II was associated with the microsomal fraction and cinnamate 4-hydroxylase activity (Alibert et al., 1972). The fraction of PAL activity associated with chloroplasts may in some cases be large. In 5-day-old barley shoots the chloroplasts appeared to contain almost one-third of the PAL in the whole shoot (Saunders and McClure, 1975). In the algae D. marina about one-fifth of the PAL activity was associated with the chloroplasts, and perhaps one-half of this was retained in thylakoid preparations (Löffelhardt et al., 1973).

Microsomal and thylakoid preparations from several sources have yielded evidence for the metabolic channeling of cinnamate (Kindl, 1979). Cinna-

TABLE III

Subcellular Localization of Phenylalanine Ammonia-Lyase in Selected Plant Species[a]

Species	Organelle	Reference[b]
Plastids[c]		
Barley, *Hordeum vulgare*	C, E	1
Castor bean, *Ricinis communis*	C, P, TC	2, 3
Corn, *Zea mays*	C, TC	3
Nasturtium, *Tropaeolum majus*	C, TC	3
Oat, *Avena sativis*	C, E	4
Petunia, *Petunia hybrida*	C	5
Spinach, *Spinacea oleracea*	C	6
Sunflower, *Helianthus annus*	C, TC	3
Watercress, *Nasturtium officinale*	C, TC	3
Marine algae, *Dunaliella marina*	C, TC	3, 7
Mitochondria		
Castor bean, *R. communis*		8
Oak, *Quercus pendunculata*		9
Microbodies[d]		
Castor bean, *R. communis*	G	2
Spinach, *S. oleracea*	P	10
Sunflower, *H. annuus*	P	10
Microsomes		
Buckwheat, *Fagopyrum esculentum*		11
Gherkin, *Cucumis sativus*		12
Oak, *Q. pendunculata*		9
Potato, *Solanum tuberosum*		13, 14
Sorghum, *Sorghum bicolor*		16

[a] Other examples are given by Alibert *et al.* (1977), and McClure (1979).

[b] 1) Saunders and McClure (1975), 2) Ruis and Kindl (1970), 3) Löffelhardt *et al.* (1973), 4) Weissenböck *et al.* (1976), 5) Ranjeva *et al.* (1977), 6) Nishizawa *et al.* (1979), 7) Czichi and Kindl (1975a), 8) Gregory (1976), 9) Alibert *et al.* (1972), 10) Ruis and Kindl (1971), 11) Amrhein and Zenk (1971), 12) Czichi and Kindl (1977), 13) Camm and Towers (1973), 14) Czichi and Kindl (1975b), 15) Stafford (1969).

[c] C, Chloroplasts; E, etioplasts; TC, thylakoid membranes.

[d] G, Glyoxysomes; P, peroxisomes.

mate formed from phenylalanine is further metabolized in preference to free cinnamate through microsomal hydroxylation. The monoxygenase systems for hydroxylating cinnamate in the ortho and para positions are membrane-bound. Some of the more detailed studies have been concerned with microsomal fractions from sorghum seedlings (Saunders *et al.*, 1977), castor bean endosperm (Young and Beevers, 1976), and Jerusalem artichoke tubers (Benveniste *et al.*, 1977), and also with thylakoid membranes (Czichi and

Kindl, 1975a). In one set of channeling experiments microsomal fractions of green cucumber cotyledons were treated with [³H]phenylalanine and [¹⁴C]-cinnamate. After 10 min, the remaining cinnamate and the o-and p-coumaric acids formed were isolated. A fivefold increase in the $^3H/^{14}C$ ratio implied that cinnamate formed from phenylalanine was preferentially hydroxylated (Czichi and Kindl, 1977). Similar results were obtained with microsomal membranes from potato (Czichi and Kindl, 1975b) and thylakoid membranes from *D. marina* (Czichi and Kindl, 1975a). In the first example, saturation with ethylene abolished channeling but did not increase cinnamate hydroxylase activity, as would be expected if permeability to the inside of a vesicle were being increased. The ethylene could produce a physical effect on membrane structure, which results in a reorientation of the membrane-bound PAL. As cinnamate is both lipophilic and hydrophilic, it is possible that cinnamate could pass from PAL to the hydroxylase system via the lipid surface of the membrane. Similar metabolic channeling has been demonstrated for benzoate formation. In *Nasturtium, Astilbe,* and *Hydrangae* species cinnamate was converted to benzoate by thylakoid-bound enzymes (Löffelhardt and Kindl, 1975). In thylakoid membranes from chloroplasts of *Astilbe chinensis* cinnamate formed from phenylalanine was more accessible for conversion to benzoate than free cinnamate.

The above results suggest that association of PAL with the hydroxylase or benzoate synthase systems is significant, but that the association need not be as precise as that implied by the term "multienzyme complex" (Stafford, 1974a). The phenomenon of channeling has been observed in cyanogenic glycoside formation (Conn *et al.,* 1979). The enzymes converting L-tyrosine to p-hydroxymandelonitrile in sorghum seedlings are membrane-associated. The C-hydroxylation of p-hydroxyphenylacetonitrile formed from the corresponding aldoxime occurs more effectively than from added nitrile (Shimada and Conn, 1977). The importance of the link between PAL and hydroxylase levels in some tissues is indicated by studies on their coordinate induction in excised hypocotyls of buckwheat (Amrhein and Zenk, 1970) and in cell suspension cultures of parsley (Hahlbrock and Wellmann, 1973); however, lack of coordination of these activities need not be an artifact of the isolation methods (see next section).

A number of reasons can be advanced for believing that compartmentalization of phenylpropanoid metabolism is to the advantage of the cell: (1) routing of different metabolites to different destinations is assisted, e.g., lignin formation in the cell walls and storage of chlorogenic acid and flavonoids in vacuoles; (2) the need for complex kinetic regulation is reduced; (3) effective pool sizes are minimized so that the system is more responsive to regulation; (4) cinnamic acid is confined and its concentration minimized. Cinnamate does not accumulate within the cell, and when cinnamate is introduced conversion to the abnormal metabolite 1-

TABLE IV

Changes in Phenylalanine Ammonia-Lyase Activity

Systems studied		Some factors influencing PAL levels[a]	Reference[b]
Barley (*Hordeum vulgare*)	Shoots	Light (phytochrome)[c]	1, 2
Bean (*Phaseolus vulgaris*)	Pods	Fungal infection, psoralen plus uv light	3
Buckwheat (*Fagopyrum esculentum*)	Hypocotyls, leaf discs	Light, sugars	4, 5
Cocklebur (*Xanthium pennsylvanicum*)	Leaf discs	Light, sugars	6, 7, 8
Duckweed (*Lemna perpusilla*)	Water culture	Circadian rhythm (also *Spirodela polyrhiza*)	9
Gherkin (*Cucumis satisvus*)	Hypocotyls	Light (blue), Mn^{2+}, temperature, cinnamate	10, 11, 12
Grapefruit (*Citrus paradisi*)	Peel	Ethylene	13
Jerusalem artichoke (*Helianthus tuberosus*)	Tuber slices	Mn^{2+}, cinnamate, oxygen, aging, wounding	14
Lettuce (*Lactuca saliva*)	Midrib	Ethylene	15
Maize (*Zea mays*)	Shoots	Gibberellic acid	16
Mustard (*Sinapis alba*)	Cotyledon, seedlings	Light (phytochrome)	17, 18, 19
Parsley (*Petroselinum hortense*)	Cell suspension cultures	Light, dilution, medium	20, 21, 22
Pea (*Pisum sativum*)	Pods, seedlings	Light, fungal infection, ethylene	23, 24, 25, 26
Potato (*Solanum tuberosum*)	Tuber slices	Light, wounding, cinnamate	27, 28, 29
Radish (*Raphanus sativus*)	Cotyledons	Light, cinnamate, D-phenylalanine	30, 31
Soybean (*Glycine max*)	Cell suspension cultures	Fungal elicitor	32, 25
Sunflower (*Helianthus annus*)	Leaves	Light, wounding, DCMU	8, 33, 34
Sycamore (*Acer pseudoplatanus*)	Cell suspension cultures	Medium	35, 32
Sweet potato (*Ipomoea batatus*)	Tuber slices	Wounding, fungal infections	36, 37, 38
Tobacco (*Nicotiana tobacum*)	Leaves, cell suspension cultures	TMV, mutant selection	39, 40, 41
Microbial sources[d]			

O-cinnamoyl-β-D-glucose takes place (e.g., potato tuber slices: Hanson, 1966; Harborne and Corner, 1961). The metabolic channeling discussed above ensures that the concentration of cinnamate and therefore the regulation of PAL activity by product inhibition is minimized. As cinnamate, like other unsaturated acids, disrupts membrane permeability (Freese *et al.*, 1973), a major reason for confining cinnamate to certain parts of the cell may be its toxicity.

C. Regulation of Enzyme Levels

1. Explanatory Schemes

Table IV lists most of the systems in which PAL levels have been studied as a function of external stimuli. Although PAL as isolated is a relatively stable enzyme, its activity is lost in tissues and crude extracts. In most systems this loss of activity may be attributed to one or more enzymes hereafter referred to as PAL-IS, where IS stands for inactivating system. The main agent is probably a proteolytic enzyme, and such an enzyme could well show specificity toward PAL. A series of schemes of increasing complexity can be envisioned to explain experiments in which PAL activity rises to a maximum (at which point synthesis is equal to breakdown) and then falls to a lower value:

1. The stimulus initiates transcription and translation of the PAL gene. PAL-IS is already present at a steady level, so that a first-order decay of PAL activity takes place as soon as it is formed. A steady state

---

[a] DCMU, Dicholoromethylurea; TMV, tobacco mosaic virus.

[b] The following citations are intended to provide entries into the extensive literature rather than to indicate all relevant studies. 1) Saunders and McClure (1975), 2) Blume and McClure (1978), 3) Hadwiger (1972), 4) Amrhein and Gödeke (1977), 5) Amrhein and Zenk (1971), 6) Zucker (1971), 7) Amrhein and Zenk (1970), 8) Creasy and Zucker (1974), 9) Gordon and Koukkari (1978), 10) Engelsma (1974), 11) Iredale and Smith (1973), 12) Attridge and Smith (1973, 1974), 13) Riov *et al.* (1969), 14) Durst (1976), 15) Hyodo *et al.* (1978), 16) Cheng and Marsh (1968), 17) Johnson and Smith (1978), 18) Tong and Schopfer (1978), 19) Acton and Schopfer (1975), 20) Schröder *et al.* (1977), 21) Betz *et al.* (1978), 22) Hahlbrock *et al.* (1971), 23) Hadwiger and Schwochau (1971), 24) Hadwiger *et al.* (1970), 25) Grisebach and Ebel (1978), 26) Hyodo and Yang (1971), 27) Lamb (1979), 28) Lamb and Rubery (1976), 29) Zucker (1968), 30) Huault and Klein-Eude (1975), 31) Klein-Eude *et al.* (1974), 32) Ebel *et al.* (1976), 33) Creasy (1976), 34) Wong *et al.* (1974), 35) Westcott and Henshaw (1976), 36) Minamikawa and Uritani (1965), 37) Tanaka and Uritani (1977), 38) Tanaka *et al.* (1977), 39) Duchesne *et al.* (1977), 40) Legrand *et al.* (1976), 41) Berlin and Widholm (1977), 42) Towers *et al.* (1974), 43) Kalghatgi and Subba Rao (1976), 44) Fritz *et al.* (1976), 45) Parkhurst and Hodgins (1971), 46) Bezanson *et al.* (1970), 47) Subba Rao *et al.* (1967).

[c] For references on phytochrome and PAL see Schopfer (1977, p. 231).

[d] See also *Polyporus hispidus* (42), *Rhizoctonia solani* (43), *Rhodotorula glutinis* (44), *Sporobolomyces pararoseus* (45), *Streptomyces verticillatus* (46), *Ustilago hordei* (47)

plateau of activity is reached when PAL synthesis is maintained. In most experimental systems, however, the rate of PAL mRNA formation probably falls off as a peak of activity rather than a plateau is observed.

2. PAL-IS is unstable or subject to inactivation. There is coinduction of both PAL and PAL-IS, and production of mRNA for both falls off.

3. In addition to the above, cinnamate and other metabolite levels may influence directly the rate of translation on cytoplasmic ribosomes of mRNA for PAL and, perhaps, PAL-IS. Metabolite control of transcription is less likely to occur, as transcription takes place in the nucleus.

4. The above considerations apply also to other enzymes associated with phenylpropanoid metabolism such as cinnamate 4-hydroxylase (C4H), flavanone synthase, and the transferase leading to chlorogenic acid formation. Each of these could have a distinct inactivating system. A scheme of this type for PAL and C4H is represented in Fig. 5.

Some additional possibilities must be considered:

5. PAL mRNA may be stored in a protected form for emergency use.

6. A precursor form of PAL may accumulate and be converted to PAL at a fixed or variable rate. (A precursor form is implied by the nature of the

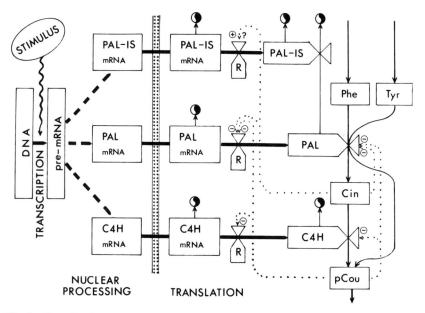

Fig. 5. Postulated control of PAL, cinnamate 4-hydroxylase, and the PAL inactivation system. No distinction is made in this scheme between different regions of the cytoplasm, although differences between chloroplast and extrachloroplast PAL have been reported (Nishizawa *et al.*, 1979). Dotted lines indicate positive or negative modulation of ribosome or enzyme activity.

prosthetic group, but this does not require the accumulation of a significant amount of precursor).

7. PAL may be stored bound to a proteinaceous inhibitor. An increase in PAL activity could occur through slow dissociation followed by destruction of the inhibitor. A loss of PAL activity could be the result of synthesis of more inhibitor.

2. Transcription and Translation in Eukaryotes

A major part of probing PAL biosynthesis has involved manipulations of the stages between stimulation of gene transcription and the actual appearance of PAL activity. Nothing is known about the mechanisms by which light, Mn^{2+}, fungal elicitors, virus infection, and wounding stimulate transcription. After a certain time these stimuli fail to maintain transcription.

Work on various eukaryotic systems (Sheinin and Humbert, 1978) indicates that DNA genes are first transcribed to a high-MW pre-mRNA (heteronuclear RNA). This is "processed" by a series of reactions, although the exact order is uncertain. The final mRNA produced may derive from several sections of the pre-mRNA, but code for a single protein. Besides cleavage and splicing, processing includes 3'-end polyadenylation, 5'-end capping and methylation, and internal methylation. The poly A and cap probably protect the mRNA against endo- and exonuclease degradation. A poly-A tail may assist the mRNA in escaping from the nucleus. Once in the cytoplasm the poly-A tail may be degraded to an oligo-A tail without loss of mRNA function. The cap is probably required for the initiation of translation (Shatkin, 1976). The complexity of this sequential process means that it takes about $\frac{3}{4}$–$2\frac{1}{2}$ h before sufficient PAL mRNA reaches the ribosomes for PAL synthesis to be detected. A simple switching on of transcription should eventually generate a steady state level of PAL with the rate of enzymatic mRNA degradation in the cytoplasm equal to the rate of export from the nucleus. The rate of transcription could fall off as a result of a time-delayed negative feedback from processing.

Translation of the relatively stable mRNA in eukaryotic systems is probably under various types of control (Ochoa and de Haro, 1979). The best understood mechanism is a blocking of polypeptide chain initiation by phosphorylation of an initiation factor subunit. In reticulocytes hemoglobin synthesis cannot take place unless free heme is available to inhibit the kinase that otherwise converts initiation factor to its blocking form. This example of positive feed-forward regulation by a metabolite is not necessarily a model for the negative feedback regulation by cinnamate and p-coumarate suggested in Fig. 5, but it makes the idea of such regulation plausible.

In studying PAL production in various systems the most popular inhibitors have been actinomycin D, 3'-deoxyadenosine (cordycepin), 7-methylguanosine triphosphate, and cycloheximide. These are believed to block transcription, mRNA transport, cap formation, and ribosomal release in translation (Agutter and McCaldin, 1979; Grollman, 1966). As indirect and differential effects of these compounds are possible, results must be interpreted with care. The failure to produce an effect may be more significant than a positive response.

3. Experimental Systems

Tobacco leaves infected by tobacco mosaic virus provide an example of a system in which PAL-IS is of minor importance. The host begins to produce local necrotic lesions after 30 h, with PAL rising from a low background level to a high plateau at about 45 h. The half-life of the enzyme is very long: ~30 h. Evidence for *de novo* synthesis was obtained from D_2O density labeling experiments (Duchesne *et al.*, 1977).

A more studied system is the phytochrome control of PAL synthesis in mustard cotyledons. The PAL from dark-grown and light-stimulated plants appeared to have the same properties (Gupta and Acton, 1979). The apparent lag after placing in continuous far-red light was 40 min, and a sharp peak of activity was reached after 20 h. The half-life of PAL was calculated to be 3–4 h. The half-life was influenced by the age of the seedlings but not by the light treatment. Given that there is a constant first-order decay rate, the actual rate of PAL synthesis may be calculated (Schopfer, 1977). The rate of PAL synthesis was a function of the $P_r \rightleftharpoons P_{fr}$ equilibrium established by special light treatments (Tong and Schopfer, 1978). The possibility that PAL is not synthesized but rather released from storage has been considered. Attempts to establish *de novo* synthesis of PAL by D_2O density-labeling experiments have generated opposing interpretations. The difficulties in reaching agreement center on the short half-life of PAL and the uncertainty as to what fraction of the amino acids used in PAL synthesis derives from a far-red light-stimulated breakdown of storage proteins and what fraction derives from amino acids whose nitrogen has been provided by exogenous nitrate. Nitrate reductase, a phytochrome-controlled enzyme, could be a rate-limiting factor in the latter process. The synthesis position championed by Acton and Schopfer (1975; Schopfer, 1977) has been opposed by Attridge *et al.* (1974), and Johnson and Smith (1978). It appears that the system is too complex for the D_2O density labeling method to lead to a definitive answer. Clarification should be possible when antiserum to the purified mustard PAL becomes available (Gupta and Acton, 1979). A proteinaceous PAL inhibitor was not detected in mustard (Billett *et al.*, 1978), but mustard PAL appeared to be more sensitive to the PAL inhibitor from gherkin than the gherkin PAL was (Gupta and Acton, 1979). As PAL-IS is presumed to be active, it is puzzling that, when seedlings were transferred to darkness after several hours in far-red light followed by light treatment to reduce the P_{fr} level and therefore turn off PAL synthesis, a plateau of PAL activity was observed rather than rapid decay (Tong and Schopfer, 1978).

The clearest example of a system conforming to possibility 1 in Section III,C,1 is provided by the light-induced formation of PAL in parsley cell suspension cultures. Prior to induction the enzyme level is undetectable. After a 2-h apparent lag a peak is reached at 15 h. The half-life of degradation is constant over a period of 30 h, encompassing the peak, but varies from 5 to 10 h depending on the age of the culture. Various labeling experiments were used to demonstrate *de novo* synthesis of PAL. At specific times the actual rate of PAL synthesis in cell samples was calculated using a radioactive amino acid in the medium and isolating PAL by immunoprecipitation. The amount of PAL accumulating could then be calculated assuming a constant rate of degradation by PAL-IS. Excellent agreement between the calculated and observed peaks in PAL activity was obtained (Betz *et al.*, 1978). The mRNA associated with polyribosomes was also assayed for its ability to synthesize PAL protein at various times (Ragg *et al.*, 1977; Schröder, 1977). The amount of mRNA synthesis was found to pass through a peak (Schröder *et al.*, 1979). Both sets of experiments were performed with continuous and short-term irradiation by light. The appearance and disappearance of PAL in soybean cell suspension cultures has not been studied in the above detail. The main emphasis has been on the role of fungal polysaccharides in eliciting PAL synthesis (Ebel *et al.*, 1976) and on the effect of nitrogen in the medium (Havir, 1981a).

Potato tuber slices (the first system to be studied: Zucker, 1968) and sweet potato slices (Tanaka *et al.*, 1977; Tanaka and Uritani, 1976) exemplify possibility 2 above. Slicing and light or slicing alone initiates PAL synthesis. Addition of cycloheximide a few hours after initiation leads to constant activity, implying that PAL-IS is not yet present, whereas addition just prior to the normal peak in activity results in a premature decay of activity. The potato tuber system has recently been reexamined using a novel D_2O pulse-labeling procedure (Lamb *et al.*, 1979). From the density labeling results the net rate of synthesis and rate constants for synthesis and degradation were calculated at eight points during a 39-h period ($dE/dt = k_s - k_d E$). Degradation was initially zero. After the peaks in PAL and PAL-IS activity a plateau was reached with about half the maximum PAL activity, and, as expected, this represented a state of low synthesis and low degradation.

For sweet potato slices, PAL activity and the titer of PAL protein determined by immunochemical procedures appeared and decreased in the same pattern (Tanaka and Uritani, 1976). Crude enzyme solutions prepared from fresh sweet potato tissue or after 5 h showed little tendency to inactivate PAL, whereas PAL activity was lost in extracts from tissue maintained for 19 h. The inactivating enzyme may be membrane-bound. Inactivation *in vitro* did not reduce the immunoprecipitate obtained with PAL antiserum. It seems probable that *in vivo* initial cleavage by PAL-IS is followed by degradation by constitutive proteinases (Tanaka *et al.*, 1977). In extracts of sunflower leaves PAL-IS activity was also associated with particulate fractions (Creasy, 1976). The delayed induction of PAL-IS has also been observed with sunflower leaf discs maintained on sucrose (Creasy and Zucker, 1974). It has been suggested that in cocklebur leaves the control of PAL activity is primarily through control of PAL-IS levels (Zucker, 1971). A further study on the potato system (Smith and Rubery, 1979) has shown that discs returned to a host tuber do not increase in the PAL activity but, if after at least 4 h in the host the discs are transferred to the usual dark environment, two peaks of PAL activity are observed within 48 h. This oscillation may reflect delayed consequences of metabolite feedback regulation of PAL synthesis. [Circadian rhythmicity in PAL and TAL activity has been observed with *Lemna perpusilla* and *Spirodela polyrhiza* maintained in continuous light (Gordon and Koukkari, 1978). Such rhythms are probably associated with general metabolic changes in the plants rather than with regulatory loops in phenylpropanoid metabolism.]

The modulation of PAL synthesis by cinnamate (possibility 3 above) was first postulated by Zucker (1965). The decrease in PAL accumulation in potato tuber discs could be attributed to effects on PAL synthesis or inactivation. Recent evidence indicates that cinnamate and *p*-coumarate modulate translation, whereas light stimulates transcription. The most useful approach has employed delayed transfer experiments (Lamb, 1979). The total time of the experiment (e.g., 9 h) remains constant, but at regular intervals a sample of the tissue (or cell suspension) is transferred from light to dark or to the inhibitor solution (inhibitor may be added instead). If a PAL precursor accumulates subsequent to translation (possibility 6 above) and the conversion to PAL is slow enough, then blocking further precursor synthesis with cycloheximide during the last time intervals will have little effect on the final PAL level and a falloff in the curve will result. No such falloff was observed for parsley cell suspension cultures (Halbrock and Ragg, 1975) or potato tubers (Lamb, 1979; Lamb and Rubery, 1976) when cycloheximide was the inhibitor. The time course showed instead the initial lag of about 2 h observed when PAL activity is sampled at regular intervals. On the other hand, actinomycin D reversed the picture for both systems: The longer the delay in blocking transcription, the greater the PAL accumulation, but blocking during the last 2 h produced a falloff in the curve. 3'-Deoxyadenosine produced an intermediate initial lag and no terminal falloff for potato tuber slices, suggesting that mRNA becomes available to cytoplasmic ribosomes soon after transport from nucleus to cytoplasm (Lamb, 1979; Agutter and McCaldin, 1979).

The above effects act as controls for experiments with other agents. Transfer from light to darkness for potato tuber slices was equivalent to transfer to actinomycin D, thus the effect is on transcription not translation. Transfer to cinnamate or p-coumarate was equivalent to transfer to cycloheximide (Lamb, 1979). Whereas PAL and hydroxycinnamoyl-CoA : quinic acid transferase are under photocontrol in potato, cinnamate 4-hydroxylase is not. p-Coumarate but not cinnamate appeared to modulate translation of cinnamate 4-hydroxylase activity. This difference rules out the possibility that a single receptor is responsible for the posttranscriptional regulation of PAL and cinnamate hydroxylase. In each case receptor and enzyme must be coded for by the same gene. Although control could be on chain initiation, as in hemoglobin biosynthesis (Ochoa and de Hara, 1979), a more economical hypothesis would be for the metabolites to bind to the nascent proteins and prevent release from the ribosome (cf. cycloheximide and emitine; Grollman, 1966).

The concept of cinnamate modulation of PAL synthesis allows observations on other systems to be explained. Exogenous D-phenylalanine may be assumed to inhibit PAL activity. The cinnamate pool is reduced, and therefore PAL synthesis should increase. This stimulation has been observed with Jerusalem artichoke tubers (Durst, 1976), radish cotyledon (Huault and Klein-Eude, 1978), and germinating mustard, rape, and oat seeds (Szkutnicka and Lewak, 1975). Transfer of cultured Jerusalem artichoke tuber discs to darkness and water or darkness and D-phenylalanine resulted in the same inactivation curves, indicating that D-phenylalanine did not influence PAL-IS levels. Anaerobiosis of the discs increased the cinnamate pool size and decreased PAL levels; on restoring oxygen the cinnamate pool rose and PAL synthesis was resumed. Whereas Mn^{2+} stimulated PAL activity fivefold, cinnamate and Mn^{2+} reduced PAL levels below the steady state level.

To test the above explanation the same system should be treated with other inhibitors of PAL. One would expect the aminooxy superinhibitors (Section II,D) to produce effects similar to those of D-phenylalanine. The PAL activity accumulating in buckwheat hypocotyls after 12 h of light in the presence and absence of L-AOPP was the same. The compound entered the tissue and was shown to inhibit PAL activity *in vivo* and thus block anthocyanin formation (Amrhein and Gödeke, 1977; Amrhein et al., 1976). These experiments point to an absence of modulation of PAL synthesis by cinnamate or any later product in the pathway. On the other hand, PAL formation in gherkin hypocotyls was inhibited by cinnamate and p-coumarate and stimulated in illuminated tissue by α-aminooxyacetic acid and L-AOPP (Amrhein and Gerhardt, 1979). Soybean seedlings also showed an increase in PAL activity when root-fed L-AOPP (Duke et al., 1980) but, for reasons that are not yet clear, there was no overproduction of PAL in soybean cell suspension cultures and 2-aminooxyacetic acid greatly reduced PAL formation in such cultures (Havir, 1981b). These compounds may have effects other than blocking cinnamate formation by PAL inhibition. The influence of cinnamate on PAL formation in a number of systems has been reviewed (Engelsma, 1979). If cinnamate modulates translation, it should be possible to demonstrate the accumulation of polyribosomal mRNA for PAL in the presence of cinnamate. Accumulation might not be observed if cinnamate prevents mRNA binding to the ribosomes or stimulates mRNA degradation.

Most of the evidence for control of PAL activity through combination with a proteinaceous inhibitor (possibility 7 above) derives from studies on gherkin seedlings by Engelsma and later by Smith and associates. The response of the system to blue light (Engelsma, 1967, 1974), Mn^{2+} (Engelsma, 1972), and temperature (Attridge and Smith, 1973; Engelsma, 1969, 1970) has been studied. The postulate that a stored form of PAL accumulates in the tissue has been supported by density labeling experiments (Attridge and Smith, 1974). A small (19,000) thermolabile hydrophobic protein that binds tightly to PAL has recently been isolated (Billett et al., 1978). It is present in the soluble fraction and is also released from microsomes by detergent treatment. The inhibitor is sensitive to proteolytic enzymes, and therefore its removal by proteolysis in vivo would result in an apparent activation of PAL. Gherkin PAL shows negative cooperativity. In the presence of inhibitor V_{max} was unchanged, but the entire curve was shifted, indicating competitive inhibition of the enzyme. On separation of inhibitor and enzyme all the PAL activity was recovered. The inhibition equilibrium appears to be reached very slowly (15–30 min, depending on the temperature) with dissociation favored at the higher temperatures (French and Smith, 1975). The protein also inhibits cinnamate 4-hydroxylase, thus it could serve to coordinate the activities of the two enzymes. The properties of the inhibitor as isolated have been said to account for observations concerning the cycloheximide-insensitive "reactivation" of PAL observed with gherkin seedlings. Seedlings during a 25-h period in blue light at 25°C gave a peak in PAL activity at 6 h, followed by a plateau. This level of PAL fell only slightly during 24 h at 4°C in the dark. On transfer in the dark to 25°C a peak in PAL activity was observed after 5 h. When cycloheximide was added at the time of transfer, the peak was enhanced, but there was no peak in the presence of chloramphenicol (Attridge and Smith, 1973). Nothing is known about the stability of the inhibitor in vivo. Although the presence of such an inhibitor has been suggested in connection with other systems, it is possible that it is important only in certain plants or at certain stages of development. Its function could be to increase the sensitivity of phenylpropanoid metabolism to changes in temperature. Some observations attributed to the presence in extracts of a proteinaceous inhibitor may be explicable in terms of complexes of lipoproteins and phenolic compounds (Gupta and Acton, 1979).

The radish cotyledon system mentioned above in connection with the effects of cinnamate and D-phenylalanine on PAL levels (Huault and Klein-Eude, 1978) has also provided evidence for PAL activation (Klein-Eude *et al.*, 1974). Radish cotyledons were exposed to far-red light for 12 h. On transfer to darkness the PAL half-life was 24 h, but on transfer to cycloheximide at a level sufficient to block PAL synthesis a plateau of activity was observed. This suggested that cycloheximide turned off both the PAL-IS system and PAL synthesis. On transfer to cycloheximide in the dark PAL levels began to decay after a 6-h plateau. Reexposure to far-red light, however, caused a swift restoration of the lost activity. There is no reason to suppose that a direct action of light on PAL is responsible for this change, and the proteinaceous inhibitor hypothesis does not seem to be relevant. It is possible the radish PAL in the dark eventually loses activity by oxidation and regains the activity when light makes reduced thioredoxin available, as discussed in Section III,D. The report (Blondel *et al.*, 1973) that an inactive form of PAL can be isolated from light-grown radish cotyledons by affinity chromatography has not been substantiated in other laboratories. The prospects for developing an affinity column for PAL seem to be poor (Gupta and Acton, 1979).

D. Kinetic Regulation of Enzyme Activity

In order to achieve a significant flux to phenylpropanoid metabolism an adequate supply of phenylalanine and, in many tissues, tyrosine must be maintained (Creasy and Zucker, 1974). The common precursor of these amino acids and of L-tryptophan is chorismic acid. In tobacco, tryptophan exerts negative feedback control on anthranilate synthase which acts on this substrate (Widholm, 1972). In a number of systems L-phenylalanine and L-tyrosine exert negative feedback control on one or more isozymes of chorismate mutase, whereas L-tryptophan may activate the same isozyme [in alfalfa (Woodin and Nishioka, 1973), oak (Gadal and Bouyssou, 1973), pea (Cotton and Gibson, 1968), mung bean (Gilchrist *et al.*, 1972), and tobacco (Widholm, 1972), but not in carrot cell suspension cultures (Palmer and Widholm, 1975; Widholm, 1974)]. Chorismate mutase leads to prephenate, the common precursor of phenylalanine and tyrosine. Whereas feedback inhibition by these amino acids serves to maintain the pool sizes of the amino acid end products at a steady level despite variable demands, cross-activation by tryptophan serves to relate the phenylalanine and tyrosine pool sizes to that of tryptophan.

The relative pool sizes of phenylalanine and tyrosine should be regulated at steps subsequent to prephenate formation. In addition to phenyl- and *p*-hydroxyphenylpyruvate the role of arogenate must be considered. This compound (the transamination product of prephenate) may serve as an alternative intermediate in tyrosine, and possibly phenylalanine, formation (Rubin and Jensen, 1979; Zamir *et al.*, 1980). As phenylalanine and tyrosine may be utilized in alternative ways and the total demand may change appreciably, the system should normally operate to maintain their pool sizes irrespective of the demand (Amrhein and Zenk, 1971; Creasy and Zucker, 1974). In a wide range of very different plant tissues the phenylalanine concentration remains relatively low (<0.1–0.2 μmol/g fr wt), stays practically constant, and varies little in response to mild stress (for references, see Margna, 1977). If the synthesis increases to the point at which transamination from aspartate or glutamate cannot keep pace, phenylpyruvate, *p*-hydroxyphenylpyruvate, and indolylpyruvate will accumulate (Wightman and Forest, 1978). The first two ketones are inhibitors of PAL, thus the demand on the phenylalanine and tyrosine pools will be lessened. Computer simulation will be required in order to clarify the function of such a negative feed-forward control.

The above discussion assumes that the enzymes and products of aromatic amino acid biosynthesis occupy a single compartment. Reversible storage in vacuoles could serve to buffer changes in the pools that operate in feedback inhibition or cross-activation. This possibility has been raised in connection with studies on wild-type and *p*-fluorophenylalanine-resistant lines of tobacco cell suspension cultures (Section III,E; Berlin and Vollmer, 1979). When AOPP was used to block PAL activity, the total phenylalanine levels increased 6- and 17-fold, respectively, but the flux from shikimate to aromatic amino acids was not impeded by these increases.

The pattern of cross-activation observed for aromatic amino acid biosynthesis has also been observed for PAL. The two forms of PAL isolated from the leaves and roots of oak were found to be activated by gallic acid (Boudet *et al.*, 1971; Alibert *et al.*, 1972, Section II,B). The V_{max} increased by ~20% (PAL I) or ~40% (PAL II). Gallic acid also derives from the shikimate pathway. The effect of this activation is to link phenylpropanoid metabolism to the pool size of gallic acid. The response to gallic acid passes through a maximum at ~1 mM (Section II,E). Benzoic acid and substituted benzoic acids were found to lower V_{max} for PAL I, which was associated with benzoate synthase, whereas caffeic and ferulic acids did not inhibit at the concentration tested. The pool sizes of benzoate, salicylate, etc., are thus under negative feedback control. Caffeic and ferulic acids inhibited PAL II, which was associated with cinnamate 4-hydroxylase, again suggesting feedback control. The products of this pathway were presumably not converted to substituted benzoates because of compartmentation and thus did not lead to inhibition of PAL I. The kinetics of the above effects have not been studied in detail.

The cooperative interactions shown by PAL from many sources probably play a role in PAL regulation (Section II,E). As isolation and purification can change K_m values, ligand dissociation constants, and cooperativity effects, important properties of the enzyme may escape detection. Negative cooperativity with respect to phenylalanine tends to minimize the effect of fluctuations in the phenylalanine pool on phenylpropanoid metabolism. The K_m values for the isolated enzyme credibly indicate the typical concentrations of L-phenylalanine present in tissues. The consequences of inhibitors or effectors acting on the cooperative enzyme are more likely to be of interest. Product inhibition by cinnamate may be negligible if cinnamate concentrations are in the range 0.2–20 μM. This seems an appropriate figure, as the K_m for cinnamate 4-hydroxylase from sorghum is about 2 μM (Potts *et al.*, 1974). For the wheat enzyme a significant concentration of benzoate increases $[S]_{0.5}$ and converts negative to positive cooperativity (Nari *et al.*, 1974, Section II,E). It is not known whether such accumulations occur in wheat tissue or whether some other compound accumulates that inhibits in this way. Homogentisic acid, a product of tyrosine degradation in plants, was found to be a noncompetitive inhibitor of PAL from barley. Negative cooperativity with respect to tyrosine was observed in the presence or absence of the inhibitor (Kindl, 1970). Flavanoids from peas were reported to inhibit pea PAL, apparently by binding to a regulatory site (Attridge *et al.*, 1971, Section II,B).

Although the extent to which PAL activity is inhibited by metabolites *in vivo* is uncertain, there appears to be good evidence that extensive regulatory inhibition occurs. In 15 out of 20 examples surveyed the isolated PAL was capable of supplying the carbon for many times the quantity of phenolic compounds that accumulated, of swiftly depleting the phenylalanine pool to zero and even of converting the entire dry matter of the tissue to cinnamic acid in a short time (Margna, 1977).

It remains to consider further the slow threefold activation of one form of PAL from spinach chloroplasts brought about by thioredoxin (Nishizawa *et al.*, 1979, Section II,B). Since disulfide bond formation is probably involved, it falls marginally outside the group of enzymes discussed by Frieden (1979) as exhibiting hysteretic behavior. It seems partinent, however, that hysteretic behavior is typically a property of regulatory enzymes. Photosynthetic electron transport sequentially brings about the reduction of ferredoxin, ferredoxin-thioredoxin reductase, thioredoxin-f, and PAL. This light effect is blocked by the photosynthetic electron transport inhibitor dichlorodiphenylmethylurea (DCMU) (Creasy, 1968). In the absence of light PAL loses its activity as a result of oxidation (probably by oxidized glutathione or dehydroascorbate). The activity of several enzymes associated with the Calvin cycle is modulated by the same system. The existence of regulated and nonregulated forms of PAL in the chloroplast parallels the existence of regulated and nonregulated forms of NADP-glyceraldehyde-3-phosphate dehydrogenase and phosphoribulokinase. Activation is fast compared to *de novo* synthesis; thus phenylpropanoid metabolism may increase a short time after the chloroplast is exposed to light, provided substrate is available. As the substrate derives, via the shikimate pathway, from Calvin cycle intermediates, a common regulation of both areas of metabolism is desirable.

E. Modification of Metabolism

As phenylpropanoid metabolism plays a key role in host–parasite interactions, and the end products of metabolism influence the food values and structural properties and durability of crop plants, an important goal of the studies discussed here is to obtain mutant plants in which the pathway has been altered in ways that cannot be obtained by conventional selection methods. The utility of such mutant genes could then be examined in different genetic contexts by standard crop-breeding techniques. The problem and its difficulties may be discussed by reference to p-fluorophenylalanine resistance.

Cell suspension cultures of tobacco (*Nicotiana tabaccum* L.) that are resistant to p-fluorophenylalanine have been isolated by Palmer and Widholm (1975). There is some uncertainty as to the fundamental difference between wild-type and resistant lines. A double mutation is intrinsically unlikely, thus it is necessary to decide where the site of mutation is and how this change produces multiple effects. The initial hypothesis was that the resistant lines were defective in the feedback control of chorismate mutase. Overproduction of L-phenylalanine would thus dilute out the harmful effects of the p-fluorophenylalanine taken up by the cells. Further study has shown that the difference in kinetic properties of chorismate mutase from the two lines is very small (Berlin and Vollmer, 1979). It seems unlikely therefore that there is in fact a structural mutation of this enzyme—the small differences must reflect secondary factors. The resistant line overproduces PAL by a factor of 10. Such a phenotypic change should provide a selective advantage in that p-fluorophenylalanine is more swiftly converted to p-fluorocinnamic acid (Hanson and Havir, 1977). This compound could then be either degraded or stored as a harmless conjugate. Mutation could involve various loci. For example, the affinity for cinnamate in the modulation of PAL synthesis could be reduced and thus PAL synthesis would increase. Overproduction of PAL might, however, be generated by lowering cinnamate levels. The mutant strain produces 6–10 times higher levels of phenolic compounds, 85% of which is caffeyoyl- and feruloylputrescine (Berlin and Widholm, 1979). This could occur because cinnamate hydroxylase levels are enhanced and cinnamate levels therefore lowered. Hydroxylase overproduction in turn might occur because p-coumarate levels are lowered. This type of conjecture can be extended. The mutation probably lies in the phenylpropanoid pathway, but it need not have a direct effect on PAL synthesis. Another observation to be explained is that the levels of all amino acids are 3–5 times lower in the mutant strain (Berlin and Vollmer, 1979).

The above analysis suggests that a desired change in a complex regulated system may be achieved by altering the system in various ways. As more p-fluorophenylalanine mutants are examined, mutants in different genes may be obtained. Selection might be achieved for less PAL production if a suicide compound were available, i.e., if PAL action converted an L-phenylalanine analogue to a substance much more toxic to the cell than the analogue itself. Even if putative mutants cannot be converted to viable plants, their study should bring about a better understanding of the interrelationship between the metabolism of aromatic amino acid and that of phenylpropanoid compounds.

IV. CONCLUDING REMARKS

One of the more encouraging aspects of the studies discussed in this review is the extent to which enzymology and physiology have moved forward together. Substantial advances in enzymology are likely to occur when it is economically possible to prepare from plant sources highly purified PAL suitable for detailed structural studies. The regulatory controls influencing PAL activity are complex, and it seems evident that no one approach can clarify fully the metabolic role of PAL. Fortunately the primary interests of investigators differ: Some are concerned with understanding the transmission of light effects, some with the integration of chloroplast metabolism or the entire phenylpropanoid pathway, some with host–parasite interactions, and some with the selection of mutant plants for potential crop improvements. This is a partial list. All these aspects of study have contributed to our present understanding, and it is important that such diversity continue.

REFERENCES

Acton, G. J., and Schopfer, P. (1975). *Biochim. Biophys. Acta* **404**, 231–242.
Agutter, P. S., and McCaldin, B. (1979). *Biochem. J.* **180**, 371–378.
Alibert, G., Ranjeva, R., and Boudet, A. (1972). *Biochim. Biophys. Acta* **279**, 282–289.
Alibert, G., Ranjeva, R., and Boudet, A. M. (1977). *Physiol. Veg.* **15**, 279–301.
Amrhein, N., and Gerhardt, J. (1979). *Biochim. Biophys. Acta* **583**, 434–442.
Amrhein, N., and Gödeke, K-H. (1977). *Plant Sci. Lett.* **8**, 313–317.
Amrhein, N., and Holländer, H. (1979). *Planta* **144**, 385–389.
Amrhein, N., and Zenk, M. H. (1970). *Naturwissenschaften* **57**, 312.
Amrhein, N., and Zenk, M. H. (1971). *Z. Pflanzenphysiol.* **64**, 145–168.
Amrhein, N., Gödeke, K.-H., and Gerhardt, J. (1976). *Planta* **131**, 33–40.
Attridge, T. H., and Smith, H. (1973). *Phytochemistry* **12**, 1569–1574.
Attridge, T. H., and Smith, H. (1974). *Biochim. Biophys. Acta* **343**, 452–464.

Attridge, T. H., Stewart, G. R., and Smith, H. (1971). *FEBS Lett.* **17**, 84–86.

Attridge, T. H., Johnson, C. B., and Smith, H. (1974). *Biochim. Biophys. Acta* **343**, 440–452.

Bandoni, R. J., Moore, K., Subba Rao, P. V., and Towers, G. H. N. (1968). *Phytochemistry* **7**, 205–207.

Bartl, K., Cavalar, C., Krebs, T., Ripp, E., Rétey, J., Hull, E. W., Günther, H., and Simon, H. (1977). *Eur. J. Biochem.* **72**, 247–250.

Beneveniste, I., Salaun, J.-P., and Durst, F. (1977). *Phytochemistry* **16**, 69–73.

Berlin, J., and Vollmer, B. (1979). *Z. Naturforsch., Teil C* **34**, 770–775.

Berlin, J., and Widholm, J. M. (1977). *Plant Physiol.* **59**, 550–553.

Betz, B., Schafer, E., and Hahlbrock, K. (1978). *Arch. Biochem. Biophys.* **190**, 126–135.

Bezanson, G. S., Desaty, D., Emes, A. V., and Vining, L. C. (1970). *Can. J. Microbiol.* **16**, 147–151.

Billett, E. E., Wallace, W., and Smith, H. (1978). *Biochim. Biophys. Acta* **524**, 219–230.

Blondel, J.-D., Huault, C., Faye, L., and Rollin, P. (1973). *FEBS Lett.* **36**, 239–243.

Blume, D. E., and McClure, J. W. (1978). *Phytochemistry* **17**, 1545–1547.

Boudet, A., Ranjeva, R., and Gadal, P. (1971). *Phytochemistry* **10**, 997–1005.

Brand, L. M., and Harper, A. E. (1976). *Biochemistry* **15**, 1814–1821.

Camm, E. L., and Towers, G. H. N. (1969). *Phytochemistry* **8**, 1407–1413.

Camm, E. L., and Towers, G. H. N. (1973). *Phytochemistry* **12**, 961–973.

Chasin, L. A., and Magasanik, B. (1968). *J. Biol. Chem.* **243**, 5165–5178.

Cheng, C. K.-C., and Marsh, H. V., Jr. (1968). *Plant Physiol.* **43**, 1755–1759.

Cleland, W. W. (1967). *Adv. Enzymol.* **29**, 1–32.

Cleland, W. W. (1977). *In* "Isotope Effects on Enzyme-Catalyzed Reactions" (W. W. Cleland, M. H. O'Leary, and D. B. Northrop, eds.), pp. 153–175. University Park Press, Baltimore, Maryland.

Conn, E., McFarlane, I. J., Moeller, B. L., and Shimada, M. (1979). *FEBS-Symp.* **55**, 63–71.

Cotton, R. G. H., and Gibson, F. (1968). *Biochim. Biophys. Acta* **156**, 187–189.

Creasy, L. L. (1968). *Phytochemistry* **7**, 441–446.

Creasy, L. L. (1976). *Phytochemistry* **15**, 673–675.

Creasy, L. L., and Zucker, M. (1974). *Recent Adv. Phytochem.* **8**, 1–19.

Czichi, U., and Kindl, H. (1975a). *Hoppe-Seyler's Z. Physiol. Chem.* **356**, 457–485.

Czichi, U., and Kindl, H. (1975b). *Planta* **125**, 115–125.

Czichi, U., and Kindl, H. (1977). *Planta* **134**, 133–143.

Dixon, N. E., Gazzola, C., Blakeley, R. L., and Zerner, B. (1976). *Science* **191**, 1144–1150.

Duchesne, M., Fritig, B., and Hirth, L. (1977). *Biochim. Biophys. Acta* **485**, 465–481.

Duke, S. O., Hoagland, R. E., and Elmore, C. D. (1980). *Plant Physiol.* **65**, 17–21.

Durst, F. (1976). *Planta* **132**, 221–227.

Ebel, J., Ayers, A. R., and Albersheim, P. (1976). *Plant Physiol.* **57**, 775–779.

Egan, R. M., and Phillips, A. T. (1977). *J. Biol. Chem.* **252**, 5701–5707.

Emes, A. V., and Vining, L. C. (1970). *Can. J. Biochem.* **48**, 613–623.

Engelsma, G. (1967). *Planta* **75**, 207–219.

Engelsma, G. (1969). *Naturwissenschaften* **56**, 563.

Engelsma, G. (1970). *Planta* **91**, 246–254.

Engelsma, G. (1972). *Plant Physiol.* **50**, 599–602.

Engelsma, G. (1974). *Plant Physiol.* **54**, 702–705.

Engelsma, G. (1979). *FEBS-Symp.* **55**, 163–172.

Erez, A. (1973). *Plant Physiol.* **51**, 409–411.

Faudin, A. S., and Macko, V. (1974). *Phytopathology* **64**, 990–993.

Freese, E., Shen, C. W., and Galliers, E. (1973). *Nature (London)* **241**, 321–325.

French, C. J., and Smith, H. (1975). *Phytochemistry* **14**, 963–966.

Frieden, C. (1979). *Annu. Rev. Biochem.* **48**, 471–489.

Fritz, R. R., Hodgins, D. S., and Abell, C. W. (1976). *J. Biol. Chem.* **251**, 4646–4650.
Gadal, P., and Bouyssou, H. (1973). *Physiol. Plant.* **28**, 7–13.
Gilchrist, D. G., Woodin, T. S., Johnson, M. L., and Kosuge, T. (1972). *Plant Physiol.* **49**, 52–57.
Givot, I. L., Smith, T. A., and Abeles, R. H. (1969). *J. Biol. Chem.* **23**, 6341–6353.
Gordon, W. R., and Koukkari, W. L. (1978). *Plant Physiol.* **62**, 612–615.
Gregor, H. D. (1976). *Z. Pflanzenphysiol.* **77**, 454–463.
Grisebach, H., and Ebel, J. (1978). *Angew. Chem., Int. Ed. Engl.* **17**, 635–647.
Grollman, A. P. (1966). *Proc. Natl. Acad. Sci. U.S.A.* **56**, 1867–1874.
Gupta, S., and Acton, J. (1979). *Biochim. Biophys. Acta* **570**, 187–197.
Hadwiger, L. A. (1972). *Plant Physiol.* **49**, 779–782.
Hadwiger, L. A., and Schwochau, M. E. (1971). *Plant Physiol.* **47**, 588–590.
Hadwiger, L. A., Hess, S. L., and von Broembsen, S. (1970). *Phytopathology* **60**, 332–336.
Hahlbrock, K., and Grisebach, H. (1979). *Annu. Rev. Plant Physiol.* **30**, 105–130.
Hahlbrock, K., and Ragg, H. (1975). *Arch. Biochem. Biophys.* **166**, 41–46.
Hahlbrock, K., and Wellmann, E. (1973). *Biochim. Biophys. Acta* **304**, 702–706.
Hahlbrock, K., Ebel, J., Ortmann, R., Sutter, A., Wellmann, E., and Grisebach, H. (1971). *Biochim. Biophys. Acta* **244**, 7–15.
Hanson, K. R. (1966). *Phytochemistry* **5**, 491–499.
Hanson, K. R. (1979). *Fed. Proc., Fed. Am. Soc. Exp. Biol.* **38**, 722.
Hanson, K. R. (1981a). *Arch. Biochem. Biophys.* (in press).
Hanson, K. R. (1981b). *Arch. Biochem. Biophys.* (in press).
Hanson, K. R., and Havir, E. A. (1969). *Fed. Proc., Fed. Am. Soc. Exp. Biol.* **28**, 602.
Hanson, K. R., and Havir, E. A. (1970). *Arch. Biochem. Biophys.* **141**, 1–17.
Hanson, K. R., and Havir, E. A. (1972a). *Recent Adv. Phytochem.* **4**, 46–85.
Hanson, K. R., and Havir, E. A. (1972b). *In* "The Enzymes" (P. D. Boyer, ed), 3rd ed., Vol. 7, pp. 75–166. Academic Press, New York.
Hanson, K. R., and Havir, E. A. (1977). *Arch. Biochem. Biophys.* **180**, 102–113.
Hanson, K. R., and Rose, I. A. (1975). *Acc. Chem. Res.* **8**, 1–10.
Hanson, K. R., Ling, R., and Havir, E. A. (1967). *Biochem. Biophys. Res. Commun.* **29**, 194–197.
Hanson, K. R., Wightman, R. H., Staunton, J., and Battersby, A. R. (1971). *Chem. Commun.* pp. 185–186.
Hanson, K. R., Havir, E. A., and Ressler, C. (1979). *Biochemistry* **18**, 1431–1438.
Harborne, J. B., and Corner, J. J. (1961). *Biochem. J.* **81**, 242.
Hassall, H., and Soutar, A. K. (1974). *Biochem. J.* **137**, 559–566.
Havir, E. A. (1979). *Fed. Proc., Fed. Am. Soc. Exp. Biol.* **38**, 354.
Havir, E. A. (1981a). *Arch. Biochem. Biophys.* (in press).
Havir, E. A. (1981b). *Planta* (in press).
Havir, E. A., and Hanson, K. R. (1968a). *Biochemistry* **7**, 1896–1903.
Havir, E. A., and Hanson, K. R. (1968b). *Biochemistry* **7**, 1904–1914.
Havir, E. A., and Hanson, K. R. (1973). *Biochemistry* **12**, 1583–1591.
Havir, E. A., and Hanson, K. R. (1975). *Biochemistry* **14**, 1620–1626.
Havir, E. A., Reid, P. D., and Marsh, H. V., Jr. (1971). *Plant Physiol.* **48**, 130–136.
Hodgins, D. S. (1968). *Biochem. Biophys. Res. Commun.* **32**, 246–253.
Hodgins, D. S. (1971). *J. Biol. Chem.* **266**, 2977–2985.
Hodgins, D. S. (1972). *Arch. Biochem. Biophys.* **149**, 91–96.
Hofinger, M., Monseur, X., Pais, M., and Jarreau, F. X. (1975). *Phytochemistry* **14**, 475–477.
Huault, C., and Klein-Eude, D. (1978). *Plant Sci. Lett.* **13**, 185–192.
Huffaker, R. C., and Peterson, L. W. (1974). *Annu. Rev. Plant Physiol.* **25**, 363–392.
Hyodo, H., and Yang, S. F. (1971). *Plant Physiol.* **47**, 765–770.

Hyodo, H., Kuroda, H., and Yang, S. F. (1978). *Plant Physiol.* **62**, 31–35.
Innerarity, L. T., Smith, E. C., and Wender, S. H. (1972). *Phytochemistry* **11**, 83–88.
Iredale, S. E., and Smith, H. (1973). *Phytochemistry* **12**, 2145–2154.
Iredale, S. E., and Smith, H. (1974). *Phytochemistry* **13**, 575–583.
Jangaard, N. O. (1974). *Phytochemistry* **13**, 1765–1768.
Johnson, C. B., and Smith, H. (1978). *Phytochemistry* **17**, 667–670.
Kalghatgi, K. K., and Subba Rao, P. V. (1975). *Biochem. J.* **149**, 65–72.
Kalghatgi, K. K., and Subba Rao, P. V. (1976). *J. Bacteriol.* **126**, 568–578.
Kamel, M. Y., and Maksoud, S. A. (1978). *Z. Pflanzenphysiol.* **88**, 255–262.
Kenyon, G. L., and Hageman, G. D. (1979). *Adv. Enzymol.* **50**, 325–360.
Kindl, H. (1970). *Hoppe-Seyler's Z. Physiol. Chem.* **351**, 792–798.
Kindl, H. (1979). *FEBS-Symp.* **55**, 49–61.
Klee, C. B., and Gladner, J. A. (1972). *J. Biol. Chem.* **24**, 8051–8057.
Klee, C. B., Kirk, K. L., Cohen, L. A., and McPhie, P. (1975). *J. Biol. Chem.* **250**, 5033–5040.
Klein-Eude, D., Rollin, P., and Huault, C. (1974). *Plant Sci. Lett.* **2**, 1–8.
Koshland, D. E., Jr., Néméthy, G., and Filmer, D. (1966). *Biochemistry* **5**, 365–385.
Koukol, J., and Conn, E. E. (1961). *J. Biol. Chem.* **236**, 2692–2698.
Lamb, C. J. (1979). *Arch. Biochem. Biophys.* **192**, 311–317.
Lamb, C. J., and Rubery, P. H. (1976). *Phytochemistry* **15**, 665–668.
Lamb, C. J., Merrit, T. K., and Butt, V. S. (1979). *Biochim. Biophys. Acta* **582**, 196–212.
Legrand, M., Fritig, B., and Hirth, L. (1976). *Phytochemistry* **15**, 1353–1359.
Levitzki, A., and Koshland, D. E., Jr. (1976). *Curr. Top. Cell. Regul.* **10**, 1–40.
Lienhard, G. E. (1973). *Science* **180**, 149–154.
Löffelhardt, W., and Kindl, H. (1975). *Hoppe-Seyler's Z. Physiol. Chem.* **356**, 487–493.
Löffelhardt, W., Ludwig, B., and Kindl, H. (1973). *Hoppe-Seyler's Z. Physiol. Chem.* **354**, 1006–1012.
McClure, J. (1979). *Recent Adv. Phytochem.* **12**, 525–556.
Mack, J. (1979). *Abstr., Int. Congr. Biochem., 11th, 1979* p. 295.
Maggio, E. T., Kenyon, G. L., Mildvan, A. S., and Hegeman, G. D. (1975). *Biochemistry* **14**, 1131–1139.
Margna, U. (1977). *Phytochemistry* **16**, 419–426.
Mildvan, A. S. (1971). *Adv. Chem. Ser.* **100**, 390–412.
Minamikawa, T., and Uritani, I. (1965). *J. Biochem. (Tokyo)* **57**, 678–688.
Monod, J., Wyman, J., and Changeux, J. P. (1965). *J. Mol. Biol.* **12**, 88–118.
Müller, U., and Crout, D. H. G. (1975). *Phytochemistry* **14**, 859–860.
Murphy, B. J., and Stutte, C. A. (1978). *Anal. Biochem.* **86**, 220–228.
Nagaiah, K., Kumar, S. A., and Mahadevan, S. (1977). *Phytochemistry* **16**, 667–671.
Nari, J., Mouttet, C., Pinna, M. H., and Ricard, J. (1972). *FEBS Lett.* **23**, 220–224.
Nari, J., Mouttet, C., Fouchier, F., and Ricard, J. (1974). *Eur. J. Biol. Chem.* **41**, 499–515.
Neish, A. C. (1961). *Phytochemistry* **1**, 1–24.
Nishizawa, A. N., Wolosiuk, R. A., and Buchanan, B. B. (1979). *Planta* **145**, 7–12.
Ochoa, S., and de Haro, C. (1979). *Annu. Rev. Biochem.* **48**, 569–580.
O'Neal, D., and Keller, C. J. (1970). *Phytochemistry* **9**, 1373–1383.
Palmer, J. E., and Widholm, J. (1975). *Plant Physiol.* **56**, 233–238.
Parkhurst, J. R., and Hodgins, D. S. (1971). *Phytochemistry* **10**, 2997–3000.
Parkhurst, J. R., and Hodgins, D. S. (1972). *Arch. Biochem. Biophys.* **152**, 597–605.
Peterkofsky, A. (1962). *J. Biol. Chem.* **237**, 787–795.
Potts, J. R. M., Weklych, R., and Conn, E. E. (1974). *J. Biol. Chem.* **249**, 5019–5026.
Ragg, H., Schroder, J., and Hahlbrock, K. (1977). *Biochim. Biophys. Acta* **474**, 226–233.
Ranjeva, R., Alibert, G., and Boudet, A. M. (1977). *Plant Sci. Lett.* **10**, 225–234.
Ranjeva, R., Boudet, A. M., and Alibert, G. (1979). *FEBS-Symp.* **55**, 91–100.
Reid, P. D., Havir, E. A., and Marsh, H. V., Jr. (1972). *Plant Physiol.* **50**, 480–484.
Ricard, J., Nari, J., and Mouttet, C. (1972). *Fed. Eur. Biochem. Soc. Meet. [Proc.]* **25**, 375–386.

Ricard, J., Mouttet, C., and Nari, J. (1974). *Eur. J. Biochem.* **41**, 479–497.
Riov, J., Monselise, S. P., and Kahan, R. S. (1969). *Plant Physiol.* **44**, 631–635.
Rubin, J. L., and Jensen, R. A. (1979). *Plant Physiol.* **64**, 727–734.
Ruis, H., and Kindl, H. (1970). *Hoppe-Seyler's Z. Physiol. Chem.* **351**, 1425–1427.
Ruis, H., and Kindl, H. (1971). *Phytochemistry* **10**, 2627–2631.
Sanwal, B. D. (1970). *Bacteriol. Rev.* **34**, 20–39.
Saunders, J. A., and McClure, J. W. (1975). *Phytochemistry* **14**, 1285–1289.
Saunders, J. A., Conn, E. E., Lin, C. H., and Shimada, M. (1977). *Plant Physiol.* **60**, 629–634.
Schopfer, P. (1971). *Planta* **99**, 339–346.
Schopfer, P. (1977). *Annu. Rev. Plant Physiol.* **28**, 223–252.
Schröder, J. (1977). *Arch. Biochem. Biophys.* **182**, 488–496.
Schröder, J., Betz, B., and Hahlbrock, K. (1976). *Eur. J. Biochem.* **67**, 527–541.
Schröder, J., Betz, B., and Hahlbrock, K. (1977). *Plant Physiol.* **60**, 440–445.
Schröder, J., Kreuzaler, F., Schäfer, E., and Hahlbrock, K. (1979). *J. Biol. Chem.* **254**, 57–65.
Shatkin, A. J. (1976). *Cell* **9**, 645–653.
Sheinin, R., and Humbert, J. (1978). *Annu. Rev. Biochem.* **47**, 277–316.
Shimada, M., and Conn, E. E. (1977). *Arch. Biochem. Biophys.* **180**, 199–207.
Smith, B. G., and Rubery, P. H. (1979). *Plant Sci. Lett.* **15**, 29–33.
Stafford, H. A. (1969). *Phytochemistry* **8**, 743–752.
Stafford, H. A. (1974a). *Recent Adv. Phytochem.* **8**, 53–79.
Stafford, H. A. (1974b). *Annu. Rev. Plant Physiol.* **25**, 459–486.
Subba Rao, P. V., Moore, K., and Towers, G. H. N. (1967). *Can. J. Biochem.* **45**, 1863–1872.
Szkutnicka, K., and Lewak, K. (1975). *Plant Sci. Lett.* **5**, 147–156.
Tanaka, Y., and Uritani, I. (1976). *J. Biochem. (Tokyo)* **79**, 217–19.
Tanaka, Y., and Uritani, I. (1977). *J. Biochem. (Tokyo)* **81**, 963–970.
Tanaka, Y., Matsushita, K., and Uritani, I. (1977). *Plant Cell. Physiol.* **18**, 1209–1216.
Tong, W. F., and Schopfer, P. (1978). *Plant Physiol.* **61**, 59–61.
Towers, G. H. N., Vance, C. P., and Namboodiri, A. M. D. (1974). *Recent Adv. Phytochem.* **8**, 81–94.
Wat, C.-K., and Towers, G. H. N. (1979). *Recent Adv. Phytochem.* **12**, 371–432.
Weissenböck, G., Plesser, A., and Trinks, K. (1976). *Ber. Dtsch. Bot. Ges.* **89**, 457–472.
Westcott, R. J., and Henshaw, G. G. (1976). *Planta* **131**, 67–73.
Widholm, J. M. (1972). *Biochim. Biophys. Acta* **261**, 52–58.
Widholm, J. M. (1974). *Physiol. Plant.* **30**, 13–18.
Wightman, F., and Forest, J. C. (1978). *Phytochemistry* **17**, 1455–1471.
Wightman, R. H., Staunton, J., Battersby, A. R., and Hanson, K. R. (1972). *J. Chem. Soc., Perkin Trans. 1* pp. 2355–2364.
Wolfenden, R. (1972). *Acc. Chem. Res.* **5**, 10–18.
Wolff, J. B. (1962). *J. Biol. Chem.* **237**, 874–881.
Wong, J. T.-F. (1975). "Kinetics of Enzyme Mechanisms," pp. 45–46. Academic Press, New York.
Wong, P. P., Zucker, M., and Creasy, L. L. (1974). *Plant Physiol.* **54**, 659–665.
Woodin, T. S., and Nishioka, L. (1973). *Biochim. Biophys. Acta* **309**, 211–223.
Young, M. R., and Neish, A. C. (1966). *Phytochemistry* **5**, 1121–1132.
Young, M. R., Towers, G. H. N., and Neish, A. C. (1966). *Can. J. Bot.* **44**, 341–349.
Young, O., and Beevers, H. (1976). *Phytochemistry* **15**, 359–362.
Zamir, L. O., Jensen, R. A., Arison, B. H., Douglas, A. W., Albers-Schönberg, G., and Bowen, J. R. (1980). *J. Am. Chem. Soc.* **102**, 4499–4504.
Zimmermann, A., and Hahlbrock, K. (1975). *Arch. Biochem. Biophys.* **166**, 54–62.
Zucker, M. (1965). *Plant Physiol.* **40**, 779–784.
Zucker, M. (1968). *Plant Physiol.* **43**, 365–374.
Zucker, M. (1971). *Plant Physiol.* **47**, 442–444.
Zucker, M. (1972). *Annu. Rev. Plant Physiol.* **23**, 133–156.

Oxygenases and the Metabolism of Plant Products

21

V. S. BUTT

C. J. LAMB

I. OXYGENASES AND OXIDASES

Oxygenases catalyze the incorporation of one or both of the atoms of molecular oxygen, O_2, into organic compounds usually, though not invariably, with the formation of hydroxyl groups. Monooxygenases (E.C. 1.14.13.- to 1.14.18.-) introduce only one atom, and dioxygenases (E.C.

The Biochemistry of Plants, Vol. 7
Copyright © 1981 by Academic Press, Inc.
All rights of reproduction in any form reserved.
ISBN 0-12-675407-1

1.14.11.- to 1.14.12.-) both. Some monooxygenases and dioxygenases require two substrates; the second substrate may be oxidized without oxygen incorporation (mixed oxidase–monooxygenase reactions) or incorporate the second oxygen atom (dioxygenase reactions). The immediate products of oxygenase action may be unstable, especially those of dioxygenation, and rapidly undergo further nonenzymatic reaction leading, for example, to dehydration or carbon-chain fission.

Oxygenases are distinct from oxidases (E.C. 1.-.3.-) in that the latter catalyze only the transfer of electrons from their substrate to O_2 with the formation of a reduced oxygen species such as superoxide anion, peroxide, or water but without the incorporation of oxygen atoms into the substrate. Plant phenolases are peculiar in their ability to act as both oxygenases and oxidases, the balance between the two activities depending on the substrate and the conditions employed.

Metabolically, oxidases are involved in the terminal oxidation stage of hydrogen or electron transport chains at least as much as in the formation of intermediates in biosynthetic pathways. Oxygenases do not take part in terminal oxidations but in the intermediary reactions of biosynthetic processes and the destruction of natural products. The products, usually with hydroxyl groups introduced at specific positions, are often peculiar to specific plants or organs in which their biological function can be seen as a consequence of the specific oxygenation that was a feature of their biogenesis.

II. OXYGENASES AND HYDROXYLATION

Under monooxygenase action, one of the two atoms of O_2 is used in hydroxylation and the second is reduced to water by a reducing agent, such as NADPH, NADH, ascorbate, tetrahydrofolate, or another tetrahydropteridine derivative, according to the general equation

$$RH + O_2 + H_2A \rightarrow ROH + H_2O + A$$

in which H_2A is the reducing agent. These monooxygenases are therefore sometimes referred to as mixed-function oxidases, in comparison with oxidases catalyzing the reduction of both atoms of O_2. With the use of ^{18}O-labeled O_2, the origins of the oxygen of the hydroxyl group can be demonstrated with isolated enzymes or in intact tissues (for example, Mason *et al.,* 1955).

These enzymes function primarily as activators of molecular oxygen. Although hydroxylation reactions under oxygenase action are virtually irreversible, the activation energies are high because, in the ground state, O_2 possesses two unpaired electrons (triplet state) and attacks substrates with paired electrons (singlet state); the unpaired electrons of O_2 must first be delocalized either by complexing with a transition metal ion, usually iron or copper, or by the generation of a radical species such as HO_2^- (Hamilton, 1974). The mechanisms of hydroxylation and oxygenase action are fully discussed by West (this series, Vol. 2, Chapter 8).

Two types of oxygenase will attract special attention in this chapter. Cytochrome P-450-dependent oxygenases are membrane-bound monooxygenases which utilize specifically NADPH as reducing agent; the iron of the cytochrome combines with O_2 which then reacts with the substrate and is hydroxylated under the catalytic action of a specific protein; the iron undergoes redox changes and interacts with the reducing agent supplied (see Hanson and Havir, 1979). The hydroxyl groups are characteristically introduced at specific carbon atoms in the substrate molecule. Thus specific monooxygenases exist for the hydroxylation of cinnamic acid to 4-hydroxycinnamic acid, kaurene to 7β-hydroxykaurene, and geraniol to 10-hydroxygeraniol. Although the enzymes engaged in these hydroxylations have only rarely been fully characterized, the introduction of hydroxyl groups into a wide range of compounds by membrane-bound or microsomal enzymes dependent on NADPH and O_2 has been demonstrated.

The second group of monooxygenases employed by plants in hydroxylation are phenolases. These are copper-containing enzymes, sometimes firmly held in membranes but usually readily solubilized. They catalyze the introduction of a second hydroxyl group into a monophenol, ortho to the hydroxyl group already present and usually meta to a carbon side chain:

$$+ O_2 + H_2A \longrightarrow \qquad\qquad + H_2O + A \qquad (1)$$

Then O_2 combines with cuprous copper, which undergoes oxidation in the course of hydroxylating the phenol (see Vanneste and Zuberbühler, 1974). The cupric copper is subsequently reduced, ultimately by the reducing agent, H_2A (Fig. 1); there is no direct reaction between H_2A and either the substrate or O_2 (McIntyre and Vaughan, 1975). The reaction must be initiated by at least a trace of the o-diphenol (Vaughan and Butt, 1970), which reduces the cupri- to the cupro-enzyme prior to oxygenation; the reducing agent serves only to reduce nonenzymatically the o-quinone so produced, and consequently a wide range of hydrogen donors can be used. The catalysis of this reaction by phenolase is variously described as its hydroxylase, monophenolase, tyrosinase, or cresolase activity, the choice of term depending often on the substrate hydroxylated. The specificity of most phenolases is rather broad, but some preparations have been resolved into fractions with different relative specificities (Stafford, 1974b; Schill and Grisebach, 1973).

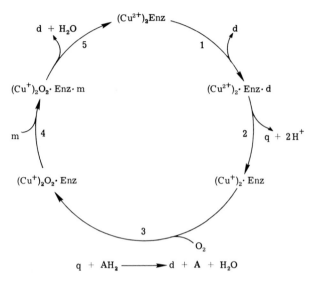

Fig. 1. Catalytic reaction cycle for hydroxylase activity of phenolase. (1) The cupri-enzyme combines with the o-diphenol, d; (2) the o-diphenol is oxidized to its o-quinone, q, which is released with generation of the cupro-enzyme; (3) the cupro-enzyme combines reversibly with O_2 to form the cupro-oxyenzyme; (4) the cupro-oxyenzyme combines with a monophenol, m; (5) the monophenol is oxygenated to an o-diphenol, which is released and the cupri-enzyme regenerated. The o-quinone released in stage (2) is reduced to the o-diphenol by reducing agent, H_2A, present; so that an added o-diphenol can act catalytically (McIntyre and Vaughan, 1975). The cupro-oxyenzyme generated in stage (4) may alternatively oxidize o-diphenol to o-quinone to regenerate the cupri-enzyme. The model assumes a single active center on a dimeric functional unit. After Vanneste and Zuberbühler (1974).

As well as participating in hydroxylase reactions, most phenolases catalyze the further oxidation of o-diphenols to o-quinones:

$$\text{(2)}$$

The accumulation of o-quinone may be suppressed by substrate quantities of reducing agent but, when these are exhausted, the o-quinone undergoes a variety of polymerization reactions (see Pierpoint, 1970) which not infrequently inactivate the enzyme. The oxidase activity of phenolase is variously referred to as polyphenol oxidase, o-diphenol oxidase, DOPA oxidase, catechol oxidase, or catecholase activity, again depending on the substrate used; it is discussed in detail by Butt (this series, Vol. 2, Chapter 3). The balance between hydroxylase and oxidase activities varies widely between enzymes in which oxidase activity is relatively low (Stafford and Dresler, 1972; Schill and Grisebach, 1973) and those in which only the oxidase activity can be detected (see Stafford, 1974a); a wide variation may be found even among fractions from the same plant organ. The hydroxylase activity is usually more labile than the oxidase activity, but both may be necessary for phenolase to exert its full biological function in some plant organs. Phenolase activity of either type may sometimes be difficult to demonstrate because of its so-called latency, from which the enzyme is released by detergent action on membranes (Robb et al., 1963; Vaughan et al., 1975) or proteolytic action (Tolbert, 1973); the latency may also be due to inhibition by substances in the extract (Satô, 1977) or to conformational change induced by the extractants employed (Robb et al., 1963).

Cytochrome P-450-dependent monooxygenases and phenolases appear to be the major hydroxylating agents in plants, but plant peroxidases can also catalyze the hydroxylation of aromatic compounds, utilizing O_2 (for which H_2O_2 cannot be substituted) and added dihydroxyfumarate (for which neither ascorbate nor isoascorbate can be substituted). The reaction gives rise to a range of isomers; e.g., salicylic acid gives 2,3- and 2,5-dihydroxybenzoic acids, suggesting random insertion of the hydroxyl group by a radical substitution mechanism (Mason et al., 1957; Buhler and Mason, 1961). Peroxygenase utilizes H_2O_2 or an organic peroxide, such as linoleic acid hydroperoxide, in the hydroxylation of phenol to p-hydroquinone and of 1-naphthol to naptho-1,4-hydroquinone; O_2 is not utilized by this enzyme (Ishimaru and Yamazaki, 1977).

In intact tissues, the role of oxygenases in hydroxylation can be assessed by the incorporation of [18]O from labeled O_2. Because most plant tissues reduce O_2 to H_2O at least as rapidly as they incorporate it into organic compounds, it is essential to demonstrate that the specific activity of the oxygen incorporated is at least comparable to that supplied, so as to eliminate the possibility of incorporation from $H_2{}^{18}O$. Use may be made of the ^{3}H shift to the carbon ortho to the carbon hydroxylated in P-450-dependent hydroxylations (Guroff et al., 1967; Zenk, 1967; Sutter and Grisebach, 1969; Amrhein and Zenk, 1969) when ^{3}H-substituted substrate is supplied, and of the release of ^{3}H into water from the carbon hydroxylated by phenolase action (Pomerantz, 1964).

III. POSITIONS AND PATTERNS OF HYDROXYLATION IN PLANT PRODUCTS

Oxygenases are the major hydroxylating agents in plants. The hydroxyl groups found in plant products vary in number from one to seven or eight per molecule. They are not randomly distributed, but their specific positions and pattern reflect the specificity of the enzymes responsible for their introduction. This review of the structural features of hydroxylated compounds in plants and the evidence for their biosynthetic origin allows us to assess the extent to which oxygenases are known to be involved in their biogenesis.

In this discussion, methoxylation and hydroxylation patterns are not distinguished but considered to reflect merely the presence and specificities of the appropriate O-methyltransferases (see Poulton, this volume, Chapter 22). Epoxidation involves the direct introduction of oxygen from O_2; dihydroxylation at adjacent carbon atoms followed by the elimination of water need not be considered (Weedon, 1970).

A. Monohydroxylation

1. Phenolic Compounds

Three positions for single hydroxyl groups in naturally occurring phenolic compounds will be discussed here: para to a carbon side chain, ortho to a carbon side chain, and the 3-hydroxyl group of flavonols and related flavonoids.

Tyrosine is the major compound with hydroxylation para to a carbon side chain. In plants, tyrosine is synthesized through the shikimate pathway, chorismate, and mainly by the amination of p-hydroxyphenylpyruvic acid. An enzyme similar to the phenylalanine oxygenase of mammalian liver (see Massey and Hemmerich, 1975) in its requirement for tetrahydrofolate or similar pteridine electron donors and its sensitivity to aminopterin has been reported in spinach leaves (Nair and Vining, 1965); this has not since been substantiated, but evidence for phenylalanine hydroxylation comes from the high specific activity of tyrosine in leaves of various species after feeding L-[^{14}C]phenylalanine (McCalla and Neish, 1959; Gamborg and Neish, 1959) and from the biogenesis of labeled hordenine when [^{14}C]phenylalanine was fed to barley roots (Massicot and Marion, 1957).

p-Coumaric (4-hydroxycinnamic acid) acid is found in plant tissues as its O-glucoside and glucose ester but is also an important intermediate in the biosynthesis of lignin and flavonoids. It is the precursor of the para-substitution in 4-hydroxycinnamyl residues of angiosperm lignin and the 4'-hydroxy group in ring B of flavonoids, sometimes subsequently methylated as in formononetin (Wengenmayer et al., 1974); these patterns stem from the high activity of hydroxycinnamoyl-coenzyme A (CoA) ligase with p-coumarate and other substituted cinnamic acids with a nonmethylated 4-hydroxyl group (see Zenk, 1979). In some grasses, p-coumaric acid is produced by the deamination of tyrosine under the so-called tyrosine ammonia-lyase activity of phenylalanine ammonia-lyase (Camm and Towers, 1977), but it is more usually synthesized from cinnamic acid by the action of cinnamate 4-hydroxylase, a microsomal cytochrome P-450-dependent monooxygenase that utilizes NADPH as hydrogen donor. This enzyme was first discovered in pea seedlings (Russell and Conn, 1967; Russell, 1971) and is the best characterized of plant monooxygenases. It has been demonstrated and assayed in many plants, special interest being attached to the increase in its activity when plants, excised organs, or cell suspensions are subjected to environmental change or stress (Section IV). 4-Hydroxybenzoic acid, found in many plant tissues, arises from *p*-coumaric acid by β-oxidation (El-Basyouni *et al.*, 1964).

Phenolic compounds with hydroxylation ortho to a side chain are also widespread in plants. Besides salicylic acid formed by the β-oxidation of *o*-coumaric acid, coumarin is synthesized by the lactonization of *cis-o*-coumaric acid, which arises from *trans-o*-coumaric acid by isomerization (Kosuge and Conn, 1961; see Brown, 1979). The enzyme catalyzing the ortho-hydroxylation of cinnamic acid has not been well characterized, but active preparations from chloroplast membranes and microsomes utilizing NADPH and O_2 have been reported (Gestetner and Conn, 1974; Czichi and Kindl, 1975).

The 3-hydroxyl group is typical of flavonols, anthocyanidins, and many other flavonoids. It appears to be introduced into flavanones at the first stage after their formation by the cyclization of malonyl-CoA and 4-hydroxycinnamoyl-CoA (Kreuzaler and Hahlbrock, 1975). This is indicated by the efficient conversion of dihydrokaempferol and dihydroquercetin into their respective flavonols in buckwheat (Patschke and Grisebach, 1968). The conversion of the chalcone, isoliquiritigenin, to the flavonol, garbanzol, and to 7,4'-dihydroxyflavonol is catalyzed by the peroxidase from *Cicer arietinum* (Wong and Wilson, 1972, 1976) and by horseradish peroxidase (Rathmell and Bendall, 1972); these flavonols are accompanied by aurones. The more likely involvement of an oxygenase is suggested by the conversion of the flavanone, naringenin, to the corresponding flavone, apigenin, by an enzyme preparation from young parsley leaves, which requires O_2, Fe^{2+}, and an unidentified thermostable cofactor for activity (Sutter *et al.*, 1975); hydroxylation at the 3-position is presumably followed by dehydration. Specific oxygenases catalyzing the introduction of the 3-hydroxyl group into the already desaturated flavones or flavyliums would be expected *in vivo*.

2. Alkaloids

Most of the alkaloids derived from aromatic amino acids and many of those from tryptophan are hydroxylated or methoxylated. Among the simple alkaloidal amines, tyramine and its *O*- and *N*-methyl derivatives, which can accumulate in some plants or act as intermediates in the biosynthesis of more complex alkaloids such as annuloline, show the hydroxylation pattern of tyrosine, from which they have been derived by deocarboxylation (see Smith, 1977). The tetrahydroisoquinolines, though derived from tyrosine, are usually further hydroxylated (see Leete, 1967b).

5-Hydroxytryptamine (serotonin) and its N-methyl derivatives are found in plants. As in mammalian tissues, serotonin is formed by decarboxylation of 5-hydroxytryptophan; there is evidence for a tryptophan 5-hydroxylase in leaf discs of *Griffonia simplicifolia*, which requires O_2 and, like the enzyme from brain tissue, is sensitive to *p*-chlorophenylalanine (Fellows and Bell, 1970), but no cell-free preparation has been described. The indole alkaloids derived from tryptophan are not usually hydroxylated; reserpine and vindoline, however, have methoxyl groups in the anomalous 6-position.

In tropane alkaloids, the —CH_2OH side group of atropic acid is derived from phenylalanine by migration of the —$CH_2 \cdot NH_2$ (see Leete, 1967a); no oxygenation is involved.

3. Fatty Acids

There exist in plants three main groups of hydroxylated fatty acids: 2-hydroxy fatty acids, ω-hydroxy fatty acids, and ricinoleic acid (D-12-hydroxyoleic acid). D-2-Hydroxy fatty acids are synthesized in young leaves and peanut cotyledons by a modification of the initial stage of the α-oxidation pathway of fatty acids, in which the δ-2-position is hydroperoxylated by a reaction dependent on FAD and O_2 (Shine and Stumpf, 1974); the hydrogen is thereby directly replaced by hydroxyl without loss of configuration (Morris and Hitchcock, 1968).

ω-Hydroxylation and the biosynthesis of ricinoleic acid both involve conventional oxygenase action. The ω-hydroxylation of palmitic acid in cutin biosynthesis has been achieved with a microsomal preparation from seedling shoots of *Vicia faba*, requiring O_2 and NADPH (or NADH) and sufficiently sensitive to CO to suggest the participation of cytochrome P-450 (Soliday and Kolattukudy, 1977). The synthesis of ricinoleic acid from oleyl-CoA required a microsomal fraction, O_2, and NADH (Galliard and Stumpf, 1966). In this reaction, CO was not inhibitory and it seemed unlikely that cytochrome P-450 was involved; both cyanide and azide inhibited the reaction, and Fe^{2+} could restore the activity of the enzyme after dialysis against *o*-phenanthroline. Direct oxygenation was implied by the observation that only the D-12 hydrogen of oleyl-CoA was displaced during the reaction (Morris, 1967). In contrast, in mycelial cultures of *Claviceps purpurea* and immature sclerotia from infected rye plants, [1-^{14}C]linoleate was efficiently converted to ricinoleate even under anaerobic conditions, by specific hydration of the $\Delta^{6,7}$-double bond, but labeled oleate was desaturated without any formation of ricinoleate (Morris *et al.*, 1966).

cis-12,13-Epoxyoleic (vernolic) acid and *cis*-9,10-epoxystearate are synthesized by the introduction of oxygen from O_2 across the double bonds of linoleic and oleic acids, respectively (Knoche, 1968; Morris, 1970); the enzyme systems have not been isolated.

4. Terpenes

Hydroxylation patterns in simple terpenes are best illustrated by monoterpenes. Two groups can be identified, neither synthesized by oxygenase action (see Charlwood and Banthorpe, 1977). Members of the first group, which includes menthol and oxidized derivatives such as menthone and pulegone, are derived directly from geranyl pyrophosphate or, more probably, neryl pyrophosphate by dephosphorylation and cyclization. However, ring closure with the elimination of pyrophosphate gives a series of cyclic compounds with nucleophilic centers available for attack by hydroxyl ion, leading to α-terpineol, borneol, fenchol, and, after oxidation, thujone (Ruzicka *et al.*, 1953; Loomis, 1967). The latter mechanism has been examined in detail in the biogenesis of isothujone in shoots of *Tanacetum vulgare* (Banthorpe *et al.*, 1977, 1978). Hydroxylation of monoterpenes by oxygenase action has been observed only in the enzymatic hydroxylation of geraniol and nerol to 10-hydroxygeraniol and 10-hydroxynerol, respectively, as intermediates in the biosynthesis of the iridoid alkaloids in *Catharanthus roseus;* a membranous fraction required O_2 and NADPH with the probable participation of cytochrome P-450 (Meehan and Coscia, 1973; Madyastha *et al.*, 1976).

Analogous hydroxylation patterns found in cyclic sesquiterpenes and diterpenes can be similarly explained. However, one group of cyclic diterpenes that has received by far the closest investigation presents a different and more complex picture. Gibberellins, over 50 of which have now been structurally identified, show a range of hydroxylation patterns based on the *ent*-gibberellane skeleton (Fig. 2). Two hydroxylations are involved in the initial stages of biosynthesis common to all gibberellins. The hydroxylations of *ent*-kaurene to *ent*-kaurenol and of *ent*-kaurenoic acid to *ent*-7α-hydroxy-kaurenoic acid are catalyzed by microsomal systems from immature seeds of *Marah macrocarpus,* endosperm of *Cucurbita maxima,* and seeds of *Pisum sativum;* O_2 and NADPH (rather than NADH) are required, and cytochrome P-450 is implicated (see Hedden *et al.*, 1978). The intermediate conversion of *ent*-kaurenol through *ent*-kaurenal to *ent*-kaurenoic acid is also catalyzed by monooxygenases and not, as might be expected, by dehydrogenases. The participation of cytochrome P-450 in the oxidation of kaurenal to kaurenoic acid is CO-sensitive, and the photochemical action spectrum for reversal of the inhibition shows a maximum at 450 nm (Murphy and West, 1969). The enzymatic conversion of kaurenol to kaurenal has been less completely characterized but probably also utilizes cytochrome P-450 (Coolbaugh *et al.*, 1978). The reaction sequence can be visualized as the successive hydroxylation of the α carbon of kaurenol.

The conversion of GA_{12} aldehyde, the intermediate common to the biogenesis of all gibberellins, to GA_{43} by a 200,000 g supernatant from

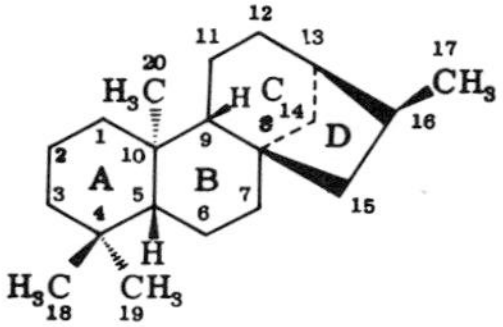

IV: GA$_{34}$

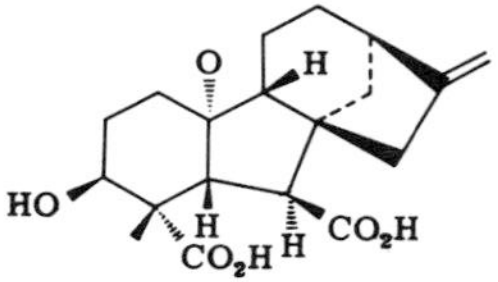

VIII: GA$_3$

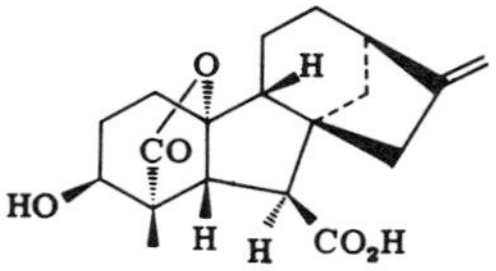

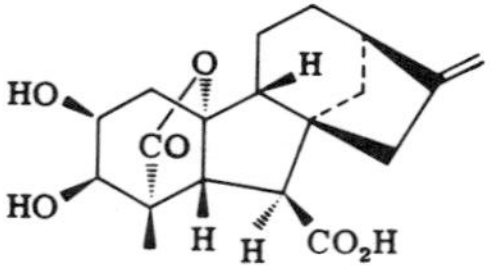

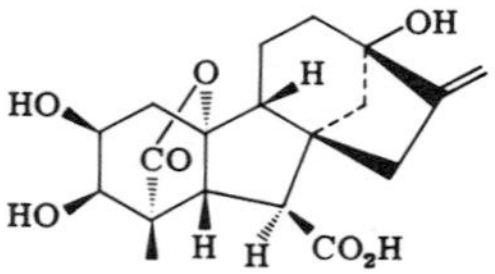

XII: GA$_8$

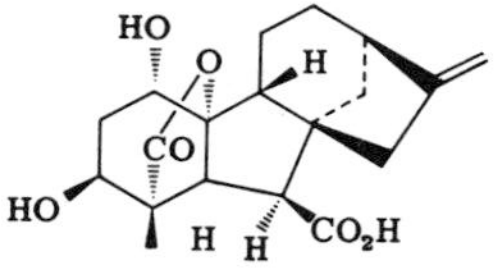

III: GA$_4$

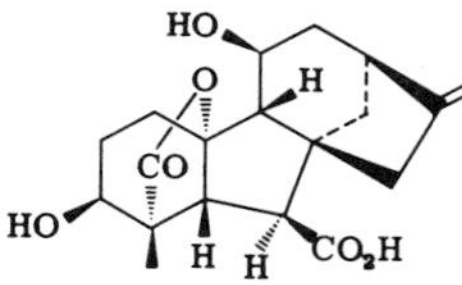

VII: GA$_{30}$

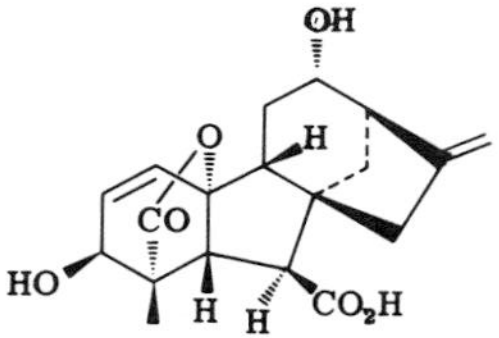

XI: GA$_{48}$

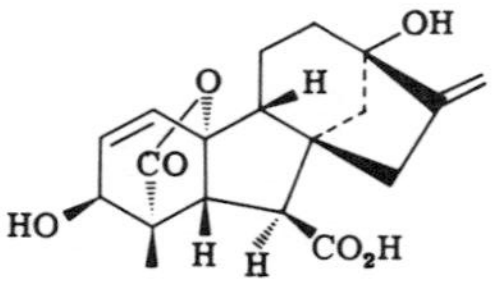

II: GA$_{14}$

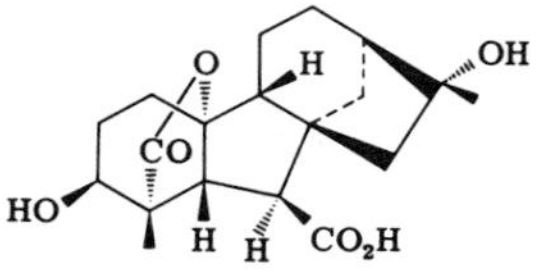

VI: GA$_{35}$

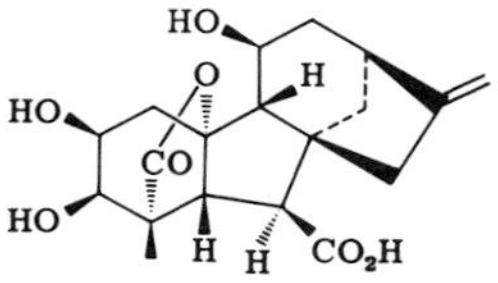

X: GA$_{50}$

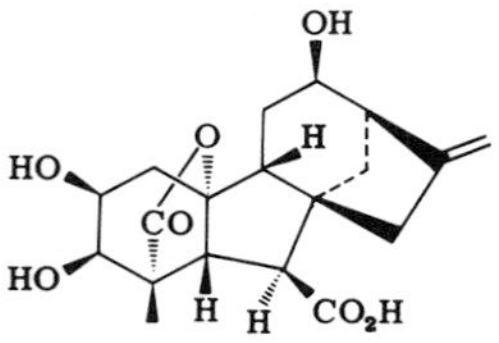

I

V: GA$_{16}$

IX: GA$_2$

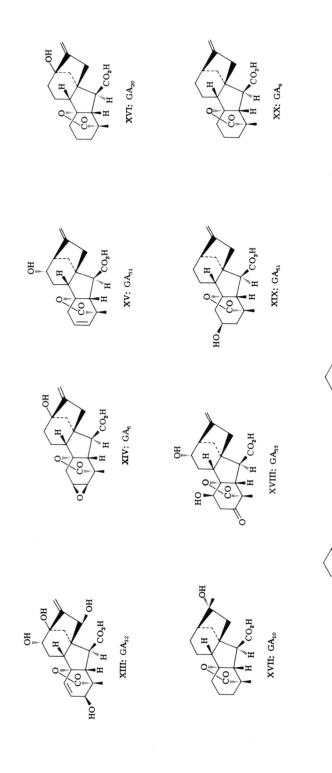

Fig. 2. Some hydroxylation patterns in natural gibberellins. I, Carbon numbering of the kaurene skeleton; II–XIII, 3β-hydroxylated gibberellin and hydroxylated derivatives; XIV–XXI, gibberellins with hydroxylation at other positions; XXII, 2α-hydroxylated gibberellin.

Cucurbita can be achieved by a network of alternative reactions, which includes at least two hydroxylations. This soluble hydroxylation system was inactivated by dialysis against EDTA, and its activity restored by Fe^{2+} (Graebe and Hedden, 1974). A soluble hydroxylase catalyzing the 2β-hydroxylation of GA_1 from cotyledons of germinating *Phaseolus vulgaris* to GA_8 was similarly affected; it was inhibited under anaerobic conditions and required ascorbate or NADPH, presumably to reduce Fe^{3+} to Fe^{2+}. A soluble hydroxylation system has also been reported in *Pisum sativum* (Ropers *et al.*, 1978). It seems likely that, with the increasing polarity of intermediates as synthesis proceeds, membrane-associated hydroxylations involving cytochrome P-450 are replaced by hydroxylations in the cytosol employing nonheme iron.

5. Steroids

Cyclic triterpenes and sterols are characterized by the 3-hydroxyl group present in all members of the series. This is an immediate consequence of the initiation of squalene cyclization by the formation of 2,3-epoxysqualene. This compound has been found in plants (see Goodwin, 1979) and is formed by the action of a microsomal monooxygenase requiring O_2 and NADPH. The subsequent cyclization is proton-initiated and does not require O_2, but ^{18}O from labeled O_2 is shown to be retained in the 3-hydroxyl group of the sterols subsequently produced (see Grunwald, 1975).

Most of the sterols and triterpenes found in plants carry no other hydroxyl group, but some further hydroxylation must occur in the synthesis of many biologically interesting derivatives such as ecdysone and other insect hormones, sapogenins, steroidal alkaloids, and the C_{19} and C_{21} steroids related to animal hormones (see Heftmann, 1973, 1977). Little is known of the nature of these hydroxylation reactions.

6. Carotenoids

Although much emphasis is often placed on hydrocarbon carotenoids, these in fact represent only about 10% of the carotenoids known. Most carotenoids are xanthophylls containing two or more hydroxyl groups. Since these hydroxyl groups are disposed symmetrically at the two ends of the carotenoid molecule, they do not interact, and these molecules are therefore functionally monohydroxylated. Lutein and zeaxanthin, which occur most widely in higher plants, are the 3,3'-dihydroxy derivatives of α- and β-carotene, respectively. With the use of *Chlorella* mutants in light and darkness (Claes, 1959) and by inhibitor studies with nicotine employing *Flavobacterium* sp. (McDermott *et al.*, 1973), it has been established that the hydroxyl groups are introduced after ring formation; hydroxylation requires O_2 since, in the absence of air, the corresponding hydrocarbons were found to accumulate but disappeared again upon aeration with formation of the corresponding xanthophylls. By ^{18}O experiments, the incorporation of oxygen from O_2 into the hydroxyl group was established (Yamamoto et al., 1962), but no enzyme was identified. Hydroxylation at C-1, -2, and -4 is found in carotenoids isolated from bacteria, fungi, and algae; in photosynthetic bacteria, hydroxylation at C-1 is an anaerobic process probably involving hydration (Singh et al., 1973), but mechanisms for these hydroxylations are otherwise not known. Hydroxylation of one or both of the geminal methyl groups and of the chain methyl groups occurs in some higher plant and algal carotenoids (see Britton, 1976).

5,6-Epoxyxanthophylls are widespread and include violaxanthin and neoxanthin, both present in chloroplasts, and antheraxanthin. The incorporation of ^{18}O from labeled O_2 into the epoxides (Yamamoto and Chichester, 1965) and the stereospecificity of the epoxide imply enzymatic oxygenation (Weedon, 1970). 5,8-Epoxyxanthophylls are also found. 1,2-Epoxylycopene occurs in plants but, despite the analogy with 2,3-epoxysqualene, there is no evidence that it participates in cyclization reactions.

B. Dihydroxylation

The presence of two hydroxyl groups in sufficient proximity for one to have affected the introduction of the other is a feature only of phenolic compounds and alkaloids, though the range of compounds may be stretched further to include quinones. By far the most typical dihydroxylation pattern in phenolic compounds is the o-diphenol grouping, but m- and p-diphenols are also found. The p-diphenols are usually, but not always, directly related to the corresponding p-quinones which accumulate in many plants.

The simplest and perhaps the key o-diphenol in phenolic metabolism is caffeic acid, which occurs widely in a variety of compounds including its 4-O-glucoside, glucose ester, and 3-quinate ester (chlorogenic acid). Caffeic acid shows the two hydroxyl groups typically ortho-related and one of them para to a carbon side chain. The hydroxyl groups may be partially methylated, in which case the para hydroxyl group is usually free, as in ferulic acid. This pattern of hydroxylation and methoxylation is repeated in compounds derived from caffeic and ferulic acids through their CoA esters, e.g., in the coniferyl residues in lignin and in the substitution patterns of ring B in many flavones, flavonols, catechins, anthocyanidins, and allied compounds.

Despite its ubiquity, the enzymatic origin of this ortho-dihydroxylation pattern is still in question. The ability of mushroom tyrosinase to catalyze the hydroxylation of monophenols, especially tyrosine, to o-diphenols was discovered early and an additional reductant shown to be involved, either the o-diphenol itself (see Mason, 1957) or added ascorbate (Satô, 1969). Small traces of o-diphenol were shown to sensitize the reaction by reducing the lag period (Mason, 1957). Oxygen from O_2 was reported to be incorporated into the hydroxyl group (Mason *et al.*, 1955). Phenolase preparations from spinach-beet leaves were found to catalyze the hydroxylation of *p*-coumaric acid to caffeic acid with similar characteristics (Vaughan and Butt, 1969, 1970), with the additional observation that the further oxidation of caffeic acid to its *o*-quinone was suppressed when dimethyltetrahydropteridine was used as reductant for the hydroxylation (Vaughan and Butt, 1972). With the exception of this latter reductant, the *o*-diphenol oxidase activity of most phenolase preparations is sufficient to oxidize rapidly the *o*-diphenol produced by hydroxylation as soon as the hydrogen donor is exhausted. However, preparations from *Sorghum* internodes have shown relatively low oxidase activity (Stafford and Dresler, 1972), and evidence for the photoinduction of an enzyme that catalyzes *p*-coumarate hydroxylation with no accompanying oxidase activity in bean leaves and spinach-beet seedlings has been found (Bolwell, 1974; Butt, 1976, 1979). Besides the problem of product oxidation, many plant phenolases have a surprisingly broad specificity, though they may show different activities and characteristics with different substrates (Roberts and Vaughan, 1971; Schill and Grisebach, 1973).

Attempts to separate and purify monophenol hydroxylases with greater specificity and without *o*-diphenol oxidase activity have so far failed. The efficient conversion of caffeic acid to ferulic acid by a highly active *O*-methyltransferase in the tissue (Shimada *et al.*, 1973; Poulton and Butt, 1975) may obviate the need for the latter. On the other hand, with the exception of the phenolase of melanomas, mammalian tissues engaged in catecholamine synthesis, such as adrenal medulla and brain, contain soluble tyrosine hydroxylase highly specific for its substrate, sensitized by Fe^{2+}, and utilizing tetrahydrofolate as reductant (Shiman *et al.*, 1971); a similar enzyme has yet to be identified in plant tissues active in phenolic synthesis.

In contrast to the situation in lignin biogenesis, the *o*-dihydroxy grouping in ring B of flavonoids persists through several intermediates, though partial or complete methylation of the product may also be found. From studies on the biogenesis of flower pigments in *Petunia hybrida*, Hess (1967, 1968) proposed that the hydroxylation pattern of ring B was established at the cinnamic acid stage (the cinnamic acid starter hypothesis); the specificities of the hydroxycinnamoyl-CoA ligase and flavanone synthase (Kreuzaler and Hahlbrock, 1975) may allow this. However, further hydroxylation of ring B at or after the flavonone stage was suggested in *Haplopappus* cell suspensions by a more efficient incorporation of naringenin (5,7,4′-trihydroxyflavonone) into cyanidin relative to *p*-coumaric acid and by the demonstration of a microsomal enzyme catalyzing the 3′-hydroxylation of naringenin and dihydrokaempferol, which was inactive with *p*-coumarate and required NADPH and O_2 (Fritsch and Grisebach, 1975).

Protocatechuic acid occurs widely in plants and can be formed from caffeic acid by β-oxidation in wheat seedlings (El-Basyouni et al., 1964). Apparently this is not the only source, for protocatechuic acid can be produced from the shikimic acid pathway without oxygenase or oxidase action by the enolization and dehydration of 5-dehydroshikimic acid. The reaction was first observed in fungi and bacteria and since extended to cell suspensions from mung bean roots (Gamborg, 1967).

Both classes of betalains, betacyanins and betaxanthins, show o-dihydroxyphenyl substitution and are biogenetically derived from 3,4-dihydroxyphenylalanine (DOPA; see Piatelli, 1976). Similar patterns are observed in many alkaloids, in which the hydroxyl groups may be partly or wholly methylated, from simple molecules such as belladine and norbelladine to the very complex such as the quinolizidine cryptopleurine; many tetrahydroisoquinoline alkaloids, such as anhalonidine, papaverine, norlaudanosine, and isothebaine show a similar hydroxylation pattern. When only one hydroxyl group is methylated, it is usually, though not invariably, the hydroxyl group meta to the carbon side chain in the DOPA precursor.

The initial stage in the biosynthesis of these compounds is the 3-hydroxylation of tyrosine. In view of the extensive investigations of tyrosinases engaged in the production of DOPA melanins in insect and animal tissues and able to catalyze this reaction, it is disappointing that our knowledge of the enzymatic reaction in plants is very limited. DOPA has been found in legumes (Nagatsu et al., 1972) and bananas (Rehr et al., 1973), the latter also accumulating dopamine (Palmer, 1963). The oxidation of DOPA through dopamine to dopachrome provides a convenient colorimetric assay for the oxidase activity of phenolase preparations (Horowitz and Shen, 1952). The spinach-beet enzyme (V. S. Butt, unpublished results) and a phenolase from the pulp of banana fruit (Deacon and Marsh, 1971) can both catalyze the hydroxylation of tyrosine and tyramine, the enzymes requiring O_2 and ascorbate as reducing agent. In a biosynthetic sequence, further oxidation of the diphenol may be prevented by active methylation. The browning reaction of banana fruit is due to the phenolase-catalyzed oxidation of dopamine (Griffiths, 1959).

While ring B of the flavonoids shows dihydroxylation patterns similar to those of the related cinnamic acids, ring A is hydroxylated characteristically in the 5- and 7-positions; that is, meta-hydroxylation occurs, which may be extended to the oxygen of the pyrone or flavylium ring. These hydroxyl groups arise not by oxygenation but during the condensation of malonyl-CoA in the initial synthesis of flavanones, in which a potential —CH_2—CO— sequence enolizes spontaneously to this pattern. Monohydroxylation at the 5- or 7-position in ring A must be due to a reduction reaction which takes place in association with, or immediately following, the condensation reactions (Wong, 1976).

Para-dihydroxylation patterns also occur. These are found in hydroquinone, in reduced ubiquinone or plastoquinone, and in more complex fungal metabolites such as griseofulvin and helminthosporin. The simplest benzoquinones and hydroquinones are formed by β-oxidation of the corresponding substituted cinnamic acid and oxidative decarboxylation of the benzoic acid (Bolkart and Zenk, 1968). A similar origin for the ring system of ubiquinone in maize plants has been postulated (Whistance *et al.*, 1967), but the ring and hydroxylation patterns of plastoquinone, tocopherols, and tocopherol-quinones, are more likely to be derived from the oxidative conversion of p-hydroxyphenylpyruvic acid to homogentisic acid (Whistance and Threlfall, 1968). A variety of mechanisms including the oxidative removal of side chains (Zenk, 1964) and polyketide formation, are involved in the biosynthesis of naphthoquinones and anthraquinones (see Leistner, 1975), but none of these requires the direct incorporation of O_2 by oxygenase action.

C. Trihydroxylation and More Complex Patterns

Among trihydroxylation patterns in plant products, by far the commonest is the simple 1,2,3-trihydroxybenzenoid pattern, with the central hydroxyl usually para to a carbon side chain. This is found in gallic acid, gallotannins, and ring B of a number of flavonoids, such as delphinidin. With partial methylation leaving the para hydroxyl group unsubstituted, the pattern is found in sinapic acid, in the sinapyl residues of dicotyledon lignin, and in ring B of flavonoids such as the anthocyanidins, petunidin, and malvidin.

The widespread occurrence of sinapyl residues in angiosperm lignin and of an O-methyltransferase that catalyzes the methylation of 5-hydroxyferulic acid specifically associated with the occurrence of these residues (Shimada *et al.*, 1973) suggests the existence of a hydroxylase active in the introduction of a hydroxyl group into ferulic acid. Neither this enzyme nor a caffeic acid 5-hydroxylase has been demonstrated, although labeled 5-hydroxyferulic and 3,4,5-trihydroxycinnamic acids are incorporated into petunidin and delphinidin, respectively, by petals of *P. hybrida* (Hess, 1967; cf. Meier and Zenk, 1965). Nor has evidence for the further hydroxylation of cyanidin, peonidin, or their biosynthetic precursors been found. The biosynthetic origin of these compounds poses a major problem for the plant enzymologist.

Among these compounds, only for gallic acid does the problem of origin appear open to solution. Although it can be formed by the β-oxidation of 3,4,5-trihydroxycinnamic acid in wheat seedlings (El-Basyouni *et al.*, 1964), it is also produced from shikimic acid by extracts of mung bean cells in the presence of NADP (Gamborg, 1967). In support of this mechanism, shikimic acid was shown to be a more effective precursor of gallic acid than glucose or phenylalanine in *Rhus typhina* (Cornthwaite and Haslam, 1965) and leaves of *Geranium pyranacium* (Conn and Swain, 1961), and especially of gallic acid esterified in the gallocatechins of the tea plant (Zaprometov, 1962). There is, however, no evidence for chain extension from gallic acid to a trihydroxy-cinnamic acid, and tracer evidence has shown that the sinapyl residues in lignin arise from phenylalanine and cinnamate (Neish, 1960).

Among alkaloids, the same trihydroxylation pattern is found in mescaline. Evidence that it is synthesized in the peyote plant from tyramine by succes-sive hydroxylation and O-methylation, followed by N-methylation (Smith, 1977), again points to enzymes catalyzing the hydroxylation of o-diphenols and their methylated derivatives.

Hydroxylation or methoxylation in ring B of the flavonoids is not necessar-ily confined to C-3', -4', and -5'. 2'-Hydroxylation is occasionally found in flavones and flavonols and must be assumed to be necessary in the biogenesis of rotenoids, but neither the stage at which it is introduced nor its mechanism is known. Anomalies in the pattern in ring A are more common, with hydrox-ylation at C-6, C-8, or both (see Harborne, 1967); it seems likely that these hydroxyl groups are introduced before or during ring closure in the course of flavanone synthesis (Wong, 1976), but no mechanism has been suggested.

D. Side-Chain Hydroxylations in Phenolic Compounds

A range of β-hydroxyphenylethylamines is found in plants (Smith, 1977). In mammalian tissues, the β-hydroxylation of dopamine to noradrenaline is catalyzed by a mixed-function copper oxygenase which requires ascorbate as hydrogen donor and incorporates ^{18}O from labeled O_2 into the product (Mason, 1965). Although the enzyme has not been reported in plants, the occurrence together of phenylethylamines and their β-hydroxylated deriva-tives in some plants points to a similar reaction.

Cyanogenic glycosides, such as amygdalin and dhurrin, are synthesized from the corresponding amino acids by N-oxidation, decarboxylation, de-hydration, and finally hydroxylation at the original β carbon adjacent to the ring. The complete sequence from tyrosine to dhurrin was catalyzed by microsomal preparations from *Sorghum bicolor;* both N-oxidation and β-hydroxylation required NADPH and O_2 (Conn *et al.,* 1979), but there was no evidence of cytochrome P-450 participation nor for the incorporation of oxygen from O_2 into the hydroxynitrile. There is, however, extensive ^{18}O incorporation into the analogous hydroxynitrile in the biosynthesis of linamarin (Zilg *et al.,* 1972).

It is thus evident that, although the introduction of hydroxyl groups into organic compounds is catalyzed in many cases by oxygenases, usually membrane-bound cytochrome P-450-dependent enzymes or phenolases, a number of alternative mechanisms exist. Hydroxylation by other metal-containing oxygenases occurs, but the addition of water across a double bond, enolization of the sequence —CH₂CO—, and reduction of the carbonyl group can also provide hydroxyl groups. Finally, enzymatic demethylation of methoxy-substituted benzoic acids has been demonstrated in cell suspensions (Harms *et al.*, 1972) but, though specific in action, the reaction is more likely to be concerned with product breakdown than with biosynthesis and accumulation.

With these alternative mechanisms possible, a range of oxygenases known, some with wide and variable specificity (e.g., phenolases), and the possibility of one of a range of isoenzymes catalyzing the hydroxylation of a specific substrate, the oxygenase active in the biogenesis of a hydroxylated product is difficult to identify. This is perhaps best achieved by relating its activity to the rate of accumulation of the product, especially under conditions that increase enzyme activity or product accumulation. These changes are discussed later in relation to the physiology of intact plant organs, excised tissues, and cell suspensions.

IV. OTHER OXYGENASE REACTIONS

The initial oxygenase reaction is the introduction of one or both of the oxygen atoms of O_2 into an organic combination, and thus the formation of hydroxylated derivatives. These may be stable, but in some instances, they rapidly undergo further reaction. Oxygenase action can therefore give rise metabolically to epoxidation, desaturation, oxidative demethylation or ring cleavage subsequent to the initial introduction of the oxygen atom or atoms.

A. Epoxidation

The production of an epoxide intermediate in microsomal hydroxylation was first demonstrated by the isolation of naphthalene 1,2-epoxide and its subsequent utilization in the synthesis of naphth-1-ol from naphthalene using preparations from rat liver (Jerina *et al.*, 1970). Epoxides formed from such substances as benzo(α)pyrene and aflatoxin are important not only as intermediates in oxidative detoxification but also because of the readiness with which they bind covalently with nucleic acids and proteins in the cell, leading to chemical carcinogenesis (Heidelberger, 1975).

In higher plants, cinnamic acid has been shown to be converted to cinnamic acid 3,4-epoxide in the course of its hydroxylation to *p*-coumaric acid by cinnamate 4-hydroxylase (Sandermann *et al.*, 1977). An enzyme catalyzing the epoxidation of 18-hydroxyoleic acid to 18-hydroxy-*cis*-9,10-epoxystearic acid has been demonstrated in a particulate preparation from spinach (Croteau and Kolattukudy, 1975a). The product is a major component of cutin, with which the enzyme is closely associated. There is stringent substrate specificity; the enzyme is inactive with oleic acid, 18-acetoxyoleic acid, and even 18-hydroxyelaidic acid which is the trans isomer of 18-hydroxyoleic acid. Since the enzyme is photoreversibly inhibited by CO, cytochrome P-450 may be involved. In addition to NADPH and O_2, ATP and CoA are required for maximum activity, suggesting that the immediate substrate is 18-hydroxyoleyl-CoA. A further enzyme has been demonstrated in apple, which catalyzes hydration of the epoxide to *threo*-9,10,18-trihydroxystearic acid, another important component of cutin (Croteau and Kolattukudy, 1975b).

The production of squalene 2,3-epoxide by a monooxygenase, which has not been characterized in higher plants, initiates its cyclization to cycloartenol, the first step in sterol biosynthesis from squalene (Goodwin, 1979).

Detoxication of insecticides by epoxidation has been demonstrated. Thus extracts of pea and bean roots have been shown to catalyze epoxidation of the organochlorine insecticide aldrin. The pea root enzyme is particulate (Mehendale, 1973), stimulated by NADPH, and inhibited by cytochrome c and menadione (Earl and Kennedy, 1975), suggesting a role for NADPH–cytochrome c reductase but offering no evidence that cytochrome P-450 is involved. Extracts of bean roots also convert *trans*-stilbene to *trans*-stilbene 1,2-epoxide (Ross *et al.*, 1978). This epoxide is hydrated to *meso*-1,2-diphenyl-1,2-ethanediol which in turn is metabolized to benzoin, benzil, and benzoic acid; in contrast, aldrin epoxide is not further metabolized despite the presence of the hydratase.

Epoxidation can thus be seen to play an important metabolic role in the biosynthesis of hydroxy compounds and essential hydroxylated intermediates and, when coupled with a hydratase (as in mammalian tissues), in the disposal of potentially toxic substances.

B. Desaturation

Desaturation occurs when hydroxylation is followed by dehydration:

$$-CH_2-CH_2- \rightarrow -CH_2-CHOH- \rightarrow -CH=CH-$$

The potential role of monooxygenases in desaturation is illustrated by oleyl-CoA desaturase from maturing seeds of *Carthamnus tinctorius* (Vijay and Stumpf, 1972). The enzyme converts oleyl-CoA to linoleyl-CoA in a reaction requiring NADPH or NADH, and O_2. NAD(P)H can be replaced by photochemically reduced ferredoxin, indicating that these compounds supply electrons to the system rather than react directly with O_2. The enzyme is highly specific; alterations in chain length or position of the double bond and substitution of trans for the cis isomer or of acyl carrier protein for CoA lead to loss of activity. The enzyme is membrane-bound and inhibited by —SH reagents, metal chelating agents, and cyanide. However, carbon monoxide is ineffective, precluding any involvement of cytochrome P-450. Oleyl-CoA desaturase has also been detected in aged slices of potato tuber (Ben-Abdelkader *et al.*, 1973).

C. Oxidative Demethylation

Oxidative demethylation of *O*- and *N*-methyl compounds involves hydroxylation followed by loss of formaldehyde:

$$-O-CH_3 \rightarrow -O-CH_2OH \rightarrow -OH + HCHO$$

$$\diagdown N-CH_3 \rightarrow \diagdown N-CH_2OH \rightarrow \diagdown NH + HCHO$$

O-Demethylation has been reported in the metabolism of methoxylated benzoic acids by mung bean and soybean cell suspension cultures (Harms *et al.*, 1972), and N-demethylation occurs during the metabolism of hordenine and *N*-methyltyramine (Meyer and Barz, 1978).

Special attention has been paid to the N-demethylation of foreign substances, including herbicides and growth substances, by plant enzymes. It has been shown that microsomal preparations from plantain, buckwheat, broad bean, and cotton oxidatively demethylate substituted 3-phenyl-1-methylureas, the most active preparations coming from hypocotyls of young etiolated cotton seedlings (Frear *et al.*, 1969). One mole of formaldehyde is produced for each mole of substrate demethylated. NADH or NADPH, and O_2, were essential for the reaction. The participation of cytochrome P-450 is inferred by the sensitivity of the system to CO and by the cosedimentation of the demethylase with NADPH–cytochrome c reductase, NADH–cytochrome c reductase, and cytochrome b_5 in microsomal preparations.

The enzyme is specific for substituted 3-phenyl-1-methylureas, among which the best substrate is the herbicide monuron, 3-(4-chlorophenyl)-1,1-dimethylurea (K_m, 29 μM), demethylated according to the equation

$$Cl-\langle\bigcirc\rangle-N-CO-NMe_2 + \tfrac{1}{2}O_2 \longrightarrow Cl-\langle\bigcirc\rangle-N-CO-NHMe + HCHO$$

The monomethyl derivative is not then further demethylated, yet when it is supplied directly to the enzyme, 4-chlorophenylurea is produced along with a second unidentified product. In this latter reaction, 3-(4-chlorophenyl)-1-hydroxymethylurea was found as an unstable intermediate (Tanaka *et al.*, 1972a), though no equivalent hydroxymethyl intermediate from the dimethyl substrate was trapped, presumably because it was even less stable. The cotton demethylase also demethylated 3-(3,4-dichlorophenyl)-1,1-dimethylurea (K_m, 1.5 mM) and 3-(3-trifluoromethylphenyl)-1,1-dimethylurea (K_m, 13 μM), but other compounds such as *N,N*-dimethyl-2,2-diphenylacetamide and 2-chloro-4-ethylamino-6-isopropylamine-5-triazine, although demethylated by plants *in vivo,* were not attacked by this enzyme.

Several *N*-methylcarbamates were competitive inhibitors, e.g., 1-naphthylmethylcarbamate (K_i, 1.5 μM); high electron density at the position ortho to the carbamate group was necessary for effective inhibition (Tanaka *et al.,* 1972b). The role of these *N*-demethylases in the detoxication of 1,1-dimethylurea herbicides is indicated by reported instances of extensive crop damage when the herbicides were used in conjunction with *N*-methylcarbamate insecticides, and furthermore the taxonomic distribution of the *N*-demethylase activity correlates well with resistance to these herbicides (Sandermann *et al.,* 1977).

Specific demethylase activity for *N*-methylarylamines has been reported in microsomes from castor bean endosperm (Young and Beevers, 1976). The system was inactive with nicotine, *N*-methyltyramine, and *N*-methyltryptamine and did not catalyze O- or S-demethylation. *p*-Chloro-*N*-methylaniline was most readily demethylated, yielding formaldehyde and 4-chloroaniline. The reaction required O_2 and NADPH or, at a much lower rate, NADH. In the absence of O_2 and NADPH, cumene hydroperoxide could support the reaction. The presence of cytochromes P-450 and b_5 in the microsomes and the demonstration of photoreversible inhibition by CO implicated cytochrome P-450 in the reaction. Demethylation of *p*-chloro-*N*-methylaniline and *p*-nitroanisole by microsomal fractions of avocado pear has been reported (McPherson *et al.,* 1975b).

D. Ring Cleavage

Whereas monooxygenases introduce a single hydroxyl group into compounds which may subsequently undergo the reactions described in the previous section, dioxygenases introduce hydroxyl groups on adjacent carbon atoms, forming diols which readily undergo ring fission. Compared with bacteria and fungi, relatively few dioxygenases have been characterized in higher plants (for review, see Sugumaron *et al.,* 1977). A number of *o*-diphenols have been shown unequivocally to be degraded by cell suspension cultures, and the *o*-diphenol has been found in each case to be a better substrate for ring cleavage than the corresponding monophenol (Barz, 1977a). Three basic types of dihydroxy ring cleavage have been described: (1) intradiol ring fission, (2) extradiol cleavage between a hydroxylated carbon and an adjacent unsubstituted atom, and (3) the homogentisic acid pathway (Barz, 1977b).

2,3-Dihydroxybenzoic acid 2,3-dioxygenase, which catalyzes the intradiol fission of 2,3-dihydroxybenzoic acid to α-carboxy-*cis,cis*-muconic acid from which the dilactone, 2,6-dioxo-3,7-dioxobicyclo[3.3.0]octane-8-carboxylic acid, accumulates, has been partially purified from leaves of *Tecoma stans*

(Sharma *et al.*, 1972; Sharma and Vaidyanathan, 1975a,b). The enzyme is found in both the soluble and chloroplast fractions from leaves; it appears to be a Cu^{2+}-requiring dioxygenase distinct from the polyphenoloxidase also present. The reaction has special interest because 2,3-dihydroxybenzoic acid undergoes extradiol cleavage in bacteria and is decarboxylated in fungi to catechol prior to ring fission (Sharma *et al.*, 1972).

Enzymes catalyzing the conversion of DOPA to stizolobinic acid (α-amino-6-carboxy-2-oxo-2*H*-pyran-3-propionic acid) and stizolobic acid (α-amino-6-carboxy-2-oxo-2*H*-pyran-4-propionic acid) have been detected and partially purified from etiolated seedlings of *Stizolobium hassjoo* (Saito and Komamine, 1976, 1978). Oxygenative extradiol ring cleavage of DOPA would produce α-hydroxy-δ-alanine muconic ϵ-semialdehyde and α-hydroxy-γ-alanine muconic ϵ-semialdehyde, and the recyclization to the α-pyronyl amino acid products might occur spontaneously on the enzyme. $NADP^+$ or NAD^+ is required for the final dehydrogenation, and it has been shown that this activity copurifies with ring fission and recyclization activities in an enzyme of 45,000 MW. Oxygenative extradiol ring cleavage of DOPA to α-hydroxy-γ-alanine muconic ϵ-semialdehyde is also involved in betalain synthesis. The key intermediate, betalamic acid, could be derived from DOPA either (1) by intradiol cleavage followed by the bonding of nitrogen to C-5', or (2) by extradiol cleavage between C-4' and C-5' and subsequent bonding of nitrogen to C-3'. However, studies on the incorporation of L-[1-^{14}C,3,5-^{3}H]tyrosine into betanin in the fruit of *Opuntia decumbens* (Fischer and Dreiding, 1972) and into indicaxanthin in the fruit of *Opuntia ficus-indica* (Impellizeri and Piatelli, 1972) are consistent only with an extradiol cleavage.

Degradation of tyrosine through the homogentisic acid ring cleavage pathway has been detected in a number of cell cultures and appears to be widely distributed in higher plants (Durand and Zenk, 1974a,b). Homogentisic acid oxygenase activity has been demonstrated in a cell-free extract of cultured cells of *Drosophyllum lusitanicum* (Durand and Zenk, 1974b). α,α'-Bipyridyl (5 mM) completely inhibited the partially purified enzyme and *in vivo* caused the accumulation of [^{14}C]homogentisic acid from [β-^{14}C]tyrosine.

It should be pointed out that not all ring cleavage reactions involve dioxygenation. For example, in the release of $^{14}CO_2$ from [^{14}C]catechol fed to soybean cell cultures, there is no evidence for such putative direct ring cleavage products as *cis,cis*-muconic acid and γ-oxalocrotonic acid; these compounds were not labeled, nor were they able to trap the release of $^{14}CO_2$ (Prasad and Ellis, 1978). Interestingly, H_2O_2 in the presence of culture filtrates or horseradish peroxidase could effect catechol ring cleavage, leading to the conclusion that, in these cases, fission occurred by an undefined mechanism involving H_2O_2 and extracellular peroxidases. As in the general study of the metabolism of natural products, the investigation of the role of oxygenases *in vivo* is made difficult here by uncertainty in comparing the metabolism of exogenous substances with that of endogenously generated compounds; Ellis (1971), for instance, observed that catechol generated internally from salicylic acid did not undergo ring fission.

V. REGULATION OF EXPRESSION OF OXYGENASE ACTIVITY

Many natural products accumulate only under precise physiological conditions in specific tissues and at particular stages in the development of the plant. The expression of oxygenase activities involved in the synthesis and breakdown of natural products must therefore be highly regulated and integrated into the overall metabolic and developmental processes.

A. Physiological Factors Affecting Oxygenase Levels

Considerable attention has been paid to the physiological factors affecting the levels of cinnamate 4-hydroxylase in the biosynthesis of phenolic compounds from phenylalanine. In the first detailed characterization of the enzyme, its activity in light-grown pea seedlings was found to vary with age; it was present only in young developing tissue, highest in the apical bud, and decreased with maturation (Russell, 1971). Enzyme levels vary during the growth of cell suspension cultures, in which transient increases in enzyme activity in soybean cultures shortly before attaining the stationary phase are correlated with accumulation of the flavonoid apigenin (Ebel *et al.*, 1974) and during the lage phase growth of carrot cell cultures are correlated with rapid phenolic accumulation (Sugano *et al.*, 1978).

The level of hydroxylase activity is increased transiently by a number of physiological stimuli, generally correlated with the accumulation of phenolics. For example, wounding or excision of tissue from Jerusalem artichoke tubers (Benveniste *et al.*, 1977), potato tubers (Lamb and Rubery, 1976a), and sweet potato roots (Tanaka *et al.*, 1974) stimulates enzyme activity concomitant with the accumulation of hydroxycinnamic esters. Ethylene further enhances this response in surface-cut sweet potato roots (Tanaka *et al.*, 1974; Rhodes *et al.*, 1976) and induces the enzyme in excised pea epicotyls (Hyodo and Yang, 1971). In contrast, cinnamate 4-hydroxylase in *Sorghum* is constitutive and not induced by wounding (Saunders *et al.*, 1977).

Light stimulates cinnamate 4-hydroxylase in pea seedlings, the response being mediated at least in part by phytochrome (Benveniste *et al.*, 1978). There is a difference in the kinetics of the response to illumination in the stem and apical bud, associated with the production of lignin and flavonoid accumulation, respectively (Butt and Wilkinson, 1979). The induction of oxygenase activity by any given stimulus usually correlates more or less precisely with other enzymes in the phenylpropanoid sequence and may be related to the accumulation of different end products in different tissues of the same plant; specification of the metabolic program is determined by the cells or tissue and its realization triggered by the environmental stimulus (Schopfer, 1977). The accumulation of flavonoids and hydroxycinnamic esters in buckwheat seedlings (Amrhein and Zenk, 1970) and of flavonoids in parsley cell cultures (Hahlbrock et al., 1971) is triggered by illumination and correlated with increases in cinnamate 4-hydroxylase activity.

However, in parsley cells, enzyme activity is induced not only by light but also by dilution of cultures into a new medium (Hahlbrock and Wellman, 1973). In this case, esters of hydroxycinnamic acids rather than flavonoids accumulate, and therefore within a tissue oxygenase activity associated with more than one metabolic program may be triggered by different stimuli to produce different substances. Cinnamate 4-hydroxylase is also involved in the biosynthesis of phenolic phytoalexins which arise notably in the Leguminosae, such as pisatin in pea and phaseollin in French beans (Vanetten and Pueppke, 1976). Production of phytoalexins is stimulated by living fungi or by elicitor molecules released from fungal cell walls (Albersheim and Valent, 1978). In addition to these natural glycan elicitors, phytoalexins also accumulate in response to a number of structurally unrelated compounds such as autoclaved RNase A and heavy metal salts (Vanetten and Pueppke, 1976). Cinnamate 4-hydroxylase shows a transient increase in activity in cultures of French bean during the induction of phaseollin accumulation by autoclaved RNase A (Dixon and Bendall, 1978) and would be expected to respond to a natural elicitor.

Relatively little is known about the physiological factors influencing the expression of other oxygenases in plants. Phenolase activity is increased in a number of tissues on illumination (Butt, 1979) and in sweet potato roots after excision (Hyodo and Uritani, 1966), but the relationship between these increases and the accumulation of polyphenols has not been traced in detail (see Butt, this series, Vol. 2 Chapter 3). Oleyl-CoA desaturase is induced in sliced potato tubers, the increase in activity coinciding with a period of rapid phospholipid biosynthesis and membrane proliferation (Ben-Abdelkader et al., 1973). The monooxygenase, which catalyzes the conversion of oleyl-CoA to ricinoleic acid in Ricinus communis seeds, is found only at certain stages of development, probably related to changes in fatty acid metabolism during seed growth (Galliard and Stumpf, 1966).

Expression of the activity of oxygenases engaged in the metabolism of xenobiotics raises particularly interesting problems, since these compounds are not found in vivo. By analogy with systems for the detoxication of foreign substances by oxidative metabolism in liver microsomes and in bacteria, substrate induction would be expected. The N-demethylases for substituted 3-phenyl-1,1-dimethylureas in cotton and other plants (Frear et al., 1969) and for N-methylarylamines in castor bean endosperm (Young and Beevers, 1976), and the aldrin and *trans*-stilbene epoxidases in pea and bean roots (Earl and Kennedy, 1975; Mehendale, 1973; Ross *et al.*, 1978), are all active in plants not previously exposed to these substances, though the aldrin epoxidase activity is found only in seedlings at least 5 days after germination. Similarly, biphenyl 2- and 4-hydroxylases, aniline hydroxylase, *p*-chloromethylaniline N-demethylase, and *p*-nitroanisole O-demethylase are active in the microsomes of avocado pear (*Persea americana*) without prior treatment with substrate. However, pretreatment of microsomes *in vitro* with safrole or 3,4-benzopyrene, but not with phenobarbitone, selectively elevates biphenyl 2-hydroxylase activity without changes in the amount of spectrophotometrically detectable cytochrome P-450 or the activity of cytochrome P-450 reductase (McPherson *et al.*, 1975a,b). It has been concluded that the enzyme is either not entirely cytochrome P-450-dependent or that a change in the biphenyl binding is promoted.

Manganese salts, ethanol, phenobarbitol, and the herbicides 3-(4-chlorophenyl)-1,1-dimethylurea and 2,6-dichlorobenzonitrile all stimulate the induction of cytochrome P-450 in sliced Jerusalem artichoke tubers, whereas N-(3-chlorophenyl)isopropyl carbamate inhibits the induction (Reichhart *et al.*, 1979). The ratio of cytochrome P-450 to cinnamate 4-hydroxylase is not altered by Mn^{2+} or N-(3-chlorophenyl)isopropyl carbamate, whereas 2,6-dichlorobenzonitrile increases P-450 more than the enzyme activity; ethanol, phenobarbitol, and 3-(4-chloro-phenyl)-1,1-dimethylurea have no effect on hydroxylase activity. The elevation of cytochrome P-450 by Mn^{2+} is fully accounted for and that by 2,6-dichlorobenzonitrile partially accounted for by specific mechanisms shown to control the expression of cinnamate 4-hydroxylase in the wound-induced biosynthesis of hydroxycinnamate esters (Section IV,B). However, the further induction of cytochrome P-450 by ethanol, phenobarbitol, and 3-(4-chlorophenyl)-1,1-dimethylurea, apparently independently of the cytochrome P-450-dependent hydroxylase, must be ascribed to the substrate induction of cytochrome P-450 involved in the oxidative metabolism of these three compounds. Similarly the component of cytochrome P-450 induced by 2,6-dichlorobenzonitrile and not accounted for by the increase in cinnamate 4-hydroxylase activity might be engaged in the metabolism of this xenobiotic. However, it has still to be established that any one of these compounds is oxidatively metabolized by cytochrome P-450-dependent enzymes in Jerusalem artichoke tubers.

A picture emerges from these studies on physiological factors affecting oxygenase levels, in which the expression of oxygenase activity is an integral part of specific metabolic programs. The temporal and spatial integration of oxygenase activity with other components of these programs is discussed in the next two sections.

B. Temporal Integration of Oxygenase Activity

The temporal integration of oxygenase activity in metabolic programs has been most extensively studied with respect to the expression of cinnamate 4-hydroxylase activity in the biosynthesis of phenylpropanoid compounds. The coinduction of the initial enzymes in this pathway from phenylalanine, namely, phenylalanine ammonia-lyase, cinnamate 4-hydroxylase, and hydroxycinnamate : CoA ligase, has been reported in a number of plant cells and tissues (Smith et al., 1977). In the apical buds and stems of pea seedlings, illumination induced precisely coordinated but transient increases in the activities of these three enzymes (Butt and Wilkinson, 1979), but such exact correlation is not generally observed. For example, illumination of parsley cell cultures caused the activities of the three enzymes to increase, but the maximum activities were attained after different periods, depending on the rates of their subsequent decay (Hahlbrock et al., 1976). Of the three enzymes, the kinetics of phenylalanine ammonia-lyase induction were most closely correlated with the accumulation of flavonoids. In dark-incubated potato tuber discs, the three enzymes underwent transient increases in activity concomitant with the accumulation of hydroxycinnamic esters. Illumination further stimulated product accumulation, but only the activity of phenylalanine ammonia-lyase was further enhanced (Lamb and Rubery, 1976a). Thus the primary control site in parsley cell cultures and potato tuber discs appears to be phenylalanine ammonia-lyase rather than cinnamate 4-hydroxylase. Nevertheless, the induction of cinnamate 4-hydroxylase must be part of the overall control system, since its basal activity in both these systems is insufficient to support the subsequent flux through the pathway. In carrot cell cultures growing in early log phase, the levels of cinnamate 4-hydroxylase activity are sufficient for the rapid accumulation of phenolic acids observed. As the cultures age, hydroxylase activity declines and the accumulation of phenolics ceases despite the persistence of high levels of phenylalanine ammonia-lyase (Sugano et al., 1978), suggesting that the hydroxylase activity may here be the primary controlling element. Clearly, within the overall pattern of coinduction of these three enzymes there is considerable flexibility presumably related to the biological requirements of specific metabolic programs.

The molecular mechanisms underlying transient increases in cinnamate 4-hydroxylase activity are not well understood, but de novo synthesis of the enzyme is likely to be involved. Induction is inhibited by cycloheximide in potato tuber discs (Lamb and Rubery, 1976b; Lamb, 1977), sliced sweet potato root (Tanaka et al., 1974; Rhodes et al., 1976), and excised pea epicotyls (Hyodo and Yang, 1971). This cannot be regarded as rigorous evidence because of the total and nonspecific inhibition of protein synthesis by cycloheximide and its well-established side effects (Smith *et al.*, 1977). Isotope incorporation studies have not been reported for the enzyme because of its insolubility, but such studies have been successful in establishing the *de novo* synthesis of phenylalanine ammonia-lyase in a number of systems in which the hydroxylase is coinduced (Hahlbrock, 1976; Hahlbrock and Grisebach, 1979; Lamb *et al.*, 1979). Antimetabolite studies show that the concomitant increases in enzyme activity originate from a coordinated stimulation of transcriptional activity (Lamb, 1977), and increased levels of translatable mRNA for phenylalanine ammonia-lyase and hydroxycinnamate : CoA ligase have been demonstrated in illuminated parsley cells (Hahlbrock and Grisebach, 1979). In potato tuber discs, the cinnamate 4-hydroxylase activity accounts for almost all of the cytochrome P-450 present, and parallel increases in enzyme activity and cytochrome P-450 content after slicing is consistent with *de novo* synthesis of the P-450 component of the hydroxylation system (Rich and Lamb, 1977). Similar findings have been reported with respect to the phytochrome-mediated light stimulation of hydroxylase activity in terminal pea buds (Benveniste *et al.*, 1978).

Product repression is also involved in the temporal integration of cinnamate 4-hydroxylase activity in phenylpropanoid synthesis. Thus both cinnamic and *p*-coumaric acids inhibit the development of phenylalanine ammonia-lyase activity in illuminated potato tuber discs, whereas exogenous *p*-coumaric acid but not cinnamic acid inhibits the development of hydroxylase activity (Lamb and Rubery, 1976c). Such feedback regulation operates after transcription (Lamb, 1979), and evidence for its operation *in vivo* is provided by the observation that treatment with D-phenylalanine and α-aminooxy-β-phenylpropionic acid, both inhibitors of phenylalanine ammonia-lyase, decrease flux through the pathway and reduce the accumulation of chlorogenic acid despite elevated levels of cinnamate 4-hydroxylase (C. J. Lamb, unpublished). Similarly, in Jerusalem artichoke tubers, either Mn^{2+} or 2,6-dichlorobenzonitrile reduces the accumulation of hydroxycinnamic acids, at the same time raising hydroxylase activity in extracts (Reichhart *et al.*, 1979).

Much less is known of the regulation of hydroxylase activity that introduces a second hydroxyl group. This reaction is probably catalyzed by the monophenolase activity of the phenolase complex and is usually measured as *p*-coumarate 3-hydroxylase. Early work with excised sweet potato roots related an increase in caffeic acid oxidase activity to the accumulation of polyphenols (Hyodo and Uritani, 1966), but an increase in the oxidase activity appears to follow rather than precede phenolic synthesis. With certain assumptions, an increase in *p*-coumarate hydroxylase has been shown to be coordinated with raised levels of phenylalanine ammonia-lyase in broad bean seedlings and of both phenylalanine ammonia-lyase and cinnamate

4-hydroxylase in spinach-beet seedlings (Bolwell, 1974; Butt, 1979); in neither case has an increase in o-diphenol oxidase been observed. Nevertheless, the demonstration of a temporal integration of increased p-coumarate hydroxylase activity with earlier enzymes in the phenylpropanoid pathway would go far to identify it with the biosynthetic process and possibly distinguish it from a more general, less specific phenolase activity perhaps exerting some other function, which is generally present in plant tissues.

C. Spatial Organization of Oxygenases

It has already been noted that two broad classes of enzyme systems are responsible for hydroxylation reactions in gibberellin biosynthesis, namely, those that are membrane-bound and cytochrome P-450-dependent and those that are soluble, possibly utilizing nonheme iron (see Hedden *et al.*, 1978). The substrates for the membrane-bound enzymes are more lipophilic and less hydroxylated than those for the soluble enzymes. Clearly the introduction of one or more oxygen atoms may have profound effects on the polarity of an intermediate in, or the product of, a biosynthetic sequence. The intracellular, or spatial, organization of oxygenase activity may therefore be critical, and it is not surprising that evidence for the microcompartmentation of these enzymes is accumulating. The importance and exclusiveness of such compartmentation is shown by the inability of microsomal preparations of pea seeds to metabolize exogenous kaurene; only kaurene synthesized *in situ* from mevalonic acid in the presence of a soluble enzyme system (Coolbaugh and Moore, 1971) can be utilized because of the requirement for a noncatalytic carrier protein readily washed out of the microsomes during preparation (Moore *et al.*, 1972). On the other hand, exogenously supplied kaurenaldehyde is metabolized by the microsomal system, so that the precise nature of the organization of the different oxygenases within the membrane is not simple and has still to be established.

The N-hydroxylation and methylene hydroxylation systems involved in biosynthesis of the cyanogenic glucoside dhurrin from tyrosine in *S. bicolor* seedlings are both localized in microsomal membranes and cosediment under a variety of conditions (Saunders *et al.*, 1977). Comparison of the $^3H/^{14}C$ ratio in the mandelonitrile produced by microsomes supplied with [^3H]tyrosine and ^{14}C-labeled intermediates such as p-hydroxyphenylacetaldoxime indicates only very limited exchange between exogenous pools and intermediates generated *in situ* (Møller and Conn, 1980). Similar metabolic channeling has been observed in the biosynthesis of o- and p-coumaric acids from phenylalanine in microsomal membranes from potato tuber discs (Czichi and Kindl, 1975) and cucumber cotyledons (Czichi and Kindl, 1977). In the latter, prolonged irradiation increases the cooperation between membrane-bound phenylalanine ammonia-lyase and cinnamate 2- and 4-hydroxylases; coupling is more marked at high phenylalanine concentrations and reduced by treating extracts with ethylene. Cinnamate hydroxylase activities are to some extent latent and can be increased by treatment with low concentrations of Triton X-100 or 8-anilinonaphthalene sulfate. Nevertheless, although substantial proportions of phenylalanine ammonia-lyase may in some cases be found associated with plastids or other membrane fractions (McClure, 1979), this enzyme is more usually found in the soluble fraction. It is suggested that, to account for this soluble lyase activity and the metabolic channeling observed, the hydroxylases are integral components of membranes with which phenylalanine ammonia-lyase forms some type of nonstoichiometric association (Czichi and Kindl, 1977). The role of these associations and their regulatory significance are open to more detailed investigation.

This is no less true of phenolases, the oxygenase action of which gives rise to even more polar compounds. Although a proportion of these enzymes can be recovered from the soluble fraction of extracts, a major component is usually associated with membranes, and not infrequently, extracts have to be treated with salts or detergents to release these membrane-bound enzymes if their full activity is to be determined. For example, the phenolase of spinach-beet leaves is wholly associated with chloroplast membranes (Bartlett *et al.*, 1972), from which it can be partially released by prolonged treatment with $(NH_4)_2SO_4$ and detergents (Vaughan *et al.*, 1975). In untreated extracts, the activity left in the supernatant after sedimenting mitochondria and chloroplasts is almost completely precipitated by prolonged centrifugation at $100,000\,g$, suggesting an association with fragmented membranes from these organelles (Parish, 1972b). Fractionation procedures have demonstrated a similar attachment of phenolases to chloroplasts and mitochondria from apple fruit (Mayer *et al.*, 1964), apple peel (Harel *et al.*, 1965), spinach-beet stems (Parish, 1972a), and tobacco leaves (Nye and Hampton, 1966) and to other subcellular particles in poppy latex (Roberts, 1971), *Sorghum* leaves (Stafford and Bliss, 1973), and potato tubers (Ruis, 1972). The hydroxylation of monophenols so increases their polarity that they are likely to penetrate membranes with difficulty and utilize more easily a surface enzyme than a deeply embedded system. The important role of membranes in the encapsulation of biosynthetic intermediates has been demonstrated in the biosynthesis of chlorogenic acid within *Petunia* chloroplasts (Alibert *et al.*, 1977; Ranjeva *et al.*, 1979); with such an organization, membrane-bound phytochrome can play an important controlling role in phenolic biosynthesis (Pecket and Bassim, 1974).

Even though the precise molecular arrangement of oxygenase enzymes within membranes has still to be established, membrane-bound compartments provide a spatial control which can direct the biosynthesis of hydroxy-lated compounds and reduce the interaction of physically and chemically active intermediates with other cellular components.

VI. BIOLOGICAL FUNCTIONS OF OXYGENATION AND HYDROXYLATION

Natural selection for specific oxygenase activity in plants can be assumed to be determined by the nature of the hydroxylated compounds produced. In some cases, these are unstable and undergo spontaneous transformation by ring cleavage, demethylation, or desaturation; the consequences of these transformations will not be considered here. The introduction of hydroxyl groups retained in the final product of the biosynthetic sequence is more usual, and the profound effects of this on the general physical properties and specific chemical binding merit some discussion. The number of hydroxyl groups in molecules affects especially their association with and penetration through membranes, while the specific position of these groups determines their likely attachment to binding sites associated with biological action or their participation in further reactions, such as oxidation.

A. Membrane Association and Penetration

The association of molecules with membranes is determined by the relative disposition of polar and nonpolar groups. Directive polarity is shown in the steroids, in which the 3-hydroxyl group, especially when glycosylated, is retained in the aqueous phase. Models in which the rest of the molecule is fitted in among the lipophilic chains of phospholipids and glycerides show how the interaction between these chains may be reduced, leading to increased membrane fluidity or a lowering of the temperature at which liquid crystals may be formed. Models of chloroplast membranes have been proposed in which the terminal 3-hydroxyl groups of xanthophylls are juxtaposed with the carotene hydrocarbons and the phytyl groups of chlorophyll. At present too little is known of the molecular structure of suberin to guess the role of the 9,10-epoxy and 9,10-dihydroxy groups in the derivatives of 18-hydroxystearic acid (see Kolattukudy, 1978) in the insoluble suberin complex. A special problem is raised concerning the biogenesis of many phenolic substances, in which successive intermediates become more polar with the introduction of two, three, and more hydroxyl groups. The products are stored largely in the vacuoles, almost invariably as glycosides, hence at some stage must pass through the tonoplast. It may be significant that glycosylation, which greatly increases the polarity of flavonoids, is the terminal stage in their synthesis, but the enzymes responsible have not been found associated with membranes nor shown to play any part in facilitating transport into vacuoles.

B. Specific Binding and Biological Activity

Hydroxylation at specific sites giving rise to characteristic and recurring hydroxylation patterns implies more than a general effect upon physical properties such as lipid solubility. Moreover, the hydroxyl group, especially in phenolic groups, is both physically and chemically highly active, readily forming associations with other molecules by hydrogen bonding or reacting with them by esterification; in alkaline solution, the hydroxyl groups ionize with a dissociation constant (pK_a) of 9.2 or less (Thompson, 1964). There is therefore strong reason to relate the hydroxylation patterns found in natural products to specific interaction with receptor sites, either within the plant producing them or, by mimicking other compounds, within other organisms; these patterns may affect the survival of the plant under conditions of predation, infection, or competition.

The wide range of hydroxylation patterns found in the gibberellins synthesized by higher plants and fungi (Fig. 2) illustrates these structure–activity relationships. Comparative studies on 9 gibberellins and over 100 allied compounds in four different biological test systems have revealed considerable differences in the sensitivities of different plant organs to these compounds, in part at least because of different rates of penetration and different rates of internal metabolic conversion (Brian et al., 1967). Only the 10-carboxyl group, which may bind to a receptor, and possibly the lactone ring appear to be essential. Molecules with two hydroxyl groups are generally active, but the introduction of a third hydroxyl group eliminates any growth response. 2β-Hydroxylation inactivates gibberellins, but 2α-hydroxy-gibberellins are as active as their nonhydroxylated analogues (see Hedden et al., 1978), suggesting that not only the polarity but the precise orientation of the molecule with its receptor presumably within a membrane determines its growth-promoting activity; 2β-hydroxygibberellins, or their glycosides, often occur in the same tissue as their nonhydroxylated analogues, suggesting that here 2β-hydroxylation plays some role in regulating the concentration of active gibberellin.

In contrast with those of the gibberellins, the hydroxylation patterns of the isoflavonoids in Trifolium sp. and alfalfa appear directly related to their estrogenic activity. The 7,4'-hydroxyl groups of genistein, daidzein, formononetin, and coumestrol are essential and related spatially to the 3-hydroxyl group and substitution in the D ring of estrone and estriol (Harborne, 1977). An ecological role for these substances in regulating the population size of quails feeding on legume crops has been suggested, the concentration of these substances in plants, hence the amounts taken in during feeding, being controlled by rainfall (Leopold *et al.*, 1976).

Flavor is of course a major determinant of diet. Hydroxylation patterns and the sugars present in phenolic glycosides are important in controlling taste. The bitterness of grapefruit is due to naringenin 7-neohesperidoside (naringin), but structural alteration to the corresponding dihydrochalcone, especially if the 4'-hydroxyl group is replaced by a 3'-hydroxy-4'-methoxy pattern, gives compounds very much sweeter than sugar (Horowitz, 1964). The aromatic phenolics vanillin and eugenol contribute to the flavor of vanilla and banana fruit, respectively, each molecule having a characteristic 3-methoxy-4-hydroxy substitution, and the same substitution pattern in gingerol and paradol in the root of ginger or in capsaicin in *Capsicum* is responsible for their pungent taste (Rohan, 1972). Feeding attraction or deterrence due to specific methoxylation–hydroxylation patterns must play an important role in the survival and dispersal of plants, as well as in their economic exploitation.

The resistance of plants to disease due to the presence of compounds in the tissues or to their production after infection is well-established (see Harborne, 1977). Among the former, hydroxycinnamic acids (especially chlorogenic acid) and hydroxylated coumarins are of major importance, and some of the fungitoxins arising in plants after infection show hydroxylation patterns remarkably similar to those with estrogenic activity already described. The patterns found in these phytoalexins are characteristic of the species producing them and directly limit the range of pathogenic fungi to which these plants are susceptible. Similar to the control of pathogen infection by higher plants is the control of competition among higher plants by the diffusion of toxic soluble substances (allelopathy). This is found in its clearest form in desert plants of the Californian chapparal, where a variety of neutral phenols, hydroxybenzoic acids, and hydroxycinnamic acids in the leaves of *Adenostoma* and *Arctostaphylos* are leached into the soil and suppress the germination of grasses and other herbs in their vicinity (Muller and Chou, 1972).

Although not strictly within the purview of this section, one property of hydroxylated compounds that does not involve specific binding within the plant itself or in other organisms, yet which affects the interaction between plants and some animals, deserves mention. The colors of fruits and flowers and of some other organs are food attractants on the one hand and facilitate dispersal and fertilization on the other. Hydroxyl groups associated with conjugated systems are chromophores modifying light absorbance, often by shifting absorbance bands from the uv region to the visible range. In the carotenoid series, 3-hydroxylation at one or both ends of the molecule has no significant effect on the absorbance in organic solvents (see Karrer and Jucker, 1950; Moss and Weedon, 1976). Among flavonoids, the optical absorbance is largely determined by the hydroxylation pattern in ring B and is surprisingly similar to that of the corresponding simple phenolics (e.g., the absorption spectra of catechin and catechol) unless the ring is conjugated to 3-hydroxyl groups, when the absorption maxima shifts to longer wavelengths (cf. apigenin and quercetin). The anthocyanin salts in acid solution, in which the electrons of the heterocyclic oxygen atom are involved in π bond formation, absorb at longer wavelengths still, giving red, mauve, and blue colors which are, however, modified by an increase in pH because of pseudobase formation and ionization of the hydroxyl groups (see Swain, 1976). The color is further modified by chelation of o-dihydroxy groups with metal ions or copigmentation with other flavonoids. The selection of chemical structure for coloration in relation to pollination and dispersal is related to color sensitivity in different groups and species of insects and other animals (see Harborne, 1977).

C. Oxidation and Quinone Production

Among the chemical properties of hydroxylated compounds, besides desaturation and cleavage as already discussed, susceptibility to further oxidation is the most important. With aliphatic alcohols, the oxidation of primary alcohols to aldehydes and carboxylic acids (as in gibberellin biosynthesis) and of secondary alcohols to ketones (as in the conversion of menthol to menthone) gives rise to compounds with identifiable and distinctive functions. However, the oxidation processes most thoroughly studied have been those of mono- and diphenols, including hydroxycinnamic acids, catechins, flavones and flavonols, anthocyanidins, and leucoanthocyanidins. The oxidation of these compounds is usually induced by cell damage arising from senescence, physical shearing, or pathogen attack. In the simplest instances, cell damage brings together o-diphenols and o-diphenol oxidases following the release of vacuolar contents into the protoplasm. Exogenously supplied o-diphenols undergo similar oxidations (Barz, 1975). The o-quinones produced then undergo rapid polymerization with further oxidation and the formation of relatively insoluble brown polymers (Mason, 1955; Pierpoint, 1970; Van Sumere et al., 1975; Wong, 1976).

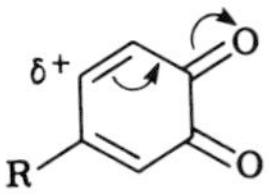

The reactivity of o-quinones is due to the readiness with which they are attacked by nucleophilic reagents. Self-polymerization, with C—O and C—C bond formation, readily occurs. Such polymers protect the surface temporarily against penetration during the period before the newly exposed cells can protect themselves by suberization (Dean and Kolattukudy, 1976; Kolattukudy, 1978) or lignification (Rhodes and Wooltorton, 1978). However, *o*-quinones may also be attacked by other nucleophilic compounds and groups, such as the ε-amino groups of lysyl residues and the thiol groups of cysteinyl residues in proteins (Pierpoint, 1969). In consequence, proteins released by the destruction of cell organization may be tanned, affording further physical protection, or during pathogen attack the surface proteins of fungal mycelia or the coat proteins of viruses may be incorporated into these polymers. The synthesis of hydroxycinnamic acids by plant tissue in response to damage (Section V) can be regarded as the production of reserve protective substances in which the second hydroxyl group introduced by phenolase may be readily oxidized further when cellular conditions change, either by the *o*-diphenol oxidase activity of the phenolase itself or by a similar oxidase which may itself have little or no hydroxylase activity (Butt, 1979).

This response to tissue damage may not depend on the presence of *o*-diphenols but requires at least substances that can be converted to them. For instance, it has been observed that the germination of spores of apple scab, *Venturia inequalis,* on apple leaves is countered by the hydroxylation of phloridzin and its aglycone, phloretin, to the *o*-dihydroxy derivatives, the further oxidation of which yields *o*-quinones active against the fungus (Overeem, 1976). Both hydroxylase and oxidase activities of the phenolase are necessary for the enzyme to exert its function *in vivo*.

The complexing of leaf protein by *o*-quinones is thought in some cases to affect the feeding habits of animals. The mountain gorilla of the African Congo has been found to have a highly selective diet, confining its attention to 29 plant species which have in common the absence of condensed tannins (Bate-Smith, 1972). The choice may be determined by the taste of these tannins, which also complex with the leaf protein to render it partly or wholly indigestible (Harborne, 1977). Similarly, the tannins in oak trees limit the number of Lepidoptera feeding on them to those that can both deal with the toughness of the leaf and also prevent reaction between the tannins and protein by, for example, having an alkaline pH in the gut (Feeny, 1970, 1976).

REFERENCES

Albersheim, P., and Valent, B. S. (1978). *J. Cell Biol.* **78,** 627–643.
Alibert, G., Ranjeva, R., and Boudet, A. M. (1977). *Physiol. Veg.* **15,** 279–301.
Amrhein, N., and Zenk, M. H. (1969). *Phytochemistry* **8,** 107–113.
Amrhein, N., and Zenk, M. H. (1970). *Naturwissenschaften* **57,** 312.
Banthorpe, D. V., Mann, J., and Poots, I. (1977). *Phytochemistry* **16,** 547–550.
Banthorpe, D. V., Ekundayo, O., and Rowan, M. G. (1978). *Phytochemistry* **17,** 1111–1114.

Bartlett, D. J., Poulton, J. E., and Butt, V. S. (1972) *FEBS Lett.* **23**, 265–267.

Barz, W. (1975). *Ber. Dtsch. Bot. Ges.* **88**, 71–81.

Barz, W. (1977a). *In* "Plant Tissue Culture and its Biotechnological Application" (W. Barz, E. Reichard, and M. H. Zenk, eds.), pp. 153–171. Springer-Verlag, Berlin and New York.

Barz, W. (1977b). *Physiol. Veg.* **15**, 261–277.

Bate-Smith, E. C. (1972). *In* "Phytochemical Ecology" (J. B. Harborne, ed.), pp. 45–56. Academic Press, New York.

Ben-Abdelkader, A., Cherif, A., Demandre, C., and Mazliak, P. (1973). *Eur. J. Biochem.* **32**, 155–165.

Benveniste, I., Salaün, J.-P., and Durst, F. (1977). *Phytochemistry* **16**, 69–73.

Benveniste, I., Salaün, J.-P., and Durst, F. (1978). *Phytochemistry* **17**, 359–364.

Bolkart, K. H., and Zenk, M. H. (1968). *Z. Pflanzenphysiol.* **59**, 439–444.

Bolwell, G. P. (1974). The control of enzyme levels in the biosynthesis of plant phenolics. D.Phil. Thesis, Oxford University.

Brian, P. W., Grove, J. F., and Mulholland, T. P. C. (1967). *Phytochemistry* **6**, 1475–1499.

Britton, G. (1976). *In* "Chemistry and Biochemistry of Plant Pigments" (T. W. Goodwin, ed.), 2nd ed., Vol. 1, pp. 262–327. Academic Press, New York.

Brown, S. A. (1979). *Recent Adv. Phytochem.* **12**, 249–286.

Buhler, D. R., and Mason, H. S. (1961). *Arch. Biochem. Biophys.* **92**, 424–437.

Butt, V. S. (1976). *Perspect. Exp. Biol.* **2**, 357–367.

Butt, V. S. (1979). *Recent Adv. Phytochem.* **12**, 433–456.

Butt, V. S., and Wilkinson, E. M. (1979). *FEBS-Symp.* **55**, 147–154.

Camm, E. L., and Towers, G. H. N. (1977). *Prog. Phytochem.* **4**, 169–188.

Charlwood, S. V., and Banthorpe, D. V. (1977). *Prog. Phytochem.* **5**, 65–125.

Claes, H. (1959). *Z. Naturforsch., Teil B* **14**, 4–17.

Conn, E. E., and Swain, T. (1961). *Chem. Ind. (London)* pp. 592–593.

Conn, E. E., McFarlane, I. J., Møller, B. L., and Shimada, M. (1979). *FEBS-Symp.* **55**, 63–71.

Coolbaugh, R. C., and Moore, T. C. (1971). *Phytochemistry* **10**, 2401–2412.

Coolbaugh, R. C., Hirano, S. S., and West, C. A. (1978). *Plant Physiol.* **62**, 571–576.

Cornthwaite, D., and Haslam, E. (1965). *J. Chem. Soc.* pp. 3008–3011.

Croteau, R., and Kolattukudy, P. E. (1975a). *Arch. Biochem. Biophys.* **170**, 61–72.

Croteau, R., and Kolattukudy, P. E. (1975b). *Arch. Biochem. Biophys.* **170**, 73–81.

Czichi, U., and Kindl, H. (1975). *Planta* **125**, 115–125.

Czichi, U., and Kindl, H. (1977). *Planta* **134**, 133–143.

Deacon, W., and Marsh, H. V. (1971). *Phytochemistry* **10**, 2915–2924.

Dean, B. B., and Kolattukudy, P. E. (1976). *Plant Physiol.* **58**, 411–416.

Dixon, R. A., and Bendall, D. S. (1978). *Physiol. Plant Pathol.* **13**, 295–306.

Durand, R., and Zenk, M. H. (1974a). *Phytochemistry* **13**, 1483–1492.

Durand, R., and Zenk, M. H. (1974b). *FEBS Lett.* **39**, 218–220.

Earl, J. W., and Kennedy, I. R. (1975). *Phytochemistry* **14**, 1507–1512.

Ebel, J., Schaller-Hekeler, B., Knobloch, K. H., Wellmann, E., Grisebach, H., and Hahlbrock, K. (1974). *Biochim. Biophys. Acta* **362**, 417–424.

El-Basyouni, S. Z., Chen, D., Ibrahim, R. K., Neish, A. C., and Towers, G. H. N. (1964). *Phytochemistry* **3**, 485–492.

Ellis, B. E. (1971). *FEBS Lett.* **18**, 228–230.

Feeny, P. (1970). *Ecology* **51**, 565–581.

Feeny, P. (1976). *Recent Adv. Phytochem.* **10**, 1–40.

Fellows, L. E., and Bell, E. A. (1970). *Phytochemistry* **9**, 2389–2396.

Fischer, N., and Dreiding, A. S. (1972). *Helv. Chim. Acta* **55**, 649–658.

Frear, D. S., Swanson, H. R., and Tanaka, F. S. (1969). *Phytochemistry* **8**, 2157–2169.

Fritsch, H., and Grisebach, H. (1975). *Phytochemistry* **14**, 2437–2442.

Galliard, T., and Stumpf, P. K. (1966). *J. Biol. Chem.* **241**, 5806–5812.
Gamborg, O. L. (1967). *Phytochemistry* **6**, 1067–1073.
Gamborg, O. L., and Neish, A. C. (1959). *Can. J. Biochem. Physiol.* **37**, 1277–1285.
Gestetner, B., and Conn, E. E. (1974). *Arch. Biochem. Biophys.* **163**, 617–624.
Goodwin, T. W. (1979). *Annu. Rev. Plant Physiol.* **30**, 369–404.
Graebe, J. E., and Hedden, P. (1974). *In* "Biochemistry and Chemistry of Plant Growth Regulators" (K. Schreiber, H. R. Schütte, and G. Sembdner, eds.), pp. 1–16. Acad. Sci. DDR, Berlin.
Griffiths, L. A. (1959). *Nature (London)* **184**, 58–59.
Grunwald, C. (1975). *Annu. Rev. Plant Physiol.* **26**, 209–236.
Guroff, G., Daly, J. W., Jerina, D. M., Renson, J., Witkop, B., and Udenfriend, S. (1967). *Science* **157**, 1524–1530.
Hahlbrock, K. (1976). *Physiol. Veg.* **14**, 207–213.
Hahlbrock, K., and Grisebach, H. (1979). *Annu. Rev. Plant Physiol.* **30**, 105–130.
Hahlbrock, K., and Wellmann, E. (1973). *Biochim. Biophys. Acta* **304**, 702–706.
Hahlbrock, K., Ebel, J., Ortmann, R., Sutter, A., Wellmann, E., and Grisebach, H. (1971). *Biochim. Biophys. Acta* **244**, 7–15.
Hahlbrock, K., Knobloch, K.-H., Kreuzaler, F., Potts, J. R. M., and Wellmann, E. (1976). *Eur. J. Biochem.* **61**, 199–206.
Hamilton, G. A. (1974). *In* "Molecular Mechanisms of Oxygen Activation" (O. Hayaishi, ed.), pp. 405–451. Academic Press, New York.
Hanson, K. R., and Havir, E. A. (1979). *Recent Adv. Phytochem.* **12**, 91–137.
Harborne, J. B. (1967). "Comparative Biochemistry of the Flavonoids." Academic Press, New York.
Harborne, J. B. (1977). "Introduction to Ecological Biochemistry." Academic Press, New York.
Harel, E., Mayer, A. M., and Shain, Y. (1965). *Phytochemistry* **4**, 783–790.
Harms, H., Haider, K., Berlin, J., Kiss, P., and Barz, W. (1972). *Planta* **105**, 342–351.
Hedden, P., MacMillan, J., and Phinney, B. O. (1978). *Annu. Rev. Plant Physiol.* **29**, 149–192.
Heftmann, E. (1973). *In* "Phytochemistry" (L. P. Miller, ed.), Vol. II, pp. 171–226. Van Nostrand Reinhold, New York.
Heftmann, E. (1977). *Prog. Phytochem.* **4**, 257–276.
Heidelberger, C. (1975). *Annu. Rev. Biochem.* **44**, 79–121.
Hess, D. (1967). *Z. Pflanzenphysiol.* **56**, 12–19.
Hess, D. (1968). "Biochemische Genetik." Springer-Verlag, Berlin and New York.
Horowitz, N. H., and Shen, S.-C. (1952). *J. Biol. Chem.* **197**, 513–520.
Horowitz, R. M. (1964). *In* "Biochemistry of Plant Phenolics" (J. B. Harborne, ed.), pp. 545–572. Academic Press, New York.
Hyodo, H., and Uritani, I. (1966). *Plant Cell. Physiol.* **7**, 137–144.
Hyodo, H., and Yang, S. F. (1971). *Arch. Biochem. Biophys.* **143**, 338–339.
Impellizerri, G., and Piatelli, H. (1972). *Phytochemistry* **11**, 2499–2502.
Ishimaru, A., and Yamazaki, D. (1977). *J. Biol. Chem.* **252**, 6118–6124.
Jerina, D. M., Daly, J. W., Witkop, B., Zaltzman-Nirenberg, P., and Udenfriend, S. (1970). *Biochemistry* **9**, 147–156.
Karrer, P., and Jucker, E. (1950). "Carotenoids." Am. Elsevier, New York.
Knoche, H. W. (1968). *Lipids* **3**, 163–169.
Kolattukudy, P. E. (1978). *In* "Biochemistry of Wounded Plant Tissues" (G. Kahl, ed.), pp. 43–84. de Gruyter, Berlin.
Kosuge, T., and Conn, E. E. (1961). *J. Biol. Chem.* **236**, 1617–1621.
Kreuzaler, F., and Hahlbrock, K. (1975). *Eur. J. Biochem.* **56**, 205–213.
Lamb, C. J. (1977). *Planta* **135**, 169–175.
Lamb, C. J. (1979). *Arch. Biochem. Biophys.* **192**, 311–317.

Lamb, C. J., and Rubery, P. H. (1976a). *Phytochemistry* **15**, 665–668.
Lamb, C. J., and Rubery, P. H. (1976b). *Plant Sci. Lett.* **7**, 33–37.
Lamb, C. J., and Rubery, P. H. (1976c). *Planta* **130**, 283–290.
Lamb, C. J., Merritt, T. K., and Butt, V. S. (1979). *Biochim. Biophys. Acta* **582**, 196–212.
Leete, E. (1967a). *Annu. Rev. Plant Physiol.* **18**, 179–196.
Leete, E. (1967b). *In* "Biogenesis of Natural Compounds" (P. Bernfeld, ed.), pp. 953–1023. Pergamon, Oxford.
Leistner, E. (1975). *Ber. Dtsch. Bot. Ges.* **88**, 163–178.
Leopold, A. S., Erwin, M., Oh, J., and Browning, B. (1976). *Science* **191**, 98–99.
Loomis, W. D. (1967). *In* "Terpenoids in Plants" (J. B. Pridham, ed.), pp. 59–82. Academic Press, New York.
McCalla, D. R., and Neish, A. C. (1959). *Can. J. Biochem. Physiol.* **37**, 531–536.
McClure, J. W. (1979). *Recent Adv. Phytochem.* **12**, 525–556.
McDermott, J. C. B., Brown, D. J., Britton, G., and Goodwin, T. W. (1973). *Biochem. J.* **132**, 649–652.
McIntyre, R. J., and Vaughan, P. F. T. (1975). *Biochem. J.* **149**, 447–461.
McPherson, F. J., Markham, A., Bridges, J. W., Hartman, G. C., and Parke, D. V. (1975a). *Trans. Biochem. Soc.* **3**, 281–283.
McPherson, F. J., Markham, A., Bridges, J. W., Hartman, G. C., and Parke, D. V. (1975b). *Trans. Biochem. Soc.* **3**, 283–285.
Madyastha, K. M., Meehan, T. D., and Coscia, C. J. (1976). *Biochemistry* **15**, 1097–1102.
Mason, H. S. (1957). *Adv. Enzymol.* **19**, 79–233.
Mason, H. S. (1965). *Annu. Rev. Biochem.* **34**, 595–634.
Mason, H. S., Fowlks, W. L., and Peterson, E. (1955). *J. Am. Chem. Soc.* **77**, 2914–2915.
Mason, H. S., Onopryenko, I., and Buhler, D. (1957). *Biochim. Biophys. Acta* **24**, 225–226.
Massey, V., and Hemmerich, P. (1975). *In* "The Enzymes" (P. D. Boyer, ed.), 3rd ed., Vol. 12, Part B, pp. 191–252. Academic Press, New York.
Massicot, J., and Marion, L. (1957). *Can. J. Chem.* **35**, 1–4.
Mayer, A. M., Harel, E., and Shain, Y. (1964). *Phytochemistry* **3**, 447–451.
Meehan, T. D., and Coscia, C. J. (1973). *Biochem. Biophys. Res. Commun.* **53**, 1043–1048.
Mehendale, H. M. (1973). *Phytochemistry* **12**, 1591–1594.
Meier, H., and Zenk, M. H. (1965). *Z. Pflanzenphysiol.* **53**, 415–421.
Meyer, E., and Barz, W. (1978). *Planta Medica* **33**, 336–344.
Møller, B. L., and Conn, E. E. (1980). *J. Biol. Chem.* **255**, 3049–3056.
Moore, T. C., Barlow, S. A., and Coolbaugh, R. C. (1972). *Phytochemistry* **11**, 3225–3233.
Morris, L. J. (1967). *Biochem. Biophys. Res. Commun.* **20**, 340–345.
Morris, L. J. (1970). *Biochem. J.* **118**, 681–693.
Morris, L. J., and Hitchcock, C. (1968). *Eur. J. Biochem.* **4**, 146–148.
Morris, L. J., Hall, S. W., and James, A. T. (1966). *Biochem. J.* **100**, 29C–30C.
Moss, G. P., and Weedon, B. C. L. (1976). *In* "Chemistry and Biochemistry of Plant Pigments" (T. W. Goodwin, ed.), 2nd ed., Vol. 1, pp. 149–261. Academic Press, New York.
Muller, C. H., and Chou, C.-H. (1972). *In* "Phytochemical Ecology" (J. B. Harborne, ed.), pp. 201–216. Academic Press, New York.
Murphy, P. J., and West, C. A. (1969). *Arch. Biochem. Biophys.* **133**, 395–407.
Nagatsu, I., Sudo, Y., and Nagatsu, T. (1972). *Enzymologia* **43**, 25–31.
Nair, P. M., and Vining, L. C. (1965). *Phytochemistry* **4**, 401–411.
Neish, A. C. (1960). *Annu. Rev. Plant Physiol.* **11**, 55–80.
Nye, T. G., and Hampton, R. E. (1966). *Phytochemistry* **5**, 1187–1189.
Overeem, J. C. (1976). *In* "Biochemical Aspects of Plant-Parasite Relationships" (J. Friend and D. R. Threlfall, eds.), pp. 195–206. Academic Press, New York.
Palmer, J. K. (1963). *Plant Physiol.* **38**, 508–513.
Parish, R. W. (1972a). *Z. Pflanzenphysiol.* **66**, 176–188.

Parish, R. W. (1972b). *Eur. J. Biochem.* **31**, 446–455.

Patschke, L., and Grisebach, H. (1968). *Phytochemistry* **7**, 235–237.

Pecket, R. C., and Bassim, T. A. H. (1974). *Phytochemistry* **13**, 815–821.

Piatelli, M. (1976). *In* "Chemistry and Biochemistry of Plant Pigments" (T. W. Goodwin, ed.), 2nd ed., Vol. 1, pp. 560–596. Academic Press, New York.

Pierpoint, W. S. (1969). *Biochem. J.* **112**, 619–629.

Pierpoint, W. S. (1970). *Annu. Rep. Rothamstead Exp. St., Part II* pp. 199–218.

Pomerantz, S. H. (1964). *Biochem. Biophys. Res. Commun.* **16**, 188–194.

Poulton, J. E., and Butt, V. S. (1975). *Biochim. Biophys. Acta* **403**, 301–314.

Prasad, S., and Ellis, B. E. (1978). *Phytochemistry* **17**, 187–190.

Ranjeva, R., Boudet, A. M., and Alibert, G. (1979). *FEBS-Symp.* **55**, 91–100.

Rathmell, W. G., and Bendall, D. S. (1972). *Biochem. J.* **127**, 125–132.

Rehr, S. S., Janzen, D. H., and Feeny, P. P. (1973). *Science* **181**, 81–82.

Reichhart, D., Salaün, J.-P., Benveniste, I., and Durst, F. (1979). *Arch. Biochem. Biophys.* **196**, 301–303.

Rhodes, M. J., and Wooltorton, L. S. C. (1978). *In* "Biochemistry of Wounded Plant Tissues" (G. Kahl, ed.), pp. 243–286. de Gruyter, Berlin.

Rhodes, M. J., Hill, A. C. R., and Wooltorton, L. S. C. (1976). *Phytochemistry* **15**, 707–710.

Rich, P. R., and Lamb, C. J. (1977). *Eur. J. Biochem.* **72**, 353–360.

Robb, D. A., Mapson, L. W., and Swain, T. (1963). *Nature (London)* **201**, 503–504.

Roberts, M. F. (1971). *Phytochemistry* **10**, 3021–3027.

Roberts, R. J., and Vaughan, P. F. T. (1971). *Phytochemistry* **10**, 2649–2652.

Rohan, T. A. (1972). *In* "Phytochemical Ecology" (J. B. Harborne, ed.), pp. 57–69. Academic Press, New York.

Ropers, J.-J., Graebe, J. E., Gasken, P., and MacMillan, J. (1978). *Biochem. Biophys. Res. Commun.* **80**, 690–697.

Ross, M. S. F., Lines, D. S., Stevens, R. G., and Brain, K. R. (1978). *Phytochemistry* **17**, 45–48.

Ruis, H. (1972). *Phytochemistry* **11**, 53–58.

Russell, D. W. (1971). *J. Biol. Chem.* **246**, 3870–3878.

Russell, D. W., and Conn, E. E. (1967). *Arch. Biochem. Biophys.* **122**, 256–258.

Ruzicka, L., Eschenmoser, A., and Heusser, H. (1953). *Experientia* **9**, 357–367.

Saito, K., and Komamine, A. (1976). *Eur. J. Biochem.* **68**, 237–243.

Saito, K., and Komamine, A. (1978). *Eur. J. Biochem.* **82**, 385–392.

Sandermann, H., Diesperger, H., and Scheel, D. (1977). *In* "Plant Tissue Culture and its Biotechnological Application" (W. Barz, E. Reinhard, and M. H. Zenk, eds.), pp. 178–196. Springer-Verlag, Berlin and New York.

Satô, M. (1969). *Phytochemistry* **8**, 353–362.

Satô, M. (1977). *Phytochemistry* **16**, 1523–1525.

Saunders, J. A., Conn, E. E., Lin, C. H., and Shimada, M. (1977). *Plant Physiol.* **60**, 629–634.

Schill, L., and Grisebach, H. (1973). *Hoppe Seyler's Z. Physiol. Chem.* **354**, 1555–1562.

Schopfer, P. (1977). *Annu. Rev. Plant Physiol.* **28**, 223–252.

Sharma, H. K., and Vaidyanathan, C. S. (1975a). *Phytochemistry* **14**, 2135–2139.

Sharma, H. K., and Vaidyanathan, C. S. (1975b). *Eur. J. Biochem.* **68**, 237–243.

Sharma, H. K., Jamaluddin, M., and Vaidyanathan, C. S. (1972). *FEBS Lett.* **28**, 41–43.

Shimada, M., Kuroda, H., and Higuchi, T. (1973). *Phytochemistry* **12**, 2873–2875.

Shiman, R., Akino, M., and Kaufman, S. (1971). *J. Biol. Chem.* **246**, 1330–1340.

Shine, W. E., and Stumpf, P. K. (1974). *Arch. Biochem. Biophys.* **162**, 147–157.

Singh, R. K., Ben-Aziz, A., Britton, G., and Goodwin, T. W. (1973). *Biochem. J.* **132**, 649–652.

Smith, H., Billett, E. E., and Giles, A. B. (1977). *In* "Regulation of Enzyme Synthesis and Activity in Higher Plants" (H. Smith, ed.), pp. 93–127. Academic Press, New York.

Smith, T. A. (1977). *Prog. Phytochem.* **4**, 27–81.

Soliday, C. L., and Kolattukudy, P. E. (1977). *Plant Physiol.* **59**, 1116–1121.

Stafford, H. A. (1974a). *Annu. Rev. Plant Physiol.* **25**, 459–486.

Stafford, H. A. (1974b). *Plant Physiol.* **54**, 686–689.

Stafford, H. A., and Bliss, M. (1973). *Plant Physiol.* **52**, 453–458.

Stafford, H. A., and Dresler, S. (1972). *Plant Physiol.* **49**, 590–595.

Sugano, N., Koide, K., Ogawa, Y., Morija, Y., and Nishi, A. (1978). *Phytochemistry* **17**, 1235–1237.

Sugumaron, M., Kishore, G., Subramanian, V., Mohan, V. P., and Vaidyanathan, C. S. (1977). *J. Indian Inst. Sci., Sect. C* **59**, 1–37.

Sutter, A., and Grisebach, H. (1969). *Phytochemistry* **8**, 101–106.

Sutter, A., Poulton, J., and Grisebach, H. (1975). *Arch. Biochem. Biophys.* **170**, 547–556.

Swain, T. (1976). *In* "Chemistry and Biochemistry of Plant Pigments" (T. W. Goodwin, ed.) 2nd ed., Vol. 1, pp. 425–463. Academic Press, New York.

Tanaka, F. S., Swanson, H. R., and Frear, D. S. (1972a). *Phytochemistry* **11**, 2701–2708.

Tanaka, F. S., Swanson, H. R., and Frear, D. S. (1972b). *Phytochemistry* **11**, 2709–2715.

Tanaka, F. S., Kojima, M., and Uritani, I. (1974). *Plant Cell. Physiol.* **15**, 843–854.

Thompson, R. H. C. (1964). *In* "Biochemistry of Phenolic Compounds" (J. B. Harborne, ed.), pp. 1–32. Academic Press, New York.

Tolbert, N. E. (1973). *Plant Physiol.* **51**, 234–244.

Vanetten, H. D., and Pueppke, S. G. (1976). *In* "Biochemical Aspects of Plant-Parasite Relationships" (J. Friend and D. R. Threlfall, eds.), pp. 239–289. Academic Press, New York.

Vanneste, W. H., and Zuberbühler, A. (1974). *In* "Molecular Mechanisms of Oxygen Activation" (O. Hayaishi, ed.), pp. 371–404. Academic Press, New York.

Van Sumere, C. F., Albrecht, J., Dedonder, A., de Pooter, H., and Pé, I. (1975). *In* "The Chemistry and Biochemistry of Plant Proteins" (J. B. Harborne and C. F. Van Sumere, eds.), pp. 211–264. Academic Press, New York.

Vaughan, P. F. T., and Butt, V. S. (1969). *Biochem. J.* **113**, 109–115.

Vaughan, P. F. T., and Butt, V. S. (1970). *Biochem. J.* **119**, 89–94.

Vaughan, P. F. T., and Butt, V. S. (1972). *Biochem. J.* **127**, 641–647.

Vaughan, P. F. T., Eason, R., Paton, J. Y., and Ritchie, G. A. (1975). *Phytochemistry* **14**, 2383–2386.

Vijay, I. K., and Stumpf, P. K. (1972). *J. Biol. Chem.* **247**, 360–366.

Weedon, B. C. L. (1970). *Rev. Pure Appl. Chem.* **20**, 51–66.

Wengenmayer, H., Ebel, J., and Grisebach, H. (1974). *Eur. J. Biochem.* **50**, 135–143.

Whistance, G. R., and Threlfall, D. R. (1968). *Biochem. J.* **109**, 577–595.

Whistance, G. R., Threlfall, D. R., and Goodwin, T. W. (1967). *Biochem. J.* **105**, 145–154.

Wong, E. (1976). *In* "Chemistry and Biochemistry of Plant Pigments" (T. W. Goodwin, ed.), 2nd ed., Vol. 1, pp. 464–526. Academic Press, London.

Wong, E., and Wilson, J. M. (1972). *Phytochemistry* **11**, 875.

Wong, E., and Wilson, J. M. (1976). *Phytochemistry* **15**, 1325–1332.

Yamamoto, H. Y., and Chichester, C. O. (1965). *Biochim. Biophys. Acta* **109**, 303–305.

Yamamoto, H. Y., Chichester, C. O., and Nakayama, T. O. M. (1962). *Arch. Biochem. Biophys.* **96**, 645–649.

Young, O., and Beevers, H. (1976). *Phytochemistry* **15**, 379–385.

Zaprometov, M. N. (1962). *Biokhimiya* **27**, 366–377.

Zenk, M. H. (1964). *Z. Naturforsch., Teil B* **19**, 856–857.

Zenk, M. H. (1967). *Z. Pflanzenphysiol.* **57**, 477–478.

Zenk, M. H. (1979). *In* "Biochemistry of Plant Phenolics" (T. Swain, J. B. Harborne, and C. F. Van Sumere, eds.), pp. 139–176. Plenum, New York.

Zilg, H., Tapper, B. A., and Conn, E. E. (1972). *J. Biol. Chem.* **247**, 2384–2386.

Transmethylation and Demethylation Reactions in the Metabolism of Secondary Plant Products

22

JONATHAN E. POULTON

The Biochemistry of Plants, Vol. 7
Copyright © 1981 by Academic Press, Inc.
All rights of reproduction in any form reserved.
ISBN 0-12-675407-1

I. METHYLATION REACTIONS IN THE METABOLISM OF SECONDARY PLANT PRODUCTS

A. Introduction

The occurrence of a vast multitude of secondary plant products possessing one or more methyl groups in their structure provides an indication as to the frequency and importance of transmethylation reactions in this area of plant metabolism. These reactions, catalyzed by a class of enzymes called methyltransferases (E.C. 2.1.1.-), involve the transfer of an intact methyl group from a donor to a nitrogen, oxygen, sulfur, or other atom of a suitable acceptor molecule. Transmethylation reactions of this nature should be clearly distinguished from processes involving the transfer of some other molecular structure which becomes secondarily modified to form the methyl moiety. The latter situation will not be considered here.

Investigations probing the nature of the major physiological methyl donors in transmethylation reactions culminated in the discovery of S-adenosyl-L-methionine (SAM), formerly called "active methionine" (Cantoni, 1953). S-Adenosyl-L-methionine is an energy-rich sulfonium compound derived from L-methionine and ATP in a reaction catalyzed by methionine adenosyltransferase (E.C. 2.5.1.6) (see Cossins, this series, Vol. 2, Chapter 9). In transmethylation reactions, the positive charge on the sulfonium atom of SAM tends to withdraw electrons from the adjacent methyl carbon, thus promoting attack by the oxygen, sulfur, or nitrogen atom of the attacking nucleophile. This chapter is primarily concerned with transmethylations involving attack by the oxygen of phenols and carboxylate structures, and by the nitrogen of primary, secondary, and tertiary amines and of heterocyclic compounds. The thioether S-adenosyl-L-homocysteine (SAH) is a common product of these reactions.

Following elucidation of the structure of SAM there was a dramatic increase in the number of reported transmethylation reactions in which this compound acts as methyl donor (see Table VII). The view that transmethylation reactions in secondary plant metabolism are catalyzed by only a few enzymes having rather broad substrate specificities may well have been based on early studies which often involved crude plant extracts containing mixtures of enzymes. This opinion is, however, no longer tenable. In the past two decades, successful attempts to characterize these methyltransferases have been made. Although purification has not been undertaken to the extent that amino acid composition or major physicochemical properties can be compared, numerous methylating enzymes are now recognized, which show surprisingly strict specificity toward only one class of substrate, for example, toward lignin precursors, flavonoids, or furanocoumarins. Activity may even be restricted to one or a limited number of closely related compounds within these classes. Another important feature of these enzymes is their ability to methylate specific positions in the substrate molecule.

Methylation fulfills many physiological functions in plants. The methylation of polyphenolic compounds reduces the chemical reactivity of the phenolic groups. There is increasing evidence that transmethylations may play a crucial role in directing intermediates toward specific biosynthetic pathways. For example, in the biosynthesis of norlaudanosoline-derived alkaloids, the sites of O-methylation determine where oxidative phenol coupling reactions will occur; complete methylation would prevent any coupling. The various alkaloid patterns of different plants may well reflect differences in the specificity of their methylating systems. O-Methylation generally increases the lipid solubility and volatility of polyphenolics. The latter property, in addition to possible changes in absorption spectra occurring on methylation (e.g., in anthocyanins), may be important in attracting or deterring pollination vectors.

In this chapter an attempt has been made to collect and interpret information describing methyltransferases involved in secondary metabolism, paying special attention to the enzymology of significantly purified preparations. Data concerning their substrate and position specificities and other important kinetic properties are reviewed in relation to the metabolic role of these enzymes in plants. The major part of this chapter concerns the metabolism of phenolic compounds, since more information is currently available about this subject. However, methylation and demethylation reactions of alkaloid metabolism are being increasingly studied at the enzymatic level.

B. Methylation of Lignin Precursors by Meta-Directing O-Methyltransferases

Lignin may constitute 20–35% of the cell walls of angiosperms and conifers, the remainder being composed of cellulose (40–50%) and hemicelluloses (20–30%). This amorphous heteropolymer is formed by the random oxidative copolymerization of certain substituted cinnamyl alcohols (Fig. 1). Ac-

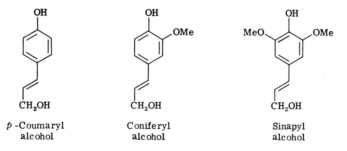

Fig. 1. Structures of lignin monomers.

cordingly, the residues of p-coumaryl, coniferyl, and sinapyl alcohols may be recognized within its structure. Multiple analytical techniques, including ^{13}C-nmr spectroscopy, have confirmed that significant differences exist among lignins isolated from various sources. Whereas angiosperm (hardwood) lignins generally consist of guaiacyl and syringyl units, gymnosperm (softwood) lignins contain mostly guaiacyl units with small amounts of p-hydroxyphenyl units. For example, the ratios of p-hydroxyphenyl, guaiacyl, and syringyl units for beech (angiosperm) and spruce (gymnosperm) lignins are approximately 5:49:46 and 14:80:6, respectively (Freudenberg 1965). Grass lignin represents a special case possessing up to 30% p-coumaryl residues; it is a mixed dehydrogenation polymer of coniferyl, sinapyl, and p-coumaryl alcohols, which additionally contains p-coumaric acid esterified with the γ-hydroxyl group of side chains of the lignin polymer.

Two transmethylation reactions (Fig. 2) are involved in lignin biosynthesis; the latter process is described fully by Grisebach (this volume, Chapter 15). Caffeic acid is methylated to ferulic acid which is thought to be hydroxylated *in vivo* to 5-hydroxyferulic acid. Subsequent methylation of this intermediate at the 5-position yields sinapic acid. The origin of the methyl ether groups of lignin was successfully investigated using doubly labeled methionine (Byerrum *et al.*, 1954). It was found that at least some of the methyl groups of this precursor were transferred as an intact unit (Mudd, 1973). In other experiments, methyl-labeled ferulate and sinapate were isolated from several plant species following the administration of [Me-^{14}C]methionine (Hess, 1964; Shimada and Higuchi, 1970).

The discovery in apple cambium tissue (Finkle and Nelson, 1963) and pampas grass (Finkle and Masri, 1964) of catechol O-methyltransferases, which methylate caffeic acid specifically at the meta position, represented a very significant advance toward a greater understanding of lignin biosynthesis. Since then, further enzymes of this type have been isolated, whose enzymological properties are in agreement with their involvement in this process. Table I summarizes the general features of these methyltransferases including their pH optima MW, and the effect of Mg^{2+} ions and EDTA. An

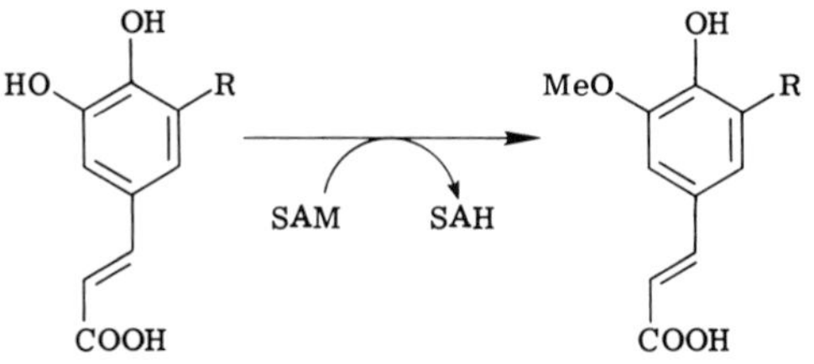

Fig. 2. Enzymatic formation of ferulic acid from caffeic acid (R = H) and of sinapic acid from 5-hydroxyferulic acid (R = OMe), respectively, by meta-specific O-methyltransferases.

TABLE I

Properties of Meta-directing O-Methyltransferases Involved in Lignin Biosynthesis

Organism	Degree of purification	pH Optimum	Substrate specificity[a]	Inhibitors	Effect of Mg^{2+} ions and EDTA	Remarks	References
Angiosperms Apple tree and *Pittisporum crassifolia* cambial tissue	Crude extract	7–8	Meta-methylation of CA	—	—	—	Finkle and Nelson, 1963
Cortaderia selloana (pampas grass) shoots	n.d.	n.d.	Predominantly meta-methylation of vicinal polyphenolic compounds including CA, but with protocatechuic aldehyde, 23% para-methylation	—	—	—	Finkle and Masri, 1964
Bamboo (*Phyllostachys pubescens* and *P. reticulata* shoots	100-fold	8.0	Meta-methylation of o-dihydric phenols including CA, 5-HFA, THC, and 5-hydroxyvanillin	Crude enzyme preparations partially inhibited by PCMB but not IAA	With crude preparations, Mg^{2+} stimulated activity; 20% inhibition by 10 mM EDTA	Enzyme stabilized by sulfhydryl reagents	Shimada, 1972; Shimada et al. 1973b

(continued)

TABLE I (Continued)

Organism	Degree of Purification	pH Optimum	Substrate Specificity	Inhibitors	Effect of Mg²⁺ ions and EDTA	Remarks	References
Beta vulgaris L. (spinach beet) leaves	75-fold	6.5	Meta-methylation of *o*-dihydric phenols including CA, protocate-chuic aldehyde, and catechol	SAH[a]	None	Enzyme stable in the absence of sulfhydryl reagents	Poulton and Butt, 1975
Glycine max L. cell cultures	60-fold	6.5–7.0	Meta-methylation of *o*-dihydric phenols including CA, 5-HFA, THC, and proto-catechuic aldehyde	SAH, quercetin, and luteolin	None	Final preparation stable in the absence of sulfhydryl reagents	Poulton *et al.*, 1976b
Tulipa anthers	n.d.	7.6	Meta-methylation of *o*-dihydric phenols CA and 5-HFA	SAH, PCMB	Mg²⁺ stabilized enzyme slightly	Slight stimulation by sulfhydryl reagents at 3 mM, but inhibitory at 30 mM	Sütfeld and Wiermann, 1978
Brassica napo-brassica (swede) root tissue	Crude extract	7	Meta-methylation of CA and 5-HFA	Some inhibition by *p*-coumaric, ferulic, and sinapic acids, and by coniferyl alcohol	—	—	Rhodes *et al.*, 1976

Species	Purification	pH optimum	Reaction	Inhibitors	Metal/cofactor effects	Stability	Reference
Populus nigra (poplar) shoots	n.d.	n.d.	Meta-methylation of *o*-dihydric phenols including CA, 5-HFA, THC, chlorogenic acid, and 5-hydroxyvanillin	—	—	—	Higuchi et al., 1977
Viscum album (mistletoe) shoots	11-fold	Approximately 7.2	Meta-methylation of CA and 5-HFA	Endogenous inhibitor of CA methylation in crude extracts	—	—	Kuroda and Higuchi, 1976
Gymnosperms *Pinus thunbergii* (Japanese black pine) seedlings	4-fold	7.5	Meta-methylation of *o*-dihydric phenols including CA, THC, protocatechuic aldehyde, and 5-HFA (poor)	PCMB, IAA	Slight stimulation by Mg^{2+}; 92% inhibition by 5 mM EDTA	Enzyme stabilized by 5 mM sulfhydryl reagents	Shimada et al., 1972b
	90-fold	—	—	CA competitively inhibits 5-HFA methylation	Mg^{2+} not required for activity, but stabilized enzyme; enzyme activated by 0.5 mM EDTA	Sulfhydryl reagents (10 mM) necessary for maximum rates	Kuroda et al., 1975
Ginkgo biloba shoots	Crude extract	n.d.	Meta-methylation of *o*-dihydric phenols including CA, THC, and 5-HFA (poor)	—	—	BSA required in homogenization buffer for active preparations	Shimada, 1972

[a] CA, caffeic acid; 5-HFA, 5-hydroxyferulic acid; THC, 3,4,5-trihydroxycinnamic acid; PCMB, *p*-chloromercuribenzoate; IAA, iodoacetic acid; SAH, *S*-adenosyl-L-homocysteine.

initial comparison indicates that these enzymes methylate 1-substituted 3,4-dihydric phenols using SAM as methyl donor; monophenols are not accepted as substrates. Generally, methylation takes place specifically at the meta position, correlating well with the m-methoxyl pattern of guaiacyl and syringyl units in lignin. The phenylpropanoids caffeic, 5-hydroxyferulic, and 3,4,5-trihydroxycinnamic acids tend to be the best substrates. However, the enzymes are not absolutely specific, since considerable activity may also be shown toward simple phenolic compounds such as protocatechuic aldehyde (Poulton and Butt, 1975; Poulton et al., 1976b; Higuchi et al., 1977), 5-hydroxyvanillin (Higuchi et al., 1977), and pyrocatechol (Higuchi et al., 1977; Poulton and Butt, 1975), and toward chlorogenic acid (Higuchi et al., 1977). Where tested, flavonoid compounds, such as luteolin and quercetin, were not substrates. Indeed, these compounds, when supplied at low concentrations (5–20 μM), strongly inhibited the methylation of cinnamic acids by soybean caffeic acid O-methyltransferase (CMT) (Poulton et al., 1976b). The failure of pine O-methyltransferase to methylate the dimeric lignin model compound catechylglycerol-β-guaiacyl ether supports the assumption that the methoxylation pattern of lignins is determined at the level of the hydroxycinnamate monomers prior to the polymerization of coniferyl and sinapyl alcohols (Higuchi et al., 1977).

Several factors affect the rate of methylation of cinnamic acids. The enzymes involved exhibit maximum activity in the range between pH 6.0 and 8.0; by comparison, flavonoid-specific methyltransferases from parsley, soybean, and Cicer cell cultures have pH optima within a much higher pH range (pH 8.7–9.7). With rat liver catechol O-methyltransferase, the pH value of the reaction medium can also influence the ratio of para- and meta-methylated products obtained (Senoh et al., 1959), but this phenomenon is unknown in plant systems. Animal catechol O-methyltransferases often require divalent metal ions, such as Mg^{2+}, for activity (Axelrod and Tomchick, 1958), but this appears not to be the case for lignin-specific O-methyltransferases. Only a slight stimulation of activity and/or stabilization of the enzyme by Mg^{2+} ions was reported for preparations from bamboo, tulip, and Pinus thunbergii, whereas they did not affect the activity of more highly purified enzymes from Glycine max cell cultures and Beta vulgaris (Poulton and Butt, 1975; Poulton et al., 1976b). Further evidence that Mg^{2+} ions are not required here as cofactor was shown by the failure of EDTA to inhibit the latter enzymes. Caution should be exercised in the interpretation of data involving the effect of divalent ions or EDTA on crude enzyme preparations. For example, whereas the activity of crude pine O-methyltransferase was almost completely inhibited by EDTA (Shimada et al., 1972b), the purified enzyme was actually activated by this compound (Kuroda et al., 1975). Considerable variation exists in the literature concerning the effect of sulfhydryl reagents and sulfhydryl inhibitors on these enzymes. Sulfhydryl reagents, such as β-mercaptoethanol, dithiothreitol, and reduced glutathione, are frequently included in extraction buffers, during purification procedures, and in assay mixtures, but no clear generalizations about their usage can be made here. The presence of sulfhydryl groups at the active site of the tulip, bamboo, and *Pinus* enzymes was suggested by inhibitor studies with *p*-chloromercuribenzoate and/or monoiodoacetate (Sütfeld and Wiermann, 1978; Higuchi *et al.*, 1967; Shimada *et al.*, 1972b).

Two biosynthetic pathways are theoretically possible for the production of sinapic acid from caffeic acid; these involve 5-hydroxyferulic acid and 3,4,5-trihydroxycinnamic acid as intermediates (see Grisebach, this volume, Chapter 15, Fig. 5). As yet, neither the *in vitro* hydroxylation of ferulic acid at the 5-position nor the presence of either of these intermediates in plant tissues has been demonstrated. These compounds would presumably be turned over rapidly in view of their high chemical reactivity. It is generally assumed that the biosynthesis of syringyl lignin involves ferulate and 5-hydroxyferulate as intermediates. The incorporation of [Me-^{14}C]ferulic acid into syringyl lignin components (Shimada, 1972) and the rapid methylation of 5-hydroxyferulate to sinapate in angiosperms (Higuchi *et al.*, 1977; Poulton *et al.*, 1976b; Sütfeld and Wiermann, 1978) support this view. However, the slow methylation of 3,4,5-trihydroxycinnamate is also known; in gymnosperms, the reaction rate with this substrate exceeds that of 5-hydroxyferulate, whereas the reverse is true in angiosperms (Higuchi *et al.*, 1977). The very low V_{max}/K_m ratio for 3,4,5-trihydroxycinnamate, especially in relation to that of 5-hydroxyferulate, has been taken as an indication that this acid is not a physiological substrate for soybean CMT. (Poulton *et al.*, 1976b).

The reported difference in composition between angiosperm and gymnosperm lignins could arise if gymnospermous plants lacked an enzyme or enzyme systems necessary for the formation of syringyl lignin. The excellent investigations of Higuchi and co-workers have revealed an extremely significant difference between the catechol *O*-methyltransferases from angiosperms and gymnosperms. Angiospermous catechol *O*-methyltransferases (e.g., from bamboo, soybean, and poplar) are called "difunction" *O*-methyltransferases in view of their capacity to methylate both caffeic and 5-hydroxyferulic acids, thereby yielding ferulic acid (a guaiacyl lignin precursor) and sinapic acid (a syringyl lignin precursor), respectively. In sharp contrast, *O*-methyltransferases from gymnospermous plants are termed "monofunction" *O*-methyltransferases since they show a preference for caffeic acid and are almost inactive with 5-hydroxyferulate (Kuroda *et al.*, 1975). The scarcity of syringyl lignin in gymnosperms could then be attributed at least in part to the very low level of 5-hydroxyferulate-methylating activity (i.e., to the monofunctional nature of the methyltransferases) found in gymnospermous plants. Higuchi *et al.* (1977) have therefore proposed that these methyltransferases are key enzymes in lignin biosynthesis, influencing the syringaldehyde-to-vanillin (S/V) ratios (obtained on alkaline nitrobenzene oxidation) of angiosperm and gymnosperm lignins. In accordance with their proposals, Table II shows the excellent correlation demonstrated among various species for the ratio of sinapic acid to ferulic acid (SA/FA) formed by the O-methyltransferases, the S/V ratio obtained on nitrobenzene oxidation of the lignins and the Mäule reaction (due to syringyl groups). In general, plants possessing higher SA/FA ratios, such as angiosperms, have a greater S/V ratio and give a positive Mäule reaction. On the other hand, gymnosperms and ferns, which show a much lower SA/FA ratio, usually possess far smaller S/V ratios and give a negative Mäule reaction.

These observed substrate specificities and their effect on the S/V ratio might be further accentuated *in vivo* by other factors. Higuchi *et al.* (1977) suggested that the feedback inhibition of bamboo O-methyltransferase by 5-hydroxyferulic acid might operate at the ferulic acid formation stage in favor of the synthesis of syringyl lignin in angiosperms (Shimada *et al.*, 1973b). Similarly, the slow methylation of 5-hydroxyferulate to sinapate by the gymnospermous O-methyltransferase from *P. thunbergii* was competi-

TABLE II

Relationship between the SA/FA ratio, the S/V Ratio, and the Mäule Reaction of Lignins[a]

Plant species	O-Methyltransferase SA/FA ratio	Lignin S/V ratio	Mäule reaction
Pteridophyta			
Psilotum nudum	0.2	0	—
Angiopteris lygodifolia	0.3	—	—
Gymnospermae			
Ginkgo biloba	0.1	0	—
Pinus thunbergii	0.1	0	—
Podocarpus macrophylla	0.0	0	—
Taxus cuspidata	0.3	0	—
Angiospermae (Dicotyledonae)			
Magnolia grandiflora	3.0	2.2	+
Cercidiphyllum japonicum	3.2	2.9	+
Populus nigra	3.0	2.0	+
Pueraria thunbergiana	2.5	—	+
Angiospermae (Monocotyledonae)			
Oryza sativa	0.9	1.0	+
Triticum aestivum	1.0	1.0	+
Phyllostachys pubescens	1.3	1.1	+

[a] Reprinted with permission from Higuchi *et al.*, *Wood Sci. Technol.* **11**, 153–167 (1977). Copyright by Springer-Verlag, New York Inc.

tively inhibited by caffeic acid; this could act *in vivo* to reduce the observed S/V ratio still further (Kuroda *et al.,* 1975).

The pronounced substrate specificities of these methyltransferases clearly could explain the phylogenic differences between gymnospermous and angiospermous lignins. That this may not be the sole factor, however, is indicated by several lines of evidence (Shimada *et al.,* 1973a). First, although angiospermous callus tissues (e.g., *Salix caprea*) exhibited an SA/FA ratio of 1.5–1.9 for their *O*-methyltransferases, their lignins possessed an S/V ratio of merely 0.10–0.12, a value more characteristic of gymnospermous lignins. Second, even within a single hardwood, softwood and hardwood lignin may be located in different morphological regions. Third, variations have been observed in the S/V ratio during the maturation of an angiospermous plant, although the SA/FA ratio of the *O*-methyltransferases remained constant. The activities of "ferulic acid 5-hydroxylase" (still hypothetical), 4-coumarate : coenzyme A (CoA) ligase, and cinnamoyl-CoA oxidoreductase may be other factors that might be expected to affect the observed S/V ratios in lignins. Later stages in lignin biosynthesis, catalyzed by cinnamyl alcohol dehydrogenases and peroxidases, are unlikely to be important factors controlling preferential formation of syringyl lignin in hardwoods (Nakamura *et al.,* 1974). Pertinent information concerning this subject is reviewed elsewhere (see Grisebach, this volume, Chapter 15).

C. Methylation of Phenolic Compounds at the Para Position

The previous section described the *O*-methyltransferases involved in lignin biosynthesis, which generally exhibit absolute specificity toward the meta hydroxyl group of 1-substituted 3,4-dihydric phenols. Other enzyme preparations are known, which methylate phenolic compounds either specifically at the para position or at both meta and para positions. In the latter case, considerable effort is required to determine whether a single protein or a mixture of methyltransferases is responsible for the observed methylation patterns.

A partially purified preparation from pampas grass (*Cortaderia selloana*) shoot tissue readily catalyzed the O-methylation of several vicinal polyphenolic compounds, including dihydroxy and trihydroxy aromatic acids, pyrogallol, esculetin, and eriodictyol (Finkle and Masri, 1964). Methylation occurred predominantly meta to the aromatic side chain, but protocatechuic acid and gallic acid were para-methylated to the extent of 23 and 7%, respectively. Purification steps and heat treatment of this preparation affected the meta and para activities differentially, as evidenced by a reduction in the meta/para methylation ratio. The data were in agreement with the presence of a labile *m-O*-methyltransferase together with a more stable *p-O*-methyltransferase (Finkle and Kelly, 1974). This situation contrasts with that in animal liver catechol O-methyltransferase, which retains a constant meta/para ratio during successive purification steps (Ball *et al.,* 1971).

A more complex picture was reported in leaves of tobacco, *Nicotiana tabacum* var. Samsun NN. Three distinct *o*-diphenol O-methyltransferases (namely, OMT I, II, and III) have been recognized in leaf extracts following partial resolution by DEAE-cellulose chromatography and by differing specificity toward 16 diphenolic substrates (Legrand *et al.,* 1978). Most substrates, including several cinnamic acids and flavonoids, were methylated almost exclusively in the meta position. In contrast, both meta- and para-methylation of protocatechuic acid, protocatechuic aldehyde, and esculetin were observed. The para-methylation of protocatechuic acid was catalyzed by only one enzyme (OMT I), whereas all three methyltransferases were responsible for para-methylation of esculetin and protocatechuic aldehyde. Purification of each protein to homogeneity and subsequent analysis of their kinetic properties would confirm whether these enzymes indeed catalyze methylation at both positions. In addition, it might allow the assignment to each methyltransferase of a metabolic role in the phenolic metabolism of tobacco leaves.

A partially purified O-methyltransferase from tobacco cell cultures catalyzed both meta- and para-methylation of caffeic acid, 5-hydroxyferulic acid, and esculetin with meta/para ratios of 9.4, 1.9, and 1.3, respectively. Further purification of this enzyme resolved two forms: One form catalyzed the O-methylation of caffeic acid almost exclusively at the meta position, and the other, quercetin at the 7-position, though each form exhibited some activity against other substrates (Tsang and Ibrahim, 1979).

An excellent system with which to investigate both para-O-methylation and methyl ester formation of hydroxycinnamic acids is afforded by the fungus *Lentinus lepideus,* which produces crystalline methyl *p*-methoxycinnamate in stationary cultures. In addition, it contains the methyl esters of cinnamic, *p*-coumaric, isoferulic, and anisic acids. Watt and Towers (1975) demonstrated the efficient para-methylation of methyl *p*-coumarate to methyl *p*-methoxycinnamate with an enzyme extract from this fungus. The highest catalytic rates were observed at pH 7 and were not stimulated by Mg^{2+} ions. The enzyme exhibited strict para specificity, failing to methylate methyl isoferulate but rapidly converting methyl caffeate and methyl ferulate to their para-O-methylated derivatives. Among the free acids tested, only cinnamic acid was methylated, albeit slowly. Since benzoic acids and their methyl esters were not utilized by this enzyme, an additional methyltransferase presumably exists in *Lentinus* showing specificity toward these substrates. The fungal enzyme preparation also contained a transmethylating enzyme for alkylation of the carboxyl group of cinnamic acid using SAM as methyl donor, thereby yielding methyl cinnamate.

Para-O-methylation of phenolic substrates has also been demonstrated with extracts from *Nerine bowdenii* (Section I,H,9), peyote (Section I,H,3), chick-pea (Section I,G), *Foeniculum vulgare* (Kaneko, 1962), and yeasts (Müller-Enoch *et al.*, 1976).

D. O-Methylation of Coumarins

Coumarins are a group of lactones derived from benzopyrone. More than 500 naturally occurring coumarins are distributed among hundreds of plant species. Some of the most common simple coumarins include coumarin itself, umbelliferone, herniarin, and scopoletin (Fig. 3). Biosynthetic studies have confirmed that simple coumarins, which possess a phenylpropanoid skeleton, are derived from shikimate and L-phenylalanine via pathways described by Brown (this volume, Chapter 10).

Scopolin (scopoletin) biosynthesis in tobacco (*N. tabacum*) has received special attention. Scopoletin is believed to arise from cinnamic acid via the intermediates *p*-coumaric and caffeic acids (Fig. 4). Methylation of the latter at the meta position yields ferulic acid which, following ortho-hydroxylation and lactonization, gives rise to scopoletin. Scopolin would be produced by glucosylation, presumably involving uridine diphosphoglucose. These data are therefore in agreement with the general belief that the full substitution pattern of the benzene ring of simple coumarins is established before lactonization, at least in the compounds thus far investigated. The failure of cell-free extracts from tobacco cell cultures to methylate esculetin to scopoletin was cited as additional evidence that methylation preceded lactonization, especially since such extracts efficiently catalyzed the production of ferulic and feruloylquinic acids from caffeic and chlorogenic acids, respectively (Fritig *et al.*, 1970).

Subsequent work has indicated that esculetin may indeed be a substrate for *O*-methyltransferases from tobacco tissue. Three distinct *O*-methyltransferases have been identified in tobacco leaf extracts, which methylate

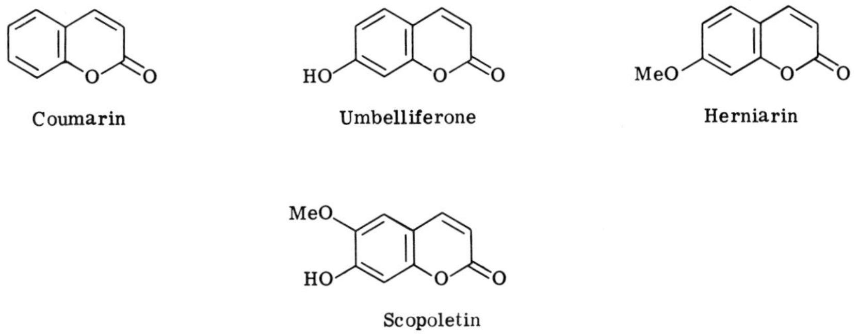

Coumarin   Umbelliferone   Herniarin

Scopoletin

Fig. 3.   Structures of naturally occurring simple coumarins.

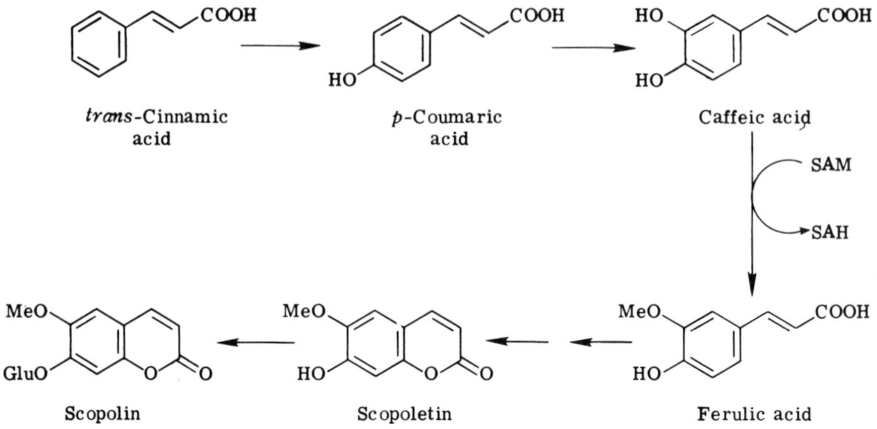

Fig. 4. The proposed biosynthetic pathway of scopolin from *trans*-cinnamic acid.

esculetin at both the meta and para positions *in vitro* (Legrand *et al.*, 1978). Similarly, two forms of SAM: *o*-dihydric phenol *O*-methyltransferase have been purified recently from tobacco cell suspension cultures (Tsang and Ibrahim, 1979). One form exhibited the highest activity with the lignin precursors caffeic and 5-hydroxyferulic acids, though it also catalyzed the methylation of esculetin and daphnetin. Tsang and Ibrahim suggest that *O*-methylation in tobacco cell cultures could still follow lactone ring formation in scopolin biosynthesis. Curiously, however, whereas daphnetin gave exclusively 7-hydroxy-8-methoxycoumarin, methylation of all other substrates including esculetin occurred at both the meta and para positions. Such *in vitro* position specificity is surprising, since isoscopoletin and isoscopolin are not reported constituents of tobacco cell cultures. Further investigation is necessary to clarify when the substitution pattern of methylated simple coumarins is determined.

It should be noted that methylation of simple coumarins has been reported also with crude extracts from pampas grass (Finkle and Masri, 1964), petunia (Hess, 1965), and yeasts (Müller-Enoch *et al.*, 1976).

Major advances have recently been made in our knowledge of the biosynthesis of linear furanocoumarins (for details, see Brown, this volume, Chapter 10). In several species, formation of the furan ring of furanocoumarins has been shown to precede methoxyl group formation. The unsubstituted linear furanocoumarin psoralen may readily be converted to methoxy derivatives via the naturally occurring intermediates bergaptol and xanthotoxol. Methylation of these intermediates has been elegantly studied by Brown and co-workers using cell cultures of *Ruta graveolens* L. Cell extracts contain a soluble *O*-methyltransferase system which methylates the 5- and 8-hydroxyl groups of linear furanocoumarins using SAM as methyl donor (Thompson *et al.*, 1978). Many approaches, including mixed-substrate experiments and MW determination, pointed to the existence of two discrete enzymes acting at these different positions. Confirmation was obtained by their clear resolution using adsorption on the general-affinity ligand 5-(3-carboxypropanamido)xanthotoxin followed by specific desorption by either bergaptol or xanthotoxol (Sharma *et al.*, 1979). Both proteins were electrophoretically homogeneous. The SAM : 8-hydroxyfuranocoumarin *O*-methyltransferase specifically methylated the 8-hydroxyl (meta) groups of xanthotoxol and 8-hydroxybergapten, yielding xanthotoxin and isopimpinellin, respectively (Fig. 5). The other enzyme, a SAM : 5-hydroxyfuranocoumarin *O*-methyltransferase, specifically methylated the 5-hydroxyl groups of bergaptol and 5-hydroxyxanthotoxin and therefore represents a novel type of methyltransferase acting ortho to the side chain.

It is clear that dimethoxypsoralen (isopimpinellin) can be derived from psoralen via several routes. Two routes involving the hydroxylation of either bergapten or xanthotoxin para to the methoxyl group were detected by *in vivo* tracer experiments with leaves of *Heracleum lanatum* and cell cultures of *R. graveolens*. The major synthetic pathway *in vivo* is presumably via xanthotoxin, since it is a better precursor than bergapten (Brown and Sampathkumar, 1977).

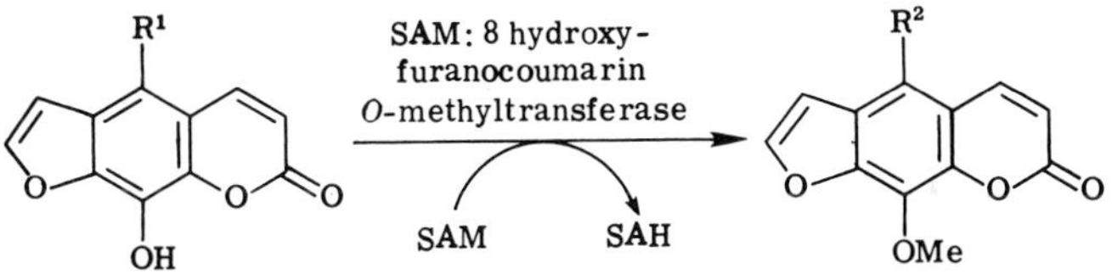

I. R¹ = H
II. R¹ = OMe

III. R² = H
IV. R² = OMe

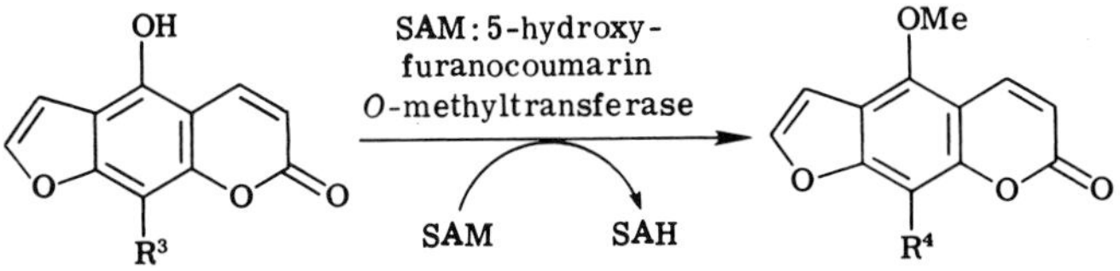

V. R³ = H
VI. R³ = OMe

VII. R⁴ = H
VIII. R⁴ = OMe

Fig. 5. Methylation of xanthotoxol (I) and 8-hydroxybergapten (II) by SAM : 8-hydroxyfuranocoumarin *O*-methyltransferase yields xanthotoxin (III) and isopimpinellin (IV), respectively. Bergapten (VII) and isopimpinellin (VIII) are formed by SAM : 5-hydroxyfuranocoumarin *O*-methyltransferase from bergaptol (V) and 5-hydroxyxanthotoxin (VI), respectively.

E. O-Methylation of the B Ring of Flavonoid Compounds

The general structure of flavonoid compounds consists of two benzene rings (namely, A and B) linked by a three-carbon unit formed into a γ-pyrone ring (Fig. 6). Various classes of flavonoids, which include flavanones, flavones, flavonols, and anthocyanins, may be differentiated by the state of oxidation of the heterocyclic (or C) ring. Within such classes, compounds are distinguished from each other primarily by the nature and degree of substitution of the A and B rings. As described by Hahlbrock (this volume, Chapter 14), the C_6–C_3 skeleton of the B and C rings is derived from certain activated cinnamic acids via the action of flavanone synthase. The substitution patterns of the B ring generally resemble those of common cinnamic acids and coumarins; C-4' usually possesses a hydroxyl group and is rarely methylated, whereas C-3' and C-5' may bear either hydroxyl or methoxyl groups. The A ring, formed by the head-to-tail condensation of three "acetate units," commonly bears hydroxyl groups at both C-5 and C-7. Alternative hydroxylation patterns are represented in compounds lacking the 5-hydroxyl group or in which extra hydroxyl groups are present at the 6- and 8-positions. Current interest is being focused also on the biosynthesis of flavonoids possessing methoxyl groups at various positions (C-5, -6, -7, and -8) of the A ring and at C-3 of the heterocyclic ring.

A central problem in the biochemistry of flavonoids is to determine at which stage B-ring substitution occurs. Two alternatives have been proposed. In 1964, Hess advanced his "cinnamic acid starter hypothesis" in which it was recognized that the start of anthocyanin synthesis from cinnamic acids was genetically controlled (Hess, 1968). For every cinnamic acid, an enzyme system should be present that uses this particular substrate to initiate synthesis of the correspondingly substituted anthocyanin. In an alternative proposal, the "substitution hypothesis," it is thought that the B ring is brought from a lower to a higher substitution level by a succession of gene-controlled substitution steps. These modifications were originally thought to involve the anthocyanins themselves, but, more recently, the substitutions have been imagined to take place in the C_{15} or cinnamic acid precursors of the anthocyanins.

Evidence favoring Hess' hypothesis included the relatively high incorpo-

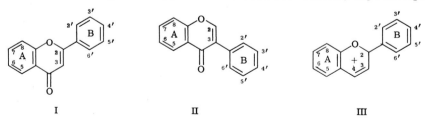

Fig. 6. Skeletal structures of flavones (I), isoflavones (II), and anthocyanidins (III).

ration of 3,4,5-trihydroxycinnamate into delphinidin in *Campanula medium* (Meier and Zenk, 1965) and the stimulation of anthocyanin synthesis in *Petunia hybrida* petals by the correspondingly substituted cinnamic acids (Hess, 1967). Furthermore, Hess showed with *P. hybrida* buds that methyl-labeled ferulate and sinapate were best incorporated into anthocyanins having identical or closely related B-ring substitution patterns. However, the value of these data must be questioned, since only 12–65% of the anthocyanin radioactivity was located in the methyl group, indicating that extensive precursor demethylation had occurred (Hess, 1964). Enzyme extracts from different lines of *P. hybrida* showed the greatest *O*-methyltransferase activity toward cinnamic acids among the *o*-diphenolic substrates tested. Caffeic and 5-hydroxyferulic acids were methylated up to approximately 100 times faster than anthocyanins (Hess, 1965, 1966). The insignificant rates obtained with anthocyanins after 6-h incubation periods were interpreted in favor of Hess' hypothesis. It remains uncertain whether methylation may nevertheless occur at the anthocyanin level, and this subject deserves further investigation, especially with cell-free extracts. The unglucosylated anthocyanidins may well be preferred substrates for these transmethylations, in view of data obtained with soybean flavonoid *O*-methyltransferase for flavonol 3-*O*-glycosides (Poulton *et al.*, 1977). Shorter incubation periods are recommended because of the lability of both substrates and products. Alternatively, the possibility should not be discounted that the substitution pattern of anthocyanins is determined at an earlier C_{15} stage.

Considerable evidence is indeed now available supporting the substitution hypothesis. Further hydroxylation of the flavonoid C_{15} skeleton was demonstrated by isotopic tracer techniques and enzyme studies (Hahlbrock and Grisebach, 1975). Moreover, in contrast to the slow methylation of anthocyanins by *Petunia* extracts, extensive methylation of other flavonoid substrates by rather specific enzymes is now recognized (Table III).

Irradiation of parsley (*Petroselinum hortense*) cell suspension cultures with uv light leads to the rapid production of flavone and flavonol glycosides, principally of apiin and graveobioside B (3'-methoxyapiin), following the coordinated induction of several enzymes involved in flavonoid biosynthesis (see Hahlbrock, this volume, Chapter 14).

One of these, a SAM : *o*-dihydric phenol *O*-methyltransferase, was purified 82-fold from this source (Ebel *et al.*, 1972). The enzyme was specific for *o*-dihydric phenols, showing no activity toward monophenols. It catalyzed the methylation of various 3',4'-dihydroxyflavonoids specifically at the meta position, correlating well with the presence of glycosides of chrysoeriol and isorhamnetin in these cells. Luteolin and its 7-*O*-glucoside served as the best substrates, having the lowest K_m and highest V_{max}/K_m values (Table IV). The flavanone eriodictyol was only weakly methylated, suggesting a necessity for

TABLE III

Properties of *O*-Methyltransferases Catalyzing Methylation of Flavonoid and Isoflavonoid Substrates in the B Ring[a]

Organism	Degree of purification	Molecular weight	pH optimum	Substrate specificity	Mg^{2+} ion requirement	Effect of EDTA	Inhibition by SAH
Petroselinum hortense cell cultures[b]	82-fold	48,000	9.7	Methylation exclusively in meta position of *o*-dihydric phenols including flavonoids and caffeic and protocatechuic acids. See Table IV for details.	Required for full activity; 10–20% of maximum activity observed in the absence of Mg^{2+}	Complete inhibition	n.d.
Glycine max cell cultures[c]	38-fold	n.d.	8.6–8.9	Methylation exclusively in meta position of *o*-dihydric phenols including flavonoids and caffeic and 5-hydroxyferulic acids. See Table IV for details	Required for full activity; 4–6% of maximum activity observed in the absence of Mg^{2+}	Complete inhibition by 1 mM	Yes
Tulipa anthers[d]	n.d.	35,000	7.6	Meta-methylation of flavonoids quercetin and luteolin	Required for full activity; 36% of maximum activity observed in the absence of Mg^{2+}		Yes
Cicer arietinum cell cultures[e]	360-fold	111,000	8.6–9.6	Para-methylation of isoflavones daidzein and genistein	None	No inhibition by 1 mM	Yes

[a] n.d., Not determined.
[b] Ebel *et al.*, 1972.
[c] Poulton *et al.*, 1977.
[d] Sütfeld and Wiermann, 1978.
[e] Wengenmayer *et al.*, 1974.

TABLE IV

Substrate Specificities of Parsley and Soybean Flavonoid O-Methyltransferase[a]

Substrate	Parsley K_m (μM)	Relative V_{max}	FMT[b] Relative V_{max}/K_m	Soybean K_m (μM)	V_{max} (μkat/kg)	FMT[c] $10^{-6} \times (V_{max}/K_m)$ (μkat/kg/mol)
Luteolin	46	88	1.9	16, 16.7*	33.7, 46.5*	2.09, 2.78*
Quercetin	n.d.	n.d.[d]	n.d.	35	70.9	2.02
Luteolin 7-O-glucoside	31	100	3.2	28	13.4	0.47
Eriodictyol	1200	15	0.012	75	11.7	0.16
Dihydroquercetin	n.d.	n.d.	n.d.	435	32.7	0.08
Texasin	n.d.	n.d.[d]	n.d.	35	10	0.29
Caffeic acid	1600	148	0.092	770*	30.9*	0.04*
5-Hydroxy ferulic acid	n.d.	n.d.	n.d.	227*	42.1*	0.19*

[a] n.d., Not determined.
[b] Data from Ebel *et al.*, 1972. Rates measured at pH 9.2.
[c] Data from Poulton *et al.*, 1977. Rates measured at pH 7.8, except for those values indicated by asterisk at pH 8.7.
[d] Substrate undergoes O-methylation, but values are unavailable.

the quasi-planar nature of the heterocyclic ring in the flavonoid substrate. Similarly, caffeic acid was only poorly methylated, as shown by its high K_m and low V_{max}/K_m values. The enzyme differed from lignin-specific methyltransferases in several respects (Table III). Magnesium ions were required for the reaction, which proceeded maximally at pH 9.7. The enzyme was completely inhibited by EDTA, whereas p-chloromercuribenzoate and iodoacetamide had no effect. Several lines of evidence strongly suggest that parsley flavonoid O-methyltransferase (FMT) is directly related to flavone glycoside biosynthesis. First, it is induced under conditions that specifically induce all other enzymes in the flavonoid biosynthetic pathway (Hahlbrock, this volume, Chapter 14). Second, it exhibits a pronounced preference for flavonoid substrates (Table IV) as compared with phenylpropanoids and phenolic acids. Moreover, the purified flavanone synthase from parsley cell cultures accepts as substrate only p-coumaroyl-CoA and caffeoyl-CoA, and not feruloyl-CoA (Saleh *et al.*, 1978). No evidence for isoenzymes in parsley has been reported. It seems certain that the 3'-methoxy-4'-hydroxy-substituted flavonoids in parsley cell cultures must therefore arise by substitution of the B ring at the C_{15} stage. The foregoing discussion indicates that the enzymatic properties of parsley FMT are well suited to fulfill this metabolic role.

An analogous FMT was partially purified from soybean (*G. max* L.) cell suspension cultures and was clearly distinguishable from the CMT present in these cells (Poulton *et al.,* 1976b, 1977). It closely resembled parsley FMT in several respects. *o*-Dihydric phenols were methylated specifically at the meta position with SAM as methyl donor. The enzyme, which had a pH optimum of 8.6–8.9, required Mg^{2+} ions for maximum activity and could be totally inhibited by EDTA. A pronounced preference was shown for flavonoid substrates, as demonstrated by their high V_{max}/K_m values; in contrast, substituted cinnamic acids had much lower values (Table IV). Flavones and flavonols were more efficient substrates than flavonones and flavanonols, again indicating a requirement for the quasi-planar structure of the heterocyclic ring. This also suggested that the *in vivo* methylation of flavonoid compounds occurs at the flavone/flavonol level of oxidation. The question whether glucosylation may precede methylation in the biosynthesis of methylated flavonoid glycosides was reinvestigated. Both luteolin and its 7-*O*-glucoside were good substrates for the soybean enzyme but, in contrast, quercetin 3-*O*-glucoside and rutin were poorly methylated relative to quercetin itself. Thus it appears that, at least in the biosynthesis of methylated flavonol 3-*O*-glucosides, methylation must precede glucosylation.

The biosynthesis of flavonoids in soybean cultures is currently under investigation. The flavone apigenin has been detected in these cells, as have also various flavonoid-specific enzymes (see Hahlbrock, this volume, Chapter 14). Unfortunately, the substrate specificity of soybean flavanone synthase remains unknown. However, it is concluded, in view of the low K_n values for flavonoid substrates and from analogy with already characterized parsley FMT, that soybean FMT is involved in the methylation of flavonoid compounds *in vivo*.

Secondary product biosynthesis is usually restricted to specific developmental stages either of the organism or of specialized cells (Luckner *et al.,* 1977). An excellent example of this phenomenon is the production of secondary compounds during development of *Tulipa* 'Apeldoorn' anthers. During microsporogenesis, these organs successively accumulate simple cinnamic acid derivatives, chalcones, flavonols, and finally anthocyanins (Wiermann, 1970). The anthers provide an excellent system for studying methylation processes, since both methylated phenylpropanoids (ferulic acid) and flavonoids (isorhamnetin) are involved. Although only partial resolution and purification were achieved, two distinct SAM:3,4-dihydric phenol 3-*O*-methyltransferases were recognized in cell-free extracts from this source, whose properties correlated very closely with those of CMT and FMT from soybean cultures (Sütfeld and Wiermann, 1978). Localization studies demonstrated the presence of these enzymes in the tapetum fraction of the anthers (Herdt *et al.,* 1978). Caffeic and 5-hydroxyferulic acids were substrates for one enzyme (CMT), whereas the other methyltransferase (FMT) methylated quercetin and luteolin but not rutin or eriodictyol. Both enzymes exhibited optimum activity at pH 7.6 and were inhibited by SAH and *p*-chloromercuribenzoate. Sulfhydryl compounds stimulated reaction rates, especially for the flavonoid-specific enzyme. In conclusion, the presence of two specific methyltransferases points to the possibility that methylation could occur at both the C_9 and C_{15} stages. Interestingly, the flavanone synthase from tulip anthers possesses a wider substrate specificity *in vitro* than the parsley enzyme, accepting feruloyl-CoA as well as *p*-coumaroyl-CoA and caffeoyl-CoA with comparable efficiency (Sütfeld *et al.*, 1978). This specificity parallels that of *Tulipa* hydroxycinnamate : CoA ligase and also agrees with the substitution pattern of the chalcones accumulated in the natural system. Thus enzymatic equipment exists for the transfer of three different cinnamic acids into the correspondingly substituted flavonoids. These data appear to support determination of the flavonoid substitution pattern at the cinnamic acid stage as postulated in the cinnamic acid starter hypothesis.

General conclusions concerning the timing of substitution of the flavonoid B ring cannot be drawn at the present time. It is conceivable that both the substitution and cinnamic acid starter mechanisms are operative, depending on the nature of the plant.

F. O-Methylation of the Flavonoid A Ring

The frequent occurrence in nature of flavonoids possessing methoxyl-substituted A rings provides direct evidence for the methylation of A-ring hydroxyl groups *in vivo*. However, *in vitro* methylation of the A ring is not well documented. Texasin (6,7-dihydroxy-4'-methoxyisoflavone) was efficiently methylated by FMT from soybean cell cultures, but the exact position of methylation remains unknown (Poulton *et al.*, 1977). Interesting in this connection is the identification of 6,7,4'-trihydroxyisoflavone 4'-glucoside and the 7-O-glucoside of 7,4'-dihydroxy-6-methoxyisoflavone in soybean. Texasin methylation was also observed with parsley FMT but not with soybean CMT.

Insight has recently been gained into the enzymes and biosynthetic pathways involved in the formation of polymethylated flavones in citrus plants and tissue cultures (Brunet *et al.*, 1978). Cell-free extracts of peel, root, and callus tissues of citrus catalyzed the methylation of quercetin to rhamnetin (3,5,3',4'-tetrahydroxy-7-methoxyflavone) and isorhamnetin (3,5,7,4'-tetrahydroxy-3'-methoxyflavone). Both products were further methylated to rhamnazin (3,5,4'-trihydroxy-7,3'-dimethoxyflavone) and possibly even to a trimethyl ether derivative. It was concluded that citrus tissue contained at least two distinct O-methyltransferases; one catalyzes the meta-methylation of the B ring, and the second introduces a methyl group at the 7-position of the A ring. Further work is in progress to characterize these enzymes more fully. Methylation of quercetin and luteolin at both the 7- and 3'-positions by a partially purified O-methyltransferase from tobacco (*N. tabacum* L.) cell cultures was reported recently (Tsang and Ibrahim, 1979). Further purification demonstrated the existence of two forms, one of which catalyzed the methylation of quercetin almost exclusively at the 7-position. The common occurrence in plants of 7-O-methoxy derivatives of these flavonoids would be consistent with the physiological role of this enzyme. In contrast to the situation in tobacco cell cultures, the three o-diphenol O-methyltransferases from leaves of *N. tabacum* var. Samsun NN methylated quercetin specifically at the 3'-position (Legrand *et al.*, 1978).

Current research is aimed toward determining whether A-ring methylation is undertaken by distinct O-methyltransferases possessing absolute specificity toward A-ring hydroxyl groups.

G. O-Methylation of Isoflavonoids

Isoflavonoids differ from other flavonoid compounds in having a branched C_6–C_3–C_6 skeleton as a basic structural feature (Fig. 6). They include isoflavones, isoflavanones, rotenoids, pterocarpans, and coumestans. Their biosyntheses must involve a rearrangement step in which the aromatic B ring migrates from C-2 to C-3 of the heterocyclic ring. Isoflavones and isoflavanones may possess methoxyl groups at C-5, -6, -7, and -8 of the A ring and at C-2', -3', -4', and -5' of the B ring. Whereas flavonol- and flavone-specific O-methyltransferases have been thoroughly investigated, isoflavonoid methylation is less well understood. Texasin (6,7-dihydroxy-4'-methoxyisoflavone) was methylated in the A ring by flavonoid-specific O-methyltransferases from soybean and parsley cell cultures (Poulton *et al.*, 1977). As in the case of flavonoids (Section I,E), an intriguing problem exists regarding the stage of isoflavonoid biosynthesis at which the methoxylation pattern of the B ring is determined. This could occur (1) at the phenylpropanoid stage, (2) at the flavonoid stage, prior to aryl migration, or (3) at the isoflavonoid stage, after aryl migration has occurred. This question has been addressed for compounds bearing a 4'-methoxyl group, one of the more common substitution patterns. The isoflavones formononetin (7-hydroxy-4'-methoxyisoflavone) and biochanin A (5,7-dihydroxy-4'-methoxyisoflavone) occur in chick-pea, *Cicer arietinum* L. (Fig. 7). Tracer experiments were carried out with chick-pea seedlings using p-[^{14}C,^{3}H]-methoxy-[β-^{14}C]cinnamic acid, p-[Me-^{2}H$_3$]methoxycinnamic acid, and 4-[Me-^{3}H]methoxy-2',4'-dihydroxy-[β-^{14}C]chalcone as potential iso-

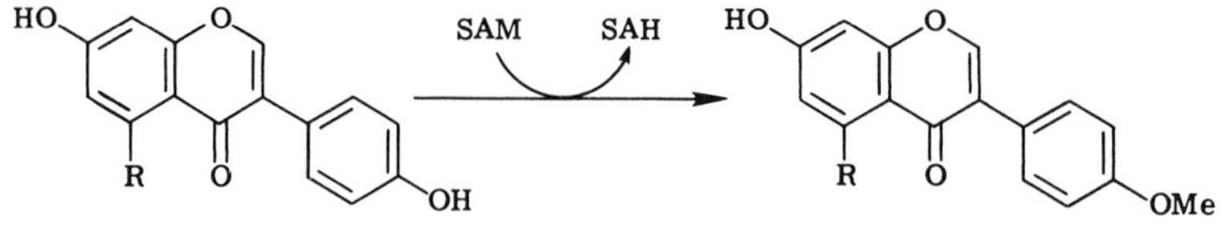

Fig. 7. Enzymatic formation of formononetin from daidzein (R = H) and of biochanin A from genistein (R = OH) by the para-specific SAM : isoflavone O-methyltransferase from *C. arietinum* seedlings and cell cultures.

flavone precursors (Ebel *et al.*, 1970a; Barz and Grisebach, 1967). However, conclusive interpretation of the results proved impossible because of the rapid demethylation reactions. An enzymatic approach to this problem resulted in the 360-fold purification of SAM:isoflavone 4'-O-methyltransferase from *Cicer* cell cultures (Wengenmayer *et al.*, 1974). This enzyme, which was also detected in *Cicer* seedlings, catalyzed the 4'-O-methylation of daidzein (7,4'-dihydroxyisoflavone) and genistein (5,7,4'-trihydroxyisoflavone) to formononetin and biochanin A, respectively, at an optimum pH of 8.6–8.9 (Fig. 7). Divalent cations, such as Mg^{2+} ions, were not required for the reaction, which was not significantly affected by EDTA (Table IV). Sulfhydryl reagents were necessary for catalytic activity. Importantly, many phenolic compounds, including several potential intermediates in the isoflavone biosynthetic pathway (e.g., *p*-hydroxycinnamic acid, apigenin), were not methylated. Therefore the high specificity toward 4'-hydroxyisoflavones suggests that methylation is the last step in the biosynthesis of these compounds. This assumption is supported by studies on mutants of *Trifolium subterraneum* L.; in one mutant, inhibition of biochanin A and formononetin formation with concurrent accumulation of genistein and daidzein was observed (Wong and Francis, 1968). Furthermore, methylation at the isoflavone level is in agreement with the hypothesis that a free hydroxyl group is required at the 4'-position for the 1,2-aryl migration occurring during isoflavone biosynthesis (Pelter *et al.*, 1971). Detailed kinetic analysis of chick-pea methyltransferase provided information as to the reaction mechanism (Wengenmayer *et al.*, 1974). *S*-Adenosyl-L-homocysteine was a competitive inhibitor ($K_i = 30 \mu M$) of methylation vs SAM ($K_m = 160 \mu M$) but exhibited noncompetitive inhibition ($K_i = 300 \mu M$) vs daidzein. The kinetic data were consistent with an ordered Bi-Bi reaction mechanism with SAM and SAH as leading reaction partners.

H. Transmethylation Reactions Involved in Alkaloid Biosynthesis

Alkaloids constitute an extremely heterogeneous class of secondary plant products, whose more than 6000 members are distributed among almost 4000 plant species. All contain nitrogen, frequently in a heterocyclic ring, and the majority are basic. These compounds have been the focus of extensive research, since many exhibit important pharmacological properties. A vast wealth of information accumulates annually, describing the biosynthesis of these secondary products from α-amino acids, acetate, and steroid or terpene precursors. Gross precursor–product relationships have now been traced for most of the structural groups by isotope incorporation studies on intact systems. Only recently, however, have enzymologists begun to take up this interesting problem, thereby confirming or eliminating proposed biosynthetic pathways.

As in many other natural products, N- and O-methyl groups are of common occurrence in alkaloids. In numerous cases, it has been demonstrated that these groups are derived from L-methionine (Table V). The transfer of an intact doubly labeled methyl group from methionine to nicotine (Dewey *et al.*, 1954) and several ergot alkaloids (Baxter *et al.*, 1964) has also been established. S-Adenosyl-L-methionine is presumed to be the immediate methyl donor in these transmethylation reactions. This concept has been strengthened by the recent identification of many SAM-dependent transmethylases specific for alkaloids; the general properties of these enzymes are presented in Table VI. Consideration is given here only to transmethylation reactions in alkaloid biosynthesis where the enzymology involved has been investigated. Readers are referred to Waller and Dermer (this volume, Chapter 12) for a complete description of the enzymatic pathways of alkaloid biosynthesis and catabolism.

TABLE V

Examples of Alkaloids Whose Methyl, Methylenedioxy, or Other Groups Are Derived from the Methyl Group of L-Methionine

Alkaloid	Nature of group[a]	Reference[b]
Hordenine	N-Me	Matchett *et al.*, 1953
Gramine	N-Me	Mudd, 1960
Stachydrine	N-Me	Robertson and Marion, 1959
Hyoscyamine	N-Me	Marion and Thomas, 1955
Ephedrine	N-Me	Shibata *et al.*, 1957
Nicotine	N-Me	Brown and Byerrum, 1952
Ricinine	N-Me and O-Me	Dubeck and Kirkwood, 1952
Mescaline	O-Me	Agurell *et al.*, 1967
N-Methylconiine	N-Me	Roberts, 1974a
Lycorine	Methylenedioxy	Archer *et al.*, 1963
Colchicine	O-Me	Leete and Nemeth, 1961
Codeine	N-Me and O-Me	Battersby and Harper, 1958
Thebaine	N-Me and O-Me	Battersby and Harper, 1958
Morphine	N-Me	Battersby and Harper, 1958
Pellotine	N-Me and O-Me	Battersby *et al.*, 1967
Protopine	N-Me, methylenedioxy	Sribney and Kirkwood, 1953
Berberine	O-Me	Gupta and Spenser, 1965
	Methylenedioxy	
	Berberine bridge	
Hydrastine	N-Me and O-Me, methylenedioxy	Gupta and Spenser, 1965
	Lactone-carbonyl carbon	

[a] N-Me, N-methyl group; O-Me, O-methyl group.

[b] References cited in Mothes and Schütte (1969, pp. 123–167), with the exception of Mudd (1960), Roberts (1974a), and Sribney and Kirkwood (1953), which are given in reference list for this chapter.

TABLE VI

Properties of Methyltransferases catalyzing O-, N-, and Carboxyl Methylations of alkaloids[a]

Organism	Degree of purification	Type of methylation	pH Optimum	Substrate specificity	Inhibitors	Remarks	References
Hordeum vulgare L. roots	13-fold	N	Approximately 8.4	Tyramine, phenethylamine, α-methyltyramine, and N-methyltyramine (poor)	SAH, PCMB	No inhibition by EDTA. Enzyme stabilized by β-mercaptoethanol	Mann and Mudd, 1963
Phalaris tuberosa L. shoots	80-fold	N	8.5	PIM: tryptamine, serotonin, and 5-methoxytryptamine. SIM: N-methyl derivatives of tryptamine, serotonin, and 5-methoxytryptamine	SAH	Resolution of PIM and SIM achieved by affinity chromatography	Mack and Slaytor, 1978
Pisum sativum L. seedlings	Crude extract	N	6–7	Nicotinic acid	—	No rate stimulation by glutathione	Joshi and Handler, 1960
Nicotiana tabacum L. roots	30-fold	N	8–9	Putrescine and N-methylputrescine	PCMB, Ag+ ions	Enzyme slightly activated by cysteine and β-mercaptoethanol. β-Mercaptoethanol stabilized enzyme on storage	Mizusaki et al., 1971

(continued)

TABLE VI (Continued)

Organism	Degree of purifi-cation	Type of methyl-ation	pH Op-timum	Substrate specificity	Inhibitors	Remarks	References
Conium maculatum L. unripe fruits	3-fold	N	8.2	Coniine	PCMB, NMM, and iodoacetamide	Sodium metabisul-fite, Polyclar AT, and DTT required during extraction. DTT included dur-ing purification and assay. Activity not affected by Mg^{2+} or EDTA	Roberts, 1974b
Thermopsis chinensis and *T. fabacea* seedlings	n.d.	N	8.5	Cytisine	*N*-ethyl-maleimide, Mg^{2+} ions	Sodium metabisul-fite, β-mercapto-ethanol, and Polyclar AT used in extraction procedure	Murakoshi *et al.*, 1977
Camellia sinensis L. leaves	n.d.	N	8.4	7-Methylxanthine, 3,7-dimethylxan-thine, and 1,7-methylxanthine	No significant inhibition by IAA and PCMB	Essential to use Polyclar AT and β-mercaptoethanol during extraction	Suzuki and Takahashi, 1975

Plant source	Purification	Position	Substrates	Inhibitors	Comments	Reference
Coffea arabica L. unripe fruits	n.d.	N	7-Methylxanthine, 3,7-dimethylxanthine, and 1,7-dimethylxanthine	PCMB, IAA, and NMM	Extraction procedures required Polyclar AT, β-mercaptoethanol, ascorbate, and rigorous exclusion of oxygen	Roberts and Waller, 1979
Lophophora williamsii (peyote)	220-fold	O	Several mono- and diphenolic phenethylamines and tetrahydroisoquinolines. Meta- and/or para-methylation, depending on nature of substrate	PCMB, Ag^+ and Cu^{2+} ions	DTT and β-mercaptoethanol stimulated activity. No effect of Mg^{2+}, EDTA, or thiourea	Basmadjian and Paul, 1971; Basmadjian *et al.*, 1978
Nerine bowdenii bulbs	5-fold	O	Norbelladine, dopamine, and 2,3-napthalenediol	8-hydroxyquinoline, β-mercaptoethanol	No stimulation by Mg^{2+}. Complete inhibition by 1 mM EDTA	Mann *et al.*, 1963
Vinca rosea seedlings	5-fold	Carboxyl methylation	Loganic and secologanic acids	PCMB, NMM, and iodoacetamide	Extracted in presence of PVP, sodium metabisulfite, and DTT. DTT enhanced activity	Madyastha *et al.*, 1973

[a] SAH, S-adenosyl-L-homocysteine: PCMB, p-chloromercuribenzoate; PIM, primary indolethylamine N-methyltransferase; SiM, secondary indolethylamine N-methyltransferase; DTT, dithiothreitol; NMM, N-methylmaleimide; IAA, iodoacetate.

1. Biosynthesis of N-Methyltyramine and Hordenine

N-Methyltyramine and hordenine occur in the roots but not in the shoots of germinating barley (*Hordeum vulgare* L.). Tracer studies indicate that they originate from L-tyrosine via tyramine as intermediate (Fig. 8). Mann and Mudd (1963) partially purified an N-methyltransferase (E.C. 2.1.1.27; trivial name, tyramine methylpherase) from barley roots, which catalyzed the methylation of tyramine at an optimum pH of 8.4, yielding N-methyltyramine and SAH. No cofactor requirement was observed. This study represents one of the few instances where the substrate specificity of the enzyme toward the methyl donor was rigorously tested. Whereas SAM ($K_m = 30 \ \mu M$) was effective as methyl donor, methionine, methionine sulfoxide, and S-methylmethionine were not utilized. Two steric configurations are possible at the sulfur atom of SAM. Methionine adenosyltransferases from rabbit liver, yeast, and barley roots synthesize only the (−)-diastereoisomer of SAM. Tyramine methylpherase showed absolute specificity toward this diastereoisomer, a property shared by other methyltransferases tested (Mudd, 1973). As observed with catechol O-methyltransferases, the enzyme was severely inhibited by SAH at extremely low concentrations ($K_i = 5 \ \mu M$). The partially purified enzyme was not completely specific for

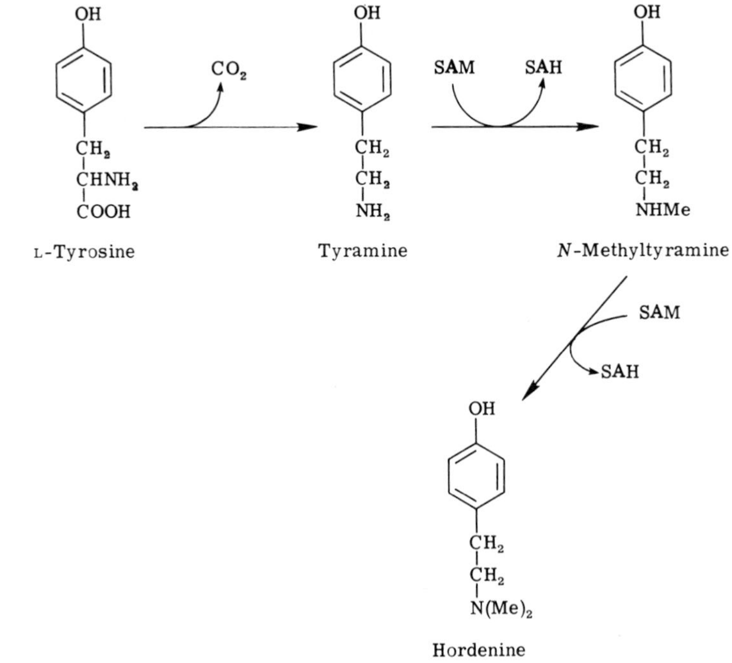

Fig. 8. The biosynthesis of hordenine from L-tyrosine in *H. vulgare* L.

tyramine, showing activity toward α-methyltyramine, phenylethylamine, and phenylethylamine analogues. N-Methyltyramine was itself poorly methylated but acted as a good competitive inhibitor ($K_i = 30$–60 μM) of tyramine methylation.

The subsequent methylation of N-methyltyramine to hordenine may also be catalyzed by barley extracts. The question remains open whether this reaction is catalyzed by tyramine methylpherase or by another distinct enzyme. Although the two activities have not been resolved by DEAE-cellulose chromatography, other lines of evidence argue in favor of separate enzymes. First, the ratio of the two methyltransferase activities may vary considerably depending on the age of the seedling. Second, methyltransferase activity toward N-methyltyramine is lost more rapidly than that toward tyramine during storage or purification procedures.

Mann et al. (1963) demonstrated the excellent correlation between the appearance and subsequent disappearance of tyramine methylpherase activity and the levels of hordenine and N-methyltyramine during the first month of germination. Additionally, evidence was obtained indicating that the methylpherase functioned at nearly its maximal capacity during the first week. The appearance of the enzyme during germination probably results from de novo synthesis, as shown by inhibitor studies with puromycin and amino acid analogues.

Hordenine biosynthesis appears to be organ-specific also in millet, *Panicum miliaceum* L. However, in contrast to the situation in barley, it is the shoots that contain both tyramine methylpherase activity and the products N-methyltyramine and hordenine (Brady and Tyler, 1958). Finally, tyramine methylpherase activity was detected in extracts from the cactus *Trichocereus spachianus;* the enzyme was specific for SAM, failing to utilize methionine, methionine sulfoxide, and S-methylmethionine as donors (Mann and Mudd, 1963).

2. Biosynthesis of Simple Indole Derivatives

Barley shoots contain a wide range of indole derivatives originating from L-tryptophan. Labeled gramine [3-(dimethylaminomethyl)tryptophan] was produced when cell-free extracts from this tissue were incubated with [β-^{14}C]tryptophan, methionine, and ATP (Breccia et al., 1966). The proposed biosynthetic pathway of this alkaloid was confirmed by Mudd (1960, 1961), who demonstrated that crude barley shoot homogenates catalyzed the transfer of a methyl group from ($-$)-SAM to 3-aminomethylindole, yielding initially 3-methylaminomethylindole and then gramine (Fig. 9).

Phalaris tuberosa, a pasture grass of southeastern Australia, contains a number of tryptamine alkaloids, principally N,N-dimethyltryptamine and 5-methoxy-N,N-dimethyltryptamine. The presence of two indolethylamine N-methyltransferases in *Phalaris* extracts has been deduced by their differen-

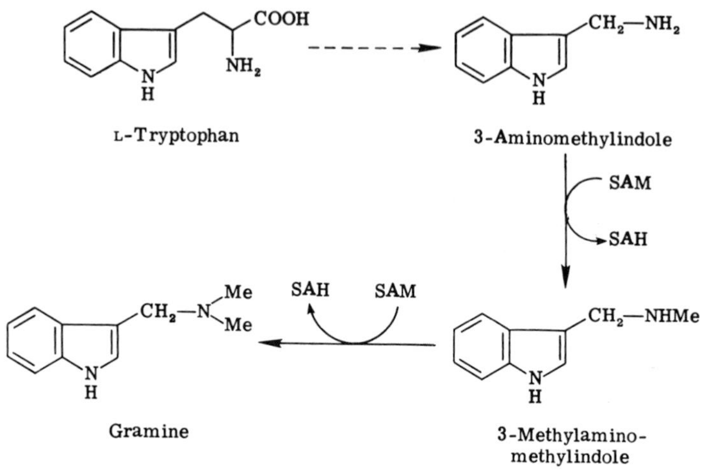

Fig. 9. The biosynthesis of gramine from L-tryptophan in *H. vulgare* L.

tial behavior toward various environmental effects (Mack, 1974). The primary indolethylamine *N*-methyltransferase catalyzes the N-methylation of tryptamine and its 5-hydroxy and 5-methoxy derivatives (Fig. 10). The corresponding N-methylated products are further methylated by the secondary indolethylamine *N*-methyltransferase to the respective *N,N*-dimethyltryptamines. The enzymes were not separated by normal chromatographic techniques but could be partially resolved by affinity chromatography (Section I,J) (Mack and Slaytor, 1978).

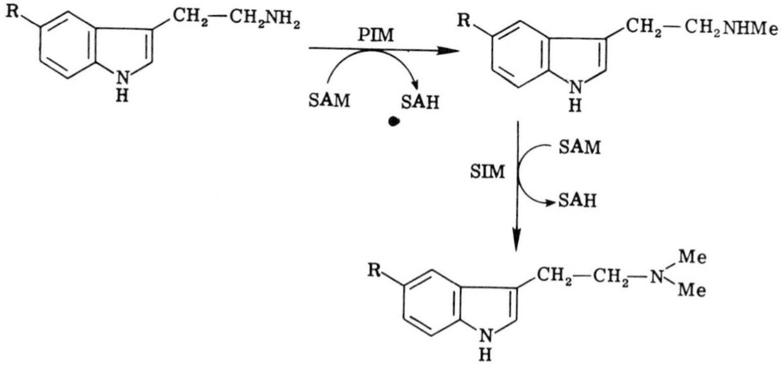

Fig. 10. The biosynthesis of tryptamine alkaloids by the primary indolethylamine *N*-methyltransferase (PIM) and secondary indolethylamine *N*-methyltransferase (SIM) from *Phalaris tuberosa*. (R = H, OH, or OMe).

3. Biosynthesis of Mescaline and Related Tetrahydroisoquinolines

Peyote (*Lophophora williamsii*) and certain other cacti contain two biosynthetically related classes of alkaloids, phenethylamines and tetrahydroisoquinolines. The proposed biosynthetic pathway (Fig. 11) from dopamine to these alkaloids, outlined by isotopic tracer studies, has been largely confirmed by probing the substrate specificity of an *O*-methyltransferase purified 220-fold from this cactus (Basmadjian and Paul, 1971; Basmadjian *et al.*, 1978). In addition to methylating dopamine specifically at the meta position, *O*-methyltransferase activity was operative toward many mono- and diphenolic phenethylamines and several tetrahydroisoquinolines. Both para and meta substitutions were exhibited, depending on the nature of the substrate. Significantly, however, once the central pyrogallol hydroxyl group (the 4-hydroxyl of phenethylamines or the 7-hydroxyl of tetrahydroisoquinolines) was O-methylated, the enzyme failed to further methylate either of the remaining hydroxyl groups. This property effectively eliminates 3,4,5-trihydroxyphenethylamine as a direct precursor of mescaline or, following cyclization to trihydroxytetrahydroisoquinoline, of the tetrahydroisoquinolines occurring naturally in peyote. In contrast, 4,5-

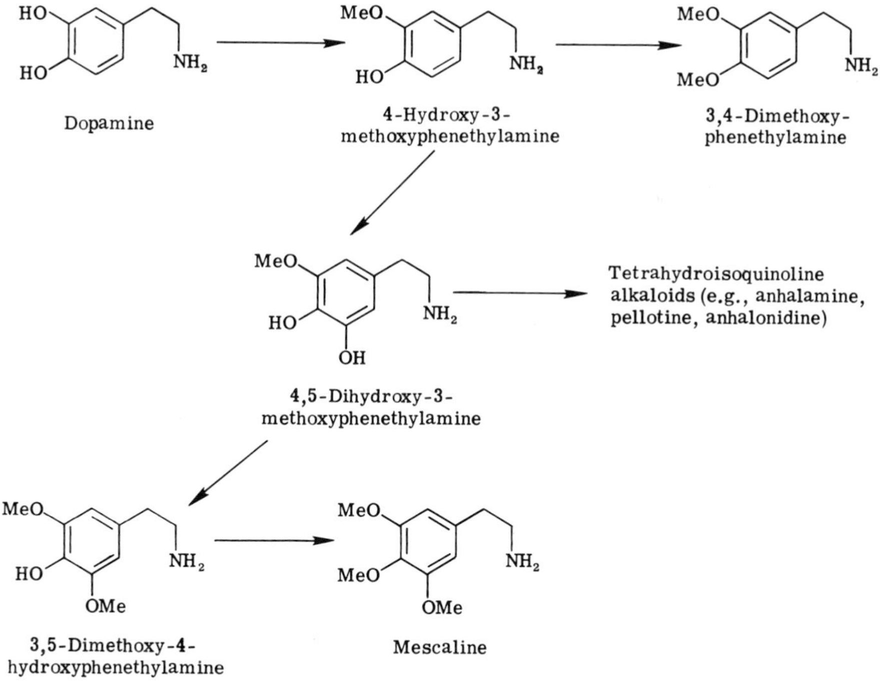

Fig. 11. Proposed biosynthesis of the peyote alkaloids.

dihydroxy-3-methoxyphenethylamine and 7,8-dihydroxy-6-methoxytetra-hydroisoquinoline were efficiently methylated, thereby strengthening conclusions from *in vivo* labeling data that the former metabolite is a common precursor of mescaline and tetrahydroisoquinolines in peyote (Paul, 1973). This enzyme differs from the alkaloid-specific O-methyltransferase from *Nerine* in being unaffected by EDTA, cyanide, and thiourea (Table VI).

4. Biosynthesis of Trigonelline

Trigonelline (N-methylnicotinic acid) occurs widely in plants and plant cell cultures and is believed to be a reservoir form of nicotinic acid and therefore of pyridine nucleotides (Willeke *et al.,* 1979). Nicotinic acid N-α-L-arabinoside fulfills the same function. Joshi and Handler (1960) reported the synthesis of SAM from ATP and L-methionine by extracts of pea (*Pisum sativum* L.) seedlings. In the presence of these extracts, SAM donated its methyl group to nicotinic acid, yielding trigonelline (Fig. 12). The enzyme

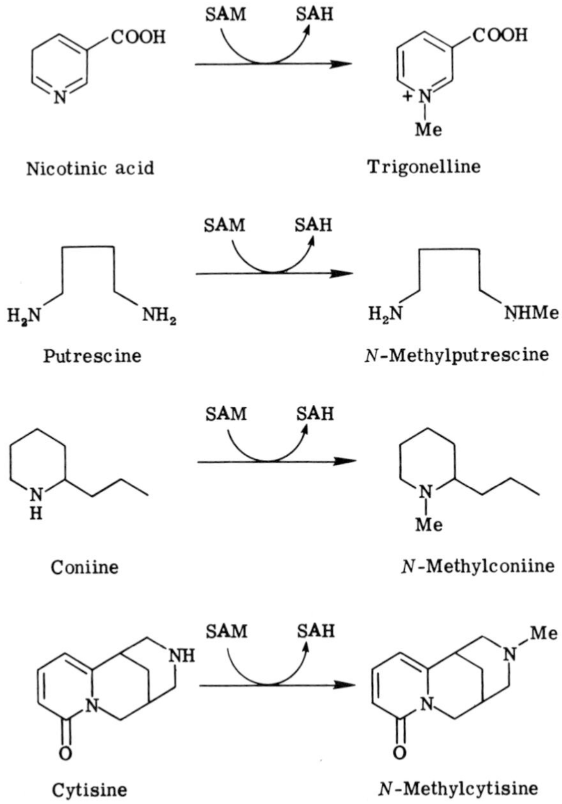

Fig. 12. Several transmethylation reactions with alkaloids as methyl acceptors.

responsible, nicotinic acid methylpherase, exhibited maximum activity between pH 6.0 and 7.0. No evidence was found for nicotinamide methylation; this contrasts with the behavior of the enzyme from mammalian liver (Cantoni, 1951) and is difficult to reconcile with the report of N-methylnicotinamide synthesis from nicotinamide and methionine in bean, wheat, and corn seedlings (Stul'nikova, 1959). Nicotinic acid methylation to trigonelline has also been reported with extracts from castor bean seedlings (Jindra, 1967) and root callus cultures of fenugreek (*Trigonella foenum graecum* L.) (Antony *et al.*, 1975). Nicotinamide is not methylated by *Trigonella* extracts.

5. Biosynthesis of Tobacco Alkaloids

Nicotine biosynthesis in *Nicotiana* (tobacco) is certainly one of the most studied pathways in alkaloid metabolism (for details, see Waller and Dermer, this volume, Chapter 12). A central problem was to determine at which stage the methyl group of nicotine was introduced. Isotopic labeling studies indicated that this group was derived from L-methionine. The enzymes ornithine decarboxylase, putrescine N-methyltransferase (PMT), and N-methylputrescine oxidase were recognized in tobacco root extracts and could convert ornithine to N-methylpyrrolinium salt, the immediate precursor of the pyrrolidone ring of nicotine. Putrescine N-methyltransferase, which catalyzes the methylation of putrescine with SAM as methyl donor, forming N-methylputrescine and probably SAH (Fig. 12), was purified 30-fold (Mizusaki *et al.*, 1971). No cofactors were required for this reaction, which proceeded best at pH 8–9. The enzyme was remarkably specific for putrescine; slight activity was observed with N-methylputrescine, but 1,3-propanediamine, 1,5-pentanediamine, Δ^1-pyrroline, nornicotine, and ornithine were not methylated. Interestingly, neither nicotine nor N-methylpyrrolinium salt affected the rate of putrescine methylation. 2,4-Dichlorophenoxyacetic acid and β-indoleacetic acid, which reduce nicotine production in tobacco plants, had no effect on PMT activity *in vitro*. Demonstration that PMT resided solely in the roots, where active nicotine biosynthesis occurs, favored the conclusion that this enzyme was specifically concerned with this process. The decapitation of shoots of hydroponically grown tobacco plants promoted simultaneous increases in the activity of the above three enzymes over normal levels (Mizusaki *et al.*, 1973). Activities reached a maximum about 24 h after decapitation before declining again. The nicotine content of root preparations correlated well with these activity changes. Additional data supported the conclusion that all three enzymes were controlled by a common regulatory system in which auxin and nicotine could be important factors.

Putrescine methyltransferase activity was absent in tobacco cell suspension cultures. The occurrence of aromatic amides such as *p-*

coumaroylputrescine, caffeoylputrescine, and feruloylputrescine in these cells suggests that the biosynthetic pathway of nicotine is blocked at the methylation step. Root preparations of *Datura* and *Atropa* plants, which incorporated putrescine into the N-methylpyrrolidone ring of tropane alkaloids *in vivo*, possessed PMT and N-methylputrescine oxidase activities, indicating that this moiety of the alkaloid molecule could be formed by a pathway similar to that of nicotine biosynthesis in *Nicotiana* (Mizusaki *et al.*, 1973).

6. Biosynthesis of Hemlock Alkaloids

Considerable evidence is now available demonstrating the probable biosynthetic interconversion of the propylpiperidine bases γ-coniceine, coniine, and N-methylconiine present in the poisonous hemlock (*Conium maculatum* L.) (Waller and Dermer, this volume, Chapter 12). The N-methyl group of N-methylconiine is derived from L-methionine, again suggesting the intermediacy of SAM in this transmethylation (Roberts, 1974a). Indeed, crude preparations of hemlock fruits catalyzed the N-methylation of coniine, using either L-methionine and ATP or SAM alone (Fig. 12) (Roberts, 1974b). The enzyme responsible, coniine N-methyltransferase, was detected only in extracts of plant parts containing methylconiine, namely, unripe fruits and, to a lesser extent, leaves of flowering plants. The use of a strong reducing agent (sodium metabisulfite) together with Polyclar AT and dithiothreitol was essential during the extraction procedure. Dithiothreitol was included in subsequent purification steps for maximum enzymatic activity.

7. Biosynthesis of Quinolizidine (Lupine) Alkaloids

Murakoshi *et al.* (1977) described the variations in lupine alkaloid content at different stages of seedling growth of *Thermopsis chinensis* (Leguminosae); during the first week of growth, the levels of N-methylcytisine increased rapidly, whereas the concentration of cytisine decreased. Cell-free extracts of seedlings at this growth stage catalyzed the N-methylation of cytisine to N-methylcytisine at an optimum pH of 8.5. S-Adenosyl-L-methionine, but not 5-methyltetrahydrofolic acid, acted as methyl donor (Fig. 12).

8. Biosynthesis of Caffeine

With the use of intact tea (*Camellia sinensis* L.) plants and callus tissue, early work established that caffeine was synthesized from the same precursors utilized for purine and methyl group synthesis in other systems. Further advances in our knowledge came from the demonstration of two N-methyltransferase activities involved in caffeine biosynthesis in extracts from tea leaves (Suzuki and Takahashi, 1975). This represented a notable achievement, considering the difficulties encountered in extracting active enzymes from tissues containing high concentrations of phenolic compounds; Polyclar AT was absolutely required in the extraction procedure. Methyl groups were transferred from SAM to 7-methylxanthine, yielding theobromine which underwent further methylation to caffeine (1,3,7-trimethylxanthine) (Fig. 13). Xanthine, xanthosine, xanthosine monophosphate, and hypoxanthine were totally inactive in this system but, unexpectedly, paraxanthine (1,7-dimethylxanthine) was the most active among methylxanthines. However, paraxanthine is not believed to be the normal precursor of caffeine *in vivo*, since its formation from 1- and 7-methylxanthine was negligible *in vitro*. Data obtained were compatible with the pathway for caffeine biosynthesis involving theobromine shown in Fig. 13. A similar conclusion was recently reached by Roberts and Waller (1979) with cell-free extracts from unripe green fruits of coffee (*Coffea arabica* L.), which had almost identical properties and substrate specificity (Table VI). Although not proven, it seems probable that these two consecutive methylation reactions are catalyzed by the same enzyme, since they display similar pH optima and are influenced almost identically by various inhibitors. Furthermore, the reported kinetic data are compatible with this viewpoint (Suzuki and Takahashi, 1975; Roberts and Waller, 1979).

9. Biosynthesis of Amaryllidaceae Alkaloids

Amaryllidaceae alkaloids constitute a large group of naturally occurring compounds of widely diverse functions and structural type, all formally derived from norbelladine. Isotopic tracer studies indicate that both tyrosine and derivatives of norbelladine serve as precursors in their biosynthesis. For example, doubly labeled [1,1'-^{14}C]norbelladine itself was incorporated into belladine, lycorine, and crinamine in *Nerine bowdenii* without randomization of activity, indicating that the precursor entered as a unit (Wildman *et al.*, 1962). Analogous feeding experiments by Barton *et al.* (1963) with multiply labeled methylated derivatives of norbelladine have concluded that (1) these compounds enter Amaryllidaceae alkaloids without scrambling of label, proving conclusively that the precursor enters as a unit, (2) the methylenedioxy function of alkaloids, such as in haemanthamine, originates from an o-methoxyphenol group [cf. the methyl carbon of L-methionine acting as a

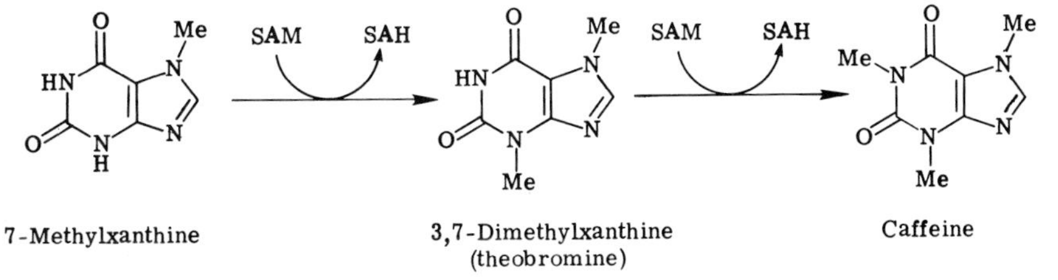

Fig. 13.　The biosynthesis of caffeine from 7-methylxanthine.

precursor of the methylenedioxy group of *Papaver* alkaloid protopine (Sribney and Kirkwood, 1953)], (3) the methylation pattern of the precursor may
determine at least partially its subsequent fate in alkaloid biosynthesis
(Spenser, 1970).

The biosynthetic route to these alkaloids was further elucidated by the
isolation of an enzyme from bulbs of *N. bowdenii* that catalyzed the
O-methylation of norbelladine to *N*-isovanillyltyramine and *N*-
vanillyltyramine (Fig. 14) (Fales *et al.*, 1963; Mann *et al.*, 1963). Methylation
occurred primarily at the para position, with a para/meta ratio of 22 : 1. In
contrast, the action of rat liver catechol-*O*-methyltransferase on norbelladine
resulted predominantly in meta-methylation. Like barley tyramine methylpherase, the enzyme appeared specific for the (−) diastereoisomer of SAM.
The D isomer of SAM was only one-third as effective as the L isomer. A
variety of other potential methyl donors, including *S*-methylmethionine,
L-methionine sulfoxide, and betaine were ineffective. In addition to norbelladine, the enzyme also methylated dopamine, 2,3-naphthalenediol, 2,4-
diphenyldiol, and catechol, but *N*-vanillyltyramine, *N*-isovanillyltyramine,
and *N*-veratryltyramine were not substrates. Norbelladine methylation occurred maximally at pH 8.1 and was not stimulated by Mg^{2+} ions (Table VI).

10. Biosynthesis of Alkaloids Related to Norlaudanosoline

The pioneering work of Winterstein, Trier, and Robinson led to the recognition that the structures of a whole series of alkaloids, including the hy-

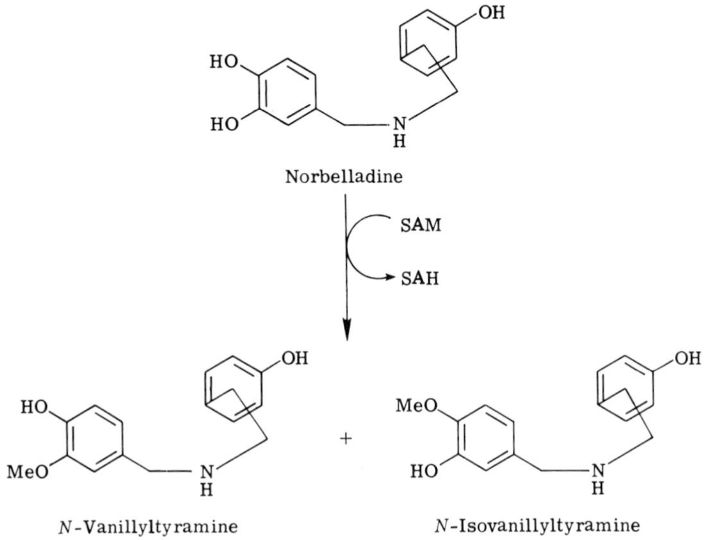

Fig. 14. O-Methylation of norbelladine by an *O*-methyltransferase from *Nerine bowdenii*.

drophenanthrene, aporphine, protoberberine, phthalideisoquinoline, and benzophenanthridine types, could be formally derived from the benzyl-tetrahydroisoquinoline skeleton of norlaudanosoline (Waller and Dermer, this volume, Chapter 12, Fig. 15). Accumulated tracer evidence is in remarkable agreement with their proposals, but knowledge about the enzymology involved is limited. It is clear, however, that if so many alkaloid types are indeed derived from the common precursor norlaudanosoline by various oxidative coupling reactions, some form of control mechanism must exist *in vivo* to direct its entry into these competing biosynthetic pathways. Both the methylation pattern of methylated derivatives of norlaudanosoline and their stereochemistry are thought to play active roles as directing factors, as described by Spencer (1970). For example, only reticuline out of the four

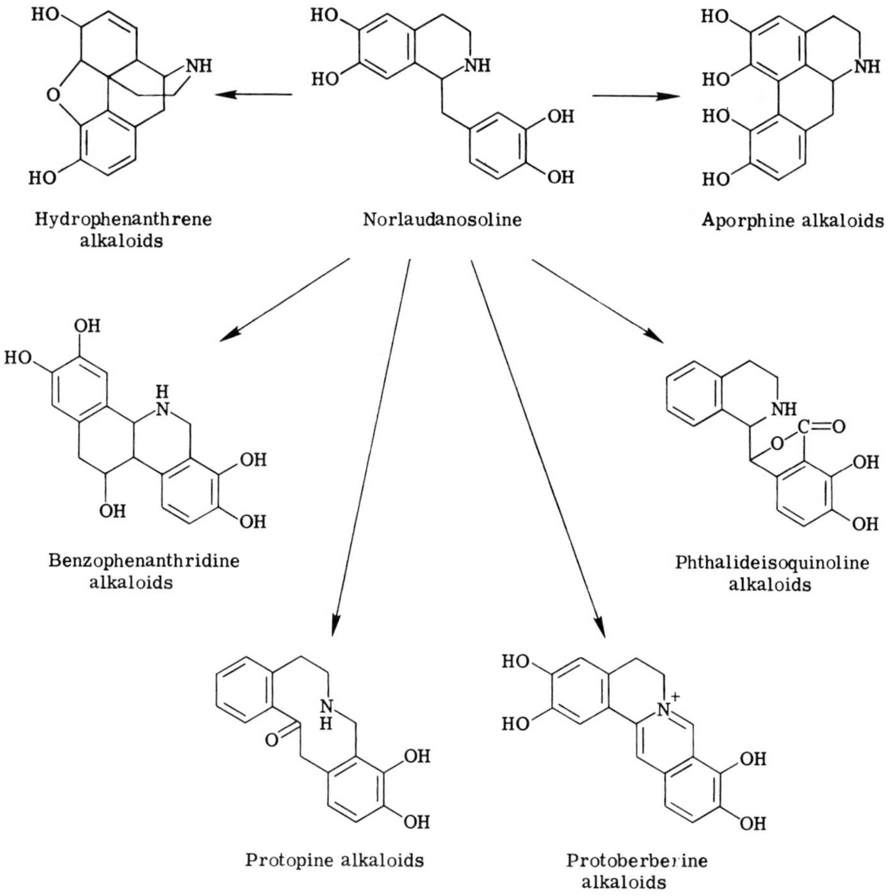

Fig. 15. Alkaloid skeletons related to norlaudanosoline.

isomeric *O,O,N*-trimethylnorlaudanosolines served as an *in vivo* precursor of thebaine and other hydrophenanthrene alkaloids of *Papaver somniferum* L. (Battersby *et al.*, 1965). O-Methylation is believed to direct cyclization by protecting, and thereby inactivating, specific phenolic groups in the nor-laudanosoline molecule. The position of the remaining unprotected hydroxyl groups would subsequently determine the mode of cyclization of the precursor. This concept presupposes the existence in plants of distinct *O*- and *N*-methyltransferases that catalyze the methylation of norlaudanosoline at specific positions, a hypothesis confirmable by the isolation of such enzymes. Recently, Antoun and Roberts (1975) demonstrated that the SAM-dependent methylation of norlaudanosoline, catalyzed by the "supernatant" fraction of *P. somniferum* latex, yielded reticuline, papaverine, codeine, and thebaine, but not narcotine. Poppy latex therefore possesses not only the methyl-transferases necessary for reticuline and papaverine biosynthesis but also demethylases converting thebaine to codeine and possibly morphine, as originally detected by tracer studies (Battersby *et al.*, 1965). Isolation of these enzymes would greatly increase our understanding of the control mechanisms involved in hydrophenanthrene alkaloid biosynthesis in *Papaver*.

11. Biosynthesis of Loganin and Secologanin

Loganin, the methyl ester of loganic acid, serves as a precursor of the nontryptophan moiety of three different classes of indole alkaloids in *Vinca rosea* (see Waller and Dermer, this volume, Chapter 12). Methionine is presumably the source of the *O*-methyl group of loganin, as shown in the related species *Menyanthes trifoliata* L. A methyltransferase partially purified from *V. rosea* seedlings by Madyastha *et al.* (1972, 1973) catalyzes a rather uncommon type of O-methylation in which the methyl group of SAM is transferred to the carboxyl group of loganic acid, forming the methyl ester (Fig. 16). The reaction proceeded optimally at pH 6.9–7.4 and showed neither divalent metal ion dependency nor inhibition by EDTA. Complex requirements for reducing agents were exhibited, suggesting an active role of reduced sulfhydryl groups of the enzyme. In confirmation, the enzyme was inhibited by sulfhydryl reagents. Both loganic and secologanic acid were

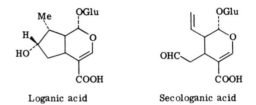

Loganic acid   Secologanic acid

Fig. 16. Substrates accepted by loganic acid methyltransferase (carboxyl-alkylating) from *Vinca rosea.*

readily methylated at the carboxyl group, but no significant methylation of 7-deoxyloganic acid, geniposidic acid, 7-epiloganic acid, or palmitic acid occurred. These results implicate loganic and secologanic acids but not 7-deoxyloganic acid as active metabolites in indole alkaloid biosynthesis.

I. Major Factors Affecting the Activity of Methyltransferases *in Vivo* and *in Vitro*

1. Factors Affecting Tissue Levels of Methyltransferases

It has become evident that the overall rates of biosynthesis and degradation of many secondary plant products are closely related to the stage of development either of the whole organism or of individual organs, tissues, or specialized cells (Wiermann, this volume, Chapter 4; Luckner *et al.*, 1977). Thus particular enzymes associated with such pathways may be present only for restricted time periods in the growth cycle of the plant or plant cell culture. This situation certainly applies to methyltransferases concerned with secondary metabolism where high activities are exhibited only at certain periods in the life cycle of cell cultures (Poulton *et al.*, 1976a; Ebel *et al.*, 1972; Wengenmayer *et al.*, 1974), whole plants (Glass and Bohm, 1972), and plant organs (Mann *et al.*, 1963; Roberts, 1974b). In contrast, the CMT activity of soybean cell cultures is independent of the age of the cells (Poulton *et al.*, 1976a). The mechanisms underlying these regulatory phenomena remain unclear.

The biosynthesis and accumulation of secondary products may be additionally influenced by physical stimuli (e.g. light), aging of tissues *in vitro*, injury, infection, and hormone treatment (Luckner *et al.*, 1977). These factors may affect the level of individual enzymes including several methyltransferases. For example, whereas light had no significant effect on the levels of CMT and FMT activity in soybean cell cultures (Hahlbrock, 1977), it caused the dramatic and specific induction of all enzymes involved in flavonoid glycoside biosynthesis in *Petroselinum* cell cultures, including FMT (Hahlbrock *et al.*, 1971). The aging of potato (Camm and Towers, 1973) and swede discs (Rhodes *et al.*, 1976) *in vitro* led to increased levels of CMT; in the latter case, ethylene stimulated this effect, whereas cycloheximide prevented these activity increases. Ethylene treatment enhanced CMT activity in postharvest lettuce also (*Lactuca sativa*) (Hyodo *et al.*, 1978).

Infection of hypersensitive *N. tabacum* plants by tobacco mosaic virus resulted in pronounced increases in CMT activity, correlating with the enrichment in *m*-methoxy, *p*-hydroxyphenyl compounds beginning about 3 days after inoculation (Legrand *et al.*, 1976). The CMTs of infected tissue were the same as those found in healthy leaves but had higher activities, especially of OMTs II and III (Legrand *et al.*, 1978). Similarly, the deposition of lignin around wounds in wheat leaves infected with the nonpathogenic fungus *Botrytis cinerea* was preceded by increases specifically within the lignifying tissues of PAL and caffeate- and 5-hydroxyferulate-methylating activities (Maule and Ride, 1976). Comparable data have been obtained following the infection of tuber slices of resistant (Orion-R_1) potatoes by *Phytophthora infestans* (Friend and Thornton, 1974). In contrast, however, infection of reed canary grass (*Phalaris arundinacea* L.) by *Helminthosporium avenae* led to a higher PAL activity and lignin content but did not influence CMT levels (Vance and Sherwood, 1976).

Tobacco cell cultures have been successfully used to study the effect of hormones on lignin synthesis (Yamada and Kuboi, 1976). Lignification was observed only in kinetin cultures and not in auxin (2,4-dichlorophenoxyacetic acid or indolebutyric acid) cultures. Levels of CMT were much higher under the former conditions, and the rise in enzyme activity coincided with the onset of this process. Activities of the enzymes shikimate dehydrogenase, cinnamate 4-hydroxylase, and caffeate and 5-hydroxyferulate *O*-methyltransferases were the greatest in large cell aggregates, promoting the extensive lignification observed in these cell clusters. Additional data suggested that enzymes in the shikimate and cinnamate pathways were coordinately enhanced during differentiation of tracheary elements (Kuboi and Yamada, 1978).

In alkaloid metabolism, β-indoleacetic acid (IAA) and 2,4-dichlorophenoxyacetic acid, which reduce nicotine content in tobacco plants, did not affect the activity of PMT *in vitro*. In contrast, administration of IAA in concentrations of 2.5–5 μM significantly raised PMT activity in roots of decapitated plants, but higher concentrations prevented the rise in enzyme activities promoted by decapitation (Section I,H,5). Yoshida (1973) demonstrated how the levels of PMT and *N*-methylputrescine oxidase in roots of young tobacco plants were affected by deprivation of nutrient substances.

2. In Vitro *Inhibition of Methyltransferases by S-Adnosyl-L-Homocysteine and Its Possible Significance in* in Vivo *Regulation*

S-Adenosyl-L-homocysteine is generally considered to be the common product of transmethylation reactions involving SAM. It is a typical feature of the majority of these reactions, whether catalyzed by an *O*-, *C*-, *N*- or *S*-methyltransferase, to be strongly inhibited by low concentrations of this product, thereby causing nonlinearity in the time course of the reaction (Table VII). Inhibition is usually of a competitive nature vs SAM, and the K_i for SAH may be severalfold lower than the K_m for SAM (Poulton and Butt, 1975; Mann and Mudd, 1963; Wengenmayer *et al.*, 1974). These features could well point to a general mechanism for the regulation of methyltransferase activities. Poulton and Butt (1975) suggested that in *B. vulgaris* the rate of *in vivo* caffeic acid methylation, and perhaps of other transmethylation reactions too, might be controlled by the intracellular SAM/SAH con-

TABLE VII

Some Transmethylation Reactions in Which SAH Acts as Inhibitor of SAM-Dependent Methyltransferases *in Vitro*

Enzyme source	Substrate	Reference
Rat liver	Catechol	Deguchi and Barchas, 1971*
Rabbit adrenal gland	Phenylethanolamine	
Bovine pineal gland	Acetylserotonin	
Hordeum vulgare	Tyramine	Mann and Mudd, 1963
Rabbit lung	*N*-Methyltryptamine	Krause and Domino, 1974*
Phalaris tuberosa	Indolethylamine	Mack and Slaytor, 1978
Guinea pig brain	Histamine	Zappia *et al.*, 1969*
Saccharomyces cerevisiae	Homocysteine	
Beta vulgaris	Caffeic acid	Poulton and Butt, 1975
Glycine max	Caffeic acid	Poulton *et al.*, 1976b
Glycine max	3′,4′-Dihydroxyflavonoids	Poulton *et al.*, 1977
Cicer arietinum	4′-Hydroxyisoflavones	Wengenmayer *et al.*, 1974
Rat liver	Arginine	Baxter and Byvoet, 1974*
Rat liver	Lysine	
Calf thymus	Protein I	Jamaluddin *et al.*, 1975*
Rabbit liver	N^2-guanine	Hildesheim *et al.*, 1973*
Wheat germ	Virus mRNA	Both *et al.*, 1975*
Mycobacterium phlei	Fatty acid (carboxyl)	Akamatsu and Law, 1970*

* References cited in Mack and Slaytor, 1978.

centration ratio. However, whatever the precise control mechanism involved, it is in any case clear that the powerful inhibition of transmethylation reactions by SAH requires the removal of this product or at least the strict regulation of its intracellular concentration for the continuation of these reactions. The biological significance of SAH inhibition is further suggested by the observation that the intracellular levels of this compound, although low, are of the same order of magnitude as those of SAM (Dodd and Cossins, 1968). *S*-Adenosyl-L-homocysteine may be cleaved in plants by SAH hydrolase (E.C. 3.3.1.1) which catalyzes the reversible synthesis of SAH from adenosine and L-homocysteine. The kinetic properties of the spinach-beet and lupine enzymes have been reported (Poulton and Butt, 1976; Guranowski and Pawelkiewicz, 1977). Since the equilibrium lies heavily in favor of synthesis, sustained SAH hydrolysis is especially governed by the disposal of the products. Accumulation of either of the products inhibits the enzyme. Cleavage of SAH by SAH hydrolases from spinach beet and *Lupinus luteus* is facilitated *in vitro* by the action of adenosine nucleosidase (E.C. 3.2.2.7) and adenosine deaminase (E.C. 3.5.4.4) (Poulton and Butt, 1975; Guranowski and Pawelkiewicz, 1977). However, it seems more likely that adenosine is conserved *in vivo* by its conversion to ADP through the successive action of adenosine and adenylate kinases. L-Homocysteine can undergo methylation to L-methionine with N^5-methyltetrahydropteroyl-monoglutamate acting as methyl donor, as demonstrated in extracts of pea seedlings (Dodd and Cossins, 1970) and spinach and barley leaves (Burton and Sakami, 1969). The rate of this reaction is expected to be closely integrated with the flow of C_1 units arising in the folate pool. Therefore SAH hydrolase plays a crucial role in metabolism by controlling the rate of SAH turnover and thereby indirectly regulating much of methyl group biogenesis and utilization.

J. Affinity Chromatography in the Purification and Characterization of S-Adenosylmethionine-Dependent Methyltransferases

Demonstration that the product SAH is a potent inhibitor of SAM-dependent transmethylases suggests its possible use as a ligand in affinity chromatography. Mack and Slaytor (1978) deduced, from available data on the binding of SAM and SAH analogues to various methyltransferases, that linkage of SAH through the carboxyl group to an immobilized support would lead to a more general affinity adsorbent for such enzymes than linkage through other functional groups. This affinity adsorbent (Fig. 17A) was used to purify two indolethylamine N-methyltransferases from *Phalaris tuberosa*. Sharma and Brown (1978), employing essentially the same affinity system (Fig. 17B), reported the 50-fold purification in a single step of SAM : 5-

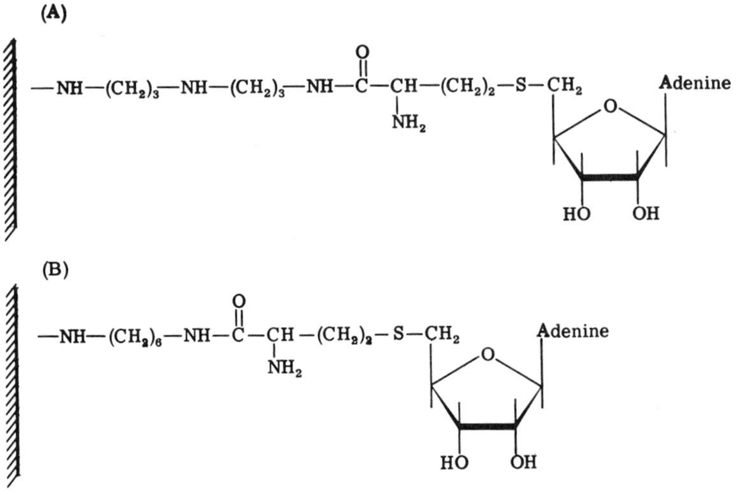

Fig. 17. The structures of SAH affinity adsorbent used in purification of SAM-dependent transmethylases. A, adapted from Mack and Slaytor (1978); B, adapted from Sharma and Brown (1978).

hydroxyfuranocoumarin O-methyltransferase from *R. graveolens* cell cultures. The 5- and 8-hydroxyfuranocoumarin methyltransferases and the CMT from these cultures became bound to this immobilized ligand. Bioelution by the addition of specific substrates to the irrigant buffer failed to displace them. Instead, coelution of all three enzymes was accomplished either by decreasing the pH of the elution buffer to 3 or, more effectively, by the addition of SAM to the irrigant buffer (Sharma and Brown, 1979). Since the methyltransferases were not resolved by this system, Sharma *et al.* (1979) investigated the use of the general affinity ligand 5-(3-carboxypropanamido)xanthotoxin followed by specific desorption by bergaptol and xanthotoxol. In addition to achieving complete separation of all three proteins, this technique provided valuable information concerning the mechanism of O-methyltransferase action, which was further supported by studies employing a ferulic acid affinity adsorbent. The findings were in agreement with a compulsory ordered kinetic mechanism for CMT and both furanocoumarin O-methyltransferases, whereby the enzyme must first bind SAM or SAH before it can bind the phenolic substrate. Similar conclusions were reached by Wengenmayer *et al.* (1974) for the isoflavone 4'-O-methyltransferase from *C. arietinum*. In contrast, kinetic data for rat liver catechol O-methyltransferase (Coward *et al.,* 1973) and the secondary indolethylamine N-methyltransferase from *P. tuberosa* (Mack, 1974) have been interpreted in terms of a random Bi-Bi mechanism involving formation of a dead-end complex.

An SAH affinity system was utilized for the purification of a SAM : protein carboxyl O-methyltransferase from calf brain (Kim *et al.,* 1978).

II. DEMETHYLATION REACTIONS IN THE METABOLISM OF SECONDARY PLANT PRODUCTS

A. Introduction

Demethylation reactions, processes whereby methyl groups are removed from methylated derivatives, have been described in mammals, microorganisms, and plants (Mason *et al.,* 1965). Considerable attention has been given to animal systems, especially in relation to drug metabolism (Testa and Jenner, 1976), but comparatively little is known about the frequency of occurrence, mechanism, or physiological significance of such processes in higher plants. N-Demethylation of certain alkaloids (e.g., ricinine, nicotine, hyoscyamine) is often connected with senescence of plants and may be involved in the transport of alkaloids within the plant. For many secondary compounds including alkaloids, demethylation generally serves to increase their water solubility and reactivity, and it probably initiates more extensive processes of degradation of the molecules.

Our knowledge of demethylation reactions in plants is based almost entirely upon data derived during the past three decades from *in vivo* radiotracer studies. Several authors have pointed to the need for caution when interpreting such data and also to various problems encountered when intact plants are employed as experimental systems (Waller and Nowacki, 1978; Harborne, 1967; Barz and Hösel, 1975). One serious concern is to prove beyond all doubt that the observed metabolic reactions result from the normal metabolism of the plant and not from microbial contamination or injury of the plant material. Early investigations generally involved intact plants grown under normal laboratory (i.e., asterile) conditions.

Barz *et al.* (1970) have elegantly demonstrated both the unsuitability of such systems for studying turnover of phenolic compounds and the absolute necessity for testing the experimental material for microbial contamination. Extensive degradation of both [Me-^{14}C]formononetin and [4-^{14}C]daidzein was observed when these isoflavones were fed to intact plants and to separated or sliced plant parts of *C. arietinum* and *Phaseolus aureus*. Making use of aseptically grown plants as controls, they showed the involvement of rhizosphere microorganisms which rapidly degraded [Me-^{14}C]formononetin to $^{14}CO_2$, part of which reentered the plant via fixation processes and led to strong labeling of carotenoid components. Moreover, isolation of labeled biochanin A and homoferreirin from *C. arietinum* indicated that methyl ether groups removed from the fed isoflavone could reenter the C_1 pool and be reutilized for further methylation reactions. Additional problems encountered during studies with intact plants included the limited transport of polyphenols and the binding of phenolic substrates to polymeric structures (Berlin, 1972).

More recently, plant cell suspension cultures were shown to be an extremely suitable system for studying the turnover of so-called secondary plant compounds (Street, 1973; Barz and Hösel, 1975). The problem of microbial contamination is avoided, and relatively large amounts of material at apparently similar developmental stages may be obtained from the same inoculum. Higher concentrations of metabolites can be expected in cell cultures than in intact plants because of easier substrate uptake. It is hoped that the use of cell cultures and aseptically grown plants and the increasing availability of suitably labeled substrates (especially doubly labeled compounds) will accelerate studies concerning the enzymatic nature of demethylation reactions.

The following sections review demethylation reactions in secondary plant metabolism, focusing attention on the enzymology involved. The numerous tracer studies in this area have been covered in depth elsewhere (Waller and Nowacki, 1978; Freudenberg and Neish, 1968). Demethylation reactions could conceivably proceed via two general mechanisms: (1) by transmethylation involving the transfer of an intact methyl group to another acceptor, or

(2) by an oxidative process releasing formaldehyde as second product. In the latter case, NADPH and oxygen-dependent N-, S- and O-demethylations may be viewed as examples of hydroxylation reactions (Brodie *et al.,* 1958). By analogy with animal systems, the second mechanism for demethylation appears more likely.

B. Demethylation Reactions Involving Simple Phenolic Compounds and Flavonoids

During the period 1950–1970, considerable advances were made in our knowledge of the biosynthetic pathways leading to lignin, flavonoids, and anthocyanins using radiotracer labeling techniques (Shimada, 1972; Hahlbrock and Grisebach, 1975). Additionally, these investigations demonstrated for the first time that phenolic compounds could undergo demethylation and even "demethoxylation" reactions in plant tissues (Table VIII). For example, ferulic acid was converted to caffeic acid, chlorogenic acid, *p*-hydroxybenzaldehyde, and *p*-coumaric acid, and sinapic acid yielded coniferyl lignin components (isolated as vanilloyl methyl ketone), ferulic acid, vanillin, and *p*-coumaric acid. The enzymology of these processes remains unknown, and it has yet to be shown whether the observed demethoxylation reactions in fact result from a demethylation followed by removal of a hydroxyl group.

C. O- and N-Demethylation of Alkaloids

The original view that alkaloids are final waste products of nitrogen metabolism is no longer tenable. The alkaloid content of many plants shows both diurnal and developmental variations, implying involvement of catabolic as well as biosynthetic processes. The surprisingly rapid turnover rates for alkaloids confirm that these compounds are not inert end products but active participants in plant metabolism. Demethylation, which often initiates more extensive breakdown processes, is the most frequently described degradative step in alkaloid catabolism. In several cases, both methylated and demethylated alkaloids exist in the same plant. However, the mechanisms for the introduction and removal of methyl groups appear to be dissimilar and are generally irreversible. Our information concerning alkaloid demethylation stems mainly from *in vivo* radiotracer studies, since most attempts to demonstrate this process with cell-free systems have been unsuccessful. The following sections introduce several demethylating systems where further research at the enzymatic level would lead to a greater understanding of the mechanisms involved.

The gradual disappearance of hordenine and N-methyltyramine from barley roots beginning 7–10 days after germination is supposedly due to

TABLE VIII

Selected Examples of 0-Demethylation of Simple Phenolic Compounds and Flavonoids by Intact Plants or Plant Cell Cultures

Organism	Compound administered	Substances isolated	Authors
Picea excelsa (spruce)	Syringin	Guaiacyl components of lignin	Kratzl and Billek, 1953*
Acer negundo (maple)	[3-¹⁴C]Sinapic acid	Guaiacyl components of lignin	Brown and Neish, 1956*
Helianthus annuus, Triticum vulgare, Zea mays	[3-¹⁴C]Ferulic acid	Caffeic acid, chlorogenic acid, scopolin, *p*-coumaric acid	Reznik and Urban, 1957a*
Brassica oleracea (red cabbage)	[3-¹⁴C]Ferulic acid	Chlorogenic acid and cyanidin glycoside	Reznik and Urban, 1957b*
Acer negundo (maple)	[3-¹⁴C]Sinapic acid	Guaiacyl components of lignin	Brown and Neish, 1959*
Triticum aestivum (wheat)	[2-¹⁴C]Sinapic acid	Vanilloyl methyl ketone isolated from lignin	Higuchi and Brown, 1963
Pinus strobus (white pine) cultures	[2-¹⁴C]Sinapic acid	Vanilloyl methyl ketone isolated from lignin	
Triticum aestivum (wheat)	[3-¹⁴C]Ferulic acid	Soluble and insoluble esters of ferulic and *p*-coumaric acids	El-Basyouni *et al.*, 1964*
	[3-¹⁴C]Sinapic acid	Soluble and insoluble esters of ferulic and *p*-coumaric acids	

Organism	Precursor	Notes	Reference
Brassica oleracea (red cabbage)	[Me-^{14}C]Ferulic acid	Sinapic acid, but radioactivity not exclusively present in the methyl group	Hess, 1964
Petunia hybrida	[Me-^{14}C]Sinapic acid	Petunidin 3-monoglucoside (in Cya genotypes)	Barz and Grisebach, 1967
Cicer arietinum	p-[^{14}C,^{3}H]Methoxy[β-^{14}C]cinnamic acid and 4-[^{14}C,^{3}H]methoxy-2',4'-dihydroxy[β-^{14}C]chalcone	Isoflavones with far smaller ^{3}H/^{14}C ratio as precursors	Ebel et al., 1970a
Cicer arietinum	p-[Me-^{2}H$_3$]methoxycinnamic	No isoflavones isolated possessing trideuterated methyl group	Ebel et al., 1970b
Robinia pseudoacacia	p-[^{14}C,^{3}H]Methoxy[β-^{14}C]cinnamic acid	Flavonoid acacetin possessed only 23% ^{3}H/2-^{14}C ratio of precursor	Steiner, 1970*
Petunia hybrida	[2-^{14}C]Sinapic acid	In Cya genotypes, the 3-monoglucosides of cyanidin, peonidin, and delphinidin. In Del genotypes, the three monoglucosides of delphinidin, petunidin, and malvidin	Berlin et al., 1971
Phaseolus aureus, Glycine max cell cultures	Methoxyl-labeled[^{14}C]p-methoxycinnamic, syringic, and 3,4,5-trimethoxybenzoic acids	Radioactive CO_2	Harms et al., 1972
Phaseolus aureus, Glycine max cell cultures	Several para-methoxylated compounds including p-methoxycinnamic, anisic, veratric, and 3,4,5-trimethoxybenzoic acids ([Me-^{14}C]labeled)	Radioactive CO_2	
Phyllostachys pubescens (bamboo)	[2-^{14}C]Ferulic acid	[2-^{14}C]p-Coumaric acid	Shimada et al., 1972a

* References cited in Shimada et al., 1972a.

catabolism (Mann *et al.*, 1963). [α-^{14}C]Hordenine fed to sprouting barley underwent partial demethylation in the roots, yielding labeled *N*-methyltyramine and probably lignin (Frank and Marion, 1956). Both tyramine and *N*-methyltyramine were products of hordenine demethylation in other studies (Rabitzsch, 1959). In addition to forming polymers, hordenine and tyramine were catabolized in barley cell cultures to 3,4-dihydroxyphenylacetic acid, which was further metabolized by ring fission reactions (Barz, 1977). *p*-Hydroxybenzoic and *p*-hydroxymandelic acids were identified as intermediates. High incorporation of label from [Me-^{14}C]-hordenine into insoluble cell residues was noted; in contrast, methyl groups resulting from O-demethylation of several substrates in cell cultures were quantitatively oxidized to carbon dioxide.

During seed germination and early growth of several plants, trigonelline disappears and pyridine nucleotides appear in the tissue. Nicotinic acid is the presumed intermediate. In *P. sativum* seedlings, N-demethylation of trigonelline to nicotinic acid, which proceeds extremely slowly by an unknown mechanism, is apparently rate-limiting in the conversion of trigonelline to pyridine nucleotides (Joshi and Handler, 1962). The yeast *Torula cremoris* utilizes trigonelline for growth and NAD biosynthesis. Joshi and Handler demonstrated that the methyl group of trigonelline was apparently removed by oxidative demethylation rather than transmethylation and then entered the C_1 pool. This situation resembles the demethylation of sarcosine to formaldehyde and glycine by rat liver mitochondria. Attempts to demonstrate trigonelline demethylation by cell-free extracts or particles from *Torula* were unsuccessful.

Trigonelline may undergo partial N-demethylation in *Melilotus officinalis, Nicotiana rustica,* and other plants. It does not function as a methyl donor in *C. arabica*. The role of trigonelline and nicotinic acid *N*-α-L-arabinoside in nicotinic acid metabolism is under investigation using plant cell cultures (Barz, 1977).

Demethylation of hyoscyamine to norhyoscyamine was observed in excised alkaloid-free shoots, but not in intact plants, of several *Datura* species (Romeike, 1964). The presence of norhyoscyamine exclusively in older plants further implied that demethylation may somehow be associated with senescence or tissue damage. Indeed, the pronounced stimulation of demethylation by a heterogeneous group of inhibitors, which included azide, fluoride, and hydroxylamine, was accompanied by marked plant damage. Demethylation seems therefore to be involved in general degradation during the course of death of the tissues.

Ricinine administered to excised old, yellow (senescent) leaves of castor bean (*Ricinus communis* L.) was partially demethylated to *N*-demethylricinine, *O*-demethylricinine, and small amounts of carbon dioxide (Waller and Nowacki, 1978). High levels of administered ricinine appeared to inhibit these demethylations. Crude homogenates of yellow leaves were inactive in demethylation. Evidence has accumulated supporting the conclusion that ricinine and/or N-demethylricinine or O-demethylricinine in yellow leaves are translocated from senescent tissues to other parts of the plant, especially to the growing apex. Both N- and O-demethylricinine undergo remethylation in green leaves to ricinine. Such interconversions of ricinine and its derivatives have been reviewed in relation to the possible physiological and metabolic functions of these alkaloids in *Ricinus*.

The N-demethylation of nicotine to nornicotine in tobacco plants and cell cultures has been extensively investigated (Mothes and Schütte, 1969). Interpretation of the literature is difficult, however, since many species of *Nicotiana* have been studied by different workers using widely dissimilar experimental conditions and approaches. Nicotine demethylation during curing and fermentation probably results from the participation by microorganisms and will not be considered here. The main features of nicotine demethylation are outlined here; readers should consult recent reviews for details of radiotracer studies (Mothes and Schütte, 1969; Waller and Nowacki, 1978; Waller and Dermer, this volume, Chapter 12).

Nicotine demethylation occurs primarily in leaves of tobacco, although the roots and stems of certain *Nicotiana* species are also active. The process may be related to leaf senescence in view of the higher demethylating activity in homogenates from older leaves. The mechanism of demethylation remains unclear. Leete and Bell (1959) administered [Me-^{14}C]nicotine to *Nicotiana tabacum* and after 1 week isolated 90% of the radioactivity in choline. Their conclusion that nicotine acted as a methyl donor received some support from cell-free studies (Bose *et al.*, 1956). Homogenates from leaves, stems, and roots of *Nicotiana glauca* and *N. tabacum* demethylated nicotine to nornicotine in the presence of the methyl group acceptor ethanolamine. Further investigations of the mechanism of demethylation pointed to an oxidative process rather than transmethylation (Mothes and Schütte, 1969). Demethylation of administered nicotine was coupled with oxygen uptake in nornicotine-containing species. Oxidation of the methyl group to carbon dioxide was also demonstrated. The proposal that nicotine 1'-oxide acted as an intermediate in this process was supported by an observed increase in nornicotine concentration following administration of nicotine 1'-oxide to tobacco leaves. Other authors, however, have eliminated nicotine 1'-oxide and N-formylnicotine as intermediates. In *N. tabacum*, nicotine demethylation appeared stereospecific, the unnatural (+) form being demethylated faster than the (−) form. (−)-Nicotine demethylated by excised *Nicotiana glutinosa* leaves yielded partially racemized nornicotine, suggesting that the pyrrolidone ring opened during demethylation. Leete and Chedekel (1974) proposed a mechanism involving (−)-nicotine N'-oxide as an intermediate to explain this phenomenon.

The demethylation system in excised leaves of *N. glutinosa* did not exhibit absolute specificity toward nicotine, since N-ethylnicotine, N-methylanabasine, and N-ethylanabasine were dealkylated at comparable rates. Nicotine demethylation was not restricted to tobacco; activity was demonstrated in several plants of the Solanaceae, which normally lack nicotine.

Early hypotheses concerning tobacco alkaloid biosynthesis assumed that nornicotine was the immediate precursor of nicotine or at least that the process was reversible. Experimental data on this subject remain controversial (Alworth and Rapoport, 1965; Mothes and Schütte, 1969; Schröter, 1966). Convincing evidence against nornicotine as a nicotine precursor was provided by Alworth and Rapoport (1965), who showed that the specific radioactivity of nicotine isolated from *N. glutinosa* plants administered $^{14}CO_2$ was always higher than that of nornicotine.

In the biosynthesis of *Papaver* opium alkaloids, Stermitz and Rapoport (1961) provided unequivocal evidence in *Papaver somniferum* of the successive O-demethylations of thebaine to codeine and finally to morphine. This sequence, supported by Kleinschmidt (1960), was irreversible. The most probable route for the conversion of thebaine to codeine involves initial demethylation to neopinone, followed by rearrangement to codeinone which is reduced to codeine. The biosynthetic enol ether cleavage of thebaine proceeds by methyl cleavage with retention of the 6-oxygen (Horn *et al.,* 1978). This mechanism appears comparable to that established in mammals (Renson *et al.,* 1965) and microorganisms (Bernhardt *et al.,* 1975) for aromatic methyl ether cleavage, in which O-demethylation proceeds by oxidation to the hydroxymethyl ether (hemiacetal) which then reverts to the phenol and formaldehyde. In *P. somniferum,* morphine itself may undergo irreversible N-demethylation to normorphine which is subsequently degraded to nonalkaloidal metabolites. In view of their high turnover rates, it has been suggested that morphine alkaloids play an active metabolic role, perhaps as specific methylating agents.

These studies clearly establish O-demethylation of alkaloids as a definite pathway of metabolism, in contrast to the original view that O-methylated alkaloids were metabolic end products.

D. Demethylation Reactions in the Metabolism of Xenobiotics

Xenobiotics may be transformed into less toxic metabolites in animals, higher plants, and insects by oxidation, reduction, hydrolysis, and conjugation reactions (McEwen and Stephenson, 1979). Investigation of dealkylation processes has been very successful, providing us with the clearest picture to date of demethylation systems at the enzymatic level in higher plants.

Phenylurea herbicides, which are potent photosynthetic inhibitors, are among a wide range of xenobiotics whose oxidative N-demethylation has been demonstrated in higher plants. N-Demethylation of phenylurea herbicides to less phytotoxic monomethylated and unmethylated derivatives is recognized as the initial step in their detoxication. Dealkylation rates have been directly correlated with the recovery rates from inhibition of photosynthesis (Swanson and Swanson, 1968). The participation of plant microsomal systems in xenobiotic detoxication might well be expected in view of the known metabolism of foreign compounds by animal and insect microsomal preparations (Shuster, 1964; Rockstein, 1978). In recent years, direct evidence supporting this concept has been accumulating.

Frear *et al.* (1972) isolated a microsomal mixed-function oxidase system from leaves and etiolated hypocotyls of cotton plants (*Gossypium hirsutum* L.) that N-demethylated monuron [3-(4'-chlorophenyl)-1,1-dimethylurea] to monomethylmonuron [3-(4'-chlorophenyl)-1-methylurea]. Active enzyme preparations were also isolated from leaves of other species. Molecular oxygen and reduced pyridine nucleotides were required as cofactors. N-Demethylation was observed with a number of closely related *N*-methylphenylurea herbicides, including diuron, monomethylmonuron, fluometuron, and fenuron, but not with the methoxymethylphenylurea linuron (Fig. 18). The system appeared quite specific for dimethyl- and monomethylphenylureas, since other compounds such as diphenamid, atrazine, and prometryne, which undergo N-demethylation, N-dealkylation, and sulfoxidation reactions *in vivo*, were not substrates. The products of monomethylmonuron N-demethylation in the cotton system were formaldehyde and 4-chlorophenylurea, produced in equimolar quantities. An

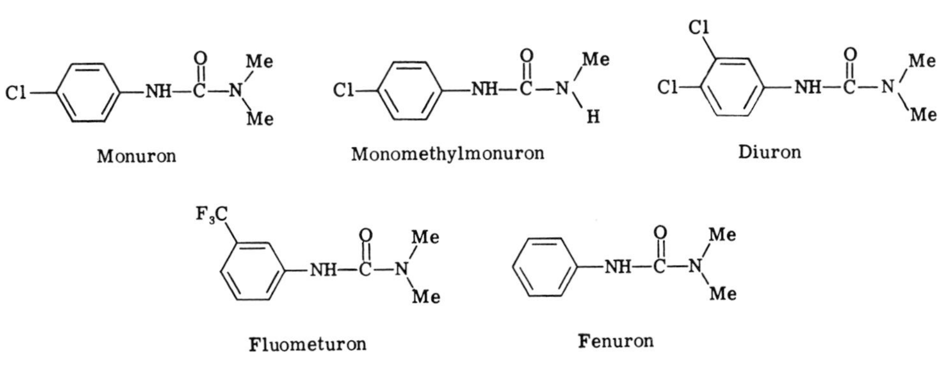

Fig. 18. Structures of the *N*-methylphenylurea herbicides demethylated by the cotton *N*-demethylase system.

unstable intermediate in this reaction was identified as 3-(4-chlorophenyl)-1-hydroxymethylurea, indicating that substrate oxidation occurred at the methyl position (Tanaka *et al.*, 1972). Evidence for the participation of this intermediate was strengthened by isolation of the β-D-glucosides of 3-(4-chlorophenyl)-1-hydroxymethyl-1-methylurea and 3-(4-chlorophenyl)-1-hydroxymethylurea from excised cotton leaves treated with monuron and monomethylmonuron.

Many similarities exist between the cotton *N*-demethylase and related animal microsomal oxidase systems (Frear *et al.*, 1969). Demethylase activity was localized in the microsomal fraction and was primarily associated with the "smooth" fraction of the endoplasmic reticulum. Its sensitivity to detergents, carbon monoxide, thiol reagents, and electron acceptors clearly resembles that reported for animal mixed-function oxidase systems (Mason *et al.*, 1965) and may indicate the involvement of a similar electron transport system in plants and animals. Indeed, NADPH–cytochrome c reductase activity and cytochromes of the b_5, P-450, and P-420 types were detected in microsomal preparations from etiolated cotton hypocotyls.

Sandermann *et al.* (1977) recently reviewed the properties of plant and animal mixed-function oxidases and the participation of cytochrome P-450 in these reactions.

The mechanism of N-demethylation of substituted 3-(phenyl)-1-methylureas by isolated cotton microsomes was examined by inhibition studies with *N*-methylcarbamates, semicarbazide analogues of diuron, substituted anilides, and thio analogues of linuron, diuron, and monomethyldiuron (Frear *et al.*, 1972). Moreover, studies with diuron, monuron, and their N-demethylated products showed that progressive N-demethylation resulted in increasing product inhibition with each N-demethylation step. Product inhibition of this detoxication pathway might be prevented by rapid removal of the demethylated products from the enzyme active site by the formation of polar metabolites and insoluble plant residues.

A comparable *N*-demethylase system was recently described by Young and Beevers (1976), who fractionated homogenates from germinating castor bean endosperm by density gradient centrifugation. Like cotton *N*-demethylase, activity was principally located in the endoplasmic reticulum fraction, was dependent upon reduced pyridine nucleotides and molecular oxygen, and responded to potential inhibitors in ways typical of mixed-function oxidases. In contrast to the wide substrate specificity of animal mixed-function oxidases, the castor bean system resembled cotton *N*-demethylase and *Sorghum* cinnamic acid 4-hydroxylase in being active with only a limited range of substrates. The enzyme appeared specific for *N*-methylarylamines, such as *p*-chloro-*N*-methylaniline, *N*-methylaniline, and *N,N*-dimethylaniline. It failed to dealkylate other potential substrates having at least one *N*-, *S*-, or *O*-methyl group. Ricinine, a compound native to castor bean, also was not demethylated, and thus the physiological substrate for this enzyme remains unknown.

III. CONCLUSIONS

Our knowledge of methylation and demethylation processes in secondary plant metabolism has expanded greatly during recent years. It is now clear that a multiplicity of distinct methyltransferases exists in plant tissues. Study of their characteristics has shown that these enzymes possess a much narrower substrate specificity than previously imagined. In order to understand fully how these methyltransferases operate *in vivo*, future research should be designed to address the following questions. What are the tissue and subcellular localizations of these enzymes? What is the nature of the kinetic and physical relationships existing between the methyltransferases and enzymes associated with them within their respective metabolic pathways? How are tissue methyltransferase levels regulated in relation to the stage of development of the organism or organs? Does product inhibition by SAH play a role in controlling the rate of transmethylation reactions *in vivo*?

Whereas the biochemistry of N-, O-, and S-methylation reactions has been well documented, little is known about C-methylations (Lederer, 1977; Speedie *et al.*, 1975). It will be of great interest to compare the latter with transmethylation systems described in this chapter when sufficient enzymological data become available.

Demethylation reactions, although well documented by *in vivo* radiotracer studies, have not lent themselves easily to investigation at the enzymatic level. It is hoped that more attention will be focused on this subject whose physiological and biochemical significance is comparable with that of methylation reactions in the metabolism of secondary plant products.

REFERENCES

Alworth, W. L., and Rapoport, H. (1965). *Arch. Biochem. Biophys.* **112**, 45–53.

Antony, A., Gopinathan, K. P., and Vaidyanathan, C. S. (1975). *Indian J. Exp. Biol.* **13**, 39–41.

Antoun, M. D., and Roberts, M. F. (1975). *Planta Med.* **28**, 6–11.

Axelrod, J., and Tomchick, R. (1958). *J. Biol. Chem.* **233**, 702–705.

Ball, P., Knuppen, R., and Breuer, H. (1971). *Eur. J. Biochem.* **21**, 517–525.

Barton, D. H. R., Kirby, G. W., Taylor, J. B., and Thomas, G. M. (1963). *J. Chem. Soc.* pp. 4545–4558.

Barz, W. (1977). *In* "Plant Tissue Culture and its Biotechnological Application" (W. Barz, E. Reihard, and M. H. Zenk, eds.), pp. 153–171. Springer-Verlag, Berlin and New York.

Barz, W., and Grisebach, H. (1967). *Z. Naturforsch., Teil B* **22**, 627–633.

Barz, W., and Hösel, W. (1975). *In* "The Flavonoids" (J. B. Harborne, T. J. Mabry, and H. Mabry; eds.), pp. 916–969. Academic Press, New York.
Barz, W., Adamek, C., and Berlin, J. (1970). *Phytochemistry* **9,** 1735–1744.
Basmadjian, G. P., and Paul, A. G. (1971). *Lloydia* **34,** 91–93.
Basmadjian, G. P., Hussain, S. F., and Paul, A. G. (1978). *Lloydia* **41,** 375–380.
Battersby, A. R., Foulkes, D. M., and Binks, R. (1965). *J. Chem. Soc.* pp. 3323–3332.
Baxter, R. M., Kandel, S. I., Okany, A., and Pyke, R. G. (1964). *Can. J. Chem.* **42,** 2936–2938.
Berlin, J. (1972). Dissertation, University of Freiburg/Br., Germany.
Berlin, J., Barz, W., Harms, H., and Haider, K. (1971). *FEBS Lett.* **16,** 141–146.
Bernhardt, F.-H., Pachowsky, H., and Staudinger, H. (1975). *Eur. J. Biochem.* **57,** 241–256.
Bose, B. C., De, H. N., and Mohammad, S. (1956). *Indian J. Med. Res.* **44,** 91–97.
Brady, L. R., and Tyler, V. E. (1958). *Plant Physiol.* **33,** 334–338.
Breccia, A., Crespi, A. M., and Rampi, M. A. (1966). *Z. Naturforsch., Teil B* **21,** 1243–1245.
Brodie, B. B., Gillette, J. R., and La Du, B. N. (1958). *Annu. Rev. Biochem.* **27,** 427–454.
Brown, S. A., and Sampathkumar, S. (1977). *Can. J. Biochem.* **55,** 686–692.
Brunet, G., Saleh, N. A. M., and Ibrahim, R. K. (1978). *Z. Naturforsch., Teil C* **33,** 786–788.
Burton, E. G., and Sakami, W. (1969). *Biochem. Biophys. Res. Commun.* **36,** 228–234.
Byerrum, R. U., Flokstra, J. H., Dewey, L. J., and Ball, C. D. (1954). *J. Biol. Chem.* **210,** 633–643.
Camm, E. L., and Towers, G. H. N. (1973). *Phytochemistry* **12,** 1575–1580.
Cantoni, G. L. (1951). *J. Biol. Chem.* **189,** 203–216.
Cantoni, G. L. (1953). *J. Biol. Chem.* **204,** 403–416.
Coward, J. K., Slisz, E. P., and Wu, F. Y.-H. (1973). *Biochemistry* **12,** 2291–2297.
Dewey, L. J., Byerrum, R. U., and Ball, C. D. (1954). *J. Am. Chem. Soc.* **76,** 3997–3999.
Dodd, W. A., and Cossins, E. A. (1968). *Phytochemistry* **7,** 2143–2145.
Dodd, W. A., and Cossins, E. A. (1970). *Biochim. Biophys. Acta* **201,** 461–470.
Ebel, J., Achenbach, H., Barz, W., and Grisebach, H. (1970a). *Biochim. Biophys. Acta* **215,** 203–205.
Ebel, J., Barz, W., and Grisebach, H. (1970b). *Phytochemistry* **9,** 1529–1534.
Ebel, J., Hahlbrock, K., and Grisebach, H. (1972). *Biochim. Biophys. Acta* **268,** 313–326.
Fales, H. M., Mann, J., and Mudd, S. H. (1963). *J. Am. Chem. Soc.* **85,** 2025–2026.
Finkle, B. J., and Masri, M. S. (1964). *Biochim. Biophys. Acta* **85,** 167–169.
Finkle, B. J., and Nelson, R. F. (1963). *Biochim. Biophys. Acta* **78,** 747–749.
Finkle, B. J., and Kelly, S. H. (1974). *Phytochemistry* **13,** 1719–1725.
Frank, A. W., and Marion, L. (1956). *Can. J. Chem.* **34,** 1641–1646.
Frear, D. S., Swanson, H. R., and Tanaka, F. S. (1969). *Phytochemistry* **8,** 2157–2169.
Frear, D. S., Swanson, H. R., and Tanaka, F. S. (1972). *Recent Adv. Phytochem.* **5,** 225–246.
Freudenberg, K. (1965). *Science* **148,** 595–600.
Freudenberg, K., and Neish, A. C. (1968). *Mol. Biol. Biochem. Biophys.* **2,** 17–26.
Friend, J., and Thornton, J. D. (1974). *Phytopathol. Z.* **81,** 56–64.
Fritig, B., Hirth, L., and Ourisson, G. (1970). *Phytochemistry* **9,** 1963–1975.
Glass, A. D. M., and Bohm, B. A. (1972). *Phytochemistry* **11,** 2195–2199.
Guranowski, A., and Pawelkiewicz, J. (1977). *Eur. J. Biochem.* **80,** 517–523.
Hahlbrock, K. (1977). *In* "Plant Tissue Culture and its Biotechnological Application" (W. Barz, E. Reinhard, and M. H. Zenk, eds.), pp. 95–111. Springer-Verlag, Berlin and New York.
Hahlbrock, K., and Grisebach, H. (1975). *In* "The Flavonoids" (J. B. Harborne, T. J. Mabry, and H. Mabry, eds.), pp. 866–915. Academic Press, New York.
Hahlbrock, K., Ebel, J., Ortmann, R., Sutter, A., Wellman, E. and Grisebach, H. (1971). *Biochim. Biophys. Acta* **244,** 7–15.
Harborne, J. B. (1967). "Comparative Biochemistry of the Flavonoids," p. 267. Academic Press, New York.

Harms, H., Haider, K., Berlin, J., Kiss, R., and Barz, W. (1972). *Planta* **105**, 342–351.
Herdt, E., Sütfeld, R., and Wiermann, R. (1978). *Cytobiologie* **17**, 433–441.
Hess, D. (1964). *Planta* **60**, 568–581.
Hess, D. (1965). *Z. Pflanzenphysiol.* **53**, 1–18.
Hess, D. (1966). *Z. Pflanzenphysiol.* **55**, 374–386.
Hess, D. (1967). *Z. Pflanzenphysiol.* **56**, 12–19.
Hess, D. (1968). "Biochemische Genetik." Springer-Verlag, Berlin and New York.
Higuchi, T., and Brown, S. A. (1963). *Can. J. Biochem. Physiol.* **41**, 613–620.
Higuchi, T., Shimada, M., and Ohashi, H. (1967). *Agric. Biol. Chem.* **31**, 1459–1465.
Higuchi, T., Shimada, M., Nakatsubo, F., and Tanahashi, M. (1977). *Wood Sci. Technol.* **11**, 153–167.
Horn, J. S., Paul, A. G., and Rapoport, H. (1978). *J. Am. Chem. Soc.* **100**, 1895–1898.
Hyodo, H., Kuroda, H., and Yang, S. F. (1978). *Plant Physiol.* **62**, 31–35.
Jindra, A. (1967). *Acta Fac. Pharm. Bohemoslov.* **31**, 23.
Joshi, J. G., and Handler, P. (1960). *J. Biol. Chem.* **235**, 2981–2983.
Joshi, J. G., and Handler, P. (1962). *J. Biol. Chem.* **237**, 3185–3188.
Kaneko, K. (1962). *Chem. Pharm. Bull.* **10**, 1085–1087.
Kim, S., Nochumson, S., Chin, W., and Paik, W. K. (1978). *Anal. Biochem.* **84**, 415–422.
Kleinschmidt, G. (1960). *Pharmazie* **15**, 663–664.
Kuboi, T., and Yamada, Y. (1978). *Biochim. Biophys. Acta* **542**, 181–190.
Kuroda, H., and Higuchi, T. (1976). *Phytochemistry* **15**, 1511–1154.
Kuroda, H., Shimada, M., and Higuchi, T. (1975). *Phytochemistry* **14**, 1759–1763.
Lederer, C. (1977). *In* "The Biochemistry of Adenosylmethionine" (F. Salvatore, E., Borek, V. Zappia, H. G. Williams-Ashman, and F. Schlenk, eds.), pp. 89–126. Columbia Univ. Press, New York.
Leete, E., and Bell, V. M. (1959). *J. Am. Chem. Soc.* **81**, 4358–4359.
Leete, E., and Chedekel, M. R. (1974). *Phytochemistry* **13**, 1853–1859.
Legrand, M., Fritig, B., and Hirth, L. (1976). *Phytochemistry* **15**, 1353–1359.
Legrand, M., Fritig, B., and Hirth, L. (1978). *Planta* **144**, 101–108.
Luckner, M., Nover, L., and Böhm, H. (1977). *Mol. Biol., Biochem. Biophys.* **23**.
McEwen, F. L. and Stephenson, G. R. (1979). "The Use and Significance of Pesticides in the Environment." Wiley, New York.
Mack, J. P. G. (1974). The Indolethylamine N-methyltransferases of *Phalaris tuberosa*. Ph.D. Thesis, Sydney University, Sydney, Australia.
Mack, J. P. G., and Slaytor, M. B. (1978). *J. Chromatogr.* **157**, 153–159.
Madyastha, K. M., Guarnaccia, R., and Coscia, C. J. (1972). *Biochem. J.* **128**, 34P.
Madyastha, K. M., Guarnaccia, R., Baxter, C., and Coscia, C. J. (1973). *J. Biol. Chem.* **248**, 2497–2501.
Mann, J. D., Fales, H. M., and Mudd, S. H. (1963). *J. Biol. Chem.* **238**, 3820–3823.
Mann, J. D., and Mudd, S. H. (1963). *J. Biol. Chem.* **238**, 381–385.
Mann, J. D., Steinhart, C. E., and Mudd, S. H. (1963). *J. Biol. Chem.* **238**, 676–681.
Mason, H. S., North, J. C., and Vanneste, M. (1965). *Fed. Proc., Fed. Am. Soc. Exp. Biol.* **24**, 1172–1180.
Maule, A. J., and Ride, J. P. (1976). *Phytochemistry* **15**, 1661–1664.
Meier, H., and Zenk, M. H. (1965). *Z. Pflanzenphysiol.* **53**, 415–421.
Mizusaki, S., Tanabe, Y., Noguchi, M., and Tamaki, E. (1971). *Plant Cell. Physiol.* **12**, 633–640.
Mizusaki, S., Tanabe, Y., Noguchi, M., and Tamaki, E. (1973). *Plant Cell. Physiol.* **14**, 103–110.
Mothes, K., and Schütte, H. R. (1969). "Biosynthese der Alkaloide." VEB Dtsch. Verlag Wiss., Berlin.
Mudd, S. H. (1960). *Biochim. Biophys. Acta* **37**, 164–165.
Mudd, S. H. (1961). *Nature (London)* **189**, 489.

Mudd, S. H. (1973). *In* "Metabolic Conjugation and Metabolic Hydrolysis" (W. H. Fishman, ed.), Vol. 3, pp. 297–350. Academic Press, New York.

Müller-Enoch, D., Thomas, H., Streng, W., Wildfeuer, W., and Haferkamp, O. (1976). *Z. Naturforsch., Teil C* **31**, 509–513.

Murakoshi, I., Sanda, A., Haginawa, J., Suzuki, N., Ohmiya, S., and Otomasu, H. (1977). *Chem. Pharm. Bull.* **25**, 1970–1973.

Nakamura, Y., Fushiki, H., and Higuchi, T. (1974). *Phytochemistry* **13**, 1777–1784.

Paul, A. G. (1973). *Lloydia* **36**, 36–45.

Pelter, A., Bradshaw, J., and Warren, R. F. (1971). *Phytochemistry* **10**, 835–850.

Poulton, J. E., and Butt, V. S. (1975). *Biochim. Biophys. Acta* **403**, 301–314.

Poulton, J. E., and Butt, V. S. (1976). *Arch. Biochem. Biophys.* **172**, 135–142.

Poulton, J. E., Grisebach, H., Ebel, J., Schaller-Hekeler, B., and Hahlbrock, K. (1976a). *Arch. Biochem. Biophys.* **173**, 301–305.

Poulton, J. E., Hahlbrock, K., and Grisebach, H. (1976b). *Arch. Biochem. Biophys.* **176**, 449–456.

Poulton, J. E., Hahlbrock, K., and Grisebach, H. (1977). *Arch. Biochem. Biophys.* **180**, 543–549.

Rabitzsch, G. (1959). *Planta Med.* **7**, 268–297.

Renson, J., Weissbach, H., and Udenfriend, S. (1965). *Mol. Pharmacol.* **1**, 145–148.

Rhodes, M. J. C., Hill, A. C. R., and Wooltorton, L. S. C. (1976). *Phytochemistry* **15**, 707–710.

Roberts, M. F. (1974a). *Phytochemistry* **13**, 1841–1845.

Roberts, M. F. (1974b). *Phytochemistry* **13**, 1847–1851.

Roberts, M. F., and Waller, G. R. (1979). *Phytochemistry* **18**, 451–455.

Rockstein, M. (1978). "Biochemistry of Insects." Academic Press, New York.

Romeike, A. (1964). *Flora (Jena)* **154**, 163–173.

Saleh, N. A. M., Fritsch, H., Kreuzaler, F., and Grisebach, H. (1978). *Phytochemistry* **17**, 183–186.

Sandermann, H., Diesperger, H., and Scheel, D. (1977). *In* "Plant Tissue Culture and its Biotechnological Application" (W. Barz, E. Reinhard, and M. H. Zenk, eds.), pp. 178–196. Springer-Verlag, Berlin and New York.

Schröter, H. B. (1966). *Abh. Dtsch. Akad. Wiss. Berlin, Kl. Chem., Geol. Biol.* **3**, 157–160.

Senoh, S., Daly, J., Axelrod, J., and Witkop, B. (1959). *J. Am. Chem. Soc.* **81**, 6240–6245.

Sharma, S. K., and Brown, S. A. (1978). *J. Chromatogr.* **157**, 427–431.

Sharma, S. K., and Brown, S. A. (1979). *Can. J. Biochem.* **57**, 986–994.

Sharma, S. K., Garrett, J. M., and Brown, S. A. (1979). *Z. Naturforsch., Teil C* **34**, 387–391.

Shimada, M. (1972). *Wood Res.* **53**, 19–65.

Shimada, M., and Higuchi, T. (1970). *Wood Res.* **50**, 19–28.

Shimada, M., Fushiki, H., and Higuchi, T. (1972a). *Phytochemistry* **11**, 2247–2252.

Shimada, M., Fushiki, H., and Higuchi, T. (1972b). *Phytochemistry* **11**, 2657–2662.

Shimada, M., Fushiki, H., and Higuchi, T. (1973a). *Mokuzai Gakkaishi* **19**, 13–21.

Shimada, M., Kuroda, H., and Higuchi, T. (1973b). *Phytochemistry* **12**, 2873–2875.

Shuster, L. (1964). *Annu. Rev. Biochem.* **33**, 571–596.

Speedie, M. K., Hornemann, U., and Floss, H. G. (1975). *J. Biol. Chem.* **250**, 7819–7825.

Spenser, I. D. (1970). *In* "Chemistry of the Alkaloids" (S. W. Pelletier, ed.), pp. 669–718. Van Nostrand-Reinhold, Princeton, New Jersey.

Sribney, M., and Kirkwood, S. (1953). *Nature (London)* **171**, 931–932.

Stermitz, F. R., and Rapoport, H. (1961). *J. Am. Chem. Soc.* **83**, 4045–4050.

Street, H. E. (1973). *In* "Biosynthesis and Its Control in Plants" (B. V. Milborrow, ed.), p. 93. Academic Press, New York.

Stul'nikova, R. I. (1959). *Int. Abstr. Biol. Sci.* **14**, 13.

Sütfeld, R., and Wiermann, R. (1978). *Biochem. Physiol. Pflanz.* **172**, 111–123.

Sütfeld, R., Kehrel, B., and Wiermann, R. (1978). *Z. Naturforsch., Teil C* **33**, 841–846.
Suzuki, T., and Takahashi, E. (1975). *Biochem. J.* **146**, 87–96.
Swanson, C. R., and Swanson, H. R. (1968). *Weed Sci.* **16**, 137–142.
Tanaka, F. S., Swanson, H. R., and Frear, D. S. (1972). *Phytochemistry* **11**, 2701–2708.
Testa, B., and Jenner, P. (1976). "Drug Metabolism: Chemical and Biochemical Aspects." Dekker, New York.
Thompson, H. J., Sharma, S. K., and Brown, S. A. (1978). *Arch. Biochem. Biophys.* **188**, 272–281.
Tsang, Y.-F., and Ibrahim, R. K. (1979). *Phytochemistry* **18**, 1131–1136.
Vance, C. P., and Sherwood, R. T. (1976). *Plant Physiol.* **57**, 915–919.
Waller, G. R., and Nowacki, E. K. (1978). "Alkaloid Biology and Metabolism in Plants." Plenum, New York.
Watt, C.-K., and Towers, G. H. N. (1975). *Phytochemistry* **14**, 663–666.
Wengenmayer, H., Ebel, J., and Grisebach, H. (1974). *Eur. J. Biochem.* **50**, 135–143.
Wiermann, R. (1970). *Planta* **95**, 133–145.
Wildman, W. C., Fales, H. M., Highet, R. J., Breuer, S. W., and Battersby, A. R. (1962). *Proc. Chem. Soc., London* pp. 180–181.
Willeke, U., Heeger, V., Meise, M., Neuhann, H., Schindelmeiser, I., Vordemfelde, K., and Barz, W. (1979). *Phytochemistry* **18**, 105–110.
Wong, E., and Francis, C. M. (1968). *Phytochemistry* **7**, 2131–2137.
Yamada, Y., and Kuboi, T. (1976). *Phytochemistry* **15**, 395–396.
Yoshida, D. (1973). *Bull. Hatano Tob. Exp. Stn.* **73**, 239–244.
Young, O., and Beevers, H. (1976). *Phytochemistry* **15**, 379–385.

Glycosylation and Glycosidases 23

WOLFGANG HÖSEL

I. GENERAL INTRODUCTION

Many secondary plant products are accumulated and stored as glycosides within plants. Specific examples are found in nearly all groups of secondary

The Biochemistry of Plants, Vol. 7

ISBN 0-12-675407-1

compounds, as can be seen in Chapters 10–17 of this volume and in other reviews (Courtois and Percheron, 1970; Miller, 1973). An overwhelming number of different glycosides are found in nature, since glycosidic linkages between the oxygen, carbon, nitrogen, or sulfur atoms of different aglycones can be linked to any one sugar. Since several sugar molecules are often present within glycosidic compounds and may be linked in various fashions, the possibility of variety is further greatly enhanced (Courtois and Percheron, 1970). Formation and breakdown, usually initiated by hydrolysis of these different secondary glycosides, are a consequence of the appropriate enzymes present in plants and can be represented in terms of enzymatic substrate specificities. The aim of this chapter is to summarize and to synopsize knowledge about the enzymes responsible for the transfer or the removal of sugars in different types of secondary plant products.

From a chemical point of view, glycosylated secondary products differ from the free aglycones in two main properties: They show enhanced water solubility and decreased chemical reactivity. This may explain why glycosylated compounds, rather than the free aglycones, are accumulated in the plant kingdom; they are better stored within the plant vacuole and are less reactive toward other cellular components than the aglycones (Pridham, 1965). Glycosylated compounds are therefore often thought of as excretion products or as physiologically inactive plant storage forms. Hydrolysis of secondary plant glycosides, on the other hand, would release the physiologically active aglycones.

Synthesis and hydrolysis of glycosides can be described by the general transglycosylation scheme depicted in Fig. 1 with glucose as an example.

From a general point of view, the transglycosylation reaction can be catalyzed by enzymes of the transferase (E.C. 2.4) and the glycosidase (E.C. 3.2) class. Hydrolytic reactions catalyzed by the latter type can be considered transferase reactions with water as the acceptor.

Nearly all biosynthetic glycosylation reactions involving secondary plant products so far investigated are catalyzed by transferases utilizing nucleotide-activated sugars. Consequently, glycosidases might be regarded as being responsible for the hydrolysis of secondary plant glycosides acting

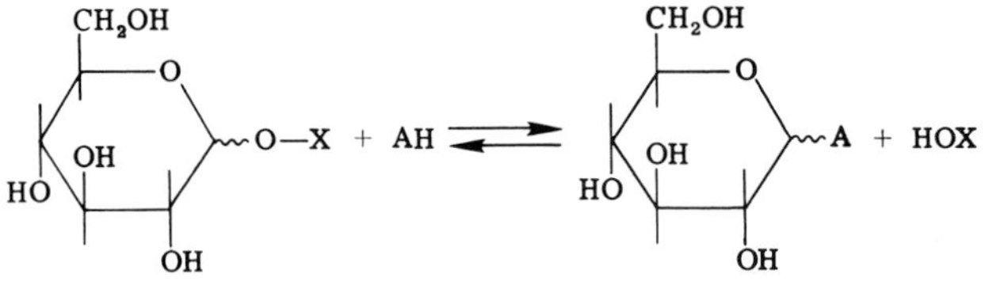

Fig. 1. General transglycosylation scheme. X, Phosphate, pyrophosphate, nucleoside-diphosphate, sugar, aromatic compound, alcohol, or hydrogen; A, sugar acceptor.

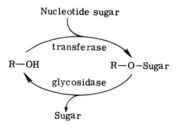

Fig. 2. Concept of synthesis and hydrolysis of secondary plant glycosides.

either within the intact plant or after destruction of the tissue. This concept, illustrated in Fig. 2, is generally accepted.

It is also recognized that characterization of the enzymes involved often provides valuable information about these processes of biosynthesis or degradation and often is the key to resolving unsolved problems defined with other methods, e.g., radioactive tracers. Therefore, in this chapter emphasis will be on describing the enzymatically well-characterized reactions of biosynthesis and hydrolysis of secondary plant glycosides. Since much information was gained in the field of plant phenolics, especially flavonoids, glycosylation reactions and glycosidases involving these compounds will be extensively discussed here. However, reference should also be made to Chapters 3, 10, and 14–17 for additional information.

II. GLYCOSYLATION

The enzymes responsible for glycosylating secondary products are transferases that utilize nucleotide sugars. The situation therefore resembles the biosynthesis of di- and oligosaccharides (starch, cellulose, hemicellulose, etc.). No effort has been made here to treat the biosynthesis or other metabolism of nucleotide sugars; see Chapter 4 in Vol. 3 of this treatise. Since the free energy of hydrolysis of nucleotide sugars in most cases by far exceeds that of secondary plant glycosides, biosynthesis of the glycoside is favored. One known exception, described in detail later, is UDP-glucose : flavonol 3-glucosylation, where a measurable reverse reaction was observed (Sutter and Grisebach, 1975).

In the following section the different groups of secondary plant products will be treated separately, and most of the emphasis will be placed on describing the substrate specificity of the glycosyltransferases with respect to the aglycone acceptors. The substrate specificity with respect to the sugar donors will be described for all transferases in a separate section, since the degree of substrate specificity is the same for most of these enzymes.

A. Glycosylation of Simple Phenolics and Phenylpropanes

Simple phenolics (e.g., hydroquinone and resorcinol) were among the first substrates used in studying glycosylation. Information initially came from feeding experiments in which various phenolics were added exogenously to higher plants, resulting in formation of the respective glucosylated compounds (Pridham, 1965, and references cited therein). The ability to glucosylate foreign phenols was observed in nearly all higher plants tested but was apparently lacking in algae and fungi (Pridham, 1964). These glucosylations were generally interpreted as detoxification processes (Pridham, 1965), although this has not been proven in any case. The glucosylations may also have resulted from the action of transferases responsible for the glucosylation of widely distributed endogenous phenolic substrates, either because of structural similarities between foreign and endogenous compounds or of low substrate specificities of the transferases involved (cf. Barz and Köster, this volume, Chapter 3).

In vitro glucosylation of simple phenolics has been demonstrated several times, the most extensive studies being devoted to the wheat germ system (Yamaha and Cardini, 1960a,b; Trivelloni *et al.,* 1962). Here an enzyme that glucosylates di- or triphenols in the presence of UDPG or ADPG and yields the respective monoglucoside has been described. Monophenols cannot serve as substrates for the enzyme. However, additional glucosyltransferase activities acting on salicylic alcohol and anthranilic acid are also present in the crude protein extract of wheat germ. Wheat germ therefore undoubtedly contains several different glucosyltransferases for phenolics (Yamaha and Cardini, 1960a). An additional enzyme has been detected and separated that catalyzes a glucosyl transfer from UDP-glucose to phenol β-glucosides, yielding phenyl β-gentiobiosides (Yamaha and Cardini, 1960b). The apparent K_m values for the phenolic substrates and UDP-glucose are in the range of 1–2 mM for the two transferases described.

The physiological aglycone substrate in wheat germ seems to be methoxyhydroquinone, for the β-glucoside of this substance occurs in this plant (cf. Yamaha and Cardini, 1960a). Similar transferases responsible for the synthesis of phenolic monoglucosides have been described in *Vicia faba* (Pridham and Saltmarsh, 1963), *Impatiens balsamina* (Miles and Hagen, 1968), and cell cultures of *Datura innoxia* (Tabata *et al.,* 1976). In the latter system only cultures pretreated with hydroquinone showed glucosylating activity. With the enzyme extracts from *V. faba* and *D. innoxia,* glucosylation of the alcoholic rather than the phenolic group of salicyclic alcohol was observed, and isosalicin rather than salicin was formed. Like results from tracer studies on *Salix* sp. (Zenk, 1967), this suggests that salicylic alcohol is not a precursor of salicin.

Glucosylation of *trans-o*-hydroxycinnamic acid by an enzyme extract from

Melilotus alba in the presence of UDP-glucose has also been demonstrated (Kleinhofs *et al.,* 1967). The product of the reaction has been tentatively identified as *trans*-β-D-glucosyl-*o*-hydroxycinnamic acid, a glucoside present in large amounts in this plant.

Synthesis of glucose esters of hydroxycinnamic acids by means of UDP-glucose-dependent transferases was observed with enzyme preparations from leaves of *Geranium zonale* (Corner and Swain, 1965) and from young, unripe apples (Macheix, 1977). In various cell cultures, esterification of other aromatic acids (e.g., benzoic and naphthylacetic acids) by means of UDP-glucose and transferases was also observed (N. Amrhein *et al.,* unpublished; Schlepphorst and Barz, 1979) (cf. Barz and Köster, this volume, Chapter 3).

An interesting glucosyltransferase, which catalyzed the synthesis of coniferin from UDP-glucose and coniferyl alcohol, was identified and purified from suspension cultures of 'Paul's Scarlet' rose (Ibrahim and Grisebach, 1976). This glucosyltransferase is thought to take part in lignin biosynthesis; for further details of the role of cinnamyl alcohol glucosides in lignin biosynthesis, the reader is referred to Grisebach (this volume, Chapter 15). Of all the different phenolic substrates tested, coniferyl alcohol and sinapyl alcohol were by far the best substrates, with apparent K_m values of 3.3 and 5.6 μM, respectively. Optimum activity occurred at pH 7.5–8, and the MW of the 120-fold purified transferase was 52,000. A similar transferase has been partially purified and characterized from lignifying stem segments of *Forsythia ovata,* and this enzyme activity seems to be widely distributed throughout the higher plant kingdom (Ibrahim, 1977).

B. Flavonoids and Anthocyanins

Figure 3 outlines different structural possibilities for flavonoid glycosides. Since all these types of glycosides may be found in each group of flavonoids (e.g., chalcones, flavones, flavonols, isoflavones, anthocyanins) and since different sugars may take part in glycoside formation, an overwhelming number of different flavonoid glycosides are theoretically possible and are indeed found in nature (cf. Wagner, 1974). Therefore it is to be expected that

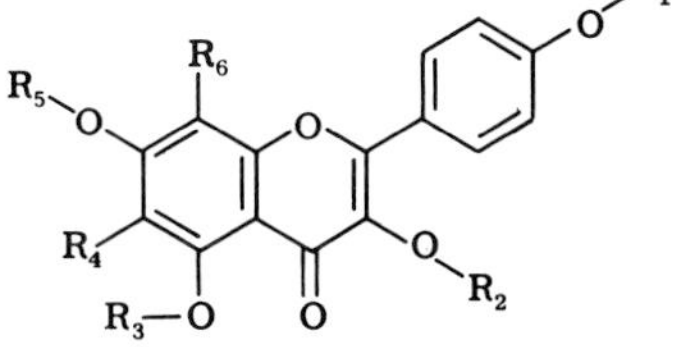

Fig. 3. Schematic outline of different types of flavonoid glycosides. Monoglycoside: Any R = sugar. Di- or oligoside: (1) Any R contains more than one sugar; (2) two different R groups (e.g., R_2 and R_5) contain one or more sugar moieties. C-glycoside: Sugar at R_4 or R_6.

the glycosyltransferases involved in biosynthesis of this large variety of glycosidic compounds will differ greatly in their substrate specificity.

The glycosyltransferases involved can be divided into two different groups: (a) those that glycosylate a phenolic hydroxyl of the flavonoid nucleus and (b) those that glycosylate a hydroxyl group of a sugar already linked to the aglycone.

1. Glycosylation of Phenolic Hydroxyls of Flavonoids

The earliest studies in this area established that transferases acting on simple phenolics (cf. Section II,A) were not able to glycosylate flavonoids in an effective manner. It was further shown quite early that flavonoid-specific glycosyltransferases showed a strict specificity for the position of the hydroxyl group to which the sugar was transferred (Barber, 1962). In most incubation mixtures tested so far, only one type of synthesized glycoside has been found, and if glycosylation occurred at different sites on the flavonoid molecule, different enzyme activities would be detected. Table I lists specific enzymatic glycosylation reactions that have been observed and the plant sources involved.

A clear-cut demonstration of the position specificity of flavonoid glycosyltransferases was achieved by separating a glucosyltransferase specific for the 3-position of flavonoids from a 7-glucosylating enzyme in parsley cell cultures (Sutter and Grisebach, 1973). All kinetic data gained so far show that glycosylation at position 7 occurs very weakly, if at all, with purified 3-glycosyltransferases, and vice versa. Another example of strict position specificity can be seen from investigations of *Silene dioica*. Among the different glycosylating activities present within petals of this plant, one is responsible for glucosylation of the 3-position and another for glucosylation of the 5-position of cyanidin (Van Brederode *et al.,* 1975). In addition, the specificity of glycosyltransferases may be restricted by the type of flavonoid (e.g., flavones, flavonols, anthocyanins) and the substitution pattern of the flavonoid. But the degree of specificity for the type of flavonoid does not seem to be as strict as that for the position of the hydroxyl group on the flavonoid nucleus. Thus the 7-glucosyltransferase from parsley cell cultures (Table I) was thoroughly investigated with respect to the structure of the flavonoid acceptor (Sutter *et al.,* 1972). The best substrates by far were flavones (luteolin, apigenin, and chrysoeriol), but the flavanone naringenin was found to be glucosylated at a remarkable rate also. Flavonols were glucosylated at a rate 10-fold lower than flavones, and cyanidin, isoflavones, and *p*-coumaric acid could not function as acceptors.

With regard to glycosylation at the 3-position, only a limited number of flavonoids (e.g., flavonols, dihydroflavonols, and anthocyanins) bear a hydroxyl at this particular position. All 3-glycosyltransferases examined to date fail to accept dihydroflavonols as substrates (Sutter and Grisebach,

TABLE I

List of *in Vitro* Glycosylations at Hydroxylic Groups of Flavonoids

Acceptor	Nucleotide-activated sugar	Product	Plant source	References
Flavonols (e.g., quercetin)	Glucose	Flavonol 3-glucoside (e.g., quercetin 3-glucoside)	*Phaseolus aureus* (leaves)	Barber (1962)
			Impatiens balsamina (flower petals)	Miles and Hagen (1968)
			Zea mays (pollen)	Larson and Lonergan (1972)
			Zea mays (endosperm)	Dooner and Nelson (1977a,b)
			Petroselinum hortense (cell cultures)	Sutter and Grisebach (1973)
			Glycine max (cell cultures)	Poulton and Kauer (1977)
			Tulipa 'Apeldoorn' (anthers)	R. Wiermann (unpublished)
	Glucuronic acid	Quercetin 3-glucuronide	*Phaseolus vulgaris* (leaves)	Marsh (1960)
	Rhamnose	Quercetin 3-rhamnoside	*Leucaena glauca* (leaves)	Barber and Chang (1968)
Anthocyanins (e.g., cyanidin)	Glucose	Cyanidin 3-glucoside	*Brassica* (seedlings)	Saleh et al. (1976a)
			Haplopappus gracilis (cell cultures)	Saleh et al. (1976b)
			Petunia hybrida	Kho et al. (1978)
			Silene dioica (petals)	Van Brederode et al. (1975)
			Hippeastrum tulipa (petal protoplasts)	Hrazdina et al. (1978)
Flavones	Glucose	Flavone 7-glucosides	*Petroselinum hortense* (cell cultures, seedlings)	Sutter et al. (1972)
Apigenin 6-C-glucoside (= Isovitexin)	Glucose, xylose	Isovetexin 7-glucoside (xyloside)	*Melandrium album* (petals)	Van Brederode and Van Nigtevecht (1973, 1974b)
			Silene dioica and *S. alba* (petals)	Van Nigtevecht and Van Brederode (1975)
			Hordeum vulgare (primary leaves, protoplast)	Blume et al. (1979)
Cyanidin 3-O-rhamnosylglucoside	Glucose	Cyanidin 3-O-rhamnosyl-glucoside 5-O-glucoside	*Silene dioica* (petals)	Van Brederode et al. (1975)
Isoflavones	Glucose	Isoflavone 7-glucosides	*Cicer arietinum* L. (roots)	Köster and Barz (submitted)

1973; Poulton and Kauer, 1977). This observation points strongly to glycosylation as being the last stage in the biosynthesis of flavonoid glycosides. However, the situation is not clear with respect to whether or not flavonols and anthocyanins are glycosylated by the same enzyme. Thus a flavonol 3-position-specific glucosyltransferase from soybean cell cultures does not glucosylate anthocyanin (Poulton and Kauer, 1977). On the other hand, cyanidin-specific glucosyltransferases from red cabbage (*Brassica oleracea*) seedlings (Saleh *et al.,* 1976a) and from *Haplopappus gracilis* cell cultures (Saleh *et al.,* 1976b) utilize flavonols at remarkable rates, in some cases as high as that of the anthocyanin itself.

It is not possible at present to generalize regarding the influence of the flavonoid substitution pattern (e.g., hydroxyl or methoxyl groups) on the substrate specificity of glycosyltransferases. In most cases examined thus far, all compounds of a given flavonoid group (e.g., flavonols) are accepted as substrates by a particular transferase for the group, although at greatly different conversion rates. The substitution pattern influences reaction velocity as well as K_m values, and in all cases reported so far the physiologically occurring substrates have been among the best acceptors (Sutter *et al.,* 1972).

There are only a few reports regarding glycosylation of a second hydroxyl of a flavonoid when a sugar moiety is already present (Table I). It has been observed that the flavonol 3-glucosylating enzyme of parsley accepts flavonol 7-glucoside as a substrate, resulting in the synthesis of flavonol 3,7-diglucosides, but the reverse sequence does not occur (Sutter and Grisebach, 1973). The affinity of flavonoid substrates for glycosyltransferases is indicated by the K_m values of different substrates and enzymes in Table II.

2. Glycosylation of Flavonoid Glycosides at Sugar Hydroxyls

In spite of the wide occurrence of flavonoid glycosides with more than one sugar unit linked in sequence (e.g., flavonol and anthocyanin 3-diglycosides; flavone 7-diglycosides) only a few transferases that introduce a second or a third sugar moiety have been described. However, it is obvious from present data that the synthesis proceeds in a sequential manner. Enzymatic steps that have been characterized are described in Table III.

Barber (1962) synthesized rutin [quercetin 3-rhamnosyl (1 → 6)-glucoside] sequentially by means of an enzyme preparation from *Phaseolus aureus* seedlings in the presence of UDP-glucose and UDP-rhamnose, but separation of the two enzymes presumably involved was not reported. A separation of these two activities, however, was achieved in the case of apiin [apigenin 7-apiosyl (1 → 2)-glucoside] synthesis. The UDP-glucose transferase (cf. Section II,B,1) was separated from the UDP-apiose transferase by means of several protein purification techniques (Sutter *et al.,* 1972; Ortmann

TABLE II

Apparent K_M Values of Some Flavonoid-glucosylating Transferases from Higher Plants

Catalyzed step	K_M (mM) of aglycone	K_M (mM) UDP-glucose	Plant source	Reference
7-Glucosylation	0.002–0.01 (flavones)	0.12–0.26	Petroselinum hortense cell cultures	Sutter et al. (1972)
	0.46 (true K_m) (vitexin)	0.12 (true K_m)	Melandrium album petals	Van Brederode and Van Nigtevecht (1973)
3-Glucosylation	0.12–0.27 (flavonols)	0.3	Glycine max cell cultures	Poulton and Kauer (1977)
	0.06 (quercetin)	0.74	Zea mays pollen	Larson and Lonergan (1972)
	0.001 (quercetin)	0.5	Petroselinum hortense cell cultures	Sutter and Grisebach (1973)
	0.4 (cyanidin)	0.5	Brassica oleracea seedlings	Saleh et al. (1976a)
	0.33 (cyanidin)	0.5	Haplopappus gracilis cell cultures	Saleh et al. (1976b)
	2.5 (K') (cyanidin) 5 (K') (delphinidin)	1.7 (true K_m)	Petunia hybrida petals	Kho et al. (1978)

TABLE III

List of *in Vitro* Glycosylations of Flavonoid Glycosides at Sugar Sites

Acceptor	Nucleotide-activated sugar	Product	Plant source	References
Quercetin 3-O-glucoside	Rhamnose	Quercetin 3-O-rhamnosyl (1 → 2)-glucoside	*Phaseolus aureus* (leaves)	Barber (1962)
Apigenin 7-O-glucoside	Apiose	Apigenin 7-O-apiosyl (1 → 2)-glucoside	*Tulipa* 'Apeldoorn' (anthers) *Petroselinum hortense* (leaves and cell cultures)	R. Wiermann (unpublished) Ortmann et al. (1972); Grisebach and Ortmann (1973)
Apigenin 6-C-glucoside	Glucose	Apigenin 6-C-glucosyl (1 → 2)-glucoside	*Melandrium album* (petals)	Van Brederode and Van Nigtevecht (1974a)
Apigenin 6-C-glucoside	Arabinose	Apigenin 6-C-arabinosyl (1 → 2)-glucoside	*Melandrium album* (petals)	Van Brederode and Van Nigtevecht (1974b)
Apigenin 6-C-glucoside	Glucose Rhamnose Arabinose	Apigenin 6-C-glucosyl (rhamnosyl, arabinosyl) (1 → 6)-glucoside	*Silene dioica* (petals) *Silene alba* (petals)	Van Nigtevecht and Van Brederode (1975)
Cyanidin 3-O-glucoside	Rhamnose	Cyanidin 3-O-rhamnosyl (1 → 6)-glucoside	*Silene dioica* (petals)	Van Nigtevecht and Van Brederode (1975)
Kaempferol and quercetin 3-O-glucoside	Glucose	Kaempferol and quercetin 3-O-triglucoside	*Pisum sativum* (seedlings)	Kamsteeg et al. (1978) Shute, et al. (1979)

et al., 1972). The substrate specificity of the UDP-apiose transferase was examined in detail (Ortmann *et al.,* 1972). Of the various UDP-activated sugars tested, only apiose could be transferred; the activity of the other sugar nucleotides was below the detection limit of 0.1%. The specificity with respect to the sugar acceptor was not as narrow as for the first glucosylation step of the flavonoid aglycone (cf. Section II,B,1). Besides flavone 7-glucosides, the following compounds were accepted as substrates: isoflavone 7-glucosides, flavanone 7-glucosides, apigenin 7-glucuronide, 4-vinylphenyl glucoside, and 4-nitrophenyl glucoside. The ability to serve as substrate decreased in the order listed. But neither flavonol 7- nor 3-glucosides were accepted. The K_m values for flavone 7-glucosides varied between 0.06 and 0.1 mM when measured at an UDP-apiose concentration of $2.7 \times 10^{-7} M$.

A glucosyltransferase from *Melandrium album,* which makes 6-glucosylvitexin, exhibited a K_m value of 0.08 mM for isovitexin as substrate (measured at a 1.8 mM UDPG concentration) (Table III) (Van Brederode and Van Nigtevecht, 1974a). 7-Glycosylated isovitexin derivatives (glucosyl or xylosyl substituents) were not substrates for this particular enzyme. Similar enzymatic activities occur in petals of *S. dioica,* where glucose, rhamnose, or arabinose is attached at the C-6 glucosyl moiety. Transfer of each sugar is catalyzed by a different enzyme. Recently the *in vitro* synthesis of a flavonol 3-triglucoside was accomplished by an enzyme preparation from *Pisum sativum* seedlings (Shute *et al.,* 1979). The transfer of the three glucose units seemed to occur stepwise, leading to the triglucoside as the main product accumulated, but the individual transferases have yet to be characterized.

With regard to the sequence of glycosylation steps leading to flavonoid triglycosides of the type flavonoid *x*-mono *y*-diglycoside, available information suggests that the second glycosylation step occurs at the sugar moiety of the acceptor molecule, forming the disaccharide rather than glycosylating a second hydroxylic group on the flavonoid nucleus (Van Brederode and Van Nigtevecht, 1974a; Saleh *et al.,* 1976b).

3. Further Characteristics of Flavonoid-Specific Glycosyltransferases

Most of the glycosyltransferases listed in Tables I and III are well characterized only with respect to their substrate specificity and the kinetics of the enzyme reaction. Other data regarding protein properties, i.e., MW, subunits, and amino acid composition, are largely absent, primarily because no transferase has been purified to homogeneity. The maximal reaction velocity of nearly all the transferases described in Tables I and III was found to occur at pH 7.5–8.5. In the four cases where MWs of transferases have been determined (see Tables I and III; Sutter *et al.,* 1972; Grisebach and Ortmann, 1973; Van Brederode and Van Nigtevecht, 1973; Dooner and Nelson, 1977a) values in the range of 50,000–55,000 have been estimated by gel chromatography (see also Section II,A). Inhibition by sulfhydryl blocking reagents was reported in several cases (Ortmann *et al.*, 1972; Larson and Lonergan, 1972). In many cases, the addition of bovine serum albumin stabilized or enhanced the transferase activity (e.g., Sutter and Grisebach, 1973; Ortmann *et al.*, 1972; Saleh *et al.*, 1976a,b). The activity of some transferases was stimulated by divalent cations such as Ca^{2+} (Larson and Lonergan, 1972) or Mn^{2+}, and Mg^{2+} or Co^{2+} (Van Brederode and Van Nigtevecht, 1974a). Strong substrate inhibition by cyanidin concentrations above 0.25 mM has been observed for cyanidin: UDPG glycosyltransferase from red cabbage seedlings (Saleh *et al.*, 1976a) and *H. gracilis* cell cultures (Saleh *et al.*, 1976b). In addition, product inhibition of both enzymes was observed, the transferase from red cabbage being more influenced (70% inhibition at 24 μM cyanidin 3-glucoside concentration) than the one from cell cultures (30% at the same concentration). Kho *et al.* (1978) did not observe a strong inhibition by cyanidin of UDP-glucose: cyanidin 3-*O*-glucosyltransferase from *Petunia hybrida*. Rather, they observed an allosteric behavior of this enzyme for the substrates cyanidin and delphinidin and measured a Hill coefficient of 1.9 for cyanidin. The enzyme behaved in a pure Michaelis–Menten fashion with respect to UDPG as substrate.

The glucosyl transfer catalyzed by UDP-glucose: flavonol 3-*O*-glucosyltransferase from parsley cell cultures (cf. Table I) was shown to be a freely reversible reaction (Sutter and Grisebach, 1975). Glucosyl transfer from either kaempferol or quercetin 3-*O*-glucoside to UDP to form UDP-glucose was observed. A value of 6.4 kcal/mol for the free energy of hydrolysis for both these flavonol 3-*O*-glucosides was calculated.

4. Physiological and Genetic Control of Flavonoid Glycosyltransferases

The appearance of flavonoid-specific glycosyltransferases in parsley (cf. Tables I and III) has been followed together with that of other flavonoid biosynthetic enzymes during the course of seedling development and also during light-stimulated flavonoid biosynthesis in cell cultures. These investigations are comprehensively reviewed by Hahlbrock (this volume, Chapter 14). Studies on the subcellular localization of UDP-glucose: anthocyanidin 3-*O*-glucosyltransferase in protoplasts of *Hippeastrum* and *Tulipa* petals (Hrazdina *et al.*, 1978) and of UDP-glucose: isovitexin 7-*O*-glucosyltransferase in *Hordeum vulgare* protoplasts (Blume et al., 1979) showed that both transferases were associated with the soluble (cytosol) fraction. Less than 2% of the transferase activity was found in any particulate fraction.

Genetic control of flavonoid biosynthesis has been extensively investigated by recording the levels and types of flavonoids and flavonoid glycosides that accumulate in plants (cf. Harborne, 1967). In the last few years, this topic has been increasingly investigated at an enzymatic level, in particular by looking for the glycosyltransferases. Thus, in *Melandrium* sp.

nearly all the glycosyltransferases leading to the different isovitexin glycosides present in this plant can be identified and correlated with specific genes (Fig. 4).

The scheme indicated in Fig. 4 is consistent with the substrate specificities of the glycosyltransferases that have been characterized (cf. Section II,B,2), with the genetic data obtained from crosses and with the flavonoid pattern found in the petals of *Melandrium* plants. The scheme outlined in Fig. 4 also applies to the flavonoids found in *Silene alba* and *S. dioica* (Van Nigtevecht and Van Brederode, 1975). In *Melandrium* plants of the type $g^G g^X$ only isovitexin 7-glucoside (and no isovitexin 7-xyloside) could be detected, indicating that the allele g^G was dominant over g^X. However, the 7-xylosyltransferase was present within this plant in the same amount as the 7-glucosyltransferase, showing that the dominance was not a consequence of transcriptional and/or translational control. It was suggested that the dominance of g^G over g^X was probably a consequence of the much higher V_{max} the 7-glucosyltransferase possesses compared to that of the 7-xylosyltransferase when operating at saturating isovitexin concentrations *in vivo* (Van Bre-

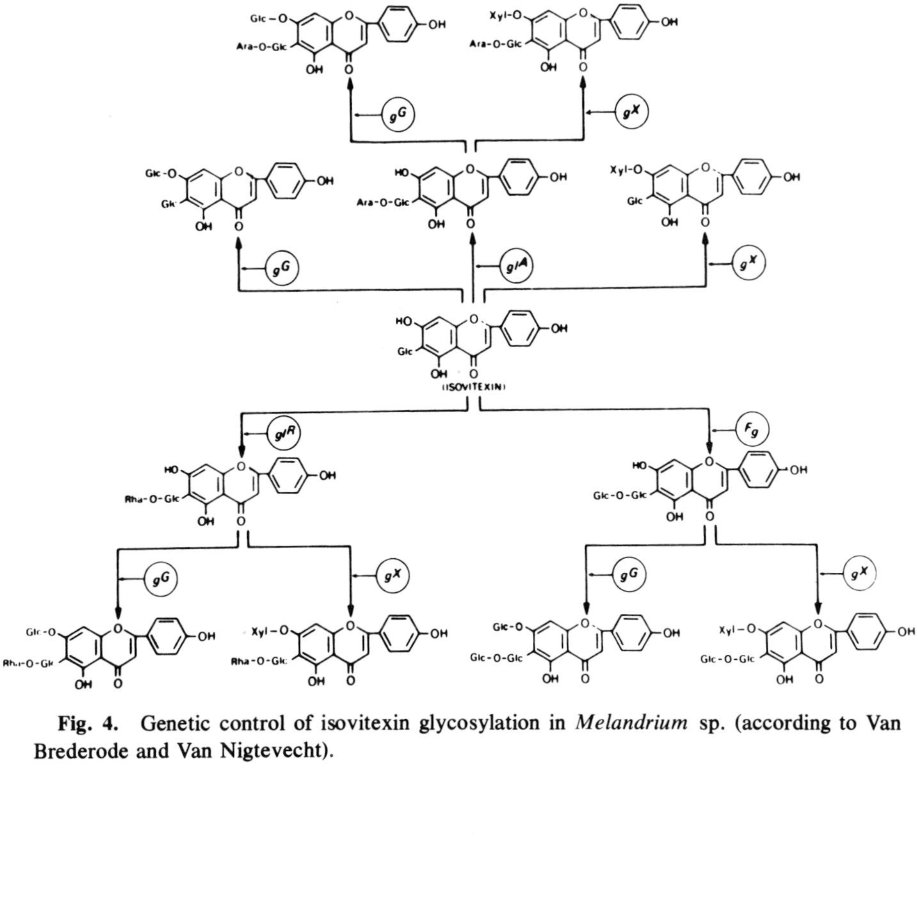

Fig. 4. Genetic control of isovitexin glycosylation in *Melandrium* sp. (according to Van Brederode and Van Nigtevecht).

derode and Van Nigtevecht, 1974c). This hypothesis was supported by re-sults from a *Melandrium* hybrid ($g^Gg^{X'}$) which contained both the 7-glucoside and the 7-xyloside in the same amount and where the 7-xylosyltransferase showed an elevated V_{max} compared to the g^X type (Van Brederode *et al.,* 1974).

Further study of the genetic control of cyanidin 3-glucosyltransferase has involved *S. dioica* (Van Brederode *et al.,* 1975; Kamsteeg *et al.,* 1978), *P. hybrida* (Kho *et al.,* 1978), and the endosperm of maize (Larson and Coe, 1977; Dooner and Nelson, 1977b). In the latter case, a structural gene (*Bz*) for the enzyme UDP-glucose: flavonol (anthocyanin) 3-*O*-glucosyltransferase and two other genes (*C* and *R*), which produce substances for the induction of this enzyme, were detected. Expression of the *Bz* locus of maize was investigated by the transposition of controlling elements (Dooner and Nelson, 1977a).

C. Glucosylation of Cyanogenic Glucosides and Glucosinolates

Cyanogenic glucosides and glucosinolates are two other groups of second-ary plant products in which glycosyltransferases are well characterized. Thus Hahlbrock and Conn (1970) isolated an enzyme from flax seedlings that catalyzed the transfer of glucose from UDP-glucose to α-hydroxyisobutyronitrile, forming linamarin, one of the two cyanogenic glucosides produced by the flax plant. Reay and Conn (1974) characterized an enzyme from *Sorghum bicolor* seedlings that brought about the synthesis of dhurrin from UDP-glucose and (*R*,*S*)-*p*-hydroxymandelonitrile. These en-zymes did not differ basically from the transferases described above. Both showed a pH optimum between 8 and 9 and a high degree of specificity for UDP-glucose as donor. In each case, the transferase was separated from β-glucosidase(s) that act on the particular cyanogenic glucoside present in the plant. With respect to the acceptor molecule, the flax transferase showed a high degree of specificity for the two aliphatic side chains of acetone and butanone cyanohydrins, the butanone cyanohydrin being more readily glucosylated than the acetone derivative. The nitrile group was shown not to be required in the acceptor molecule, since the corresponding alcohols lack-ing a nitrile group (e.g., isopropanol, 2-butanol) were also glucosylated, but not to the same degree as the cyanohydrins. The question whether or not the formation of linamarin and lotaustralin *in vivo* was catalyzed by the same enzyme could not be conclusively answered, but during the purification pro-cedure no indication of the presence of two different enzymes was found. Aromatic cyanohydrins turned out to be extremely poor substrates. On the other hand, a second glucosyltransferase seemed to be present in the prep-aration that glucosylated *p*-hydroxybenzaldehyde at the phenolic hydroxyl.

The sorghum transferase exhibited a high specificity for an aromatic group in the acceptor substrate, since no glucosylation occurred with the cyanohydrins of acetone or acetaldehyde. A 4-hydroxy substitution seems not to be necessary for the aromatic acceptor, since mandelonitrile was converted as fast as 4-hydroxymandelonitrile. On the other hand, the latter substance was converted better than 4-hydroxybenzyl alcohol, indicating some specificity for the α-nitrile group. In neither case were the K_m values for the cyanohydrin substrates determined, since the concentration-dependent dissociation of these substances prevented exact measurement. The sorghum glucosyltransferase possessed strict stereospecificity for p-hydroxy-(S)-mandelonitrile, forming dhurrin exclusively from a racemic mixture of the substrate. The same degree of stereospecificity has been observed for a glucosyltransferase from *Triglochin maritima* seedlings, which glucosylates only the p-hydroxyl-(R)-mandelonitrile from racemic mixtures to produce taxiphyllin, the enantiomer of dhurrin (Hösel and Nahrstedt, 1980). Such stereospecificity was not observed with the flax transferase that glucosylated both enantiomers of the racemic cyanohydrin mixture (R,S)-2-hydroxy-2-methylbutyronitrile. This transferase formed equal amounts of the epimeric glucosides, lotaustralin and (S)-epilotaustralin, since it is not able to distinguish between the methyl and ethyl groups attached to the chiral carbon atom of the cyanohydrin.

The involvement of a specific glucosyltransferase was also demonstrated in a related group of compounds, glucosinolates (cf. Larsen, this volume, Chapter 17). An enzyme that catalyzes the formation of desulfobenzylglucosinolate by glucosyl transfer from UDP-glucose to phenylacetothiohydroximate has been isolated from leaves of *Tropaeolum majus* and purified 20-fold (Matsuo and Underhill, 1971). As in the synthesis of cyanogenic glucosides, this glucosyltransferase also showed a high specificity for both the sugar donor (UDPG) and the acceptor (thiohydroximate). Acceptors with a hydroxyl group instead of a thiol group were not found to be glucosylated. Sulfhydryl, chelating, and reducing compounds increased enzymatic activity. Interestingly, a MW of 50,000 was determined for this transferase, as well as for all transferases in the flavonoid group (cf. Section II,B,3). Similar glucosyltransferases were also found in other glucosinolate-containing plants (Matsuo and Underhill, 1971).

D. Glycosylations of Other Secondary Compounds

There are a large number of secondary heterosides about which nothing is known regarding their enzymatic *in vitro* glycosylation (e.g., cardiac glycosides, saponin glycosides, glycoalkaloids, most quinone glycosides, C-glycosides, and most secondary N-glycosides). It is reasonable to anticipate that no basic difference exists in the glycosylation of these secondary glycosides, but it would be interesting to learn more about the enzymes that glycosylate structurally exceptional aglycones and synthesize tri- or tetraglycosides of C-glucosides. A start in this area has been made in that preliminary information concerning the *in vitro* glycosylation of gibberellins (Müller *et al.*, 1974) and quinones (Müller and Leistner, 1978) is available.

E. Substrate Specificity with Respect to the Sugar Donor

As stated earlier, nucleotide-activated sugars have generally proved to be the sugar donors for the glycosylation of all groups of secondary products investigated so far. In each case when a "low-energy" donor such as α-glucose-1-phosphate was used, it failed to show any activity (Marsh, 1960; Barber, 1962; Sutter *et al.*, 1972; Ortmann *et al.*, 1972).

Among the nucleotide sugars UDP derivatives have proven to be the best substrates by far. Exceptions are the glucosylation of quercetin by a mung bean transferase, where TDP-glucose is equally as active as UDP-glucose (Barber, 1962), and the glucosylation of diphenols by wheat germ transferases, where ADP-glucose seems to be preferred over UDP-glucose (Trivelloni *et al.*, 1962). TDP sugars are used by other transferases in preference to UDP derivatives in several cases (Barber, 1962; Sutter *et al.*, 1972; Larson and Lonergan, 1972; Matsuo and Underhill, 1971; Barber and Chang, 1968). In all other cases examined, nucleotide sugars with adenine, cytosine, or guanine bases have shown little or no activity. The specificity of transferases with regard to the sugar residue transferred seems to be very high too, since no activity is found when the proper UDP sugar is replaced by other UDP sugars (Matsuo and Underhill, 1971; Sutter *et al.*, 1972; Grisebach and Ortmann, 1973; Van Brederode and Van Nigtevecht, 1973). Since the sugars are linked in an α configuration to the nucleotide and the resulting products of transferase activity have a β-type linkage, a Walden conversion is likely to occur during enzymatic synthesis.

III. GLYCOSIDASES

If transferases are the biosynthetic enzymes of secondary heterosides, the glycosidases found within plants are therefore seen as being responsible for the hydrolysis of such compounds (Fig. 1). This viewpoint is strongly supported by one observation: In spite of the fact that the enzymatic hydrolysis of most glycosides by glycosidases is a reversible reaction, hydrolysis is by far the favored reaction in most cases. In addition, glycosyltransferases and glycosidases active on the same type of secondary compound have repeatedly been demonstrated in the same plant tissue.

Although these facts strongly favor the view that the primary role of glycosidases is the hydrolysis of secondary plant glycosides, the exact physiological role is in most cases far from understood. A specific example is the β-glucosidases that act on cyanogenic glucosides. The first such β-glucosidase was demonstrated more than 150 yr ago (Liebig and Wöhler, 1837), and these enzymes have been the subject of thorough investigations (cf. Nisizawa and Hashimoto, 1970). Nevertheless the exact role of these β-glucosidases remains unknown (cf. Conn, this volume, Chapter 16).

Elucidation of the exact role of glycosidases in plant secondary metabolism has been hampered until now for several reasons. First of all, the main emphasis of most investigations has been directed toward the sugar moiety of the substrates, and the influence of the aglycones has hardly been evaluated. This is reflected in the wide and almost exclusive use of synthetic and easily measured substrates such as nitrophenyl and methylumbelliferyl sugars instead of physiological substrates. Thus information about one group of glycohydrolases (e.g., β-glucosidases, β-galactosidases) is obtained, but no distinction within the group is made as regards the aglycone. Even the information about the sugar specificity is of limited value, since absolute sugar specificity of glycosidases seldom exists (cf. Nisizawa and Hashimoto, 1970). With our present state of knowledge, an essential prerequisite for the evaluation of glycohydrolases is thorough purification of the enzymes. Since different activities of a single type (e.g., β-glucosidase) may occur in a single plant tissue, often in very different quantities, a clear-cut separation of the enzyme under investigation from all other glycohydrolases, especially other β-glucosidases, has to be achieved before determination of its properties is possible. It should be emphasized, moreover, that physiological substrates must be used for monitoring the purification procedures. Synthetic substrates are of limited value, since the hydrolytic activity of a particular glycosidase for these substrates cannot be known beforehand and may be very low. Furthermore, in many investigations, the hydrolysis of secondary glycosides has been followed only qualitatively (e.g., by chromatography) without measuring reaction velocities, a procedure that also hampers the exact evaluation of substrate specificity. It is the aim of this section to concentrate on glycohydrolases that show a clear specificity for secondary plant glycosides. Hence glycosidases specific for carbohydrates lacking a secondary aglycone (e.g., oligo- and polysaccharides) are not dealt with here and the reader is referred to other reviews (Vol. 3 of this series; Nisizawa and Hashimoto, 1970). Since nearly all naturally occurring plant heterosides are β-linked (Courtois and Percheron, 1970), and since glycohydrolases consistently show an absolute specificity with regard to the configuration of the linkage hydrolyzed, only β-glycosidases will be discussed. Emphasis will be

placed on a description of the specificity of purified glycosidases with respect
to the sugar *and* the aglycone parts of the substrate, since both moieties of a
secondary glycoside are of importance.

A. Glycosidases Specific for Phenolic and Flavonoid Glycosides

It has long been known that glucosides of simple phenolics such as arbutin
(hydroquinone β-D-glucoside) and salicin [2-(hydroxymethyl)phenyl β-
D-glucoside] can be hydrolyzed by certain plant extracts (cf. Nisizawa and
Hashimoto, 1970). Emphasis was placed on the transferase activity of these
enzymes, i.e., their ability to catalyze transfer of the glucose moiety to
acceptors other than water, e.g., alcohols (Rabaté, 1935). The specific
glucosidases involved apparently have not been further investigated.

Studies in the early 1950s suggested that cinnamyl alcohol glucosides (e.g.,
coniferin, syringin), together with a specific β-glucosidase, might be involved in lignification (Freudenberg *et al.*, 1952; Reznik, 1955; cf. Grisebach,
this volume, Chapter 15). Only very recently have β-glucosidases with preferential specificity for coniferin been purified to apparent homogeneity from
Picea abies seedlings (Marcinowski and Grisebach, 1978) and *Cicer arietinum*
cell cultures (Hösel *et al.*, 1978). Both enzymes showed a Michaelis constant
near 1 mM for coniferin, and the V_{max} values for this substrate were among
the highest for all substrates tested. Other β-glucosidases, which were present in both plants and which were able to hydrolyze aromatic β-glucosides,
actually hydrolyzed coniferin only poorly. Both β-glucosidases shared additional common features such as an isoelectric point at pH 9–10 and a glycoprotein character. Likewise, syringin was less effective than coniferin as a
substrate. Another β-glucosidase, purified from *Glycine max* plants and cell
cultures, possessed a high specificity for these two substrates, with a K_m of
0.25 mM for syringin and 0.5 mM for coniferin and identical V_{max} values with
both substrates (Hösel and Todenhagen, 1980). Most interestingly, all other
β-glucosides tested were poor substrates; for example, 4-nitrophenyl
β-glucoside was hydrolyzed at only 5% of the rate of the cinnamyl alcohol
glucosides. These data strongly reinforce the statement made above, that
physiological substrates must be used for the purification and characterization of glycosidases.

In healthy, intact sweet clover leaves, essentially all the coumarin is present in "bound" form, specifically as glucoside of *o*-hydroxycinnamic acid
(Haskins and Gorz, 1961; Kosuge, 1961). Upon disruption of the tissue,
extensive conversion of the bound coumarin to free coumarin occurs (cf.
Brown, this volume, Chapter 10). Kosuge and Conn (1961) and Schön (1966)
have demonstrated a specific β-glucosidase activity in sweet clover for *cis-
o*-coumaric acid β-glucoside. Whereas the glucoside of melilotic acid was

hydrolyzed at half the rate of the *cis*-glucoside, *trans-o*-coumaric acid β-glucoside was hydrolyzed rather slowly. The specificity of the clover β-glucosidase was quite different from that of almond β-glucosidase, which hydrolyzed all three substrates at approximately the same rate (Kosuge and Conn, 1961). A K_m value of 2.2 mM was determined for *cis-o*-coumaric acid β-glucoside for a 40-fold-enriched clover enzyme.

A fairly interesting β-glucosidase specific for coumarin glucosides has been described in *Daphne odora* (Sato and Hasegawa, 1971). The substrate specificity of this glucosidase seems to be rather narrow, since only coumarin glucosides among numerous aromatic and aliphatic substrates tested were hydrolyzed. This particular glucosidase also showed substantial transferase activity, since daphnetin 7-glucoside was extensively converted into daphnetin 8-glucoside. The reverse transfer was not observed. Both the glucosidase and transferase activities behaved similarly on gel electrophoresis, but the pH optima of the two activities at 5.3 and 6.3, respectively, were clearly different. Since daphnetin 8-glucoside accumulates at the end of plant development and is preceded by daphnetin 7-glucoside, the transferase activity is believed to catalyze the interconversion *in vivo*.

An important example of a β-glucosidase with a high degree of specificity for a secondary phenolic glucoside is the enzyme that occurs in stratified apple seeds. Such seeds contain a β-glucosidase specific for phloridzin (4,2′,4′,6′-tetrahydroxydihydrochalcone 2′-β-glucoside), which was separated from two other activities that hydrolyzed this glucoside only slowly, if at all (Podstolski and Lewak, 1970). A similar activity was detected in apple leaves. This glucosidase hydrolyzed phloridzin 370 times more rapidly than 4-nitrophenyl β-glucoside under standard conditions, and the K_m value for phloridzin was 14 μM.

It is interesting to note that another β-glucosidase has been purified from apple seeds that preferentially hydrolyzes amygdalin, a cyanogenic glucoside that also occurs in this tissue. The amount of phloridzin-specific β-glucosidase increases in seed up to 90 days during stratification, whereas the β-glucosidase activity for amygdalin decreases after 60 days. These alterations in enzyme levels showed a good correlation with the amounts of the respective substrates in the seeds (Podstolski and Lewak, 1973; Bogatek *et al.*, 1976).

That β-glucosidases having high specificity for flavonoid glycosides also occur in higher plants can be seen from investigations of the isoflavone 7-glucoside-specific β-glucosidases of *C. arietinum* seedlings. These enzymes have been purified to apparent homogeneity from roots and leaves of this plant (Hösel and Barz, 1975). Among the many substrates investigated, the isoflavones formononetin and biochanin A 7-glucosides exhibited the highest V_{max} and by far the lowest K_m values (0.02–0.03 mM). All kinetic data indicate that the substrate specificity of these β-glucosidases is determined by the following features: β linkage, sugar (glucose) moiety, aglycone (isoflavone), the 7-position of the sugar–aglycone linkage. The structure of the aglycone obviously contributes in a major way to the substrate affinity. The development and the organ distribution of this β-glucosidase activity correlate very well with the isoflavone content in chick pea seedlings (Hösel, 1976a). While β-glucosidases with a high specificity for phloridzin or isoflavone 7-glucoside are known, enzymes with similar specificity have not been described for other flavonoid glycosides. It can be predicted that these β-glucosidases exist, but further work is needed to demonstrate it unequivocally. There are several reports of the hydrolysis of flavonoid glycosides, especially flavonol and anthocyanin 3-glycosides by protein extracts from higher plants (Suzuki, 1962; Boylen *et al.*, 1969; Bourbouze *et al.*, 1974). However, these studies in most cases have not involved purified enzymes and thus lack adequate quantitation of enzymatic activity. Since many β-glycosidases do not show absolutely strict substrate specificity for either the glycosidic or the aglycone moiety, hydrolysis of a wide range of substrates might be possible to some extent without showing a *preferential* specificity for these substrates. Thus when a large number of higher plant tissues and plant cell cultures known to possess or to metabolize flavonol 3-glycosides were screened with a sensitive photometric assay for flavonol 3-glycoside β-glycosidases, no clear indication of specific β-glycosidases was obtained (Surholt and Hösel, 1978). This problem therefore still awaits a solution, but preliminary results by Zaprometov (1978) indicate that an active, specific glycosidase for the hydrolysis of rutin occurs in *Bupleurum* plants.

B. Glucosidases Specific for Cyanogenic Glucosides

The presence of enzymes in higher plants hydrolyzing cyanogenic glucosides was discovered nearly as early as the cyanogenic glucosides themselves when it became apparent that injured or crushed plants released hydrogen cyanide. The hydrolysis of cyanogenic glucosides is brought about by β-glucosidases, and the liberated α-hydroxynitrile either decomposes spontaneously or its dissociation is catalyzed by a second enzyme (α-hydroxynitrile lyase) present in the plant. Only β-glucosidases involved will be discussed here.

The β-glucosidases from almonds were first detected by Liebig and Wöhler (1837) and are by far the most intensively studied β-glucosidases from higher plants (cf. Nisizawa and Hashimoto, 1970). Nevertheless, even now the relationships between β-glucosidases and the endogenous substrates amygdalin and prunasin are not fully understood. This is again due to the fact that synthetic substrates rather than physiological ones have been used for the enzymatic investigations (cf. Helferich, 1933; Nisizawa and Hashimoto, 1970).

The hydrolysis of amygdalin by β-glucosidases present in the enzyme preparation from almonds called "emulsin" is postulated to occur stepwise (Haisman and Knight, 1967; Haisman et al., 1976; Lalégérie, 1974). One β-glucosidase first removes the terminal glucose unit and gives rise to prunasin and glucose, and a second enzyme hydrolyzes the prunasin, yielding mandelonitrile and a second glucose molecule. The intact gentiobioside unit does not seem to be formed during hydrolysis by almond enzymes. The K_m values for amygdalin and prunasin are in the range of 1–2 mM (Haisman and Knight, 1967; Lalégérie, 1974).

The β-glycosidases from almonds have long been the subject of intensive investigations. Separation of the different glycosidases present in the crude protein extract emulsin, examination of the substrate specificity of the β-glucosidases with respect to the sugar moiety, and characterization of the enzymes' properties including mechanism have been described. Since these studies are adequately reviewed (Helferich, 1933; Pigman, 1944; Veibel, 1951; Nisizawa and Hashimoto, 1970), no attempt is made to deal with them here. I will discuss only some more recent investigations. Grover and Cushley (1977) and Grover et al. (1977) have reinvestigated the question of whether the β-glucosides and β-galactosides hydrolyzed by almond β-glucosidase are acted upon at the same catalytic site. They have proposed that both substrates are hydrolyzed at different catalytic sites by a bifunctional enzyme. These results have been seriously questioned by Walker and Axelrod (1978), whose results are in accordance with earlier investigations by Heyworth and Walker (1962) claiming one catalytic site. By means of the covalent inhibitor conduritol B epoxide Legler and Harder (1978) characterized a short amino acid sequence at the catalytic site of bitter almond β-glucosidase. Like other glycosidases, the inhibitor was bound at an aspartic acid residue involved in enzymatic catalysis.

There is no doubt that the β-glucosidases that occur together with cyanogenic glucosides in the same plants possess preferential specificity for these particular substrates. In a survey of a number of plants with β-glucosidase activity, Butler et al. (1965) showed that only extracts from plants that contain linamarin or lotaustralin were able to hydrolyze these substrates effectively. With partially purified β-glucosidases from Trifolium repens (Hughes and Maher, 1973) and Manihot esculenta (cassava) leaves (Cooke et al., 1978) K_m values of 0.7 and 1.45 mM have been determined for linamarin. The results from Hughes (1968) and Hughes and Maher (1973) suggest that a great deal of the β-glucosidase activity extracted from white clover plants is related to linamarin hydrolysis and is governed by the Li genetic locus within this plant. Immunological tests have shown that plants homozygous for the recessive Li allele do not contain an enzymatically inactive protein antigenically related to the normal enzyme (Hughes and Maher, 1973).

That cyanogenic plants contain β-glucosidases with a distinct specificity for particular cyanogenic glucosides was further demonstrated by the discovery and characterization of a β-glucosidase from *Alocasia macrorrhiza* that is highly specific for triglochinin, the cyanogenic glucoside found in this plant species (Hösel and Nahrstedt, 1975; Hösel and Klewitz, 1977). The enzyme was purified to apparent electrophoretic homogeneity by only a 9-fold enrichment procedure, suggesting that $\frac{1}{10}$ of the extracted protein consisted of this β-glucosidase. Moreover, nearly all the β-glucosidase activity extracted was specific for triglochinin. A similar observation was made for the linamarin-specific β-glucosidase from *T. repens* (Hughes and Maher, 1973). Thorough investigations of substrate specificity revealed triglochinin to be by far the best substrate, with a K_m of 0.3 mM and a V_{max} 10-fold higher than that of the next best hydrolyzed substrate, 4-nitrophenyl β-glucoside. Of all the other cyanogenic glucosides investigated (aliphatic and aromatic), only dhurrin and sambunigrin were hydrolyzed, and these to only a small extent. A β-glucosidase with protein properties and catalytic activity similar to that from *A. macrorrhiza* is also present in *T. maritima* (Nahrstedt *et al.*, 1979), the plant in which triglochinin was first detected (Eyjolfssón and Ettlinger, 1972). Interestingly, a second cyanogenic glucoside, taxiphyllin, and a β-glucosidase that shows good activity toward this substance, are also found in seedlings of this plant. Dhurrin, the enantiomer of taxiphyllin, is also a substrate for this β-glucosidase, showing the same K_m value as taxiphyllin (0.4 mM) but a 6-fold lower V_{max}. A crude protein preparation of *Sorghum vulgare*, a plant in which dhurrin is the cyanogenic compound, showed β-glucosidase activity toward several substrates, of which taxiphyllin and 4-nitrophenyl β-glucoside were much better hydrolyzed than dhurrin (Mao and Anderson, 1967). Yet a recent reinvestigation and thorough separation of the β-glucosidases of *Sorghum* showed that this plant contained two β-glucosidases that were nearly entirely specific for dhurrin and could not hydrolyze any other cyanogenic or synthetic β-glucoside to a significant extent (S. H. Eklund and E. E. Conn, unpublished). This observation again clearly demonstrates that information on substrate specificities obtained with a mixture of β-glycosidases can be highly misleading.

An investigation of the distribution of the dhurrin hydrolyzing β-glucosidase in the tissues of *Sorghum* seedlings showed that this activity was present almost entirely in the mesophyll cells. The dhurrin, in contrast, was located only in the vacuoles of the epidermal cells, thus providing a clear separation between the substrate and the hydrolyzing enzyme within the intact plant tissue (Kojima *et al.*, 1979). Hence it seems likely that the β-glucosidase acts only after disruption of the plant tissue, probably providing a defense mechanism against herbivores by liberating the toxic cyanide.

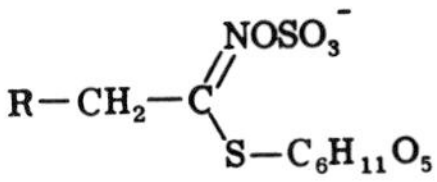

$$R-CH_2-C\begin{matrix} NOSO_3^- \\ \\ S-C_6H_{11}O_5 \end{matrix}$$

Fig. 5. General structure of glucosinolates.

C. Enzymatic Hydrolysis of Glucosinolates

The enzymatic hydrolysis of glucosinolates (Fig. 5) in many ways resembles that of cyanogenic glucosides. Thus it has long been known that glucosinolates are accompanied by specific hydrolytic enzymes, called myrosinases, within plants (cf. Ettlinger and Kjaer, 1968).

A major problem here concerns the substrate specificity of thioglucosidases with respect to the hydrolysis of O-glucosides. Whereas earlier investigations had indicated some hydrolysis of O-glucosides by thioglucosidases (Reese *et al.*, 1958; Gaines and Goering, 1962), more recent investigations with purified enzymes showed that O-glucosides were not hydrolyzed by these enzymes (Snowden and Gaines, 1969; Lein, 1972). It was further observed that isoenzymes of these thioglucosidases occur (Björkman and Lönnerdal, 1973, and literature cited therein). Björkman and Janson (1972) separated three isoenzymes from *Sinapis alba* seeds and purified one to apparent homogeneity. Studies by Lönnerdal and Janson (1973) on the thioglucohydrolases in *Brassica napus* seeds resulted in four isoenzymes, one of which was also purified to homogeneity. The isoenzymes were characterized with regard to different protein and catalytic properties. They differed mainly in isoelectric point and carbohydrate content, but all showed nearly the same substrate specificity with respect to different glucosinolates (Björkman and Lönnerdal, 1973). Hence it was concluded that myrosinases from different species were not specifically adapted to a particular set of glucosinolates. The kinetic constants of the isoenzymes were influenced by ascorbic acid, an effect that had been detected some years ago (Ettlinger *et al.*, 1961). The K_m values increased from 0.05 mM up to 0.4 mM, and the V_{max} values increase 20- to 40-fold when assayed in the presence of 1 mM ascorbic acid.

D. Hydrolysis of Intact Disaccharide Units from Various Secondary Diglycosidic Substances

Some plant β-glucosidases, which are active on monoglucosides and have low substrate specificity, are able to release disaccharide units from diglycoside substrates (cf. Hösel and Barz, 1975). On the other hand, there are data indicating that this type of hydrolysis is catalyzed by specific glycohydrolases. Thus β-glycosidases responsible for the release of primverose (6-O-D-xylosyl-D-glucose) (Bridel, 1926), vicianose (6-O-α-L-arabinosyl-D-glucose) (Bertrand and Rivkid, 1907), and rutinose (6-O-α-L-rhamnosyl-D-glucose) (Bridel and Charaux, 1926), linked to different aglycones, have been known for a long time. Bourbouze *et al.* (1974) purified a glycosidase from seeds of *Fagopyrum esculentum* that specifically released rutinose from different rutinosides. A similar β-glycosidase acting on furcatin [p-vinylphenylapiosyl (1 → 6)-β-glucoside] was detected in leaves of *Viburnum furcatum* and partially characterized (Imaseki and Yamamoto, 1961). This enzyme required the apiosyl (1 → 6)-glucoside unit for activity, since neither the apiosyl (1 → 2)-glucoside nor the rhamnosyl (1 → 6)-glucoside moiety was liberated from substrates containing these disaccharides. Inhibitor experiments indicated that a SH group was involved in the catalytic action. It was further established that the enzyme could transfer the intact apiosylglucose unit from furcatin to other acceptors, a fact that was also observed for the primverose-releasing glycosidase (Rabaté, 1935). Another β-glycosidase showing properties similar to those of the enzyme in *Viburnum* leaves was purified to apparent homogeneity from leaves of *C. arietinum* plants (Hösel, 1976b). This enzyme preferentially acted on isoflavone 7-apiosyl (1 → 2)-glucosides but could also hydrolyze monoglycosides. Enzyme properties (the SH group involved in catalytic action, transferase activity, substrate specificity) were similar to those of the furcatin-hydrolyzing β-glycosidase. The *Cicer* glycosidase shows additionally a rather narrow affinity for the isoflavone aglycone, since flavone 7-glucosides and flavonol 3-apiosyl (1 → 2)-glucosides were not hydrolyzed.

E. Glycosidases with Specificity for Other Heterosides

Although many other β-glycosidases are known to occur in plants and to be involved in secondary plant metabolism, they cannot be dealt with here in detail because of limited space. However, two aspects should be stressed. The glycosidases that hydrolyze plant steroid glycosides (cardiac glycosides) occurring in species such as *Digitalis* and *Strophantus* have long been known (Stoll *et al.,* 1935; Stoll and Renz, 1939). It appears that all plants containing such cardiac glycosides also possess β-glucosidases that hydrolyze these glycosides at the positions indicated by an arrow in the scheme below.

$$\downarrow$$
Steroid aglycone—deoxy sugar$_x$—glucose$_y$
$$\downarrow$$
Steroid aglycone —glycose$_y$

Although no thorough purification of these β-glucosidases has been carried out, studies employing crude plant extracts suggest a fairly high specificity for the enzymes present. This is true also for an enzyme found in various Cucurbitaceae that specifically hydrolyzes a tetracyclic terpenoid glucosidic compound, elaterinide, which is the bitter principle of these plants (Enslin *et al.*, 1956; Joubert, 1960). It might well be that these β-glucosidases possess the same degree of specificity as was recently established for a steryl β-D-glucoside hydrolase from *S. alba* seedlings (Kalinowska and Wojciechowski, 1978). A highly purified form of this β-glucosidase, which was separated from other β-glucosidases present in the seedlings, hydrolyzes only steryl glucosidic substrates such as cholesteryl and sitosteryl glucoside. A comparable situation may hold for gibberellin *O*-glucosides, since the occurrence of these substances and their hydrolysis by endogenous glucosidases, present within the plants, have been observed (Knöfel *et al.*, 1974). Trim (1953) investigated the β-glycosidase "galionase," which occurs in the Rubiaceae, and found it to be specific for the two anthraquinone glycosides purpurin-3-carboxylic acid primveroside and glucoside. Other anthraquinone glycosides such as ruberythric acid and rubiadin primveroside, known to occur also in roots of the Rubiaceae, are not hydrolyzed by galionase but by another β-glycosidase called erythrozyme which is present in the roots. Both glycosidases have been separated from each other, but a thorough description of substrate specificities has not yet appeared. The concentrations of galionase and its substrates seems to be approximately proportional in certain tissues of *Galium* sp., and both increase after excision of parts of the plants. The localization of galionase in the shoots of several Rubiaceae has also been studied at the tissue level (Trim, 1953). It has yet to be determined what kind of specificity is shown by a β-glucosidase that brings about hydrolysis of the biosynthetic intermediate strictosidine in the biosynthesis of indole alkaloids (Scott *et al.*, 1977; Treimer and Zenk, 1978). However, a very recent report from Hemscheidt and Zenk (1980) shows that this β-glucosidase displays high specificity for its physiological substrate 3β-(*R*)-strictosidine. Finally, it should be added that hydrolysis of *C*- and *N*-glycosides by glycosidases from higher plants has not been investigated to my knowledge.

F. β-Glycosidases Specific for Secondary Plant Products from Fungi

No attempt has been made to discuss the β-glycosidases in fungi that act on secondary plant products. There is ample knowledge of such enzymes, and the reader is referred to other reviews covering this subject (e.g., Nisizawa and Hashimoto, 1970).

IV. CONCLUSIONS

It is generally accepted that the synthesis of most, if not all, secondary plant glycosides is brought about by glycosyltransferases acting with nucleotide-activated sugars. The glycosylation steps are usually at the end of the particular biosynthetic pathways. Consequently, the role of the glycohydrolases found in plants can be seen, from my point of view, to be responsible for hydrolysis of the secondary heterosides. This view is supported by the examples outlined above in which distinct enzymatic specificities of many glycohydrolases for particular secondary heterosides were demonstrated. It has been one of the main objectives of this chapter to emphasize and to show that plant glycohydrolases possess the same degree of specificity as other enzymes involved in the metabolism of secondary compounds. It might be predicted that the list of plant glycohydrolases showing a clear specificity for secondary plant glycosides will be further increased drastically if a thorough purification and characterization of glycosidases using the proper endogenous instead of synthetic substrates occurs. The well-known transferase activity of glycohydrolases when the sugars are transferred to acceptors other than water (Rabaté, 1935; Dedonder, 1961; Nisizawa and Hashimoto, 1970) can be seen as an *in vitro* reaction occurring under certain conditions. There is no proof that this type of activity is realized *in vivo,* and only a few cases exist where transferase activity of glycosidases might be a rationale for explaining the biosynthetic processes of secondary glycosides in higher plants (Sato and Hasegawa, 1971).

However, the exact physiological role of most glycohydrolases, in contrast to that of glycosyltransferases, has yet to be determined. It has been suggested for a long time that glycosidases are localized within plants separately from their particular substrates and act only after destruction of the plant tissue (Czapek, 1921), but this has been proved so far only for the dhurrin-specific β-glucosidase in *S. bicolor* seedlings (Kojima *et al.,* 1979). There are indications that this situation may generally hold for glycosidases with preferential specificity for some groups such as glucosinolates (Tang, 1973; Pihakaski and Iversen, 1976), cyanogenic glucosides (Conn, 1980; cf. Conn, this volume, Chapter 16), and coumarinic acid glucosides (Schön, 1966), but it is possible that they are also involved to some extent in the endogenous catabolism of secondary plant glucosides, releasing the free aglycone which is further degraded (cf. Barz and Köster, this volume, Chapter 3). It is further a matter of speculation whether glucosylation and deglycosylation play a role in inactivation or activation of physiologically active compounds such as plant hormones, but there is growing information that they do (cf. Knöfel *et al.,* 1974; Müller *et al.,* 1974).

REFERENCES

Barber, G. A. (1962). *Biochemistry* **1,** 463–468.
Barber, G. A., and Chang, M. T. Y. (1968). *Phytochemistry* **7,** 35–39.

Bertrand, G., and Rivkid, L. (1907). *Bull. Soc. Chim. Biol.* **4**, 497.

Björkman, R., and Janson, J.-C. (1972). *Biochim. Biophys. Acta* **276**, 508–518.

Björkman, R., and Lönnerdal, B. (1973). *Biochim. Biophys. Acta* **327**, 121–131.

Blume, D. E., Jaworski, J. G., and McClure, J. W. (1979). *Planta.* **146**, 199–202.

Bogatek, R., Podstolski, A., Ostaszewska, A., and Lewak, S. (1976). *Biol. Plant.* **18**, 241–250.

Bourbouze, R., Pratviel-Sosa, F., and Percheron, F. (1974). *Biochimie* **56**, 1305–1313.

Boylen, C. W., Hagen, C. W., Jr., and Mansell, R. L. (1969). *Phytochemistry* **8**, 2311–2315.

Bridel, M. (1926). *C.R. Hebd. Seances Acad. Sci.* **180**, 1421.

Bridel, M., and Charaux, C. (1926). *Bull. Soc. Chim. Biol.* **8**, 40–49.

Butler, G. W., Bailey, R. W., and Kennedy, L. D. (1965). *Phytochemistry* **4**, 369–381.

Conn, E. E. (1980). *Encycl. Plant Physiol., New Ser.* **8**, 461–492.

Cooke, R. D., Blake, G. G., and Battershill, J. M. (1978). *Phytochemistry* **17**, 381–383.

Corner, J. J., and Swain, T. (1965). *Nature (London)* **207**, 634.

Courtois, J. E., and Percheron, F. (1970). *In* "The Carbohydrates" (W. Pigman and D. Horton, eds.), Vol. 2, Chapter 32, pp. 213–240. Academic Press, New York.

Czapek, F. (1921). *In* "Biochemie der Pflanzen," 2nd ed., Chapter III, pp. 205–220. Fischer, Jena.

Dedonder, R. A. (1961). *Annu. Rev. Biochem.* **30**, 347–382.

Dooner, H. K., and Nelson, O. E., Jr. (1977a). *Proc. Natl. Acad. Sci. U.S.A.* **74**, 5623–5627.

Dooner, H. K., and Nelson, O. E., Jr. (1977b). *Biochem. Genet.* **15**, 509–519.

Enslin, P. R., Joubert, F. J., and Rehm, S. (1956). *J. Sci. Food Agric.* **7**, 646–655.

Ettlinger, M. G., and Kjaer, A. (1968). *Recent Adv. Phytochem.* **1**, 59–144.

Ettlinger, M. G., Dateo, G. P., Harrison, B. W., Mabry, T. J., and Thompson, C. P. (1961). *Proc. Natl. Acad. Sci. U.S.A.* **47**, 1875–1880.

Eyjolfssón, R., and Ettlinger, M. G. (1972). *J. Chem. Soc., Chem. Commun.* pp. 572–573.

Freudenberg, K., Reznik, H., Boesenberg, H., and Rasenack, D. (1952). *Chem. Ber.* **85**, 641–647.

Gaines, R. D., and Goering, K. J. (1962). *Arch. Biochem. Biophys.* **96**, 13–19.

Grisebach, H., and Ortmann, R. (1973). *In* "Methods in Enzymology" (V. Ginsburg, ed.), Vol. 28, Part B, Chapter 61, pp. 473–477. Academic Press, New York.

Grover, A. K., and Cushley, R. J. (1977). *Biochim. Biophys. Acta* **482**, 109–124.

Grover, A. K., MacMurchie, D. D., and Cushley, R. J. (1977). *Biochim. Biophys. Acta* **482**, 98–108.

Hahlbrock, K., and Conn, E. E. (1970). *J. Biol. Chem.* **245**, 917–922.

Haisman, D. R., and Knight, D. J. (1967). *Biochem. J.* **103**, 528–534.

Haisman, D. R., Knight, D. J., and Ellis, M. J. (1967). *Phytochemistry* **6**, 1501–1505.

Harborne, J. B. (1967). "Comparative Biochemistry of Flavonoids," Chapter 8, pp. 251–266. Academic Press, New York.

Haskins, F. A., and Gorz, H. J. (1961). *Crop Sci.* **1**, 320–323.

Helferich, B. (1933). *Ergeb. Enzymforsch.* **2**, 74–89.

Hemscheidt, T., and Zenk, M. H. (1980). *FEBS Lett.* **110**, 187–191.

Heyworth, R., and Walker, P. G. (1962). *Biochem. J.* **83**, 331–335.

Hösel, W. (1976a). *Planta Med.* **30**, 97–103.

Hösel, W. (1976b). *Hoppe-Seyler's Z. Physiol. Chem.* **357**, 1673–1680.

Hösel, W., and Barz, W. (1975). *Eur. J. Biochem.* **57**, 607–616.

Hösel, W., and Klewitz, O. (1977). *Hoppe-Seyler's Z. Physiol. Chem.* **358**, 959–966.

Hösel, W., and Nahrstedt, A. (1975). *Hoppe-Seyler's Z. Physiol. Chem.* **356**, 1265–1275.

Hösel, W., and Nahrstedt, A. (1980). *Arch. Biochem. Biophys.* **203**, 753–757.

Hösel, W., and Todenhagen, R. (1980). *Phytochemistry* **19**, 1349–1353.

Hösel, W., Surholt, E., and Borgmann, E. (1978). *Eur. J. Biochem.* **84**, 487–492.

Hrazdina, G., Wagner, G. J., and Siegelman, H. W. (1978). *Phytochemistry* **17**, 53–56.

Hughes, M. A. (1968). *J. Exp. Bot.* **19**, 427.

Hughes, M. A., and Maher, E. P. (1973). *Biochem. Genet.* **8**, 1–12.

Ibrahim, R. K. (1977). *Z. Pflanzenphysiol.* **85**, 253–262.

Ibrahim, R. K., and Grisebach, H. (1976). *Arch. Biochem. Biophys.* **176**, 700–708.

Imaseki, H., and Yamamoto, T. (1961). *Arch. Biochem. Biophys.* **92**, 467–474.

Joubert, F. J. (1960). *Arch. Biochem. Biophys.* **91**, 11–16.

Kalinowska, M., and Wojciechowski, A. Z. (1978). *Phytochemistry* **17**, 1533–1537.

Kamsteeg, J., Van Brederode, J., and Van Nigtevecht, G. (1978). *Biochem. Genet.* **16**, 1045–1058, 1059–1071.

Kho, K. F. F., Kamsteeg, J., and Van Brederode, J. (1978). *Z. Pfanzenphysiol.* **88**, 449–464.

Kleinhofs, A., Haskins, F. A., and Gorz, H. J. (1967). *Phytochemistry* **6**, 1313–1318.

Knöfel, H.-D., Müller, P., and Sembdner, G. (1974). *In* "Biochemistry and Chemistry of Plant Growth Regulators" (K. Schreiber, H. R. Schütte, and G. Sembdner, eds.) pp. 121–124. Halle/Saale, GDR.

Kojima, M., Poulton, J. E., Thayer, S. E., and Conn, E. E. (1979). *Plant Physiol.* **63**, 1022–1028.

Kosuge, T. (1961). *Arch. Biochem. Biophys.* **95**, 211–218.

Kosuge, T., and Conn, E. E. (1961). *J. Biol. Chem.* **236**, 1617–1621.

Lalégérie, P. (1974). *Biochimie* **56**, 1297–1303.

Larson, R. L., and Coe, E. H., Jr. (1977). *Biochem. Genet.* **15**, 153–156.

Larson, R. L., and Lonergan, C. M. (1972). *Planta* **103**, 361–364.

Legler, G., and Harder, A. (1978). *Biochim. Biophys. Acta* **524**, 102–108.

Lein, K.-A. (1972). *Angew. Bot.* **46**, 137–159.

Liebig, J., and Wöhler, F. (1837). *Annalen* **22**, 11.

Lönnerdal, B., and Janson, J.-C. (1973). *Biochim. Biophys. Acta* **315**, 421–429.

Macheix, J.-J. (1977). *C.R. Hebd. Seances Acad. Sci.* **284**, 33–36.

Mao, C.-H., and Anderson, L. (1967). *Phytochemistry* **6**, 473–483.

Marcinowski, S., and Grisebach, H. (1978). *Eur. J. Biochem.* **87**, 37–44.

Marsh, C. A. (1960). *Biochim. Biophys. Acta* **44**, 359–361.

Matsuo, M., and Underhill, E. W. (1971). *Phytochemistry* **10**, 2279–2286.

Miles, C. D., and Hagen, C. W., Jr. (1968). *Plant Physiol.* **43**, 1347–1354.

Miller, L. P. (1973). *In* "Phytochemistry" (L. P. Miller, ed.), Vol. 1, chapter 11. Van Nostrand Reinhold Company, New York.

Müller, P., Knöfel, H.-D., and Sembdner, G. (1974). *In* "Biochemistry and Chemistry of Plant Growth Regulators" (K. Schreiber, H. R. Schütte, and G. Sembdner, eds.), pp. 115–120. Halle/Saale, GDR.

Müller, W.-U., and Leistner, E. (1978). *Phytochemistry* **17**, 1739–1742.

Nahrstedt, A., Hösel, W., and Walther, A. (1979). *Phytochemistry* **18**, 1137–1141.

Nisizawa, K., and Hashimoto, Y. (1970). *In* "The Carbohydrates" (W. Pigman and D. Horton, eds.), 2nd ed., Vol 2A, Chapter 33, pp. 241–299. Academic Press, New York.

Ortmann, R., Sutter, A., and Grisebach, H. (1972). *Biochim. Biophys. Acta* **289**, 293–302.

Pigman, W. W. (1944). *Adv. Enzymol.* **4**, 41–74.

Pihakaski, K., and Iversen, T.-H. (1976). *J. Exp. Bot.* **27**, 242–258.

Podstolski, A., and Lewak, S. (1970). *Phytochemistry* **9**, 289–296.

Podstolski, A., and Lewak, S. (1973). *Acta Soc. Bot. Pol.* **42**, 193–199.

Poulton, J. E., and Kauer, M. (1977). *Planta* **136**, 53–59.

Pridham, J. B. (1964). *Phytochemistry* **3**, 493–497.

Pridham, J. B. (1965). *Annu. Rev. Plant Physiol.* **16**, 13–36.

Pridham, J. B., and Saltmarsh, M. J. (1963). *Biochem. J.* **87**, 218–224.

Rabaté, J. (1935). *Bull. Soc. Chim. Biol.* **17**, 572.

Reay, P. F., and Conn, E. E. (1974). *J. Biol. Chem.* **249**, 5826–5830.

Reese, E. T., Clapp, R. C., and Mandels, M. (1958). *Arch. Biochem. Biophys.* **75**, 228–242.

Reznik, H. (1955). *Planta* **45**, 455–469.

Saleh, N. A. M., Poulton, J. E., and Grisebach, H. (1976a). *Phytochemistry* **15**, 1865–1868.

Saleh, N. A. M., Fritsch, H., Witkop, P., and Grisebach, H. (1976b). *Planta* **133**, 41–45.

Sato, M., and Hasegawa, M. (1971). *Phytochemistry* **10**, 2367–2372.

Schlepphorst, R., and Barz, W. (1979). *Planta Med.* **36**, 333–342.

Schön, W. J. (1966). *Angew. Bot.* **40**, 38–54.

Scott, A. I., Lee, S. L., and Wan, W. (1977). *Biochem. Biophys. Res. Commun.* **75**, 1004–1009.

Shute, J. L., Jourdan, P. S., and Mansell, R. L. (1979). *Z. Naturforsch., Teil C* **34**, 738–741.

Snowden, D. R., and Gaines, R. D. (1969). *Phytochemistry* **8**, 1649–1654.

Stoll, A., and Renz, J. (1939). *Enzymologia* **7**, 1362–382.

Stoll, A., Hoffmann, A., and Kreis, W. (1935). *Hoppe-Seyler's Z. Physiol. Chem.* **235**, 249–260.

Surholt, E., and Hösel, W. (1978). *Phytochemistry* **17**, 873–877.

Sutter, A., and Grisebach, H. (1973). *Biochim. Biophys. Acta* **309**, 289–295.

Sutter, A., and Grisebach, H. (1975). *Arch. Biochem. Biophys.* **167**, 444–447.

Sutter, A., Ortmann, R., and Grisebach, H. (1972). *Biochim. Biophys. Acta* **258**, 71–87.

Suzuki, H. (1962). *Arch. Biochem. Biophys.* **99**, 476–483.

Tabata, M., Ikeda, F., Hiraoka, N., and Konoshima, M. (1976). *Phytochemistry* **15**, 1225–1229.

Tang, C. H. (1973). *Phytochemistry* **12**, 769–773.

Treimer, J. F., and Zenk, M. H. (1978). *Phytochemistry* **17**, 227–231.

Trim, A. R. (1953). *J. Exp. Bot.* **6**, 100–125.

Trivelloni, J. C., Recondo, E., and Cardini, C. E. (1962). *Nature (London)* **195**, 1202.

Van Brederode, J., and Van Nigtevecht, G. (1973). *Mol. Gen. Genet.* **122**, 215–229.

Van Brederode, J., and Van Nigtevecht, G. (1974a). *Biochem. Genet.* **11**, 65–81.

Van Brederode, J., and Van Nigtevecht, G. (1974b). *Phytochemistry* **13**, 2763–2766.

Van Brederode, J., and Van Nigtevecht, G. (1974c). *Genetics* **77**, 507–520.

Van Brederode, J., Van Wielink-Hillebrands, G. H., and Van Nigtevecht, G. (1974). *Mol. Gen. Genet.* **130**, 307–314.

Van Brederode, J., Van Nigtevecht, G., and Kamsteeg, J. (1975). *Heredity* **35**, 429–430.

Van Nigtevecht, G., and Van Brederode, J. (1975). *Heredity* **35**, 429.

Veibel, S. (1951). *In* "The Enzymes" (J. B. Summer and K. Myrbäck, eds.), 1st ed., Vol. 1, pp. 584–634. Academic Press, New York.

Wagner, H. (1974). *Fortschr. Chem. Org. Naturst.* **31**, 153–216.

Walker, D. E., and Axelrod, B. (1978). *Arch. Biochem. Biophys.* **187**, 102–107.

Yamaha, T., and Cardini, C. E. (1960a). *Arch. Biochem. Biophys.* **86**, 127–132.

Yamaha, T., and Cardini, C. E. (1960b). *Arch. Biochem. Biophys.* **86**, 133–137.

Zaprometov, M. H. (1978). *12th FEBS Meet.* Abstract No. 311.

Zenk, M. H. (1967). *Phytochemistry* **6**, 245–252.

Index

Contents of Other Volumes

VOLUME 1—THE PLANT CELL

VOLUME 2—METABOLISM AND RESPIRATION

VOLUME 3—CARBOHYDRATES: STRUCTURE AND FUNCTION

VOLUME 4—LIPIDS: STRUCTURE AND FUNCTION

VOLUME 5—AMINO ACIDS AND DERIVATIVES